Geological Disposal of Carbon Dioxide and Radioactive Waste: A Comparative Assessment

ADVANCES IN GLOBAL CHANGE RESEARCH

VOLUME 44

For other titles published in this series, go to
www.springer.com/series/5588

Ferenc L. Toth
Editor

Geological Disposal of Carbon Dioxide and Radioactive Waste: A Comparative Assessment

Springer

Editor
Ferenc L. Toth
Department of Nuclear Energy
International Atomic Energy Agency
Vienna, Austria

Cover: CO2 disposal scheme; Illuminated Tokyo City in Japan at night from high above © Achim Baqué [Image provider: Fotolia.co.uk]; The KBS-3 radioactive waste disposal scheme © Svensk Kärnbränslehantering AB (SKB), Sweden. KBS-3 is a registered trade mark of SKB.

ISBN 978-90-481-8711-9 e-ISBN 978-90-481-8712-6
DOI 10.1007/978-90-481-8712-6
Springer Dordrecht London Heidelberg New York

Library of Congress Control Number: 2011922503

Printed on acid-free paper

Springer is part of Springer Science+Business Media (www.springer.com)

Preface

Energy is key to life and a primary engine for socio-economic development. The Commission on Sustainable Development (CSD) in its 9th session specifically recognized that 'Energy is central to achieving the goals of sustainable development'. Access to energy opens up many new opportunities, while lack thereof is one of the contributing factors to persistent poverty afflicting individuals, communities, nations and regions. Indeed, unless there is universal access to clean and affordable energy services, the United Nations Millennium Development Goals will not be achieved.

Improving access to energy is a multi-faceted challenge with far-reaching implications and long-lasting obligations. Delivering energy services involves several steps (resource extraction/harvesting, conversion/transformation, transmission, distribution and service production at the point of use) and many players from both the public and private sectors.

Technology is the critical link between access, affordability and environmental compatibility of energy services. But technology is more than power plants, motor vehicles and appliances; it includes infrastructures such as buildings, settlement patterns, road and transportation systems, industrial plants and equipment and, of course, the production of goods and services. Each step along the different energy service delivery chains is subject to investment and operating costs—hence, the competition and the choice between technologies and fuels that provide the same energy service. Technology choices are also subject to laws and regulations that reflect national capabilities, social preferences and cultural backgrounds.

Energy extraction, conversion and service production always generate undesirable by-products and waste—far more, in fact, than any other process chain. The careless use of energy can have devastating effects on ecosystems and life on planet Earth. Most energy plants, equipment and infrastructure have long operating lives (25–50 years or more) and, in some cases, require special management long after their operational lives have ceased. Today's choices about how energy services are produced will determine the sustainability of the future energy system and thus of socio-economic progress as a whole.

Dangerous anthropogenic interference with the climate system has emerged as the main global environmental challenge for a global energy system that is 80% reliant for its energy supply on fossil fuel combustion. At the 15th Conference of the Parties to the United Nations Framework Convention on Climate Change held in Copenhagen in December 2009, the international community agreed that the threshold for dangerous interference would be a 2°C rise in global mean temperature: approximately equivalent to a maximum atmospheric greenhouse gas (GHG) concentration of 450 ppm. Since pre-industrial times carbon dioxide (CO_2) from fossil fuel combustion has been the main cause of increased GHG concentrations. Without a drastic shift to an energy system that minimizes GHG emissions in the production of energy services, the 450 ppm threshold will probably be reached within a few decades. GHG mitigation—how best to reconcile the dilemma of continued reliance on (still relatively cheap and plentiful) fossil energy and associated (long-lived) infrastructures while protecting the climate system and still providing affordable energy services—has thus become a major challenge.

Based on the recognition that, on a full life-cycle basis, no technology can provide energy services without interaction with the environment in terms of emissions or waste, a crucial question must be posed: What is the most efficient and cost-effective approach to the decarbonization of the global energy system? The options are known and range from efficiency improvements (not really an option for a quarter of world population without access to modern energy services such as provided by electricity) to the use of renewables, nuclear energy and CO_2 capture and disposal (CCD); in other words, continued use of fossil fuels, but alongside technological measures that prevent the majority of combustion products reaching the atmosphere. CCD, involving geological disposal of captured CO_2, has been advanced as a way of giving fossil fuels a new lease on life in a heavily carbon emission constrained future. Likewise, there is renewed interest in nuclear energy for the generation of low-GHG-emitting electricity. While in the past fossil fuel combustion and nuclear energy had little in common, the advent of CCD may change this and commonalities in the area of waste disposal could emerge. To date, there has been little experience of large-scale CO_2 disposal, geological retention times, leakage rates, etc. Disposal in geological repositories of high-level nuclear waste from reprocessing or spent nuclear fuel is, however, a generally favoured approach, and several countries have embarked upon the development of such repositories.

The question then arises as to how fossil and nuclear fuels stack up against each other in terms of the final waste disposal strategies applied in their respective cases. The Special Report on Carbon Capture and Storage of the Intergovernmental Panel on Climate Change provides a useful synthesis of the knowledge available from a fast-evolving research field. Compared with carbon, geological storage of radioactive waste has a much longer history of research and technology development; however, there has been no recent international research synthesis on the geological storage of radioactive waste, like that of the IPCC report on carbon. Are there lessons to be learned from the much longer R&D work regarding nuclear waste repositories that can be useful for CCD? Is public acceptance of CCD greater than of nuclear waste disposal? How can long-term leakage of CO_2 and ionized radiation be minimized? All these questions need solid answers if informed decision making is to take place.

Effective decision making with respect to the appropriate energy technologies to use, taking into account climate, cost or other considerations, requires comprehensive energy systems analysis and planning at the national level. This type of analysis helps policymakers to study the costs and effectiveness of different GHG mitigation options and to chart long-term scenarios of sustainable energy development. It also helps them test various climate change policies and response strategies, including the flexible mechanisms of the Kyoto Protocol, such as the Clean Development Mechanism (CDM), Joint Implementation (JI) and emissions trading.

The IAEA assists Member States in building national capacity to conduct independent energy and environmental assessments and to develop national strategic energy plans. One cornerstone of this capacity-building effort is comparative assessment of different demand and supply options. This type of planning approach prevents a situation arising where one technology option is rejected (for whatever reason) without an alternative solution having been specified that provides the same energy service in terms of quality and reliability. IAEA assistance involves transferring analytical and planning tools, and training of national experts in hands-on use of these tools to conduct energy and electricity demand and supply studies. A fast-growing planning tool application has been the analysis of least-cost GHG mitigation options. Through these and other activities, the IAEA advises and helps countries to identify the most suitable and feasible national energy mix, irrespective of whether or not this includes nuclear power.

This book must be seen in the context of capacity building and comparative assessment. Its objective is to summarize the state of the art in the fields of CO_2 and nuclear waste disposal by providing an in-depth comparative assessment of their similarities and differences, of related issues that have already been resolved and of the key challenges that remain; it also evaluates the policy implications for moving the process further. It is the product of the first close collaboration between leading scientists involved in the comparative assessment of various aspects of the geological disposal of CO_2 and radioactive waste. The contributors come from a broad range of scientific disciplines, including geology, geography, environmental sciences, engineering, economics, psychology and political science.

I believe the comparative assessment presented here to be of interest to a wide audience. The greatest effort was made by the authors and the editor to ensure the neutrality and objectivity of the comparative technology analyses. Considering the ample opportunities for knowledge transfer and learning between the CCD and radioactive waste management research communities, this book can be expected to trigger more collaborative projects to explore the open issues still further. On the policy side, the insights presented by the authors are likely to provide useable knowledge to assist policymakers in resolving major challenges encountered during the formulation of national energy strategies.

H.-Holger Rogner
Section Head, Planning and Economic Studies Section
Department of Nuclear Energy
International Atomic Energy Agency (IAEA)
Vienna, Austria

Contents

Contributors

Stefan Bachu
Alberta Innovates—Technology Futures, Edmonton, Canada,
stefan.bachu@arc.ab.ca

Nadira Barkatullah
International Atomic Energy Agency, Wagramer Str. 5, P.O. Box 100,
1400 Vienna, Austria
N.Barkatullah@iaea.org

Sara Bergan
University of Minnesota Law School, 229 19th Avenue South, Minneapolis,
MN 55455, USA

Jens T. Birkholzer
Earth Sciences Division 90-1116, Lawrence Berkeley National Laboratory,
Berkeley, CA 94720, USA
jtbirkholzer@lbl.gov

Donald A. Brown
Program on Science, Technology, and Society, Penn State University,
201 Willard Building, University Park, PA 16802, USA
dab57@psu.edu

Brian Brunskill
Helix Geological Consultants Ltd., Regina, Canada
brianbrunskill@sasktel.net

Alexey Cherepovitsyn
Department of Organisation and Management, Saint Petersburg State
Mining Institute, St Petersburg, Russia
alekseicherepov@inbox.ru

Marthinus Cloet
Council for Geoscience, 280 Pretoria St, Pretoria 012 841 1911, South Africa
mcloete@geoscience.org.za

Fanny Descamps
Mining Engineering Department, Faculty of Engineering, University of Mons, rue du Joncquois 53, 7000 Mons, Belgium
fanny.descamps@umons.ac.be

György Falus
Eötvös Loránd Geophysical Institute, Kolumbusz utca 17-23, 1145 Budapest, Hungary
falus@elgi.hu

Darío R. Gómez
National Atomic Energy Commission, Gerencia Química, Avenida General Paz 1499, B1650KNA, San Martín, Argentina
dgomez@cnea.gov.ar

Judith I. M. de Groot
School of Design, Engineering and Computing, Bournemouth University, Psychology, Poole House, Fern Barrow, Poole BH12 5BB, United Kingdom
jdgroot@bournemouth.ac.uk

Zsuzsanna Hódossyné Hauszmann
Department of Strategy, Nuclear Safety Directorate, Hungarian Atomic Energy Authority, Fényes Adolf utca 4, 1539 Budapest, Hungary
hodossyne@haea.gov.hu

Roberto Heemann
João Marcelo Ketzer, Brazilian Carbon Capture and Storage Research Center (CEPAC), Porto Alegre, Brazil

Goro Hiromoto
Brazilian Energy and Nuclear Research Institute (IPEN), São Paulo, Brazil

Alexander Ilinsky
All-Russia Petroleum Research Exploration Institute (VNIGRI), St Petersburg, Russia
alex.ilinsky@bk.ru

Hitoshi Koide
CC Geosystem Laboratory, 2-6-1-602 Higashi-Ikebukuro, Toshima, Tokyo 170-0013, Japan
MXE02263@nifty.com

Kinichiro Kusunose
Geological Survey of Japan, AIST, 1-1-1, Higashi, Tsukuba 305-8567, Japan
k.kusunose@aist.go.jp

Phillip J. Lloy
Cape Peninsula University of Technology, Symphony Way, Cape Town 021 959 6911, South Africa
lloydp@cput.ac.za

Philip Maul
Quintessa Limited, The Hub, 14 Station Road, Henley-on-Thames, Oxfordshire, United Kingdom RG9 1AY
philipmaul@quintessa.org

Tim McEwen
McEwen Consulting, Melton Mowbray, United Kingdom
tmcewen@globalnet.co.uk

Asami Miketa
International Atomic Energy Agency, Wagramer Str. 5, P.O. Box 100, 1400 Vienna, Austria
A.Miketa@iaea.org

William J. Nuttall
Judge Business School, University of Cambridge, Trumpington Street, Cambridge CB2 1AG, United Kingdom
wjn21@cam.ac.uk

Curtis M. Oldenburg
Earth Sciences Division 90-1116, Lawrence Berkeley National Laboratory, Berkeley, CA 94720, USA
cmoldenburg@lbl.gov

Zhonghe Pang
Institute of Geology and Geophysics, Chinese Academy of Science, Beijing 100029, China
z.pang@mail.iggcas.ac.cn

Jonathan M. Pearce
British Geological Survey, Keyworth, Nottingham NG12 5GG, United Kingdom
jmpe@bgs.ac.uk

David M. Reiner
Judge Business School, University of Cambridge, Trumpington Street, Cambridge CB2 1AG, United Kingdom
dmr40@cam.ac.uk

Richard A. Roehrl
International Atomic Energy Agency, Wagramer Str. 5, P.O. Box 100, 1400 Vienna, Austria
aroehrl@gmail.com

Péter Scholtz
Eötvös Loránd Geophysical Institute, Kolumbusz utca 17–23, 1145 Budapest, Hungary
scholtz@elgi.hu

Alexandro Scislewski
Brazilian National Nuclear Energy Commission (CNEN), Rio de Janeiro, Brazil

Richard P. Shaw
British Geological Survey, Keyworth, Nottingham NG12 5GG, United Kingdom
rps@bgs.ac.uk

Linda Steg
Department of Psychology, Faculty of Social and Behavioural Sciences,
University of Groningen, Grote Kruisstraat 1/2, 9712 TS Groningen,
The Netherlands
E.M.Steg@rug.nl

Anthony D. Surridge
South African National Energy Research Institute, 158 Jan Smuts Ave,
Sandton 011 280 0300, South Africa
tonys@saneri.org.za

Ferenc L. Toth
International Atomic Energy Agency, Wagramer Str. 5, P.O. Box 100,
1400 Vienna, Austria
F.L.Toth@iaea.org

Jean-Pierre Tshibangu K
Faculty of Engineering, Mining Engineering Department, University of Mons,
rue du Joncquois 53, 7000 Mons, Belgium
jean-pierre.tshibangu@umons.ac.be

Michael Tyacke
Idaho National Laboratory, 1625, Idaho Falls, ID, USA
michael.tyacke@inl.gov

Ju Wang
Beijing Research Institute of Uranium Geology, China National
Nuclear Corporation, 9818, Beijing 100029, China
wangju9818@163.com

Julia M. West
British Geological Survey, Keyworth, Nottingham NG12 5GG, United Kingdom,
jmwes@bgs.ac.uk

Elizabeth J. Wilson
Humphrey Institute of Public Affairs, University of Minnesota,
301 19th Avenue South, Minneapolis, MN 55455, USA
ewilson@umn.edu

Malcolm Wilson
University of Regina, Regina, Canada
malcolm.wilson@uregina.ca

Comparing the Geological Disposal of Carbon Dioxide and Radioactive Waste: Introduction and Overview

Ferenc L. Toth

Abstract Fossil fuels will remain the backbone of the global energy economy for the foreseeable future. The contribution of nuclear energy to the global energy supply is also expected to increase. With the pressing need to mitigate climate change and reduce greenhouse gas emissions, the fossil energy industry is exploring the possibility of carbon dioxide disposal in geological media. Geological disposal has been studied for decades by the nuclear industry with a view to ensuring the safe containment of its wastes. Geological disposal of carbon dioxide and that of radioactive waste gives rise to many common concerns in domains ranging from geology to public acceptance. In this respect, comparative assessments reveal many similarities, ranging from the transformation of the geological environment and safety and monitoring concerns to regulatory, liability and public acceptance issues. However, there are profound differences on a broad range of issues as well, such as the quantities and hazardous features of the materials to be disposed of, the characteristics of the targeted geological media, the site engineering technologies involved and the timescales required for safe containment at the disposal location. There are ample opportunities to learn from comparisons and to derive insights that will assist policymakers responsible for national energy strategies and international climate policies.

Keywords Geological disposal • Carbon dioxide • Radioactive waste • Comparative analysis • Climate change mitigation • Sustainable energy development

F.L. Toth (✉)
Department of Nuclear Energy, International Atomic Energy Agency, Wagramer Str. 5, 1400 Vienna, Austria
e-mail: F.L.Toth@iaea.org

F.L. Toth (ed.), *Geological Disposal of Carbon Dioxide and Radioactive Waste: A Comparative Assessment*, Advances in Global Change Research 44,
DOI 10.1007/978-90-481-8712-6_1, © International Atomic Energy Agency, 2011

1 Introduction

In the second half of the last century the use of fossil fuels, especially coal and oil, has gone through two major technology transformations, first in the developed countries, followed by the economies in transition as well as in more affluent developing countries. The first transformation was triggered by local/regional air pollution problems (urban smog with severe visibility degradation and human health impacts) and entailed the removal or wider dispersion of heavy hydrocarbons (C_xH_y) from stack gases. The second change was prompted by continental-scale pollution problems that involved long-range transport of air pollutants (mainly SO_x) causing material corrosion, forest degradation and the acidification of water bodies. The response to both transitions encompassed a set of technologies ranging from pre-combustion fuel treatment to flue gas scrubbing to reduce the emissions of pertinent compounds as well as fuel and technology switching. The increasing concern over anthropogenic climate change and the need to reduce greenhouse gas (GHG) emissions poses the next challenge to the fossil fuel industry. If large reductions of GHG emissions are necessary over the next few decades, the viability of fossil fuels will depend on the possibility and prospects of preventing the release of carbon dioxide (CO_2) into the atmosphere by capturing and disposing of it in geological formations. CO_2 capture and disposal (CCD) has emerged as one of the principal fields of scientific research and technological R&D.

Over the same time frame, nuclear energy has been pursued by many countries for a variety of reasons, ranging from fast growing energy demand to energy supply security and, more recently, as part of climate change mitigation strategies. The safe disposal of the resulting radioactive waste (RW) has been one of the main predicaments from the beginning, and it remains an issue that the nuclear industry needs to resolve in order to improve the prospects for nuclear energy to contribute to resolving the enormous energy challenges the world faces in this century. Over the past 2 decades, major scientific and technological advances have been made towards the safe temporary storage and final disposal of RW. The disposal of RW in geological media is considered by most scientists and engineers engaged in the issue to be a safe and viable method for isolating it from the hydrosphere, the atmosphere and the biosphere.

Geological disposal of the waste products (CO_2 and RW) establishes a curious link between the fossil fuel and nuclear energy industries. The question arises whether, despite the profound differences, at least at first sight, there is any chance to learn from comparing the diverse array of issues involved and what the possibilities are for sharing experience and transferring lessons between the two fields. This chapter introduces a book that is intended to explore these questions across relevant thematic areas and in selected geographical regions. It presents the broader context of global energy challenges and the potential role of fossil fuels (combined with CCD) and of nuclear power (combined with RW disposal) in long-term climate change mitigation and sustainable energy development.

It is important to clarify the terminology used here right at the outset. The emplacement of CO_2 in geological formations is widely called ‘storage’. This is a somewhat misleading euphemism because the primary meaning of the word 'storage' is the action of putting something away for future use whereas it is not foreseen to use the disposed CO_2 ever again. Therefore, most chapters in the book use 'disposal', but some authors prefer to adhere to 'storage' or 'sequestration' (a widely used term especially in the North American literature) and these preferences are respected. Hence, in connection with CO_2 the three terms, geological disposal, storage and sequestration are used interchangeably throughout the book. With regard to RW, there is more clarity: 'storage' is used for the keeping of spent fuel and other RW in temporary storage facilities even if such arrangements last for decades in many cases, while the term 'disposal' is used for permanent emplacement in geological formations, even if it is intended to leave open the option of retrievability for 100 years or longer.

The next section presents a short summary of the global energy challenges for the twenty-first century as the broader context for this book. This is followed in Sect. 3 by an outline of the key issues pertinent to the comparative assessment of CO_2 and RW disposal. An overview of the thematic and regional chapters (all peer-reviewed by at least three referees) is presented in Sect. 4. Section 5 summarizes the most important points raised in this chapter.

2 Energy Challenges for the Twenty-First Century

Energy is generally recognized as a central issue in sustainable development. Several high-level conferences and declarations have emphasized that the provision of adequate energy services at affordable costs, in a secure and environmentally benign manner and in conformity with social and economic developmental needs is an essential element of sustainable development. Reliable energy services are an important precondition for investments that bring about economic development. Among other things, they facilitate the learning and study and improved health care that are crucial for developing human capital. They also promote gender equity by allowing women to use their time for more productive activities than collecting firewood, and social equity by giving the less well-off the chance to study, thus providing a possible escape from poverty. Energy is therefore vital for alleviating poverty, improving human welfare and raising living standards. Yet, worldwide, 2.4 billion people rely on traditional biomass as their primary source of energy and 1.6 billion people do not have access to electricity (UNDP 2005), and this severely hampers socioeconomic development.

All recent socioeconomic development studies forecast major increases in energy demand, driven largely by demographic and economic growth in today's developing countries. Of the world's 6.8 billion people, about 82% live in non-OECD countries and consume only 53% of global primary energy. Alleviating this energy inequity will be a major challenge. A growing global population will compound the problem. The medium variant of the latest projection by the United Nations estimates an additional 1.5 billion people by 2030, and another 840 million by 2050, bringing the world's population to about 9.15 billion by the middle of this century (UN DESA 2009).

It is also anticipated that the rising population will enjoy increasing economic welfare over the next decades. According to the World Bank (2009a), after the projected meagre 0.9% global GDP growth in 2009, it is expected to rebound to 2% in 2010 and 3.2% in 2011. Developing countries are projected to expand by 4.4% (2010) and 5.7% (2011). Over the long term, the World Bank (2009b) projects a 3.1% average annual growth rate for the world economy up to 2015 and 2.5% between 2015 and 2030. Developing countries will grow fastest, while OECD countries will grow at the slowest rate. Per capita incomes in developing countries are projected to triple from US$1,550 in 2004 to US$4,650 in 2030.

In its World Energy Outlook (WEO) 2008 (IEA 2008a), the OECD International Energy Agency (IEA) adopts the population projection developed by the United Nations Department of Economic and Social Affairs (UN DESA) and makes similar assumptions as the World Bank about longer term economic development. World population is estimated to increase to 8.2 billion by 2030, while the global economy is assumed to grow at an annual average rate of 4.2% up to 2015 and 2.8% between 2015 and 2030. Based on these two main drivers of energy demand and additional assumptions about technological development and resource availability for the energy sector, the IEA projects in its Reference Scenario that world total primary energy demand will grow to over 17 gigatonnes of oil equivalent (Gtoe) by 2030 (IEA 2008a) and, according to the extended Reference Scenario presented in Energy Technology Perspectives (ETP) 2008 (IEA 2008b), it will exceed 23 Gtoe in 2050 (see Fig. 1).

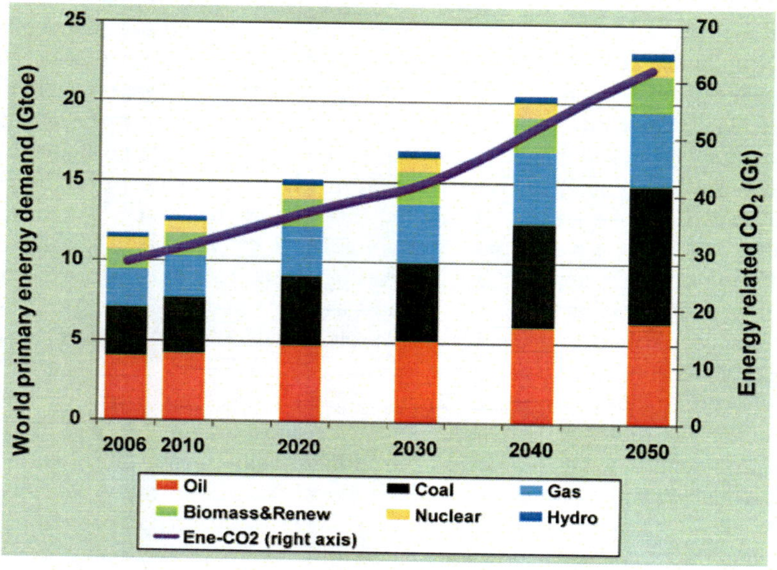

Fig. 1 Global primary energy sources (*left axis*) and energy-related CO_2 emissions (*right axis*) in the IEA's reference scenarios (Based on IEA 2008a, b) (*see* Colour Plates)

The ETP study (IEA 2008b) presents the global energy prospects up to the middle of the century. The most notable changes anticipated for the next half century in the IEA Reference Scenario include the following:

- Coal is expected to surpass oil as the largest primary energy source by 2040 due to the persistent strong growth in demand for electricity in coal-rich countries such as China and India.
- Gas is estimated to level out at around 4.5 Gtoe by the middle of the century.
- Despite a 31% increase in volume between 2005 and 2050, the nuclear share in the global primary energy balance is projected to decline from 6.3% in 2005 to 4.8% by 2030 and to 4% by 2050.

The climate change implications of the Reference Scenario are severe. Energy-related CO_2 emissions, the largest component of global GHG emissions, will have increased by 55% in 2030 and by 130% in 2050 relative to 2005. Assuming that other GHGs increase at comparable rates, this would put the Earth on track towards atmospheric GHG concentrations on the order of 800 ppm CO_2 equivalent and an equilibrium warming of over 5°C in terms of global mean temperature increase above the pre-industrial level (IPCC 2007a). Thus these trends stand in sharp contradiction to the declaration issued by the Group of Eight (G8) summit in 2009 on the need to keep global mean temperature increase below 2°C, and point to the urgent requirement for deploying low-carbon technologies.

In addition to the staggering increases in demand for all forms of energy, particularly electricity, and the need to reduce GHG emissions, there are several other issues on the current energy policy agendas of many countries that nuclear power and coal-based electricity using CCD might contribute to resolving.

The first factor is the price of oil and gas energy sources. The rate of infrastructure development in resource extraction and delivery in key supply regions is lagging behind the fast growing energy needs. This exerts a sustained upward pressure on international oil and gas prices even if one takes into account the speculative bubble that affected commodity prices and culminated in mid-2008. This in itself is a strong motivation for countries that depend on high shares of imported fuels for their electricity generation to look for substitutes. Political conflicts in key supply regions exacerbate the price pressure and raise severe concerns over the security of supply per se, even at high prices. This is yet another reason for considering alternative electricity sources.

Energy importing developing countries tend to be more concerned about the sustained high price level because of the prospect of its severely increasing their energy import bills, affecting their current account balances and undermining the competitiveness of their export industries. In most developed countries (except those with very small energy resource endowments) energy is a relatively smaller fraction of the total import bills and the energy content of exports is lower. These countries are more concerned about direct losses due to supply disruptions, especially if these might render expensive capital and labour capacities idle for some time.

Another, but closely related, factor is price volatility. All elements of the energy supply infrastructure are long lived. Energy intensive industries base their investment decisions on cautious expectations about future energy and electricity prices. A reasonable degree of stability and predictability of resource prices is crucial for such decisions because hedging against large price fluctuations might be vastly expensive.

In many countries continued reliance on large and cheap domestic coal reserves could help alleviate energy security fears, but the use of currently prevailing technologies would aggravate the climate problem. In other countries nuclear power could help mitigate supply security concerns and reduce GHG emissions at the same time. The choice between establishing or expanding coal-based power generation combined with CCD, or nuclear electricity accompanied by the need to find a means of safe disposal for the resulting RW will be influenced by many factors and will depend on natural resource and environmental endowments as well as on social, economic and political preferences.

3 Why Compare CO_2 and Radioactive Waste Disposal?

This section delineates the considerations that motivated the initiation of the comparative assessments and the preparation of this book. It also highlights the broader linkages, similarities and differences between the two areas, some of which will be explored in more detail in subsequent chapters.

3.1 Objectives

Fossil fuels (mainly coal but also natural gas and to some extent oil) provide the bulk of electricity generated in the world today, and they are projected to dominate the power sector up to 2030 (IEA 2008a) and beyond (IEA 2008b). Fossil-based electricity sources are under increasing pressure to reduce their GHG (mainly CO_2) emissions in order to mitigate climate change. This requirement has accelerated technological R&D efforts to capture CO_2 and dispose of it in geological formations.

Another important source of electricity is nuclear power. The emissions of GHGs and other air pollutants are very low even if one considers the indirect emissions arising from the construction to the decommissioning of power plants and all activities in the nuclear fuel cycle, from uranium mining to enrichment and fuel fabrication to final disposal of the RW. This last item has been a conundrum for the nuclear industry for decades but there is now a general consensus that disposal of high-level RW in suitable geological formations is the ultimate solution and that it is technically viable.

The Intergovernmental Panel on Climate Change (IPCC) Special Report on Carbon Dioxide Capture and Storage (IPCC 2005) provided a useful synthesis of the then available knowledge from a fast evolving research field. Research and technological development related to geological disposal of RW has a somewhat longer history but no recent international synthesis has been published. Except for a few sporadic efforts dealing with selected topics, no systematic comparison has been prepared so far about the issues involved in the geological disposal of CO_2 and RW. This book intends to fill the gap by reviewing the state of the art in these two fields, preparing an in-depth comparative assessment of the similarities and differences, the already resolved issues and the remaining key challenges, and by evaluating the policy implications emerging from the comparative study.

Accordingly, the main objective of this book is to present a comparative assessment of CO_2 and RW disposal. Information from such an assessment is expected to foster future scientific research and to become a useful component of the knowledge base for policymakers when considering various options for the future energy supply in their countries or regions.

The main *scientific* objectives of the study are to explore:

- The main issues/challenges in the geological disposal of CO_2 and RW;
- The state of the art in these two fields: issues already resolved, those remaining open, unknown or uncertain;
- The common issues in and the main similarities and differences between CO_2 and RW;
- The possibilities regarding what scientists working in these two fields can learn and/or adopt from each other.

The main *policy-relevant* objectives are to examine:

- The key factors to consider in domestic decision making (especially in formulating long-term energy strategies);
- The relative benefits and drawbacks of geological disposal of CO_2 and RW;
- The issues/aspects requiring international coordination and treaties;
- The main domestic regulatory requirements for implementation.

Implementing these ambitious objectives is not a simple task. According to the experience gained from this project, the links between the two communities working on CO_2 and RW disposal, in terms of sharing knowledge and experience, are rather sparse (limited to a few special aspects) in the area of the natural sciences (e.g. geology) and the environmental and engineering sciences as well as in the social sciences (ranging from legal to economic and public acceptance issues). However, the results indicate that there are many similarities between these areas and that one can derive useful information from the differences as well.

It is important to note that utmost attention has been devoted to keeping this comparative assessment neutral and non-adversary. It is an explicit objective of this book to avoid any comparison, let alone conclusion, as to the superiority of one technology over the other. In any case, this would be a futile exercise since the numerous local and nation-specific factors will ultimately determine the relative importance, advantages and shortcomings of each technology in accordance with national energy strategy priorities.

3.2 Shared Issues, Similarities and Differences

Over its long history, mankind has been changing the environment at increasing temporal, spatial and complexity scales. Already in the nineteenth century George Perkins Marsh recorded the transformation of several components of the natural environment through human activities (Marsh 1874). Since the 1980s several publications have documented the human-induced changes in land cover and soils, the biosphere and the atmosphere (see, for example, Turner et al. 1990). Beneath the surface, deep mining has been going on for a long time and has also clearly impacted at depth (e.g. gold mines or drilling for oil and gas extending to a depth of 3,500 m). However, with the introduction of geological disposal of CO_2 and RW, humanity is entering a new phase in transforming the Earth, this time impacting on the deep underground in a different way.

Both CO_2 and RW disposal involve what might be called 'inverse geological transformation'. As opposed to traditional geological exploration that looks for underground space from which to extract material and remove what is useful, the objective in the case of disposal is to look for underground space in which to deposit something. RW research started doing this decades ago and CO_2 disposal has triggered a new upswing more recently. RW disposal will affect relatively small tracts for a very long time while CCD will spread over large expanses under the terrestrial and oceanic surface for considerably shorter periods of time, except in such cases as that of depleted oilfields, in which pressurized CO_2 could remain in place for very long time as well. This also implies a reversal of concerns at the surface regarding the hazards associated with removing material from beneath the ground surface as opposed to those associated with placing substances there.

A good understanding of geological formations and processes is a prerequisite for geological disposal of CO_2 and RW. Cross-learning between the fossil resources sector and the area of RW disposal has been going on for decades in both directions in a few very specific areas. In the exploration stage of an RW disposal site, geophysical methods and other techniques that were invented by the oil industry are used. Several organizations working on RW disposal have used the know-how of the oil and gas industry. The transfer of knowledge in the other direction is more recent. Although the main technical aspects concerning scale, risks and scope are different, the scientific advances made in RW disposal research over the past 3 decades in simulating multiphase flows and reactive transport processes in deep geological systems is valuable for research on CO_2 disposal. Various concepts, methods and tools developed in establishing the scientific foundations for RW disposal have been adopted in the geological research related to CO_2 disposal.

Looking at the geological aspects first, we find interesting similarities as well as major differences between the geological disposal of CO_2 and that of RW. Both substances require reasonable tectonic stability, and locations with at least one natural barrier against migration. The principal geological formation for CO_2 disposal is certain types of sedimentary (soft) rocks while radioactive wastes can be disposed of in hard rock as well. Both substances will trigger local effects on the geological environment as a result of the emplacement, although the nature of the effects (e.g. thermal cooling versus heating, different geochemical and geomechanical effects, etc.) differs.

Post-emplacement monitoring is usually required in both cases, although RW disposal should be passively safe and not have to rely on monitoring or any other action. Therefore, unless monitoring takes place very close to the disposal site, it is very unlikely that any releases of radioactivity will be detected for a very long time. One obvious area of joint interest is risk assessment methods: how best to evaluate long-term risks and prove the security cases. RW has a long history that CCD can learn from.

Perhaps the largest differences are related to the volume and toxicity of the waste products for disposal. Gigatonnes of fluid CO_2 will need to be injected into the disposal media whereas the volume of high-level radioactive waste accumulated so far amounts to a few hundred thousand tonnes. In contrast, the direct environmental and health hazards of CO_2 are relatively modest (except in extreme cases of seepage in valleys with human settlements), while high-level waste contains radioisotopes which emit alpha, beta, gamma and neutron radiation. External exposure to high levels of gamma radiation or neutrons is harmful and can be fatal to most species, including humans.

Another important difference is in the disposal technologies. CO_2 disposal is carried out through wells that extend to great depths and is based on oil/gas drilling techniques in terms of engineering, while for RW, mining techniques are used to create the tunnels and vaults at a depth of a few hundred metres. The latter technology uses a combination of engineered and natural barriers.

Both substances undergo long-term decay: CO_2 will be bound and absorbed by the host media through chemical processes, while the radiotoxicity of RW will decline as well. The timescales and containment period may be significantly shorter for CO_2, ranging from centuries to millennia, whereas RW may require safety assessment timescales to cover at least 10,000 and possibly as much as 100,000 or a million years. Yet the timescales for both are long enough for these to become a public liability if remediation of leakage, compensation of victims or repair and rehabilitation of the affected area is required.

Alternative solutions to underground disposal exist for both substances to reduce the time until toxicity levels or hazards are acceptable. They could be transformed into less harmful or totally harmless matter, at least partially. Partitioning and transmutation of RW would reduce its volume, radiotoxicity and the duration of the hazard. Chemical mineralization of CO_2 would immediately eliminate both atmospheric and geological hazards. However, both methods have their drawbacks.

A comparable variety of similarities and differences can be observed in the issues concerning the implementation of CO_2 and RW disposal. The timing of the disposal activity relative to the time of the waste generation has several implications. CO_2 will require disposal within a short time after it has been captured because temporary storage, albeit in principle possible, would be very expensive considering the huge volumes involved. In contrast, RW has been safely stored for decades in the past and this practice could continue for decades into the future before emplacement in a final repository. This means that CO_2 disposal will require an upfront investment in exploration, site assessment, licensing, infrastructure, equipment, etc., which will be recovered during the operation time of the disposal site through the avoided CO_2 emission costs (tax or tradable permits), while nuclear reactor operators can set aside a small fraction of their per kWh sales revenues for establishing the ultimate disposal site at a later time.

A broader economic aspect in which the management of CO_2 and RW become similar with the advent of CCD is the internalizing of the costs. This has largely been the case for RW, while CCD involves bringing home in two ways what has so far been a global externality: economically, by paying for the costs of separating CO_2 from the biogeochemical cycle and keeping it away from the atmosphere, and geographically, by keeping the waste within or relatively close to the region of its origin.

At the boundary between economics and law the question arises as to the ownership of the underground space in which these waste products will be disposed of. Some legal systems (e.g. that of the USA) grant property rights (including the right to extract resources) to the owner of the surface area. In most cases, however, the underground space is in public (government) ownership. In either case, securing the right to use this space for disposal involves contentious issues. The case of CO_2 is somewhat more complicated because it can migrate underground to large distances from the injection wells, depending on the geological formation, while RW will stay at the location of the engineered barrier system for a thousand years or longer.

Another legal issue is liability. With the introduction of the geological disposal of CO_2, the fossil power industry enters new legal terrain on account of the need to deal with the liability associated with the CO_2 disposal sites for possibly hundreds of years. The final solution for the extremely long liability period is likely to be similar in both domains: transfer of responsibility and liability to a state or government entity. The nature and magnitude of the payment by the operator of the disposal sites for the virtually infinite public liability will need to be resolved in both cases.

The lack of public acceptance or outright public opposition can prevent the implementation of any project irrespective of the actual and proven risks and benefits. Energy infrastructure, industrial sites and hazardous material are particularly exposed to the vagaries of public sentiments that can be easily manipulated by interest groups whose stakes or political agendas are at odds with the proposed project. These tendencies have long been observed for RW and are emerging for CO_2 as well. An unequivocal similarity between fossil electricity with CCD and nuclear power with RW disposal is that both are condemned and campaigned against by most environmental non-governmental organizations (NGOs).

The long struggle and many failures in various countries in earlier attempts to search for, characterize and select sites for RW repositories, and the experience from more recent and successful site selection procedures, could be a valuable source of information for those working on CO_2 disposal. The importance of openness and transparency, public information and public participation during not only site selection but all phases of decision making during RW disposal programmes cannot be overemphasized. The experience with such procedures could well be beneficial for all phases of CO_2 disposal programmes (capture facilities, transport routes, disposal sites). At this stage it is not clear what will be easier, organizing information campaigns and public dialogues to foster public acceptance at a few potential RW disposal sites or in many potentially affected communities for large-scale CO_2 disposal programmes.

A related issue is the possible link between liability, compensation and public acceptance relevant for both CO_2 and RW disposal. Willingness to accept (WTA) studies in economics indicate that people are willing to accept some level of environmental menace if they feel properly compensated. The unresolved question is whether very large compensation schemes would really increase public acceptance or not. Astronomic compensation schemes might lead to diverging public reactions. They might increase trust ('there must be a very high level of confidence that nothing will go wrong') or might undermine it ('it will be such a big disaster that no one will be left to compensate or to be compensated'). This is possibly a cultural issue that cannot be resolved in a general way.

Even if CO_2 and RW disposal are demonstrated to be safe, economically efficient (in terms of preserving the economic competitiveness of the related energy technology) and acceptable to the current generation, there are still some concerns that could be raised and should be discussed from the perspective of environmental ethics. Intergenerational equity and the concept of sustainability imply two important principles: first, the present generation should properly take care of its wastes and not leave them and the resulting burden to future generations; second, the present generation should leave all options (including technologies) open for future generations to the largest possible extent. In addition to other concerns, opposition by environmentalists against nuclear power and fossil-based electricity stems to a large extent from the alleged inability of the nuclear industry to dispose of RW safely and on the fossil fuel industry's dumping its CO_2 into the atmosphere and both thus potentially harming future generations. However, they ignore the value future generations might attach to the availability of these technological options for serving their own energy needs.

One option to be considered for reducing the risk of geological disposal in the case of both CO_2 and RW is siting disposal facilities in distant, possibly unpopulated, areas. Although long distance transport of electricity is possible, it is practical to have power plants relatively close to large population centres. This will involve transport of CO_2 and RW to the disposal sites. Transport of both substances is technically feasible. It seems to be easier and less expensive for the relatively small volume of RW to go by road, rail or sea. CO_2 will need pipelines, possibly with boosters, and this might become a more significant cost factor. Multinational (i.e. joint) disposal sites shared by small countries would make a lot of sense economically for both substances, especially for RW from countries with few nuclear reactors, high population density or an unsuitable environment for disposal, but they may prove politically impossible.

A possibly serious disturbance that might affect both CO_2 and RW disposal is 'remote infection', where remote can be just a few hundred kilometers or continents away. As examples of nuclear power accidents or, more recently, the offshore oil industry disaster in the Gulf of Mexico (spill after the explosion of a drilling platform) indicate, remote events can trigger profound changes in policy, regulation, public acceptance and other conditions anywhere in the world. Distant incidents might lead to much more stringent safety standards (irrespective of whether they are justified under the local conditions) with severe cost implications. The nuclear industry, and thus RW, has long been globalized in this respect. CO_2 disposal might be more heavily exposed to the risk of remote infections because dozens to hundreds of sites will be established and operated in a country compared to one or at most two RW disposal facilities.

Another important similarity between CO_2 and RW disposal is the prominent role of international coordination. In connection with RW disposal, the International Atomic Energy Agency (IAEA) has been supporting its Member States and the international community through scientific and technical information (IAEA 1989, 2007, 2009) and management and safety guides (IAEA 2006, 2008a, b). Work on CCD has become an increasingly important area of activity of the IEA Greenhouse Gas R&D Programme, which is an international collaborative research programme established as an Implementing Agreement under the IEA. Workshops, conferences and web-based seminars provide forums for information exchange; general and technical publications serve the CCD community (see, for example, IEA GHG 2007, 2008, 2009).

In relation to international climate change negotiations under the United Nations Framework Convention on Climate Change (UNFCCC), the two technologies seem to face the same problem. In the Marrakesh Accords (specifying the detailed rules for the implementation of the Kyoto Protocol), nuclear energy was excluded as a GHG mitigation technology eligible to earn Certified Emission Reductions (CERs) in connection with international mitigation activities like the Clean Development Mechanism (CDM) or Joint Implementation (JI). No explicit exclusion exists for CCD but it is not a recognized technology either. Negotiations texts discussed in 2009–2010 list various options regarding the role of both nuclear energy and CCD in the flexibility mechanisms, ranging from exclusion to full recognition. De Coninck (2008) presents the diversity of stakeholders' convictions about CCD that influence the outcome of the negotiations. Although scientific assessments by the IPCC (2007b) and the IEA (2008a, b) as well as others clearly demonstrate the importance of both technologies in climate change mitigation, the outcome of the negotiations concerning their inclusion in flexibility mechanisms under future protocols to the UNFCCC is difficult to predict.

4 Comparative Assessments Across Themes and Regions

This section provides a succinct overview of the chapters that follow and indicates the logic behind the order in which they are arranged. This overview is explicitly not intended to steal the thunder by presenting results of individual chapters. They are all worth reading for their own merit. It is hoped that this summary will be useful for readers by providing an overall framework and some background information about each paper.

4.1 Thematic Chapters

The first part of the book explores selected aspects of the geological disposal of CO_2 and RW. It is remarkable to note the number of disciplines that are needed to contribute towards resolving the issues associated with the various steps of the disposal process, from early site exploration to post-closure liability regulation.

The starting point for the comparative assessment across the many complex issues involved in geological disposal is geology itself. The bedrock for the whole book is the chapter by Bachu and McEwen (2011). They provide a superb overview of the issues to be taken into account when searching for appropriate geological formations for disposing of CO_2 and RW. They start with the properties of these waste materials and compare the resulting essential requirements for the geological media and the emplacement as well as the impacts of emplacement on the geological environment and the related site activities. This chapter is not only an excellent comparative study, it is also valuable for scientists working on specific issues of CO_2 or RW disposal but who would like to have a state-of-the-art overview of the broad range of relevant geological topics. The comparison table developed by the authors has served as a starting point for many regional chapters.

Once the deep geological factors have been clarified, the next step is to assess possible implications for humans and the environment near and above the surface. Numerous environmental issues arise during the operation of the disposal sites for CO_2 and RW, and after these are closed. They will need to be taken into consideration in selecting, designing, establishing and closing the sites. West et al. (2011) consider the main environmental and human health hazards, their essential features and impact mechanisms. The different properties of CO_2 and RW give rise to rather different kinds of hazards; however, the authors identify interesting similarities in the approaches to addressing the related environmental issues.

Addressing the environmental and human health risks properly requires their in-depth assessment and management. Maul (2011) explores the related methodological issues in these two fields and compares them in the context of the risk assessment process, from basic principles to analysing uncertainties by using scenarios and conceptual models. He observes that many tools developed for risk assessment in RW disposal, especially generic databases and computer models, can likewise be used for assessing the risk involved in CO_2 disposal.

In order to minimize the potential for detrimental health and environmental impacts and to support the pertinent risk management and remediation activities outlined in the chapters discussed above, extensive monitoring is required. Monitoring activities track changes in the geological media and follow the fate of the disposed material from site selection through operation to long after the closure of the disposal facilities. Brunskill and Wilson (2011) provide an overview of the applicable monitoring methods for CO_2 and RW disposal, compare their relevant aspects and provide examples of the adoption of certain methods employed in one domain for use in the other.

A major step between capturing CO_2 at the power plant or storing RW at temporary facilities and their geological disposal is transport, which can involve long distances in both cases. Gómez and Tyacke (2011) present the transport systems for transferring CO_2 and RW to the disposal site. The profound differences in the volume and key properties of these materials require completely different transport techniques (pipelines for CO_2; rail, ship or truck for RW), thus the associated safety standards also differ. Yet there are some commonalities as well: concerns about routing, the need for a clear regulatory framework and public perception of the transport-related risks, and thus the acceptance of the transport schemes, are examples of these.

Establishing the disposal sites involves rather different kinds of engineering activities for CO_2 and RW. The former involves deep-wellbore technologies with a long history of technological development in the oil and gas industry, whereas the latter can rely on an even longer history and experience in mining. Tshibangu and Descamps (2011) explore these aspects. Given the differences in the required properties of the geological media and in the volume and properties of the waste material, the comparative analysis mostly reveals obvious differences in site engineering but also finds some interesting similarities.

Suitable geological formations, reassuring risk assessment results and monitoring concepts, safe transport and site engineering schemes are all important prerequisites for geological disposal of both CO_2 and RW. Whether and to what extent it will be used also depends on the costs and the resulting competiveness of the electricity generated. Toth and Miketa (2011) present an overview of recent disposal cost estimates for CO_2 and RW and analyse the repercussions of the disposal costs on the total electricity costs. Their results indicate that the costs of RW disposal amount to a small fraction of the cost of electricity and in many countries have long been considered in the costing of nuclear power in one way or another while CO_2 disposal cost is a new element in costing fossil fuel-based electricity and, together with capture and transport, can increase the total electricity cost significantly. The alternative to CCD is continued CO_2 emission and either paying the applicable carbon tax or buying emission permits, both of which also lead to an increase in power cost.

A diverse range of legal and regulatory issues arise in the geological disposal of both CO_2 and RW. One of the major concerns, liability issues, is addressed by Wilson and Bergan (2011). The authors take case studies from several countries on managing liability for RW, on the one hand, and compare the current proposals regarding liability for CO_2 in the USA and the European Union, on the other. They present a matrix of seven liability-related questions and pertinent features of geological disposal for analysing similarities and differences between the CO_2 and RW cases. The key similarity is the following: owing to the very long time horizons (a few hundred to tens of thousands of years), industry and government will be jointly responsible for managing liability over the short term but liability will eventually be transferred to the government over the long term.

The obvious and considerable risks involved in the disposal of CO_2 and RW make public acceptance a particularly sensitive issue. In their chapter, which takes the form of a detailed overview rather than a systematic comparison, Reiner and Nuttall (2011) identify many factors that influence public perception. They take a closer look at the drivers of public acceptance related to CO_2 and RW disposal and conclude that it is difficult to separate the perception of disposal risks from the fears engendered by the images and associations of the related technologies, like power plant accidents and nuclear weapons in the case of RW.

While it is relatively easy to solicit and measure the views of the current generation, the situation is much more difficult when today's actions have implications for future generations over a very long time horizon. Brown (2011) explores the ethical principles of intergenerational equity involved in the conundrum of changing the Earth's climate by emitting CO_2 versus mitigating CO_2 emissions but leaving behind CO_2 and/or RW in geological formations. The ethical dilemmas are complicated by the state of science and the magnitude of uncertainties associated with the various options because improving knowledge and reduced uncertainties can change the ethical conclusions within the same ethical framework while the same level of knowledge and uncertainty can lead to different conclusions in different ethical frameworks.

Ethical considerations are one of the psychological factors determining people's perceptions and eventual acceptance of a technology. De Groot and Steg (2011) analyse five psychological factors driving acceptability judgements: the dreaded or unknown character of the technology in question, the related affect, the moral aspects, fairness and trust. The relative importance of these factors varies somewhat between CO_2 and RW, the latter being a better known substance but its disposal technology less known. The authors argue that reasoning and understanding play an important role; therefore it is possible to influence and improve acceptability by public information campaigns that deliver clear and objective information about the risks. However, it seems to be more difficult to overcome emotional barriers stemming from hunch-based attitudes.

4.2 Regional Chapters

We are also seeking here to learn lessons from a series of case studies that look at geological disposal of CO_2 and RW in a regional or national context. The second part of the book presents comparative assessments for selected regions. The early availability of some of the thematic chapters, particularly that on geological foundations by Bachu and McEwen (2011), was very helpful in preparing some of the regional case studies. The thematic chapters summarized above present comparative assessments in general; the regional chapters focus on region-specific issues, particularly the prevailing geological and environmental conditions relevant for CO_2 and RW disposal. They also highlight socioeconomic issues (economic, legal, public acceptance, etc.) to the extent that these aspects have already been addressed in a given country or region.

We start our world tour in North America and proceed eastward. Oldenburg and Birkholzer (2011) review the current status of CO_2 and RW disposal in the USA and Canada. Their comparative analysis surveys the targeted geological formations in this region and observes that the disposal of both CO_2 and RW is believed to be technically feasible. The authors also look at the opportunities identified and the remaining challenges within a comparative framework.

There have been long-established RW disposal programmes for decades in many Western European countries while CCD research projects emerged more recently. The European Commission supports and coordinates research in both areas. Toth et al. (2011) investigate three large countries in the region that have significant shares of both nuclear and fossil electricity in their national generation mixes: Germany, France and the UK. They focus on the comparative analyses between CO_2 and RW within the three countries and intentionally avoid the comparison of CO_2 and RW programmes across the three countries, this being beyond the scope of this book.

Several countries in Eastern Europe also rely on a combination of nuclear and fossil sources for their electricity generation. The search for RW disposal solutions has been going on for some time at varying levels of intensity in most of these countries. CCD is being increasingly considered as well because these countries are listed in Annex I of the UNFCCC (United Nations 1992), and as such they are obliged to reduce their GHG emissions, although their mitigation commitments under the Kyoto Protocol are well above their current emissions. Hódossyné Hauszmann et al. (2011) present a regional overview across eight countries in Central and Eastern Europe. Their analysis highlights the challenges that small countries are facing in the geological disposal of both CO_2 and RW.

The next country on the journey towards the east is the Russian Federation. Cherepovitsyn and Ilinsky (2011) focus on the European part of the country, where most of the population, economic activities and energy use are located. The authors observe that the disposal of RW accumulated from civilian and military nuclear programmes is an increasingly pressing task, and identify several suitable formations in the region under consideration. Since GHG emissions in the Russian Federation are also well below the Kyoto Protocol commitment, CO_2 disposal is less urgent, on account of which the authors concentrate on lucrative options like enhanced oil recovery (EOR).

As the world's largest CO_2 emitter, albeit without any legally binding mitigation commitment so far, and also a country with a very ambitious nuclear power expansion programme, China is a particularly interesting case for comparing the geological disposal of CO_2 and RW. Wang and Pang (2011) point out that currently both substances are considered as a resource in China: CO_2 for EOR, enhanced gas recovery (EGR) and enhanced coalbed methane (ECBM) recovery, and spent nuclear fuel for its uranium and plutonium content accessible by reprocessing. The search for possible disposal sites seems to follow similar patterns for both CO_2 and RW, from national screening to gradually zooming in on promising areas and then increasing the depth of the investigation. The authors compare a broad range of issues involved in geological disposal in the Chinese context.

Any underground activity creates special challenges in a region close to plate boundaries, crustal movements and the resulting active faults. The preference for tectonic stability for the disposal of both CO_2 and RW appears to be difficult to satisfy in such cases. Koide and Kusunose (2011) summarize relevant elements of the immense knowledge base accumulated in Japan about the long-term stability of the various formations and the prediction of crustal movement relevant for RW disposal as well as the impact of large-scale CO_2 disposal on the geological environment under such circumstances. They draw on an impressive diversity of studies and experiments. They present solid scientific foundations for analysing geological formations with regard to their suitability for disposal of CO_2 and RW, but they also raise a series of open questions and uncertainties due to the complex geological characteristics of the Japanese archipelago.

The next step in our journey takes us to the southern hemisphere. Surridge et al. (2011) describe the current status and future plans for the geological disposal of CO_2 and RW in South Africa. This country is characterized by the common problem of countries with only a few nuclear power plants: the small amount of RW that accumulates even over decades of operation and the high fixed costs of establishing a geological repository makes the latter economically unattractive. Nonetheless, South Africa is also exploring final disposal options while at the same time establishing safe RW storage facilities. As a developing country, it is not yet committed to reducing its GHG emissions but the CCD option is being seriously investigated because of the country's increasing reliance on its abundant and cheap coal resources for power generation and because CO_2 mitigation may be required under future global climate change agreements.

Ending our world tour by returning to the American continent, the study by Heemann et al. (2011) compares the region-specific issues of geological disposal of CO_2 and RW in two Latin American countries utilizing nuclear power in their electricity mix, namely Brazil and Argentina. Brazil is apparently blessed with huge and diverse energy resources: hydropower dominates the electricity sector, abundant uranium reserves have been identified and the CO_2 disposal potential assessed so far is also vast. A reasonably good understanding of suitable CO_2 and RW disposal options seems to be emerging in both countries.

The regional assessments indicate the availability of huge geological capacities for CO_2 disposal and also of suitable geological formations for RW disposal in several regions (North America, Latin America, South Africa, Russian Federation). This allows a great deal of flexibility in choosing the most suitable energy sources based on other important decision criteria specified for the energy strategies of these regions.

5 Concluding Remarks

Fossil fuel-based electricity and nuclear power remain two key energy supply technologies to satisfy the fast increasing energy demand under increasing GHG emissions constraints and other energy policy concerns. This book is the first attempt to provide a comprehensive comparative assessment of these two technologies, explore their relative merits and shortcomings and identify opportunities for learning and transferring experience between them.

We have shown here that there are several reasons originating in scientific research and technological development that make such a comparative assessment a meaningful and promising exercise. The value of the insights gained from evaluating the two technologies in a comparative framework is also obvious for policymakers.

The thematic chapters in this book indicate that the balance of similarities and differences as well as the mutual learning opportunities vary across the topical areas. More differences have been detected in the domains of the natural and environmental sciences like geology and environmental impacts as well as in engineering. The number of similar features is higher in the areas primarily addressed by the social sciences, like public acceptance, legal and liability issues, etc. The regional chapters demonstrate that the relative importance of these similarities and differences varies depending on the broader context and the prevailing geological, geographical and socioeconomic conditions of a given country or region.

The benefits and drawbacks of introducing/expanding nuclear power as well as of continued reliance on fossil energy sources with CO_2 capture need to be systematically assessed with a view to the geological disposal of the waste products (RW and CO_2) across a wide range of issues and against numerous criteria in order to make informed choices. Such assessments require input from a large and diverse array of scientific disciplines as well as innovative approaches to integrate the disciplinary findings for decision making. The comparative assessments presented in this book represent a first but hopefully useful step in this process.

Acknowledgements I am indebted to the authors of this book for their commitment and creative thinking in engaging in this first-of-a-kind exercise and for their patience in the long production process. I am also grateful to more than 70 reviewers who devoted their time to help the authors improve the chapters. The book has also benefited greatly from comments received from colleagues at the IAEA: Hans-Holger Rogner, Irej Jalal, Romain Boniface, Hans Forsström, Paul Degnan, Asami Miketa, Jan-Marie Potier and Bernard Neerdael. Special thanks are due to Vivien Landauer for her unrelenting commitment to perfection in the many rounds of editorial corrections and improvements.

References

Bachu S, McEwen T (2011) Geological media and factors for the long-term emplacement and isolation of carbon dioxide and radioactive waste. In this volume

Brown DA (2011) Comparative ethical issues entailed in the geological disposal of radioactive waste and carbon dioxide in the light of climate change. In this volume

Brunskill B, Wilson M (2011) Monitoring methods used to identify the migration of carbon dioxide and radionuclides in the geosphere. In this volume

Cherepovitsyn A, Ilinsky A (2011) Comparison of the geological disposal of carbon dioxide and radioactive waste in European Russia. In this volume

de Coninck H (2008) Trojan horse or horn of plenty? Reflections on allowing CCS in the CDM. Energ Policy 36:929–936

de Groot JIM, Steg L (2011) Psychological perspectives on the geological disposal of radioactive waste and carbon dioxide. In this volume

Gómez DR, Tyacke M (2011) Transport of carbon dioxide and radioactive waste. In this volume
Heemann R, Ketzer JM, Hiromoto G, Scislewski A (2011) Assessment of the geological disposal of carbon dioxide and radioactive waste in Brazil, and some comparative aspects of their disposal in Argentina. In this volume
Hódossyné Hauszmann Z, Scholtz P, Falus Gy (2011) Carbon dioxide and radioactive waste in Central and Eastern Europe: a regional overview of geological storage and disposal potential. In this volume
IAEA (International Atomic Energy Agency) (1989) Safety Principles and Technical Criteria for the Underground Disposal of High Level Radioactive Wastes, vol 99, IAEA Safety. IAEA, Vienna
IAEA (International Atomic Energy Agency) (2006) Geological Disposal of Radioactive Waste, vol WS-R-4, IAEA Safety Standards. IAEA, Vienna
IAEA (International Atomic Energy Agency) (2007) IAEA Safety Glossary: Terminology Used in Nuclear Safety and Radiation Protection: 2007 edition. IAEA, Vienna
IAEA (International Atomic Energy Agency) (2008a) The Management System for the Disposal of Radioactive Waste, vol GS-G-3.4, Safety Standards. IAEA, Vienna
IAEA (International Atomic Energy Agency) (2008b) The Management System for the Processing, Handling and Storage of Radioactive Waste, vol GS-G-3.3, Safety Standards. IAEA, Vienna
IAEA (International Atomic Energy Agency) (2009) Classification of Radioactive Waste, vol GSG-1, IAEA Safety Standards. IAEA, Vienna
IEA (International Energy Agency) (2008a) World Energy Outlook 2008. OECD/IEA, Paris
IEA (International Energy Agency) (2008b) Energy Technology Perspectives 2008: Scenarios & Strategies to 2050. OECD/IEA, Paris
IEA GHG (International Energy Agency Greenhouse Gas R&D Programme) (2007) Storing CO_2 Underground. IEA Greenhouse Gas R&D Programme, Cheltenham
IEA GHG (International Energy Agency Greenhouse Gas R&D Programme) (2008) Geologic Storage of Carbon Dioxide: Staying Safely Underground. IEA Greenhouse Gas R&D Programme, Cheltenham
IEA GHG (International Energy Agency Greenhouse Gas R&D Programme) (2009) Natural and Industrial Analogues for Geological Storage of Carbon Dioxide. IEA Greenhouse Gas R&D Programme, Cheltenham
IPCC (Intergovernmental Panel on Climate Change) (2005) IPCC Special Report on Carbon Dioxide Capture and Storage. Metz B, Davidson OR, de Coninck HC, Loos M, Meyer LA (eds) Cambridge University Press, New York/Cambridge, UK
IPCC (Intergovernmental Panel on Climate Change) (2007a) Climate change 2007: synthesis report. Pachauri RK, Reisinger A (eds) Cambridge University Press, New York/Cambridge, UK
IPCC (Intergovernmental Panel on Climate Change) (2007b) Climate change 2007: mitigation. Contribution of Working Group III to the Fourth Assessment Report of the Intergovernmental Panel on Climate Change. Metz B, Davidson OR, Bosch PR, Dave R, Meyer LA (eds) Cambridge University Press, New York/Cambridge, UK
Koide H, Kusunose K (2011) Geological disposal of carbon dioxide and radioactive waste in the geotectonically active country of Japan. In this volume
Marsh GP (1874) The Earth as Modified by Human Action. Scribner, New York
Maul P (2011) Risk assessment, risk management and remediation for the geological disposal of radioactive waste and storage of carbon dioxide. In this volume
Oldenburg CM, Birkholzer JT (2011) Comparative assessment of status and opportunities for carbon dioxide capture and storage and radioactive waste disposal in North America. In this volume
Reiner DM, Nuttall WJ (2011) Public acceptance of geological disposal of carbon dioxide and radioactive waste: similarities and differences. In this volume
Surridge AD, Cloete M, Lloyd PJ (2011) The geological storage of carbon dioxide and disposal of nuclear waste in South Africa. In this volume
Toth FL, Miketa A (2011) The costs of the geological disposal of carbon dioxide and radioactive waste. In this volume
Toth FL, Roehrl RA, Miketa A, Barkatullah N (2011) Comparing the geological disposal of carbon dioxide and radioactive waste in Western Europe. In this volume

Tshibangu KJ-P, Descamps F (2011) Engineering challenges in the geological disposal of radioactive waste and carbon dioxide. In this volume
Turner BL II, Clark WC, Kates RW, Richards JF, Mathews JT, Meyer WB (eds) (1990) The Earth as Transformed by Human Action: Global and Regional Change in the Biosphere over the Past 300 Years. Cambridge University Press, New York
UN DESA (United Nations Department of Economic and Social Affairs, Population Division) (2009) World population prospects: the 2008 revision. http://esa.un.org/unpd/wpp2008/peps_documents.htm
UNDP (United Nations Development Programme) (2005) Energizing the Millennium Development Goals: A Guide to Energy's Role in Reducing Poverty. UNDP, New York
United Nations (1992) United Nations Framework Convention on Climate Change. United Nations, New York
Wang J, Pang Z (2011) Comparison between geological disposal of carbon dioxide and radioactive waste in China. In this volume
West JM, Shaw RP, Pearce JM (2011) Environmental issues in the geological disposal of carbon dioxide and radioactive waste. In this volume
Wilson EJ, Bergan S (2011) Managing liability: comparing radioactive waste disposal and carbon dioxide storage. In this volume
World Bank (2009a) Global Development Finance: Charting a Global Recovery, vol I, Review, analysis, and outlook. The World Bank, Washington, DC
World Bank (2009b) Global Economic Prospects 2009: Commodities at the Crossroads. The World Bank, Washington, DC

Part I
Thematic Assessments

Geological Media and Factors for the Long-Term Emplacement and Isolation of Carbon Dioxide and Radioactive Waste

Stefan Bachu and Tim McEwen

Abstract A review is presented of the factors considered important in the selection of environments and sites for the geological storage of carbon dioxide (CO_2) and the disposal of radioactive waste (RW)—with a focus on those of a geological nature. The distinction between the terms *storage* for CO_2 and *disposal* for RW is not significant in this regard. The relevant properties of the two product types are presented, as are the desirable characteristics and types of geological environments that are considered suitable for disposal purposes. The role that the geological barrier plays in trapping the disposed substance, in the case of CO_2, and in containing and slowly releasing the waste, in the case of RW, is explained. The comparative roles played by the geological barrier and the engineered barrier system of a repository for RW is also outlined—although the emphasis of the discussion is on the geological barrier itself. The status and challenges associated with the storage of CO_2 are presented, together with a discussion of the geographic distribution of areas of the world potentially suitable for its storage and the criteria for site selection that could be applied. A discussion is also presented of the geological environments that are most likely to be used for the disposal of RW.

A considerable part of the chapter presents a comparison between the storage or disposal of the two types of disposed substances, discussing their similarities and differences. This comparison is considered under the four subject headings: Characteristics of the Geological Media, Emplacement Characteristics, Effects of Emplacement and Potential Migration from the Disposal Site, and Site Activities.

Keywords Radioactive waste disposal • CO_2 storage • Trapping mechanisms • Migration mechanisms and pathways • Repository • Geological/natural barrier • Engineered barrier system

S. Bachu (✉)
Alberta Innovates—Technology Futures, Edmonton, Canada
e-mail: stefan.bachu@albertainnovates.ca

T. McEwen
McEwen Consulting, Melton Mowbray, United Kingdom
e-mail: tmcewen@globalnet.co.uk

F.L. Toth (ed.), *Geological Disposal of Carbon Dioxide and Radioactive Waste: A Comparative Assessment*, Advances in Global Change Research 44,
DOI 10.1007/978-90-481-8712-6_2, © International Atomic Energy Agency, 2011

1 Why Geological Storage of CO_2 and Disposal of Radioactive Waste?

1.1 Introduction

The emplacement in geological media of radioactive waste (RW) and carbon dioxide (CO_2) is considered to be a safe method for isolating these substances from the hydrosphere, the atmosphere and the biosphere. The disposal of long-lived RW, e.g. spent fuel (SF), long-lived intermediate-level waste (ILW-LL), etc., currently takes place at only one location, at the Waste Isolation Pilot Plant (WIPP) in the USA, although plans to dispose of SF are well advanced in several countries and the disposal of this type of waste is likely to be taking place at several sites over the next few decades. Investigations and research programmes concerning the disposal of RW have, however, been taking place since the 1970s or 1980s in many countries. The recent increased interest in the use of nuclear power for electricity generation has provided a greater focus on developing long-term management solutions for the waste that is inevitably produced.

In contrast, the storage of CO_2 is a relatively recent consideration, although there has been injection of approximately one million tonnes per year of CO_2 at Sleipner in the North Sea since the mid-1990s and, similarly, at In Salah in Algeria since the mid-2000s. The importance of CO_2 storage has risen rapidly up the political agenda over the last decade as representing a climate change mitigation strategy with significant potential, in particular as the significance of the effects of global warming has been appreciated.

1.2 Carbon Dioxide

The widening gap, on the one hand, between the increase in CO_2 emissions due to the expected increase in population, global standards of living and carbon intensity of the energy system, and, on the other hand, the decrease in CO_2 emissions due to the increase in energy efficiency and conservation, can be partially or totally covered by artificially increasing the capacity and uptake rate of CO_2 sinks through CO_2 storage or sequestration. This involves either the diffuse removal of CO_2 from the atmosphere after its release through terrestrial and marine photosynthesis, with subsequent storage of the carbon-rich biomass (natural sinks), or the capture of CO_2 emissions prior to their potential release and their storage in deep oceans or geological media, or through surface mineral carbonation (known collectively as carbon capture and storage, or CCS).

In contrast to natural sinks, CCS is a process that consists of separating and capturing CO_2 from large stationary sources, transporting it to a storage site, and isolating it from the atmosphere for very long periods of time, in the order of several centuries to millions of years. Three processes have been considered: surface mineral carbonation, ocean storage and geological storage (IPCC 2005). Surface mineral carbonation consists of converting CO_2 into solid, inorganic carbonates by chemical reactions, but requires the use of certain minerals such as olivine and serpentine, mining on a large scale, large amounts of energy for crushing, milling and heating the minerals, and the transportation and disposal of very large amounts of the resulting carbonate rock, thus excluding this process as a viable option for reducing atmospheric CO_2 emissions (IPCC 2005). Ocean storage consists of injecting CO_2 at great depths, where it will dissolve or form hydrates or heavier-than-water plumes that will sink to the bottom of the ocean (Aya et al. 1999), thus removing CO_2 from the atmosphere for several hundreds of years. However, ocean CO_2 storage would result in a measurable change in ocean chemistry, with corresponding consequences for marine life (IPCC 2005), notwithstanding issues of ocean circulation, storage efficiency, technology, cost, technical feasibility, international limitations regarding dumping at sea, and strong public opposition.

Geological storage of CO_2 thus currently represents the best and likely only short- to medium-term option for significantly enhancing CO_2 sinks. The technology exists today and can be applied immediately, being based on experience to date from the oil and gas industry, from the deep disposal of liquid wastes and from water resources management (IPCC 2005), and is forecasted to play an important role in reducing anthropogenic CO_2 emissions into the atmosphere in the first part of this century and beyond (IEA 2004, 2006). The storage of CO_2 in geological media shares many similar features with oil and gas accumulations in hydrocarbon reservoirs and methane in coalbeds, whilst the capture, transportation, injection and monitoring of CO_2 in the subsurface has already been practised for a few decades in enhanced oil recovery, acid gas disposal and CO_2 storage (IPCC 2005). However, although the individual components of this technology all exist separately, they have not yet been implemented on a large scale in an integrated system because of significant challenges and barriers of an economic, legal and regulatory nature and due to public attitudes to large-scale deployment (Bachu 2008a).

Although various climate change mitigation options have different spatial and temporal ranges of applicability and timing of deployment, it is clear that the reduction in atmospheric CO_2 emissions needed for stabilizing the climate can be achieved through the application of a portfolio of measures, which includes energy efficiency and conservation, increasing the share of non-fossil fuel energy sources and carbon capture and storage (Pacala and Socolow 2004; Socolow 2005). The latter could provide 15–43% of the emissions reduction needed to stabilize atmospheric greenhouse gas levels at 550 ppm CO_2 equivalent (Pacala and Socolow 2004), compared to 380 ppm today and 280 ppm in the mid-nineteenth century.

In this context, carbon capture and storage means the removal of CO_2 directly from anthropogenic sources and its emplacement in geological media for long periods of time. From an engineering point of view, this is a geological disposal operation, similar to acid gas (CO_2 and hydrogen sulphide (H_2S)) disposal at more than 70 sites in North America (e.g. Bachu and Gunter 2005) and to other fluid-waste disposal operations, albeit on a much larger scale. However, for various reasons the term CO_2 disposal has been avoided, and various terms have been used historically such as CO_2 removal, CO_2 sequestration and CO_2 storage. The term *CO_2 sequestration* continues to be used preferentially in the USA, where it is defined as the long-term isolation of CO_2 from the atmosphere through physical, chemical, biological or engineered processes. Geological CO_2 sequestration refers specifically to the emplacement of CO_2 deep underground. The term *CO_2 storage* is sanctioned by UN agencies and is used, particularly in Europe, to indicate CO_2 underground emplacement, the term CO_2 sequestration in these countries being reserved for other processes that reduce atmospheric CO_2 emissions. For consistency with the purpose of this book and with other chapters, the term *CO_2 disposal* will be used from now on in this chapter, the meaning, nevertheless, being the same as that of CO_2 sequestration and carbon capture and storage, or CCS, namely the injection of CO_2 into geological media in order to isolate it from the atmosphere and biosphere for long periods of time—at least several centuries to millennia.

1.3 Radioactive Waste

Deep geological disposal (generally at hundreds of metres depth) is the option favoured internationally for the long-term management of heat generating RWs (i.e. SF and high-level waste (HLW)) and RWs with a considerable content of long-lived radionuclides, such as ILW-LL, which produce only negligible amounts of heat. Countries that possess these waste types typically have significant active programmes aimed at developing suitable geological repositories.

Direct experience of the geological disposal of HLW does not yet exist, as the only operating repository is the WIPP in New Mexico, USA, which has been licensed to dispose of transuranic RW (i.e. intermediate-level waste (ILW)) derived from the research and production of nuclear weapons. Several countries' disposal programmes for SF and HLW are, however, nearing fruition: what will be the access route to a repository for SF at Olkiluoto, Finland, is currently under construction; Sweden has recently chosen a preferred site for an SF repository at Forsmark; and France is investigating a potential disposal area on the border of the Departments of Meuse and Haute Marne, around the Bure site where the Underground Research Laboratory is located, to take all wastes not acceptable for surface disposal. In addition to the waste disposal programmes in these and other countries, international organizations such as the International Atomic Energy Agency (IAEA) and the OECD Nuclear Energy Agency (NEA) are contributing towards developing confidence in relevant technologies, approaches and concepts for the geological disposal of RW. These same organizations, in addition to others, such as the European Union, are also supporting international projects on training and demonstration, in line with the general principles defined in the IAEA Safety Fundamentals (IAEA 1995) and with the principle of sustainability. This has been defined by the Brundtland Commission as: 'development that meets the needs of the present without compromising the ability of future generations to meet their own needs' (WCED 1987).

The fundamental principles involved in geological disposal are discussed in, for example, Chapman and McKinley (1987), Savage (1995), Chapman and McCombie (2003) and Alexander and McKinley (2007). A key concept in this disposal is the multi-barrier principle, in which long-term safety is assured by a series of engineered and natural barriers that act in tandem (Fig. 1)—geological repositories are designed to be passively safe. These barriers prevent or reduce the transport of radionuclides in groundwater, which is generally the most important transport mechanism. The barriers may also influence the migration of gas, which will be evolved in RW repositories by chemical and biochemical reactions and by radioactive decay (e.g. Rodwell et al. 2003). For example, some radionuclides (such as ^{14}C) may be transported in the gaseous phase, being subject to many of the same transport processes as CO_2.

The long-term safety of a deep geological repository for RW will be strongly dependent on the performance of the geosphere. The geosphere potentially isolates the RW from possible future intrusions by humans; provides a stable physical and chemical environment for the engineered barriers within the repository, insulating against external perturbations such as earthquakes and climate change; and prevents, delays and attenuates radionuclide transport by virtue of its hydraulic and sorptive properties.

A safety case for a deep geological repository typically makes use of geoscientific information within a long-term safety assessment that evaluates potential impacts. These studies require a conceptual model of the geosphere that quantifies, for instance, groundwater flow rates and consequent radionuclide transport (as, eventually, the RW will come into contact with, and dissolve in, the groundwater—although this process may take place many thousands of years in the future). Geoscientific information can, however, play a larger role in the development of a safety case; in particular, geoscience can offer multiple and independent lines of evidence (both qualitative and quantitative) to support a safety case. Moreover, it can play an important role in other repository activities that bear on safety, such as site selection and repository design.

2 Current Status of CO_2 Disposal in Geological Media

2.1 *Relevant CO_2 Properties*

The concept of disposing of anthropogenic CO_2 by injecting it deep underground is based on the properties and behaviour of CO_2 at the conditions found at depth and on the physical and chemical properties of the rocks. At normal atmospheric conditions CO_2 is an odourless, colourless gas, slightly heavier than air, which is present in the atmosphere at concentrations of ~0.4%. Its density is 1.872 kg/m^3 at standard conditions of temperature and pressure. Like any substance, CO_2 changes phase from gaseous to liquid, solid or supercritical, depending on pressure and temperature (Fig. 2a). At very low temperatures CO_2 is a solid (dry ice), and is used as such

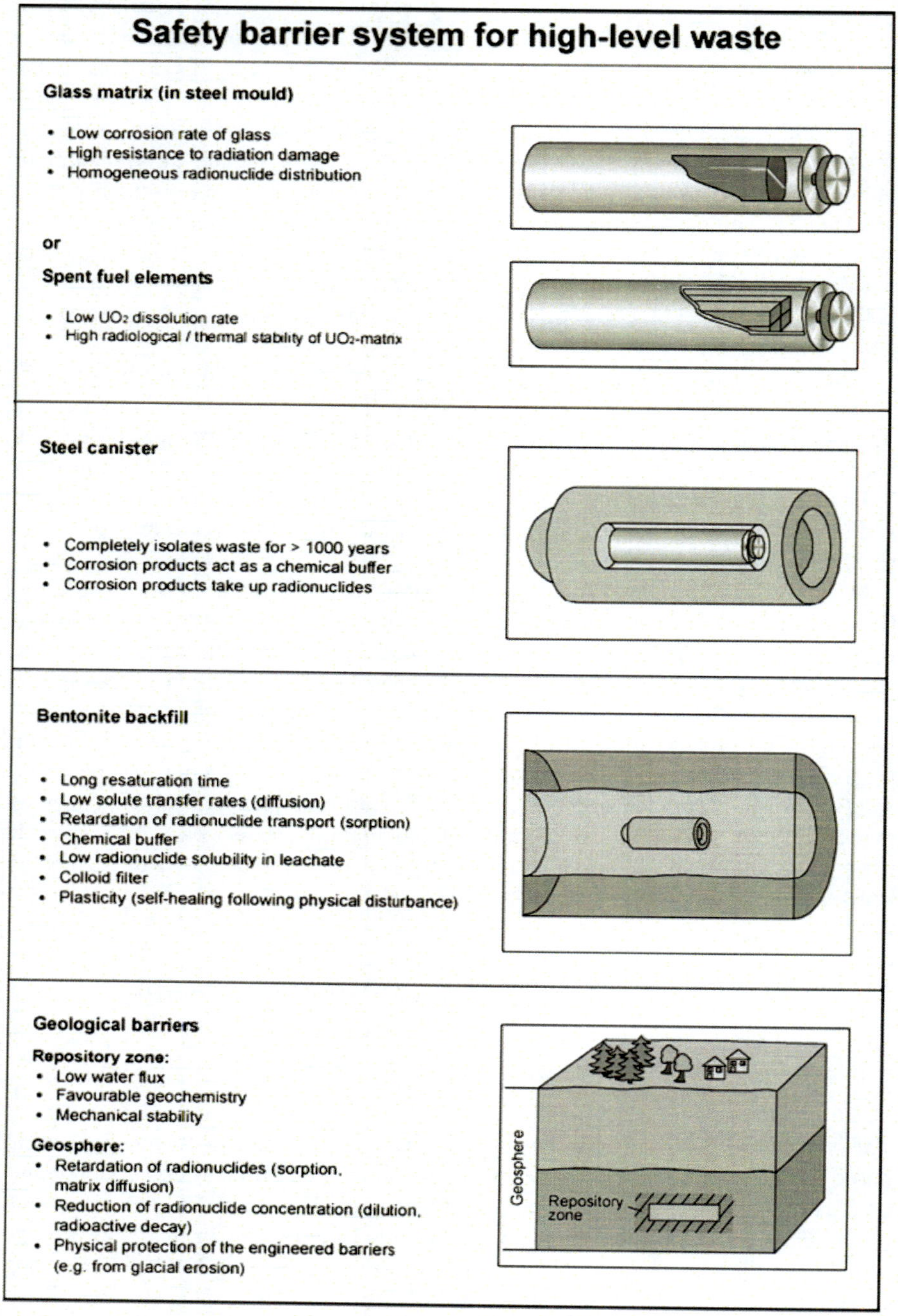

Fig. 1 The safety barriers for high-level waste, based on Nagra's disposal concept for use in Switzerland (From Nagra 2002)

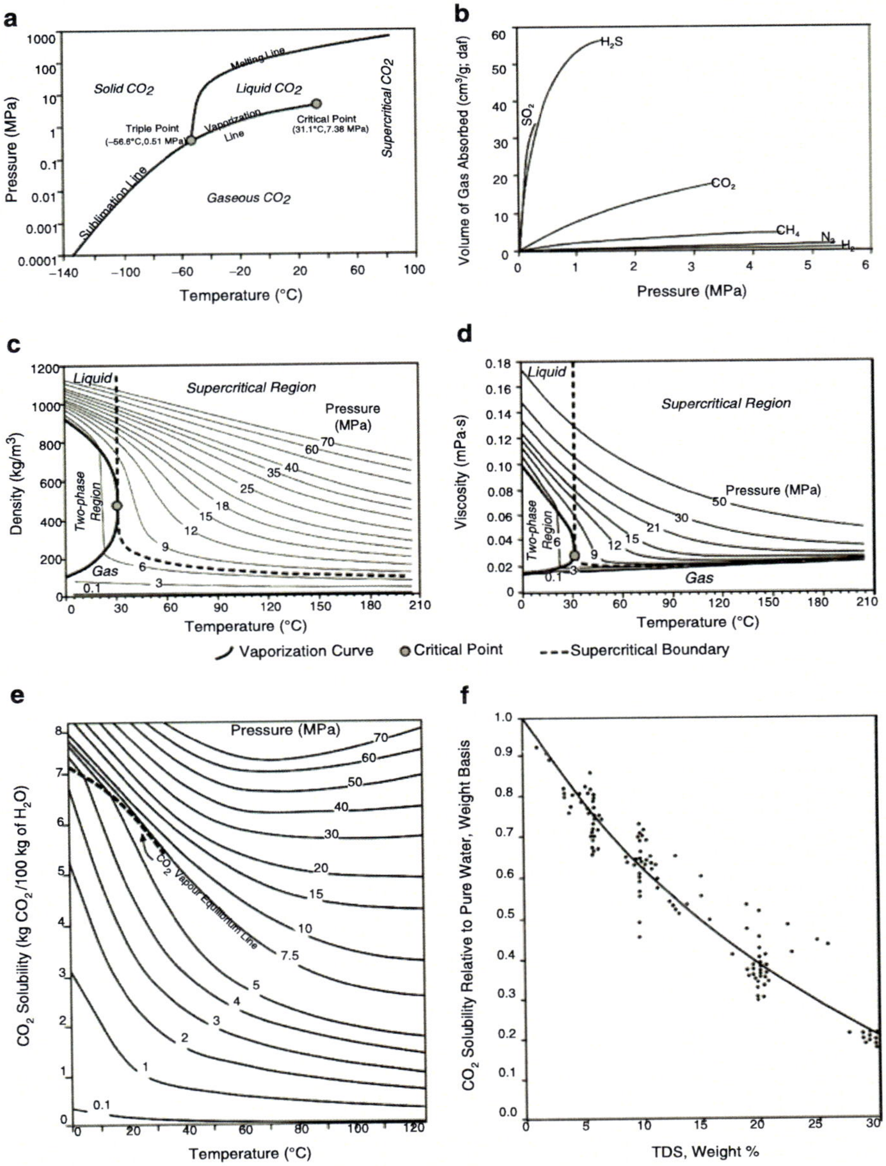

Fig. 2 Relevant CO_2 properties: (**a**) phase diagram; (**b**) adsorption capacity of various gases on coal (From Chikatamarla and Bustin 2003); (**c**) density variation with pressure and temperature (From IPCC 2005); (**d**) viscosity variation with pressure and temperature (From IPCC 2005); (**e**) solubility in water as a function of pressure and temperature (From Kohl and Nielsen 1997); and (**f**) decrease in solubility with increasing water salinity (From Enick and Klara 1990)

in industrial processes. However, except at shallow depths in Arctic and Antarctic regions and at high altitudes, where temperatures may be below 0°C, temperatures in the ground are always greater than zero and increase with depth according to the local geothermal gradient, whose global average is ~30°C/km, but which can vary widely, particularly in areas of active tectonics (e.g. volcanic regions and along the margins of tectonic plates).

At temperatures less than 31.1°C (the critical temperature, T_c) an increase in pressure will result in CO_2 changing phase from gaseous to liquid once it reaches the vaporization line (Fig. 2a). The pressure needed for CO_2 to change phase from gaseous to liquid increases with increasing temperature, reaching 7.38 MPa (the critical pressure, P_c) at the critical temperature, T_c (Fig. 2a). For reference, this pressure is equal to the hydrostatic pressure exerted at the bottom of a column of pure water at a depth of 738 m. For temperatures greater than the critical temperature, gaseous CO_2 becomes supercritical for pressures greater than the critical pressure. The characteristics of a supercritical fluid that are relevant for CO_2 disposal in geological media are that its density is comparable to that of the liquid phase (Fig. 2c) whereas it retains gas-like behaviour by filling the entire volume available and by mixing with other gases according to gas mixing rules. For temperatures below the critical point, CO_2 condensation from gas to liquid across the vaporization line (Fig. 2a) takes place gradually in the so-called 'two-phase' region (Fig. 2c) where the two phases coexist until all the gaseous CO_2 liquefies. The density difference at the vaporization line between gaseous and liquid CO_2 is sharp and significant, although decreasing along the vaporization line (Fig. 2c). For temperatures greater than the critical temperature, the transition from gaseous CO_2 to supercritical and the associated increase in density are gradual (Figs. 2a, c). The viscosity of CO_2, which depends strongly on its density (Fenghour et al. 1998), displays a similar behaviour (Fig. 2d). Notably, in the supercritical region CO_2 viscosity is closer to the viscosity of the gaseous phase than to that of the liquid phase (Fig. 2d).

The significance of this phase behaviour and of the variation of density and viscosity with temperature and pressure can be understood in the context of the increase with depth in the Earth's crust of both pressure and temperature. Broadly, pressure increases hydrostatically with depth (i.e. with a gradient of ~10 kPa/m), although lower (sub-hydrostatic) gradients have been documented, and overpressurized zones have been identified, where pressure gradients approach lithostatic (20 kPa/m and higher). Thus, the increase in pressure with depth would normally lead to a continuous increase in CO_2 density. However, the corresponding increase in temperature associated with the same increase in depth leads to a decrease in density such that, after a significant increase in density with depth in the first few hundreds of metres, at a certain depth the two factors (pressure and temperature) balance each other, leading to a marginal increase in density, a constant value or even a decrease in density, depending on the interplay between mean long-term surface temperature, geothermal gradient and pressure (Bachu 2003). Assuming a hydrostatic pressure gradient, the density of CO_2 would be higher in regions characterized by a low mean long-term surface temperature and low geothermal gradient (up to 800 kg/m^3) than at the same depth (up to only 500 kg/m^3) in a region characterized by a high mean long-term surface temperature and/or high geothermal gradient (i.e. 'cold basin' versus 'warm basin' (Bachu 2003)). Correspondingly, the volume occupied by the same mass of CO_2 emplaced underground at the same depth will be smaller in the 'cold basin' than in the 'warm basin' case.

The void space in rocks at depth, in the form of pores or fractures, is saturated with fluids, the great majority of which is water, with oil and hydrocarbon gases accumulated in oil and gas reservoirs. Many gas reservoirs naturally contain CO_2 in various proportions, with several giant pure CO_2 reservoirs in the USA that are used to produce CO_2 for enhanced oil recovery (Stevens 2005). CO_2 dissolves in water, with its solubility increasing with increasing pressure and decreasing with increasing temperature (Kohl and Nielsen 1997; Fig. 2e); however, the presence of other dissolved substances reduces significantly the CO_2 solubility in water, by a factor of up to 5 (Enick and Klara 1990; Fig. 2f). Once dissolved in water, CO_2 forms a weak carbonic acid that, depending on the mineralogy of the rock, may lead to CO_2 precipitation in the form of carbonate minerals (Gunter et al. 2004). CO_2 has a greater solubility in oil and, depending on oil gravity and reservoir temperature, at pressures greater than a minimum miscibility pressure it mixes with oil (Holm and Josendal 1982). CO_2 mixes with other gases in gas reservoirs and with air in the unsaturated or vadose zone, although in the latter case, being heavier than air, it tends to accumulate at the bottom of the zone (Oldenburg and Unger 2003).

Finally, coal has variable adsorption affinity for various gases, including CO_2 (Chikatamarla and Bustin 2003; Fig. 2b). Coal has a higher affinity for CO_2 than for methane (CH_4) (also a greenhouse gas, which, for a given quantity, has 25 times greater global warming potential than CO_2 over a time horizon of 100 years) by a factor of 2–8, and for nitrogen (N_2), a gas that forms the majority of flue gases in fossil fuel power plants. Conversely, H_2S, found in gas reservoirs, and sulphur oxides (SO_x), found in flue gases, have greater affinities for coal than CO_2 (Fig. 2b). These adsorption properties are important because: (1) injecting CO_2 into coalbeds would replace methane, which should be recovered and used as a clean fossil fuel (it has the lowest carbon/hydrogen ratio), and (2) the CO_2 stream will most likely contain impurities in various proportions, and these, except for N_2, would preferentially adsorb onto the coal surface, with the advantage of retaining toxic substances such as H_2S and SO_x, but with the associated disadvantage of reducing the disposal capacity available for CO_2.

The properties of CO_2 on which its disposal is based are, therefore, its increased density with depth, its solubility in water and oil (with the associated potential mineral reactions) and its higher adsorption affinity onto coal than that of methane.

2.2 *Geological Media for CO_2 Disposal*

CO_2, being a fluid, will be disposed of at depth in rocks via well injection, and will retain its fluid characteristics and ability to flow as long as it does not precipitate as a carbonate mineral or adsorb onto coal. A decrease in pressure in coal will result in CO_2 being desorbed, with subsequent flow through any fractures present. Where it is dissolved in formation water or oil, CO_2 can be transported by the movement of the fluid and may exsolve when pressure and temperature conditions change, thereby regaining its free-phase form and its ability to flow. The geological disposal of CO_2 therefore needs to meet three requirements:

1. *Capacity*: the disposal unit has to have sufficient capacity to receive and retain the intended volume of CO_2;
2. *Injectivity*: which is the ability to inject CO_2 deep into the ground at the rate that it is supplied from the CO_2 source;
3. *Confinement*: if CO_2 is not confined, then, due to its buoyancy (being lighter than water, see Fig. 2c) it will flow upwards, ultimately entering the shallow hydrosphere (including potable groundwater), the biosphere and the atmosphere.

The first condition for CO_2 disposal requires the availability of large volumes of suitable rock (*capacity*). As an example, a coal-fired power plant that emits five million tonnes of CO_2 per year (Mt CO_2/year) would require a disposal volume of 10×10^6 m^3/year, or 0.4 km^3 over 40 years lifetime of emissions and an in situ CO_2 density of 500 kg/m^3. The volumes required for CO_2 disposal can be provided by the pore volume of the rocks or by mined caverns. At a porosity of 10%, the volume of rock needed to store the previously quoted storage volume is 4 km^3. Crystalline and metamorphic rocks have very low porosities unless they are fractured, and only sedimentary rocks, such as sandstones and carbonates, have generally sufficient connected porosity to provide the space needed for CO_2 disposal.

The second condition, *injectivity*, depends on the fluid viscosity and the permeability of the rock. CO_2 is less viscous than water by a factor of 10–20 and much less viscous than oil, which means that it is easier to inject CO_2 than water into the same rock, but, conversely, CO_2 is more mobile and may escape more easily than the other two fluids. Rocks that allow the production or injection of fluids (water, oil, gas) through wells are considered as permeable and, if they are saturated with water, are known as aquifers or, if they contain oil and/or gas, as reservoirs. Such rocks vary from unconsolidated gravel and sands to their lithified equivalents (conglomerates and sandstones) and also include carbonates. Other rocks, such as clays and shales and evaporites (such as halite), generally have such low permeabilities that they are referred to as aquitards or aquicludes in hydrogeology, and form caprocks, because they cap oil and gas reservoirs, impeding the flow of hydrocarbons out of the reservoir.

Capacity and injectivity are not completely independent of each other. Whilst the volumetric capacity, known also as static capacity (i.e. the necessary pore space), may exist, limitations in injection rates due to low injectivity (i.e. maintaining the maximum pressure below a certain limit imposed by safety measures) may reduce the amount of CO_2 that can be safely injected during the active injection period (this actual capacity is referred to as dynamic capacity).

The third condition for CO_2 disposal, *confinement*, requires the existence of impermeable rock units that would impede the upward migration and leakage of the injected CO_2. Sedimentary basins characterized by layered sequences of permeable and impermeable rocks, such as sandstone, carbonate, shale/claystone and evaporite, provide the type of geological environment that might prove suitable. In contrast, crystalline and metamorphic rocks do not meet any of the requirements for CO_2 disposal because of their lack of suitable porosity and permeability. Some volcanic rocks (e.g. basalts) may possess the required porosity and permeability, but generally they lack the necessary confinement properties. Mined caverns in soft or hard rock are also unsuitable for CO_2 disposal for a variety of reasons, including their low capacity (CO_2 would have low density because of the low pressures at the relatively shallow depths of such caverns) and the likely lack of confinement (which would have to be provided by engineered seals). Salt caverns formed via solution mining could allow the necessary pressurization through well injection, and confinement of the CO_2 would be ensured by the low permeability and plastic properties of the salt; however, such caverns would have only relatively low capacities (typically a fraction of 1 Mt CO_2 (Dusseault et al. 2004)), which would be insufficient for their use on the scale needed. Salt caverns mined through solution mining, not through regular shaft and tunnel systems, may, however, be used for the temporary disposal of CO_2, or as a buffer element in a CO_2 collection and distribution (i.e. transportation) system.

The conditions of capacity and injectivity are somewhat flexible, in the sense that some measures can be taken if any of these criteria are not being met. For example, injectivity can be increased by drilling more wells and/or drilling long horizontal wells and/or stimulating the wells whilst maintaining caprock integrity. Or, if capacity is insufficient, either several sites may be considered (e.g. store in the first site whilst the search and/or the preparation for another site is being pursued), or a smaller amount of CO_2 will ultimately be stored. But if the third condition, that of confinement, which basically relates to the safety and security of CO_2 disposal, is not being met, then that site will definitely not be considered and approved.

The above considerations indicate that the vast majority of crystalline, metamorphic and volcanic rocks are not suitable for large-scale CO_2 disposal; in addition, many sedimentary rocks also do not meet all three conditions. Sedimentary rocks that are faulted, folded and fractured generally do not meet the condition of confinement because CO_2 may escape along transmissive faults and fractures. For example, the Rocky Mountains in North America, which were formed by the compression and uplifting of sedimentary strata, are, generally, unsuitable because of their faulted and fractured nature, although storage structures can be found locally (e.g. oil and gas reservoirs in the foothills). Sedimentary basins, preferably with relatively simple geological histories and displaying minimal faulting and with successions containing at least one, if not several, low permeability confining units, are, thus, most likely to be suitable for CO_2 disposal (Bachu 2003, 2010; Bradshaw and Dance 2005; IPCC 2005).

Within sedimentary basins, aquitards and aquicludes (e.g. shales and evaporitic rocks such as salt and anhydrite) do not meet the requirement of injectivity and constitute barriers to the upward migration and leakage of CO_2. For reasons explained in more detail in the next section, the environments most suitable for CO_2 disposal are deep saline aquifers, oil and gas reservoirs and coalbeds.

In contrast to water supply aquifers that are normally relatively shallow, with low groundwater salinities (e.g. water with a salinity of less than 4,000 or 5,000 ppm for protected groundwater), deep saline aquifers are defined here as aquifers whose groundwater salinity makes them unfit for human consumption and that meet the necessary conditions for CO_2 disposal. Groundwater salinity may be in excess of 400,000 ppm, particularly in the vicinity of evaporitic beds (by comparison, seawater has a salinity of ~33,000 ppm), and in some places minerals dissolved in formation water are extracted for industrial purposes. In such cases, the respective aquifers constitute an economic resource that would be sterilized if used for CO_2 disposal.

Oil and gas reservoirs have properties similar to those of confined aquifers (i.e. permeable porous reservoir rocks capped by impermeable strata), but are saturated with hydrocarbon fluids (oil and/or gas) rather than water. The oil and gas would be produced first before any consideration could be given to the disposal of CO_2. In many cases oil and gas reservoirs are underlain by aquifers with which they are in hydraulic communication, and this factor would need to be taken into account. Coalbeds retain CO_2 as a result of a different process, but they too may constitute a resource that could be mined (or in which in situ combustion could be employed), or may represent aquifers by themselves due to their relatively high permeability, in which case they are not suitable for CO_2 disposal. Figure 3 diagrammatically illustrates the geological conditions and emplacement system for CO_2 disposal.

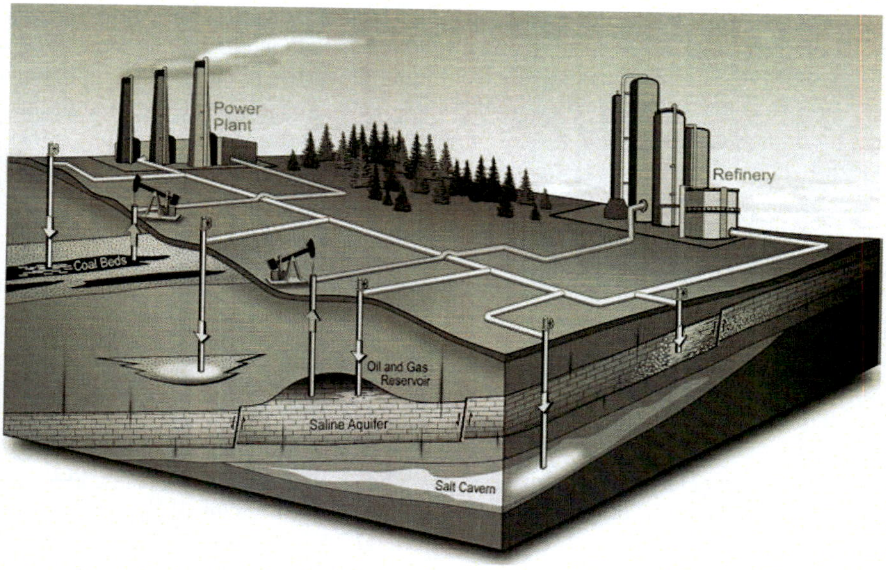

Fig. 3 Diagrammatic representation of the geological media and the transportation and injection system for onshore CO_2 disposal

2.3 Trapping Mechanisms for CO_2 in Geological Media

Long-term geological processes can result in the formation of oil and gas from organic rich shales, from which they are expulsed (primary migration) into adjacent aquifers. Once in aquifers, hydrocarbons flow updip along bedding and upwards, driven by their buoyancy (secondary migration), until they are trapped in geological regions in an aquifer, where changes in permeability impede any upward and lateral flow. This leads to oil and/or gas accumulation in what then become hydrocarbon reservoirs. The changes in permeability that form the trap for buoyant fluids (in this case oil or gas) are due to depositional and/or diagenetic changes (stratigraphic traps) or to the development of structural traps (due to folding and faulting) (Gunter et al. 2004). It is important to note that there are many such stratigraphic and structural traps in sedimentary basins that are not charged with oil or gas because they were not located along the hydrocarbon migration path. These stratigraphic and structural traps, saturated initially either with water (aquifers) or hydrocarbons (reservoirs), constitute the main targets for CO_2 disposal. Obviously oil and gas reservoirs, because of their economic value, may or will be used for CO_2 disposal only after their depletion. These traps can be very large in size (up to hundreds of square kilometers in areal extent and tens to hundreds of metres thick). CO_2 injected into these traps forms a continuous phase and can flow through the pore space, and actually will flow throughout the trap until steady state conditions are achieved, but it will not flow out of the trap. This type of trapping is called *stratigraphic and structural trapping*.

CO_2 is a non-wetting fluid that may flow through the rock pore space where it is continuous. However, when water (a wetting fluid) invades the rock previously saturated with CO_2, disconnected gas bubbles are caught in the pore space due to capillary snap-off, losing their ability to flow and becoming immobile at residual gas saturation. This is due to the hysteretic nature of the relative permeability of the two fluids, water and CO_2. Significant amounts of CO_2 can be trapped this way in the pore space in the wake of a migrating stream or plume of CO_2 (Kumar et al. 2005; Juanes et al. 2006; Ide et al. 2007). In this case there is no need for a stratigraphic or structural trap because the CO_2 is immobilized in the pore space. This type of trapping is called *residual gas trapping*.

As mentioned before, CO_2 in contact with water, either at the interface between a stream or plume of CO_2, or in each pore (non-wetting CO_2 against wetting water), will dissolve in water over a timescale of years to centuries (Gunter et al. 2004). Once dissolved, CO_2 loses its free-phase buoyant properties and will flow with the natural flow of water in the aquifer. Because CO_2-saturated water is heavier by approximately 1% than unsaturated water, if certain instability requirements are met, the heavier water will flow in a cellular pattern (free convection), dropping to the bottom of the aquifer, thus removing the CO_2-saturated water from the CO_2–water interface and moving it downwards whilst unsaturated water replaces it, in this way accelerating the process of dissolution (Ennis-King and Paterson 2003). This process is called *dissolution trapping*.

The weak carbonic acid formed by CO_2 dissolution reacts with rock minerals and may precipitate as carbonate rocks in what is called *mineral trapping*

(Bachu et al. 1994), in a process that usually takes centuries to millennia to deposit significant amounts of CO_2 as solid rock (Xu et al. 2003; Perkins et al. 2005).

If CO_2 is injected outside stratigraphic or structural traps in deep, regional-scale saline aquifers, whose size is in the order of tens to hundreds of kilometres and where formation water usually flows with velocities in the order of millimetres to centimetres per year, CO_2 will form a plume that will migrate updip along the strata but still below the caprock until it is immobilized through the combined effects of residual gas trapping, dissolution and mineral precipitation, regardless of the presence or absence of stratigraphic and/or structural traps along the migration pathway. This combined trapping mechanism is called *hydrodynamic trapping* (Bachu et al. 1994) lately known also as Migration Assisted Storage (MAS).

Finally, if injected into coalbeds, CO_2 will flow through the coal's natural system of fractures (cleats), diffuse through the coal's micropores, and adsorb onto the surface of the coal, displacing methane, in a process called *adsorption trapping*. It is desirable that the coalbeds into which CO_2 is injected be themselves overlain by impermeable strata to impede the upward flow of any excess CO_2 that is not adsorbed by the coal. Coal's permeability depends on the effective stress, which increases with depth and closes the coal cleats. Thus, coals tend to lose injectivity with increasing depth (McKee et al. 1988) such that coals at depths greater than 800–1,200 m cannot be used for CO_2 disposal because of lack of injectivity. In addition, CO_2 has the effect of swelling the coal (Cui et al. 2007), further closing the cleats and reducing permeability, and hence its injectivity.

The various CO_2 trapping mechanisms identified above can be variously classified as physical and chemical, or as primary and secondary. Physical trapping mechanisms are those where CO_2 retains its chemical composition: structural and stratigraphic trapping, and residual gas trapping. Dissolution, mineral and adsorption trapping are chemical trapping mechanisms. Hydrodynamic trapping is based on both physical and chemical trapping processes.

More important is the evaluation of CO_2 trapping mechanisms in relation to the duration of injection, which for a power plant or industrial process would be in the order of several decades (Fig. 4a). Primary trapping mechanisms are those whose timescale is comparable with that of the CO_2 injection, namely the emplacement of CO_2 in the trapping geological medium (Fig. 4a). These are structural and stratigraphic trapping, adsorption trapping and hydrodynamic trapping. A key characteristic of the disposal unit in all these cases is that it must have the necessary capacity to take all the CO_2 that is injected during the active disposal period. Residual gas trapping, dissolution and mineralization are secondary trapping mechanisms because they are dependent on the primary trapping (CO_2 emplacement) occurring first; they depend on CO_2 and water movement, and they operate on longer timescales, from centuries to millennia (Fig. 4a). On the other hand, the secondary trapping mechanisms contribute to increasing disposal security and a reduction in the risk with increase in time because, through CO_2 immobilization (residual gas trapping), dissolution and mineralization, less free-phase mobile CO_2 is left that may migrate and leak to the shallow hydrosphere, biosphere and atmosphere (Fig. 4b). The security of CO_2 disposal broadly increases, and hence the risk also

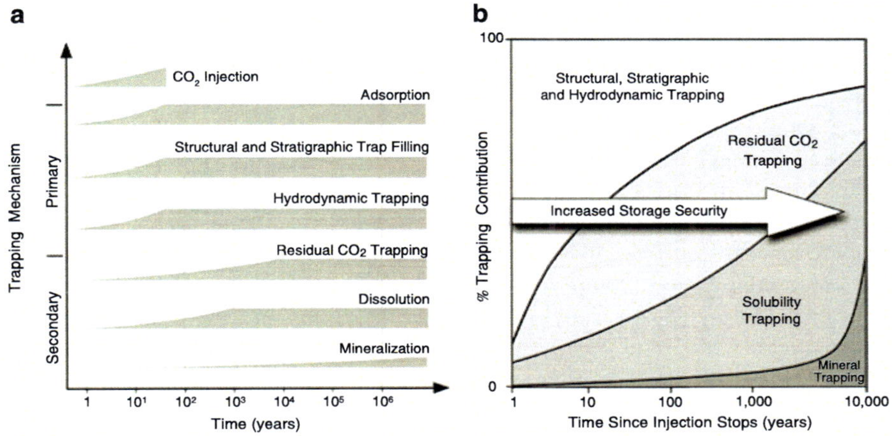

Fig. 4 Diagrammatic representation of the characteristics of CO_2 trapping mechanisms in geological media: (**a**) timescales for achieving full efficiency; and (**b**) variation in time of the amount of CO_2 trapped by various mechanisms when injected in deep saline aquifers (From IPCC 2005)

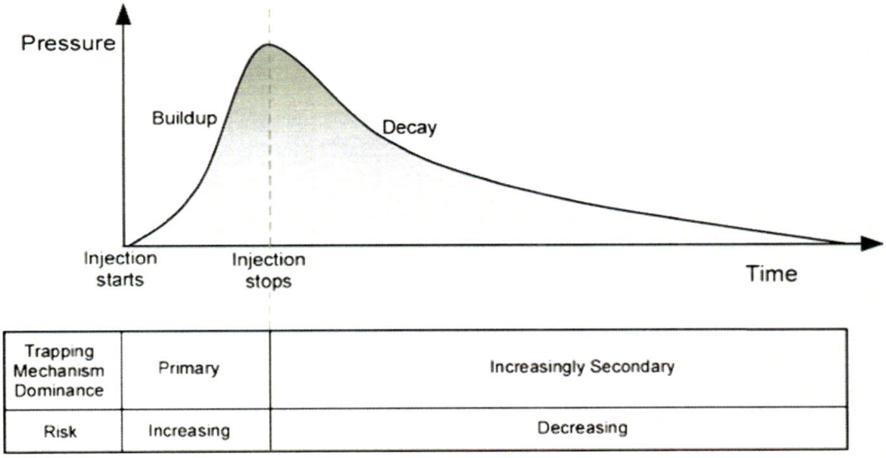

Fig. 5 Diagrammatic representation of the pressure variation with time in a CO_2 disposal operation, of risk and of dominance of trapping mechanisms (After Bachu 2008a)

decreases, after cessation of CO_2 injection because, after injection ceases, the pressure, which increases continuously during injection, decays, thus reducing the driving force acting on the injected CO_2. The combination of pressure decay and the increasing role of secondary CO_2 trapping mechanisms leads to a decrease in the risk associated with CO_2 disposal after injection has ceased (Fig. 5). This scenario is generally true unless the plume of migrating CO_2 encounters a leaky well or an open fracture or fault, in which case the risk may locally increase as a result of leakage along this newly found leakage pathway.

2.4 *Long-Term Fate and Potential Migration Mechanisms and Pathways*

As discussed previously, CO_2 injected in deep saline aquifers or depleted oil and gas reservoirs may retain its form or may dissolve in aquifer brine or reservoir oil, or may precipitate as a carbonate mineral due to time-dependent processes. CO_2 injected into coalbeds will adsorb onto the coal surface. As long as CO_2 remains in, or, through exsolution or desorption, regains its original state, regardless of the phase (gaseous, liquid or supercritical), it will be subjected to hydrodynamic and buoyancy forces. The hydrodynamic forces are the result of injection (pressure forces) and of the natural flow systems in the injection aquifer. The buoyancy force is due to the in situ density difference between CO_2 and the groundwater or oil. If injected into porous rocks (deep saline aquifers or depleted gas reservoirs), CO_2 will, in addition, be subjected to viscous and capillary forces whereas, if injected into coalbeds, it will be subjected to molecular bonding forces. If the hydrodynamic and buoyancy forces are stronger than the capillary or adsorption forces, CO_2 will flow upwards if a pathway is available.

The transport mechanisms for free-phase CO_2 in porous media are diffusion and advection accompanied by dispersion. The former dominates in low permeability rocks such as shales, whilst the latter dominates in permeable aquifer and reservoir rocks and in fractures. Since the whole concept of CO_2 disposal is predicated on the existence of low permeability barriers that impede upward CO_2 flow, the issue is under what conditions these barriers could be breached, allowing upward CO_2 leakage. There are two possible mechanisms for the failure of the confining caprock caused by the injection of CO_2. Mechanical failure may take place due to hydraulic fracturing, the opening of pre-existing fractures or due to fault reactivation. This occurs when the injection pressure, which is highest at the injection well, exceeds a certain value P_m, equal to the minimum horizontal stress, if pre-existing fractures normal to the minimum stress direction are present or, in their absence, equal to the rock fracturing pressure. In general, mechanical failure is unlikely to occur because, during the injection stage, regulatory agencies limit the maximum bottom hole pressure at the injection well to values below the pressure corresponding to mechanical failure, and because of pressure decay in the post-injection stage (Fig. 5).

The other case of caprock failure occurs when the pressure at the interface between the CO_2 and the caprock exceeds the displacement pressure P_d (known also as the capillary entry pressure), above which water that saturates the caprock is displaced by the intruding gas (CO_2) phase. The capillary entry pressure depends on the interfacial tension (IFT) between CO_2 and water, which in turn depends on the in situ pressure, temperature and salinity conditions (Bachu and Bennion 2008) and is about half of that between methane and water (Chiquet et al. 2007). Usually P_m is smaller than P_d, particularly for low permeability rocks, whose capillary entry pressure is very high (Bennion and Bachu 2007), such that the integrity of the caprock is maintained by keeping the injection pressure below the threshold for mechanical failure. However, gas migration from gasfields has been documented (Gurevich et al. 1993). It is possible to have gas reservoirs that are overpressurized close to the displacement pressure

P_d corresponding to methane–water systems and, if these reservoirs are filled instead with CO_2 up to their initial pressure, it will exceed the displacement pressure for the CO_2–water system because of the lower IFT for the latter than for the former, resulting in CO_2 migration through the caprock. Even in such extreme cases, the timescale for leakage to occur will be very large (centuries to millennia and longer) because of the low permeability of the caprock and of relative permeability effects. The duration of CO_2 migration through the caprock depends not only on the caprock flow characteristics, but also on the caprock thickness.

Notwithstanding the possibility of CO_2 upward flow due to caprock failure, which has a very low probability, wells represent the most significant potential pathway for free-phase CO_2 leakage (Bachu and Celia 2009), as shown by documented natural gas leakage along wells in Alberta, Canada (Watson and Bachu 2007). The potential for leakage through wells is enhanced by the presence of CO_2, either in direct contact with well cement and casing, or dissolved in water, although under certain conditions well cement degradation is halted by the chemical reactions taking place in the presence of CO_2 (Scherer et al. 2005; Kutchko et al. 2007). Work to date seems to indicate that, depending on the type of cement used, if wells are properly drilled, constructed, completed and abandoned, the potential for leakage, including that of CO_2, is quite low due to the protective carbonate layer that forms when the CO_2-saturated brine reacts with well cement (Kutchko et al. 2007). However, preferential flow paths may be present due to pre-existing well defects, particularly in older wells, such as an annular space between the cement and the casing, poor bonding between the cement and the rock and cement fractures, which may be enhanced by the presence of CO_2 (Carey et al. 2007; Watson and Bachu 2009). Similarly, wells with cements that contain additives such as bentonite, or that have been stimulated through fracturing or acidizing, or that were abandoned with bridge plugs containing elastomers, will be more susceptible to CO_2 leakage (Watson and Bachu 2008).

In the case of free-phase CO_2 leakage through faults, fractures and wells, CO_2 will decompress relatively quickly as it flows upwards (due to the Joule–Thompson effect) and three-phase conditions will form, self-limiting the CO_2 flow rate due to three-phase relative permeability effects (Pruess 2004, 2005). On the other hand, in the case of diffusive transport across a caprock, or if CO_2 is dissolved in formation water that reaches the surface through faults, fractures and wells, the movement of CO_2 is extremely slow such that temperatures equalize and Joule–Thompson effects are avoided. In the case of CO_2 transport in solution, as the pressure decreases and the solubility drops, CO_2 will exsolve. The leakage rates in such degassing cases are very low and do not pose a significant risk (Shipton et al. 2005).

2.5 *Geographic Distribution and Criteria for Site Selection*

The selection of sites for CO_2 disposal has to consider the disposal requirements: confinement, capacity and injectivity. The confinement requirement implicitly includes an assessment of the long-term fate of the injected CO_2 and an assessment of the potential for leakage. From the analysis of the geological environments suitable

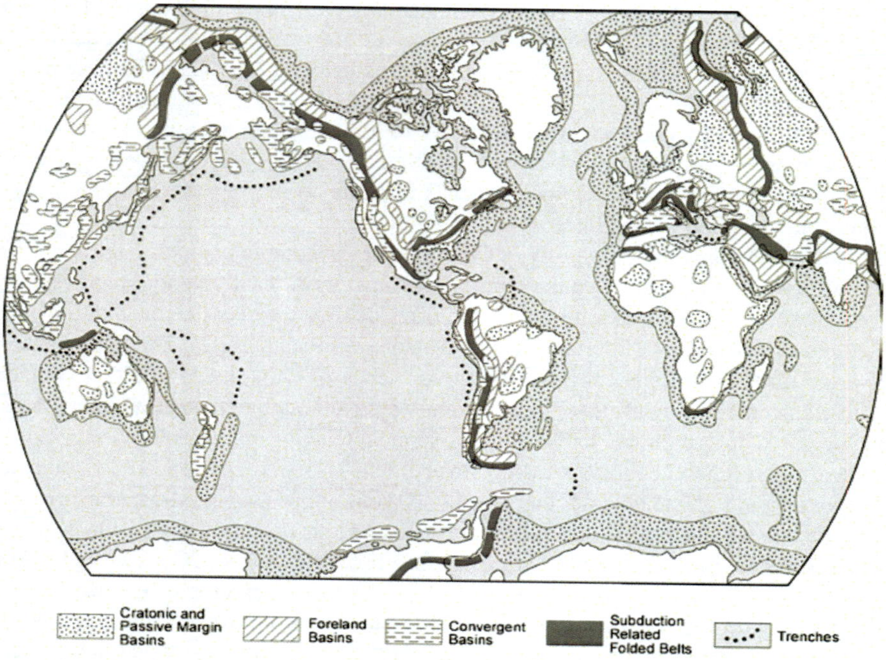

Fig. 6 Global distribution of sedimentary basins and their main types

for CO_2 disposal it is evident that only sedimentary basins could be considered, but even within these there are basins which are less favourable for CO_2 disposal, such as those located in areas of tectonic plate convergence, and basins better suited for CO_2 disposal, such as intracratonic and passive margin basins (Hitchon et al. 1999; Bachu 2003). Figure 6 shows the distribution and type of sedimentary basins around the world. It is instructive to note that circum-Pacific basins are of the convergent type, and hence are likely to be faulted and prone to tectonic activity and also tend to be comparatively small whereas circum-Atlantic basins and those around the Indian Ocean are large and of the passive margin type, which are more favourable to CO_2 disposal due to their simpler geological histories and more stable natures.

There are few sedimentary basins in Africa and Asia relative to their size, population and CO_2 emissions. In North America, foreland and intracratonic basins are found between the Rocky and Appalachian mountains, whilst in South America they are found east of the Andes Mountains. In Europe, foreland basins are found north of the Alps and the Carpathian Mountains and west of the Urals in Russia, but the sedimentary basin with the greatest potential is the prolific North Sea basin. Mediterranean basins are located in an area of plate convergence and possess all the associated unfavourable characteristics. Foreland basins in south-west and southern Asia are located south of the Zagros Mountains in Iran, where the major Middle East oil and gas resources are found, and south of the Himalayas in the Indian subcontinent. The main sedimentary basins in Australia with the largest CO_2 disposal potential are offshore.

Other criteria for assessing the suitability of a sedimentary basin for CO_2 disposal are its size, depth, geology and degree of faulting and fracturing, hydrogeological and geothermal regimes, and the presence of coals, oil and gas reservoirs, salt beds and deep saline aquifers (Bachu 2003, 2010). For example, a 'warm' sedimentary basin is less suited for CO_2 disposal than a 'cold' basin because, for the same depth, temperatures will be higher in the former, hence CO_2 density will be lower by a factor of up to two, leading to higher CO_2 buoyancy and lower efficiency in terms of the utilization of the pore volume. Other considerations are basin maturity (degree of exploration and production of oil and gas reservoirs, if present), accessibility and existence of infrastructure (e.g. roads, pipelines).

In terms of the potential for CO_2 disposal, another major element in site selection is the location of major stationary CO_2 sources (emitters) in relation to possible disposal sites (also known as source–sink matching). For example, there are many Arctic, sub-Arctic and Antarctic basins, many offshore basins, intracratonic basins in Africa or in the Amazon in Brazil (Fig. 6) that are too far from any significant CO_2 source amenable to capture and disposal; transportation of CO_2 by ship and/or pipeline to disposal sites in these basins would be uneconomic.

Even in countries that, overall, have sufficient CO_2 disposal potential, it may, in some cases, be located too far from large CO_2 sources. For example, in Australia the major CO_2 sources are located along the coast in the southeast (mainly coal-fired power plants), whilst the best sites for CO_2 disposal are offshore in the north-west (Bradshaw et al. 2002). In Canada, the capacity and potential for CO_2 disposal lies in the western provinces of Alberta and Saskatchewan, whilst the major sources of CO_2 in central Canada (Ontario) have no conveniently located disposal sites (Bachu 2003). In the USA, major CO_2 sources in the north-east and the Midwest (Ohio Valley) do not have sufficient CO_2 disposal capacity within an economic distance. Even if a sedimentary basin meets the criteria for CO_2 disposal in general terms, there will be regions within the basin that do not meet these criteria, particularly along the shallow edge of the basin or in faulted and folded regions. Such is the case of the Alberta Basin in western Canada—where major CO_2 sources related to the production of synthetic oil from tar sands are located in the north-east close to the basin edge, where there is no CO_2 disposal potential—or in south-western Ontario, where major coal-fired power plants and refineries are located on a sedimentary wedge less than 1,000 m deep that separates the Michigan and Appalachian Basins in the USA. In these cases, CO_2 captured at these large sources would have to be transported by pipeline, several hundred kilometres in length, to appropriate disposal sites.

Yet another consideration in the selection of CO_2 disposal sites is the type of geological medium. Countries with significant oil and gas reserves and in an advanced stage of exploration and production will most likely consider oil and gas reservoirs for CO_2 disposal, either at reservoir depletion or to increase production through CO_2 enhanced oil recovery (EOR). This is the case of countries in the Middle East and around the North Sea, and Indonesia and Mexico, but this is also a viable option in the USA and Canada. On the other hand, sedimentary basins in China and southern Africa are rich in coal, which puts them at a disadvantage because CO_2 disposal in coalbeds is an immature technology (IPCC 2005) and because coal is used for energy production and hence will not be available for disposal.

On a global basis and considering the major world CO_2 emitters, the distribution and type of sedimentary basins and the main disposal media, it seems that Asian countries along the Pacific Rim (i.e. Japan, South Korea, China) do not have sufficient CO_2 disposal capacity (Newlands et al. 2006), neither do India (Holloway et al. 2009) or South Africa. Middle Eastern countries (e.g. Saudi Arabia, United Arab Emirates) and European countries around the North Sea (e.g. Germany, UK, Norway) are likely to have sufficient CO_2 disposal capacity, although an extensive pipeline infrastructure would have to be built. Continental-size countries like the USA, Canada, Australia, Russia and Brazil appear likely to possess the necessary CO_2 disposal capacity, but in some cases there is a mismatch between the location of major CO_2 sources and disposal sites. In countries such as the USA, Canada and Russia and in the Middle East, CO_2 disposal will most likely be implemented onshore, whilst in northern Europe, Brazil and Mexico it is more likely to be implemented offshore.

On a local scale, site selection has to be based on the same criteria of confinement, capacity and injectivity. Additional criteria are protection from possible contamination of other energy and mineral resources and of groundwater, land ownership and rights of access, ownership of the 'pore space' (i.e. the right to inject CO_2), and infrastructure (roads, pipelines and wells). In some countries the subsurface is owned by the state, in others by both freeholders (individuals or private companies) and the state. Specific selection criteria for the case of oil and gas reservoirs are the degree of depletion, their suitability for EOR, reservoir heterogeneity, and the individual reservoir capacity (i.e. it is not economic to build the necessary CO_2 disposal infrastructure for reservoirs that will be quickly filled up). In the case of coalbeds, in addition to the standard criteria, the lack of any economic potential for the coal, now and in the foreseeable future, is a major consideration in site selection. If the coal is likely to be mined for power generation or for industrial use (e.g. steel making), or could be used for gasification or coal liquefaction to increase energy security and sustainability, then the coalbeds will not be used for CO_2 disposal. This is particularly important for countries endowed with large coal resources and with major energy needs such as the USA, China and India. Also, unlike deep saline aquifers and oil and gas reservoirs, the use of coalbeds for CO_2 disposal is limited to a narrow depth range because of their loss of permeability with increasing depth and in the presence of CO_2 and because shallow coalbeds are likely to have already been mined or lie in the depth range where the protection of groundwater resources is an issue.

A very preliminary estimate of the worldwide capacity for CO_2 disposal suggests that coals have the lowest potential at 15–200 gigatonnes of CO_2 (Gt CO_2), oil and gas reservoirs have ultimately a capacity of 675–900 Gt CO_2, and deep saline aquifers have the largest capacity at more than 1,000 Gt CO_2 (IPCC 2005). This should be compared with global annual emissions from fossil fuel use of approximately

25 Gt CO_2/year, of which emissions from large stationary sources (each greater than 0.1 Mt CO_2/year) constitute approximately 60%, or 15 Gt CO_2/year. The latter are clustered mainly in the midwestern and eastern USA, in central and northern Europe, eastern Asia (China, Korea and Japan), India and South Africa. If, in addition to the criteria of confinement, capacity and injectivity, other considerations for site selection (such as individual site size, access, economics, land ownership and use, and population distribution) are applied, the worldwide CO_2 disposal capacity is likely to become smaller by probably an order of magnitude.

2.6 *Status and Challenges*

CO_2 disposal in geological media has not yet been implemented as a mitigation measure for climate change, although CO_2 injection and disposal has occurred for different reasons in the last 3 decades.

The most significant experience with CO_2 transportation and injection exists in the Permian basin in west Texas, USA, where there are more than 90 CO_2 EOR projects, injecting approximately 30 Mt CO_2/year (Moritis 2006). Of the amount injected, approximately 60% is produced together with oil, and is captured and recirculated, whilst the other 40% remains in the ground. The oldest CO_2 EOR scheme in west Texas has been in operation since 1974. There are a few other CO_2 EOR operations in the world, the most notable being at Weyburn in south-eastern Saskatchewan, where approximately 5,000 t CO_2/day are injected. The Weyburn operation is a CO_2 EOR scheme, like all the others, except that it has been accompanied by a monitoring research programme (Wilson and Monea 2004).

The other important experience with CO_2 disposal has occurred in conjunction with the production of sour natural gas, which is natural gas that contains CO_2 and/or H_2S (both CO_2 and H_2S form a corrosive acid in the presence of water, hence the industry designation as 'acid gas' once these are stripped of the natural gas to meet pipeline and market specifications). As a result of regulatory requirements in western Canada that do not allow venting and/or flaring of H_2S, and because incineration or desulphurization of the acid gas are uneconomic, operators are increasingly turning to the geological disposal of acid gas in depleted hydrocarbon reservoirs and deep saline aquifers. Consequently, in 2007 there were close to 50 such operations in western Canada that have injected more than 6 Mt of acid gas since 1990, approximately half of which is CO_2 (Bachu and Gunter 2005). There are more than 20 such operations in the USA, mostly in Texas, Oklahoma and Wyoming, and new operations are currently being built in Iran and Kazakhstan. The main driver for these disposal operations is the need to deal with H_2S, which is a toxic hazardous substance.

Also worthy of note are two CO_2 disposal operations where CO_2 is stripped of natural gas that contains approximately 9–10% CO_2 and is injected on site into deep saline aquifers, with the gas being sent to markets in Europe. Both operations inject in the order of 1 Mt CO_2/year. The first one is at Sleipner in the North Sea, where the CO_2 has been injected into the Utsira formation since the mid-1990s, approximately 800 m below the seabed, and where a project for monitoring the fate of the injected CO_2 has been in operation (Torp and Gale 2003). The driver for the Sleipner operation is a carbon tax imposed by the Norwegian government on CO_2 emissions from offshore gas production, and in this regard this project can be considered as being a mitigation measure for climate change. The second operation is at In Salah in Algeria, which started in the mid-2000s and where the CO_2 is injected in the downdip water leg of the gas reservoirs that produce the gas containing CO_2 (Riddiford et al. 2003). A third operation started in 2008 at Mongstad, offshore Norway in the Norwegian Sea.

With regard to the injection of CO_2 into coalbeds, the only successful operation to date was run between 1995 and 2001 at the Allison Unit in the San Juan Basin, New Mexico, USA, as a pilot for enhanced coalbed methane (ECBM) production (Reeves 2003); however, no monitoring project was run in conjunction.

In addition to these commercial scale projects, there are a number of demonstration and pilot operations, mostly funded by governments, mainly for testing and developing technology for monitoring the fate of the injected CO_2 and developing monitoring techniques in the case of CO_2 injected into deep saline aquifers and depleted gas reservoirs (e.g. van der Meer et al. 2005; Hovorka et al. 2006; Förster et al. 2006). Pilot operations run to test CO_2 disposal in coalbeds in Canada, Poland, China and Japan have been less successful, mainly because of coal swelling in the presence of CO_2 (e.g. van Bergen et al. 2006; Yamaguchi et al. 2008; Wong et al. 2007).

These commercial and pilot scale operations indicate that CO_2 injection through wells does not pose any particular technological challenge. Generally, except for CO_2 disposal in coalbeds, the technology is mature and can be deployed immediately, at least on a demonstration scale (i.e. several large-scale operations, greater than 1 Mt CO_2/year each). However, there are still a few geoscientific and technical challenges that need addressing before the large-scale deployment of CO_2 disposal as a mitigation measure for climate change. These are:

1. *Resource mapping*: If the disposal volume that would be required for large-scale deployment is defined as a resource, there is a need to implement a sustained geoscience programme for the definition, identification, mapping and characterization of this resource.
2. *Timescale and effect of geochemical reactions*: If geochemical reactions between CO_2 and in situ fluids and rocks are likely to have a discernible effect over a time frame of millennia, then it may be possible to neglect them from a disposal point of view (where time frames of the order of a few centuries are likely to be more significant). Currently there is a divergence of opinion with regard to the geochemical effects associated with CO_2 disposal, particularly with respect to mineral trapping.
3. *Predictive modelling*: In order to properly predict the fate of the injected CO_2 over periods of time measured in centuries to millennia, there is a need to develop comprehensive mathematical and numerical models that couple multi-phase fluid flow, heat transfer and phase change(s), reactive geochemistry and geomechanical

effects of CO_2 disposal. Currently there are sophisticated models that treat one or two of these processes (e.g. flow and geomechanical, flow and geochemistry, flow and heat transfer, geomechanical and heat transfer), but there are no models that can treat three or more of these processes, because of the complexities involved, the nonlinearity of the system, and limitations in computing capabilities.

4. *Data collection*: There are insufficient physical and geochemical data, such as relative permeability and reaction kinetics, to characterize and model the fate of the injected CO_2 for the pressure, temperature and salinity conditions found at the disposal depths in various geological environments.
5. *Fate of wells*: Although wells have been drilled for more than 100 years with improving technology, there is no experience with the 'thousand year well', i.e. there is no experience with wells that should last as long as the CO_2 disposal operations should retain their effectiveness. This concerns existing wells, some from the nineteenth century, and new wells, both for CO_2 disposal and for other uses, mainly oil and gas exploration and production. This is essential for maintaining disposal efficacy (i.e. avoiding or minimizing CO_2 leakage). The magnitude of the problem is best illustrated by the following facts: there are more than 1,000,000 wells in Texas alone; there are more than 350,000 wells in Alberta, Canada, and new wells are being drilled at a rate of approximately 20,000 per year; generally there are no records about the completion and abandonment of old wells, particularly those drilled in the nineteenth century and early in the twentieth century. The fate of cement and casing in a CO_2-rich environment has to be understood and remediation measures have to be developed.
6. *Applicability of CO_2 disposal in coalbeds*: The loss of permeability due to coal swelling, and coal plasticization in the presence of CO_2 under certain conditions of temperature and pressure, severely limit the potential of coal to be used as a medium for CO_2 disposal. Coal is a brittle (glassy) material that becomes plastic at high temperatures and pressures. In the presence of CO_2 the temperature at which coal becomes plastic drops dramatically to around 30°C for pressures above 5 MPa (Larsen 2003).
7. *Effect of impurities*: CO_2 streams from power generation, energy production and industrial processes will contain various impurities, such as H_2S, SO_x and nitrogen oxides (NO_x), whose effects in the long term are not well understood. There is a trade-off between the increasing cost of purification and the fact that these impurities reduce the available disposal volume and may have a negative effect in the long term.
8. *Fate of displaced water*: Injecting such large volumes of fluid (liquid or supercritical CO_2) which are required to achieve climate stabilization targets would displace very large volumes of saline water, whose fate needs to be determined because they may have adverse impacts on potable groundwaters and the surface ecology if they migrate into shallow aquifers or to the surface.

There are other challenges facing the large-scale development of the geological disposal of CO_2, but they are of an economic, financial, legal and regulatory nature and are also likely to be linked to the attitude of the public to such developments (Bachu 2008a). These subjects are considered in other chapters of this volume.

3 Current Status of Radioactive Waste Disposal

3.1 What Are Long-Lived Radioactive Wastes?

Radioactive waste is defined by the IAEA (1994) as 'material that contains or is contaminated with radionuclides at concentrations or activities greater than the clearance levels as established by the regulatory body, and for which no use is foreseen.' National policy may consider some of the potential RW to be a resource, but this is likely to apply only to SF, which can be recycled to produce reusable plutonium and uranium for possible reuse in nuclear reactors. In other countries SF is considered a waste and is disposed of directly, although whether the SF is considered a resource or a waste is not necessarily based on an economic assessment, but often on political considerations. In this respect, RW is treated differently from other forms of hazardous and/or toxic waste.

RW is classified so as to determine how it should be handled and how suitable disposal options can be identified. The classification of the different types of RW varies from country to country and, as such, makes comparison difficult (see Vankerckhoven and Mitchel 1998). The IAEA has, however, implemented the Net Enabled Waste Management Database (NEWMDB) (www-newmdb.iaea.org), which attempts to harmonize waste definitions (Table 1) and these are used in this chapter.

The RW that is of interest here is the long-lived waste derived from the following sources that will require disposal in a geological disposal facility or repository:

- SF from reactors (which is heat emitting);
- Reprocessed SF, which results in the formation of HLW (which is also heat emitting) and other by-products, which are classified mainly as long-lived low- and intermediate-level waste (LILW-LL);

Table 1 Details of the waste classes defined by the IAEA

Waste class	Typical characteristics	Possible disposal options
Short-lived (L/ILW-SL)	Restricted long-lived radionuclide concentrations, e.g. long-lived α-emitters average <400 Bq/g or 4000 Bq/g maximum per package	Near-surface or (in some countries) geological disposal facility
Long-lived (L/ILW-LL)	Long-lived radionuclide concentrations exceeding limitations for short-lived wastes	Geological disposal facility
High-level waste (HLW)	Thermal power greater than about 2 kW/m^3 and long-lived radionuclide concentrations exceeding limitations for short-lived wastes (includes SF and HLW)	Geological disposal facility

L/ILW-LL long-lived low/intermediate-level waste, *L/ILW-SL* short-lived low/intermediate-level waste, *SF* spent fuel

- ILW from other sources such as reactor operations and decommissioning;
- Some countries, such as the UK, may also require the disposal of some long-lived low-level waste in a geological facility;
- Waste derived from military sources in countries that have nuclear weapons (this waste can be of a variety of types);
- Medicine and industry (although, again, the majority of this waste is not long-lived).

RWs need to be treated and conditioned to convert the waste materials into a form that is suitable for subsequent management, such as transportation, storage and disposal. The principal aims are to minimize the volumes requiring management via optimized treatment processes and to reduce the potential hazard of the waste by conditioning it into a stable, solid form that immobilizes it and provides containment. This is to ensure that the waste can be safely handled during its management. The processes used in this treatment and containment depend on the level of activity of the waste, with each country having its own waste management policy that influences the approach taken.

Many of the treatment methods, such as compaction and incineration, are applicable only to the shorter-lived wastes. Conditioning methods include cementation, bituminization and vitrification. Whilst the first two of these are applicable to ILW, vitrification is most commonly used for conditioning the highly radioactive liquors that result from reprocessing (where SF is dissolved in concentrated nitric acid to recover the uranium and plutonium, which can be reused), with the resulting glass being cast into stainless steel containers and then stored. SF is already in a reasonably stable waste form and its conditioning consists of placing it inside a metal canister. Canister designs vary, with existing designs including a copper canister with a cast iron insert (to be used in Sweden and Finland, e.g. SKB (2004) and Fig. 7) and a titanium–carbon steel equivalent, e.g. JNC (2000). Further information on waste sources and classification can be found in McGinnes (2007).

After nuclear fuel has been involved in the nuclear fission process, the fuel becomes intensely radioactive, largely as a result of the formation of new radionuclides, known as fission products, which reduces the efficiency of the reactor. After a few years the fuel needs to be removed from the reactor and becomes SF and, after some period of surface storage so as to reduce its heat output, it is normally placed in canisters. The storage time depends on the disposal concept considered, which in turn will determine the maximum acceptable temperature in the near field of a repository. It may also depend on other factors such as the regulations in the country in question.

HLW originates as a liquid residue from reprocessing SF to extract the uranium and plutonium for reuse, with the liquid containing most of the radioactivity from the original SF. It is commonly then evaporated to dryness and the residue containing the radionuclides then melted with a much larger volume of inert borosilicate glass-forming material to produce a homogeneous, solid, vitreous waste form. The glass is cast into stainless steel containers that are sealed and may be placed in an additional metal container for emplacement in a repository.

Fig. 7 The repository design proposed by SKB (Sweden) for the disposal of spent fuel in steel canisters sheathed with copper and emplaced within a bentonite buffer in disposal holes drilled into the floor of horizontal disposal tunnels (an alternative, but similar system has the waste canisters emplaced in horizontal disposal holes). The repository would be located at a depth of approximately 500 m in hard, fractured rock. A similar repository concept is being developed in Finland (Picture courtesy of SKB)

ILW can come in many forms. It arises principally from reactor operations, from reprocessing of SF and from decommissioning nuclear facilities. It is also derived from the production and decommissioning of nuclear weapons—and this is the primary source of wastes that are being disposed of in the WIPP repository in New Mexico, USA.

The volume of RW produced by the nuclear industry is very small compared with the other wastes generated. For example, in the OECD countries some 300 million tonnes of toxic wastes are produced each year, compared with 81,000 m^3 of conditioned RWs. In countries with nuclear reactors, RWs comprise less than 1% of total industrial toxic wastes. The volumes of RWs worldwide, as taken from the NEWMDB database (which includes the majority of the installed nuclear power capacity worldwide) were last updated in 2007, and are listed in Table 2.

Figure 8 shows a curve of relative radioactivity (compared with the radioactivity of the mined uranium ore) for typical SF (Swedish boiling water reactor fuel) as a function of time after discharge from the reactor, showing the early contribution of

Table 2 Volumes of declared waste arising worldwide

Class	In storage (m^3)	Disposed wastes (m^3)
L/ILW-SL	2,222,980	23,777,710
L/ILW-LL	3,127,681	10
HLW	365,404	0

From the NEWMDB database (www-newmdb.iaea.org) as of 2007. A much more detailed breakdown of the available data can be found on the NEWMDB website
L/ILW-LL long-lived low/intermediate-level waste, *L/ILW-SL* short-lived low/intermediate-level waste, *HLW* high-level waste

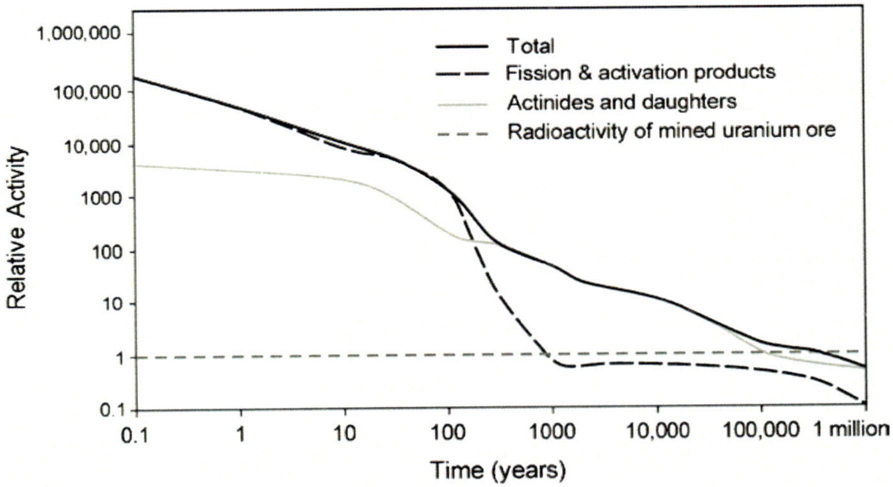

Fig. 8 The relative activity of spent fuel over time for SKB spent fuel, compared with the radioactivity of the mined uranium ore. (After Hedin 1997)

the fission and activation products. The sharp decline in fission product activity between 100 and 1,000 years is largely a result of the decay of ^{90}Sr and ^{137}Cs, both with half lives of about 30 years. After a few hundred years the actinide elements become dominant. After a few hundred thousand years the total activity of the fuel is similar to that of the uranium ore from which the fuel was produced. Other waste types will have different activity-time curves.

3.2 *Geological Disposal of Long-Lived Wastes*

The discussion below refers to the common form of the geological disposal of RW, in what is often referred to as a *mined repository*, or *disposal facility*, located at depth in water-saturated rocks. The site originally proposed for a repository for

HLW in the USA, at Yucca Mountain, is different in that it is located in the unsaturated zone and so some of the statements below, for example in relation to reducing conditions at depth, are not applicable. (Following an announcement in March 2009 regarding funding for the Yucca Mountain project, it is now certain that this will not be the site of an HLW repository.) In addition to a mined repository, there are other disposal concepts for geological disposal, such as deep borehole disposal, which are potentially suitable for only certain types of waste and which are different in certain specific regards from mined repositories (see McEwen 2004).

The multi-barrier system, introduced above, consists of two main elements:

- The engineered barrier system (EBS), which comprises the solid waste matrix and the various containers and backfills used to immobilize the waste inside the repository;
- The natural barrier (also referred to as the geosphere), which is principally the rock and groundwater system that isolates the repository and the EBS from the biosphere. The host rock is the part of the natural barrier in which the repository is located. In some cases the host rock is effectively equivalent to the geosphere, e.g. in the situation where the crystalline rock, in which the repository is located, extends to the surface.

The extent to which these two principal components act to provide containment, the way in which the different parts of the EBS control the behaviour of individual radionuclides, and the relative importance of the natural and engineered barriers at different times in the future evolution of the disposal system, constitute what is known as the *safety concept*, with what are referred as *safety functions* allocated to the different components of the system. The safety functions of the host rock are, according to Posiva (2008): (a) to isolate the repository from the biosphere and normal human habitat, (b) to provide favourable and predictable mechanical, geochemical and hydrogeological conditions for the engineered barriers, protecting them from potentially detrimental processes taking place above and near the ground surface, such that they contain the SF, and (c) to limit and retard inflow to and release of harmful substances from the repository. Similarly worded descriptions of the safety functions of the host rock or geosphere have been developed by other waste management agencies. Other safety functions are associated with the EBS (see below).

The safety concept can be different for each disposal system. Thus Fig. 7 provides a contrast with Nagra's disposal concept shown in Fig. 1.

3.2.1 The Natural Barrier or Geosphere

The natural barrier, or geosphere, is the rock that surrounds the disposal facility. As indicated in Table 1 and in the Introduction, there are certain requirements placed on the geosphere which will vary with the disposal concept considered, the geological environment chosen and with the time after waste emplacement. Emplacement of the waste in carefully engineered structures placed at depth in suitable rocks is chosen principally for the long-term stability that the geological environment provides (see item 1 in Sect. 4.1). At depths of several hundred metres in a tectonically stable environment, processes that could disrupt the repository are so slow that the rock and groundwater systems at depth will remain almost unchanged for perhaps hundreds of thousands of years, and possibly longer.

There is considerable flexibility in selecting a suitable geological environment for hosting a repository, as can be seen from the list of environments given in Table 3 for existing and proposed disposal facilities. The host rocks for disposal can vary quite widely, from hard, fractured rocks such as granite and gneiss through argillaceous rocks, mainly mudstones and clays, to evaporites, normally halite—and these rocks can be present in a variety of geological environments, from ancient basement terrains through to relatively young sedimentary basins. The argillaceous rocks and the evaporites, in particular, are chosen for their very low hydraulic conductivities, normally $<10^{-11}$ m/s (equivalent permeability in the range of 10^{-20} to 10^{-18} m^2), so that diffusive transport processes tend to dominate. Hard, fractured rocks are unlikely to have such low hydraulic conductivities but, even so, values at depths of several hundred metres and on the scale of tens of metres in suitable environments are likely to be $<10^{-10}$ to 10^{-9} m/s (equivalent permeability less than 10^{-17} to 10^{-16} m^2). All suitable disposal environments also need to possess chemically reducing conditions at depth (indicated by factors such as negative Eh and the presence of sulphides and Fe(II)). (See comment at the beginning of Sect. 3.2 with reference to Yucca Mountain, where conditions may be only locally reducing.) A useful discussion of the factors that are of greatest interest and concern regarding the properties of the rock mass and the hydrogeological and hydrogeochemical environment at depth is provided, for the case of hard, fractured rocks in Sweden, by Andersson et al. (2000). Similar considerations are likely to apply to any host rock although, of course, the strength of the sedimentary rocks, including evaporites, is considerably lower than hard, fractured rock, with the result that there will be notable differences in the repository concepts, depending on the type of host rock.

The disposal concept will, thus, vary with the type of geological environment under consideration, specifically the host rock, and also the waste forms for disposal. The relative importance of the natural barrier compared with the EBS will also vary, with host rocks in which solute transport is determined by diffusive processes (e.g. mudstones and halite) allowing the EBS to provide a more secondary, but nevertheless complementary, role (there is transport of solutes through all rocks, even halite, although at very low rates). This is in comparison with disposal in hard, fractured rocks, where the EBS, in the form of the bentonite buffer and the long-lasting canister, provides the dominant barrier to radionuclide migration (see, for example, Fig. 7).

There are important interactions between the natural and engineered barriers that lie at the heart of the multi-barrier principle. These are illustrated in Fig. 9 for the case of a KBS-3 type repository concept (as shown in Fig. 7) that is to be employed in Sweden and Finland. (KBS is an abbreviation for *Kärnbränslesäkerhet*, a Swedish term which means 'nuclear fuel safety'.) Similar interactions would exist for other disposal concepts for spent fuel or HLW. For the first 1,000 years, the EBS provides complete containment;

Table 3 Examples of the geological environments considered for hosting repositories for long-lived radioactive wastes

Geological environment	Location	Comments
Hard, fractured rocks (the geological environments of these three sites are similar –they all consist of old basement crystalline rocks)	Olkiluoto, Finland	Site of Finland's proposed spent fuel repository. Access ramp and shafts to what is planned to be the repository are currently under construction
	Forsmark or Laxemar, Sweden	Forsmark was chosen ahead of Laxemar in 2009 as the location of Sweden's spent fuel repository. Site investigations are complete at both sites
Mudstones (both the mudstones in France and Switzerland are Jurassic in age and are in structurally relatively simple geological environments)	Bure, France	A URL has been constructed and further work over the next decade is likely to result in the development of a repository for HLW and ILW-LL close, or quite close, to the URL
	Northern Switzerland	Investigations of the Opalinus Clay took place (see Nagra 2002); a site selection programme has been developed (which is currently under review by the regulatory authorities), and this formation may be chosen to host a repository for HLW and some ILW-LL
Evaporites (the host horizon at the WIPP is halite and other countries that have considered evaporites for disposal, such as Germany, have also chosen halite)	New Mexico, USA	The WIPP facility for military-derived ILW-LL (referred to as transuranic waste) has been operating for several years
		Gorleben, Germany, was for many years the proposed location for HLW and ILW-LL disposal. The German disposal programme was in abeyance for several years for political reasons, but is now active again, and the programme at Gorleben may be restarted

(continued)

Table 3 (continued)

Geological environment	Location	Comments
Volcanic tuffs (this geological environment is different from any other in the world, as the originally proposed repository was located in the unsaturated zone)	Yucca Mountain, USA	An extensive investigation programme was carried out at Yucca Mountain for many years, together with considerable safety case development and the construction of many kilometers of exploratory tunnels. It is now known that the site will not be developed as a repository

There are other geological environments being considered, such as plastic clay in Belgium—but these can be considered as subsets of the environments listed here

HLW high-level waste, *ILW-LL* long-lived intermediate-level waste, *URL* underground research laboratory, *WIPP* waste isolation pilot plant

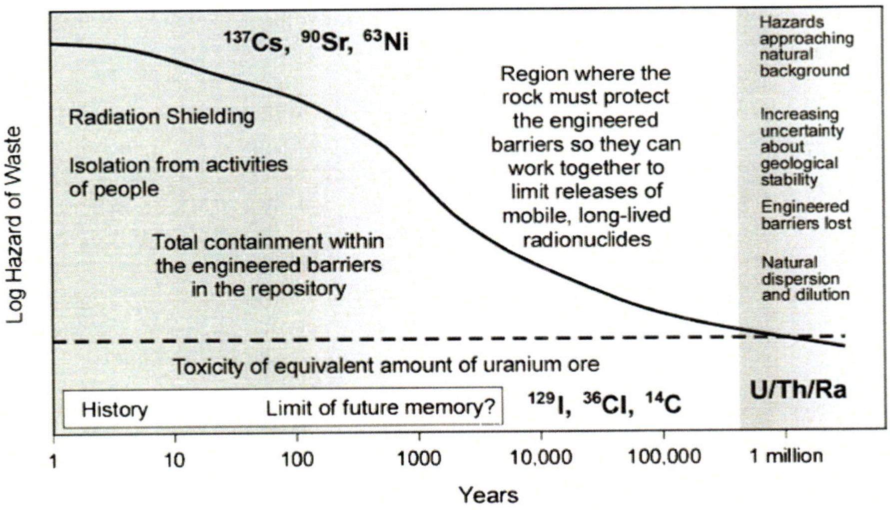

Fig. 9 The log of the hazard (relative to the hazard from the original uranium ore used to make the fuel elements) from spent fuel or high-level waste against time, illustrating the achievement of geological disposal in isolating the waste from the surface environment. The radionuclides which are of greatest significance in determining the hazard at different times in the future are also shown (From Chapman and McCombie 2003)

this period ends at approximately one to a few thousand years, following which, up to perhaps 100,000 years, the host rock is required to protect the EBS so that it can operate as planned and limit the release of the mobile, long-lived radionuclides. After this period, as discussed below, there is a gradual loss in the efficacy, or performance, of the EBS to limit the release of radionuclides.

After this period of up to 100,000 years, there are several factors that need to be taken into account when developing a safety case (see Fig. 9), as discussed below.

- *Geological stability can no longer be guaranteed*: at least not to the same extent that is possible up to this period. There are two separate components to this stability—the effects of climate change and tectonically related processes. For the first of these, regardless of the extent of future global warming, the Milankovitch forcing factors will ensure that glacial–interglacial cycling will reassert itself, with the result that ice sheets will advance and sea levels will change considerably, with a periodicity of approximately 100,000 years. For the second, it is more the effects of uncertainty as to what may take place that is important—for many geological environments, such as in the majority of Europe, tectonically related processes are unlikely to be significant for the next few million years as geological activity is relatively benign. The most important process over this period of time in Europe is likely to be uplift and erosion. For other, more tectonically active regions of the world, such as Japan, tectonic activity is likely to play a more significant role in locating a site for a nuclear waste repository and in developing a safety case. In such countries processes such as uplift and erosion, earthquakes, fault movement and volcanic activity may need to be an integral part of a safety case.
- *The hazard due to the waste is approaching the natural background*: it is also, by this period, likely to be below the toxicity of the uranium ore used to produce the fuel rods. In fact, the time after closure of the repository, when the crossover takes place with respect to the toxicity of the uranium ore, may be as little as 10,000 years. It could, therefore, be argued that it is necessary to demonstrate safety only up to this time following closure.
- *Engineered barriers are lost*: there can be no guarantee for times in excess of approximately 100,000 years that the EBS will maintain its essential functions. For example, the waste canisters will eventually degrade and allow the release of radionuclides; the compressed bentonite that might be surrounding an HLW waste canister is relatively thin and cannot be assumed to provide its diffusive barrier for ever—for example, it may be degraded by erosive processes or could undergo mineral transformations. Again, all these processes will need to be considered in a safety case.
- *Natural dispersion and dilution*: radionuclides released from the waste will be transported by flowing groundwater or diffuse away from the repository and be dispersed and diluted in the geosphere. The extent of this dispersion and dilution will depend on the types of rocks surrounding the host rock or the repository—on their porosities and permeabilities and on the hydrogeological environment in which the site lies, e.g. the hydraulic gradients and groundwater fluxes, etc.

The discussion above relates to the disposal of SF. For other waste types, in particular ILW-LL, which are likely to be surrounded by an EBS that will not restrict the release of radionuclides to the same extent, releases may occur earlier. The most likely conditioning methods for ILW-LL are cementation or bituminization, and such waste may be placed in vaults with additional cementitious backfill. In fact, it is likely to be more difficult to make a safety case for this type of waste than for the higher activity wastes, which may be present in considerably smaller volumes. This will be particularly significant in countries that carry out reprocessing of SF, such as the UK and France, where the volumes of ILW-LL are considerably in excess of those of HLW.

3.2.2 Engineered Barrier Systems

The type of EBS is linked to the type of waste, its conditioning and the disposal concept being considered. As indicated in the Introduction, this chapter concentrates on the geological aspects of waste disposal so that the discussion of the EBS is purposely limited in its extent. Examples of key engineered components of disposal systems currently being considered by waste management organizations (see, for example, Figs. 1 and 7) include:

- *Concrete or metal waste containers*: concrete and steel containers, although they may actually last for thousands of years, are generally conservatively assumed in safety analyses not to have any physical containment function after about 1,000 years. They can, however, buffer chemical conditions in the repository so as to limit the release and transport of radionuclides for very much longer times. Copper and titanium waste containers are expected to have a containment function for up to 100,000 years, although their corrosion may take even longer.
- *Backfill and buffer (around the waste)*: concretes can limit transport of radionuclides by diffusion for a long period and can also buffer the chemistry of the pore water and act as a sorbing medium for radionuclides; clays, such as bentonite, are naturally occurring materials which can provide a diffusion barrier for extremely long times.

Some disposal concepts place great emphasis on the protective roles of these EBS materials for protracted periods of time, the longest being the Scandinavian concepts for SF disposal in thick copper containers surrounded by a bentonite buffer (Fig. 7). Others rely more on the geochemical barriers in the near field of the repository and on dispersion and dilution in regions of the natural barrier system for some of the radionuclides. An example of this more chemically based approach is the phased geological repository concept developed by Nirex in the UK for the geological disposal of LILW-LL (Nirex 2005). There can be significant differences from one national programme to another, from site to site, and from one repository concept to another so that the role and the relative importance ascribed to each part of the multi-barrier system is very variable.

3.3 Implementation of Disposal Facilities

The development of a geological disposal facility, its operation and its final closure will take many decades. A proper legal and organizational framework must be established and a disposal strategy agreed with the various stakeholders before much progress can be made. The long timescales to implementation and the novel structure of the task mean that the activities themselves need to be carried out in a staged or stepwise manner.

The allocation of the functions for waste management and regulatory control is an important first step. In the majority of countries the regulatory task is left to the government and the implementation to those responsible for producing the waste, although there are exceptions regarding who is responsible for implementation as in some countries (e.g. the USA) this is also the responsibility of a government department. The nuclear power plant owners can join forces to form dedicated waste management organizations and there are many examples of these, e.g. Posiva (Finland), SKB (Sweden), Nagra (Switzerland), etc.

Following the establishment of an organizational structure within any country, it is necessary to formulate an overall waste management strategy. Such a strategy needs to include the key decision points, to decide how decisions will be taken and to ensure that sufficient resources will be available. Extensive guidance is available on such matters in international consensus documents produced by the IAEA. Any strategy, as indicated above, will need to be phased or staged: SF and HLW, for example, may need to be stored for several decades to reduce their heat outputs; waste repositories take several decades to develop (the combined effects of site selection, site characterization and construction); the repository operation is also likely to last several decades; and post-closure safety needs to be assured for many thousands of years. These extensive times have resulted in the development of a proposal for 'adaptively managing' such a staged development (National Research Council 2003). This implies adopting a flexible process in which the new knowledge gained at each stage is used to plan the content and duration of subsequent stages as opposed to defining in advance all the deadlines and milestones at the beginning of the programme. A useful review of repository implementation by McCombie (2007) includes a description of all the stages of the process, including the cost implications and the status of the disposal programmes in selected countries.

4 Comparison Between the Disposal of CO_2 and Radioactive Waste

A comparison between the disposal of CO_2 and RW is shown in Table 4, which is used to guide the discussion below regarding the similarities and differences between the disposal of the two types of substances.

4.1 *Characteristics of the Geological Media*

Four characteristics are discussed below: tectonic stability, the past stability of the site and the area in which it is located, the geological environment and the host rock type for disposal. These cover the main geological aspects considered in this chapter.

Table 4 Comparison of CO_2 and radioactive waste disposal, concentrating only on the geologically related issues considered in this chapter

Characteristic or attribute	CO_2 disposal	Radioactive waste disposal
Characteristics of the geological media		
1. Tectonic stability	Tectonically stable region preferred	Tectonic stability preferable, but still possible in tectonically active areas, such as Japan. Limitations regarding features such as active faults, Quaternary volcanoes, uplift rates, etc.
2. Past stability	Currently is not considered, or very little consideration is being given in a few projects	Important to understand and demonstrate past physical and chemical stability to increase confidence that such stability will continue into the future
3. Geological environment	In sedimentary basins in strata that • Have sufficient porosity (for capacity) and permeability (for injectivity) • Are confined by low permeability caprock (shales and/or evaporites) • Are minimally fractured, faulted or discontinuous	One in which • Groundwater fluxes at depth are sufficiently low • Reducing conditions exist within the disposal zone • Sufficient volume of host rock exists to house repository • Host rock has suitable geotechnical properties for underground construction • Geological complexity is acceptably low so that site can be adequately investigated and a convincing safety case developed
4. Rock type	Sedimentary rocks (sandstone, carbonate)	Hard, fractured (crystalline) rock, sedimentary rocks of various types (most probably mudstones and clays) and evaporites (most probably halite) (see Table 3)
Emplacement characteristics		
5. Mode of disposal	Injection through wells	Emplacement in (and from) tunnel and/or vault systems, i.e. in-tunnel, borehole/hole (both vertical and horizontal and both long and short) and vault emplacement. Considerable use of EBS, which can take a variety of forms
6. Volume	Very large (Gigatonnes, or 10^{10} m^3/year)	Comparatively very small (see Table 2) (total volume of long-lived wastes generated to date is approximately 4×10^6 m^3). A typical reactor generates approximately 30 t of packaged HLW per year

(continued)

Table 4 (continued)

Characteristic or attribute	CO_2 disposal	Radioactive waste disposal
7. Depth	>800 m up to >5,000 m	Relatively shallow: >200 m and probably <1,000 m (for the great majority of disposal concepts). One concept, deep borehole disposal, would require depths of up to 5,000 m approx.
8. Physical state	Fluid (mostly supercritical)	Solid
9. Containment mode	Natural barriers (shale and/or evaporitic layers) Most likely to have multiple natural barriers (confining strata)	Both natural and engineered barriers. Always a geological barrier acting in tandem with an EBS
10. Timescale of interest	Two timescales • Associated with global warming (greater than centuries) • Associated with local risks posed by injection and possible leakage (decades to centuries)	Widely discussed in radwaste community. Detailed, quantitative calculations required probably for at least 100,000 years, less quantitative for longer, possibly up to one million years (and possibly beyond)
11. Containment period	At least several centuries, up to millennia	Depends on disposal concept and waste types. Absolute containment, for some disposal concepts, could be $>10^4$ years and possibly as much as 10^5 years. For some waste types and disposal concepts (most likely for ILW-LL) absolute containment cannot be guaranteed for these periods. Releases are treated in a probabilistic manner and are acceptable if below the dose or risk target
Effects of emplacement and potential migration from the disposal site		
12. Direct effects of disposal	• Pressure increase • Thermal effects due to cooling • Geochemical reactions in the presence of formation water in a weak acidic environment • Geomechanical effects as a result of pressure increase and stresses • Brine displacement	• Thermal effects due to radioactive decay (for heat emitting wastes) • Geochemical reactions and processes in both the near and far fields • Biochemical processes in both the near and far fields • Geomechanical and hydrogeological effects due to repository construction and operation

(continued)

Table 4 (continued)

Characteristic or attribute	CO_2 disposal	Radioactive waste disposal
13. Effects on the natural barrier	No significant structural modifications to the geological environment caused by the engineered systems (wells), but the CO_2 itself may have effects on barrier integrity	The construction of the repository and the EBS employed will directly affect the natural barrier (although probably only locally). Also, heat emitting wastes will directly affect the natural barrier, although any effects will be limited to a few thousand years, at most
14. Transport mechanisms of CO_2 or radionuclides	The CO_2 itself (excess pressure due to its injection and also its buoyancy)	Dominantly via groundwater (advective and diffusional transport), but to a lesser extent via gas (produced by a variety of geochemical and biochemical processes in the near field and by radioactive decay). Transport of radionuclides can also take place in colloidal form
15. Return to the biosphere, hydrosphere and atmosphere	There are, in effect, no engineered barriers, and leaky wells, fractures and other local geological features may provide a pathway for the return of CO_2. Evaluating the potential impact of wells may be one of the key issues of assessing the performance of the disposal system	Considerable proportions of the long-lived wastes are encapsulated within containers that will remain intact for considerable times. Even after canister failure, the EBS will delay the release of radionuclides for further times (see Fig. 9)
Site activities		
16. Site characterization	Considerably simpler and shorter investigation programme. Considerably sparser information, based only on limited boreholes and seismic imaging	Very comprehensive and lengthy investigation programme. Eventual underground access allows considerably greater level of detail and certainty regarding the near field (a prerequisite for the development of a final safety case)
17. Monitoring	Monitoring required for baseline conditions (site selection), during injection, and decreasingly after cessation of injection for site closure and ensuring long-term safety of the system	All disposal concepts have extensive barrier systems, so, although monitoring will be a requirement, no releases are likely to be detected for a considerable time (i.e. several millennia) after closure. Monitoring is obviously linked to the possibility of waste retrieval and/or the reversibility of the disposal process

(continued)

Table 4 (continued)

Characteristic or attribute	CO_2 disposal	Radioactive waste disposal
18. Future access, intrusion or penetration	Penetration by future wells drilled for other purposes (e.g. exploration, production) is quite possible. Mining of coal is also possible	Sites will be selected only in areas where the intrusion risk is considered low (i.e. no mineral resources)
19. Retrievability of the waste	This is not an issue, except with regard to specific cases mentioned in Sect. 6	Retrievability of the waste has implications for repository operation and for the design of the monitoring programme. It does not necessarily influence the type of disposal environment selected nor does it necessarily exclude specific host rock types from consideration

EBS engineered barrier system, *HLW* high-level waste, *ILW-LL* long-lived intermediate-level waste

1. *Tectonic stability*: Relatively benign and stable tectonic conditions are preferred for both forms of disposal. (It is important to appreciate what is meant by the term *stable* in this regard. It does not imply that conditions at depth are unchanging, but that they change only slowly or, in most cases, extremely slowly. In fact, what is most significant is that they can be shown to change *sufficiently slowly* to ensure the long-term safety of the repository; see NEA 2005, 2009.) The presence of stable conditions is likely to provide an intrinsically safer disposal environment (and is also likely to make the demonstration of long-term safety an easier task); however, it has been shown that the presence of active tectonics does not preclude the possibility of disposing of RW (see also item 5 below in relation to the mode of disposal and item 10 in relation to the timescales involved of interest with regard to tectonic stability). An interesting discussion of this subject in relation to the siting of nuclear facilities (including waste repositories) is provided in Connor et al. (2009). It is not normally possible, especially in respect of waste repositories, to make specific statements regarding the required level of tectonic stability, except with regard to specific features, such as active volcanoes. It is probably better to develop preferences in this regard, for example the separation of any repository from an active fault, with any such preferences being related to the implications for the long-term safety of the facility. Japan provides an example of a country where the level of tectonic activity is of particular interest for the disposal of RW. Japan has had an RW disposal programme that has been operating for many years and the Nuclear Waste Management Organization of Japan (NUMO) is confident that it will be possible to dispose of long-lived RW in Japan, even though the entire country is tectonically active, with numerous active faults, areas of active volcanoes and areas with geologically rapid uplift rates. A series of reports produced as part of the H12 project in Japan (JNC 2000), combined with a large R&D programme, have led to the development of a structured approach to the

development of a repository for long-lived wastes (NUMO 2007), in which it is envisaged that a repository for HLW will be available for use in 2035. (The H12 project, whose full name is the Project to Establish the Scientific and Technical Basis for HLW Disposal in Japan, is named after the 12th year of the Heisei era—related to the current Emperor.) As a contrast, reference can be made to Posiva's proposed repository site in Finland, which is a site in a tectonically stable environment (see McEwen and Andersson 2009). Even in this stable environment, the possibility of future rock movements needs to be taken into account—in this case it is the effect of future glaciations and their impact on fracture displacements that is potentially important in locating waste canisters. Two recent reports on the significance of geological stability (which is a broader subject than just tectonic stability) in the context of the disposal of RW are of interest here (NEA 2005, 2009). These reports, which represent the outcomes of two workshops, discuss the significance of geological stability in all its forms, i.e. mechanical, hydrogeological, hydrogeochemical, etc., with reference to the use of argillaceous and crystalline rocks for disposal purposes. One of the conclusions of these reports is that, as long as disposal sites are selected with care, there would appear to be no significant reasons why RW cannot be disposed of safely at depth, even in areas with relatively high levels of tectonic activity. With regard to CO_2 disposal, although to date most existing pilot and commercial scale operations are in tectonically stable regions, such as the North Sea, Sahara and the Williston Basin in Canada, the Westcarb Regional Partnership on CO_2 Sequestration in the USA is looking at identifying and piloting CO_2 injection sites in California, and in Japan pilot demonstrations took place in a coalfield at Ishikari (Shi et al. 2008) and in a deep saline aquifer at Nagoaka, where an earthquake of magnitude 6.8 with an epicentre distance of 20 km occurred without adverse effects on the injected CO_2. Both California and Japan are located in tectonically active regions around the Pacific Rim. Nevertheless, as in the case of RW, stable areas are preferred and volcanic areas and sites close to active faults should be avoided.

2. *Past stability*: This is currently not explicitly considered in the disposal of CO_2, although it is suggested that this subject may become part of the assessment of a proposed CO_2 disposal facility (Maul et al. 2007; Maul 2011). Site stability in CO_2 disposal is implicitly considered under tectonic stability (see point 1 above). In the disposal of RW it is an important requirement to demonstrate such past stability, and in fact to understand the evolution, over perhaps the last 100,000–1,000,000 years, of the site and the area in which it is located, in order that a convincing account can be developed of its likely evolution, and hence its continuing stability, in the future. A very considerable amount of work is required to produce such an account. One field in particular is of interest in this regard, that of the use of palaeohydrogeology, where studies are made of the past hydrogeological evolution of a site. A useful review of the use of such data is presented in the conclusions of the PADAMOT (Palaeohydrogeological Data Analysis and Model Testing) programme in Degnan et al. (2005). The two NEA reports on geological stability (NEA 2005, 2009), referred to in item 1 above, are also of relevance here, and several of the presentations given at these workshops concern the past stability of a site.

3. *Geological environment*: The range of geological environments currently considered suitable for the disposal of CO_2 is smaller than those considered suitable for the disposal of RW. The disposal of CO_2, as currently accepted, can take place only in sedimentary basins, in environments that have the characteristics listed in Table 4, e.g. have sufficient porosity (for capacity) and permeability (for injectivity), are minimally faulted and/or discontinuous (see Sects. 2.2 and 2.3), and are confined by low permeability caprock (shales and/or evaporites) that is not fractured. Lately there has been interest in exploring the potential for CO_2 storage in basalts due to the fact that rapid geochemical reactions are likely between the injected CO_2 and the basalt, and two test sites are being pursued to explore this concept, one in the USA and another in Iceland in the Hellisheidi hydrothermal field. However, it is worth mentioning that in Iceland the concept consists of dissolving the CO_2 in water at the surface and injecting the CO_2-saturated water, rather than injecting free-phase, high density CO_2 which otherwise will most likely leak due to the characteristics of the basalt. RW can be safely disposed of in a variety of geological environments, i.e. a repository can be located in an argillaceous rock, in an evaporite (either bedded or domal in form) and in a wide range of hard, fractured rocks, any of these existing in a wide range of geological environments, from Archean basement terrains to relatively recent sedimentary basins. As long as the environments have the characteristics and properties listed in Table 4, they are likely to be suitable. Probably the most significant of these are that groundwater fluxes at depth are sufficiently low and that chemically reducing conditions are present (and that they remain reducing) (see Sect. 3.2). Another important factor is the requirement for there to be sufficient volume of suitable rock to house the repository. The location and layout of disposal tunnels or vaults may be most constrained in crystalline rock, due to the ubiquitous presence of fracture zones, which need to be avoided. The design and layout of the repository is likely to vary considerably between different types of geological environments and can be modified to take into account the specific characteristics of the site in question.
4. *Rock type*: The disposal of CO_2 is likely only in sedimentary rocks with sufficient porosity and permeability, such as sandstones and carbonates, although disposal in salt caverns is also possible (see Sects. 2.2 and 2.3). If proven successful and economic, storage of CO_2 in basalts may be considered sometime in the future. RW can be disposed of in a larger range of rock types, including many types of hard, fractured (crystalline) rocks, sedimentary rocks of various types (but probably most likely mudstones and clays, as these have the necessary low permeabilities) and in evaporites (most likely halite) (see Sect. 3.2).

4.2 Emplacement Characteristics

Seven attributes, associated with the characteristics of the emplacement of CO_2 or RW, are considered below: the mode of disposal, the volume of the waste, the disposal depth, the physical state of the waste, the mode of containment, the timescale of interest and the containment period.

5. *Mode of disposal*: The disposal of CO_2 will take place via direct injection from wells. In contrast, RW will be emplaced in or from tunnels or in vaults, e.g. in tunnels (e.g. Fig. 1), in boreholes or holes (both horizontal and vertical) drilled from disposal tunnels (e.g. Fig. 7).
6. *Volume*: The volume of CO_2 that requires disposal is extremely large, in the order of gigatonnes (equivalent to billions of cubic meters) per year whereas, in comparison, the volumes of RW, especially HLW and SF, are very small—many orders of magnitude lower. As of 2003, the total volume of all long-lived RWs, both in storage and already disposed of was approximately 650,000 m^3. Table 2 lists some figures which indicate some of the volumes of RW that require disposal. The total volume of conditioned RWs produced per year in all the OECD countries is 81,000 m^3, and much of this waste is relatively short-lived and does not require geological disposal.
7. *Depth*: The disposal of CO_2 is likely to take place in the range 800–5,000 m, for the reasons given in Sect. 2.1. The disposal of RW will almost certainly take place over a smaller and shallower depth range (probably >200 m, and more likely >400 m and <1,000 m) with the exception of the deep borehole disposal concept, where disposal could take place at depths as great as approximately 5,000 m (see McEwen 2004).
8. *Physical state*: CO_2 will be disposed of as a fluid, and most likely a supercritical fluid, whereas RW will definitely be disposed of as a solid (but see item 5 above).
9. *Containment mode*: No engineered barriers are present in the disposal of CO_2 (at least not in the disposal zone itself) and, given the size and nature of the disposal operations, none can be constructed; although, of course, the necessary boreholes are sealed so as to prevent leakage, and there has to be at least one, but preferably several, natural barriers (i.e. confining strata) in the geological succession (see Sects. 2.2 and 2.3). An important axiom of RW disposal is the necessity of having multiple barriers (normally referred to as the multi-barrier concept), in which an EBS (see Figs. 1 and 7 and discussion in Sect. 3.2) acts in tandem with the geological barrier. The EBS is likely to be composed of several components (each of which is itself a barrier). The relative significance given to the EBS and the geological barrier in the safety concept will vary with the type of repository under consideration and, in particular, with the geological environment for disposal, but both types of barrier will always be present.
10. *Timescale of interest*: There are two timescales currently considered for the disposal of CO_2. The first of these is that associated with global warming, which is likely to be in the order of centuries to millennia, and the second is that associated with local risks related to the injection of the CO_2 and to possible leakages (Figs. 4 and 5), in the order of decades to centuries. The first timescale is associated with the stabilization and subsequent decrease of CO_2 concentration in the atmosphere, whilst the second timescale is associated with the immobilization of CO_2 in the ground and the disappearance of the potential for and risk of leakage.

The timescales for the disposal of RW are considerably longer and the distinction between the different timescales considerably more complicated. Figures 8 and 9 illustrate the decreasing risk or hazard of the waste with time, so that the events that are of most interest in the short term, i.e. during the operational phase of the repository and immediately post-closure (less than about 150 years), will be different from those in the far future, and the contribution of the different radionuclides will change with time. Detailed, quantitative calculations of doses or risk will be required for at least 10,000 years, and more likely as much as 100,000 years, with more qualitative calculations and reasoned arguments being presented for times up to, and possibly exceeding, one million years. Figure 8 shows that for SF the toxicity of the waste is similar to that of the uranium ore from which the fuel was fabricated after about 100,000 years; for HLW this period will be less, of the order of 10,000 years. The timescales of interest can be related to this changing toxicity of the waste, and therefore the risk associated with its disposal, so that several waste management organizations have developed timescales of interest related to the development and evolution of the repository, for example:

- Operation (approximately 100 years);
- Near future (post-closure monitoring phase, phase of global warming, no expected release from repository via groundwater pathway, etc.) (perhaps up to 1,000 years);
- Period where the integrity of the EBS is guaranteed, where climate change does not include glacial phases, where there is considerable confidence in the behaviour of the disposal system (up to 10,000 years approximately);
- Period during which the toxicity of the waste approaches and/or equals the toxicity of the uranium ore, when there is the possibility of major climate change and when the confidence in the behaviour of the disposal system may be considerably reduced (10,000–100,000 years approximately);
- Period when the processes and events associated with the disposal system are those illustrated in Fig. 9 and when major climate change may be even more significant (>100,000 years).

11. *Containment period*: For CO_2 disposal, containment will likely be required for several centuries up to millennia, although to what extent this needs to be absolute containment, or whether such absolute containment is indeed possible, is unclear. There are some organizations that, by analogy with RW disposal, are suggesting that containment (absence of leakage) should be demonstrated for 10,000 years, but so far this view has not gained acceptance. Recognizing that some CO_2 leakage may be unavoidable, several studies have suggested that, from the point of view of climate stabilization, small global leakage rates (0.01–1%/year) would still be acceptable (Pacala 2003; Hepple and Benson 2005; IPCC 2005). However, these are peak leakage rates, and the long-term average permissible leakage rates should be much lower. The extent of absolute containment for RW (i.e. no escape from the EBS) depends on the waste type and disposal concept and could vary from less than 10,000 years (perhaps as little as approximately 1,000 years) to as much

as 100,000 years—possibly as much as one million years in some circumstances. Some disposal concepts assume that the waste containers have no containment function immediately after closure of the repository—although this does not imply that release of radionuclides takes place immediately. There may be problems with the escape of radioactive gas relatively soon after repository closure, especially in the disposal of ILW-LL, although whether this gas ever enters the biosphere depends on several factors, many of which are site specific. Releases are treated in a probabilistic manner and are acceptable if the consequences of such releases are below a risk or dose target.

4.3 *Effects of Emplacement and Potential Migration from the Disposal Site*

The subjects discussed below concern the effects of the disposal of CO_2 or RW and mechanisms by which they could migrate from the disposal site. These include: the direct effects of disposal, the effects on the natural barrier, the transport mechanisms of CO_2 or radionuclides and the possible return to the biosphere, hydrosphere and atmosphere.

12. *Direct effects of disposal*: For CO_2 disposal there are a variety of effects resulting from the emplacement and presence of CO_2 in the subsurface, such as pressure increase, stress changes and deformation, and geochemical reactions, including mineral dissolution and/or precipitation in the presence of formation water in a weakly acidic environment, and also brine displacement. Another possible effect will likely be cooling of the disposal reservoir or aquifer due to the fact that the injected CO_2 will likely be at a lower temperature than the initial formation temperature. This thermal effect may in turn affect stresses in the disposal unit and overlying confining layer. For RW the direct effects are thermal, geochemical reactions and processes, in both the near and far fields, and hydrogeological and geomechanical effects due to repository construction and operation; although the extent of any such effects depends on many factors, such as the types and characteristics of the wastes (there may be several different waste types in a single repository), the rock type, the repository design, the depth of the repository and the length of the operational period.
13. *Effects on the natural barrier*: With regard to the disposal of CO_2, no significant structural changes to the geological environment are expected due to the drilling and sealing of boreholes, although the CO_2 itself may affect the integrity of the natural barrier and there will be physicochemical changes to the rock mass in the disposal zone, which have been discussed in Sects. 2.3 and 2.4. The physical (pressure and temperature) and geochemical changes induced by CO_2 disposal may locally affect the integrity of the natural barrier at the interface between the disposal unit and the overlying natural barrier, but they should not affect the barrier's confinement/containment ability, otherwise the site should not be selected for disposal.

With regard to the disposal of RW, the construction of the repository and the EBS will directly affect the natural barrier (although mainly only locally). These changes may be chemical, due, for example, in some but not all repositories, to the effects of the alkaline plume caused by the cementitious components of the waste, the EBS and the rock support. Also, heat emitting wastes will directly affect the natural barrier, although any thermal effects will be limited to a few thousand, years, at most, with the dominant thermal phase lasting perhaps 1,000 years.

14. *Transport mechanisms of CO_2 or radionuclides*: With regard to CO_2 disposal, the mechanisms that drive the flow of free-phase CO_2 are the excess pressure, due to the injection process (Fig. 5), and its own buoyancy (see Sects. 2.1 and 2.4). In the case of CO_2 dissolved in formation water, the transport mechanism is the hydrodynamic drive of formation water and/or free convection induced by density differences in the order of 1% between the CO_2-saturated brine and unsaturated brine, if unstable conditions develop. The dominant transport mechanism for radionuclides is either advection in the groundwater or, in some host rocks with very low hydraulic conductivities such as clays and halite, diffusional transport. There are likely to be elements of both advective and diffusional transport in many repositories, e.g. perhaps diffusion in the EBS and advection in the geosphere. Transport in a gas phase is also possible—some wastes, such as LILW-LL, may produce considerable volumes of radioactive gas, due to geochemical and biochemical reactions and radioactive decay, and the potential effects of pressurization of the repository system need to be considered; other wastes, such as HLW, can produce much smaller volumes of gas, mainly due to processes such as anaerobic corrosion of the steel components of the waste form. Transport of radionuclides can also take place in colloidal form.
15. Return to the biosphere, hydrosphere and atmosphere: With regard to CO_2 disposal, the potential for leakage exists only as long as CO_2 is in free, mobile phase. There are no engineered barriers against CO_2 leakage, and leaky wells, fractures and other local geological features may provide pathways for the return of free-phase, mobile CO_2 to shallow potable groundwater, the vadose zone, soil, the biosphere and/or atmosphere (see Sect. 2.4). Evaluating the potential impact of wells and well integrity may be one of the key issues of assessing the performance of the disposal system. Such wells include both those used for CO_2 injection and those that may already be present or that may be drilled in the future and penetrate the disposal horizon. The issue of long-term cement and casing integrity is central to CO_2 disposal. With regard to RW, some of the long-lived wastes are encapsulated within containers that will remain intact for considerable times, e.g. those contained in copper or iron canisters may last for periods in excess of 100,000 years (see Figs. 7 and 9). Other wastes are encapsulated in concrete, which is designed to limit the release of the radionuclides. Even after canister failure, the EBS will delay the release of radionuclides for further times (see Fig. 9 for the situation regarding HLW and SF). The safety case for a repository for long-lived wastes needs to consider the effect of the return of any radionuclides to the biosphere. A variety of scenarios are normally modelled, in which different evolutionary paths for the repository are considered, i.e. different future

climates, premature failure of waste packages, etc. The return of radionuclides to the biosphere needs to be determined for all of these scenarios and the doses to critical groups determined for different times in the future. The processes that can take pace at the geosphere–biosphere interface can be complex, as can the processes and pathways in the biosphere. Again, as in item 14 above, both water-borne and gaseous releases need to be considered.

4.4 Site Activities

Included here are the activities that take place on and around the disposal site (except for the mode of disposal, which is considered above in Sect. 4.2). The subjects considered below are: site characterization, monitoring and future access, and intrusion or penetration of the disposal site.

16. *Site characterization*: Adequate site characterization is essential in ensuring proper site selection and performance of a CO_2 disposal site (IPCC 2005). Sites will be characterized based on well and seismic data, rock and fluid samples, laboratory analyses, and computer modelling, but a degree of uncertainty will always remain. Site characterization will likely be continuously refined after the start of injection as monitoring data are acquired, interpreted and fed back into the system. Research programmes will most likely accompany site characterization in the case of CO_2 disposal operations deployed in the near term, as is currently the case with the few pilot, demonstration and commercial operations implemented to date. With regard to RW disposal, a very comprehensive and lengthy investigation programme is inevitable—almost certainly more comprehensive and more extensive than one for CO_2 disposal. Such a programme is likely to involve a number of phases, may last many years and be accompanied by an extensive R&D programme. Eventually, underground access will be required, which will allow a considerably greater level of detail to be obtained regarding the near field (a prerequisite for the development of a final safety case). Site investigations at sites that are likely to be developed into repositories are currently taking place (or one phase of such investigations has been completed, with the investigations planned to be resumed later) in Finland, Sweden, Canada and France, and other countries are planning to carry out similar investigations starting in the next few years. Investigations are also taking place at underground research laboratories. Information on these investigation programmes can be obtained from the respective websites (e.g. www.posiva.fi, www.skb.se and www.andra.fr), whilst www.radwaste.org provides a list of the majority of websites of waste management organizations, waste repositories, regulatory organizations, etc.
17. *Monitoring*: With regard to CO_2 disposal, monitoring is likely to be required for defining the baseline conditions (before any disposal takes place), during injection, and decreasingly as the risk decreases (see Fig. 5) after the cessation of injection for site closure and for ensuring the long-term safety of the system.

With regard to RW disposal, all disposal concepts have extensive barrier systems, so, although monitoring will be a requirement (due, in part to public concern regarding the possibility and consequences of any release of radionuclides), no releases are likely to be detected for a considerable time (i.e. several millennia) after closure. Monitoring will also be required as part of the site investigation programme in order to study medium-term changes in parameters such as hydraulic head and seismicity, but a monitoring system will also be required to be in place before any underground construction takes place so as to monitor the effect of such construction on groundwater heads, groundwater chemistry and rock stress, i.e. to define the baseline conditions (EC 2004). Monitoring systems for this purpose are currently in operation at Olkiluoto, Finland (e.g. Pitkänen et al. 2007), where construction of what will become the access ramp to the SF repository is currently taking place. Monitoring will also be required in association with any requirement for the reversibility of the waste emplacement process or the retrievability of the waste.

18. *Future access, intrusion or penetration*: Sites used for CO_2 disposal may be penetrated in the future by exploration and/or production wells due to hydrocarbon exploration in, or production from, deeper strata. In the case of CO_2 disposal in coal seams, if successful, consideration should be given to coal mining or underground coal gasification at some future date. The possibility of such future intrusion appears, however, not to be an insurmountable problem, although it has been suggested that such potential intrusion be included in any future performance assessments for CO_2 disposal (Maul et al. 2007; Maul 2011). Sites for an RW repository will only be selected where the future intrusion risk, i.e. after closure of the repository, is considered to be low. This is likely to exclude all areas where there are mineral resources (of all types), and there has been extensive work in this area. The retrievability of the waste and the reversibility of the disposal operations need also to be considered—and all these subjects have been extensively discussed over recent years (see discussion in Chapman and McCombie 2003). Opponents of deep disposal would prefer to leave the wastes indefinitely in monitored surface or underground stores. Proponents argue that this is not a sustainable solution and that it is a higher risk option and that one should proceed in a stepwise manner towards eventual disposal. It may be necessary, in order to obtain public support for the disposal of RW, at least in some countries, to evolve a strategy that includes the possibility of retrieval of the wastes at all stages of the repository development programme. Such a programme is likely to last at least 100 years before final closure of the repository takes place and a stepwise approach, in which waste is slowly emplaced in the repository, combined with extensive monitoring of the performance of the repository, may be sufficient to allow the programme to proceed. The subjects of the accidental penetration of a CO_2 storage or RW disposal site and of the retrievability of RW from a repository are large subjects by themselves and much of their discussion is not geological in nature, and, being outside the scope of this chapter, are discussed elsewhere (e.g. see West et al. 2011).

5 Insights and Implications

5.1 Main Insights from the Comparative Assessment

It can be seen from the above discussion that there are similarities, but more, and sometimes significant, differences between the disposal of CO_2 and the disposal of RW. In both cases tectonic stability is preferred, and both require at least one, and preferably several, natural barriers against migration. Monitoring after emplacement will be required in both cases (for CO_2 monitoring, see Brunskill and Wilson 2011), although, in the case of RW disposal, unless monitoring takes place very close to the waste, it is very unlikely that any releases of radioactivity will be detected (and even close to many of the waste forms no releases are likely for very considerable periods in the future (EC 2004)). Also, in both cases there will be local effects on the geological environment as a result of the emplacement, although not identical (e.g. thermal cooling versus heating, different geochemical and geomechanical effects, etc.).

Whilst the current thinking is that CO_2 can be stored only in certain types of sedimentary (soft) rocks, RWs can be disposed of in hard rock as well. Very large volumes of fluid CO_2 will ultimately be disposed of through wells at great depths in natural geological media, whilst considerably smaller volumes of RW are or will be disposed of in solid form in tunnels or vaults at relatively shallow depths, using a combination of engineered and natural barriers. The timescales and containment period are significantly shorter for CO_2, in the order of centuries to millennia, whilst for RW they are in the order of at least ten thousand to possibly as much as a million years.

Whilst the main mechanism for the possible migration of CO_2 out of the disposal unit is its own buoyancy, being lighter than water (i.e. it is self-propelled), the main mechanism for the transport of RW once outside the EBS is transport by groundwater (i.e. it needs a carrier).

Site characterization in the case of CO_2 disposal is inherently simpler, but the results of such characterization are likely to be less certain because of sparser data and information. In contrast, a site characterization programme at an RW disposal site is likely to be considerably more comprehensive, lengthier and also more expensive. The programme will also include eventual access to the disposal zone via shafts or inclined tunnels, which will provide considerably more data on the rock mass (see the example of the current construction of the ONKALO at Olkiluoto, Finland (Posiva 2009; Andersson et al. 2007)), and this forms part of the site characterization programme. The extent of the safety case that will be required for obtaining permission to dispose of RW appears likely to be more comprehensive than the equivalent for CO_2 disposal (and this applies also to the associated R&D programme). Regarding the costs of characterization, they are likely to be lower for CO_2 disposal than for RW disposal, except possibly in the case of offshore CO_2 disposal in the deep sea in a region with limited data coverage.

Whilst the intent is that disposal sites for RW will never be penetrated or accessed (with disposal taking place only in areas with no mineral reserves), it is likely that in some cases the injected CO_2 will encounter existing wells, and it is possible that CO_2 disposal sites will be penetrated in the future by wells drilled for other purposes, given the very large areal footprint of CO_2 disposal operations. The potential impact of such penetration is unclear, as it may lead to CO_2 leakage into other strata, potable groundwater and even to the surface, posing local risks and also reducing the efficacy of the CCS process and limiting its usefulness. For this reason operators and regulatory agencies must take care in properly drilling, completing and abandoning such wells, and possibly new regulations will have to be developed for such situations.

There is considerably more experience in the disposal of RW, as there have been research programmes for several decades in many countries and there are now some operating repositories, although only one for long-lived RW—at the WIPP in the USA—whilst the disposal of CO_2 in geological media has been considered as a climate change mitigation measure only in the last decade. Several additional repositories for long-lived RW should, however, be developed over the next 2 decades, whilst large-scale demonstration projects for the disposal of CO_2 are under way. The science of performance assessment for RW is also well developed and there is general international consensus as to the suitability and efficacy of geological disposal for such wastes, whilst the science and criteria for performance assessment in the case of CO_2 disposal are currently under development. Consequently, the nascent CO_2 disposal industry can learn from the mistakes and successes of the RW disposal industry. Also, each national and subnational jurisdiction should develop a proper legal and regulatory framework for the selection, characterization and acquisition of CO_2 disposal sites, and a framework for the management of this new natural resource that is the CO_2 disposal pore space.

5.2 Implications

There are two areas of a geological nature where the disposal of CO_2 in geological media, which is in an incipient phase, can benefit from the experience gained to date from the disposal of RW:

- Site selection;
- Performance assessment (which itself involves three main elements and is itself part of the safety case) including the development of a good site understanding.

The mistakes made in earlier attempts at selecting sites for RW repositories, and the public opposition that was generated, made the nuclear industry in many countries realize the importance of openness and transparency during not only site selection, but in all elements of their RW disposal programmes (e.g. McEwen 2007; see Reiner and Nuttall 2011). An open and transparent approach is likely to be beneficial, not only in selecting sites for the disposal of CO_2, but during all phases of any subsequent disposal programme.

It is probably in the area of performance assessment (and the associated more comprehensive safety case) that the greatest opportunity lies for the proponents of the disposal of CO_2 to make use of what has been learned regarding the disposal of RW. As emphasized in Maul et al. (2007) and Maul (2011), the development of methods for undertaking performance assessments for the disposal of CO_2 is at an early stage. Much can, therefore, be learned with regard to the disposal of CO_2 from the experience of RW disposal over 3 decades: the need to employ a systematic and transparent methodology; the advantage of using system-level models; and the need to make maximum use of information from natural systems (i.e. natural analogues) (see also Maul 2011), where the safety case is discussed. There are, for example, both natural and engineered analogues for CO_2 disposal (Pearce et al. 2004; IPCC 2005). Natural accumulations are being studied in Australia, Europe and the US (IPCC 2005), supplying information on trapping and migration mechanisms and the potential impacts of leakage, as well as providing field-based testing grounds for deep, shallow, surface and atmospheric monitoring tools. This subject is discussed in greater detail in Maul et al. (2007, 2011).

In RW disposal, in addition to the performance assessment itself, the requirement to demonstrate a good understanding of the geology, hydrogeology and hydrogeochemistry of the site, and the region in which it lies, has grown in importance over the years, and is now a major element in developing a safety case for a potential disposal site. As outlined in items 1, 2 and 16 in Sect. 4, this site understanding needs to include an understanding of the way in which the site has developed over at least the last million years and how it is likely to develop in the future over a similar time frame. It seems unlikely that this element will be a requirement in a safety case for CO_2 disposal, at least not to the same extent as that required in the disposal of RW, as the timescales of interest are so much shorter. In all other respects, however, a similar emphasis is likely to be placed on the requirement to develop a good site understanding, and in placing the site in its regional geological context, thereby requiring more geological information than would be required for the performance assessment itself. This has important implications for the design, areal extent and operation of a site investigation programme at a potential CO_2 disposal site and is likely also to require more work in areas such as model validation, together with associated R&D, than is normally the case in the hydrocarbons industry.

Broadly speaking, CO_2 and RW disposal operations follow, or should follow, similar processes of site selection and characterization, application and permission, design and construction, disposal, site closure and post-closure monitoring, with remediation being an activity to be considered in case of leakage and/or migration out of the disposal zone. Whilst there is a large body of experience in the RW disposal industry with regard to site selection and characterization, there is no such experience yet in the incipient CO_2 disposal industry. The characteristics the geological media should possess for CO_2 disposal and, where relevant, any analogies with the RW disposal industry, should serve as a basis for developing policy and regulatory procedures for the site selection, characterization and siting of CO_2 disposal operations.

Given the large volumes of CO_2 for disposal, and the corresponding large footprint of such operations, having the necessary rock volumes available that meet the basic requirements for CO_2 disposal becomes a matter of resources, i.e. the pore space suitable for CO_2 disposal becomes a natural resource to be appropriately managed within each jurisdiction. Thus, as a matter of policy at the national and subnational level, each jurisdiction (state, province, country) should proceed with an inventory of its CO_2 storage capacity in terms of its size and distribution. Such inventories have been completed in some countries (e.g. Australia and most countries in northern Europe), or are in the process of being completed in other countries, such as Canada, the USA and other European countries. A lack of resources and data makes such inventories a challenge in developing countries. On the other hand, the lack of suitable geological environments or sufficient capacity for CO_2 disposal in some countries, as is the case in Japan and South Korea, will affect their policies with regard to a reduction of greenhouse gas emissions because carbon capture and storage, as CO_2 disposal in geological media is known, is an option with limited or no potential. Preliminary studies also indicate that large CO_2 emitters such as China and India may not have sufficient CO_2 storage capacity compared with their current and projected CO_2 emissions. The same limitations regarding the availability of potentially suitable geological disposal environments do not apply to RW, as there is greater flexibility in the use of different geological environments and the volumes of rock required for disposal are so much less than those required for CO_2 disposal.

There are certain legal implications regarding this new natural resource because currently it is not covered by existing legislative acts that would provide for acquisition of the right to dispose of CO_2 and access to the disposal site (i.e. mineral, mining, oil and gas acts, and alike, provide for access to and production of mineral and energy resources, but not for the utilization of the pore space for CO_2 disposal on a large scale). Similarly, regulatory agencies that will be mandated with regulating the CO_2 disposal industry will have to develop clear criteria for site selection, for predicting and monitoring the fate of the injected CO_2 and its effects on the subsurface environment, and for permitting CO_2 disposal projects. (See Bachu (2008b) and Wilson and Bergan (2011) for a review of legal and regulatory aspects that need to be addressed in relation to the management of this new resource.)

Finally, regulatory agencies will have to keep track, both geographically and stratigraphically, of the location of the CO_2 disposal operations and of their large footprints because of the potential for their penetration in the future by well drilling activities. Similar requirements will, naturally, also apply to RW disposal sites, although for the majority of countries there is likely to be only one deep geological disposal site. As already discussed, the likelihood of any future intrusion of such a disposal site is considerably lower than that for a CO_2 disposal site as, in contrast to perhaps the majority of CO_2 disposal sites, such sites will only be located in areas with no mineral reserves.

The potential impacts of CO_2 disposal on RW disposal programmes, or vice versa, also need to be considered. In most countries, RW repository projects are a long way from completion. On the other hand, if it is to be an effective means of mitigating climate change, CO_2 disposal will probably need to be implemented on a large scale within the next 10–20 years. Therefore, the majority of countries with both CO_2 disposal and RW management programmes will probably implement CO_2 disposal first. Thus, in these cases there is no danger that the implementation of CO_2 disposal itself will cause human intrusion of an RW repository. However, although RWs will be disposed of at shallower depths than CO_2, when siting an RW repository in sedimentary rocks, it might be necessary to determine the spatial extent of rock volumes that are likely to be affected by CO_2 disposal in the future and/or the footprint of CO_2 disposal operations. Potentially, large-scale implementation of CO_2 disposal could effectively rule out large areas of a country from consideration as possible locations for an RW repository. This factor may have to be included as part of a site selection programme, although it was not specifically included in the list of initial subsurface screening criteria for the geological disposal of RW published recently in the UK (Defra 2008). On the other hand, in the future the presence of RW disposal sites may reduce the potential for CO_2 disposal in their vicinity.

6 Conclusions

Various geological media are suitable for the disposal of the many products of human activity, among them RW and anthropogenic CO_2 captured from large stationary sources. For both these particular types of waste, the objective of disposal is their isolation from the hydrosphere, atmosphere and biosphere for very long periods of time—in the order of centuries to thousands of years for CO_2, and in the order of tens of thousands to perhaps a million years for RW. However, there are some fundamental differences between the two products that, consequently, dictate the types of geological environments that are suitable and required for their disposal. RWs are in solid form and are relatively limited in volume, but are extremely hazardous. As a result, RWs are to be emplaced in subsurface engineered systems (often known as repositories) at depths of a few hundred metres, using both tunnels and shafts to provide access to the point of waste emplacement. CO_2 is a fluid that can be emplaced in the disposal unit only by its injection through wells, usually at depths greater than 800–1,000 m. The volumes of CO_2 that need to be injected to achieve significant reductions in atmospheric CO_2 emissions are huge, with consequently a very large subsurface footprint (tens to hundreds of square kilometers at a single disposal site). As a result, no barriers to CO_2 escape can be engineered, unlike RW, and containment has to rely entirely on the natural system, or natural barrier.

For both RW and CO_2 disposal a stable geological environment is preferable, with additional safety and remediation measures possibly being necessary if the disposal site is located in a less tectonically stable region. Currently, only sedimentary rocks with suitably high porosity and permeability are considered for CO_2 disposal, although laboratory and field experiments are being carried out to assess the possibility of disposing of CO_2 in other rocks, such as basalts. Hard crystalline rocks, but also sedimentary rocks and evaporites are considered potentially suitable for RW disposal, but it is important that all such rocks should have low to very low permeability. For both forms of waste, any site should be geomechanically sound and associated with low velocities of groundwater flow, although the requirements in this regard are more onerous for RW than they are for CO_2.

The main effects of the disposal of CO_2 are an increase in pressure at depth, with accompanying changes in the stress regime at the disposal site, thermal effects as a result of the temperature difference between the injected CO_2 and the disposal zone (CO_2 being cooler), and geochemical changes as a result of the dissolution of CO_2 in formation water, forming a weak carbonic acid. In the case of RW disposal, the main effects due to waste emplacement are thermal (at least for HLW and SF), as a result of the heat emitted by the waste, geochemical and biochemical, as well as geomechanical and hydrogeological, due to the construction and operation of the repository. The natural barrier is affected, in the case of CO_2 disposal, by the drilling of wells and may be affected by the CO_2 itself and, in the case of RW disposal, mainly by the construction of the repository, but also by the boreholes required to investigate the site. CO_2 may escape from the disposal unit, if a pathway is available, as a result of its own buoyancy and due to the pressure build-up caused by injection, whereas the main escape route for radioactivity is via groundwater, although the release of radioactive gas also needs to be considered. Consequently, extensive site characterization and monitoring is or will be required in both cases, although it seems likely that any such site characterization programme is likely to be more detailed and more prolonged in the case of RW disposal.

A significant difference between the two types of disposal is retrievability, which may be a requirement in the case of RW, but which is not, and may in any case be impossible, in the case of CO_2. There may be cases in the future where wells will be drilled into a plume of CO_2 to release CO_2 and reduce the pressure, but this scenario is envisaged only as a remediation measure in case of uncontrolled leakage. In any case not all of the CO_2 will be recovered because some will be dissolved in formation water and some will be immobilized in the pore space at irreducible saturation or through mineralization. The subject of the retrievability of RW is included, although not discussed in detail, in item 18 of Sect. 4.4. Although the subject has been extensively debated, the possibility of waste retrieval does not necessarily influence the type of disposal environment selected nor does it necessarily exclude specific host rock types from consideration.

There are two areas of a geological nature where the disposal of CO_2 in geological media, which is in an incipient phase, can benefit from the experience gained to date from the disposal of RW: firstly in site selection and, perhaps more importantly, in performance assessment. It is, therefore, probably in the area of performance assessment (and the associated more comprehensive safety case, which includes the development of a good site understanding) that the greatest opportunity lies for the proponents of the disposal of CO_2 to make use of what has been learned regarding the disposal of RW. There are many areas in the development of a comprehensive safety case for a disposal site where geological input is required.

The analysis presented in this chapter indicates that there are, therefore, similarities, but also many significant differences, between the geologically related issues and requirements for the long-term emplacement and isolation of CO_2 and RW, although, in carrying out such an analysis, useful comparisons can nevertheless be drawn between the disposal of these two types of waste.

References

Alexander WR, McKinley LE (eds) (2007) Deep Geological Disposal of Radioactive Waste. Elsevier, Amsterdam

Andersson J, Ström A, Svemar C, Almén K-E, Ericsson L (2000) What requirements does the KBS-3 repository make on the host rock? Geoscientific suitability indicators and criteria for siting and site evaluation. SKB Technical Report TR-00-12. Svensk Kärnbränslehantering AB, Stockholm

Andersson J, Ahokas H, Hudson JA, Koskinen L, Luukonen A, Löfman J, Keto V, Pitkänen P, Mattila J, Ikonen A, Ylä-Mella M (2007) Olkiluoto site description 2006. Report POSIVA 2007-03. Posiva Oy, Olkiluoto

Aya I, Yamane K, Shiozaki K (1999) Proposal of self sinking CO_2 sending system: COSMOS. In: Eliasson B, Riemer PWF, Wokaun A (eds) Greenhouse Gas Control Technologies: Proceedings of the 4th International Conference on Greenhouse Gas Control Technologies. Pergamon, Amsterdam, pp 269–274

Bachu S (2003) Screening and ranking of sedimentary basins for sequestration of CO_2 in geological media in response to climate change. Environ Geol 44(3):277–289

Bachu S (2008a) CO_2 storage in geological media: role, means, status and barriers to deployment. Prog Energy Combust Sci 34(2):254–273

Bachu S (2008b) Legal and regulatory challenges in the implementation of CO_2 geological storage: an Alberta and Canadian perspective. Int J Greenhouse Gas Control 2(2):259–273

Bachu S (2010) Screening and selection criteria, and characterisation for CO_2 storage. In: Maroto-valer M (ed) Developments and Innovation in Carbon Dioxide (CO_2) Capture And Storage Technology, v. 2, Woodhead Energy Series 16, Woodhead Publishing Ltd., UK, pp. 27–56

Bachu S, Bennion B (2008) Effects of in-situ conditions on relative permeability characteristics of CO_2-brine systems. Environ Geol 54:1707–1722. doi:10.1007/s00254-007-0946-9

Bachu S, Celia MA (2009) Assessing the potential for CO_2 leakage, particularly through wells, from geological storage sites. In: McPherson BJOL, Sundquist E (eds) The Science of CO_2 Storage. AGU Monograph Series GM148. American Geophysical Union, Washington, DC, USA, pp. 203–216

Bachu S, Gunter WD (2005) Overview of acid gas injection operations in western Canada. In: Rubin ES, Keith DW, Gilboy CF (eds) Greenhouse Gas Control Technologies: Proceedings of the 7th International Conference on Greenhouse Gas Control Technologies, vol 1. Elsevier, London, pp 443–448

Bachu S, Gunter WD, Perkins EH (1994) Aquifer disposal of CO_2: hydrodynamic and mineral trapping. Energy Convers Manage 35(4):269–279

Bennion B, Bachu S (2007) Permeability and relative permeability measurements at reservoir conditions for CO_2-water systems in ultra low permeability confining caprocks. Paper SPE 106995. Europec/EAGE Annual Conference and Exhibition, London, 11–14 June 2007

Bradshaw J, Dance T (2005) Mapping geological storage prospectivity of CO_2 for the world's sedimentary basins and source-to-sink matching. In: Rubin ES, Keith DW, Gilboy CF (eds) Greenhouse Gas Control Technologies: Proceedings of the 7th International Conference on Greenhouse Gas Control Technologies, vol 1. Elsevier, London, pp 583–591

Bradshaw J, Bradshaw BE, Allinson G, Rigg AJ, Nguyen V, Spencer A (2002) The potential for geological sequestration of CO_2 in Australia: preliminary findings and implications for new gas field development. Aust Petrol Prod Explor Assoc J 42(1):25–46

Brunskill B, Wilson M (2011) Monitoring methods used to identify the migration of carbon dioxide and radionuclides in the geosphere. In this volume

Carey JW, Wigand M, Chipera SJ, Woldegabriel G, Pawar R, Lichtner PC, Wehner SC, Raines MA, Guthrie GD (2007) Analysis and performance of oil well cement with 30 years of CO_2 exposure from the SACROC Unit, West Texas, USA. In: Gale J, Bolland O (eds) 8th International Conference on Greenhouse Gas Control Technologies—GHGT-8. Int J Greenhouse Gas Control 1(1):75–85

Chapman N, McCombie C (2003) Principles and Standards for the Disposal of Long-lived Radioactive Wastes. Elsevier, Amsterdam

Chapman NA, McKinley IG (1987) The Geological Disposal of Nuclear Waste. Wiley, London

Chikatamarla L, Bustin MR (2003) Sequestration potential of acid gases in Western Canadian coals. In: Proceedings of the 2003 International Coalbed Symposium. The University of Alabama, Tuscaloosa, AL

Chiquet P, Daridon JL, Broseta D, Thibeau S (2007) CO_2/water interfacial tensions under pressure and temperature conditions of CO_2 geological storage. Energy Convers Manage 48(3):736–744

Connor CB, Chapman NA, Connor LJ (eds) (2009) Volcanic and Tectonic Hazard Assessment for Nuclear Facilities. Cambridge University Press, Cambridge

Cui X, Bustin RM, Chikatamarla L (2007) Adsorption-induced coal swelling and stress: implications for methane production and acid gas sequestration into coal seams. J Geophys Res 112:B10202. doi:10.1029/2004JB003482

Defra (Department for the Environment, Food and Rural Affairs) (2008) Managing Radioactive Waste Safely: A Framework for Implementing Geological Disposal. Stationery Office, London

Degnan P, Bath A, Cortés A, Delgado J, Haszeldine S, Milodowski A, Puigdomenech I, Recreo F, Silar J, Torres T, Tullborg E-L (2005) PADAMOT: Project overview report. PADAMOT Project Technical Report. http://www.bgs.ac.uk/padamot/documents.html

Dusseault MB, Bachu S, Rothenburg L (2004) Sequestration of CO_2 in salt caverns. J Can Pet Technol 43(11):49–55

EC (European Commission) (2004) Thematic network on the role of monitoring in a phased approach to the geological disposal of radioactive waste. Nuclear science and technology, Final Report EUR 21025 EN. European Commission, Luxembourg

NEA (Nuclear Energy Agency) (2005) Stability and buffering capacity of the geosphere for long-term isolation of radioactive waste: application to argillaceous media. In: "Clay Club" Workshop Proceedings, Braunschweig, Germany, 9–11 Dec 2003. OECD, Paris

NEA (Nuclear Energy Agency) (2009) Stability and buffering capacity of the geosphere for long-term isolation of radioactive waste: application to crystalline rock. In: Workshop Proceedings, Manchester, UK, 13–15 Nov 2007. OECD Publishing, Paris

Enick RM, Klara SM (1990) CO_2 solubility in water and brine under reservoir conditions. Chem Eng Commun 90:23–33

Ennis-King JP, Paterson L (2003) Role of convective mixing in the long term storage of carbon dioxide in deep saline formations. SPE J 10:349–356

Fenghour A, Wakeham WA, Vesovic V (1998) The viscosity of carbon dioxide. J Phys Chem Ref Data 27(1):31–44

Förster A, Norden B, Zink-Jørgensen K, Frykma P, Kulenkampff J, Spangenberger E, Erzinger J, Zimmer M, Kopp J, Borm G, Juhlin C, Cosma C, Hurter S (2006) Baseline characterization of the CO_2SINK geological storage site at Ketzin, Germany. Environ Geosci 13(3):145–161

Gunter WD, Bachu S, Benson SM (2004) The role of hydrogeological and geochemical trapping in sedimentary basins for secure geological storage of carbon dioxide. In: Baines SJ, Worden RH (eds) Geological Storage of Carbon Dioxide, vol 233, Geological society special publication. Geological Society of London, London, pp 129–145

Gurevich AE, Endres BL, Robertson JO, Chilingar GV (1993) Gas migration from oil and gas fields and associated hazards. J Petrol Sci Eng 9:223–238

Hedin A (1997) Spent nuclear fuel—How dangerous is it? SKB Technical Report TR-97-13. Svensk Kärnbränslehantering AB, Stockholm

Hepple RP, Benson SM (2005) Geologic storage of carbon dioxide as a climate change mitigation strategy: performance requirements and the implications of surface seepage. Environ Geol 47:576–585

Hitchon B, Gunter WD, Gentzis T, Bailey RT (1999) Sedimentary basins and greenhouse gases: a serendipitous association. Energy Convers Manage 40:825–843

Holloway S, Garg A, Kapshe M, Deshpande A, Pracha AS, Khan SR, Mahmood MA, Singh TN, Kirk KL, Gale J (2009) An assessment of the CO_2 storage potential of the Indian subcontinent. In: Gale J, Herzog H, Braitsch J (eds) Proceedings of the 9th International Conference on Greenhouse Gas Control Technologies. Energy procedia, vol 1, pp 2607–2613

Holm LW, Josendal VA (1982) Effect of oil composition on miscible-type displacement by carbon dioxide. SPE J 22(1):87–98

Hovorka SD, Benson SM, Doughty C, Freifeld BM, Sakurai A, Daley TM, Kharaka YK, Holtz MH, Trautz RC, Nance HS, Myer LR, Knauss KG (2006) Measuring permanence of CO_2 storage in saline formations: the Frio experiment. Environ Geosci 13(2):105–121

IAEA (International Atomic Energy Agency) (1994) Classification of Radioactive Waste, IAEA Safety Series No. 111-G-1.1. IAEA, Vienna

IAEA (International Atomic Energy Agency) (1995) The Principles of Radioactive Waste Management, IAEA Safety Series No. 111-F. IAEA, Vienna

Ide ST, Jessen K, Orr FM Jr (2007) Storage of CO_2 in saline aquifers: effects of gravity, viscous and capillary forces on amount and timing of trapping. Int J Greenhouse Gas Control 1(4):481–491

IEA (International Energy Agency) (2004) Prospects for CO_2 Capture and Storage. OECD/IEA, Paris

IEA (International Energy Agency) (2006) Energy Technology Perspectives: Scenarios and Strategies to 2050. OECD/IEA, Paris

IPCC (Intergovernmental Panel on Climate Change) (2005) IPCC Special Report on Carbon Dioxide Capture and Storage. Metz B, Davidson O, de Coninck HC, Loos M, Meyer LA (eds) Cambridge University Press, Cambridge, New York

JNC (Japan Nuclear Cycle Development Institute) (2000) H12: Project to Establish the Scientific and Technical Basis for HLW Disposal in Japan. Japan Atomic Energy Agency, Tokai

Juanes R, Spiteri EJ, Orr FM Jr, Blunt MJ (2006) Impact of relative permeability hysteresis on geological CO_2 storage. Water Resour Res 42:W12418

Kohl AL, Nielsen RB (1997) Gas Purification. Gulf Publishing Company, Houston

Kumar A, Ozah R, Noh M, Pope GA, Bryant S, Sepehrnoori K, Lake LW (2005) Reservoir simulation of CO_2 storage in deep saline aquifers. SPE J 10(3):336–348

Kutchko BG, Strazisar BR, Dzombak DA, Lowry GV, Thaulow N (2007) Degradation of well cement by CO_2 under geologic sequestration conditions. Environ Sci Technol 41(12):4787–4792

Larsen JW (2003) The effects of dissolved CO_2 on coal structure and properties. Int J Coal Geol 57(1):63–70

Maul P (2011) Risk assessment, risk management and remediation for the geological disposal of radioactive waste and storage of carbon dioxide. In this volume

Maul PR, Metcalfe R, Pearce J, Savage D, West JM (2007) Performance assessments for the geological storage of carbon dioxide: learning from the radioactive waste disposal experience. Int J Greenhouse Gas Control 1(4):444–455

McCombie C (2007) Repository implementation. In: Alexander WR, McKinley LE (eds) Deep Geological Disposal of Radioactive Waste. Elsevier, Amsterdam, pp 169–193

McEwen T (2004) A review of the deep borehole disposal concept for radioactive waste. Report to United Kingdom Nirex Limited. Nirex Report No. N/108. Harwell

McEwen T (2007) Site selection and characterisation. In: Alexander WR, McKinley LE (eds) Deep Geological Disposal of Radioactive Waste. Elsevier, Amsterdam, pp 77–111

McEwen T, Andersson J (2009) Stable tectonic settings: designing site investigations to establish the tectonic basis for design and safety evaluation of geological repositories in Scandinavia. In: Connor CB, Chapman NA, Connor LJ (eds) Volcanic and Tectonic Hazard Assessment for Nuclear Facilities. Cambridge University Press, Cambridge

McGinnes DF (2007) Waste sources and classification. In: Alexander WR, McKinley LE (eds) Deep Geological Disposal of Radioactive Waste. Elsevier, Amsterdam, pp 8–40

McKee CR, Bumb AC, Koenig RA (1988) Stress-dependent permeability and porosity of coal and other geologic formations. SPE Formation Eval 4:81–91

Moritis G (2006) CO_2 injection gains momentum. Oil Gas J 104(15):37–57

Nagra (Nationale Genossenschaft für die Lagerung radioaktiver Abfälle—National Cooperative for the Disposal of Radioactive Waste) (2002) Project Opalinus Clay: safety report—demonstration of

disposal feasibility for spent fuel, vitrified high-level waste and long-lived intermediate-level waste (Entsorgungsnachweis). Technical Report NTB 02-05. Nagra, Wettingen. http://www.nagra.ch
National Research Council (2003) One Step at a Time: The Staged Development of Geologic Repositories for High-level Radioactive Waste. The National Academies Press, Washington, DC
Newlands IK, Langford RP, Causebook R (2006) Assessing the CO_2 storage prospectivity of developing economies in APEC—applying methodologies developed in GEODISC to selected sedimentary basins in the Eastern Asian region. In: Gale JJ, Røkke N, Zweigel P, Svenson H (eds) Proceedings of the 8th International Conference on Greenhouse Gas Control Technologies. Elsevier, London, CD-ROM
Nirex (2005) The viability of a phased geological repository concept for the long-term management of the UK's radioactive waste. Nirex Report N/122. United Kingdom Nirex Limited, Harwell
NUMO (Nuclear Waste Management Organization of Japan) (2007) The NUMO structured approach to HLW disposal in Japan. Report NUMO-TR-07-02. NUMO, Tokyo
Oldenburg CM, Unger AJA (2003) On leakage and seepage from geologic carbon sequestration sites: unsaturated zone attenuation. Vadose Zone J 2:287–296
Pacala S (2003) Global constraints on reservoir leakage. In: Gale JJ, Kaya Y (eds) Greenhouse Gas Control Technologies: Proceedings of the 6th International Conference on Greenhouse Gas Control Technologies, vol 1. Pergamon, Amsterdam, pp 267–272
Pacala S, Socolow RH (2004) Stabilization wedges: solving the climate problem for the next 50 years with current technologies. Science 305:968–972
Pearce JM, Czernichowski-Lauriol I, Lombardi S, Brune S, Nador A, Baker J, Pauwels H, Hatziyannis G, Beaubien S, Faber E (2004) A review of natural CO_2 accumulations in Europe as analogues for geological sequestration. In: Baines SJ, Worden RH (eds) Geological Storage of Carbon Dioxide, vol 233, Geological society special publication. Geological Society of London, London, pp 29–41
Perkins EH, Czernichowski-Lauriol I, Azaroual M, Durst P (2005) Long term predictions of CO_2 storage by mineral and solubility trapping in the Weyburn Midale reservoir. In: Wilson M, Morris T, Gale JJ, Thambimuthu K (eds) Greenhouse Gas Control Technologies: Proceedings of the 7th International Conference on Greenhouse Gas Technologies, vol 2. Elsevier, London, pp 2093–2101
Pitkänen P, Partamies S, Lahdenperä A-M, Lehtinen A, Ahokas T, Hirvonen H, Hatanpää E (2007) Results of monitoring at Olkiluoto in 2006: hydrogeochemistry. Working Report 2007-51. Posiva Oy, Olkiluoto
Posiva (2008) Safety case plan 2008. Report POSIVA 2008-05. Posiva Oy, Olkiluoto
Posiva (2009) Olkiluoto site description 2008. Report POSIVA 2009-01. Posiva Oy, Olkiluoto
Pruess K (2004) Numerical simulation of CO_2 leakage from a geologic disposal reservoir, including transitions from super- to subcritical conditions, and boiling of liquid CO_2. SPE J 9(2):237–248
Pruess K (2005) Numerical studies of fluid leakage from a geologic disposal reservoir for CO_2 show self-limiting feedback between fluid flow and heat transfer. Geophys Res Lett 32:L14404
Reeves SR (2003) Coal-Seq project update: field studies of ECBM recovery/CO_2 sequestration in coal seams. In: Gale JJ, Kaya Y (eds) Greenhouse Gas Control Technologies: Proceedings of the 6th International Conference on Greenhouse Gas Control Technologies, vol 1. Pergamon, Amsterdam, pp 557–562
Reiner DM, Nuttall WJ (2011) Public acceptance of geological disposal of carbon dioxide and radioactive waste: similarities and differences. In this volume
Riddiford FA, Tourqui A, Bishop CD, Taylor B, Smith M (2003) A cleaner development: the In Salah gas project, Algeria. In: Gale JJ, Kaya Y (eds) Greenhouse Gas Control Technologies: Proceedings of the 6th International Conference on Greenhouse Gas Control, vol 1. Pergamon, Amsterdam, pp 601–606
Rodwell WR, Norris S, Mäntynen M, Vieno T (2003) A thematic network on gas issues in safety assessment of deep repositories for radioactive waste (GASNET). Report EUR 20620. European Commission, Luxembourg
Savage D (ed) (1995) The Scientific and Regulatory Basis for the Geological Disposal of Radioactive Waste. Wiley, Chichester
Scherer GW, Celia MA, Prévost J-H, Bachu S, Bruant R, Duguid A, Fuller R, Gasda SE, Radonjic M, Vichit-Vadakan W (2005) Leakage of CO_2 through abandoned wells: role of corrosion of cement.

In: Benson SM (ed) Carbon Dioxide Capture for Storage in Deep Geologic Formations—Results from the CO_2 Capture Project, vol 2, Geologic storage of carbon dioxide with monitoring and verification. Elsevier, London, pp 827–848
Shi J-Q, Durucan S, Fujioka M (2008) A reservoir simulation study of CO_2 injection and N_2 flooding at the Ishikari coalfield CO_2 storage pilot project, Japan. Int J Greenhouse Gas Control 2(1):47–57
Shipton ZK, Evans JP, Dockrill B, Heath J, Williams A, Kirchner D, Kolesar PT (2005) Natural leaking CO_2-charged systems as analogs for failed geologic storage reservoirs. In: Benson SM (ed) Carbon Dioxide Capture for Storage in Deep Geologic Formations—Results from the CO_2 Capture Project, vol 2, Geologic storage of carbon dioxide with monitoring and verification. Elsevier, London, pp 699–712
Socolow RH (2005) Can we bury global warming? Sci Am 293(1):49–55
Stevens SH (2005) Natural CO_2 fields as analogs for geologic CO_2 storage. In: Benson SM (ed) Carbon Dioxide Capture for Storage in Deep Geologic Formations—Results from the CO_2 Capture Project, vol 2, Geologic storage of carbon dioxide with monitoring and verification. Elsevier, London, pp 687–697
Torp TA, Gale JJ (2003) Demonstrating storage of CO_2 in geological reservoirs: the Sleipner and SACS projects. In: Gale JJ, Kaya Y (eds) Greenhouse gas control technologies: Proceedings of the 6th International Conference on Greenhouse Gas Control, vol 1. Pergamon, Amsterdam, pp 311–316
van Bergen F, Pagnier H, Krzystolik P (2006) Field experiment of enhanced coalbed methane-CO_2 in the Upper Silesian basin of Poland. Environ Geosci 13(3):210–224
van der Meer LHG, Hartman J, Geel C, Kreft EE (2005) Re-injecting CO_2 into an offshore gas reservoir at a depth of nearly 4000 metres subsea. In: Rubin ES, Keith DW, Gilboy CF (eds) Greenhouse Gas Control Technologies: Proceedings of the 7th International Conference on Greenhouse Gas Control, vol 1. Elsevier, London, pp 521–530
Vankerckhoven P, Mitchel K (1998) Radioactive waste categories: Current position (98) in the EU member states and in the Baltic and Central European countries. Report EUR 18324. European Commission, Luxembourg
SKB (Svensk Kärnbränslehantering AB—Swedish Nuclear Fuel and Waste Management Company) (2004) RD&D-Programme 2004. Programme for research, development and demonstration of methods for the management and disposal of nuclear waste, including social science research. SKB Technical Report TR-04-21. Svensk Kärnbränslehantering AB, Stockholm
Watson TL, Bachu S (2008) Identification of wells with high CO_2-leakage potential in mature oil fields developed for CO_2-enhanced oil recovery. Paper SPE 112924. SPE oil recovery symposium, Tulsa, OK, 19–23 Apr
Watson TL, Bachu S (2009) Evaluation of the potential for gas and CO_2 leakage along wellbores. Paper SPE 106817, SPE Drilling and Completion, 24(1):115–126
WCED (World Commission on Environment and Development) (1987) Our Common Future. Oxford University Press, Oxford
West JM, Shaw RP, Pearce JM (2011) Environmental issues in the geological disposal of carbon dioxide and radioactive waste. In this volume
Wilson EJ, Bergan S (2011) Managing liability: comparing radioactive waste disposal and carbon dioxide storage. In this volume
Wilson M, Monea M (eds) (2004) IEA GHG Weyburn monitoring and storage project: summary report 2000–2004. In: Proceedings of the 7th International Conference on Greenhouse Gas Control Technologies, vol III. Petroleum Technology Research Center, Regina
Wong S, Law D, Deng X, Robinson J, Kadatz B, Gunter WD, Jianping Y, Sanli F, Zhiqiang F (2007) Enhanced coalbed methane and CO_2 storage in anthracitic coals—Micro-pilot test at South Qinshui, Shanxi, China. In: Gale J, Bolland O (eds) 8th International Conference on Greenhouse Gas Control Technologies—GHGT-8. Int J Greenhouse Gas Control 1(2):215–222
Xu T, Apps JA, Pruess K (2003) Reactive geochemical transport simulation to study mineral trapping for CO_2 disposal in deep arenaceous formations. J Geophys Res 108(B2):2071–2084
Yamaguchi S, Ohga K, Fujioka M, Nako M, Muto S (2008) Field experiment of Japan CO_2 geosequestration in coal seams project (JCOP). In: Gale JJ, Røkke N, Zweigel P, Svenson H (eds) Proceedings of the 8th International Conference on Greenhouse Gas Control Technologies. Elsevier, London, CD-ROM

Environmental Issues in the Geological Disposal of Carbon Dioxide and Radioactive Waste

Julia M. West, Richard P. Shaw, and Jonathan M. Pearce

Abstract A comparative assessment of the post environmental issues for the geological disposal of carbon dioxide (CO_2) and radioactive waste (RW) is made in this chapter. Several criteria are used: the characteristics of RW and CO_2; their potential environmental impacts; an assessment of the hazards arising from RW and CO_2; and monitoring of their environmental impacts. There are several differences in the way that the long-term safety of the disposal of RW and CO_2 is regulated and evaluated. While the regulatory procedures relating to the development of a facility for the disposal of RW in many countries with nuclear power programmes are well defined having evolved over several decades, those relating to CO_2 disposal are less well developed. The results of this assessment show that, despite key differences, many of the approaches addressing environmental issues are similar. Additionally, much can be learned from the RW disposal experience which will be particularly relevant to the assessments of site performance for CO_2 within a regulatory framework, particularly in the methods and approaches to long-term site performance assessment.

Keywords Carbon dioxide storage • Environmental impacts • Radioactive waste disposal • Technology comparison

1 Introduction

This chapter provides a comparative assessment of the environmental issues surrounding the geological disposal of carbon dioxide (CO_2) and radioactive waste (RW). These are diverse and influence the entire disposal chain including transport

J.M. West (✉), R.P. Shaw, and J.M. Pearce
British Geological Survey, Keyworth, Nottingham NG12 5GG, UK
e-mail: jmwes@bgs.ac.uk; rps@bgs.ac.uk; jmpe@bgs.ac.uk

F.L. Toth (ed.), *Geological Disposal of Carbon Dioxide and Radioactive Waste: A Comparative Assessment*, Advances in Global Change Research 44,
DOI 10.1007/978-90-481-8712-6_3, © International Atomic Energy Agency, 2011

and the construction and operation of facilities. However, the issues considered here are post-closure, that is after the closure of an RW repository or after CO_2 injection has ceased and the site has been formally closed. Consideration will be made of both terrestrial and marine environments although, for RW, the focus is mainly on terrestrial environments.

In CO_2 capture and storage (CCS), the injection of CO_2 into a geological formation is known as 'storage' although there is no intention to retrieve the CO_2 once it has been injected. This is also the case for RW because it is generally envisaged that waste emplacement, or 'disposal', at depth is permanent, although there may be a long phase of active management prior to the decision to initiate repository sealing and closure. However, both RW and most CO_2 could be technically retrieved.

RW includes all waste materials that are too radioactive for disposal within an ordinary landfill facility. This will include wastes derived from nuclear power generation, including fuel reprocessing, medical wastes and laboratory wastes. It may also include naturally occurring radioactive materials (NORM) such as scale (removed from the inside of oil pipelines) which is naturally radioactive. It will include some long-lived low-level wastes (LLW), intermediate-level waste (ILW) and high-level waste (HLW), and could include spent nuclear fuel and plutonium and uranium if these materials are considered to be waste. Most LLW is disposed of to surface or shallow disposal facilities and is not considered further here. CO_2 streams will comprise almost pure CO_2 captured from large point sources such as fossil fuel-based power stations, cement and some chemical and refinery plants.

For both technologies, post-closure environmental concerns focus on the impacts of either unpredicted releases of radionuclides or leakage of CO_2 into the biosphere, which includes the shallow subsurface (the soil, vadose zone and potable aquifers), and surface ecosystem. Performance Assessment (PA) (described in the chapter by Maul (2011)) is usually used to evaluate the (post-closure) evolution of repository systems with some of the output expressed in terms of risk to human health and the environment. PAs provide a rigorous and comprehensive approach to site appraisal and, in the context of project planning and regulatory decision making, they are crucial in developing the long-term 'safety case' which, for the geological disposal of RW, is commonly extrapolated over a period in the order of 10^6 years (e.g. Nirex 1997). Currently, formal PA is not implemented in existing CO_2 storage projects because the technology is still evolving from the research and development stage. However, guidelines are being developed for the risk assessment of CO_2 storage, such as the Convention for the Protection of the Marine Environment of the North-East Atlantic (OSPAR) Guidelines for Risk Assessment and Management of Storage of CO_2 Streams in Geological Formations (OSPAR 2007), and the European Commission's (EC) draft directive on the geological storage of carbon dioxide (EC 2008) or, at an earlier stage, Strategic Environmental Assessment (SEA), which compares different management strategies. Environmental impact assessments (EIAs) are undertaken in both technologies and these give 'an evaluation of … impacts of a proposed activity, where the performance measure is overall environmental impact, including … global measures of impact on safety and the environment'

(IAEA 2003). Thus EIAs have been used for construction and operational phases where, for example, physical and ecological effects are being evaluated. However, EIAs in the oil and gas industry, on which CO_2 storage practice is based, are normally concerned with environmental impacts during construction, operation and decommissioning, and have not necessarily been used to consider potential impacts over the long term. Several CO_2 storage demonstration projects have also included an element of long-term risk assessment (e.g. Weyburn, Canada (Zhou et al. 2004; Stenhouse et al. 2005); Gorgon, Australia (Gorgon Joint Ventures (GJV) 2005a, b) and Schweinrich, Germany (Svensson et al. 2005)).

When examining the environmental issues surrounding RW disposal and CO_2 disposal, several criteria need to be examined. These include:

- The characteristics of RW and CO_2;
- The potential environmental impacts of RW and CO_2;
- Assessment of the hazards arising from RW and CO_2;
- Monitoring of environmental impacts.

The environmental issues for the two technologies are discussed in the following sections, with particular emphasis on these criteria. A comparative assessment is then made, using the above criteria, highlighting similarities and differences between the two areas. The conclusions from these comparisons are then discussed in terms of future research and policy.

2 Geological Storage of Carbon Dioxide: Environmental Issues

2.1 International Regulatory Background

Globally, emissions of CO_2 from fossil fuel use in the year 2000 totalled about 23.5 Gt with 60% attributed to large (>0.1 Mt CO_2/year) stationary emission sources such as power stations, cement production and refineries (IPCC 2005). Clusters of these sources are found in North America, Europe, East Asia and South Asia, and a variety of mitigation strategies, including CCS, will be required to reduce CO_2 emissions from these sources.

To date, the major projects demonstrating CCS at Weyburn, Canada (Wilson and Monea 2004) and Sleipner, in the North Sea (Torp and Gale 2002), have particularly focused on technological and economic viability, and whether these sites could leak. Consequently, these studies are focusing on monitoring, verification and risk assessment—it is intended that such work will assist regulators and reassure other stakeholder groups (especially the public) that the sites will not leak. These projects operate within existing oil and gas regulatory frameworks. At Weyburn, for example, injection of CO_2 is used to enhance oil recovery from an existing oil field.

However, if CCS is conducted outside hydrocarbon-related operations these existing regulations may not be appropriate.

At the time of writing, the regulatory frameworks governing geological CO_2 storage are being developed (described in the chapter by Wilson and Bergan (2011)). In general, current projects are licensed under petroleum legislation. However, OSPAR has provided guidance on the steps it requires before geological storage in reservoirs at depth below the seabed can be allowed in marine jurisdictions of contracting parties (OSPAR 2007). Further, a draft EC Directive enabling European Member States to enact legislation of the regulation of CCS is currently under discussion (EC 2008). However, within these draft regulations, it is recognised that issues of leakage and potential long-term stewardship must be addressed if the potential for CCS to provide substantial reductions in atmospheric CO_2 emissions is to be realised (Mace et al. 2007; Zakkour and Haines 2007). Additionally, studies on public perception of CCS (see, for example, Shackley et al. 2004) indicate concerns about the effect of leakages on the environment.

2.2 *Environmental Impacts of CO_2*

It can be assumed that storage sites will be selected to 'permanently' store the injected CO_2. However, if leakage from storage sites did occur after formal closure of the injection site, it could be over small areas from discrete point sources, such as abandoned wells, resulting in locally high concentrations of CO_2. This could reach tens of per cent levels in soil gas (West et al. 2005), well above any background levels, and which will impact on organisms (Table 1). Although extensive physiological research is available, the overall environmental impacts of localised elevated CO_2 concentrations on terrestrial, subsurface and marine ecosystems are still poorly understood and, as a result, are areas of active research (see following section).

Essentially, respiratory physiology and pH control are the primary physiological mechanisms controlling responses in organisms to elevated CO_2 exposures. Information is available from a diverse research base, and some examples are given in Table 1. These data, however, are mostly from studies on organisms exposed to either slightly elevated concentrations of CO_2 or the high concentrations that give a lethal response.

In economic terms, leaks from a storage site into marine and freshwater systems might affect fisheries by altering pH, with accompanying physiological effects (Turley et al. 2004). For terrestrial systems, leakages might damage crops, groundwater quality and/or human and animal health. Other concerns include acidification, changes in biological diversity and species composition, and asphyxiation at high CO_2 concentrations. In addition, biogeochemical processes may be affected as increased CO_2 concentrations could change pH, microbial populations and nutrient supply. It is also important to understand the local effects in comparison to the effects of global increases of CO_2 concentrations on the environment and habitats. In contrast to studies of the effects of elevated atmospheric CO_2 concentrations

Table 1 Examples of tolerances to CO_2 exposure in selected organisms (From West et al. 2005)

	Exposure	Effect	Reference
Humans (healthy adults)	Below 3%	No adverse effects but increased breathing, mild headache and sweating	Hepple 2005
	4–5% for 'few minutes'	Headache, increased blood pressure and difficulty in breathing	
	7–10% up to 1 h	Headache, dizziness, sweating, rapid breathing and near or full unconsciousness	
	15%+	Loss of consciousness in less than one minute. Narcosis, respiratory arrest, convulsions, coma and death	
	30%	Death in few minutes	
Terrestrial invertebrates	Insect (rusty grain beetle—*Cryptolestes ferrugineus*)		Mann et al. 1999
	15%	Death after ~42 days	Benson et al. 2002
	100%	Death after ~2 days	
	40%	Used to preserve food from microbes and fungi	
	Soil invertebrates 20%	Majority of any one species have 'behavioural changes'	Sustr and Siemk 1996
	11–50%	Lethal for 50% of species	
Terrestrial vertebrates	Rodents 2%	Observed in burrows and nests	References in Maina 1998
	Gophers 4%		
	Birds 9%		
Plants	>0.2%	Stimulation of C3 photosynthesis plants (includes temperate cereal crops such as wheat)	Hepple 2005
	15–40%	Acid tolerant grasses dominate pasture. Few dicotyledonous plants	Beaubien et al. 2008
	Trees, Mammoth Mountain, USA	Tree killed probably by suppression of root zone respiration via hypoxia	Hepple 2005
	20–90%		
Fungi	15–20%	Significant inhibition of growth of spores for two types of fungi	Haasum and Nielsen 1996
	30%	No measurable growth of spores	Tian et al. 2001
	50%	No germination of spores	
Subsurface microbes	None known	Increased concentrations (from injection) are likely to have profound effects because aerobic organisms will be inhibited but anaerobic organisms e.g. Fe (III) reducers, S reducing reducers and methanogens will respond to rock/water/CO_2 interactions and are likely to increase in population size and activity	Onstott 2005 (Discussion paper)

(continued)

Table 1 (continued)

	Exposure	Effect	Reference
Marine invertebrates	Commercial shellfish	Few data specifically on CO_2 effects. The little evidence is limited to effect of pH change on e.g. shells	Turley et al. 2004; SMR 1999
Marine vertebrates	Fish	More sensitive to hypoxia than invertebrates. Mostly unknown effects on reproduction/ development	Turley et al. 2004

(say, a rise from current levels to 550 ppm), levels of CO_2 in soils resulting from leaks from engineered storage sites underground could be enhanced by several orders of magnitude above atmospheric levels, causing damage or, in the worst case, serious damage to an ecosystem.

Organisms close to a leakage could be exposed to acute and perhaps lethal concentrations whilst those at increasing distances from the leakage could be exposed to firstly acute and then to chronic concentrations. How such exposures will influence an existing ecosystem as a whole, or the individual species within an ecosystem, is unknown and further work is required to obtain a better understanding. Thus, for all ecosystems of interest, the potential indicator groups at the different trophic levels need to be identified and effects determined. At an economic level, it can be envisaged that particular concern will lie with certain key receptors. For example, in marine environments key fishery groups and their food sources may be specific target receptors, whilst in terrestrial systems these may include humans and crop plants. However, such key receptor groups should not be seen in isolation because they will interact with other species within an ecosystem.

CO_2 leakage could also affect subsurface and surface biogeochemical processes by changing, for example, pH and possibly redox conditions. CO_2 mobilisation of trace metals is also a common geological process, albeit typically on long timescales and at slow rates. The potential for heavy metal mobilisation via leaking CO_2 has been proposed by several authors (e.g. Kharaka et al. 2006) although, as yet, little direct evidence from analogue systems has been obtained. It is also important to consider the effect of potential environmental impacts resulting from impurities (such as H_2S, SO_2 and NO_x) that may be present in leaking CO_2. Such changes could have significant implications for groundwater quality in terms of acidification of supplies and possible dissolution of minerals and mobilisation of heavy metals. Little work has been undertaken in this area, although Onstott (2005) and Stenhouse et al. (2009) have undertaken some preliminary modelling work. H_2S is a toxic gas and as such poses a hazard to humans and is closely regulated. H_2S, SO_x and NO_x could, if they were co-transported within a leaking CO_2 plume, alter pH and redox conditions in the soil environment, which could result in changes in nutrient supply, microbial and plant diversity and habitats (IEA-GHG 2004).

2.3 Current Research

At the time of writing, several projects are under way to examine the environmental impacts of CO_2 leakage into both terrestrial and marine systems. CO2GeoNet is a European Network of Excellence (http://www.co2geonet.com) for geological storage of CO_2, involving 13 partners. Some of its research activities have focused on studying the ecosystem responses to natural CO_2 leakages at sites in Italy and Germany (see e.g. Beaubien et al. 2008; Krüger et al. 2009) and a generic system model is also being developed (described in the chapter by Maul (2011) and in West et al. (2006)). Field sites are also being developed to study impacts of CO_2 leakage on agricultural crops (Artificial Soil Gassing and Response Detection (ASGARD) site, Nottingham, UK (West et al. 2009); and see http://www.nottingham.ac.uk/geography/asgard/) and to test monitoring technologies and models (Zero Emission Research and Technology Center (ZERT) site, Montana, USA (Spangler et al. 2009); and see: http://www.montana.edu/zert/). Specific work is also being undertaken on the impacts of CO_2 leakage on marine systems by the Research Institute of Innovative Technology for the Earth (RITE), Japan, with CO2GeoNet partners (Ishida et al. 2006) and by Plymouth Marine Laboratory (PML), UK (Turley et al. 2004). However, all these projects are in their early stages with only limited results currently available.

2.4 Gaps in Knowledge

As detailed above, no explicit acknowledgement or guidance is available in any existing regulations on the release and environmental impacts of CO_2 from terrestrial and marine storage sites. Additionally:

- No indicator species for specific ecosystems have been identified. While to some extent ecosystems will be site specific, basic supporting research on generic processes is still needed to build confidence.
- No data on total ecosystem responses to a CO_2 leak and their recovery times are available.
- No specific data are available on the potential impacts on groundwater or surface water quality. Although the potential for CO_2 mobilisation of trace metals, other gases and hydrocarbons has long been recognised, little data have been generated.
- Co-transported and -injected species have received little attention so far but could include low to trace concentrations of O_2, SO_2, NO, H_2S, CO, Hg, Cd, Ar, N_2, H_2O, and NH_3. Hg and Cd are likely to be at ppb levels (Aspelund and Jordal (2007) and references therein). Many of these potentially co-injected gases (e.g. O_2, SO_2, H_2S) are biogeochemically important and could alter microbial populations either in the reservoir or, if released with CO_2, in the

overburden and near-surface environment. We are not aware of any research that has determined the fate of co-injected species during CO_2 storage.
- Few data exist on impacts on the soil environment from high concentrations of CO_2 emerging from depth.
- There is currently a lack of integration between considerations of potential impacts of CO_2 leaks on terrestrial and marine ecosystems and PAs. EIAs have traditionally been used to assess the impacts of engineering schemes over the lifetime of the project, which have included legacy issues such as site abandonment, clean-up, remediation and liability following the end of the project. However, CO_2 storage projects present new challenges because of the very long timescales that need to be considered after the injection project has finished, particularly when considering performance.

3 Geological Disposal of Radioactive Waste

3.1 International Regulatory Background

RWs comprise less than 1% of total industrial toxic wastes with a total arising of 81,000 m^3/year (~210 kt/year) of conditioned wastes in the Organisation for Economic Cooperation and Development (OECD) countries (McGinnes 2007). The composition and characteristics of RWs vary and a recent summary of waste classes defined by the International Atomic Energy Agency (IAEA) is given in Table 2. In countries which use nuclear power, roughly 90% of the volume is LLW containing 1% of the total radioactivity, 7% is ILW with 4% of total radioactivity and 3% is HLW, containing 95% of the radioactivity (McGinnes 2007).

3.2 Environmental Impacts of Radionuclides in Radioactive Wastes

RW contains radioisotopes of a wide range of elements which will emit alpha, beta, gamma and neutron radiation. While minimal shielding will protect people and the environment from alpha and beta radiation, external exposure to high levels of gamma radiation or neutrons is harmful and can be fatal to some species, including humans. Internal exposure to alpha or beta radiation sources, for example through inhalation or ingestion, is also harmful at high levels and can be fatal in serious cases. Some radioactive elements are also chemically toxic. Additionally, some RWs also contain chemically toxic materials, such as lead from shielding, but these are not considered further here.

Table 2 Details of radioactive waste classes by the IAEA (From McGinnes 2007)

Waste class	Typical characteristics	Possible disposal options
Exempt waste (EW)	Activity levels at or below clearance levels	No radiological restrictions, normal landfill
Short-lived (L/ILW-SL)	Restricted long-lived radionuclide concentrations, e.g. long-lived α-emitters average <400 Bq/g or 4,000 Bq/g maximum per package	Near-surface or geological repository
Long-lived (L/ILW-LL)	Long-lived radionuclide concentrations exceeding limitations for short-lived wastes	Geological disposal facility
High-level waste (HLW)[a]	Thermal power greater than about 2 kW/m^3 and long-lived radionuclide concentrations exceeding limitations for short-lived wastes	Geological disposal facility

L/ILW-SL Short-lived low/intermediate-level waste, *L/ILW-LL* Long-lived low/intermediate-level waste
[a] If spent fuel is considered a waste, then this falls into this class

The nature of radioactive elements means that their impacts on organisms are very complex. Moreover, interpretation of data is further complicated by the debate surrounding the relationship between radiation dose and subsequent biological impacts. As a result, it is not possible to produce a definitive summary of the impacts of radionuclides on organisms (as has been given for CO_2 in Table 1).

Radioactivity is easily measured and controversy exists as to whether it is harmful at low levels. Even in regions with naturally high background radiation (e.g. uranium ore deposits in Africa (Bowden and Shaw 2007) and Brazil (Chapman et al. 1992)) it does not necessarily have any identifiable effect on the surface environment or local plant, animal or human populations. Following the Chernobyl accident in April 1986 a large amount of work has been undertaken in evaluating the environmental impact of the disaster including monitoring the response of the natural environment to radiation exposure (IAEA 2005). Within the 30 km exclusion zone, localised sites of acute adverse effects on animals and plants have been recorded in areas of higher radiological exposure. However, no adverse effects have been reported in plants and animals exposed to a cumulative dose of less than 0.3 Gy absorbed dose during the first month following the accident (IAEA 2005).

In order to isolate higher activity RW from the environment, most waste management organisations are proposing geological disposal of these wastes in deep (greater than 200 m) repositories. Wastes will be conditioned and emplaced in engineered barrier systems designed to minimise radionuclide migration, within a suitable geological environment which will isolate the waste for an extended period of time. In most geological settings it is inevitable that there will eventually be some dispersion of the radionuclides from the repository, but this will be very slow and occur only in the distant future, when the hazard from the waste has been considerably reduced by radioactive decay. The processes in the engineered and geological barriers will reduce mobility of the majority of any radionuclides that 'escape', ensuring that only a small fraction will ever reach the near-surface and surface environments. Additionally, their dispersion will ensure that they only contribute a small fraction to the doses received by plants and animals, including humans, when compared to doses received from natural radiation sources.

The IAEA specify that the annual dose to a member of the public from a closed geological repository in the future should not exceed 0.3 mSv (IAEA 2006). This compares to the global annual average effective dose from natural background radioactivity of 2.4 mSv (UNSCEAR 2000). However, regulators in many countries require a target more stringent than 0.3 mSv. For example, for land-based disposal of RW, the UK environmental agencies have defined that the assessed radiological risk to an individual of developing a fatal cancer or a serious hereditary defect should be less than one-in-a-million per year (EA 2009). This compares to the 1 in 100,000 per year risk constraint suggested by the IAEA Safety Requirement of 0.3 mSv and the 1 in 1,000 to 1 in 10,000 as a result of exposure to natural background levels (2.23 mSv (Watson et al. 2005)) in the UK. Thus the accepted dose from a repository in the UK is between 100 and 1,000 times below the radiological risk to which members of the population are exposed as a result of natural background radiation levels.

Studies of natural and anthropogenic analogues provide information on the impacts of environmental exposure to radiation; how radionuclides behave over geological timescales; and an understanding of how the materials used in an RW repository are likely to perform in the long term. Examples of work include the impacts of exposed/near-surface uranium mineralisation on the local habitat (Needles Eye, Scotland; Poços de Caldas, Brazil) (Miller et al. 2000) and the behaviour of reactor products in the geological environment produced by a natural reactor 2 billion years ago (Oklo, Gabon) (Miller et al. 2000). Such studies are important in helping to predict the future performance of a repository and also have a significant role in promoting confidence in the wider stakeholder community that a repository will provide the intended isolation of the waste.

3.3 *Examples of Current Work*

Significant effort has been directed over many years, particularly by national waste management programmes, into designing waste packaging and the engineering of a repository and its backfill to ensure optimum retention of the radionuclides within the repository, understanding the processes by which radionuclides may eventually be released from a repository and how they may migrate or be retained within the geosphere (Alexander and McKinley 2007). Extensive databases on their potential impact on reference plant and animal species and on humans in various uptake pathways have also been compiled (ICRP 2007).

Many studies are site specific, relating to particular waste types in defined geological environments. Other studies are generic and are aimed at understanding, for example, the processes that may be involved in radionuclide migration. Considerable experimental work is also being undertaken in several underground research facilities (including Äspö (Swedish Nuclear Fuel and Waste Management Company (SKB), Sweden), Bure (Agence Nationale pour la Gestion des Déchets Radioactifs (ANDRA), France), Grimsel and Mt Terri (National Cooperative for the Disposal of Radioactive Waste (NAGRA), Switzerland) and Mol (Belgian Nuclear Research Centre (SCK•CEN), Belgium)), into how repositories in different rock types will perform during operational and post-closure phases. This is supported by extensive work on natural analogue systems (Miller et al. 2000). Examples of other recent work includes the palaeohydrogeological studies carried out under the European Union Euratom funded EQUIP (Evidence from mineralogy and geochemistry for the evolution of groundwater systems during the Quaternary for use in RW repository safety assessment) and PADAMOT (Palaeohydrogeological Data Analysis and Model Testing) projects (Degnan and Bath 2005) that included mineralogical studies to elucidate the impacts of glaciations on groundwater systems. Ongoing research is examining the role that microbial activity, including biofilms, has in retarding or enhancing radionuclide migration through different geological environments (Coombs et al. 2010). The Large Scale Gas Injection Test (Lasgit) experiment in the Äspö Underground Research Laboratory (Harrington et al. 2007) is studying bentonite saturation and gas migration through the bentonite backfill of a full-scale deposition hole.

3.4 Gaps in Knowledge

Compared to CCS, RW disposal is a relatively mature science with a 50 year history (POST 1997; McKinley et al. 2007). During this time significant advances have been made in understanding and assessing the long-term performance of a repository. Appropriate sites will be selected to allow RWs to be disposed of with confidence that the impacts on the near-surface and surface environment will be minimal over very long time periods—in fact much more securely than we currently dispose of many other wastes, some of which are also highly toxic (Savage 1995). RW disposal is also highly regulated, ensuring that it is undertaken safely and appropriately.

However, there are still some issues that are not fully understood and which additional research will clarify and permit more robust predictions to be made on repository behaviour and overall performance. For the purposes of this chapter, it is relevant to note that these include:

- Gas generation within a repository and its subsequent migration through the engineered systems and into and through the geological environment;

- Understanding the processes that may help to reduce the mobility of conservative isotopes, such as ^{14}C, in the repository and geological environments, and thus mitigation to reduce their migration can be introduced;
- Further understanding of the processes of the migration of radionuclides at the interfaces between the repository and the surrounding geosphere (the rocks in which the repository is sited) and the geosphere and the biosphere (the plants and animals, including humans, in the near-surface and surface environment).

4 Technology Comparison

Having described the environmental issues surrounding both technologies, it is now possible to make comparisons between them using the criteria outlined in the introductory section above. These are also summarised in Table 3.

4.1 Characteristics of Radioactive Waste and CO_2

The nature, composition and volumes of the two wastes are very different, as detailed in previous sections, and thus are important considerations for environmental impacts. RW is toxic at high concentrations and much is long-lived, with the highest activity material being so radioactive that heat generation is a real issue when considering handling, storage and disposal. Thus the appropriate management of waste is required to ensure the safety of workers, the general public and the surrounding environment because of the radiation emitted. However, not all RW has the same level of potential hazard to the environment and so classification of waste makes it easier to determine how they can be handled and helps to identify suitable disposal options (Table 2). Additionally, repositories often have individual limits for specific radionuclides which are defined as part of the licensing of facilities. Waste inventories are also very well defined. The production of RWs is not limited to nuclear power generation but is generated wherever radioisotopes are used (e.g. nuclear medicine, military applications, research). Additionally, the use of raw materials such as rocks, soils and minerals containing NORM in certain industrial activities can concentrate their natural radioactivity e.g. oil pipeline scales, soap manufacture from phosphate.

In comparison, CO_2 is a non-radioactive, naturally occurring gas, asphyxiating at higher concentrations, which is being emitted into the atmosphere in huge volumes. CO_2 waste streams from many sources, particularly power plants, will probably also contain impurities. There is considerable uncertainty in the estimates of volumes of these impurities although it is important to note that some, for example H_2S, are in themselves toxic. Thus, in contrast to RW, the specifications of some CO_2 streams have yet to be clearly defined.

Table 3 Comparison of the environmental issues relevant to the geological disposal of CO_2 and radioactive waste (post-emplacement)

Comparison criteria	Geological storage of CO_2	Geological disposal of radioactive waste
Characteristics	Large volume/mass (emissions from fossil fuels 23.5 Gt CO_2/year (2001))	Small volume/mass (81,000 m^3/year or ~210 kt/year conditioned wastes in OECD countries)
	Naturally occurring gas. Not radioactive	Radioactive but some isotopes not found in nature
	Asphyxiant at high concentrations	Toxic at high concentrations. Some low concentrations have health hazard
	Waste streams may contain other impurities; uncertainty in estimates of volumes of impurities	Generally a very complex composition. Inventories are usually very well-defined
Environmental impacts	Many sites needed (potentially large area, kms depth)	Few sites needed (small area, 1 km depth)
	Mostly surface infrastructure	Surface and underground infrastructure
	Depends entirely on geological isolation	Geological isolation critical but complemented by engineered barriers
	CO_2 will be able to alter the geological environment	Repository barriers and gases from degradation of waste and barriers will be able to alter the geological environment
	Small research database on the impacts of CO_2 leakages from storage sites	Large research database on impacts on biological systems (particularly humans)
	No regulatory framework currently exists	Exposure and dose limitations are highly regulated
Assessment of hazards	Hazard as long as concentrated	Hazard as long as concentrated but decreases with time due to radioactive decay
	Containment using geological environment only. Likely to be tested early post-closure	Repository design tailored to waste type and will involve an engineered multi-barrier approach
	Post-closure, leakage could occur through caprock, undetected faults, fractures, abandoned, leaking wells. Risk of leakage will decrease with time because trapping mechanisms become more efficient	Post-closure, leakage could result if both the engineered barriers and geological environment failed
	Emphasis on expected post-injection performance	Emphasis on low probability, high consequence scenarios over long term

(continued)

Table 3 (continued)

Comparison criteria	Geological storage of CO_2	Geological disposal of radioactive waste
Monitoring environmental impacts	Baseline environmental information needed from undisturbed site	Baseline environmental information needed from undisturbed site
	Monitoring high profile in safety case	Monitoring, if any, depends on regulatory requirements (not in safety case)
	Range of monitoring requirements is being refined. Duration of monitoring requires regulatory framework	Technical background on monitoring available

4.2 *Environmental Impacts of Radioactive Waste and CO_2*

4.2.1 Impacts from the Disposal Facilities

The relatively low volume of RW produced by the nuclear industry means that it can be managed and disposed of in relatively small, usually national facilities—and the understanding and regulation of environmental issues can be similarly constrained. Both surface and underground infrastructure will be required to ensure isolation of the wastes. In contrast, CO_2 storage facilities will be numerous and probably large-scale. Surface infrastructure will be needed for injection with associated transport facilities. Consequently, evaluating post-closure performance will be more diverse and challenging, particularly in terms of environmental issues.

For RW disposal, it is important to recognise that all repository designs use an engineered multi-barrier system approach and these barriers, in themselves, can alter the surrounding host rock environment. An example is the generation of a hyperalkaline plume from a repository containing cementitious materials, which will alter the mineralogy and porosity of the surrounding rock. Because of radioactive decay, RWs become progressively less radioactive with time and, within a million years of its removal from a reactor, spent fuel is less radioactive than the uranium ore from which it was made (Chapman and Curtis 2006). If disposed of in a deep geological repository it is likely to be much more isolated from the near-surface environment by the intervening strata and so have much less environmental impact than the original ore deposits, many of which lie near the surface. For vitrified HLW, which has had the potentially valuable long-lived uranium and plutonium removed by reprocessing, the reduction to natural uranium ore deposit levels of radioactivity occurs within a few thousand years (Chapman and Curtis 2006).

With the exception of the well completions, no engineered barriers will be used for CO_2 storage and, as a result, it is possible for the CO_2 to change the environment both chemically (alteration of groundwater conditions through CO_2/rock interactions) and, in extreme cases, physically. However, the degree of risk to the environment from CO_2 leakage from the geological environment will significantly reduce with time from the end of injections, as a combination of initially physical (such as residual trapping and pressure decreases) and subsequent chemical trapping mechanisms become more effective, e.g. chemical reactions with minerals (Benson 2005; IPCC 2005).

4.2.2 Impacts of Leakages on Biological Systems

Radiation, from whatever source, represents a potential danger to biological systems and hence to the environment. The actual danger from RW depends on many factors such as the nature of the radionuclides in the waste and the type and energy of the radiation emitted, its rate of exposure and the type, age and health of the receiving receptor (usually human). At high radiation exposures, death will occur within months or less; at moderate levels, radiation exposure increases the chance that an individual will develop cancer; at lower levels, the cancer risk decreases although the relationship between cancer risk and the magnitude of exposure is unclear. In order to minimise and control these risks, national radiation protection agencies have issued rules with legal force on dose limitations and limits of intake of radioactivity as well as guidelines for working with radioactive substances. The International Commission on Radiological Protection (ICRP) regularly publishes recommendations and guidelines, and is currently considering a framework for assessing the impact of ionising radiation on non-human species. In this framework the ICRP proposes the use of 'reference animals and plants' because there is now an increasing need to demonstrate, directly and explicitly, that the environment is being protected even under planned radiation exposure situations (see ICRP 2007).

Although it is an asphyxiant at high concentrations, CO_2 has a fundamental role in the global biogeochemical cycle which is well recognised. This chapter has identified some of the impacts of elevated CO_2 on the environment in the context of CO_2 storage. However, no equivalent of the ICRP exists and no guidance is currently available on the release and environmental impacts of CO_2 from terrestrial and marine storage sites. No 'reference animals and plants' have been identified and, indeed, little information is available on total ecosystem responses to a CO_2 leak and their recovery times. Consequently, the scientific understanding of the environmental impacts of CO_2 leaking from a storage site, which is needed to assist in the development of regulatory guidelines, is not yet fully understood.

4.3 Assessment of the Hazards Arising from Radioactive Waste and CO_2

RW inventories vary and, consequently, so does the radiological hazard and the duration of that hazard. Thus any particular repository design will need to reflect the nature of the RWs to be emplaced, and the associated hazard. For example, waste will be emplaced in a matrix which will provide a stable waste form that is resistant to leaching and gives slow rates of radionuclide release for the long-term. This will be decades for less hazardous LLW but will need to be up to hundreds of thousands of years for very hazardous HLW. In contrast, although CO_2 could be mixed with impurities on injection, it is only hazardous in high concentrations. However, this hazard will remain constant at higher concentrations.

The risks of leakage of CO_2 from a geological storage site to the environment can be classified as either global or local. Global risks involve the release of CO_2 that may contribute significantly to climate change if there is a large leakage from a geological formation into the atmosphere—although this risk should be compared to that arising if there is no storage. This risk, although low, is higher during the injection phase when reservoir pressures are highest. With regard to local risks, these include sudden and rapid CO_2 leakage from an injection well or from abandoned wells, or gradual leakages through undetected faults, fractures, caprock or leaking wells. Risks of this type of leakage are higher early post-closure before other trapping mechanisms reduce the mass of buoyant CO_2. Consequently, much emphasis is placed on assessing post-injection performance, before formal closure. Leakage from a post-closure RW repository would also be a local risk to the environment and would include unpredicted failure of the engineered barriers coupled with subsequent migration of radionuclides through the host rock. While unlikely, much work has been undertaken to evaluate and manage risk of leakages from RW repositories using low probability/high consequence scenarios, particularly in the context of PA and the repository 'safety case', and a similar holistic system model approach is now being proposed for CO_2 (described in the chapter by Maul (2011)).

4.4 *Monitoring Environmental Impacts*

Monitoring is an important aspect of the development and operation for both technologies and will also provide confidence in successful containment of the wastes (Stenhouse and Savage 2004). It will be important to obtain baseline information on the undisturbed site and, for environmental impacts, it will be crucial to obtain near-surface and surface data using a variety of ecological, chemical and physical parameters. Subsequent operational and post-injection monitoring data can then provide meaningful inputs to assessments. It is unlikely that there will be radionuclide releases from a repository soon after closure because of the engineered barrier system, so surface monitoring will be relatively unimportant and is dependent on regulatory requirements. However, the integrity of the geological containment of CO_2 may be tested soon after closure because there are no engineered barriers in place, as is the case for an RW repository. A range of standard protocols would be needed to undertake effective environmental monitoring for CO_2 and these are currently being developed. Environmental monitoring is likely to become less important with time as retention processes become more important. However, the decision on when to cease monitoring of any kind will be one that can only be made when the necessary regulatory framework is in place.

5 Conclusions

Given the discussions and comparisons above, several key points emerge which can be summarised in two general areas: science and policy.

5.1 *Science*

Both CO_2 and RW can be hazardous to a wide range of organisms although their effects on life processes are very different. Much is known about radiological effects on organisms. In contrast, little is known about the effects of CO_2 leakages from a storage site on ecosystems and subsurface environments and this is currently an active area of research.

The volume of RW is very small when compared to CO_2 emissions from stationary sources. Consequently, the numbers and relative sizes of RW repositories and CO_2 storage sites will be very different. Moreover, this means that RW management and disposal can be tightly constrained. Additionally, repositories are usually considered as national facilities, whereas CO_2 storage projects are often considered to be regional in nature. Currently, most CO_2 emissions from stationary sources are directly into the atmosphere with no management—effectively this means that there is 100% leakage. If CCS is to be a successful mitigation technology, then it will be crucial to demonstrate that the environmental impacts of the technology, particularly in the long term, are acceptable when compared to those of global warming.

RW repositories use an engineered multi-barrier approach for containment and these barriers can alter the environment. In contrast, CO_2 storage relies on the integrity of the geological environment for containment and this is likely to be tested early post-closure. Additionally, the CO_2 itself will also alter the geological environment. Consequently, it will be important to develop protocols for monitoring environmental changes as a result of CO_2 leakage. Methods will be needed for monitoring the shallow subsurface, ecosystems and reference organisms.

Much work has been undertaken to evaluate and manage risk of leakages from RW repositories, particularly in the context of PA and the repository 'safety case' and much can be learned from this considerable experience. A similar system model approach is now being proposed for CO_2 (described in the chapter by Maul (2011)). This will help to ensure a systematic approach to assessing environmental impacts for any CO_2 storage site.

5.2 Policy

The criteria that a radioactive repository must satisfy for long-term, post-closure safety are very well defined internationally. Currently, no similar specific regulatory framework for geological CO_2 storage is in place (described in the chapter by Wilson and Bergan (2011)) although it is recognised that leakages to the environment must be addressed. Currently, most EIAs for existing CO_2 storage projects under existing oil and gas regulations have focused on the operational period, but it is increasingly recognised that long-term performance will form a critical component when assessing potential environmental impacts and site liability issues. Although the two technologies are different, an examination of the approaches used for regulating RW repositories could be very useful for the development of the CO_2 storage regulatory framework.

Regulators will also require information on impacts of CO_2 on 'yet to be defined' reference organisms in order to establish appropriate threshold and safety criteria. Recovery rates will also need to be defined. Additionally, the impacts on groundwaters will need to be assessed.

In conclusion, it is worth noting that many countries around the world continue to face difficulties with implementing programmes for geological disposal of RW. Technically speaking, although geological disposal is well understood and regulated, the general public has concerns and fears about the long-term safety of a repository which focus on the effects of leaks on human health and the environment. Clearly without addressing these concerns, the implementation of waste disposal programmes will continue to flounder and this is now being recognised by the nuclear industry. Recent studies of the public's perception of CCS have revealed the same concerns about the effects of leakages of CO_2 from a storage site on the environment (as described in the chapter by Reiner and Nuttall (2011) and by Shackley et al. 2004). The RW disposal experience strongly suggests that it is crucial that these perceived CO_2 leakage concerns are addressed if the technology is to gain public acceptance and be successfully implemented.

Acknowledgements This work is published with the permission of the Executive Director of the British Geological Survey (Natural Environment Research Council (NERC)). The authors would like to thank Jon Harrington and Gary Kirby (British Geological Survey), Ian McKinley (McKinley Consulting, Switzerland), Peter Hogarth and the anonymous reviewers for their helpful comments.

References

Alexander WR, McKinley LE (eds) (2007) Deep Geological Disposal of Radioactive Waste, vol 9. Elsevier, Oxford

Aspelund A, Jordal K (2007) Gas conditioning—the interface between CO_2 capture and transport. Int J Greenhouse Gas Control 1:343–354

Beaubien S, Ciotoli G, Coombs P, Dictor MC, Krüger M, Lombardi S, Pearce JM, West JM (2008) The impact of a naturally occurring CO_2 gas vent on the shallow ecosystem and soil chemistry of a Mediterranean pasture (Latera, Italy). Int J Greenhouse Gas Control 2(3):373–387. doi:10.1016/j.ijggc.2008.03.005

Benson SM (2005) Lessons learned from industrial and natural analogs for health, safety and environmental risk assessment for geologic storage of carbon dioxide. In: Benson SM (ed) Carbon Dioxide Capture for Storage in Deep Geologic Formations—Results from the CO_2 Capture Project, vol 2, Geologic storage of carbon dioxide with monitoring and verification. Elsevier Science, Oxford, pp 1133–1141

Benson SM, Hepple R, Apps J, Tsang CF, Lippmann M (2002) Lessons Learned from Natural and Industrial Analogues for Storage of Carbon Dioxide in Deep Geological Formations. Report LBNL-51170. Lawrence Berkeley National Laboratory, Berkeley, CA

Bowden RA, Shaw RP (2007) The Kayelekera uranium deposit, Northern Malawi: past exploration activities, economic geology and decay series disequilibrium. Appl Earth Sci Trans Inst Min Metall B 116(2):55–67

Chapman N, Curtis C (2006) Confidence in the safe disposal of radioactive waste. Geological Society of London. http://www.geolsoc.org.uk/gsl/views/policy_statements/rwd. Accessed 27 Apr 2009

Chapman NA, McKinley IG, Shea ME, Smellie JAT (eds) (1992) The Poços de Caldas Project: natural analogues of processes in a radioactive waste repository. J Geochem Explor 45(1/3): 1–603 and 46(1):1–146

Coombs P, Wagner D, Bateman K, Harrison H, Milodowski AE, Noy D, West JM (2010) The role of biofilms in subsurface transport processes. Quart J Eng Geol 43:131–139

Degnan P, Bath A (eds) (2005) PADAMOT: Project Overview Report. PADAMOT Project Technical Report. United Kingdom Nirex Limited, Harwell

EA (Environment Agency) (2009) Geological disposal facilities on land for solid radioactive wastes: guidance on requirements for authorisation. Environment Agency, Northern Ireland Environment Agency. http://www.environment-agency.gov.uk/business/sectors/99322.aspx. Accessed 27 Apr 2009

EC (European Commission) (2008) Proposal for a directive of the European parliament and of the council on the geological storage of carbon dioxide. http://ec.europa.eu/environment/climat/ccs/eccp1_en.htm. Accessed 27 Apr 2009

GJV (Gorgon Joint Ventures) (2005a) Draft environmental impact statement/Environment review and management programme for the proposed Gorgon development: executive summary. http://www.chevronaustralia.com/Libraries/Chevron_Documents/Executive_Summary.sflb.ashx. Accessed 27 Apr 2009

GJV (Gorgon Joint Ventures) (2005b) Draft environmental impact statement/Environment review and management programme for the proposed Gorgon development: main report. http://www.chevronaustralia.com/ourbusinesses/gorgon/environmentalresponsibility/environmentalapprovals.aspx. Accessed 27 Apr 2009

Haasum I, Nielsen PV (1996) Preincubation of *Penicillium commune* conidia under modified atmosphere conditions: influence on growth potential as determined by an impedimetric method. J Stored Prod Res 32(4):329–337

Harrington JF, Birchall DJ, Noy DJ, Cuss RJ (2007) Large scale gas injection test (Lasgit) performed at the Äspö Hard Rock Laboratory: summary report 2007. BGS Technical Report CR/07/211. British Geological Survey, Nottingham

Hepple RP (2005) Human health and ecological risks of carbon dioxide. In: Benson SM (ed) Carbon Dioxide Capture for Storage in Deep Geologic Formations – Results from the CO_2 Capture Project, vol 2, Geologic storage of carbon dioxide with monitoring and verification. Elsevier Science, Oxford, pp 1143–1172

IAEA (International Atomic Energy Agency) (2003) Radioactive Waste Management Glossary. IAEA, Vienna

IAEA (International Atomic Energy Agency) (2005) Chernobyl's Legacy: Health, Environmental and Socio-economic Impacts and Recommendations to the Governments of Belarus, the Russian Federation and Ukraine. The Chernobyl Forum: 2003–2005. IAEA, Vienna

IAEA (International Atomic Energy Agency) (2006) Geological disposal of radioactive waste. IAEA Safety Standards Series No. WS-R-4. IAEA, Vienna

ICRP (International Commission on Radiological Protection) (2007). Environmental protection: the concept and use of reference animals and plants. Draft report 4a. http://www.icrp.org/draft_animals.asp. Accessed 27 Apr 2009

IEA-GHG (International Energy Agency Greenhouse Gas R&D Programme) (2004) Impact of Impurities on CO_2 Capture, Transport and Storage. Report PH4/32. IEA Greenhouse Gas R&D Programme, Cheltenham

IPCC (Intergovernmental Panel on Climate Change) (2005) IPCC Special Report on Carbon Dioxide Capture and Storage. Metz B, Davidson O, de Coninck HC, Loos M, Meyer LA (eds). Cambridge University Press, Cambridge, UK, and New York NY

Ishida H, Fukuhara T, Watanabe Y, Shirayama Y, Golmen L, Kita H, Magi M, Ohsumi, T (2006) Assessing the effect of high concentration of CO_2 on deep-sea benthic microorganisms using a benthic chamber system. In: Gale JJ, Rokke N, Zweigel P, Svenson H (eds) Proceedings of the 8th International Conference on Greenhouse Gas Control Technologies, Elsevier, Oxford, CD-ROM

Kharaka YK, Cole DR, Thordsen JJ, Kakouros E, Nance HS (2006) Gas–water–rock interactions in sedimentary basins: CO_2 sequestration in the Frio Formation. J Geochem Explor 89:183–186

Krüger M, West J, Frerichs J, Oppermann B, Dictor M-C, Jouliand C, Jones D, Coombs P, Green K, Pearce J, May F, Möller I (2009) Ecosystem effects of elevated CO_2 concentrations on microbial populations at a terrestrial CO_2 vent at Laacher See, Germany. In: Gale J, Herzog H, Braitsch J (eds) Proceedings of the 9th International Conference on Greenhouse Gas Control Technologies. Energy Procedia, 1: 1933–1939

Mace MJ, Hendriks C, Coenraads R (2007) Regulatory challenges to the implementation of carbon capture and geological storage within the European Union under EU and international law. Int J Greenhouse Gas Control 1:253–260

Maina JN (1998) The Gas Exchangers: Structure, Function and Evolution of the Respiratory Processes, vol 37, Zoophysiology. Springer, Berlin

Mann DD, Jayas DS, White NDG, Muir WE (1999) Mortality of adult *Cryptolestes ferrugineus* (Stephens) exposed to changing CO_2 concentrations. J Stored Prod Res 35:385–395

Maul P (2011) Risk assessment, risk management and remediation for the geological disposal of radioactive waste and storage of carbon dioxide. In this volume

McGinnes DF (2007) Waste sources and classification. In: Alexander WR, McKinley LE (eds) Deep Geological Disposal of Radioactive Waste, vol 9. Elsevier, Oxford, pp 8–40

McKinley IG, Alexander WR, Blaser PC (2007) Development of geological disposal concepts. In: Alexander WR, McKinley E (eds) Deep Geological Disposal of Radioactive Waste, vol 9. Elsevier, Oxford, pp 41–76

Miller W, Alexander R, Chapman N, McKinley I, Smellie J (2000) Geological Disposal of Radioactive Wastes and Natural Analogues, vol 2, Pergamon waste management series. Elsevier Science, Oxford

Nirex (1997) Nirex 97*:* An Assessment of the Post-closure Performance of a Deep Waste Repository at Sellafield. Overview. Nirex Report S/97/012. United Kingdom Nirex Limited, Harwell

Onstott TC (2005) Impact of CO_2 injections on deep subsurface microbial ecosystems and potential ramifications for the surface biosphere. In: Benson SM (ed) Carbon Dioxide Capture for Storage in Deep Geologic Formations – Results from the CO_2 Capture Project, vol 2, Geologic storage of carbon dioxide with monitoring and verification. Elsevier Science, Oxford, pp 1217–1250

OSPAR (Convention for the Protection of the Marine Environment of the North-East Atlantic) (2007) OSPAR guidelines for risk assessment and management of storage of CO_2 streams in geological formations. OSPAR 07/24/1-E, Annex 7. http://www.ospar.org/v_measures/browse.asp?menu=01110305610124_000001_000000. Accessed 29 Apr 2009

POST (Parliamentary Office of Science and Technology) (1997) Radioactive Waste – Where Next? Parliamentary Office of Science and Technology, London

Reiner DM, Nuttall WJ (2011) Public acceptance of geological disposal of carbon dioxide and radioactive waste: similarities and differences. In this volume

Savage D (ed) (1995) The Scientific and Regulatory Basis for the Geological Disposal of Radioactive Waste. Wiley, Chichester

Shackley S, McLachlan C, Gough C (2004) The public perceptions of carbon capture and storage. Tyndall working paper No 44. Tyndall Centre for Climate Change Research. http://www.tyndall.ac.uk/publications/working_papers/wp44.pdf. Accessed 29 Apr 2009

SMR (Senter for Miljo-Og Ressursstudier) (1999) CO_2 and Marine life. Report from a Workshop in Bergen. SMR-Report 30/99. Senter for Miljo-Og Ressursstudier, University of Bergen, Bergen, Norway 16–18 Feb 1998

Spangler LH, Dobecka LM, Repasky K, Nehrirb A, Humphries S, Barr J, Keith C, Shaw J, Rouse J, Cunningham A, Benson S, Oldenburge CM, Lewickie J, Wells A, Diehl R, Strazisar B, Fessenden J, Rahn T, Amonette J, Barr J, Pickles W, Jacobson J, Silver E, Male E, Rauch H, Gullickson H, Trautz R, Kharaka Y, Birkholzer J, Wielopolski L (2009) A controlled field pilot for testing near surface CO_2 detection techniques and transport. In: Gale J, Herzog H, Braitsch J (eds) Proceedings of the 9th International Conference on Greenhouse Gas Control Technologies. Energy Procedia 1:2143–2150

Stenhouse MJ, Savage D (2004) Monitoring experience associated with nuclear waste disposal and its application to CO_2 sequestration projects. In: Baines SJ, Worden RH (eds) Geological Storage of Carbon Dioxide. Geological society Special Publication 233, Geological Society of London, London, pp 235–247

Stenhouse MJ, Zhou W, Savage D, Benbow S (2005) Framework methodology for long-term assessment of the fate of CO_2 in the Weyburn field. In: Benson SM (ed) Carbon Dioxide Capture for Storage in Deep Geologic Formations—Results from the CO_2 Capture Project, vol 2, Geologic storage of carbon dioxide with monitoring and verification. Elsevier Science, Oxford, pp 1251–1261

Stenhouse M, Arthur R, Zhou W (2009) Assessing environmental impacts from geological CO_2 storage. In: Gale J, Herzog H, Braitsch J (eds) Proceedings of the 9th International Conference on Greenhouse Gas Control Technologies. Energy Procedia 1:1895–1902

Sustr V, Siemk M (1996) Behavioural responses to and lethal effects of elevated carbon dioxide concentration in soil invertebrates. Eur J Soil Biol 32:149–155

Svensson R, Bernstone C, Eriksson S, Kreft E, Arts R, Obdam A, Meyer R (2005) Safety assessment of structure Schweinrich. Part of CO_2STORE case study Schwarze Pumpe. Internal unpublished CO_2STORE report

Tian S, Fan Q, Xu Y, Wang Y, Jiang A (2001) Evaluation of the use of high CO_2 concentrations and cold storage to control *Monilia fructicola* on sweet cherries. Postharvest Biol Technol 22(1):53–60

Torp TA, Gale J (2002) Demonstrating storage of CO_2 in geological reservoirs: the Sleipner and SACS projects. In: Gale JJ, Kaya Y (eds) Greenhouse Gas Control Technologies: Proceedings of the 6th International Conference on Greenhouse Gas Control Technologies, vol 1. Pergamon, Amsterdam, pp 311–316

Turley C, Nightingale P, Riley N, Widdicombe S, Joint I, Gallienne C, Lowe D, Goldson L, Beaumont N, Mariotte P, Groom S, Smerdon G, Rees A, Blackford J, Owens N, West J, Land P, Woodason E (2004) Environmental Impacts of a Gradual or Catastrophic Release of CO_2 into the Marine Environment Following Carbon Dioxide Capture and Storage. Report to the UK Department for Environment, Food and Rural Affairs. DEFRA: MARP 30 (ME2104), London, 31 Mar 2004

UNSCEAR (United Nations Scientific Committee on the Effects of Atomic Radiation) (2000) Sources and Effects of Ionizing Radiation. UNSCEAR 2000 Report to the General Assembly, with scientific annexes, vol I: sources. http://www.unscear.org/unscear/en/publications/2000_1.html. Accessed 29 Apr 2009

Watson SJ, Jones AL, Oatway WB, Hughes JS (2005) Ionising radiation exposure of the UK population: 2005 review. RPD Series Report HPA-RPD-001. Health Protection Agency. http://www.hpa.org.uk/webw/HPAweb&Page&HPAwebAutoListNameDesc/Page/1197637096018?p=1197637096018. Accessed 29 Apr 2009

West JM, Pearce JP, Bentham M, Maul P (2005) Issue profile: environmental issues and the geological storage of CO_2. Eur Environ 15:250–259

West JM, Pearce J, Bentham M, Rochelle C, Maul P (2006) Environmental issues, and the geological storage of CO_2—a European perspective. In: Gale JJ, Rokke N, Zweigel P, Svenson H (eds) Proceedings of the 8th International Conference on Greenhouse Gas Control Technologies, Elsevier, Oxford, CD-ROM

West JM, Pearce JM, Coombs P, Ford JR, Scheib C, Colls JJ, Smith KL, Steven MD (2009) The impact of controlled injection of CO_2 on the soil ecosystem and chemistry of an English lowland pasture. In: Gale J, Herzog H, Braitsch J (eds) Proceedings of the 9th International Conference on Greenhouse Gas Control Technologies. Energy Procedia 1:1863–1870

Wilson EJ, Bergan S (2011) Managing liability: comparing radioactive waste disposal and carbon dioxide storage. In this volume

Wilson M, Monea M (2004) IEA-GHG Weyburn CO_2 monitoring and storage project summary report 2000–2004. In: Rubin ES, Keith DW, Gilboy CF (eds) Proceedings of the 7th International Conference on Greenhouse Gas Control Technologies, vol 3. International Energy Agency Greenhouse Gas R&D Programme, Cheltenham

Zakkour P, Haines M (2007) Permitting issues for CO_2 capture, transport and geological storage: a review of Europe, USA, Canada and Australia. Int J Greenhouse Gas Control 1:94–100

Zhou W, Stenhouse M, Sheppard M, Walton F (2004) Theme 4: long-term risk assessment of the storage site. In: Wilson M, Monea M (eds) IEA-GHG Weyburn CO_2 Monitoring and Storage Project Summary Report 2000–2004. Petroleum Technology Research Centre, Regina, pp 211–268

Risk Assessment, Risk Management and Remediation for the Geological Disposal of Radioactive Waste and Storage of Carbon Dioxide

Philip Maul

Abstract Risk assessment, risk management and remediation in the fields of geological disposal of radioactive waste (RW) and storage of carbon dioxide (CO_2) are discussed and compared. In both fields detailed site characterization is a fundamental requirement and it is necessary to consider the evolution of the system over long timescales so that natural analogues for key processes can be valuable. Some of the most important differences are:

- In RW disposal, performance assessment methods have been developed over a period of more than 2 decades, whilst for CO_2 methods for modelling the system as a whole are still at an early stage of development.
- Similarly, mature regulatory regimes are in place in most countries with deep disposal programmes for RW, but this is not the case for the geological storage of CO_2.
- The possibility of material returning to the surface in the first few decades after operations cease is much more likely for CO_2, so that monitoring will be important. If surface leakage of CO_2 is detected during this period it should be possible to sink borehole(s) to extract some of the injected CO_2.
- For RW disposal systems, engineered barriers will inevitably degrade with time, whilst for CO_2 some of the important natural barriers may actually become more effective with time. This affects the way that risk assessments are undertaken and uncertainties managed.

Keywords Risk assessment • Risk management • Performance assessment • Remediation • systems modelling

P. Maul (✉)
Quintessa Limited, The Hub, 14 Station Road, Henley-on-Thames, Oxfordshire, United Kingdom RG9 1AY
e-mail: philipmaul@quintessa.org

F.L. Toth (ed.), *Geological Disposal of Carbon Dioxide and Radioactive Waste: A Comparative Assessment*, Advances in Global Change Research 44,
DOI 10.1007/978-90-481-8712-6_4, © International Atomic Energy Agency, 2011

1 Introduction

For any technology the associated risks have to be assessed and managed. In this chapter, risk assessment, risk management and remediation in the fields of geological disposal of radioactive waste (RW) and storage of carbon dioxide (CO_2) are discussed and compared.

Different waste management options are appropriate for different categories of RW. Here consideration is restricted to those wastes that require geological disposal, as these are of most direct interest when making comparisons with the geological storage of CO_2.

The risks considered here are post-operational, after an RW repository has been closed or after CO_2 injection has ceased. The focus is on the methods used; a detailed consideration of the potential impacts from radionuclides and CO_2 returning to the accessible environment is given in the chapter on environmental impacts.

The term 'risk assessment' is used with slightly different meanings in different fields. The term 'risk' itself is used in a number of different ways. As indicated by the IAEA (2003), when used quantitatively, risk is usually defined to be the product of the probability that a specified hazard will cause harm and the consequence of that harm. Risk assessment, as applied in major hazards industries, is generally applied to the analysis of accidental events that can occur to operational plants and facilities.

In RW disposal programmes, performance assessment (PA) is used to assess 'the performance of a system or subsystem and its implications for protection and safety at a planned or an authorized facility' (IAEA 2003). PA is usually applied to analysing the post-operational (post-closure) evolution of systems that depend on passive environmental controls for this function, and part of the output from a PA may be expressed in terms of risks (particularly to human health and the environment).

Although this chapter is concerned with 'risk assessment', the term will here be used to cover the same ground as considered in PAs. Further discussion of the use of PA in the field of RW disposal and its relevance to carbon capture and storage (CCS) is given by Maul et al. (2007). In that paper, priorities were suggested for the development of performance assessment methods for CO_2 storage based on areas where experience from RW disposal can be usefully applied. These included, inter alia, dealing with the various types of uncertainty, using systematic methodologies to ensure an auditable and transparent assessment process, developing whole system models and gaining confidence to model the long-term system evolution by considering information from natural systems.

Some of the key issues that are addressed in this chapter are:

1. What methods are available to assess risks from geological disposal?
2. What options are available for risk management and remediation?
3. How does the regulatory regime affect how risks are assessed and managed?
4. What are the key technical challenges to demonstrating safety and what are the priorities for further research and development?
5. What can workers in each field learn from experience gained in the other?

Background material for the two technologies is given in Sects. 2 and 3 and some comparisons between the two are made in Sect. 4. The conclusions drawn on these topics are then summarized in Sect. 5.

2 The Geological Storage of Carbon Dioxide

2.1 Hazards and Regulations

A detailed discussion of potential impacts is given in the chapter on environmental impacts, but it is worth noting here that little is known about the direct impacts of CO_2 at the levels that may be seen when it returns to the accessible environment from a storage facility (West et al. 2005). In addition, a number of indirect impacts may be important. These include formation water/brine displacement, with the potential for adverse impacts on the quality of drinking water supplies.

In enhanced oil recovery (EOR) schemes, CO_2 is injected into oil reservoirs to increase the amount of oil that can be extracted, so that the primary motivation is not the geological storage of the CO_2. Such schemes are undertaken under the regulatory regime applicable to the original extraction process, and there are no explicit requirements to assess potential environmental impacts over long timescales (Stenhouse et al. 2005a). General regulatory criteria for CCS have yet to be fully developed in most countries. Regulatory frameworks are at various stages of development (see, for example, EC 2008 and Forbes et al. 2009), but there is little experience with their implementation.

2.2 Status of the Technology

CO_2 has been routinely used for several decades for EOR in several countries, notably the Permian Basin in the US, where there were 80 such projects in 2006 (Moritis 2006), although most of the CO_2 used was extracted from natural accumulations. At the Weyburn oilfield in Saskatchewan (Wilson and Monea 2004), CO_2 produced from the North Dakota coal gasification plant is transported via pipeline and then injected. Other projects include the Sleipner gasfield in the Norwegian North Sea (Torp and Gale 2003), where naturally occurring CO_2 within the methane natural gas is separated and injected into a saline aquifer below the seabed. A similar project is also being carried out in the Algerian In Salah gasfield (Riddiford et al. 2005).

The use of CO_2 in EOR projects is well established, but few projects have so far been initiated where the primary motivation is the geological storage of CO_2. If CCS is to become a major contributor to climate change mitigation, CO_2 from power plants will need to be captured and stored. The European Technology

Platform on Zero Emission Fossil Fuel Power Plants (ZEP) programme is aiming at 10–12 demonstration plants by 2015 prior to commercially available 'zero emission' fossil-fired power plants in 2020 (ZEP 2006). If this technology does become extensively employed there will be a requirement for a large number of storage sites in many countries.

2.3 Natural and Industrial Analogues

There are both natural and industrial analogues for CO_2 storage (Pearce et al. 2004; IPCC 2005). Holloway et al. (2005) show that natural systems can provide important information on specific relevant processes. Natural accumulations can provide information on trapping and migration mechanisms and provide field-based testing grounds for monitoring methods. Volcanic or tectonically unstable areas can provide valuable information on leakage impacts (e.g. Beaubien et al. 2008). Natural analogues can therefore provide information that is directly relevant to risk assessments.

Industrial analogues include natural gas storage and acid gas injection, and these provide experience relevant to the risk management of the injection and closure phases of CO_2 storage schemes, although this is less relevant to the post-closure period that is the focus of this chapter.

2.4 Containment Philosophy

Figure 1 illustrates some of the general features of geologic storage systems. CO_2 is injected at depth (several hundred metres below the surface) into a reservoir formation with a caprock, which provides the most important barrier to vertical movement back towards the surface. It is possible that some projects may use a reservoir without a conventional caprock, for example CO_2 may be injected into a shallow-dipping aquifer, sufficiently far from outcrop that trapping mechanisms will prevent the CO_2 from returning to the surface. The area over which potential impacts from the injection may need to be considered could be large, with horizontal distance scales of up to about 100 km being relevant.

The principal storage reservoirs are likely to be either oil and gas reservoirs or saline aquifers. Oil- and gasfields are generally characterized by proven traps with caprocks that can retain buoyant fluids for geological timescales. As illustrated in Fig. 1, possible pathways back to the surface (indicated by the red arrows) are via a well or a fracture that passes through the caprock.

Well integrity is one of the major issues for CO_2 storage, especially in mature onshore hydrocarbon fields where the numbers of wells can be large. Particularly in the cases of old wells, records may have been lost or may be inaccurate.

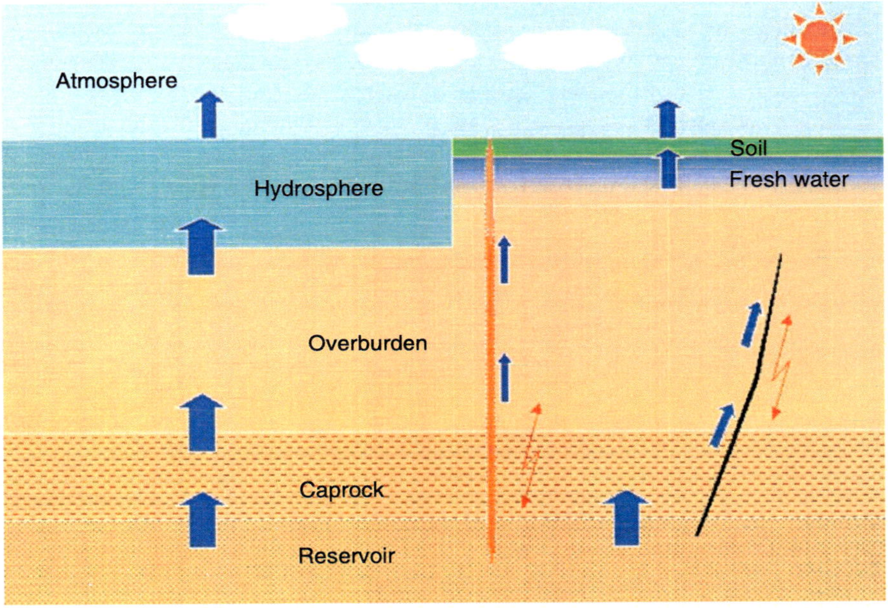

Fig. 1 Barriers and transport pathways for carbon dioxide (*see* Colour Plates)

Consequently, the existence, location or condition of wells may be unknown or uncertain.

There are a number of barriers, both physical and chemical, that can be part of the overall containment capacity of the system. In some geological settings there will be secondary seals, so that even if CO_2 is transported through the primary caprock, this may not result in transport all the way to the surface. Geochemical reactions may eventually immobilize some or all of the CO_2 and even if some CO_2 does reach the near-surface environment, there are a number of dispersive mechanisms that may result in surface fluxes being small.

2.5 *Risk Assessment*

Two different timescales of interest can be considered for risk assessments in this field. The first timescale is associated with the potential global impacts of CO_2 returning to the atmosphere. If the primary purpose of storing the CO_2 is to mitigate climate change effects, then timescales of a few centuries may be relevant (IPCC 2005), although it may be necessary to consider periods of several thousand years (Torvanger et al. 2006). The second timescale is associated with potential local impacts, which are more likely to constrain acceptable leakage rates; the relevant timescales will be determined by when such local impacts may be incurred.

Methods for assessing long-term risks are currently being developed. The Weyburn project (Wilson and Monea 2004) was amongst the first in which long-term site performance was considered (Stenhouse et al. 2005b).

Because of extensive experience in reservoir modelling in the oil and gas industries, several groups involved with assessment of the long-term fate of CO_2 have developed models based on reservoir simulation codes to investigate the transport of CO_2 (see, for example, Pruess 2004 and Rutqvist et al. 2002). There is extensive experience in this field of modelling coupled thermal, mechanical, hydraulic and chemical processes. These studies can represent the multiphase transport nature of the problem, but do not generally address in any detail the consequences in the accessible environment of potential releases from the system.

The systems approach to risk assessment is illustrated in Fig. 2, where reference is made to Features, Events and Processes (FEPs), which are different types of factors affecting the evolution of the system. It is possible to differentiate between FEPs that are external to the system (EFEPs) and those that are internal to the system. The EFEPs can combine to generate scenarios for system evolution. For example, relevant EFEPs might be associated with climate change and/or seismicity.

The system may be split up into a number of interacting subsystems and it is necessary to model all the relevant FEPs that affect the quantities of interest.

This approach has been used in other fields (particularly RW disposal) and FEP analyses have been undertaken for some CO_2 storage risk assessments (for example, Stenhouse et al. 2005b). Lewicki et al. (2007) conducted an audit of FEPs of natural systems that identified some of the key processes for CO_2 storage sites. These included secondary trapping and release in shallow reservoirs, specific events that release CO_2, faults and fractures acting as conduits for CO_2 migration and the importance of high-quality well completions.

Progress has been made in developing a generic FEP database for the geological storage of CO_2. Figure 3 shows an example entry in the FEP database described in Maul et al. (2005). The FEPs included in the database are not specific to any particular

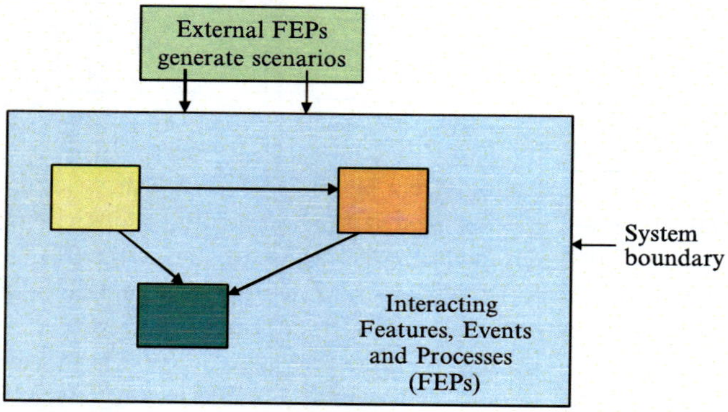

Fig. 2 Systems modelling

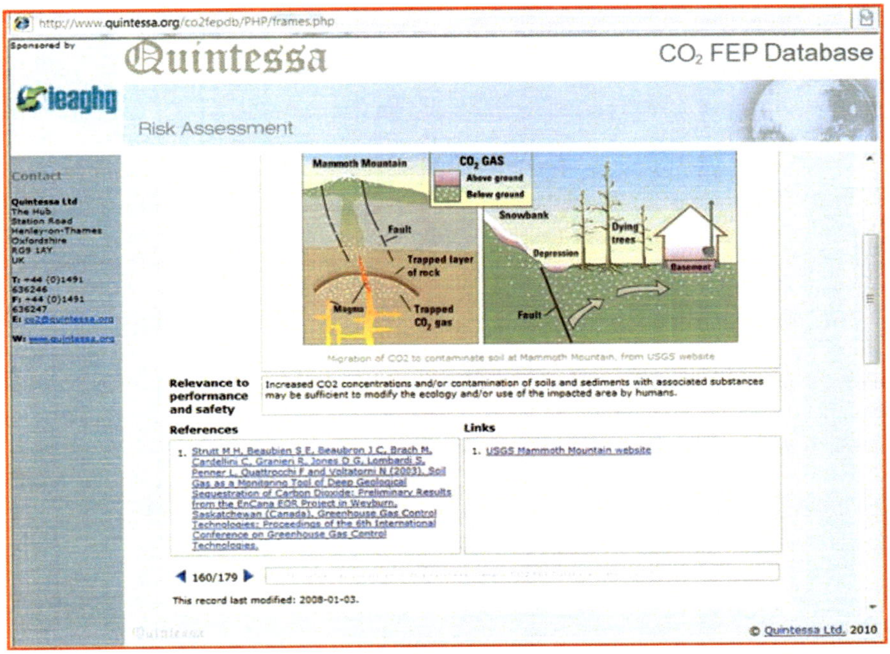

Fig. 3 An example entry in the generic FEP database (*see* Colour Plates)

model; different models will represent the FEPs in different ways, and there will not be a one-to-one correspondence between FEPs and model parameters.

This database is available through the OECD International Energy Agency (IEA) website (http://www.co2captureandstorage.info/riskscenarios/riskscenarios.htm), and has the potential to provide a basis for documenting the key sources of information. This database was originally produced in 2004, but continues to be maintained and was updated in 2008.

The development of models that satisfactorily represent the whole system remains at an early stage. With the extensive experience of detailed reservoir simulation modelling, the development of models for important specific processes, such as well leakage, and the computing power now available, most of the components required for the development of system-level models are available (see, for example, Pawar et al. 2006). The modelling and software development requirements are challenging, but not insuperable. The use of a system-level model for a natural analogue site has been demonstrated by Maul et al. (2009). Representing the potential impact of wells is one of the key challenges, and innovative methods are being developed for doing this (see, for example, Nordbotten et al. 2005).

Risk assessments may have to take account of both quantitative information from model calculations and qualitative information, for example from expert judgement. Methods for bringing these two types of information together are being developed (see, for example, Metcalfe et al. 2009).

2.6 Risk Management and Remediation

As indicated by the Intergovernmental Panel on Climate Change (IPCC 2005), risk management methods have yet to be fully demonstrated, but overall frameworks for risk management are being developed. In particular, the recent European Union Directive (EC 2008) provides such a framework, requiring, for example, that the operator should remain responsible for monitoring and undertaking any required remediation measures until responsibility for the storage site is transferred to the relevant competent authority.

The development of remote sensing techniques to detect CO_2 leakage is currently an active area of research (see, for example, Pearce et al. 2005). One way that small leakages may be detected is in changing patterns of vegetation growth. With slightly enhanced CO_2 levels crop fertilization effects may be seen, but at higher levels crop damage is seen (see, for example, Beaubien et al. 2008). There are also innovative techniques for monitoring subsurface CO_2 migration. Repeat seismic surveys have been used to monitor subsurface CO_2 at Sleipner (Arts et al. 2004) and satellite altimetry has been employed for this purpose at In Salah (Mathieson et al. 2009).

Research is also being undertaken into remediation options, including the recovery of CO_2 that has been injected if this proves to be necessary. Akervoll et al. (2009), for example, concluded that it would be possible to retrieve a significant proportion of the mobile CO_2 at Sleipner if serious problems with caprock integrity were detected.

3 Geological Disposal of Radioactive Waste

3.1 Hazards and Regulations

Despite residual uncertainties, a great deal is known about the impacts of radiation on humans and the environment (e.g. ICRP 2000), and associated regulatory criteria are well developed. Radiation doses can be calculated from human contact with radioactive materials, and a linear relationship between impacts on human beings and the radiation dose is then assumed—which is almost certainly a pessimistic assumption.

Safety criteria for RW repositories may be expressed in terms of radiation dose or risk, although the numerical values used in national regulations vary (NEA 2007). Regulatory requirements for the timescale over which quantitative PAs should be undertaken vary from country to country. There is a general acceptance that less reliance should be placed on calculations far into the future, but detailed quantitative calculations may be required for 10,000 years or longer (e.g. NEA 2007). Clearly, the long half-lives of some radioactive elements play a part in defining these assessment timescales, but long timescales are also necessary because: (1) well-located sites imply releases of contaminants only very far into the future;

and (2) ethical considerations mean that the same level of environmental protection should exist in the future as that which is applicable today.

3.2 *Status of the Technology*

RW disposal generally operates within national boundaries, with each state commissioning state-owned organizations to develop and implement the disposal plans and another agency to act as a regulator. The number of deep repositories in any country will be few. Deep geological disposal programmes are being developed in many countries. Examples include:

- In France, the Agence nationale pour la gestion des déchets radioactifs (ANDRA) is proposing a repository to be hosted in argillites in Meuse/Haute-Marne where an underground repository has been constructed. Granite has also been considered (ANDRA 2005).
- The US Department of Energy's (US DOE) Waste Isolation Pilot Plant (WIPP) commenced operations in 1999. The facility is located in rock salt (halite) in Texas. There is very little groundwater movement and the salt will flow to seal man-made structures in the rock to help isolate the waste (US DOE 2004).
- In Sweden, Svensk Kärnbränslehantering (SKB) is planning a deep repository in hard rock to be operational around 2020. A preliminary PA has recently been published (SKB 2006). Similar developments are being carried out in Finland.
- In Switzerland, the Nationale Genossenschaft für die Lagerung radioaktiver Abfälle (Nagra) is considering a repository in a low permeability sedimentary host rock environment, the Opalinus Clay (Nagra 2002).

A summary of national programmes in the OECD is given in NEA (2005).

3.3 *Natural Analogues*

The whole concept of deep geological disposal is based on an understanding of the evolution of geological systems over long timescales, and so confidence in modelling the system is increased if information from natural systems can be used (see, for example, Miller et al. 2000). Almost all national disposal programmes are involved in natural analogue studies.

An example natural analogue site is Maqarin in Jordan (Alexander and Smellie 2002). This has enabled some aspects of models for interactions between repository host rocks and alkaline pore-fluids to be tested, which is important for repositories where cement is used. The results from this study are also relevant to understanding the long-term alteration that might occur in the rock that surrounds cement used as seals or to bond casings with the rock, in wells that penetrate a CO_2 storage site. Several other analogue studies for RW also have relevance to the geological storage of CO_2.

3.4 The Multi-Barrier Concept

A key concept in RW disposal is the multiple barrier principle, in which long-term safety is assured by a series of engineered and natural barriers (see, for example, Savage 1995).

These barriers prevent or reduce the transport of radionuclides in groundwater, which is generally the most important transport mechanism. The barriers may also influence the migration of gas (e.g. Rodwell et al. 2003). Some radionuclides, such as C-14, may be transported in the gaseous phase, which will be subject to many of the same transport processes as CO_2.

The use of multiple barriers to provide a range of safety functions is one of a number of siting and design principles that are observed in order to achieve so-called 'robust' systems. For example, at any given time in the evolution of a system, some safety functions may be 'latent', i.e. they operate only if other safety functions (unexpectedly) fail to operate. Others may be 'reserve', i.e. they may contribute positively to safety, but residual uncertainties in quantitative understanding of their contributions lead to their being omitted from conservative ('worst case') safety analyses. The relative importance of the barriers may change with time.

3.5 Risk Assessments

Significant advances have been made over the last 2 decades in PAs in this field. In particular, systematic PA methodologies help to ensure that the whole process is auditable and transparent. Figure 4 shows the stages in a typical methodology, which is based on an internationally developed methodology (IAEA 2004) for near-surface repositories, although the principles apply equally to geological disposal.

Systematic analysis of FEPs (see Sect. 2.5) that can influence radionuclide transport and the impacts of radionuclides on humans and the environment has proved to be effective for documenting and auditing PA models. The Nuclear Energy Agency FEP database (NEA 2000) has been widely used in this context.

For disposal concepts that rely on the performance of engineered barriers, the evolution of the system through coupled thermal, hydraulic, mechanical and chemical (THMC) processes can be complex (see, for example, SKB 2006). Modelling the evolution of such systems remains an area of intensive research activity.

Detailed supporting models will always be needed, for example, to investigate groundwater flows in three dimensions. However, the continuing increase in modern computing power means that the distinction between systems-level and detailed models is becoming increasingly blurred.

Probabilistic assessments are one powerful tool for investigating uncertainties and are widely used in the field of RW disposal, particularly where regulatory

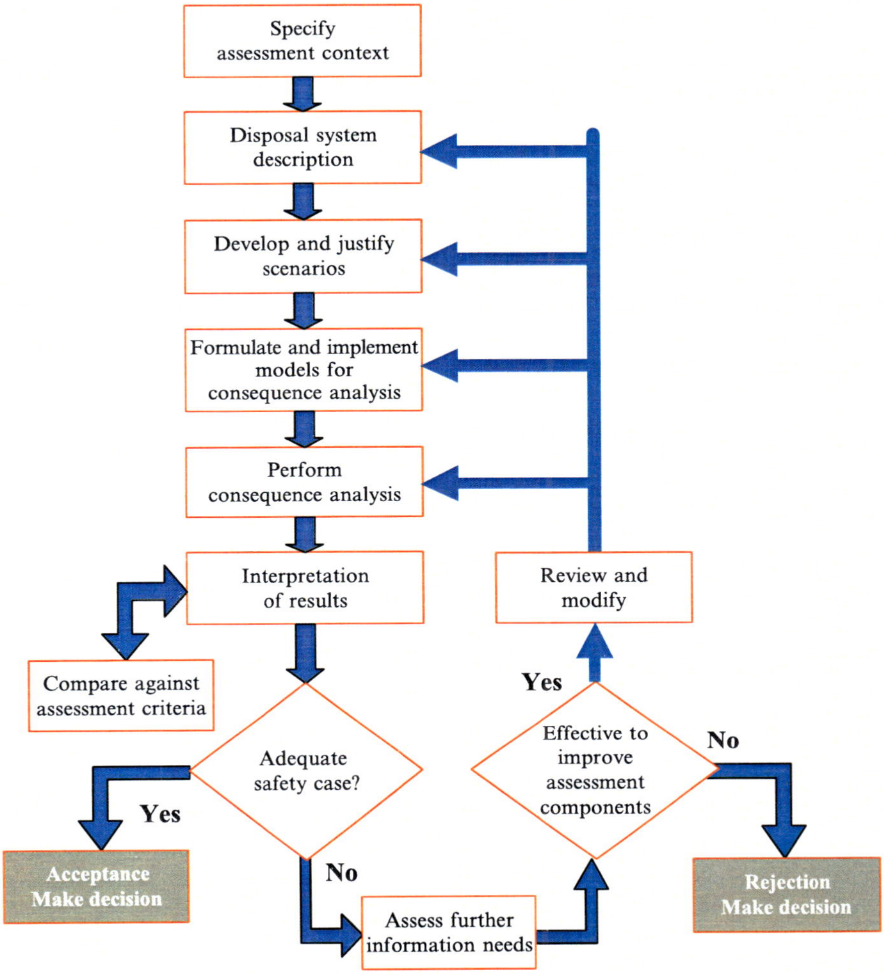

Fig. 4 A structured approach to performance assessment

criteria are expressed in terms of risk. However, experience in using these methods has highlighted a number of important problems that can arise:

- Probability density functions (PDFs) need to be defined for all input parameters, and for some of these the only way to do this is to use knowledge that experts in the field possess. Formal methods are available for using expert knowledge to elicit these (see, for example, O'Hagan et al. 2006), but this can be an extremely resource-intensive activity and it is frequently only possible to obtain such information for a few key parameters.
- The use of parameter PDFs can hide important distinctions between uncertainties due to our 'ignorance' of the system (which might change as more

information becomes available) and genuine variations due to, for example, system heterogeneities. If uncertainties and variability are not distinguished, the calculated spread in the endpoints of interest may be overestimated.

- Probabilistic assessments do not always properly represent correlations between parameters. If correlations exist, but are not properly represented, conclusions drawn from the calculated impacts may be misleading.
- Probabilistic calculations can result in counter-intuitive outputs. For example, it is possible that in admitting to a greater level of 'ignorance' in a key parameter we may actually decrease the calculated risks. This has been termed 'risk dilution' (Savage 1995).
- Probabilistic calculations can hide so much detail that the transparency of the proponent's case may be lost. The use of deterministic calculations to support conclusions drawn from probabilistic assessments can be helpful; these can be more readily reproduced by third parties and can exemplify the key features of the arguments being put forward.

3.6 Risk Management and Remediation

Some repositories are designed to facilitate the retrieval of waste over long periods, but the most important contribution to the management of risks is in the site selection process and the design of engineered barrier systems (EBSs). Measures such as restricting access to the site and the maintenance of records can be employed following repository closure, but, because of the long timescales involved, no reliance can be placed on remediation measures far into the future and the assurance of safety in regulatory criteria has to be demonstrated without human intervention.

4 Comparisons Between Technologies

4.1 Introduction

Based on the descriptions given in the previous two sections, the two technologies are compared in this section. Table 1 summarizes the key issues, and further details are then given in each case.

4.2 Basic Principles

For both RW disposal and the geological storage of CO_2 the fundamental concept is to isolate the material from the biosphere and natural resources for very long timescales. In both cases the feasibility of this approach is based on an understanding of the behaviour of natural systems, with many of the processes that affect the long-term

Table 1 Summary of technology comparisons

Issue	Geological Storage of Carbon Dioxide	Radioactive waste disposal
Basic principles	Natural processes provide isolation	Emphasis on the multi-barrier approach
Site selection and characterization	Mostly remote information, possibly supplemented by information from boreholes	Resource intensive; need to avoid natural resources
Assessment timescales	Not yet well defined, but likely to be up to several thousand years	Typically up to a million years
System evolution	Injected CO_2 may directly affect geosphere evolution	Construction of engineered barriers, but radionuclides are 'trace' contaminants
Leakage	Probability may reduce with time	Probability will generally increase with time
Risk assessments	System-level modelling methods beginning to be developed	Well established performance assessment methodologies
Regulatory regime	Generally not fully developed	Mature in most countries
Monitoring	Important for the first few decades	Required for public reassurance
Remediation	Should be feasible	Likely to be difficult

evolution of the system being the same. As discussed in Sect. 3.5, some natural analogues are relevant to both technologies.

The multi-barrier concept is emphasized in RW disposal, but can also be seen to be applicable to CO_2 storage, as a number of different barriers may operate. In the case of RW disposal, the near-field barriers are engineered, and their effectiveness will inevitably reduce with time. For CO_2 storage, borehole seals can also be considered to be engineered barriers. The respective roles of the natural and engineered barriers will depend on the type of the host rock.

4.3 *Site Selection and Characterization*

Detailed site characterization is a fundamental requirement for both concepts. In RW disposal the geosphere is an important barrier in the overall design concept and a detailed knowledge of the geology may be essential in order to make the safety case. Here the underground environment hosting the waste will be accessible via shafts, tunnels or drifts, but for CO_2 storage projects the amount of information will be much sparser, perhaps being limited to a few boreholes and indirect characterization such as seismic surveys. As discussed in Sect. 2, the most important features in the system for risk assessment may be abandoned wells, but it may simply not be possible to identify all such features in the area of interest as part of site characterization.

Selection of sites for the geological disposal of RWs includes avoiding locations with obvious natural resource potential. However, for CO_2 storage, it is almost inevitable that such regions will be utilized if CCS becomes a widely employed technology with possibly hundreds of storage sites in some countries.

This emphasizes that human intrusion scenarios are likely to be important in assessments for CO_2 storage.

As previously indicated, individual CO_2 storage projects may be significantly smaller in financial terms than national RW disposal programmes. This will directly affect the resources that will be appropriate for undertaking site characterization and risk assessments, subject to satisfying regulatory requirements.

4.4 Assessment Timescales, System Complexity and Uncertainty

In both cases it is necessary to consider the evolution of the system over long timescales, as materials may not return to the surface for many thousands of years (if at all). This issue has received detailed consideration by regulators in the field of RW disposal, but the regulatory regime is not yet fully developed in the field of CO_2 storage.

Because of the complexity of the natural system and the long assessment timescales, an integral part of any assessment is the management of uncertainties. Uncertainties can be categorized in a number of different ways, but one useful approach is to consider scenario, conceptual model and parameter uncertainties (e.g. Savage 1995). Scenario uncertainty reflects the fact that we can never know how the system is going to evolve in the future, and have to consider feasible examples of possible future evolutions. Conceptual model uncertainty reflects the fact that our models of natural processes will always be approximations, and that there may be several different models for the same process or groups of processes. For each model there will be uncertainty about the parameter values to use. This parameter uncertainty is, in a sense, the easiest to deal with, and there is often an over-emphasis on this type of uncertainty at the expense of inadequate consideration of the other sources. The assessment needs to demonstrate that all the different uncertainties have been addressed, and that the system performance remains satisfactory in the light of those uncertainties.

4.5 Modelling System Evolution and Material Transport

In the case of RW, radionuclides released from the near-field engineered barriers essentially act as 'trace' contaminants; they do not significantly affect the evolution of the system. On the other hand, an EBS employed in an RW repository may significantly modify the surrounding geological environment. The actual environmental changes that occur will depend upon the particular repository design and operation, which will in turn reflect the nature of the RWs. For example, where steel waste canisters are employed, corrosion may generate hydrogen gas, which might in turn influence groundwater pressures and hence flow (Rodwell et al. 2003). Another example is the emanation of an alkaline groundwater plume from a repository employing cementitious barriers. The mineralogy and porosity of the surrounding rock may be changed by reactions involving this plume.

In contrast, for CO_2 storage there would be no significant modifications to the geological environment caused by engineered systems other than boreholes, but the CO_2 itself could affect the environment. For example, CO_2 injected into deep geological strata could result in microseismic events or geochemical changes. The physical form of the CO_2 will vary with depth, as will its potential impact on system evolution. From this perspective, the technical challenge of modelling CO_2 transport may be considered to be more demanding (Pruess 2004).

Whilst there are more issues to address for the return of CO_2 to the surface ('leakage') on relatively short timescales, if this does not happen, the probability of leakage occurring may actually decrease with time as some of the natural barriers (e.g. dissolution into pore water, residual trapping and geochemical reactions with minerals) become more effective (see, for example, Benson 2005). There are, however, some processes that might lead to increased risk of leakage over time in some circumstances, notably degradation of borehole seals.

4.6 Risk Assessment Methods

As previously discussed, systematic PA methodologies are well established in the field of RW disposal, whilst the development of system-level models is at an early stage of development in the field of CO_2 storage.

4.7 Regulatory Regimes

In the field of radioactive disposal, regulatory regimes are well established in most countries that have a disposal programme. Some of these programmes have been in place for several decades. These regulatory regimes directly affect the type of risk assessment undertaken by the proponent, particularly through the specified safety requirements that have to be met.

Currently, CO_2 storage as part of EOR schemes is undertaken under the regulatory regime applicable to the original extraction process. If CCS becomes a widely employed technology with CO_2 from power plants being captured and stored, major developments in the regulatory regime will be required. It can be anticipated that the large number of demonstration plants currently planned will provide the impetus for this development.

4.8 Monitoring

Monitoring is an important aspect of the development and operation of both RW repositories and CO_2 storage sites. It is necessary to collect adequate baseline data representative of the undisturbed site, and operational and post-operational monitoring data can provide important inputs to the required assessments.

Because there will be extensive EBSs for RW, it is very unlikely that there will be releases from the repository soon after repository closure, and so surface-based monitoring is very unlikely to see radioactivity derived from the repository soon after repository closure. This does not apply to CO_2 storage where the natural barriers will be tested at an early stage. As discussed previously, if there is no short-term leakage from the host geology shortly after injection, retention processes may become more effective with time. Monitoring after operations cease is therefore likely to be an important feature of risk management for CO_2 storage. This monitoring will include the implementation of measures to detect surface leakage, but also surface-based monitoring of underground movements of CO_2, for example by carrying out repeated seismic surveys or even by using satellite altimetry. The length of time for which monitoring may be required has yet to be defined, but will depend upon the regulatory regime under which any particular project is undertaken. The period of monitoring could last for many decades following the end of operations.

4.9 Remediation

An important issue for risk management is remediation in the event that unacceptable levels of radionuclides or CO_2 are released at the surface. For a deep RW repository, remediation is highly unlikely to be required in the short term (few decades) after repository closure. Depending on the nature and extent of the contamination, some remediation techniques might be applicable, but the most effective response may be based on simply restricting human access to contaminated areas.

By contrast, if surface leakage of CO_2 is detected in the first few decades after injection has ceased, it would be possible to sink one or more boreholes to extract some of the injected CO_2 that had been injected at depth.

5 Conclusions

Given the discussion in Sects. 2–4, it is possible to summarize the conclusions that can be drawn for the key issues identified in Sect. 1. These are addressed in turn.

What methods are available to assess risks from geological disposal?

For RW disposal, systematic methods for PA have been developed over more than two decades. Radionuclide transport codes are well developed, although modelling the evolution of EBSs remains an active area of research. For CO_2 storage, extensive experience is available in reservoir modelling, but the development of methods to represent the evolution of the system as a whole over long timescales is at an early stage.

What options are available for risk management and remediation?

For both technologies the most important aspect of risk management is the selection of suitable sites. Surface-based monitoring in the first few decades after injection ceases is particularly useful for CO_2 storage, as remediation by, for example, removal of (some of) the CO_2 is a practical option on this timescale.

How does the regulatory regime affect how risks are assessed and managed?

In those countries where a risk-based criterion is used in regulations for RW disposal, this effectively requires the proponent to undertake probabilistic assessments. For CO_2 storage regulatory regimes are yet to be fully developed, and so there is scope for national and international authorities developing criteria that ensure that 'fit for purpose' risk assessments are undertaken.

What are the key technical challenges to demonstrating safety and what are the priorities for further research and development?

As discussed in Sect. 4, many of the technical challenges are similar in the two fields. In both cases it is necessary to model the evolution of a complex system over long timescales in the presence of inevitable uncertainties of different types. In the field of RW disposal regulatory criteria are frequently expressed in terms of risks as low as 10^{-6} per year. Demonstrating that this criterion is met over long assessment timescales may be challenging, depending on the host geology. If the host geology does not provide an effective barrier to radionuclide transport, then detailed information is needed in order to provide confidence in the performance of EBSs over thousands of years. Risk assessment in this field is a mature activity, but further research in the area of THMC modelling is needed for those disposal concepts that rely on the performance of the engineered barriers. For CO_2 storage, less detailed site characterization information may be available, and it may be necessary to demonstrate that consequences will be tolerable even if leakage occurs through unidentified abandoned wells. A key challenge is the development of methods that can represent all important processes in the system as a whole.

What can workers in each field learn from experience gained in the other?

Many tools that have been developed in the field of RW disposal either have been, or potentially could be, used in risk assessments of CO_2 storage. Examples include the use of generic FEP databases to audit assessment models and the use of general-purpose computer codes to enable systems-level modelling to be undertaken. Experience with the use of probabilistic methods (both good and bad) is a specific area where lessons learned in RW disposal are relevant to CO_2 storage assessments. Many of the techniques developed for reservoir modelling in the oil and gas industry are directly or indirectly relevant to the THMC modelling that needs to be undertaken in the field of RW disposal.

References

Akervoll A, Lindeberg E, Lackner A (2009) Feasibility of reproduction of stored CO_2 from the Utsira formation at the Sleipner gas field. In Gale J, Herzog H, Braitsch J (eds) Proceedings of the 9th International Conference on Greenhouse Gas Technologies. Energy Procedia, 1:2557–2564

Alexander WR, Smellie, JAT (2002) The maqarin natural analogue project (1989–1998). In Proceedings of the Eighth EC Natural Analogue Working Group Meeting European Commission Report EUR 19118 EN. European Commission, Luxembourg, pp 87–102

ANDRA (Agence nationale pour la gestion des déchets radioactifs—National Radioactive Waste Management Agency) (2005) Evaluation de sûreté du stockage géologique. http://www.andra.fr/interne.php3?id_rubrique=161

Arts R, Eiken O, Chadwick A, Zweigel P, van der Meer L, Zinszner B (2004) Monitoring of CO_2 injected at Sleipner using time-lapse seismic data. In Freund P, Kaya Y, Lior N (eds) Proceedings of the 6th International Conference on Greenhouse Gas Control Technologies. Energy, 29:1383–1392

Beaubien SE, Ciotoli G, Coombs P, Dictor MC, Krüger M, Lombardi S, Pearce JM, West JM (2008) The impact of a naturally occurring CO_2 gas vent on the shallow ecosystem and soil chemistry of a Mediterranean pasture (Latera, Italy). In Busch A, Kühn M, Etheridge D (eds) EGU General Assembly 2007: Advances in CO2 Storage in Geological Systems—EGU (2007). International Journal of Greenhouse Gas Control, 2(3):373–387. doi: 10.1016/j.ijggc.2008.03.005

Benson SM (2005) Lessons learned from industrial and natural analogs for health, safety and environmental risk assessment for geologic storage of carbon dioxide. In: Benson SM (ed) Carbon Dioxide Capture for Storage in Deep Geologic Formations—Results from the CO_2 Capture Project, vol 2, Geologic storage of CO_2 with monitoring and verification. Elsevier, London, pp 1133–1141

EC (European Commission) (2008) Directive on the Geological Storage of Carbon Dioxide. http://eur-lex.europa.eu/LexUriServ/LexUriServ.do?uri=COM:2008:0018:FIN:EN:PDF

Forbes SM, Verma P, Freedmann SJ, Curry TE, Wade SM, Venezia J (2009) WRI CCS guidelines and emerging geologic sequestration regulations: a comparative assessment. In Gale J, Herzog H, Braitsch J (eds) Proceedings of the 9th International Conference on Greenhouse Gas Control Technologies. Energy Procedia, 1: 1759–1766

Holloway S, Pearce JM, Ohsumi T, Hards VL (2005) A review of natural CO_2 occurrences and their relevance to CO_2 storage. International Energy Agency Greenhouse Gas R&D Programme Report 05/08. IEA Greenhouse Gas R&D Programme, Cheltenham

IAEA (International Atomic Energy Agency) (2003) Radioactive Waste Management Glossary. IAEA, Vienna

IAEA (International Atomic Energy Agency) (2004) Safety Assessment Methodologies for Near Surface Disposal Facilities. IAEA-ISAM-1. IAEA, Vienna

ICRP (International Commission on Radiological Protection) (2000) Radiation protection recommendations as applied to the disposal of long-lived radioactive waste. ICRP publication 81. Annals of the ICRP, 28(4)

IPCC (Intergovernmental Panel on Climate Change) (2005) IPCC Special Report on Carbon Dioxide Capture and Storage. Metz B, Davidson O, de Coninck HC, Loos M, Meyer LA (eds). Cambridge University Press, Cambridge, UK, and New York, NY

Lewicki JL, Birkholzer J, Tsang CF (2007) Natural and industrial analogues for leakage of CO_2 from storage reservoirs: identification of features, events, and processes and lessons learned. Environ Geol 52:457–467

Mathieson A, Wright I, Roberts D, Ringrose P (2009) Satellite imaging to monitor CO_2 movement at Krechba, Algeria. In Gale J, Herzog H, Braitsch J (eds) Proceedings of the 9th International Conference on Greenhouse Gas Control Technologies. Energy Procedia, 1:2201–2209

Maul PR, Savage D, Benbow SJ, Walke RC, Bruin R (2005) Development of a FEP database for the Geological Storage of Carbon Dioxide. In: Rubin ES, Keith DW, Gilboy CF (eds) Greenhouse Gas Control Technologies: Proceedings of the 7th International Conference on Greenhouse Gas Control Technologies, vol 1. Elsevier, London, pp 701–710

Maul PR, Metcalfe R, Pearce J, Savage D, West JM (2007) Performance assessments for the Geological Storage of Carbon Dioxide: learning from the radioactive waste disposal experience. Int J Greenhouse Gas Control 1(4):444–455

Maul PR, Beaubien SE, Bond A, Limer L, Lombardi S, Pearce J, Thorne M, West J (2009) Modelling the fate of carbon dioxide in the near-surface environment at the Latera natural analogue site. In Gale J, Herzog H, Braitsch J (eds) Proceedings of the 9th International Conference on Greenhouse Gas Control Technologies. Energy Procedia, 1:1879–1885

Metcalfe R, Maul PR, Benbow SJ, Watson CE, Hodgkinson DP, Paulley A, Limer L, Walke RC, Savage D (2009) A unified approach to performance Assessment (PA) of geological CO_2 storage. In Gale J, Herzog H, Braitsch J (eds) Proceedings of the 9th International Conference on Greenhouse Gas Control Technologies. Energy Procedia, 1:2503–2510

Miller W, Alexander R, Chapman N, McKinley I, Smellie J (2000) Geological disposal of radioactive wastes and natural analogues. Lessons from nature and archaeology. Pergamon, Amsterdam

Moritis G (2006) CO_2 injection gains momentum. Oil Gas J 104(15):37–57

Nagra (Nationale Genossenschaft für die Lagerung radioaktiver Abfälle—National Cooperative for the Disposal of Radioactive Waste) (2002) Project Opalinus Clay: safety report—Demonstration of the disposal feasibility for spent fuel, vitrified high-level waste and long-lived intermediate-level waste (Entsorgungsnachweis). Technical Report 02-05. Nagra, Wettingen

NEA (Nuclear Energy Agency) (2000) Features, Events and Processes (FEPs) for Geologic Disposal of Radioactive Waste: An International Database. OECD Nuclear Energy Agency, Paris

NEA (Nuclear Energy Agency) (2005) Radioactive Waste Management Programmes in OECD/NEA member countries. OECD Publishing, Paris

NEA (Nuclear Energy Agency) (2007) Regulating the long-term safety of geological disposal: towards a common understanding of the main objectives and bases of safety criteria. Report NEA No. 6182. OECD Nuclear Energy Agency, Paris

Nordbotten JM, Celia MA, Bachu S, Dahle HK (2005) Analytical solutions for CO_2 leakage between two aquifers through an abandoned well. Environ Sci Technol 39(2):602–611

O'Hagan A, Caitlin EB, Daneshkhah A, Eiser JR, Garthwaite P, Jenkinson D, Oakley J, Rakow T (2006) Uncertain Judgements: Eliciting Experts' Probabilities. Wiley, London

Pawar R, Carey J, Chipera S, Fessenden J, Kaszuba J, Keating G, Lichtner P, Olsen S, Stauffer P, Viswanathan H, Ziock H, Guthrie G (2006) Development of a framework for long-term performance assessment of geologic CO_2 sequestration sites. In: Røkke N, Bolland O, Gale J (eds) Proceedings of the 8th International Conference on Greenhouse Gas Control Technologies. Elsevier, London, CD-ROM

Pearce JM, Czernichowski-Lauriol I, Lombardi S, Brune S, Nador A, Baker J, Pauwels H, Hatziyannis G, Beaubien S, Faber E (2004) A review of natural CO_2 accumulations in Europe as analogues for geological sequestration. In: Baines S, Worden RJ (eds) Geological Storage of Carbon Dioxide, vol 233, Geological society special publication. Geological Society of London, London, pp 29–41

Pearce J, Chadwick A, Bentham M, Holloway S, Kirby G (2005) Technology status review – monitoring technologies for the geological storage of CO_2. Technology Status Review Report DTI/Pub URN 05/1032. UK Department of Trade and Industry, London

Pruess K (2004) Numerical simulation of CO_2 leakage from a geologic disposal reservoir, including transitions from super- to subcritical conditions, and boiling of liquid CO_2. SPE J 9(2):237–248

Riddiford F, Wright I, Bishop C, Espie A, Tourqui A (2005) Monitoring geological storage: the In Salah gas CO_2 storage project. In: Wilson M, Morris T, Gale J, Thambimuthu K (eds) Greenhouse gas control technologies: proceedings of the 7th International Conference on Greenhouse Gas Control Technologies, vol 2. Elsevier, London, pp 1353–1359

Rodwell WR, Norris S, Mäntynen M, Vieno T (2003) A thematic network on gas issues in safety assessment of deep repositories for radioactive waste *(GASNET)*. European Commission Report EUR 20620. European Commission, Luxembourg

Rutqvist J, Wu YS, Tsang CF, Bodvarsson G (2002) A modelling approach for analysis of coupled multiphase fluid flow, heat transfer, and deformation in fractured porous rock. Int J Rock Mech Min Sci 39(4):429–442

Savage D (ed) (1995) The Scientific and Regulatory Basis for the Geological Disposal of Radioactive Waste. Wiley, Chichester

SKB (Svensk Kärnbränslehantering AB—Swedish Nuclear Fuel and Waste Management Company) (2006) Long-term safety for KBS-3 repositories at Forsmark and Laxemar—A first evaluation. Main report of the SR-Can project. SKB Technical Report TR-06-09. Svensk Kärnbränslehantering AB, Stockholm

Stenhouse MJ, Wilson M, Herzog H, Cassidy B, Kozak M, Zhou W, Gale J (2005a) Regulatory issues associated with deep (geological) CO_2 storage. In: Rubin ES, Keith DW, Gilboy CF (eds) Greenhouse Gas Control Technologies: Proceedings of the 7th International Conference on Greenhouse Gas Control Technologies, vol 1. Elsevier, London, pp 961–970

Stenhouse MJ, Zhou W, Savage D, Benbow S (2005b) Framework methodology for long-term assessment of the fate of CO_2 in the Weyburn field. In: Benson SM (ed) Carbon Dioxide Capture for Storage in Deep Geologic Formations—Results from the CO_2 Capture Project, vol 2, Geologic Storage of CO_2 with Monitoring and Verification. Elsevier, London, pp 1251–1261

Torp TA, Gale J (2003) Demonstrating storage of CO_2 in geological reservoirs: the Sleipner and Sacs projects. In: Gale J, Kaya Y (eds) Proceedings of the 6th International Conference on Greenhouse Gas Control Technologies, vol 1. Pergamon, Amsterdam, pp 311–316

Torvanger A, Fuglestvedt J, Grimstadl AA, Georgescul S, Lindeberg E, Olevier ML, Rive N, Rypdal K, Skeie RB (2006) Climate impacts of leakage from geological CO_2 storage. In: Røkke N, Bolland O, Gale J (eds) Proceedings of the 8th International Conference on Greenhouse Gas Technologies. Elsevier, London, CD-ROM

US DOE (United States Department of Energy) (2004) 2004 WIPP compliance recertification Application (CRA)—Main Volume. DOE/WIPP 04-3231. Carlsbad, NM. http://www.wipp.energy.gov/library/CRA/CRA_Index.htm

West JM, Pearce JM, Bentham M, Maul PR (2005) Issue profile: environmental issues and the geological storage of CO_2. Eur Environ 15:250–259

Wilson M, Monea M (eds) (2004) IEA GHG Weyburn CO_2 monitoring and storage project: summary report 2000–2004. In Proceedings of the 7th International Conference on Greenhouse Gas Control Technologies, vol III. Petroleum Technology Research Centre, Regina

ZEP (Zero Emission Fossil Fuel Power Plants Technology Platform) (2006) A vision for zero emission fossil fuel power plants. European Commission Report EUR 22043. European Commission. Luxembourg

Monitoring Methods Used to Identify the Migration of Carbon Dioxide and Radionuclides in the Geosphere

Brian Brunskill and Malcolm Wilson

Abstract The disposal of industrial wastes in the subsurface has been ongoing for some time. Effective monitoring methods are necessary to verify both the safety of the disposed materials and the reliability of the methods used under present and future conditions. Utilizing reliable monitoring and verification methods is critical to understanding what is happening to both carbon dioxide and radioactive waste sequestered in the subsurface. Information gained while monitoring is useful to help determine what remedial action can be taken in the event of premature or unexpected escape of such geologically sequestered materials. This chapter looks at some of the general technologies used for monitoring the behaviour of these wastes in the subsurface and provides a general comparison of the methods used. An example is provided of how one method being used to monitor the behaviour of carbon dioxide in the subsurface could be adapted to monitor radioactive waste.

Keywords Radioactive waste • Carbon dioxide • Monitoring • Drilled radioactive waste repository

1 Introduction

Within the earth, locations exist that are suitable for the disposal of industrial wastes. The ultimate safety of any geological repository is dependent upon the mobility of the fluids surrounding the rocks. If these fluids are contaminated by the

B. Brunskill (✉)
Helix Geological Consultants Ltd, Regina, Canada
e-mail: brianbrunskill@sasktel.net

M. Wilson
Director, Office of Energy and Environment, University of Regina, Regina, Canada
e-mail: malcolm.wilson@uregina.ca

F.L. Toth (ed.), *Geological Disposal of Carbon Dioxide and Radioactive Waste: A Comparative Assessment*, Advances in Global Change Research 44,
DOI 10.1007/978-90-481-8712-6_5,

waste, but are relatively immobile, the escaped material will be contained. If, however, the contaminated fluids are mobile, both the surrounding geosphere and, potentially, the biosphere are at risk of contamination. The suitable geological character of the containing system is therefore critical to safely dispose of carbon dioxide (CO_2) or radioactive waste (RW) underground.

The injection of anthropogenic CO_2, captured from large, single-point industrial emitters into deep saline aquifers or depleted hydrocarbon reservoirs is one method of significantly reducing greenhouse gas emissions. When CO_2 is injected into an aquifer, it tends to rise and migrate updip due to its buoyancy. The geological and hydrogeological characteristics of a host aquifer control this migration of CO_2 while it is still buoyant, and provide the conditions necessary for its ultimate neutralization in the aquifer.

RW contains fission products, which become harmless to humans and to the environment only through their natural decay over time. Since some isotopes take hundreds of thousands of years to decay, disposal solutions must be safe and secure for very long periods of time. Although competent engineered barriers will mitigate the escape of radionuclides, the migration of contaminating material into the geosphere surrounding the repository site can be effectively mitigated only by natural barriers intrinsic to the geosphere. Disposal methods currently being considered by regulators in many countries rely heavily upon the competence of highly engineered barriers placed within a supportive geological framework. It may be desirable to use disposal methods that provide for both secure containment and for material retrievability in the event that future societies wish to retrieve the material for currently unforeseen reasons.

Monitoring and verification methodologies proposed for both CO_2 and RW disposal are designed to identify contamination of the geosphere surrounding the disposal site so that mitigating action could be taken, but if leakage into the distant geosphere occurs, remediation may be very challenging. Eventual contamination of the distant geosphere should be expected.

The nuclear energy and fossil fuel energy industries are greatly influenced by their respective experience-based knowledge and conventions. Proposed RW disposal methods are, therefore, supported by known mining methodologies and CO_2 disposal methods have been influenced by oilfield drilling experience. Although the methodologies are different, both are subject to the application of fundamental engineering and geological principles.

This chapter provides a preliminary comparison of the application of geological disposal and monitoring methods used for CO_2 and RW. Although material management is accomplished by different 'industries', there are useful analogues to share.

This chapter also presents a conceptual model for developing RW disposal repositories beneath sedimentary basins. Disposing of waste under the proposed conditions will provide: (1) an effective and reliable monitoring platform that will be available indefinitely, and (2) greater utilization of natural barriers that may provide isolation and containment for geological periods of time.

2 Geological Disposal of CO_2

In geological basins where the conditions are favourable, the distribution and character of the sedimentary rocks have provided environments in which CO_2 has naturally accumulated, and the genesis, migration and accumulation of hydrocarbons have occurred.

Our current understanding of the movement and accumulation of buoyant fluids in aquifers is largely based upon principles developed for hydrocarbon exploration. The trapping of hydrocarbons demonstrates the long-term effectiveness of overlying rocks as seals that have prevented further migration. The entrapment conditions will also apply to anthropogenic CO_2 injected into appropriate regions of the subsurface. Sophisticated reservoir modelling and simulation applications used by the petroleum industry have been adapted to model CO_2 disposal. It is important to note, however, that our detailed knowledge of the behaviour of CO_2 in the subsurface is far from complete.

2.1 Monitoring and Verification of Injected CO_2

Monitoring the performance of the CO_2 injection and disposal operation requires observations both at the surface and in the subsurface. Surface equipment monitoring involves the application of standard oilfield practices for the regular inspection of the CO_2 distribution infrastructure, injection volumes and pressures, general well performance and regular scheduled maintenance. The petroleum industry has been injecting CO_2 into hydrocarbon-bearing aquifers since the 1970s in enhanced oil recovery (EOR) projects, so many practices are well established.

Monitoring programmes will extend from the pre-operational, through operational and post-operational periods. Pre-operational monitoring activities will provide baseline data that will be useful for the disposal site characterization, developing the safety case for the disposal system and for the development of performance models. Many of the methods used will continue into the operational phase of the project. Following the closure and sealing of the disposal site, monitoring methods used would ideally not compromise the integrity of the geological container. Seismic imaging is one very useful method that can be used repeatedly for as long as this information is useful.

Observation wells can be useful throughout all the operational periods. Wells that are located distant from the injection location, and are completed in the disposal aquifer and other strata, can be used to monitor the migration of CO_2 in the subsurface and for conducting various geophysical surveys. Information gathered can be used to verify the movement of the CO_2 and of the geochemical evolution of the native brines and host rock components. If leakage is detected there may be an opportunity to mitigate its escape into the biosphere. Observation wells can remain operational for a period of time far beyond the injection period, until such time as public confidence in the disposal method used can be assured. Eventually, observation wells will be abandoned according to regulated procedures. There is risk, however, that, if abandonment materials used fail prematurely, then CO_2 could potentially escape from the disposal aquifer.

The Intergovernmental Panel on Climate Change (IPCC) report on CO_2 capture and storage (IPCC 2005) has summarized both the direct and indirect methods used to monitor the movement of CO_2 in the geosphere. The following are examples of the types of methods used

- Time-lapse, 3-D seismic imaging to identify the development and geometry of the CO_2 plume, and seismic profiling and imaging techniques that help to detect the distribution of CO_2 in the aquifer and identify potential leakage through fractures and faults.
- Hydrogeological testing to assess aquifer properties, flow directions and rates, fluid densities and hydraulic heads, and to develop both local and regional models.
- Geochemical testing of fluids from observation wells to determine the degree of fluid interaction and trapping; tracers in the injected fluids may be utilized.
- Seismic assessments to estimate the probability and magnitude of tectonic events.
- Surface soil-sampling programmes that detect leakage to the biosphere.

Understanding the behaviour of CO_2 that has been injected into the geosphere has evolved significantly with the implementation of, and experience gained from, various carbon capture and storage (CCS) projects. Two world-class projects are being conducted, one at Sleipner in the Norwegian portion of the North Sea, and the other at Weyburn, Canada.

Beginning in 1996, Statoil has been injecting about one million tonnes of CO_2 per year into a deep saline aquifer in the Sleipner Field in the Norwegian sector of the North Sea. The International Energy Agency Greenhouse Gas R&D Programme (IEA GHG), with its industry partners and several research institutes developed a Best Practice Manual (Holloway et al. 2003) to share relevant information. The use of time-lapse seismic surveying is one technique that has been a reliable tool for monitoring the development and movement of the CO_2 plume at Sleipner (Arts et al. 2004; Holloway et al. 2003).

Since 2000, EnCana Corporation (then PanCanadian Petroleum) has been injecting over 5,000 t of CO_2 per day into the Weyburn oil reservoir as an EOR solvent to extract additional crude oil. Within the geoscience framework of the IEA GHG Weyburn CO_2 Monitoring and Storage Project, research has provided abundant information regarding the injection of CO_2. This CO_2 EOR project has provided a dynamic, commercial-sized laboratory where the geochemical and physical nature of the reservoir is being observed and documented as the conditions evolve with the continual introduction of CO_2.

Before the project was initiated, a robust information baseline about the character of the reservoir was developed so that effective monitoring of the changes could be observed. These efforts were focused on the anticipated physical and chemical effects, and on the tracking of the CO_2 as it spread in the reservoir and potentially outside the intended area (White et al. 2004). The results from the monitoring efforts are providing an ongoing verification of the modelling process, reliable estimates of the distribution of the CO_2, and confidence in the effectiveness of the disposal container. Analyses from production data provide an ongoing geochemical survey of the evolving aquifer, using reservoir pressure data and analysis of injected and produced products. Understood leakage routes have been identified (there may be others) and corresponding monitoring efforts have been initiated. Although detection and remediation strategies have been developed, based largely upon common oilfield practices, more experience will provide for increasingly effective mitigation efforts.

Time-lapse seismic imaging has been a very effective monitoring tool for identifying the shape of the CO_2 plume and its movement in the reservoir, and repeated sampling of soil–gas concentrations has so far indicated that no CO_2 is escaping to surface (White et al. 2004). Risk assessments have concluded that the geological setting of the Weyburn project is well suited for the secure, long-term disposal of CO_2 (Whittaker et al. 2004).

These operations and others in various stages of implementation provide critical background experience that can lead to improved CO_2-disposal, -monitoring and -verification methods. Standard protocols to verify geological disposal have not yet been fully developed, but long-term monitoring will be a likely requirement (Benson et al. 2005).

The concept of disposing of anthropogenic CO_2 in the deep subsurface is relatively recent, so modifications to existing and proposed practices are also evolving. For example, rather than injecting a relatively pure stream of CO_2 into the aquifer, it may be beneficial to pre-mix the CO_2 with brine from the intended disposal aquifer. As a result, the development of a CO_2 plume could be avoided and the CO_2 would be more widely dispersed in the aquifer. Greater dispersion would provide greater surface-area contact between the CO_2 and the native brines and minerals, potentially accelerating the rate of CO_2 neutralization. If this method of injection was deemed suitable, then monitoring methods would also require adjustment. Coincidentally, developing a brine-premix source well may also provide a geothermal energy source.

2.2 Containment and Potential Failure of Seals

Where sequences of sedimentary rock comprise aquifers interbedded with less permeable seals or aquitards, the contrast of high lateral permeability in the aquifers with low vertical permeability in the aquitards has provided some of the fundamental conditions necessary for the lateral migration and accumulation of hydrocarbons and will provide for the safe disposal of CO_2.

CO_2 is compressed to a supercritical or 'liquid-similar' density when injected into the disposal aquifer. The pressure necessary to maintain the CO_2 in this supercritical state is usually available at depths greater than 800 m (Gunter et al. 2004).

When injected into the aquifer, the CO_2 will move away from the point of injection and, due to its buoyancy, will tend to rise and migrate updip (Flett et al. 2005) subject to controlling mechanisms such as pressure gradients, natural hydraulic gradients, buoyancy, dissolution into formation fluids, and chemical interaction with rock-forming minerals. The intended disposal aquifer must effectively contain the CO_2 until the CO_2 reacts fully with the host rock and associated formation fluids, possibly requiring a period of many thousand years (Bachu et al. 1994). These natural conditions will both contain the CO_2 and ultimately provide for its permanent sequestration.

Primary seals are composed of impermeable rocks that provide a cap directly overlying the intended disposal aquifer. For as long as injected CO_2 remains buoyant and migrates updip in the aquifer, there is a risk that it will encounter a permeable breach in the seal. This risk must be carefully assessed, for naturally occurring fractures and faults in rocks can provide potential vertical conduits between aquifers. If the vertical conduit terminates, upward flow will cease or, if the relative permeability of an overlying intersected aquifer is greater, the flow of CO_2 may be recaptured by this 'relief' aquifer. The presence of secondary seals higher in the rock sequence provide additional barriers to vertical fluid movement.

Abandoned well bores from past drilling activity which intersect the CO_2 disposal aquifer create additional risk to CO_2 containment, as they provide potential conduits for the vertical movement of CO_2. As CO_2 migrates updip in the aquifer it may encounter a hole in a previously drilled oil exploration prospect. One intention of the site selection process is to identify these conditions so that the potential for cross-formational flow and contamination within the geosphere, and for contamination of near-surface potable water aquifers or escape to the biosphere are mitigated.

Depleted oil- and gasfields may also provide secure geological containers for CO_2 disposal because the hydrocarbon trapping mechanism has contained buoyant hydrocarbons for millions of years (Gunter et al. 2004; Shaw and Bachu 2002). However, in many depleted fields, particularly older ones where there are numerous (possibly hundreds) of well casings, the potential for escape is significant. The concern resides around the ageing of the materials in the wells and the resulting possibility of providing migration paths for the CO_2.

2.3 Complete Neutralization of CO_2

The disposal potential and ultimate sequestration of CO_2 in deep saline aquifers depends to a great extent upon the degree of reactivity between the injected CO_2, the formation fluids and the host rock constituents. Geochemical reactions will vary according to differences in mineralogy, formation-fluid chemistry, pressure, pH, temperature and many other aquifer characteristics (Gunter et al. 2004).

Physical trapping occurs when the CO_2 is confined as a supercritical 'bubble' (Bachu et al. 1994). Deep saline aquifers with extremely slow flow rates provide an effective geological container which can trap injected CO_2 hydrodynamically, as it takes from hundreds of thousands to millions of years for CO_2 to travel any significant distance by buoyant forces. As CO_2 moves through the aquifer, it also experiences solubility trapping when it dissolves into the brine and no longer migrates as a separate phase, then moving at the same rate as the brine in the aquifer. With the associated changes in pH, ionic trapping may occur with the formation of ionic species. With time, CO_2 will geochemically react with rock minerals, particularly feldspars and clays, becoming permanently trapped by mineral trapping (Gunter et al. 2004). At the tail of the rising plume, residual CO_2 will 'imbibe' to the host rock (Flett et al. 2005). As the character of the CO_2 plume evolves, the various trapping mechanisms interact in a complex way, both simultaneously and at different timescales. Over time, these mechanisms lower the potential for leakage because the CO_2 becomes less mobile (Benson et al. 2005). It may require several thousands of years for mineral trapping to be effectively complete (Bachu et al. 1994), so containment must be reliable for this length of time. Monitoring changes in the geochemical nature of the CO_2 and native brines taken from observation wells provides an opportunity to evaluate the evolution of the neutralization process.

3 Geological Disposal of Radioactive Waste

An RW repository must ultimately provide safety to humans and the environment, so final disposal solutions must be secure for many thousands of years (NEA 2004). Multiple safety barriers, both engineered and geological in origin, combine to provide this assurance. The expectations for developing a repository include isolation from the biosphere, confinement of the RW in the geosphere in the near term (10,000 years) and, due to anticipated material failure, mitigated release to the geosphere in the long term (Sykes 2003). As the character of the geosphere will provide the most reliable conditions for long-term, safe isolation of the RW, repositories must be sited in stable geological environments where the geomechanical, geochemical and groundwater-flow characteristics are favourable.

3.1 Disposal Systems

Disposal systems will be inherently passive in character. Isolation from the biosphere will be maintained by conditions that are not reliant upon any active measures in the future. Based upon the timely degradation of engineered barriers, escape of the radionuclides into the geosphere surrounding the disposal site will be retarded due to the robust nature of the multiple containment design (NEA 2004). The NEA (2006) summarizes the safety functions of an RW repository as described by the European Commission 'Testing of Safety and Performance Indicators' (SPIN) Project (Becker et al. 2002), where barriers identified for saturated formations perform both individually and collectively over relative periods of time and levels of radioactivity:

- During the early post-closure history of the repository, vessels that contain the RW will provide a watertight barrier that isolates the material; this represents the most transient period due to resaturation, the greatest level of heat and radiation release and pressure rebuilding.
- When container failure occurs groundwater will eventually come in contact with the RW and various physical and chemical processes will result in the very slow leeching of radionuclides into the buffer materials surrounding the container.
- Groundwater flow rates in the rock surrounding the repository site will be very slow (relatively stagnant), so the migration of dissolved radionuclides into the distant geosphere will be retarded; migration is retarded further due to sorption of some radionuclides onto minerals in the buffer and host rock materials.
- Long-lived radionuclides will eventually be mobile in the distant geosphere and may enter surrounding aquifers; by the time these materials enter parts of the biosphere they will be widely diluted and dispersed.

It is assumed that containment will be fully satisfied through the site-selection process and applied engineered methods, and that barrier failures will occur in a timely and predictable fashion.

Several countries are in various stages of investigating and developing deep geological repositories. For example, Finland, Sweden and Canada are investigating development in crystalline rock, and in France, Belgium and Switzerland development in sedimentary rocks is being considered (McCombie 2003).

At all the sites under consideration, traditional mining methods including the creation of shafts, tunnels and rooms up to 1,000 m below the ground surface are used. Proposals for disposal in crystalline rocks in Canada, Finland and Sweden envision that the spent nuclear fuel be placed in steel (or iron) and copper containers having a predicted lifetime of at least 100,000 years, and that these containers be placed in rooms which are subsequently backfilled with chemically and physically supportive bentonitic clays (McCombie 2003).

The proposed highly engineered barriers are expected to provide the greatest blockade to material escape. For example, in Sweden canisters housing the RW will be constructed to withstand the anticipated mechanical load and potential corrosive conditions of the repository, and supportive buffer materials around the canisters will protect them and mitigate the movement of radionuclides that escape; backfill materials will stabilize the repository and are intended to prevent groundwater flow in the tunnels (SKB 2004).

Although the use of highly engineered containers is also proposed for RW disposal in sedimentary-rock repositories in France, Belgium and Switzerland, greater reliance would be placed on the hydrogeological environment to contain eventual leakage into the geosphere (Mazurek 2004).

In the USA, a repository is being developed in volcanic tuff at Yucca Mountain, Nevada. Although the porous rocks surrounding the repository are considered to be unsaturated, fractures are common and could provide conduits for groundwater movement. Highly engineered containers and barriers would be used to keep stored material permanently dry and isolated (OCRWM 2001).

3.2 *Methods Used to Monitor a Radioactive Waste Disposal Site*

The primary objective of monitoring programmes is to assess the performance of the repository site and the reliability of the barriers, and to progressively update the safety case through each evolutionary phase of the project. Monitoring activities would begin during the siting process to establish baseline information under present or unperturbed conditions and would continue into the future, ending sometime following the closure of the facility. Collected data will be useful in the development of predictive models and in the assessment of those models over time. Pre-closure activities would include site selection and characterization, repository evaluation and construction, RW placement operations, decommissioning and repository closure. Post-closure activities would follow the final sealing of the facility, during which time institutional control is maintained (Simmons 2006).

Various monitoring methods can be utilized to confirm the performance of the barriers during pre-closure activities. Results from these efforts would assist operators in proceeding from one operational stage to the next. To assess the performance of the repository, instrumentation is placed within the host rock to monitor the conditions while access to the underground is available. Methods that require the use of boreholes in the host rock will require appropriate sealing when the site is being decommissioned so that sealing systems are not compromised. Examples of the types of methods utilized (Simmons 2006) include:

- Rock-mass monitoring to assess changes in stress, displacement and micro-seismic activity;
- Temperature monitoring to assess the role of heat load in rock stress;
- Hydraulic monitoring of the excavation site to assess the development of communication pathways;
- Hydrogeological monitoring to assess changes in pressure and groundwater flow;
- Geochemical monitoring to identify changes in groundwater composition.

The duration of the monitoring efforts being used must be sufficient so that reliable information provides confidence in the performance models, possibly for a few hundred years.

Following the closure of a facility monitoring would continue for some time to support ongoing performance assessments and, ultimately, to assure public confidence in the disposal methods used. The intention of all national RW disposal programmes is to not burden future generations with having to care for the RW, so only when the long-lived safety of the repository is assured will it be sealed (Stenhouse and Savage 2004). Therefore, long-term safety and security will be achieved using disposal methods that do not require active monitoring, maintenance or institutional control (NEA 2004).

4 Comparison of CO_2 and Radioactive Waste Disposal Monitoring Techniques

The concept of using deep geological repositories to safely dispose of CO_2 and RW may be becoming both socially and politically acceptable. The physical conditions and time frame necessary for implementation are, however, broadly different. For example:

- There is significant interest in reducing global anthropogenic greenhouse gas emissions as soon as possible, and the geological disposal of CO_2 is viewed by many as capable of making a significant contribution to these reductions. Several monitoring methods used for CO_2 disposal are being 'field tested' and are evolving concurrently with active disposal. The eventual disposal of RW in geological repositories is also practical but has a much longer time horizon for its implementation. With the exception of the Waste Isolation Pilot Plant (WIPP) site in New Mexico, USA, most national facilities are utilizing underground research laboratories (URLs) to conduct in situ monitoring (Stenhouse and Savage 2004).
- The quantities of material to store are widely different. Nuclear material is solid and dense, and the amount of product to store globally can be measured in tonnes per year, whereas CO_2 is light and buoyant, the amount being measured in millions of tonnes per day. The geological characteristics of the CO_2 repository will include well developed porosity, permeability and fluid-mobility potential, whereas those of the RW disposal site will be in excavated caverns where the rocks have very limited permeability and restricted fluid-mobility potential (i.e. the characteristics are opposite). Some of the methods used for the monitoring of both products will rely upon groundwater sampling during the RW pre-closure and CO_2 operational periods.
- The area required for disposal is potentially much greater for CO_2 than for RW. Monitoring methods used will be required to accommodate these widely different spatial requirements. Injecting millions of tons of CO_2 per year for several years at a single site could, depending on the thickness of the aquifer, result in the development of a plume over 100 km^2 in size, whereas a single RW repository would likely require a significantly smaller area. Monitoring programmes for CO_2 must be able to accommodate the large areas and volumes involved, so techniques with vertical resolutions in the order of metres to tens of metres are acceptable and even lower resolution may prove adequate. With RW disposal, the resolution required will need to be much finer in order to detect changes in the stresses in rock, fractures, backfill, the disposal containers and hydraulic features.
- The sites must provide safe disposal for as long as the products are potentially mobile and/or harmful. For CO_2 disposal, the period is probably less than about 10,000 years, whereas for nuclear material, containment must be safe for a much longer period of time. These temporal differences require durable containment systems that are effective, potentially over geological periods of time. There is an inverse relationship between risk and time when comparing the safe disposal of CO_2 and RW. The risk of escape of nuclear material increases with time due

to the potential for premature degradation of the engineered barriers, whereas the risk of leakage of free CO_2 decreases with time due to the ongoing process of its neutralization.

- Once CO_2 is injected into the disposal aquifer, it is the natural character of the geosphere that will provide the conditions necessary to contain it for as long as it remains buoyant, and for several thousands of years after that until it reacts completely with the rock-forming minerals—no reliance is placed on human-made barriers. Monitoring the distant geosphere will potentially confirm the migration and behaviour of CO_2 in the subsurface during the operational and post-operational periods. Under current strategies, the safety of RW relies on highly engineered barriers that are supported by the character of the geosphere. Monitoring the distant geosphere in the post-closure period could be conducted if methods used did not affect the passive safety of the RW repository.

Programmes have been established to monitor the safe disposal of CO_2 from the pre-operational, operational and post-operational periods and for RW from the pre-closure through post-closure periods. Table 1 identifies several of these monitoring methods.

5 Knowledge Transfer Potential

Many of the principles involved with CO_2 disposal are similar to those used in hydrocarbon exploration and development. Operators have benefited from their experience which has allowed them to modify operating procedures as previously unforeseen conditions have arisen. They have also been able to modify monitoring and verification techniques as the amount of practical knowledge increases.

Since there is no immediacy for the disposal of RW, a comprehensive, cautionary approach to disposal has been taken. The consequences of nuclear leakage into the biosphere have very long-term environmental implications, whereas an unintended release of CO_2 would likely have few lasting effects once the leak was remedied. This gradual approach to RW disposal allows for the development of policies and regulatory protocols, whereas with CO_2 disposal, many of these issues have yet to be resolved and some policies are being established by precedent 'as we go'. Several RW monitoring methods have been tested for many years in separate URLs in different countries. Monitoring the behaviour of anthropogenic CO_2 in the subsurface is more recent.

The body of monitoring experience is significant for both the RW and CO_2 research communities, and some of this knowledge and experience may be transferrable. For example:

- Abandoned wellbores provide potential pathways for CO_2 to escape to the biosphere. If current abandonment methods are successfully applied, then this risk is mitigated; however, there is the potential for premature failure of the materials used. Several RW monitoring methods require the use of boreholes

Table 1 Examples of carbon dioxide and radioactive waste monitoring methods

CO_2 monitoring	Purpose	Radioactive waste monitoring	Purpose
Pre-operational period		**Pre-closure period**	
Soil gas and near-surface hydrology	Establish baseline surface characterization	Environmental and near-surface hydrology	Establish baseline surface characteristics
Use of existing local and regional subsurface data: aquifer characteristics, geochemistry, hydrogeology, seismic	Establish baseline subsurface characterization	Use of existing local and regional subsurface data: geochemistry, hydrogeology, seismic	Establish baseline subsurface characterization
Remote sensing	Identify lineaments and surface-expressed faults to predict potential escape pathways	Hydrogeological monitoring using surface and subsurface boreholes; may include use of tracers	Establish baseline conditions and identify changes in hydraulic head and groundwater geochemical properties
Operational period			
Time-lapse 3-D seismic profiling	Track CO_2 plume development and migration patterns	Overcoring with borehole deformation instrumentation	Establish in situ rock mass stability during site characterization and construction
Time-lapse gravity measurements and electrical conductivity surveys	Detect and track migration of CO_2 in disposal and other aquifers	Seismic detection (seismometers, geophones, hydrophones, accelerometers, acoustic emission, microseismic)	Determine the location of seismic activity, including events caused by mining and operational activities
Use of observation wells:			
Pressure and temperature changes, fluid sampling; may include use of tracers	Track physical conditions and geochemical evolution of CO_2 and native fluids in disposal and other aquifers; on-going hyrogeological assessment	Displacement of rock mass following excavation	Confirm mechanical properties of the host rock
		Hydraulic monitoring following excavation	Assess the influence construction has on the development of communication pathways to the more distant geosphere
Borehole geophysical techniques (seismic tomography, cross-hole tomography, vertical seismic profiling, acoustic emission, microseismic and passive seismic)	Assess geomechanical stability and structural disturbances	Temperature monitoring during construction and operation	Assess the rock mass response to temperature changes

(continued)

Table 1 (continued)

CO_2 monitoring	Purpose	Radioactive waste monitoring	Purpose
Post-operational period		**Post-closure period**	
Continuation of surface procedures as is deemed necessary; borehole monitoring	Escape of CO_2 to the biosphere	Continued non-intrusive geophysical procedures as is deemed necessary	Possibly provide greater societal assurance
Time-lapse 3-D seismic profiling	Monitor continued evolution of plume development and dissipation	Monitoring is not required for safety beyond the period of institutional control but monitoring may be conducted if desired	Methods used must be non-intrusive to avoid compromising the passive safety of the disposal system

in the excavated areas and in the surrounding geosphere. These holes will eventually be sealed during the pre-closure period of the repository. Some aspects of the sealing methods and materials used for RW borehole closure may be useful for CO_2 well abandonment procedures.

- Downhole instruments are used by both research communities. The reliability and durability of these instruments have been 'field tested' for RW monitoring for a longer period of time than for CO_2 monitoring. Some aspects of this RW monitoring experience may be useful for monitoring CO_2.

6 Application of a CO_2 Monitoring Method to Radioactive Waste Monitoring

As described previously, observation wells located strategically distant from the CO_2 injection well can be used to track the movement of CO_2 in the surrounding geosphere. These wells will eventually be abandoned, likely during the post-operational period. There remains, however, the option to develop a new observation well at any time, allowing future decision makers the ability to 'have a look' anytime, and respond to the arising of currently unforeseen circumstances. Future societies may also desire additional monitoring.

If the geological character of the RW repository site is suitable, sampling groundwater from the geosphere surrounding a repository site may be useful if it can be conducted without compromising the integrity of the containment barriers. Sampling can be conducted over the short term (less than 300 years) or, indeed, indefinitely into the future beyond the decommissioning of the repository, if either technical conditions or public demand require further sampling.

Selection of a repository site which places its greatest reliance on suitable geological systems is more likely to provide for permanent isolation of the RW, particularly in the event of premature engineered-barrier failure. It is appropriate, therefore, to develop repositories where the natural environment provides reliability for geological periods of time—for millions of years. Locating RW repositories beneath suitable intracratonic sedimentary basins may provide: (1) an opportunity to monitor the integrity of the containment system indefinitely, and (2) permanent isolation and containment.

The Williston Basin, for example, is generally located in southern Saskatchewan, Canada, and North Dakota, USA, and conditions there may provide for this reliability (Brunskill 2006). An RW repository could be developed in the Precambrian Shield beneath the stagnant, dense brines (e.g. 250–350 g/l Total Dissolved Solids) which occupy aquifers at the base of the basin. As well as great depth (e.g. 3,000–4,000 m), the hydrogeological environment of the repository site will likely inhibit the vertical migration of contaminated material because the water that would carry the contaminating material would be unable to move significant vertical distances. The dense brines will potentially provide complete isolation of any leakage for a period of time far longer than any nuclear material would be harmful. Even following a significant tectonic event, contamination would likely remain in the very deep geosphere.

The development of these repositories is technically possible and may be economically feasible if, for example, surface-drilling methods currently utilized in the petroleum industry are used. The disposal space for the RW would be developed by drilling long, small-diameter ‘rooms’ that are lined with continuous, metallurgically suitable casing. Although nuclear material placed in this lateral section of the hole would be in the abandonment position, material could potentially be retrieved and inspected as deemed necessary. With this option of being readily retrievable for some time, future decision makers would have greater flexibility as new concerns and technologies arise.

In the Williston Basin example, the presence of this overlying aquifer also provides a means to conduct reliable and timely monitoring of the repository site without compromising the integrity of the repository. Observation wells can be placed strategically in and around the disposal site and be used to circulate native brines from the overlying aquifer across the repository area to the surface where any contamination can be detected. If deemed appropriate, remedial action may be taken.

Figure 1 provides a sectional view of a model RW disposal facility. In this scenario the hole is drilled vertically from the surface through the sedimentary section of rocks to a depth of about 3,000 m, now being roughly 300 m beneath the Precambrian surface. The hole would then be drilled laterally to its maxim depth of approximately 6,400 m. RW would be repackaged and placed in this lateral section. Radionuclides that eventually escape into the overlying, brine-filled aquifer would likely remain in the very deep geosphere and be subject to detection during the monitoring programme.

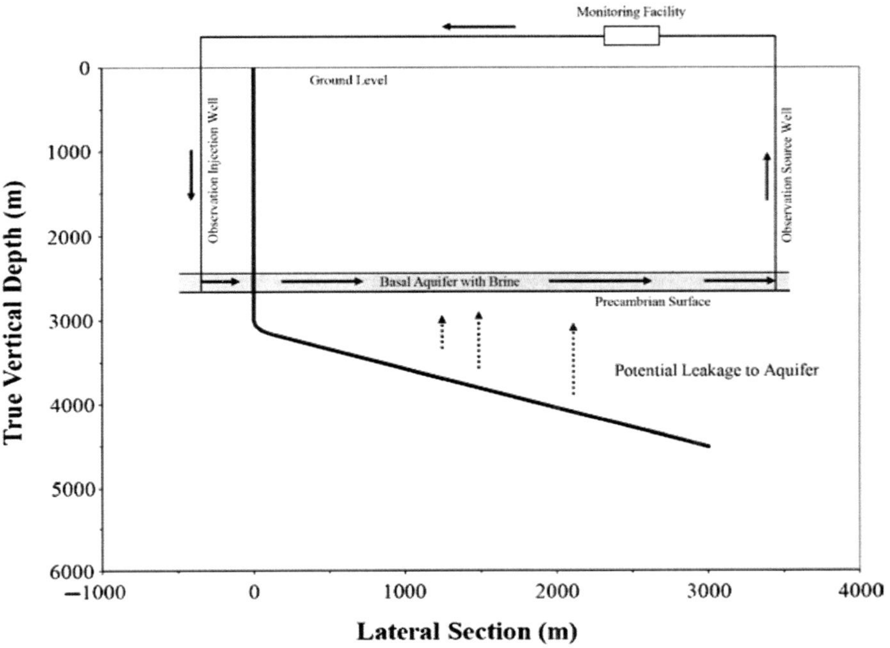

Fig. 1 Illustration of a drilled radioactive waste repository beneath the Williston Basin, Canada. Includes the brine circulation loop in the basal aquifer to monitor the migration of radioactive material that escapes into the overlying geosphere

7 Conclusions

In suitable locations the geological conditions provided by the geosphere can effectively isolate anthropogenic CO_2 and RW from the biosphere, although the conditions necessary for disposal are widely different. Once injected into the disposal aquifer, the containment of CO_2 relies upon the natural conditions provided by the geosphere. Under most current strategies, containment of RW is reliant upon highly engineered barriers that are supported by the character of the geosphere surrounding the repository.

Many of the monitoring methods used during the site-selection and geological characterization stages are similar for both RW and CO_2 disposal. Surface hydrology and subsurface geochemical and hydrogeological monitoring programmes contribute significantly to this process. During the operational stage of a CO_2 repository, geophysical evidence provided by time-lapse seismic surveys is one reliable monitoring tool and sampling aquifer fluids from observation wells support the confirmation of the geochemical evolution of CO_2 in the subsurface. Operating in RW excavations and URLs provide additional opportunities to develop effective techniques to monitor geomechanical and geochemical variations in the subsurface.

Both the RW and CO_2 disposal research communities are well experienced at 'field testing' various monitoring methods, and there is potential for a significant transfer of knowledge and experience between these communities.

Under the appropriate conditions, the geological disposal of both CO_2 and RW is a very effective way to safely and securely dispose of these products. The ongoing development of effective monitoring programmes will continue to provide both technical and societal confidence. Furthermore, the geological disposal of CO_2 is one method available today that can make a significant contribution to reductions in the emission of anthropogenic CO_2 in the very near term. Public confidence gained through the efforts of objective, 'third party' educators is critical to societal acceptance for the disposal of both CO_2 and RW.

Confidence in programmes that can effectively monitor and actively control materials like RW and CO_2 in the geosphere thousands of years from now and, indeed, over geological periods of time is unrealistic. Societies may evolve in such a way that they are no longer reliant on traditionally mined materials, and taking remedial action in response to premature leakage, for example, 2,000 years from now, may not be possible. Human understanding of highly technical issues is also very recent. To provide perspective, it has been only about 10,000–12,000 years since humans left the Paleolithic Period.

The development of very deep geological repositories for RW beneath sedimentary basins is technically possible. Great depth, the geological character and the hydrogeological environment could potentially provide the conditions necessary for safety and security for millions of years. A repository developed under these conditions would also provide for retrievability for some time and an option for future generations to conduct effective monitoring, particularly in response to currently unforeseen circumstances if they so desire. It may be beneficial to also support further investigations of this model in conjunction with continued research on current disposal strategies.

Acknowledgements The authors are very grateful to the editor and reviewers for their supportive comments, which led to considerable improvements in the clarity of the text.

References

Arts R, Eiken O, Chadwick A, Zweigel P, van der Meer L, Zinszner B (2004) Monitoring of CO_2 injected at Sleipner using time-lapse seismic data. In: Freund P, Kaya Y, Lior N (eds) Proceedings of the 6th International Conference on Greenhouse Gas Control Technologies. Energy 29(9–10):1383–1392

Bachu S, Gunter WD, Perkins EH (1994) Aquifer disposal of CO_2: hydrodynamic and mineral trapping. Energ Convers Manage 35(4):269–279

Becker DA, Buhmann D, Storck R, Alonso J, Cormenzana JL, Hugi M, van Gemert F, O'Sullivan P, Laciok A, Marivoet J, Sillen X, Nordman H, Vieno T, Niemeyer M (2002) Testing of safety and performance indicators (SPIN). Nuclear Science and Technology, Final Report EUR 19965 EN. European Commission, Luxembourg

Benson S, Cook P et al (2005) Underground geological storage. In: Metz B, Davidson O, de Coninck HC, Loos M, Meyer L (eds) IPCC Special Report on Carbon Dioxide Capture and Storage. Cambridge University Press, Cambridge and New York, pp 196–276

Brunskill B (2006) Discussion of an option for geological storage of used nuclear fuel beneath the Williston basin of southern Saskatchewan. In: Summary of Investigations 2006, 1:1–8. Misc Rep 2006–4.1. Saskatchewan Geological Survey, Sask. Industry Resources, Regina

Flett MA, Gurton RM, Taggart IJ (2005) Heterogeneous saline formations: long-term benefits for geo-sequestration of greenhouse gases. In: Rubin ES, Keith DW, Gilboy CF (eds) Greenhouse Gas Control Technologies: Proceedings of the 7th International Conference on Greenhouse Gas Control Technologies, vol 1. Elsevier, London, pp 501–509

Gunter WD, Bachu S, Benson S (2004) The role of hydrogeological and geochemical trapping in sedimentary basins for secure geological storage of carbon dioxide. In: Baines SJ, Worden RH (eds) Geological Storage of Carbon dioxide for Emissions Reduction, vol 233, Geological society special publication. Geological Society of London, London, pp 129–145

Holloway S, Chadwick A, Lindeberg E, Czernichowski-Lauriol I, Arts R (eds) (2003) Saline aquifer CO_2 storage project (SACS) best practise manual. Report Number PH4/21. International Energy Agency Greenhouse Gas R&D Programme, Trondheim

IPCC (Intergovernmental Panel on Climate Change) (2005) IPCC Special Report on Carbon Dioxide Capture and Storage. Metz B, Davidson O, de Coninck HC, Loos M, Meyer LA (eds). Cambridge University Press, Cambridge and New York

Mazurek M (2004) Long-term used nuclear fuel waste management—Geoscientific review of the sedimentary sequence in southern Ontario. Technical Report TR 04–01. Institute of Geological Sciences, University of Bern, Bern

McCombie C (2003) Status of geological repositories for used nuclear fuel. Background Paper 6–3. Nuclear Waste Management Organization, Toronto

NEA (Nuclear Energy Agency) (2004) Post-closure safety case for geological repositories: nature and purpose. Report NEA No. 3679. OECD Nuclear Energy Agency, Paris

NEA (Nuclear Energy Agency) (2006) Advanced nuclear fuel cycles and radioactive waste management. Report NEA No. 5990. OECD Nuclear Energy Agency, Paris

OCRWM (Office of Civilian Radioactive Waste Management) (2001) Yucca Mountain Science and Engineering Report. DOE/RW-0539. US Department of Energy, Office of Civilian Radioactive Waste Management, Washington, DC

Shaw J, Bachu S (2002) Screening, evaluation, and ranking of oil reservoirs suitable for CO_2-flood EOR and carbon dioxide sequestration. J Can Petrol Technol 41(9):51–61

Simmons GR (2006) A framework for the long-term monitoring of a deep geologic repository for used nuclear fuel. Report 06819-REP-01300–10119-R00. Ontario Power Generation, Toronto

SKB (Svensk Kärnbränslehantering AB—Swedish Nuclear Fuel and Waste Management Company) (2004) RD&D-Programme 2004. Programme for research, development and demonstration of methods for the management and disposal of nuclear waste, including social science research. SKB Technical Report TR-04–21. Svensk Kärnbränslehantering AB, Stockholm

Stenhouse MJ, Savage D (2004) Monitoring experience associated with nuclear waste disposal and its application to CO2 sequestration projects. In: Baines SJ, Worden RH (eds) Geological Storage of Carbon dioxide, vol 233, Geological Society Special Publication. Geological Society of London, London, pp 235–247

Sykes JF (2003) Characterizing the Geosphere in High-Level Radioactive Waste Management. Nuclear Waste Management Organization, Toronto, Background Paper 4–2

White D, Hirsche K, Davis T, Hutcheon I, Adair R, Burrows G, Graham S, Bencini R, Majer E, Maxwell S (2004) Theme 2: prediction, monitoring and verification of CO_2 movements. In: Wilson M, Monea M (eds) IEA GHG Weyburn CO_2 Monitoring & Storage Project Summary Report 2000–2004. Petroleum Technology Research Centre, Regina, pp 73–148

Whittaker S, Rostron B, Khan D, Hajnal Z, Qing H, Penner L, Maathuis H, Goussev S (2004) Theme 1: geological characterization. In: Wilson M, Monea M (eds) IEA GHG Weyburn CO_2 Monitoring & Storage Project Summary Report 2000–2004. Petroleum Technology Research Centre, Regina, pp 15–69

Transport of Carbon Dioxide and Radioactive Waste

Darío R. Gómez and Michael Tyacke

Abstract A comparative assessment of carbon dioxide (CO_2) and radioactive waste transport systems associated with electricity generation was undertaken on the basis of 15 criteria grouped under three areas, namely the transport chain, policy aspects and state of the technology. For CO_2, we considered exclusively the transport that would take place under a future large-scale capture and storage infrastructure. Our study allowed a certain hierarchy of criteria to be identified for the comparative assessment. We discovered that the physical state for transport (fluid for CO_2 and solid for radioactive waste) and the volumes involved are the key properties for determining the most suitable modes of transport. These are pipelines (on- and offshore) for liquid or supercritical CO_2, and rail, ship or truck for spent nuclear fuel and high-level waste. Ship-based transport has also been suggested for future applications of large-scale CO_2 transport. Leakage and accidental releases are the main risks underlying the safety policies of both transport systems. However, because of the large differences between transport chains, safety standards are specific to each system. Regulatory frameworks both at national and international levels are at very different stages of development. Routing is a common concern for both transport systems. In this study we cite over 90 references covering the main literature published on this topic over the last decade.

Keywords Transport • Carbon dioxide • Spent nuclear fuel • High-level waste • State of the technology • Policy aspects • Regulatory framework

---

D.R. Gómez (✉)
National Atomic Energy Commission, Gerencia Química, Avenida General Paz 1499, B1650KNA, San Martín, Argentina
e-mail: dgomez@cnea.gov.ar

M. Tyacke
Idaho National Laboratory, 1625, Idaho Falls, ID, USA
e-mail: michael.tyacke@inl.gov

F.L. Toth (ed.), *Geological Disposal of Carbon Dioxide and Radioactive Waste: A Comparative Assessment*, Advances in Global Change Research 44,
DOI 10.1007/978-90-481-8712-6_6, © International Atomic Energy Agency, 2011

1 Introduction

We aim here to compare the transport of carbon dioxide (CO_2) and that of radioactive waste (RW) from electricity generation through to their respective geological disposal. This may seem a somewhat paradoxical undertaking, considering that no facilities at either end of these transport chains have been built to date.

CO_2-rich streams, for instance, are presently being transported in the oil industry for the purpose of enhanced oil recovery (EOR) and also in a few CO_2 capture and storage (CCS) projects worldwide (DGC 2008; Maldal and Tappel 2004; Statoil 2007). This CO_2 is not captured from thermal power plants but is either of natural origin or captured from industrial facilities such as natural gas processing or chemical plants. However, it is expected that after 2010 new CCS demonstration projects worldwide will involve CO_2 capture from power plants (Gale 2009). Furthermore, our assessment presupposes that a large deployment of CCS will occur in the future at a scale estimated in the range of several hundred to several thousand million tonnes of CO_2 per year worldwide (Gale et al. 2005), with power plants being significant CO_2 sources.

With respect to RW, geological disposal has yet to occur. However, progress towards implementation is evident in a number of countries that have adopted this option as the reference long-term management solution for their high-activity, long-term RW (NEA 2008). Consequently, there is currently no transport of RW to the last step of the chain; however, transportation of spent nuclear fuel (SNF) and high-level waste (HLW) from nuclear power generation and other sources for different purposes has evolved over 4 decades.

This comparative assessment then draws on international experience in the transport of CO_2 and RW, although the existing systems do not yet connect the initial stage (for CO_2) or the final stage (for RW) of the transport chains associated with electricity generation. Within this framework, we have looked at several aspects of three broad areas, namely the transport chain itself, its associated policy aspects and the state of the art of both technologies.

For the transport chain we have considered the requirements and associated technical aspects of the conditioning process that is necessary before actual transportation of the CO_2-rich stream captured from the fossil fuel-fired power plants or the SNF or HLW from the nuclear fuel cycle can take place.

We have characterized the central transport system according to five inherent attributes: (1) the appropriate physical state of the waste for transport; (2) the volumes involved; (3) the means of transport; (4) the experience obtained thus far by industry; and (5) the energy requirements and associated environmental loads, particularly additional greenhouse gas (GHG) emissions. Finally, we have briefly looked at the ways of transferring waste from the transport system to the disposal site. We have also considered the environment, safety and risk, particularly the characterization of the main risks and the availability of statistics on incidents.

Policy issues concerning the status of the international regulatory framework have also been evaluated. The transport of hazardous goods is usually a highly political issue; therefore we have looked at public acceptance issues associated with the transport of CO_2 and RW.

The last step in characterization concerns the state of the art of the technology, the maturity of the science, engineering and regulatory aspects and the gaps in knowledge.

These three broad aspects are presented in the relevant sections below for each transport chain. The chapter concludes with a discussion regarding similarities and differences between all elements selected to characterize the transport systems of CO_2 and RW.

2 The CO_2 Transport Chain

Transport is the step that connects the first and last elements of a CO_2 capture and storage system. Presently, CO_2 transport takes place on- and offshore using several methods, including pipelines, ships, trucks and rail. Recent assessments (Berger et al. 2004; Svensson et al. 2004) have indicated that pipelines (on- and offshore), ships (offshore) and combinations of these are the most cost-effective alternatives for the bulk transport of CO_2 associated with a large-scale CCS infrastructure. The CO_2-rich stream from the capture facilities needs to be conditioned to meet the requirements of the transport alternative chosen. That is why, following Aspelund and Jordal (2007), we consider that the CO_2 transportation chain starts with the gas conditioning of the captured CO_2-rich stream and ends with its injection in a high-density phase (see Fig. 1), although conditioning has previously been considered to be part of the capture system. After gas conditioning, the captured CO_2-rich stream needs to be compressed ahead of the pipeline suction point or liquefied for ship-based transport.

2.1 Conditioning

CO_2 is transported to the storage site in liquid (ship-based) or supercritical phase (pipeline) to make the best possible use of the transport capacity. Removal of water and certain impurities is required before the captured and conditioned CO_2-rich gas is ready for transmission.

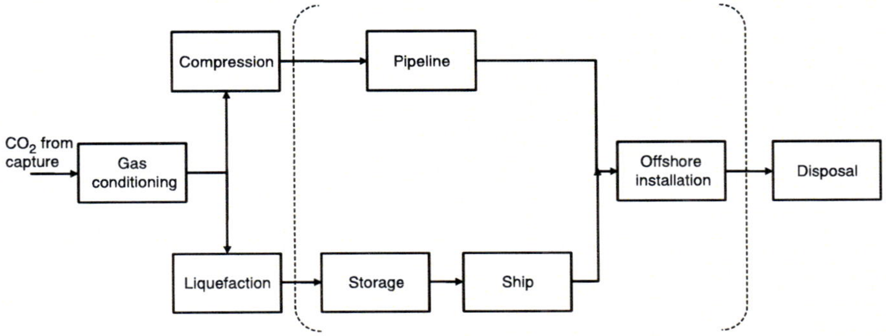

Fig. 1 CO_2 transport chain including conditioning and disposal (Based on Aspelund and Jordal 2007)

Water must be removed to avoid corrosion, freezing and the formation of solid hydrates that can block pipelines, valves or other equipment (Birkestad 2002; Heggum et al. 2005; Aspelund and Jordal 2007; Seiersten and Kongshaug 2005). Depending on the pressure of the captured CO_2-rich gas stream, three or more compression stages are typically required to reach transport conditions. The bulk of water (and other liquids) is removed in each of the compressor suction drums to prevent the ingress of liquid into the compressor. Active dehydration is generally necessary to avoid corrosion and hydrate formation and thus meet the requirements for transport.

Gas conditioning is designed so that the CO_2 stream leaving the capture process satisfies both transport and reservoir specifications. This stream contains a number of impurities in the form of non-condensable gases that differ depending on the CO_2 sources and the type of capture systems. The presence of sulphur compounds, particularly hydrogen sulphide, may raise health and safety concerns. Most of these non-condensable components must be removed for ship-based transport to avoid liquefaction temperatures that may cause the formation of dry ice. This removal is not strictly necessary for pipelines, but it is nevertheless convenient from an economic standpoint. In addition, the presence of small amounts of these non-condensable gases has a major impact on flow properties in terms of influencing the relationship between pipeline pressure drop, on the one hand, and temperature and elevation, on the other (Farris 1983).

For pipeline transport, compression of the captured CO_2 stream is the most power consuming operation of the conditioning step and involves large investment costs. For the ship-based transport chain, the liquefaction system is typically the most energy-intensive process (Aspelund et al. 2006).

Aspelund and Jordal (2007) recently authored a thorough study on the conditioning of CO_2-gas rich streams for CCS. They considered pipeline and ship transport for nine types of streams reported in the benchmark study by Kvamsdal et al. (2007) that assessed different approaches for capturing CO_2 from a reference 400 MW combined cycle plant. The authors reported that the overall energy requirements for the conditioning processes were typically between 90 and 120 kWh/t CO_2. As electricity is required for compression, average GHG emissions would depend on the primary energy supply and on the fuels used for heating purposes.

2.2 *Transport*

After the CO_2 has been conditioned, it is ready to be sent to the pipeline suction or to intermediate storage for subsequent ship loading. This section deals with the transport step itself.

2.2.1 State of Matter for Transport

The operating regions for pipeline transport and the suggested operating conditions for large-scale ship-based transport are depicted in Fig. 2. The thin triangle at the

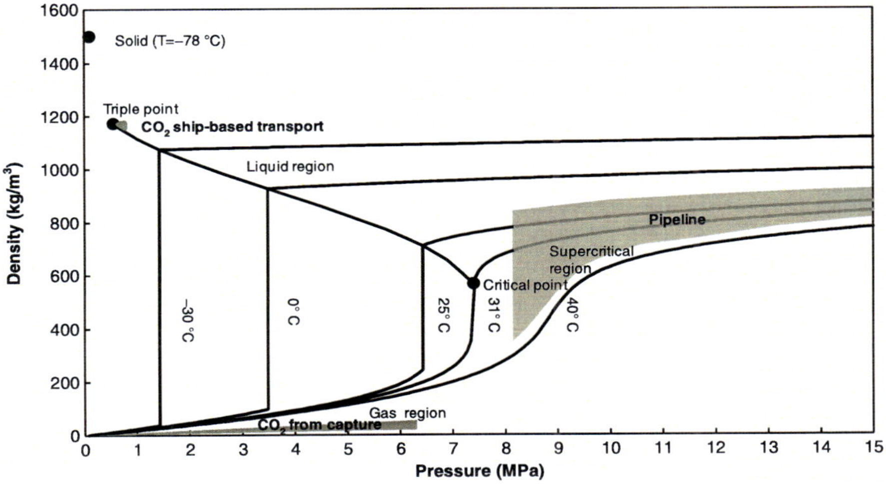

Fig. 2 Key physical CO_2 properties for pipeline and ship-based transport (Based on Aspelund et al. 2006, with data from Lemmon et al. 2005)

bottom of the phase diagram indicates the conditions under which the CO_2-rich stream is delivered from the capture system.

Pipelines operate beyond CO_2 critical pressure (7.38 MPa), mainly in the 8–10 MPa range, in which density versus compression ratio is normally optimal. Higher pressures require more energy and investment costs with little gain in density. However, higher inlet pressures (up to 20 MPa) may be required to overcome the pressure drop along the pipeline without adding intermediate booster stations. The lower pressure limit depends on the CO_2 phase behaviour and is chosen to avoid two phase mixtures. Operating temperatures are in the 4–38°C range. The upper temperature limit is set by the exit conditions of the compression unit and the maximum allowable temperature of the external pipeline coating. The lower temperature limit is determined by the winter soil temperature.

Semi-pressurized vessels at 1.4–2 MPa are presently used to transport liquid CO_2 by ship in much smaller volumes than would be expected for a large-scale implementation of CCS. Conceptual designs for future implementation recommend operating conditions near the triple point (0.52 MPa, 56.6°C) to keep CO_2 in liquid phase close to the lowest-possible pressure to allow large-volume cargo tanks (pressure vessels) to be built with practical wall thickness. In principle, CO_2 could also be transported by ship as a solid. However, Aspelund et al. (2006) have discarded this option on the basis that complex loading and unloading procedures would make it economically unfeasible.

2.2.2 Volume

Gas volumes and concentration levels of CO_2 from thermal power plants depend on the type of fuel used and the excess air level used for optimal combustion conditions. Concentration levels by volume range from 3% to 4% for natural gas-fired power plants up to 14% for coal power plants (Gale et al. 2005). The capture system produces gaseous CO_2-rich streams with typical specific volumes ranging from 300 kg CO_2/MWh for natural gas combined cycles up to more than 800 kg CO_2/MWh in the case of coal power plants (Thambimuthu et al. 2005). It is preferable to capture these CO_2-rich streams from power plants from large point sources (>100,000 t CO_2/year). In 2000, such sources worldwide numbered 4,942 and their associated emissions amounted to 10,539 million tonnes of CO_2 (Mt CO_2).

Presently, several Mt CO_2 per year are transported for EOR and ~2 Mt CO_2 per year for CCS. A large-scale deployment of CCS would require the transportation of several hundreds to thousands million tonnes of CO_2 per year worldwide.

2.2.3 Modes

Pipelines are the preferred option for the land-based transport of large quantities of CO_2 across long distances up to 1,000 km (Skovholt 1993). The pipeline structure depends on the required transport capacity, diameter, inlet pressure, route, need and location of booster pumps, pressure regulators and valves. In mountainous areas, terrain elevation is key, as the static head increases with downhill flows and decreases with uphill flows, which influences the temperature profile in the pipeline. Ideally, the simplest approach is to boost the CO_2 pressure at the suction point to drive the fluid along the whole length of the pipeline as far as the injection point. This is not always possible, and it may be necessary to include intermediate boosters and/or pressure regulators (Farris 1983). Additional considerations include special features for compressors and pumps to compensate for the poor lubricating properties of dry CO_2 and the use of sealing materials (Barrie et al. 2004; Gale and Davison 2004; DGC 2008).

The potential of a large-scale infrastructure to transport several million tonnes of CO_2 per year by ship has received attention in recent years from researchers who have proposed several conceptual designs (Aspelund et al. 2004a, b, 2006; Aspelund and Jordal 2007; Barrio et al. 2004; Berger et al. 2004; Haugen et al. 2009; Hegerland et al. 2004; Ozaki et al. 2004; Svensson et al. 2004).

Aspelund and co-workers (Berger et al. 2004; Barrio et al. 2004; Aspelund et al. 2006) developed a conceptual design for a large-scale ship-based transport of ~2 Mt CO_2 per year in the North Sea. Ozaki et al. (2004) also assessed a system for the transport of ~6 Mt CO_2 across distances in the range of 200–12,000 km. These integrated designs consider all the equipment and machinery necessary to carry out all the steps from conditioning to injection, namely intermediate storage, loading, ship-based transport to the storage site and unloading.

Intermediate storage at harbours would be required for ship transport, as CO_2 is typically captured in a continuous process whereas ships are generally loaded batch-wise. At present, steel tanks are used to store CO_2; however, it has been suggested that rock caverns could also be used for this purpose (Svensson et al. 2004).

The loading system from the onshore storage tanks to the ship includes piping between tanks and ship, pumps adapted for high pressure and low temperature CO_2 service, marine loading arm and a return line for any vaporized CO_2 generated at the ship (Aspelund et al. 2006; Barrio et al. 2004; Ozaki et al. 2004). The cargo tanks are first filled and pressurized with gaseous CO_2 to prevent contamination by humid air and the formation of dry ice (Doctor et al. 2005).

When the delivery point is onshore, the liquid CO_2 is unloaded from the ship into temporary storage tanks. For offshore delivery, the use of a submerged turret loading system has been suggested to transfer the CO_2 from the ship to a platform for further injection (Barrio et al. 2004; Aspelund et al. 2006).

2.2.4 Experience

The transport of high purity CO_2 was originally developed to supply CO_2 for injection in EOR. In the USA there are more than 6,000 km of pipelines (US DOT 2008a) that transport several million tonnes of mostly naturally occurring CO_2 annually. Industrially produced CO_2 (e.g. from gas processing, coal gasification, fertilizer and ethylene plants) is transported for use in EOR in a limited number of cases (Gozalpour et al. 2005). The oil industry has more than 37 years' experience in successfully transporting and injecting CO_2 for EOR operations.

For storage purposes only, CO_2 from natural gas processing has been transported onshore in Algeria since 2004 and also in the first long distance (170 km) offshore pipeline of the Snøhvit project at 318 m below sea level in the Norwegian North Sea (Maldal and Tappel 2004; Statoil 2007). In the well known Sleipner gasfield development in Norway there is no need for a long pipeline. Here, after CO_2 has been captured at an offshore platform, its pressure is boosted to 8 MPa and it is then piped to a nearby platform for injection (Hansen et al. 2005). In the Weyburn-Midale project, the CO_2 stream captured from the Dakota Gasification Company's synfuels plant (in North Dakota, USA) is liquefied and transported 320 km by pipeline to the Weyburn field and the Apache's Midale field (both located in Saskatchewan, Canada). This large international collaborative research programme is aimed at exploring and testing key scientific and technological aspects of the long-term storage of CO_2 used in EOR (IEA GHG 2005).

At present, CO_2 is routinely transported by tankers with a capacity of up to ~1,500 t CO_2. The much larger ships needed for a large-scale CO_2 infrastructure can be built based on experience in the construction and operation of semi-pressurized liquefied petroleum gas (LPG) and liquefied natural gas (LNG) ships (Barrio et al. 2004; Ozaki et al. 2004; Aspelund et al. 2006). CO_2 tankers of this type can be constructed in 1–2 years, depending on the ship's size, by the same shipyards currently building LPG and LNG tankers (Doctor et al. 2005).

2.2.5 Energy Requirements and Generation of Waste and/or Greenhouse Gas Emissions

The pipeline transport of the captured CO_2 generates additional emissions, as energy may be needed for intermediate boosters that compensate for pressure drops along the pipeline. The need for these boosters depends on the length of the pipeline, the characteristics of the terrain and the diameter of the pipeline. Boosters may be avoided by increasing the pipeline diameter and reducing the flow velocity. Waste generation is relatively low and disposal is readily available. Greenhouse gases may be emitted from vented streams containing CO_2 and from compressors, depending on the energy supply.

Since a transport system based on large-scale semi-pressurized ships has not been implemented to date, estimates of energy requirements are available only from design studies. Ship fuel consumption of 25 kWh/t CO_2 was reported for a 20,000 m^3 tanker by Aspelund et al. (2006). The demand for unloading is about 7 kWh/t CO_2. GHG emissions are associated with energy requirements, and the levels depend on the assumptions that the modellers have made in their design about the characteristics of the energy supply. The ratio between CO_2 emitted from ships and transported CO_2 is proportional to distance and decreases when larger and lower-speed ships are selected.

2.3 *Disposal*

At this step, the CO_2 that has been transmitted via pipeline or ship is transferred to geological storage via one or more injection wells. The design of the injection system depends on the conditions at the point at which the transported CO_2 is delivered as well as the geometry of the reservoir and its physical characteristics such as faulting, porosity and permeability, which determine the flow rate and pressure required for injection. (This is covered elsewhere and is not further discussed in this chapter.) The main design variables include: (1) number of wells required; (2) well diameter; (3) the need for additional boosters and the corresponding injection pressure; and (4) the maximum injection flow rate (Cockerill 2005).

The injection system is typically composed of a pressurized surge storage tank, injection pumps (if needed), piping to distribute CO_2 to the injection wells, and monitoring and control equipment (Smith et al. 2002). The injection well consists of two or more concentric protective casings, with the injection tube as the innermost part. The main purpose of the exterior casing is to protect aquifers and to prevent water contact with the intermediate protective casing.

For offshore CO_2 storage the injection wellheads can be located on a fixed platform above the waterline or on the seafloor and fitted with valves to control fluid distribution. Regarding the CO_2 injection developments in the North Sea, the former option has been adopted in the Sleipner field (Hansen et al. 2005) and the latter in the Snøhvit field (BERR 2007).

2.4 *Environment, Safety and Risks*

As CCS is a new technology still under development and few projects have been carried out, many of the legal and regulatory implications are not yet widely understood (Mace et al. 2007). For the same reasons, social research into public perceptions and acceptance of CCS is still at an early stage of development, with only a few finished or ongoing studies (ETP-ZEP 2006). Within this framework, it is often difficult to isolate the specific issues associated with CO_2 transport from the general context concerning regulatory requirements, public acceptance and communication of the entire CCS system. We have made an effort here, however, to discuss specific questions concerning transport; for the general framework the reader is referred to the respective background chapter.

2.4.1 Characterization of Main Risks

Leakage and accidental releases, the main risks associated with CO_2 transport and injection, are typically of a short-term and local nature. They may occur at hazard levels spanning from small leaks to major failures or ruptures of pipes, vessels, pumps or compressors. CO_2 transport safety is often likened to that of natural gas and hazardous liquid transport systems. Unlike other gases or liquids regulated as hazardous materials, pure CO_2 is neither combustible nor toxic. However, because it is heavier than air, compressed CO_2 tends to pool near the ground, displacing all the oxygen, and forming a vapour cloud that can cause respiratory problems including suffocation and even death. The US National Institute for Occupational Safety and Health (NIOSH 1995) has established a value of 40,000 ppm for the immediately dangerous to life or health concentration (IDLH) of CO_2. This is based on statements: (1) by the American Conference of Governmental Industrial Hygienists that a 30 min exposure at 50,000 ppm (5%) CO_2 produces signs of intoxication, and a few minutes of exposure between 70,000 and 100,000 ppm produces unconsciousness; and (2) by the American International Health Alliance that 100,000 ppm is the atmospheric concentration that is immediately life threatening. The consequences of a release may entail further risks if the transported CO_2 contains substantial amounts of hazardous or toxic impurities, particularly hydrogen sulphide (Doctor et al. 2005). (According to NIOSH the exposure threshold at which hydrogen sulphide is immediately dangerous to life or health is 100 ppm.)

Under pipeline conditions a large, sudden release of CO_2 could have catastrophic consequences in a populated area. Therefore pipeline routing must be carefully considered with a view to assuring the rapid dispersion of any leak to prevent CO_2 accumulation, to selecting well ventilated areas and to avoiding depressions such as valleys. Moreover, pipeline blowdowns during maintenance need to be undertaken as quickly as possible (Gale and Davison 2004). Typically, pipeline control and monitoring are performed by means of a supervisory control and data acquisition (SCADA) system. The use of emergency shutdown valves that are activated automatically is common practice to mitigate risks associated with leaks and their propagation.

The transportation of supercritical or liquid CO_2 also involves risks of long-running brittle fractures due to the effects of cooling around leaks and long-running ductile fractures due to phase changes during depressurization. Crack arrestors are normally installed along the pipeline to prevent the propagation of fractures (Race 2006).

Several pipeline risk assessments have been undertaken that consider different design and operating conditions and also several release types (Kruse and Tekiela 1995; Turner et al. 2006). For details and the main results of these risk assessments the interested reader is referred to the original publications.

Collision, foundering, stranding and fire are some of the risks involved in water-borne navigation. For CO_2 tankers, there is risk of asphyxiation if a collision causes the rupture of a tank. One way of improving safety is to adopt the high standards of construction and operation currently applied to LPG tankers. Liquid CO_2 released onto the sea surface in the event of a ship accident could lead to the formation of hydrates, with ice and temperature differences inducing strong currents. Under poor ventilation, a CO_2 cloud may form and present similar respiratory problems to those of onshore releases, possibly causing stoppage of the ship's engines (Doctor et al. 2005). Risk mitigation involves routes being carefully planned and personnel highly qualified.

Care must be taken when designing large-scale CO_2 liquefaction systems and storage tanks, especially in harbour areas, where a gas detector system is required. Procedures for loading and unloading liquid CO_2 near the triple point have been developed to avoid dry ice formation, as blockage and operational problems may occur (Aspelund et al. 2004a). During offshore unloading, the vessel should be kept at a safe distance from the platform (Barrio et al. 2004).

Risks during injection are typically associated with releases like blowouts or leakage due to mechanical failure of the injection equipment (Hendriks et al. 2005). The main reasons for these are inner and outer corrosion of tubing, outer corrosion of casings and wellbore blockage. Measures normally taken to prevent outer corrosion consist of lining the exterior of the tube with polyethylene and filling the annulus between the protective casing and tubing with a corrosion inhibitor fluid (Vendrig et al. 2003). To avoid wellbore blockage, it is essential to ensure that CO_2 stays in supercritical phase to minimize hydrate and ice formation.

In the event of leakage through the wellbore annulus, CO_2 can migrate into adjacent reservoir zones and aquifers, with the risk of contaminating underground sources of drinking water. Checks for wellbore integrity are normally undertaken by the operator to protect aquifers and prevent reservoir cross-flow. All materials used in the injection well should be designed to anticipate peak volume, pressure and temperature (Cailly et al. 2005).

2.4.2 Statistics of Incidents

Statistics on pipeline incidents are available from the US Department of Transportation (US DOT 2008b), which requires the reporting of accidents and incidents involving CO_2 and other hazardous liquid pipelines. Within these data, of the 3,695 serious accidents reported on hazardous liquid pipelines since 1994, only 36 involved CO_2 pipelines. Among the 36 incidents, only one injury, and no fatalities, were reported. It is difficult to statistically characterize the reasons for the incidents because they are so relatively few in number. Based on previous statistics, Gale and Davison (2004) have indicated that while most incidents in CO_2 pipelines were related to the pipeline itself (failures of relief valves, failures of weld/gasket/valve packing and corrosion), the principal cause of incidents for natural gas pipelines was external force, such as damage by excavator buckets. This contrast should be taken into account when estimating failure frequencies for CO_2 pipelines from the available failure databases of natural gas or hazardous liquid transmission in the context of risk assessment.

2.5 Regulatory Requirements

Development of national standards is under way in several countries, in which CO_2 transport is not specifically addressed, and where the adaptation of existing environmental rules governing drilling, injection and gas transportation is typically the favoured approach. Transport of CO_2 across national boundaries and transport by ships and via sub-sea pipelines is covered by various international legal conventions. The following features of the transport system play a role in determining the applicability and application of regulatory and liability regimes: (1) mode (pipeline, ship or a combination of both); (2) geographical location (within or across national boundaries, onshore, offshore, proximity to population centres); (3) land ownership (private, publicly owned or managed); (4) impacts (local or transboundary); (5) risks (to the public, to the terrestrial, marine or aquatic environment, to groundwater); and (6) identity of the party responsible for damages resulting from accidental release of CO_2 (pipeline owner or supervisor, ship owner).

These characteristics also influence the design of permission procedures, the identity of the relevant permit authority or authorities, responsibility for monitoring, and environmental impact assessment procedures (Hendriks et al. 2005). Existing international liability regimes may need to be extended or clarified to cover the bulk transport of CO_2 in view of the large scale envisaged for these activities and the corresponding risk levels.

The design and operation of pipelines is typically governed by national codes and standards. Our discussion focuses mainly on the regulatory status in the USA, which is the country with the most extensive pipeline network and the largest construction and operating experience. The USA currently has three different regulatory schemes for transportation of energy resources by pipeline (FERC 2008). Under the scheme governing CO_2 transport to date, pipelines are sited under state law and there is no federal role involved. Operators of interstate pipelines are free to set their own rates and terms of service. Safety standards and reporting requirements for CO_2 pipelines are aimed at ensuring safety in pipeline design, construction, testing, operation and maintenance, corrosion control and qualification of personnel. Similar regulations are in place in Canada.

The transport and injection of CO_2 in sub-seabed repositories may involve different categories of marine pollution under the relevant international conventions, namely the UN Convention on the Law of the Sea (UNCLOS), the London Convention on the Prevention of Marine Pollution by Dumping of Wastes and Other Matter, the London Protocol and the International Convention for the Prevention of Pollution from Ships (MARPOL). The categorization depends on whether the captured CO_2 is transported by ship and injected from platforms, transported and injected from land-based pipelines running across or beneath the seabed, or injected from facilities used for offshore oil and gas exploration and exploitation (Hendriks et al. 2005). Regional rules include the conventions and protocols of the United Nations Environment Programme (UNEP) Regional Seas Programme and other regional and subregional arrangements. The MARPOL Convention includes regulation of emissions from routine operations and accidental pollution associated with ships, fixed and floating platforms and mobile offshore drilling rigs that might be used to inject CO_2 into the seafloor. Annex III of the Convention, dealing with the prevention of pollution by harmful substances in packaged form and implemented through the International Maritime Dangerous Goods (IMDG) Code, is relevant for the bulk transport of liquid CO_2 for injection. Ships transporting liquefied CO_2 would be subject to the general requirements under Annex III, which lists detailed standards on packing, marking, labelling, documentation, stowage, quantity limitations, exceptions and notifications for preventing pollution by harmful substances (Hendriks et al. 2005).

2.6 Public Acceptance

Most of the available studies addressing the acceptance of CCS technology focus on its role as a GHG mitigation option and on the issues concerning CO_2 disposal (de Coninck and Huijts 2004; Gough et al. 2002; Itaoka et al. 2004; Palmgren et al. 2004; Shackley et al. 2005; Wright et al. 2007; ETP-ZEP 2006), although few of them have considered issues related to CO_2 transport (Itaoka et al. 2004; Wright et al. 2007). In general terms and with the exception of the results of Palmgren et al. (2004), these studies seem to indicate that, if given adequate information about the climate change context, the public may look favourably on CCS.

The study by Itaoka et al. (2004) detected four important factors influencing public opinion: (1) environmental impacts and risks, including the possibility of leakage; (2) the effectiveness of CCS as a GHG mitigation option; (3) societal responsibility for CO_2 mitigation; and (4) concern that CCS would allow continuation of the current levels of fossil fuel use. Concern about accidents during CO_2 transport was one of the 19 items making up the first of these factors.

Wright et al. (2007) provided a prioritized assessment of perceptions and issues affecting the deployment of CCS. It considered seven regions/countries (North America, Europe, Australia and New Zealand, Japan, China, India and South Africa) and five stakeholder groups in each region (government, industry, non-governmental organizations, the public, and research and development organizations).

Of the 27 issues included in the survey, two were specific to transport and concerned routing and safety of CO_2 pipelines. Routing was considered as a potentially negative driver of public opinion in 71% of answers and safety in 65%.

2.7 *State of the Technology*

2.7.1 Science and Engineering

The onshore transport of high purity CO_2 by pipeline is a mature technology with more than 6,000 km of pipeline worldwide and an annual capacity of several million tonnes. Most of these pipelines presently transport naturally occurring CO_2 and, to a minor extent, CO_2 extracted from natural gas processing or other industrial applications. The streams that, in the future, will originate in facilities capturing CO_2 from combustion processes will contain different types of impurities. In their recent review paper on CCS, Steeneveldt et al. (2006) have indicated that there is a need to improve understanding of the influence of such impurities on the thermophysical properties of these CO_2-rich streams and how possibly changing properties will affect the design and operating conditions of the pipeline transport system.

The information necessary to undertake environmental and health impact assessments of onshore pipeline transport is relatively well defined and does not involve significantly different requirements to those of the many impact assessments conducted every year. However, there is still a need for a comprehensive definition of exposure limits and for a deeper discussion about modelling the release, as well as about the preferred models available for CO_2 dispersion (IEA GHG 2007; Turner et al. 2006; Koornneef et al. 2009).

Leakage from offshore pipelines and wells could adversely affect large areas through CO_2 dissolution in the surrounding seawater and subsequent acidification thereof, which could detrimentally affect marine ecosystems (Chadwick et al. 2007). This reinforces the need to ensure that the risk of leakage is minimized through proper site selection, design and monitoring. Owing to gaps in knowledge regarding the effects of ocean acidification on marine ecology, these effects remain uncertain as this area of science is relatively young (IEA GHG 2007). There is also uncertainty about the impact on the onshore water environment. The 318 m pipeline of the Snøhvit project in the Norwegian North Sea is the only offshore facility that has been built to date. The learning curve concerning offshore pipeline operation and maintenance has thus only just started.

There is experience in transporting relatively small quantities of CO_2 by ship. However, large-scale ship-based transport of CO_2 has yet to occur, and only conceptual designs for this option are available. These designs rely on the experience in the construction and operation of semi-pressurized LPG and LNG tankers. It remains to be assessed if there are rock caverns close to harbours that would be suitable for the intermediate storage of hundreds of thousands of cubic metres of CO_2 in a similar manner as is done for LPG.

2.7.2 Regulatory Aspects

The regulatory framework for CO_2 transport is under way in most countries interested in the deployment of large-scale CCS. For pipeline transport an evolutionary approach based on existing environmental rules governing drilling, injection and gas transportation is the preferred option. There are presently no recognized specifications for CO_2 quality in terms of its transport for CCS purposes; however, it is likely that future specifications for the transport of CO_2 will take into consideration maximum allowable impurity content in the storage site, the local legislation governing CO_2 transportation, and the type and level of impurities that are acceptable.

Several authors (Gale and Davison 2004; Hendriks et al. 2005) consider that the substantial experience regarding the regulation of CO_2 pipelines in North America could be used by other countries as a reference. However, some key actors in the USA believe that there may be still gaps in the existing rules addressing the construction and operation of CO_2 pipeline networks required for a large-scale deployment of CCS. Furthermore, Kerr et al. (2009) have recently indicated the concern of the UK regarding uncertainties associated with CO_2 transport; the country has called for initiatives and projects to develop best practice guidelines for onshore and submarine CO_2 pipelines.

The United States Environmental Protection Agency is currently working on the regulation of CO_2 injection to ensure that this activity will not endanger underground sources of drinking water. Key components of the proposed regulation include requirements related to: geological site characterization to ensure that wells are sited in suitable areas to limit the potential for migration of injected and formation fluids into an underground source of drinking water; well construction and well operation to ensure that the wells are properly constructed and managed; well integrity testing and monitoring to ensure that the wells perform as designed; and well closure, post-closure care and financial responsibility to ensure proper plugging and abandonment of the injection wells (US EPA 2008).

No integrated international framework is yet available for ship-based transport, offshore pipelines and injection of CO_2 in sub-seabed repositories, which may involve different categories of marine pollution under the relevant international conventions.

2.7.3 Policy Aspects

Further assessment is necessary to evaluate public perception of CO_2 transport. Most of the studies available are of a general nature and only a few of them deal with specific issues of transport. It is likely that acceptance of transport in general may become more problematic since this is the most visible part of the CCS system (Coleman 2009).

There are considerable gaps in the knowledge of the effects of CO_2 release and impurities on the marine environment, both on specific organisms and on ecosystems. There is a certain amount of knowledge about the effects of CO_2 on animals and vegetation in the terrestrial environment; however, effects on smaller organisms are less well researched. Human health effects are well understood, but effects on members of the population with suboptimal health are less well understood.

2.7.4 Cost Estimates

Transport costs depend strongly on the distance and the quantity transported. Pipeline material costs are a function of the diameter and the thickness of the pipeline, the linear weight and the price of the selected steel, and the price of the external coating. The type of pipeline (onshore or offshore) and the characteristics of the route and the terrain play an important role in determining the final investment and operating and maintenance costs.

Offshore pipelines that typically operate at higher pressures and lower temperatures than onshore pipelines are generally more expensive. Doctor et al. (2005) have compiled cost estimates for both onshore and offshore pipelines that have been reported in several studies. These studies have considered the pipeline only and did not include either conditioning or compression costs. Investment costs for onshore pipelines, expressed in terms of the diameter and the length of the pipeline, were US$0.6–1/m/km for pipeline diameters in the 0.1–1.2 m range, with lower values corresponding to larger diameters. For offshore pipelines, investment costs were US$1–2/m/km for the same range of pipeline diameters. Doctor et al. (2005) also reported transport costs per mass of CO_2 for a nominal distance of 250 km as a function of both pipeline diameter and mass flow rate of CO_2. The costs decrease exponentially with either of the two variables. For a pipeline diameter of 0.3 m, transport costs related to the use of onshore pipelines are in the US$2.5–4.2/t CO_2 range and in the US$4.2–5.5/t CO_2 range for offshore pipelines. For a pipeline diameter of 1 m, costs in the US$0.7–1.4/t CO_2 range include both types of pipeline.

The cost of ship-based transport depends mainly on the ship size and the transport distance. Table 1 provides an overall picture of the results from the studies by Ozaki et al. (2004) and Aspelund et al. (2006), the main features of which were

Table 1 Summary of key parameters for the cost estimation of ship-based transport of CO_2

	Ozaki et al. (2004)	Aspelund et al. (2006)
Annual amount of CO_2 transported (Mt)	6	2–4
Distance from storage tank to unloading (km)[a]	200–12,000[b]	[c]
Ship capacity (kt CO_2)	10, 30, 50	22
Liquefaction requirements (kWh/t CO_2)	130	110
Oil consumption by ship (kWh/t CO_2)	n.a.	25
CO_2 emissions/CO_2 transported (%)[d]	12–30	1.4
Cost (US$/t CO_2) for the range of distances considered	17–58	20–30
Cost (US$/t CO_2) for a distance of 1,000 km (Ozaki et al. 2004) and 750 km (Aspelund et al. 2006)	20	25

[a]Distances correspond to one-way trips from intermediate storage to injection; costs are for the round trip journey

[b]The results of this study reflect the very wide range of selected transport distances

[c]Distances limited to the North Sea

[d]The difference in the results may be explained in part by the assumptions made by Aspelund et al. (2006) that the required power in their model has no associated CO_2 penalties since it comes from a power plant with 100% CO_2 capture while the corresponding penalty in the model by Ozaki et al. (2004) is ~10%

summarized in Sect. 2.2.3. Care is necessary when comparing the pipeline costs reported above that do not include conditioning costs with those reported in Table 1 that include the liquefaction facility, which is an important component of the investment and, particularly, the operating costs. Other cost elements are associated with storage tanks, loading and unloading facilities, the sailing route and harbour fees. For distances under 1,000 km the estimated costs in both studies are in agreement, in the US$20–25/t CO_2 range. Aspelund et al. (2006) reported the following contributions to the cost components considered in their assessment: liquefaction (42%) > ship (30%) > unloading (16%) > storage (9%) > loading (3%).

3 The Radioactive Waste Transport Chain

The life cycle transport chain for nuclear material used to generate electricity starts at the point of the raw uranium's removal from a mine and ends with the final disposal of the spent nuclear fuel (SNF) or high-level waste (HLW) in a deep geological repository. This study is limited to radioactive waste (RW) associated with nuclear reactor fuel and does not discuss the transport of raw uranium or other types of nuclear materials, such as sealed sources, medical isotopes and low-level waste.

The nuclear fuel transport begins at the fuel fabrication facility. After fabrication, the nuclear fuel is transported to a nuclear reactor site where it is placed in the reactor and burned to generate heat to make electricity. When the nuclear material in the fuel has been used up or spent, the spent fuel is removed from the reactor and placed in a storage pool for several years to allow it to cool. From the storage pool, the SNF can be transported in one of three directions. It can be sent to a dry storage facility, to a reprocessing facility, or directly to disposal (see Fig. 3). If, however, the spent fuel is reprocessed, there are two other transport considerations. These are: (1) transport of the fuel material retrieved from the spent fuel reprocessing back to fuel fabrication for use as new fuel; and (2) transport of the treated HLW (i.e. vitrified/solidified waste) to either an HLW storage facility or directly to disposal.

3.1 Conditioning

There are two types of RW that need to be conditioned for transport: SNF and HLW, both of which form in the nuclear reactor and which are segregated and recovered from the SNF reprocessing. Conditioning of the SNF is primarily done by placing the material into transport packages, also known as transport casks. Conditioning of the HLW requires the liquid radioactive material from the reprocessing process to be solidified, usually by vitrification, before being placed into the transport packages. The outer transport casks are generally intended for multiple and extended use possibly for more than 20 years.

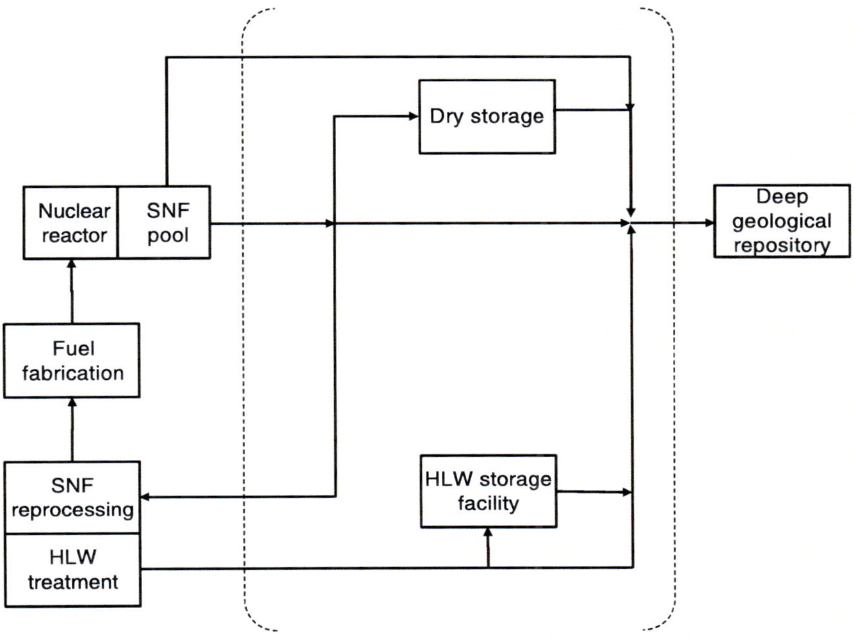

Fig. 3 Example of the nuclear fuel/material/waste transport chain

The reusable nuclear material retrieved from reprocessing requires minimal conditioning prior to transport to the fuel fabrication facility. In most cases this material is in powder form and is placed in special canisters. This material has minimal radioactivity and does not require the same rigorous transportation packaging as is needed for RW. However, this material involves extensive security requirements for transport because of its purity, its convenient handling and the ease with which it can be used for proliferation purposes.

Nuclear fuel is usually composed of fingernail-sized pellets of uranium dioxide inside hollow metal rods, typically constructed of zirconium oxide alloy (zircaloy). These fuel rods are generally between 3.5 and 4.5 m in length and are bundled together into fuel assemblies, each weighing between around 275 and 685 kg (National Research Council 2006). The assemblies are placed in commercial nuclear reactors and used to generate heat through a nuclear reaction, i.e. nuclear fission. It takes 1–2 years for the assemblies to lose their ability to produce heat or become spent; hence the term 'spent nuclear fuel'. As part of the process of expending energy during a nuclear reaction, the fuel becomes highly radioactive and thermally hot. Spent fuel emits radiation as a result of radioactive decay. The SNF is removed from the reactor and placed in specially designed storage pools near the reactors where it is cooled in preparation for transport to dry storage, reprocessing or final disposition.

Conditioning of SNF for transport from the reactor storage pool to dry storage, reprocessing facility or deep geological repository is quite an involved process. The highly radioactive nature of the material means that it must be handled with great care and with scrupulous regard for the safety of the workers, the public and the environment. The SNF must be conditioned to protect against criticality, radiation exposure and radioactive contamination under normal and hypothetical accident conditions. The first protective barrier is the cladding around the fuel meat in the fuel assemblies. The second and most important protective barrier is a specially designed, tested and licensed performance-based package. Transport packages provide protection in terms of containment, shielding, heat management and nuclear criticality safety for the radioactive material that they contain (National Research Council 2006).

Containment is provided by cladding around the nuclear fuel and/or by placing the nuclear material in canisters that are custom-designed for SNF. Specially designed transport packaging for shipment provides the final and main layer of containment.

To shield the workers, the public and the environment from the hazards of radiation, the package is enclosed in multiple layers of dense material that limit the amount of radiation that can escape from it. The structure of an SNF transport package most commonly consists of an inner and an outer stainless steel structure which enclose the materials that shield against gamma radiation; in some designs, the structure is comprised of a monolithic thick-walled steel cylinder which at the same time provides gamma shielding. Neutron shielding is usually placed over the outer cylinder enclosing the gamma shielding materials and held in place by a thin-walled stainless steel structure (EPRI 2004). Typically, for every tonne of SNF there are approximately 4 t of shielding materials in the package.

It is of utmost importance to ensure that internal nuclear reactions (i.e. self-sustaining nuclear reactions such as those that occur in the reactor) do not take place and cause criticality events while the RW is being transported. Criticality control is achieved by limiting the amount of RW in the package, minimizing nuclear moderator, and/or ensuring adequate spacing of the materials within the package. Thus, inside the package is a structure (referred to as a basket) that provides support, positioning, criticality safety and heat management.

The package is closed with one or two steel lids, which have an airtight seal to the package body. The package is also designed with impact limiters to absorb mechanical forces generated in the event of transport accidents and to provide thermal protection for the lid seals in case of fires (National Research Council 2006).

In most cases the transport of SNF and HLW is done in so-called Type B packages (see Fig. 4). These packages come in over 150 types and are built to maintain gamma and neutron radiation shielding, even under extreme conditions (WNA 2008).

The energy requirements for SNF conditioning is limited to what is needed for the nuclear material handling facility–primarily electricity for lights, cooling and heating. Conditioning of HLW requires the use of high-temperature furnaces capable of vitrifying matrices for a wide spectrum of fission products and specific elements such as sodium, phosphate, iron, molybdenum or actinides. The furnaces operate at temperatures of between 1,150°C and 1,600°C (Petitjean et al. 2002).

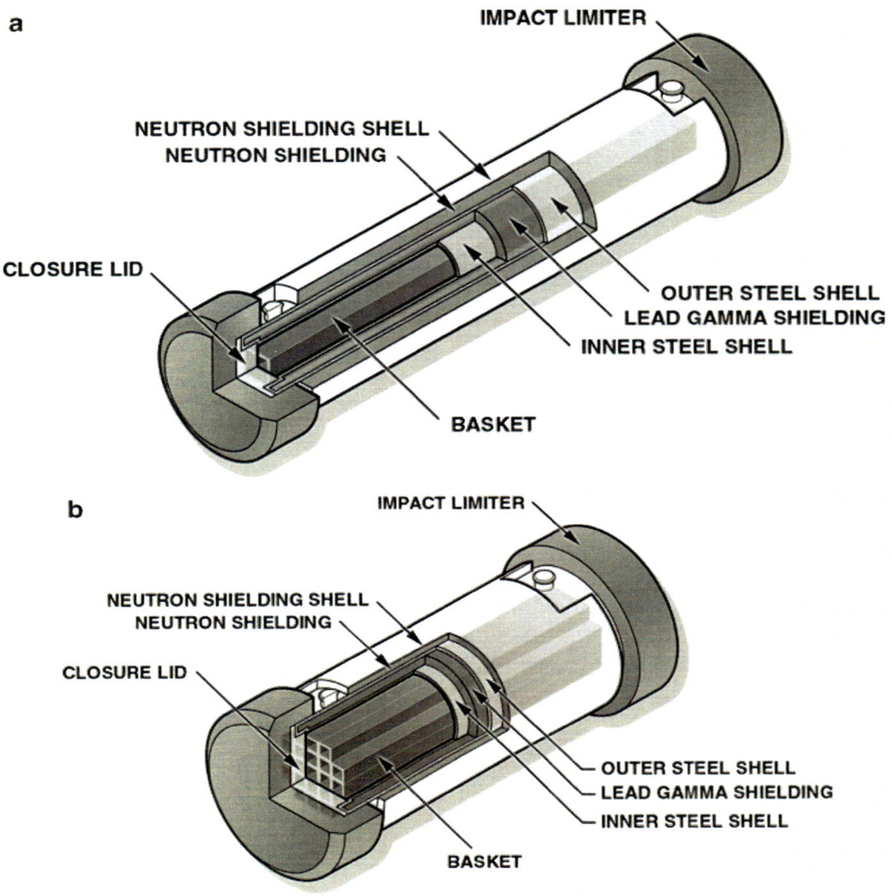

(a) Generic truck cask.

Typical specifications are:
Gross weight (including fuel): 25 t
Cask diameter: 1.2 m
Overall diameter (including impact limiters): 1.8 m
Overall length (including impact limiters): 6 m
Capacity: Up to four pressurized water reactor (PWR) or nine boiling water reactor (BWR) fuel assemblies.

(b) Generic rail cask.

Typical specifications are:
Gross weight (including fuel): 125 t cask diameter: 2.4 m
Overall diameter (including impact limiters): 3.4 m
Overall length (including impact limiters): 7.6 m
Capacity: Up to 26 PWR or 61 BWR fuel assemblies.

Fig. 4 Schematic representation of typical spent fuel transportation casks (Source: United States Nuclear Regulatory Commission website http://www.nrc.gov)

The waste generated during SNF conditioning consists of small quantities of low-level RW generated during the loading and unloading operations. High-level liquid waste is also generated during SNF reprocessing. The amount of GHG emissions depends on the energy sources supplying the facility, particularly the fuel used to generate the heat for the furnace.

3.2 *Transport*

Once the RWs have been conditioned and loaded into the appropriate transport package, they are ready for transport. This section looks at the main characteristics associated with the transport itself.

3.2.1 State of Matter for Transport

The RWs are in a solid form when transported. As previously described, nuclear fuel is originally solid and remains in that state after it has been spent. HLW is solidified through a glass-forming process that reduces its volume and eliminates the gaseous fission products that it contains (National Research Council 2006).

3.2.2 Volume

Nuclear power produces an amount of spent fuel of roughly the same mass and volume as the fuel that is fed to the reactor. This amounts to 2.7–3.6 g/MWh (Ewing 2006; Garwin 2008; EIA 2008).

SNF transport casks designed for road transport weigh normally about 25 t, however, some casks may weigh up to 40 t, not only necessitating the use of heavy trucks but also potentially requiring the consideration of routing particulars and special permits (EPRI 2004). Packages designed for railway transportation and/or intermodal barge shipping weigh up to 125 t. There is roughly a six to one fuel capacity advantage of rail casks over road casks.

Presently, the largest inventories of HLW and SNF from both defence and power production are stored in the USA and Russia. The SNF inventory of the USA was about 42,000 t in 2000 and that of Russia about 8,500 t in 1999. The worldwide SNF inventory is expected to grow significantly over the next 30 years at least. For example, the USA inventory will nearly double to about 83,800 t by 2035. Data reported to the International Atomic Energy Agency (IAEA) by 23 countries (excluding the USA and Russia) indicated that, overall, inventories of SNF through 1996 had accumulated to 42,466 t and are projected to be 90,472 t by 2014 (National Research Council 2001).

3.2.3 Modes

There are three modes of transporting SNF or HLW (i.e. road/truck, railway and ship/barge). Transport by road/truck and rail is the most likely mode for overland transport. The difference in capacity between one large rail cask that can accommodate roughly six times more SNF than a truck cask makes rail a more efficient transport mode. Both road and rail transport require specialized equipment. Road transport uses specially designed trailers that provide integral tie downs to fasten the cask to the conveyance, while a 125 t rail cask requires more than a four-axle goods wagon to transport it (EPRI 2004).

Ship/barge transport is typically used for shipments between most continents, island countries, and in situations where sea transport is easier than transport through transit countries.

3.2.4 Experience

The international community has decades of experience in the conditioning, regulating and safe handling of SNF and HLW. Some industrialized countries have considerable experience, while other less developed countries or countries without nuclear reactors have little or none. There are no complete statistics on the worldwide transport of RW. Based on a literature search and a series of informal contacts with about 25 of its member states, the IAEA was able to compile information on shipments of SNF to 2000 (National Research Council 2006). A summary of this information, as presented by Pope et al. (2001) and modified by the National Research Council (2006), is presented in Table 2. The compilers recognized the informal and incomplete nature of this information as some of the countries contacted did not respond and some respondents provided incomplete or inconsistent data.

In spite of the preliminary nature of this information, it is clear that rail has been the prevalent transportation mode and that, in general terms, the most intensive traffic has occurred within and across the borders of 11 European countries (Czech Republic, Finland, France, Germany, Hungary, Italy, Russian Federation, Slovakia, Sweden, Ukraine and the UK). The disaggregated data compiled by Pope et al. (2001) have also shown that most of the shipments are concentrated in France and the UK and that most of them are destined for the reprocessing facilities at La Hague and Sellafield, respectively. The survey also reported that SNF rail shipments within the UK are made using dedicated trains (i.e. trains carrying only one commodity from origin to destination), whereas shipments to France are made using both scheduled and dedicated trains. These trains share the rails with other freight and pass through large cities. Most of the other spent fuel shipments within or between countries are bound for interim storage (National Research Council 2006).

The sea-based transport system in Sweden, operative since 1985, uses a dedicated ship (M/S Sigyn); heavy trucks are used for complementary land transport at terminals. Dybeck (2004) has reported that up to 2004 some 1,400 transport casks

Table 2 Estimates of spent nuclear fuel shipments worldwide (Source: Pope et al. 2001; National Research Council 2006)

Mode	Europe		Japan		North America	
	Mass of SNF (tHM)	Number of packages	Mass of SNF (tHM)	Number of packages	Mass of SNF (tHM)	Number of packages
Road	81	52				
Road and rail	258	131			2,270	3,020
Mostly rail	45,702–65,142	17,565–34,065				
Sea	4,400					
Sea and land			7,821	2,130		
Unspecified	5,297–5,438	1,507–1,572			100	187
Total	55,738–75,319	19,255–5,820	7,821	2,130	2,370	3,207

Europe: Domestic or international shipments made in or between 11 European countries
Japan: Domestic shipments and international shipments from Japan to France or the UK
North America: Canada and the USA
SNF spent nuclear fuel, *tHM* tonnes of heavy metal (usually uranium)

with SNF and 130 casks with core components had been transported from Swedish reactors to the central interim SNF storage site in Sweden; these shipments amounted to 4,200 t of heavy metal, which is consistent with the information reported by Pope et al. (2001).

3.2.5 Energy Requirements: Generation of Waste and/or Greenhouse Gas Emissions

Standard fuels (mostly diesel oil and fuel oil) and/or electricity are used to supply the power needed to transport the RW. There is nearly no waste generated during the transport of RW. The only GHG emissions are from transport exhaust.

3.3 Disposal

Typically, the SNF or HLW will either be in a specially designed disposal canister when it arrives at the disposal facility, or it will be unloaded and placed into a disposal canister. The canisters will then be transported into the disposal facility using specially designed transport equipment (i.e. special fork lifts, air pallets, transfer casks, etc.).

3.4 Environment, Safety and Risks

Package safety is primarily based on robust mechanical design, the application of a substantial engineering safety margin and the use of protective features to mitigate any physical impacts that may occur during transportation (EPRI 2004).

3.4.1 Characterization of Main Risks

Risks for transporting RW arise from conventional vehicular accidents and exposure to ionizing radiation under both normal and accident conditions. Radiation risks are primarily a concern for transportation workers and for people who live near shipment routes and also for those travelling on these routes (National Research Council 2006).

Packages are effective in shielding well over 99% of the radiation emitted by the SNF or HLW. However, a small amount of radiation, primarily gamma rays, can escape from the interior of the packages and provide external doses to workers and the public (National Research Council 2006). The IAEA (2004) recently summarized the findings of several assessments of dose and risk associated with the transport of radioactive material in the nuclear fuel cycle, indicating that annual individual doses to the public are low (well below 0.1 mSv (millisievert)) and also that annual individual doses to workers are generally low (less than 1 mSv). (The sievert is a unit of equivalent dose (1 J/kg) that considers the type and effect of the radiation). Equivalent dose equals absorbed dose times Q, a quality factor (e.g. $Q=1$ for X-rays and $Q=20$ for alpha particles). These figures are below regulatory limits and also lower than the total annual global per capita effective dose due to natural radiation sources (cosmic rays, terrestrial gamma rays, inhalation and ingestion), which has been reported to be 2.4 mSv (UNSCEAR 2008).

The greatest risk arises from accidents affecting the transportation package, the likely result of which would be damage to the vehicle and/or little to no damage to the package and the RW contained in it. Degradation and/or loss of package containment have the potential to increase such radiation exposure incidents and possibly result in the release of radioactive material from the package to the environment (National Research Council 2006). However, the robust design of transportation packages makes such releases unlikely. Experience thus far indicates that no event of this type has occurred after thousands of shipments and 50 years of RW transport.

Rhoads et al. (1986) have provided a framework for the comparative assessment of the risks associated with a number of activities, including the transportation of hazardous materials such as SNF, explosives, chlorine and propane, as well as natural and man-made phenomena such as lightning, tornadoes, dam failures and industrial accidents. The results of this study showed that the individual risk (i.e. the probability of an individual at risk of dying from this cause in a given year) from SNF transport was 1 in 10^{15}. This risk was 4×10^{7} times lower than for chlorine transport, 7×10^{7} times lower than for propane transport, 4×10^{9} times lower than for railway accidents and 3×10^{11} times lower than for motor vehicle accidents.

Most regulatory bodies have relied on the operational experience of the safe transport of SNF as validation for their regulations. Since the early 1970s, some regulatory agencies such as the US Nuclear Regulatory Commission have undertaken several risk assessments, analytical studies and cask testing programmes to ensure that the regulations governing radioactive materials transport are strong enough to protect the public. EPRI (2004) has recently summarized some significant studies undertaken in the USA concerning: (1) SNF shipping response to severe road and railway accident conditions (US NRC 1987); (2) re-examination of SNF risk estimates (US NRC 2000); (3) additional assessment of SNF responses under actual road and railway transportation accidents unrelated to SNF transport (US DOE 2003); and (4) physical testing programmes of SNF shipping casks (Jefferson and Yoshimura 1978; Huerta 1981; US NRC 2003). These recent assessments have concluded that: (1) the earlier risk assessments were conservative and that the risks associated with SNF transport remain small; and (2) the probability of an accident severe enough to violate the integrity of a SNF cask was extremely small. Consequently, the risk to the general public of any credible accident is also extremely small.

3.4.2 Statistics of Incidents

Notification of accidents and incidents in transport is typically required by the regulations of most countries where competent authorities are responsible for receiving and recording these events. Individual countries keep track of accidents and incidents involving radioactive materials within their borders (Shaw et al. 2001; McClure 1997; EPRI 2004).

The IAEA maintains a database (Events in the Transport of Radioactive Material (EVTRAM)) of such information (Young 2004). However, this database has been supported only to a limited extent by IAEA member states, which report on a voluntary basis, and the experience thus far has shown that this type of reporting system leads to incomplete information (Shaw et al. 2001).

The combined information from national and IAEA data sources indicates that in spite of transportation accidents involving SNF casks in several countries, there have been no serious injuries to transport workers, emergency response personnel, or the general public from the radioactive contents of the casks (EPRI 2004).

3.5 Regulatory Requirements

The transportation of SNF is perhaps the most comprehensively regulated of all hazardous materials (EPRI 2004). The international recommended requirements for the packaging and transport of radioactive material have evolved over 4 decades, resulting in today's IAEA regulations for the safe transport of radioactive material (IAEA 2005; Pope 2004). This set of regulations includes requirements for shippers and carriers; packaging, including analysis or testing for both normal and accident conditions of transport; security and physical protection; training and emergency response; and inspection and quality assurance (EPRI 2004). In addition, each nation has developed its own requirements following, in the vast majority of cases, the IAEA advisory regulations. Adherence to these regulations ensures that the transport package: (1) is appropriate for the radioactive material to be transported; (2) is designed according to a quality assured process; (3) is properly prepared for transport; (4) is properly labelled in accordance with national and international requirements; (5) is properly operated, handled and maintained in accordance with the requirements stated in the transport package safety case; (6) has the appropriate documentation during transport to provide the necessary information to those involved in transport and those responding to any incident that may occur; and (7) performs in a predictable manner under normal transport and accident conditions.

The IAEA advisory regulations for the safe transport of radioactive materials were first published in 1961 (IAEA 1961). They are reviewed on a biennial basis and are revised as needed; this periodic review is essential to ensure safety. The IAEA regulations are now recognized throughout the world as the uniform basis for both national and international transport requirements and have been adopted by over 60 countries, the International Civil Aviation Organization (ICAO) for air transport, the International Maritime Organization (IMO) for sea transport, and regional transport organizations (Pope 2004). In addition, all of the IAEA regulatory requirements have been incorporated into the latest edition of the United Nations recommendations on the transport of dangerous goods (UN/SCETDG 2001).

The IAEA regulations for the safe transport of radioactive materials (IAEA 2005) contain requirements for both normal conditions of transport and hypothetical accident conditions. For the particular case of SNF and HLW transport,

Table 3 Tests specified by IAEA regulations for demonstrating the ability of a package to withstand normal and accident conditions of transport (Based on IAEA 2005, Section VII)

Test	Brief description
Normal conditions of transport	
Water spray	The specimen is exposed to a spray simulating an exposure to rainfall of approximately 5 cm/h for at least 1 h
Free drop	The specimen is dropped from specified heights according to the package mass, from 0.3 m (>15 t) to 1.2 m (<5 t)
Stacking	Unless the shape of the packaging effectively prevents stacking, the specimen is subjected to a compressive load of: 5 × (actual package mass) or 13 kPa × (vertically projected area of the package), whichever is greater, for a period of 24 h
Penetration	A 6 kg bar of 3.2 cm diameter with a hemispherical end is dropped from a height of 1 m and is directed to fall, with its longitudinal axis vertical, onto the centre of the weakest part of the specimen
Accident conditions of transport	
Free drop	The specimen is dropped from a height of 9 m onto a flat, essentially unyielding horizontal surface, so as to suffer maximum damage
Puncture	The specimen used in the free drop test is dropped so as to suffer maximum damage from a height of 1 m onto a solid mild steel bar of circular section (15 cm in diameter and 20 cm long), which has been rigidly mounted perpendicularly on an unyielding horizontal surface. The steel bar has a flat and horizontal upper end with its edge rounded off to a radius of not more than 6 mm
Thermal	The specimen used in the previous mechanical tests is fully engulfed in a hydrocarbon fuel/air fire for 30 min in sufficiently quiescent ambient conditions to assure a minimum average flame emissivity coefficient of 0.9 and an average temperature of at least 800°C. The specimen is subsequently exposed to an ambient temperature of 38°C, subject to specified solar insolation conditions and subject to the design maximum rate of internal heat generation within the package by the radioactive contents for a sufficient period to ensure that temperatures in the specimen are everywhere decreasing and/or are approaching initial steady state conditions
Water immersion	A separate undamaged specimen is immersed under a head of water of at least 15 m for a period of not less than 8 h in a position that will lead to maximum damage. In addition, for packages designed to contain more than $10^5 A_2$, an enhanced water immersion test is specified under which the specimen is immersed under a head of water of at least 200 m for a period of not less than 1 h

the requirements specify that Type B packages should be designed to withstand severe accident conditions without a loss of containment or an increase in external radiation to levels that would endanger emergency responders or the general public. Under normal transport conditions, the regulations require that if Type B packages are subjected to the water spray, free drop, puncture and stacking tests briefly described in Table 3, the corresponding specimens must maintain their containment effectiveness by restricting the loss of radioactive contents to not more than $10^{-6}A_2$/h. (A_2 is the activity value of radioactive material which is given in special tables in IAEA (2005) and is used to determine the activity limits for the requirements of these regulations.) Under accident conditions of transport, the IAEA regulations specify that if Type B packages were subjected to the mechanical, thermal and immersion tests presented in Table 3, they should:

(i) Retain sufficient shielding to ensure that the radiation level at 1 m from the surface of the package would not exceed 10 mSv/h with the maximum radioactive contents which the package is designed to contain; and
(ii) Restrict the accumulated loss of radioactive contents in a period of one week to not more than $10A_2$ for krypton-85 and not more than A_2 for all other radionuclides.

3.6 *Public Acceptance*

Establishing a route for a nuclear material shipment can be very political and highly emotional if the public is made aware of the shipment. Some countries (i.e. the USA and Germany) require that the public be made aware of certain nuclear material shipments, whereas other countries (i.e. the Czech Republic, Russian Federation, Slovakia and Ukraine) specifically prohibit dissemination of information to the public for security reasons. The countries that notify the public of nuclear shipments provide a significant amount of public/media awareness training and outreach before the first shipment is made. There is a significant amount of experience of effective outreach of this nature.

Although the security of radioactive materials in transport, understood as 'the protection of humankind and the environment from the potential consequence of malicious, purposeful and unlawful acts of an individual or group' (Pope and Luna 2004), is not new for the nuclear transport industry, it has received increased attention following recent world events. To meet the security needs, the IAEA began in 2002 a series of activities to provide additional guidance on the basis of model regulations developed by the United Nations Sub-Committee of Experts on the Transport of Dangerous Goods (UN/SCETDG 2001). The implementing guide on security in the transport of radioactive material (IAEA 2008) constitutes the main result of these activities. In addition to considering the quantity of the radioactive material being transported, the transport modes and the type of packages being used, the guidance requires measures: 'to deter, detect and delay unauthorized access to the radioactive material', 'to identify the actual possible malicious acts involving any consignment', and 'to provide rapid response to any… malicious acts involving radioactive material while in transport or storage incidental to such transport.' The guidance specifies that establishing 'an adequate security regime for the transport of radioactive material is the responsibility of each State', and discusses the role of the operators in implementing adequate security measures.

3.7 *State of the Technology*

3.7.1 Science and Engineering

The science and engineering for making RW shipments is well established. The engineering for the packages is fully recognized and the science for ensuring the shielding, criticality, containment, and structural integrity is well known. An important aspect of assuring safety is the graded approach to package design, whereby a proportionate robustness of packaging is required according to the materials being carried and the safety risk of individual components. There is over 50 years of experience in this area and it continues to improve as technologies and experience evolve.

Type B packages are performance-based packages; their design, licensing and fabrication require complex expertise in technical design areas such as structural engineering, heat transfer, nuclear criticality safety and radiation shielding. As discussed in Sect. 3.5, regulatory requirements impose a set of strict performance criteria on designers and manufacturers to ensure that each Type B package can withstand normal transportation and hypothetical accident conditions.

The analytical tools used for the design of any SNF transportation cask and its other transportation system components (structural and thermal computer codes, nuclear codes for criticality safety and shielding) are utilized well within their demonstrated range of benchmarked capability. Physical testing may be conducted during design in several circumstances such as when new materials are used, in cases in which numerical methods may not be fully capable of accurately predicting behaviour or where performance data are incomplete. Full-scale testing of components or partial-scale testing of components and packages is done using standardized material testing methods (EPRI 2004).

The construction of SNF and HLW packages normally follows the industrial practices used in the fabrication of large pressure vessels. Specialty materials such as lead, depleted uranium or hydrogen-containing materials are uniquely identified and specifically tested to assure compliance with the design specifications. Before its initial use, each completed cask undergoes acceptance testing that includes leak checking, hydraulic testing for integrity, shielding continuity testing and thermal testing. During the entire life of the cask, it is operated and maintained to specified requirements and under a strict quality assurance programme with approved procedures. In general terms, no other hazardous material container undergoes the same level of scrutiny (EPRI 2004).

3.7.2 Regulatory Aspects

No sector of transport is regulated more stringently than the nuclear transport industry, which has to take many actions regarding: (1) requirements for loading, stowage, carriage, handling and unloading of the package; (2) restrictions on the mode of transport and routing instructions; and (3) emergency and safety arrangements (IAEA 2005). However, the underlying philosophy, based on a set of performance criteria for packages rather than specific design specifications, requires that the package provide the primary means of ensuring the necessary safety during incident-free transport and during accidents, whatever mode of transport is used (Green 2004).

The nuclear transport industry (Green 2004) and other stakeholders (IAEA 2004) have called for greater standardization, harmonization, global application and simplification of transport safety standards. Among harmonization issues, the industry has mentioned: (1) different time schedules for introduction of new regulations in different jurisdictions; (2) different interpretation of the regulations by different competent authorities (e.g. the order in which package tests are carried out); and (3) different assumptions being used by different authorities in carrying out reviews of the criticality safety of packages. These harmonization issues may lead to considerable time intervals between the renewal of a package certificate in one country and the relevant revalidation in another country, occasioning delays in transport. One key question in the implementation concerns the independent reviews of package designs and revalidation of approved packaging carried out by various national competent authorities in the context of international shipments. As sometimes different underlying assumptions are used, a single design may require, for instance, the preparation of multiple criticality analyses to obtain base approval and foreign validation (Green 2004).

Transport security has received increasing attention, and the IAEA has recently published an implementing guide for security in the transport of radioactive material (IAEA 2008). International transport security standards have also been developed, especially by IMO. In some cases, international standards are supplemented by national requirements. However, there is a need for harmonization because differing requirements between national jurisdictions may lead to greater complexity, with the potential for confusion and misinterpretation (Green 2007). In addition, the transport industry still faces the challenge of balancing the traditional safety approach, which needs to be clearly declared, with the need to maintain security (Morgan-Warren 2003).

3.7.3 Policy Aspects

Transport of RW is very political and, although its low associated risk has been estimated based on sound science and demonstrated over 50 years of experience, the nuclear transport industry still needs to make efforts to win over the public. However, there is extensive experience showing that well planned and executed public and media training and outreach programmes, demonstrating that shipments can and will be carried out safely and securely, serves to overcome the general public's fears, resulting in minimal opposition. Although a significant amount of knowledge is required to do this effectively, issues such as denials that shipments contain nuclear waste, delays to shipments, and transport security, remain major challenges (Green 2007).

3.7.4 Cost Estimates

Costs depend on factors such as the volume of waste shipped, the origin and destination of shipments and the specific route used. However, costs for truck and rail shipments can be estimated based primarily on the weight of the load and the length of the trip (Tang and Saling 1990). Other components of the cost may involve leasing and demurrage costs, the latter being the waiting time for the cargo to be loaded or unloaded at the originating and terminating facilities.

Many studies like the recent one by the University of Chicago (2004) have adopted a reference value of US$63/kg of uranium (2003 prices) for the transportation costs of SNF. This value was selected from the report by NEA (1994), which addresses relatively short transportation distances within the European area, assuming, for sensitivity reasons, transport costs in the range US$25–100/kg of uranium.

4 Comparative Assessment of the Transport of CO_2 and Radioactive Waste Associated with Electricity Generation

All the criteria composing the three guiding principles that we proposed for this comparative assessment, namely transport chain, policy aspects and state of the technology, are summarized and addressed in Table 4, while the main findings are discussed hereafter.

4.1 Transport Chain

For large-scale operations associated with CCS, CO_2 is transported in liquid or supercritical state to make the best possible use of the transport capacity. This is totally different for SNF, which remains in the same solid state as the original

Table 4 Comparison between the transport of CO_2 and radioactive waste resulting from the generation of electricity

CO_2	RW
Transport chain	
1. Conditioning	
1.1. Type of processing up to the inlet of the transport system	
Removal of water and certain impurities. Compression before pipeline suction. Liquefaction for ship-based transport	Proper packaging for the type of material and the mode of transport
1.2. Energy requirements. Generation of waste and/or greenhouse gas emissions	
Compression and liquefaction are very energy intensive. Waste generation is relatively low. GHGs depend on the energy supply	Standard energy requirements. Small quantities of low-level radioactive waste may be generated during loading and unloading
2. Transport	
2.1. State of matter	
Supercritical or liquid for pipeline. Liquid for ship-based transport	Solid for both SNF and HLW
2.2. Volume	
~300 kg CO_2/MWh for natural gas-fired power plants 600–800 kg CO_2/MWh for coal power plants	3–4 g SNF/MWh. Typically, for every tonne of SNF there are ~4 t of protective shielding materials in the reusable package
2.3. Modes	
Pipeline (on- and offshore), ship and combinations of these are regarded as the most cost-effective alternatives for a large-scale CCS infrastructure	Rail, the dominant mode, is followed by ship for long distances involving maritime transport. Road/truck is the third mode. Air transport is unlikely
2.4. Experience	
Onshore pipeline: >6,000 km pipeline annually transporting several Mt CO_2 Offshore pipeline: The first long distance (170 km, ~2,500 t CO_2/day) pipeline has been constructed in the Norwegian North Sea Ship: tankers with capacities <1,500 t CO_2. Large-scale ship-based transport (2–6 Mt CO_2/year): only conceptual designs are available	Until 2000, 66,000–85,000 tHM, usually uranium of SNF have been transported worldwide in ~12,000 transportation casks Modes in terms of tHM transported: mostly rail (46,000–65,000) > unspecified (~5,000) > sea and land (~12,000) > road and rail (~2,500) > road (<100)
2.5. Energy requirements. Generation of waste and/or greenhouse gas emissions	
Pipeline: Intermediate boosters may be required to compensate for pressure drop along the pipeline *Ship*: Fuel consumption (~30 kWh/tCO_2, for a 20,000 m^3 tanker) > unloading (<10 kWh/t CO_2)	Standard type energy sources are used to generate the power needed to transport the nuclear material (i.e. gasoline and diesel)
Fuel combustion: GHG emissions are associated with pumping through the pipeline or with ship-based operations. *Fugitive*: CO_2 from venting. Waste generation is relatively low and disposal is readily available	There is nearly no waste generated during transport. The only GHG emissions are from the exhaust of the mode of transport

(continued)

Table 4 (continued)

CO_2	RW
3. Disposition (manner of transfer from the transport system to the disposal site)	
Injection system: pressurized surge storage tank, injection pumps (if needed), piping to distribute CO_2 to the injection wells, monitoring and control equipment	Typically, the SNF or HLW will either be in a specially designed canister for disposal when it arrives at the disposal facility or it will be unloaded and placed into a disposal canister
4. Environment, safety and risks	
4.1. Characterization of main risks	
Leakage and accidental releases are the main risks associated with CO_2 transport and injection. They are typically of short-term and local nature *Onshore pipeline*: CO_2 from leaks could accumulate near the ground *Offshore pipeline*: Leaks could adversely affect a large area because of the dissolution and acidification of the surrounding seawater *Ship*: Collision, foundering, stranding and fire are the risks involved *Injection*: Releases or leakage due to mechanical failure of the injection equipment. CO_2 could migrate into adjacent reservoir zones and aquifers	Risks arise from conventional vehicular accidents and exposures to ionizing radiation under both normal and accident conditions Radiation risks are primarily a concern for transportation workers and for people who live near shipment routes and also for those travelling on these routes
4.2. Statistics of incidents	
Data from 36 CO_2 pipeline incidents that occurred in the USA show that most incidents were related to the pipeline itself. These features are different from natural gas pipelines, for which the principal cause of incidents was outside force, such as damage by excavator buckets	Combined data from national sources and the IAEA indicate that while there have been transportation accidents involving SNF casks in several countries, there have been no serious injuries to transport workers, emergency response personnel, or the general public due to the radioactive contents of the casks
Policy aspects	
5. Regulatory requirements	
Development of national standards is under way in several countries. The transport of CO_2 across national boundaries and transport by ships and by sub-sea pipelines is covered by various international legal conventions. The applicability and application of regulatory and liability regimes depend on: transport mode, geographical location, land ownership, impacts, risks and identity of the party responsible for damage	In 1961 the IAEA started publishing advisory regulations for the safe transport of radioactive materials. Those regulations are now recognized throughout the world as the uniform basis for both national and international transport safety requirements. These regulations have been adopted by over 60 countries, the International Civil Aviation Organization (ICAO), the International Maritime Organization (IMO), and regional transport organizations

(continued)

Table 4 (continued)

CO_2	RW
6. Public acceptance	
It is difficult to isolate the specific issues associated with CO_2 transport from the general context of public acceptance and communication of the entire CCS system. For a large-scale deployment of CCS, public concerns about CO_2 transport may be a significant barrier. Public and media awareness training and outreach programmes will be required	There is a significant amount of experience in undertaking public and media awareness training and outreach programmes to demonstrate that the shipment(s) can and will be done safely and securely. Establishing a route for a nuclear material shipment can be very political and highly emotional if the public is made aware of the shipment
State of the technology	
7.1. Science and engineering	
Onshore pipeline: The transport of high purity CO_2 by pipeline is a mature technology *Offshore pipeline*: The 318 m pipeline of the Snøhvit project is the only facility that has been built. The learning curve has just started *Ship-based*: There is experience in transporting relatively small quantities of CO_2 by ship. Large-scale ship-based transport of CO_2 has yet to occur and only conceptual designs for this option are available	The science and engineering for making nuclear material shipments is well established. Type B packages are performance-based packages. The engineering for the packages is fully recognized and the science for ensuring that the shielding, criticality, containment, and structural integrity is well known. An important aspect of assuring safety is the graded approach to package design. There are over 50 years of experience in this area, and it continues to improve as technologies and experiences evolve
7.2. Regulatory aspects	
An integrated international framework is not yet available. The regulatory framework for CO_2 transport is under way in most countries interested in the deployment of large-scale CCS	No sector of transport is regulated more stringently than the nuclear transport industry. There has been a call for greater standardization, harmonization, global application and simplification of transport safety standards
7.3. Policy aspects	
Further assessment is necessary to evaluate public perception of CO_2 transport	Although low risk has been estimated based on sound science and demonstrated over 50 years of experience, the nuclear transport industry still needs to make efforts to convince people that nuclear transportation is safe

CCS carbon capture and storage, *GHG* greenhouse gas, *HLW* high-level waste, *SNF* spent nuclear fuel, *tHM* tonnes of heavy metal

nuclear fuel, or for HLW, which is solidified before transport. Therefore, the transport of CO_2 and nuclear waste essentially differ in the many aspects associated with the transportation of bulk fluids versus the transport of properly identified packages containing solid materials.

For each MWh of electricity, about 300 kg CO_2 can be captured from natural gas-fired thermal power plants and 600–800 kg of CO_2 from coal-fired power plants. These figures are five orders of magnitude higher than the amount of waste generated by nuclear power plants, which is 3–4 g SNF/MWh. This large difference in specific emissions/waste is reflected in the projected volumes that would be required for a large deployment of CCS, which are estimated to be several hundreds to thousands million tonnes of CO_2 per year worldwide (Gale et al. 2005), while the inventory of SNF worldwide would be several hundred thousand tonnes in 2030, which would definitely not all be transported in 1 year.

The physical state for transport and the volumes involved largely determine the preferred means of transport. They are pipelines (on- and offshore) for liquid or supercritical CO_2 and railways, ship or truck for SNF and HLW. Ship-based transport has also been suggested for future large-scale CCS. Ship-based transport is the only common mode for both transport systems. However, there is a difference between the state of the art for ship-based transport of CO_2 and that of RW. There is a mature market for SNF and HLW, particularly in countries such as Sweden and Japan, while large-scale transport of CO_2 (2–6 Mt CO_2) is at the research phase, with only conceptual designs available.

Conditioning is necessary for the stream of captured CO_2 and the SNF or HLW before they are actually transported. Because of the physical state and the associated risks, the type of processing for each type of material is very different. Removal of water and certain impurities, compression before pipeline suction or liquefaction for ship-based transport are required for CO_2-rich gas. This conditioning is primarily aimed at providing adequate physical properties for transport, with safety playing a secondary role in defining the characteristics of the process. Conditioning of the SNF is primarily done by placing the material into transport packages (denominated Type B packages) while the HLW is subject to a solidifying process (usually vitrification) before being placed in the transport packages. Safety is the main concern for this processing and the specially designed, tested and licensed performance-based transport packages constitute the most important barrier providing protection regarding containment, shielding, heat management and nuclear criticality safety for the radioactive material that they contain.

In spite of the differences between both transport systems, onshore pipeline transport of CO_2 has a somewhat similar level of experience to SNF and HLW transport. The transport of high purity CO_2 was originally developed to supply CO_2 for injection in EOR, and the oil industry has presently over 4 decades of experience in successfully transporting and injecting CO_2 for this purpose. In the USA more than 6,000 km of pipelines annually transport several million tonnes of mainly naturally occurring CO_2. With respect to RW transport, the international community has about 5 decades of experience in the conditioning, regulating and safe handling of SNF and HLW. By 2000, total shipments worldwide had totalled 55,000–75,000 t of heavy metal in 19,000–36,000 packages. Rail was the predominant shipping mode, followed by sea and land.

The transport systems also differ in the infrastructure they require. Pipelines (on- and offshore) must be built especially for CO_2 transport. On the other hand, trucks and trains normally share roads and rails with other vehicles without requiring the construction of a dedicated infrastructure. Maritime shipments of SNF or HLW are usually done in dedicated ships as will be the case for large-scale ship-based CO_2 transport.

In general terms, waste generation is relatively low and disposal is readily available for all the steps of both transport systems. GHG emissions would depend on the structure of the electricity supply and on the transport distance for those modes that use fossil fuels, particularly trucks and ships.

The main risks associated with both transport systems are similar in that they constitute leakage and accidental releases, typically of short-term and local nature. The main difference is the nature and impacts of these releases. Pure CO_2 is neither combustible nor toxic, unlike other gases or liquids regulated as hazardous materials. The main risk of compressed CO_2 is that, being denser than air, it tends to pool near the ground, displacing all oxygen and forming a vapour cloud that can cause respiratory problems, including suffocation and even death. The main risk associated with a damaged SNF or HLW transport package is that of an accidental release resulting in radiation exposure, contamination and/or criticality.

As a consequence of the differences in risk discussed above, the safety measures for both systems also differ. The performance-based approach for Type B packages requires that the package be the primary safety barrier during normal transport and during accidents, whatever mode of transport is used. This is the main difference not only to the transport of CO_2 but also to the transport of many hazardous cargoes where the mode of transport is the only primary safety measure.

The pipeline transport of CO_2 and the transport of SNF and HLW have a similar record regarding incidents. While there have been accidents in both transport systems, there have been no serious injuries to transport workers, emergency response personnel or the general public as a result of the radioactive contents of the packages or CO_2-rich releases.

4.2 Policy Aspects

Of all hazardous materials transport, that of SNF and HLW is perhaps the most comprehensively regulated. The IAEA regulations for the safe transport of radioactive material have evolved over 4 decades and are recognized worldwide as the uniform basis for both national and international safety standards. They include requirements for shippers and carriers; packaging, including analysis or testing for both normal and accident conditions of transport; security and physical protection; training and emergency response; and inspection and quality assurance. This status is quite different from that of CO_2 transport, which is lacking a uniform international approach. Development of national standards is at different stages in several countries ranging from an advanced regulatory scheme in the USA, which is the country with the most extensive pipeline network and the largest construction and operating experience, to countries whose legislation has not yet specifically addressed CO_2 transport. Transport of CO_2 across national boundaries and transport by ships and by sub-sea pipelines is covered by various international legal conventions.

Public acceptance of nuclear material shipments, particularly concerning aspects such as routing and hearings, can be very political and highly emotional.

The nuclear transport industry has extensive experience in providing public and media awareness training and outreach programmes on the safety and security of SNF and HLW transport. We have been unable to register any major problems with respect to the public acceptance of pipeline transport of CO_2 for EOR. This may be because these pipelines are not generally built across very populated areas. However, under a scenario of large deployment of CCS, public acceptance of CO_2 could also encounter problems similar to those involved in RW transport because CO_2 may need to be transported in large amounts over significant distances in populated areas. In that case, the number of people potentially exposed to risks of the CO_2 transport system may be larger than the number exposed to potential risks of capture and storage facilities, and public concerns about CO_2 transport may be a significant barrier.

4.3 *State of the Technology*

The science and engineering involved in SNF and HLW shipments and CO_2 pipeline transport are well established. Onshore pipeline and all modes of nuclear material shipments are mature technologies. The status is different for CO_2 transport via offshore pipeline or large-scale ships. The learning curve for offshore pipeline transport of CO_2 has recently started with the construction of a 318 km pipeline in the North Sea. Large-scale ship-based transport of CO_2 has yet to occur, and only conceptual designs for this option are available.

There is room for improvement in the regulatory framework of both transport systems. However, while the nuclear transport industry has called for greater standardization, harmonization, global application and simplification of transport safety standards, the CO_2 transport industry still lacks an integrated international approach. The regulatory framework for CO_2 transport is under way in most countries interested in the deployment of large-scale CCS. For pipeline transport an evolutionary approach based on existing environmental rules governing drilling, injection and gas transportation is the preferred option. Ship-based transport, offshore pipelines and injection of CO_2 in sub-seabed repositories may involve different categories of marine pollution under the relevant international conventions.

Further assessment is necessary to evaluate public perception of CO_2 transport. Most of the available studies addressing CCS as a GHG mitigation option are of a general nature, and only few of them deal with specific issues of transport. It is likely that acceptance of transport in general may become more problematic as this is the most visible part of the CCS system. Transport of RW is very political and, although its low associated risk has been estimated based on sound science and demonstrated over 50 years of experience, the nuclear transport industry still needs to make efforts to convince people that it is safe. Issuing denials that a shipment is carrying RW and delays to shipments, together with other problems involving transport security, remain major challenges. Transport security has received increasing attention; balancing the traditional safety approach that requires declaration with the need to maintain security poses a significant challenge.

5 Conclusions

Our discussion of the individual transport systems and the overall picture presented in Sect. 4 allows a hierarchy of criteria for the comparative assessment of the transport of CO_2 and RW associated with electricity generation to be identified. We found that the main factors determining the mode of transport to be used are the volumes involved, the physical state in which the substance will be transported and the radioactive nature of the SNF and HLW. These properties, which are listed below, show that there are more differences than similarities between the systems analysed.

- Volume: rather than being a specific property of each transport chain, the amount that needs to be transported is an inherent characteristic of each electricity generation system (i.e. 300–800 kg CO_2/MWh, depending on the fossil fuel used, versus ~0.004 kg SNF/MWh) and determines the scale of the transport system.
- Physical state: the CO_2 present in the flue gases from any thermal power plant remains gaseous after being captured and is subsequently transformed to a denser phase (liquid or supercritical) to make transport economically feasible; on the other hand, solid nuclear fuel remains in the same state after being spent, while HLW is solidified through vitrification.
- Waste radioactivity: international standards establish specific requirements for performance-based packages that provide the necessary protection for workers and the general public under normal and accident conditions.

The contrast between the most convenient systems, i.e. bulk transport of liquid CO_2 via (mainly buried) pipeline versus surface transport by rail, ship or truck of properly identified performance-based packages for solid RW singles out the main difference. Under a scenario of large-scale deployment of capture and disposal of CO_2, ship-based transport has been pointed out as a future option. In this case, the right ships for this purpose would be much closer to the tankers used for transporting LPG and LNG than to the specialist vessels carrying nuclear cargoes.

There is a similarity in that for both transport systems: (1) leakage and accidental releases are in general the main risks; and (2) the available records of incidents show that there have been no serious injuries as a consequence of any accident. But once again, the distinctive nature of these risks associated with the differences in both chains does not permit a strict comparison. Accordingly, safety standards are specific to each transport system; with the performance-based approach for Type B packages being a unique feature of the nuclear transport industry. There is a need for more exhaustive information of incidents and an effort on the part of both transport industries in this regard would be welcome because: (1) for onshore pipelines, the available specific information comes mainly from the USA and extrapolations from natural gas pipelines do not seem advisable; and (2) for RW transport, the valuable international information from the IAEA's EVTRAM database is somewhat limited on account of the voluntary basis of reporting by member states.

In spite of all the differences, both transport systems share a well established status with regard to the science and engineering aspects of the existing techno logies. Both onshore CO_2 pipelines and RW transport are mature markets and, although the learning curve has recently started for offshore CO_2 pipelines, while large-scale ship-based CO_2 transport is at a research phase, the specialists suggest that fully developed technologies and good experience of natural gas transfer by offshore pipeline or ships would be readily available when needed on a large scale.

There are also differences regarding the two main policy aspects analysed. Regulatory frameworks both at national and international levels are at very different stages of development. For CO_2 transport, some authors have considered more intensive international cooperation in this field as vital (van Alphen et al. 2009). The process followed by the IAEA in developing the regulations for the safe transport of RW may be of interest in the development of a unified international regulatory approach for the safe transport of CO_2. However, pipeline deployment in the oil and gas industries has made it necessary to contemplate a number of site-specific issues, suggesting that strict standardization such as that of RW transport may be unsuitable.

While routing is a common concern for both transport chains, RW transport has had a higher degree of visibility, particularly in countries requiring that the public be made aware of the shipment, than onshore CO_2 transport via buried pipelines, which to date have mainly occurred in areas of low population. A large deployment of CCS may make this transport more visible, and it is difficult to evaluate from the available studies how public opinion would evolve regarding CO_2 transport for disposal purposes. Public concern may focus on the risk of leakages and accidental releases irrespective of the view of CCS as a technology aimed at decarbonizing the power and industrial sectors. In any case, CO_2 transporters may learn from the experience of the nuclear transport industry in planning and executing public and media training and outreach programmes.

When we started planning this comparative assessment, we feared that a study of this kind would be like comparing apples and oranges. We then realized that our aim was to compare the transport rather than the fruits themselves; assuming that the transport of apples and oranges was comparable, we therefore decided to undertake the study. Furthermore, recent research showed that the fruits themselves were not only comparable but quite similar (Barone 2000). As for apples and oranges, the transport systems for CO_2 and RW turned out to be amenable to comparative analysis; however, similarity was not the determining feature. The distinctive nature of CO_2 and SNF or HLW largely determines the numerous differences between the two transport systems.

Acknowledgements Thanks are due to Kathryn A. McBride (Idaho National Laboratories) for her exhaustive compilation of references on radioactive waste transport.

References

Aspelund A, Jordal K (2007) Gas conditioning–the interface between CO_2 capture and transport. Int J Greenhouse Gas Control 1(3):343–354

Aspelund A, Sandvik TE, Krogstad H, de Koeijer G (2004a) Liquefaction of captured CO_2 for ship-based transport. In: Wilson M, Morris T, Gale JJ, Thambimuthu K (eds) Proceedings of the 7th international Conference on Greenhouse Gas Technologies, vol 2. Elsevier, Oxford, pp 2545–2549

Aspelund A, Sandvik TE, Wongraven LR, Krogstad H (2004b) Offshore unloading of CO_2 to an oilfield. In: Wilson M, Morris T, Gale JJ, Thambimuthu K (eds) Proceedings of the 7th International Conference on Greenhouse Gas Technologies, vol 2. Elsevier, Oxford, pp 2551–2554

Aspelund A, Mølnvik MJ, de Koeijer G (2006) Ship transport of CO_2 technical solutions and analysis of costs, energy utilization, energy efficiency and CO_2 emissions. Chem Eng Res Des 84(A9):847–855

Barone JE (2000) Comparing apples and oranges: a randomised prospective study. Br Med J 321:1569–1570

Barrie J, Brown K, Hatcher PR, Schellhase HU (2004) Carbon dioxide pipelines: a preliminary review of design and risks. In: Rubin ES, Keith DW, Gilboy CF (eds) Proceedings of the 7th international Conference on Greenhouse Gas Technologies, vol 1. Elsevier, Oxford, pp 315–320

Barrio M, Aspelund A, Weydahl T, Sandvik TE, Wongraven LR, Krogstad H, Henningsen R, Mølnvik M, Eide SI (2004) Ship-based transport of CO_2. In: Wilson M, Morris T, Gale JJ, Thambimuthu K (eds) Proceedings of the 7th International Conference on Greenhouse Gas Technologies, vol 2. Elsevier, Oxford, pp 1655–1660

Berger B, Kaarstad O, Haugen HA (2004) Creating a large-scale CO_2 infrastructure for enhanced oil recovery. In: Rubin ES, Keith DW, Gilboy CF (eds) Proceedings of the 7th International Conference on Greenhouse Gas Technologies, vol 1. Elsevier, Oxford, pp 321–328

BERR (Department for Business, Enterprise and Regulatory Reform) (2007) Development of a CO_2 transport and storage network in the North Sea. Report to the North Sea basin task force. http://www.berr.gov.uk/files/file42476.pdf

Birkestad H (2002) Separation and compression of CO_2 in an O_2/CO_2-fired power plant. Master's thesis, Department of Energy Conversion, School of Mechanical Engineering, Chalmers University of Technology, Göteborg

Cailly B, Le Thiez P, Egermann P, Audibert A, Vidal-Gilbert S, Longaygue X (2005) Geological storage of CO_2: a state-of-the-art of injection processes and technologies. Oil Gas Sci Technol 60(3):517–525

Chadwick A, Arts R, Bernstone C, May F, Thibeau S, Zweigel P (eds) (2007) Best practice for the storage of CO_2 in saline aquifers: observations and guidelines from the SACS and CO2STORE projects. European Commission. http://www.ngu.no/FileArchive/91/CO2STORE_BPM_final_small.pdf. Accessed 15 Feb 2008

Cockerill T (2005) Outline of the new carbon dioxide storage cost model. In: Gough C, Shackley S (eds) An Integrated Assessment of Carbon Dioxide Capture and Storage in the UK. Technical Report 47. Tyndall Centre for Climate Change Research, Manchester, pp 210–229

Coleman DL (2009) Transport infrastructure rationale for carbon dioxide capture & storage in the European Union to 2050. In: Gale J, Herzog H, Braitsch J (eds) Proceedings of the 9th International Conference on Greenhouse Gas Control Technologies. Energy Procedia 1:1673–1681

de Coninck HC, Huijts NMA (2004) Carbon dioxide capture and storage: public perception, policy and regulatory issues in the Netherlands. In: Wilson M, Morris T, Gale JJ, Thambimuthu K (eds) Proceedings of the 7th International Conference on Greenhouse Gas Technologies, vol 2. Elsevier, Oxford, pp 2491–2495

DGC (Dakota Gas Company) (2008) Pipeline information. http://www.dakotagas.com. Accessed 11 Feb 2008

Doctor R, Palmer A, Coleman D, Davison J, Hendriks C, Kaarstad O, Ozaki M (2005) Transport of CO_2. In: Metz B, Davidson O, de Coninck HC, Loos M, Meyer LA (eds) IPCC Special Report on Carbon Dioxide Capture and Storage. Cambridge University Press, Cambridge, pp 179–193

Dybeck PBJ (2004) Facilities and systems for transport of encapsulated spent fuel to a final repository in Sweden. Pack Transport Stor Secur Radioact Mater 15(2):113–119

EIA (Energy Information Administration) (2008) Nuclear power and the environment. US Department of Energy. http://www.eia.doe.gov/cneaf/nuclear/page/nuclearenvissues.html. Accessed 11 Apr 2008

EPRI (Electric Power Research Institute) (2004) Spent nuclear fuel transportation: an overview. Final report TR-1009226. Electric Power Research Institute, Palo Alto

ETP-ZEP (European Technology Platform for Zero Emission Fossil Fuel Power Plants) (2006) ZEFFPP–WG5 public acceptance and communication. Contribution to the strategic deployment document. Draft report. http://www.zero-emissionplatform.eu/website/library/index.html. Accessed 20 Mar 2008

Ewing RC (2006) The nuclear fuel cycle: a role for mineralogy and geochemistry. Elements 2:331–334

Farris CB (1983) Unusual design factors for supercritical CO_2 pipelines. Energy Prog 3(3):150–158

FERC (Federal Energy Regulatory Commission) (2008) Testimony of the Honorable Joseph T. Kelliher, Chairman of the Federal Energy Regulatory Commission, before the Committee on Energy and Natural Resources, United States Senate, 31 Jan. http://energy.senate.gov/public/_files/KelliherTestimony013108.pdf

Gale J (2009) Impure thoughts. Int J Greenhouse Gas Control 3:1–2

Gale J, Davison J (2004) Transmission of CO_2 – safety and economic considerations. In: Freund P, Kaya Y, Lior N (eds) Proceedings of the 6th International Conference on Greenhouse Gas Control Technologies. Energy 29:1319–1328

Gale J, Bradshaw J, Chen Z, Garg A, Gomez D, Rogner H-H, Simbeck D, Williams R (2005) Sources of CO_2. In: Metz B, Davidson O, de Coninck HC, Loos M, Meyer LA (eds) IPCC Special Report on Carbon Dioxide Capture and Storage. Cambridge University Press, Cambridge/New York, pp 75–103

Garwin RL (2008) Nuclear power in the world's energy future. Felice Ippolito Lecture, Rome, Italy, 22 May. http://www.fas.org/rlg/Ippolito_Lecture_3-1.pdf. Accessed 27 Mar 2009

Gough C, Taylor I, Shackley S (2002) Burying carbon under the sea: an initial exploration of public opinion. Energy Environ 13(6):883–900

Gozalpour F, Ren SR, Tohidi B (2005) CO_2 EOR and storage in oil reservoirs. Oil Gas Sci Technol 60(3):537–546

Green L (2004) The industry commitment to global transport safety standards. Pack Transport Stor Secur Radioact Mater 15(3–4):181–184

Green L (2007) Emerging issues in radioactive materials transport. Opening plenary presentation. 15th International Symposium on the Packaging and Transportation of Radioactive Materials (PATRAM 2007), Miami, 21–26 Oct. http://www.wnti.co.uk/UserFiles/File/public/publications/speeches/2007/lg_plenary.pdf

Hansen H, Eiken A, Aasum TA (2005) Tracing the path of carbon dioxide from a gas-condensate reservoir, through an amine plant and back into a subsurface acquifer. In: Proceedings of Offshore Europe 2005: Oil & Gas Exhibition & Conference. Paper SPE-96742-PP. Society of Petroleum Engineers, Richardson. CD-ROM

Haugen HA, Eldrup N, Bernstone C, Liljemark S, Pettersson H, Noer M, Holland J, Nilsson PA, Hegerland G, Pande JO (2009) Options for transporting CO_2 from coal fired power plants Case Denmark. In: Gale J, Herzog H, Braitsch J (eds) Proceedings of the 9th International Conference on Greenhouse Gas Control Technologies. Energy Procedia 1:1665–1672

Hegerland G, Jørgensen T, Pande JO (2004) Liquefaction and handling of large amounts of CO_2 for EOR. In: Wilson M, Morris T, Gale JJ, Thambimuthu K (eds) Proceedings of the 7th International Conference on Greenhouse Gas Technologies, vol 2. Elsevier, Oxford, pp 2541–2544

Heggum G, Weydahl T, Roald W, Mølnvik M, Austegard A (2005) CO_2 conditioning and transportation. In: Benson SM (ed) Carbon Dioxide Capture for Storage in Deep Geologic Formations–Results from the CO_2 Capture Project, vol 2, Geologic Storage of Carbon Dioxide with Monitoring and Verification. Elsevier, Amsterdam, pp 925–936

Hendriks C, Mace MJ, Coenraads R (2005) Impacts of EU and international law on the implementation of carbon capture and geological storage in the European Union. Report ECS04057. Ecofys, Utrecht, and FIELD, London

Huerta M (1981) Analysis, scale modeling, and full-scale testing of a railcar and spent-nuclear-fuel shipping cask in a high-velocity impact against a rigid barrier. Report SAND78-0458. Sandia National Laboratories, Albuquerque

IAEA (International Atomic Energy Agency) (1961) Regulations for the safe transport of radioactive materials. Safety Series No. 6. IAEA, Vienna

IAEA (International Atomic Energy Agency) (2004) Safety of transport of radioactive material. In: Proceedings of an International Conference, Vienna, 7–11 July 2003. Proceedings Series. IAEA, Vienna

IAEA (International Atomic Energy Agency) (2005) Regulations for the safe transport of radioactive material. IAEA Safety Standards Series No. TS-R-1. IAEA, Vienna

IAEA (International Atomic Energy Agency) (2008) Security in the transport of radioactive material, vol 9, IAEA Nuclear Security. IAEA, Vienna

IEA GHG (International Energy Agency Greenhouse Gas R&D Programme) (2005) IEA GHG Weyburn CO_2 monitoring and storage project. IEA Greenhouse Gas R&D Programme, Cheltenham

IEA GHG (International Energy Agency Greenhouse Gas R&D Programme) (2007) Environmental assessment for CO_2 capture and storage. IEA Greenhouse Gas R&D Programme, Cheltenham

Itaoka KA, Saito A, Akai M (2004) Public acceptance of CO_2 capture and storage technology: a survey of public opinion to explore influential factors. In: Rubin ES, Keith DW, Gilboy CF (eds) Proceedings of the 7th International Conference on Greenhouse Gas Technologies, vol 1. Elsevier, Oxford, pp 1011–1019

Jefferson RM, Yoshimura HR (1978) Crash testing of nuclear fuel containers. Report SAND77-1462. Sandia National Laboratories, Albuquerque

Kerr T, Havercroft I, Dixon T (2009) Legal and regulatory developments associated with carbon dioxide capture and storage: a global update. In: Gale J, Herzog H, Braitsch J (eds) Proceedings of the 9th International Conference on Greenhouse Gas Control Technologies. Energy Procedia 1:4395–4402

Koornneef J, Spruijt M, Molag M, Ramirez A, Faaija A, Turkenburg W (2009) Uncertainties in risk assessment of CO_2 pipelines. In: Gale J, Herzog H, Braitsch J (eds) Proceedings of the 9th International Conference on Greenhouse Gas Control Technologies. Energy Procedia 1:1587–1594

Kruse H, Tekiela M (1995) Calculating the consequences of a CO_2-pipeline rupture. Energy Convers Manage 37(6–8):1013–1018

Kvamsdal HM, Jordal K, Bolland O (2007) A quantitative comparison of gas turbine cycles with CO_2 capture. Energy 32:10–24

Lemmon EW, McLinden MO, Friend DG (2005) Thermophysical properties of fluid systems. In: Linstrom PJ, Mallard WG (eds) NIST Chemistry WebBook, NIST Standard Reference Database Number 69. National Institute of Standards and Technology. http://webbook.nist.gov. Accessed 12 Mar 2008

Mace MJ, Hendriks C, Coenraads R (2007) Regulatory challenges to the implementation of carbon capture and geological storage within the European Union under EU and international law. In: Gale J, Bolland O (eds) 8th International Conference on Greenhouse Gas Control Technologies–GHGT-8. Int J Greenhouse Gas Control – 1(2):253–260

Maldal T, Tappel IM (2004) CO_2 underground storage for Snøhvit gas field development. In: Freund P, Kaya Y, Lior N (eds) Proceedings of the 6th International Conference on Greenhouse Gas Control Technologies. Energy 29:1403–1411

McClure JD (1997) Historical safety performance of spent fuel transportation. Presentation at the Meeting of the Panel on the Waste Management System, US Nuclear Waste Technical Review Board, 19 Nov

Morgan-Warren EJ (2003) Radioactive materials transport: a summary of the institution of nuclear engineers' conference in 2002. In: Safety of transport of radioactive material. Proceedings of an International Conference, Vienna, 7–11 July 2003. Proceedings Series. IAEA, Vienna, pp 59–67

National Research Council (2001) Disposition of High-Level Waste and Spent Nuclear Fuel: The Continuing Societal and Technical Challenges. The National Academies Press, Washington, DC

National Research Council (2006) Going the Distance? The Safe Transport of Spent Nuclear Fuel and High-Level Radioactive Waste in the United States. The National Academies Press, Washington, DC

NEA (Nuclear Energy Agency) (1994) The Economics of the Nuclear Fuel Cycle. OECD Publishing, Paris

NEA (Nuclear Energy Agency) (2008) Moving forward with geological disposal of radioactive waste. A collective statement by the NEA Radioactive Waste Management Committee (RWMC). Report NEA No. 6433. OECD Nuclear Energy Agency, Paris

NIOSH (US National Institute for Occupational Safety and Health) (1995) Documentation for immediately dangerous to life or health concentrations (IDLH): NIOSH chemical listing and documentation of revised IDLH values (as of 3/1/95). http://www.cdc.gov/niosh/idlh/intridl4.html. Accessed 14 Feb 2008

Ozaki M, Davison J, Minamiura J (2004) Marine transportation of CO_2. In: Wilson M, Morris T, Gale JJ, Thambimuthu K (eds) Proceedings of the 7th International Conference on Greenhouse Gas Technologies, vol 2. Elsevier, Oxford, pp 2535–2539

Palmgren CR, Morgan MG, de Bruin WB, Keith DW (2004) Initial public perceptions of deep geological and oceanic disposal of CO_2. Environ Sci Technol 38(24):6441–6450

Petitjean V, Fillet C, Boen R, Veyer C, Flament T (2002) Development of vitrification process and glass formulation for nuclear waste conditioning. In: Proceedings of the Waste Management 2002 Symposium. WM Symposia Tucoon. http://www.wmsym.org/archives/2002/Proceedings/18/39.pdf

Pope RB (2004) Historical background to the development of various requirements in the international regulations for the safe packaging and transport of radioactive material. Pack Transport Stor Secur Radioact Mater 15(1):5–13

Pope RB, Luna RE (2004) The development of recommended requirements for security in the transport of radioactive material: a status report. In: Safety of transport of radioactive material. Proceedings of an International Conference, Vienna, 7–11 July 2003. Proceedings Series. IAEA, Vienna, pp 115–130

Pope RB, Bernard-Bruls X, Brittinger M (2001) A worldwide assessment of the transport of irradiated nuclear fuel and high-level waste. Paper Presented at the 13th International Symposium on Packaging and Transportation of Radioactive Materials (PATRAM 2001), Chicago, 3–7 Sept

Race J (2006) Designing a safe carbon dioxide transport infrastructure for the UK. Presentation to the Carbon Capture and Storage Association (CCSA) Meeting on Carbon Capture and Storage—Health, Safety and Environment Issues, London, 24 July

Rhoads RE, Franklin AL, Lavender JC (1986) Evaluation of methods to compare consequences from hazardous material transportation accidents. Report SAND86-7117. Sandia National Laboratories, Albuquerque

Seiersten M, Kongshaug KO (2005) Materials selection for capture, compression, transport and injection of CO_2. In: Benson SM (ed) Carbon Dioxide Capture for Storage in Deep Geologic Formations–Results from the CO_2 Capture Project, vol 2, Geologic Storage of Carbon Dioxide with Monitoring and Verification. Elsevier, Amsterdam, pp 937–953

Shackley S, McLachlan C, Gough C (2005) The public perception of carbon dioxide capture and storage in the UK: results from focus groups and a survey. Climate Policy 4(4):377–398

Shaw KB, Hughes JS, Ringot C, Schwarz G, Francois Y, Schmitz BM et al (2001) Notification of incidents/accidents during transport: final report. National Radiological Protection Board, Chilton

Skovholt O (1993) CO_2 transportation systems. Energy Convers Manage 34(9–11):1095–1103

Smith LA, Gupta N, Sass BM, Bubenik TA (2002) Engineering and economic assessment of carbon dioxide sequestration in saline formations. J Energy Environ Res 2:5–22

Statoil (2007) Snøhvit–a true adventure. http://www.statoil.com/STATOILCOM/snohvit/svg02699.nsf?OpenDatabase&lang=en. Accessed 28 Feb 2008

Steeneveldt R, Berger B, Torp TA (2006) CO_2 capture and storage: closing the knowing-doing gap. Chem Eng Res Des 84(A9):739–763

Svensson R, Odenberger M, Johnsson F, Strömberg L (2004) Transportation systems for CO_2–application to carbon capture and storage. Energy Convers Manage 45:2343–2353

Tang YS, Saling JH (1990) Radioactive Waste Management. Taylor and Francis, Basingstoke

Thambimuthu K, Soltanieh KM, Abanades JC, Allam R, Bolland O, Davison J et al (2005) Capture of CO_2. In: Metz B, Davidson O, de Coninck HC, Loos M, Meyer LA (eds) IPCC Special Report on Carbon Dioxide Capture and Storage. Cambridge University Press, Cambridge/New York, pp 105–178

Turner R, Hardy N, Hooper B (2006) Quantifying the risks associated with a CO_2 sequestration pipeline: a methodology and case study. In: Gale JJ, Røkke N, Zweigel P, Svenson H (eds) Proceedings of the 8th International Conference on Greenhouse Gas Technologies. Oxford, Elsevier, CD-ROM

University of Chicago (2004) The Economic Future of Nuclear Power. University of Chicago, IL

UNSCEAR (United Nations Scientific Committee on the Effects of Atomic Radiation) (2008) Effects of ionizing radiation: UNSCEAR 2006 report, vol 1: report to the general assembly, scientific annexes A and B. United Nations Office at Vienna, Vienna

UN/SCETDG (United Nations Committee of Experts on the Transport of Dangerous Goods) (2001) Recommendations on the transport of dangerous goods: model regulations, 14th revised edn. Report ST/SG/AC.10/1/Rev.14. United Nations, New York

US DOE (United States Department of Energy) (2003) Comparison of selected highway and railway accidents to the 10CFR71 hypothetical accident sequence and NRC risk. US Department of Energy National Transportation Program, Albuquerque

US DOT (United States Department of Transportation) (2008a) Written statement of Krista L. Edwards, Deputy Administrator, Pipeline and Hazardous Materials Safety Administration, before the Committee on Energy and Natural Resources, United States Senate, 31 Jan. Pipeline and Hazardous Materials Safety Administration. http://www.phmsa.dot.gov/media/testimony. Accessed 13 Feb 2008

US DOT (United States Department of Transportation) (2008b) Data & statistics. Office of pipeline safety, Pipeline and Hazardous Materials Safety Administration. http://www.phmsa.dot.gov/pipeline/library/data-stats. Accessed 12 Feb 2008

US EPA (United States Environmental Protection Agency) (2008) Testimony of Benjamin H. Grumbles, Assistant Administrator for Water, U.S. Environmental Protection Agency, before the Committee on Energy and Natural Resources, United States Senate, 31 Jan. http://www.epa.gov/ow/speeches/1-31-08_EPA_Testimony_SENR_CSS.pdf

US NRC (United States Nuclear Regulatory Commission) (1987) Shipping container response to severe highway and railway accident conditions. Report NUREG/CR-4829, UCID-20733. Lawrence Livermore National Laboratory, Livermore

US NRC (United States Nuclear Regulatory Commission) (2000) Reexamination of spent fuel shipment risk estimates. Report NUREG/CR-6672, SAND2000-0234. Sandia National Laboratories, Albuquerque

US NRC (United States Nuclear Regulatory Commission) (2003) Package performance study test protocols–draft report for comment. NUREG-1768. US Nuclear Regulatory Commission, Washington, DC

van Alphen K, Hekkert MP, Turkenburg WC (2009) Comparing the development and deployment of carbon capture and storage technologies in Norway, the Netherlands, Australia, Canada and the United States–an innovation system perspective. In: Gale J, Herzog H, Braitsch J (eds) Proceedings of the 9th International Conference on Greenhouse Gas Control Technologies. Energy Procedia 1:4591–4599

Vendrig M, Spouge J, Bird A, Daycock J, Johnsen O (2003). Risk analysis of geological sequestration of carbon dioxide. DNV Consulting Report No. R246, DTI Report DTI/Pub URN 03/1320. Department of Trade and Industry, London

WNA (World Nuclear Association) (2008) Transport of radioactive materials. http://www.world-nuclear.org/info/inf20.html. Accessed 28 Feb 2009

Wright I, Ashworth P, Xin S, Di L, Yizhong Z, Liang X, Anderson J, Shackley S, Itaoka K, Wade S, Asamoah J, Reiner D (2007) Public perception of carbon dioxide capture and storage: prioritised assessment of issues and concerns. CO2 Capture Project, Anchorage

Young CN (2004) The role of national competent authorities in facilitating regulation of the transport of radioactive material. In: Safety of transport of radioactive material. Proceedings of an International Conference, Vienna, 7–11 July 2003. Proceedings Series. IAEA, Vienna, pp 93–102

Engineering Challenges in the Geological Disposal of Radioactive Waste and Carbon Dioxide

Jean-Pierre Tshibangu K. and Fanny Descamps

Abstract This chapter deals with engineering issues related to the geological disposal of radioactive waste and carbon dioxide. An overview of the methodology for tackling these challenges is given, starting from the understanding of the geological context and the rock characterization (in laboratory and in situ) to the design and construction of the repository. We recall first the fundamentals of porous media and the transport mechanisms of solutes and gas in geological formations. Then we describe the various steps in the engineering design of underground workings, from site investigation to long-term safety and performance assessment. The particular cases of radioactive waste and carbon dioxide disposal are developed independently. Finally, we compare both types of disposal from the engineering point of view and show that, even if obvious differences exist, some requirements are similar. It is therefore valuable to develop a comparative view of the two approaches in order to benefit from the experience acquired.

Keywords Natural barrier • Engineered barrier system • Transport in porous medium • Long-term sealing • Geological disposal

1 Introduction

The main engineering challenge involved in storing waste materials in geological formations is to develop technologies that are safe enough to protect public health and avoid pollution or contamination of potential future resources (potable water, energy resources). Disposal should then be designed so as to limit the migration of pollutants from the geological formations.

J-P Tshibangu K. (✉) and F. Descamps
Mining Engineering Department, University of Mons, rue du Joncquois 53, 7000 Mons, Belgium
e-mail: jean-pierre.tshibangu@umons.ac.be; fanny.descamps@umons.ac.be

F.L. Toth (ed.), *Geological Disposal of Carbon Dioxide and Radioactive Waste: A Comparative Assessment*, Advances in Global Change Research 44,
DOI 10.1007/978-90-481-8712-6_7,

The first step in addressing geological storage is to find geological formations with secure holes (or voids) to host the waste material. The hole concept can be understood either on a small scale, in terms of pores or cracks to allow the displacement of fluid waste, or on a bigger scale, in terms of cavities (natural or man-made). By secure, we mean that the voided zone should be surrounded by a barrier that is able to considerably slow down the migration of any pollutant beyond the boundaries of the targeted disposal reservoir.

Every geological formation has its own distinctive physical properties, and all will allow chemical transfer to a certain extent. Thus, after identification of formations that are able to host waste materials, a second step would be to perform laboratory and field tests to characterize the rock materials that make up the formation.

In a third step, the engineers must design the requisite technology to allow the disposal site to be accessed and the waste to be stored securely. In fact, construction techniques must be designed in such a way that secure openings can be built that permit no or very limited leakage over time. Depending on the depth and size of a future repository, the techniques will be developed using the approaches used in the mining or petroleum industries. In general, a radioactive waste (RW) repository would be dealt with in accordance with mining technologies and carbon dioxide (CO_2) disposal in accordance with deep-wellbore petroleum technologies.

To assess the performance of a RW repository on a very long-term basis, Gomit et al. (1997) carried out an extensive study in the framework of the EVEREST project, funded by the European Union. The study described and evaluated the impact of events that can affect the quality of the repository: phenomena of natural origin (variation of Earth orbital parameters, tectonics, diapirism and meteorite impact) and phenomena of human origin (non-detected features, sealing defects, inadvertent human intrusion, human-induced climate change, voluntary human intrusion and war). Comparable approaches are being developed, though in a less detailed manner, to address long-term security issues for CO_2 disposal. These combine mechanisms of structural and stratigraphic, residual, solubility, and mineral trapping (Benson et al. 2005).

2 Theoretical Issues Related to Fluid Solutes and Gas Transport in Geological Formations

2.1 *The Porous Medium*

Every geological formation can be considered to a certain degree as a porous medium and can therefore exhibit two essential characteristics: capacity for storage and transmissibility of fluids.

A porous medium contains voids or spaces that form the porosity. Two types of porosity can be distinguished: *primary or matrix porosity*, which generally refers to void spaces in sedimentary rocks that remain after sedimentation and compaction,

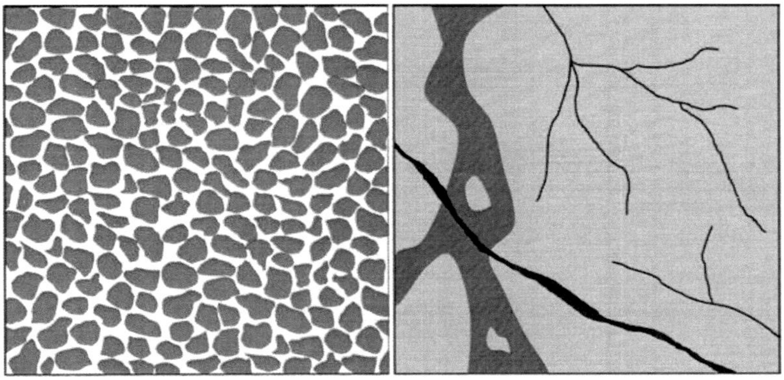

Fig. 1 Primary (matrix) and secondary (fractures) porosities

and *secondary porosity* which is due to fractures and other discontinuities in the material (Fig. 1). Because of past tectonic activities, most sedimentary rock formations exhibit both primary and secondary porosity.

Porosity (ϕ) is defined as a percentage or fraction of the void space with respect to the bulk volume of the rock. It is expressed as a percentage by:

$$\phi = \frac{V_v}{V} \cdot 100 \tag{1}$$

where V_v is the volume of voids and V the total volume.

If the porous medium is saturated by a fluid, usually only part of this fluid will flow through the medium. An effective porosity, ϕ_e, also known as the capacity to permit free flow, can then be defined by:

$$\phi_e = \frac{Volume\ free\ fluid}{Total\ volume} < \phi \tag{2}$$

ϕ_e depends on both the porosity and the grain fineness; the smaller the grain size, the smaller this quantity. For clays, effective porosity is very small compared with total porosity, whereas in sandstones, these two properties are very close in size.

When the porous medium contains more than one fluid, the saturation concept has to be defined. This is the case, for instance, when the pores contain a liquid phase, like water, and a gaseous one, like air.

Determination of porosity requires measurement of the total volume and either the pore volume or the matrix volume. The total volume of a rock sample can be measured by fluid displacement, while the pore volume can be measured according to different techniques, the most usual one being as follows: the rock sample is dried and weighed (W_d), and then saturated in brine (salt-saturated water) or another fluid, and then weighed again (W_{sat}). The connected porosity is then given by:

$$\phi = \left(\frac{W_{sat} - W_d}{\gamma_{fl} \cdot V_{sam}} \right) \cdot 100 \tag{3}$$

where γ_{fl} represents the unit weight of the injected fluid and V_{sam} is the volume of the sample.

2.2 *Transport Mechanisms in Porous Media*

Different mechanisms can be invoked to describe the transport of molecular species through porous media (Mody and Hale 1993; Horseman et al. 1996; Marivoet et al. 1997). Among them, we focus on the following two main mechanisms:

- *Hydraulic flow* or *advection*, the driving force of which is the hydraulic pressure difference and whose flow rate depends on the permeability of the porous medium;
- *Diffusion*, the driving force of which is the chemical potential or concentration difference of dissolved species contained in the pore fluid.

Other interesting mechanisms favouring the geological disposal of RW and/or CO_2 can be mentioned: retardation due to chemical sorption (reaction with minerals on the solid surface), dissolution into the formation fluid, mineralization, dispersion caused by formation heterogeneities, and buoyancy (due to the difference in density between the two fluids). Most of the mechanisms listed can be modelled mathematically using the same thermodynamics concepts presented in the subsections below (Marivoet et al. 1997; Benson et al. 2005).

2.2.1 Hydraulic Conductivity

When choosing a coordinate system such that the z axis is oriented in the direction of gravity g (i.e. downwards), and neglecting the effect of velocity (because of the very low kinetic energy involved), a fluid particle (water in the present case) having an ordinate z and a pressure p will have a hydraulic charge (h) defined by:

$$h = \frac{p}{\gamma_w} - z \tag{4}$$

where γ_w is the water unit weight.

The hydraulic charge represents a quantity that is proportional to the internal energy of a particle of mass M, and this is the main driving force in the flow of fluids through porous media. For deep reservoirs (i.e. more than 1,000 m), the hydraulic charge is given mainly by the pressure term.

The magnitude of the hydraulic flow is characterized by the permeability of the medium. The permeability of a rock is a measure of a specific flow capacity and

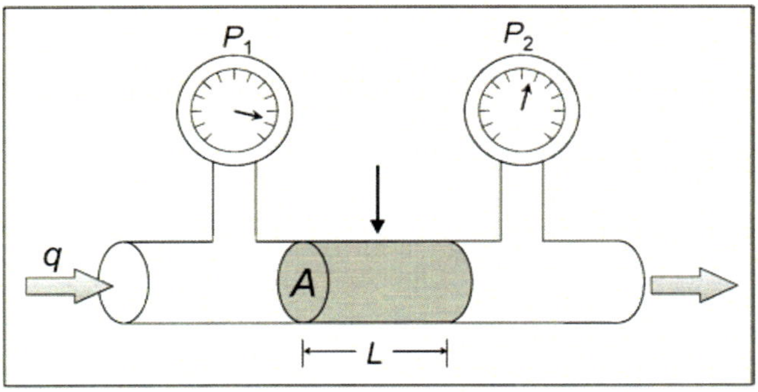

P_1 and P_2	Inlet and outlet fluid pressures
A	Cross section of the sample
L	Length of the sample
q	Flow rate

Fig. 2 Permeability test principle

can be determined only by a flow experiment. As permeability depends upon continuity of pore space, there is no unique relation between the porosity of a rock and its permeability.

The permeability can be expressed by Darcy's law as:

$$\vec{q} = -k\ \nabla p \tag{5}$$

where $\vec{q}$ is the flow rate (m³/s); k is the permeability (Darcy); p is the fluid pressure.

In this equation, the permeability depends on both the rock characteristics and the fluid viscosity. The two effects can be split and the parameter expressed independently with respect to the fluid.

Measurement of the permeability can be performed on a cylindrical sample by flushing it with water, gas (nitrogen or air) or another fluid, the viscosity of which is known. A differential pressure is applied on the two faces of the sample, and the flow rate is measured to assess the permeability (Fig. 2).

Multiphase Flow in Porous Media (Non-miscible Fluids in Saturated Media)

In oil reservoirs three different fluids can be displaced: water, oil and gas. The effective permeability for each fluid is derived from Darcy's law and is always lower than the overall true permeability of the medium, also known as absolute permeability (Dake 1978).

The *relative permeability* is defined as the ratio of effective permeability to absolute permeability. It depends on saturation and wettability (defined as the tendency of a fluid to displace another fluid from a solid surface). The relative permeability notion is very important when attempts are made to recover oil, for example by injecting CO_2.

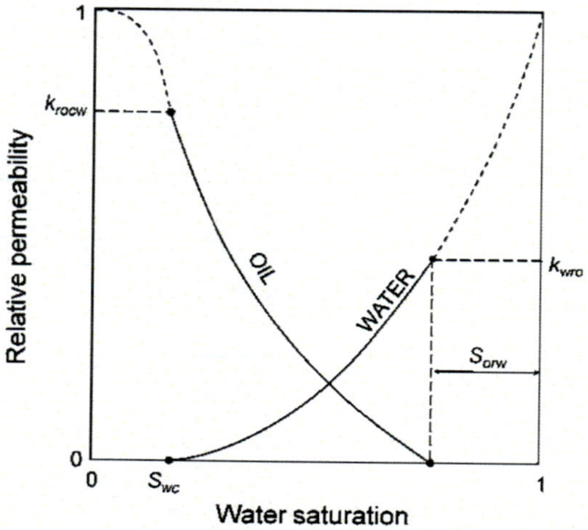

Fig. 3 Water–oil relative permeability curves in the case of oil recovery by water injection

Figure 3 shows the evolution of water–oil relative permeabilities versus water saturation. For a new petroleum reservoir (onset of production), the water saturation is minimum (S_{wc} is the connate or irreducible water saturation), and the relative permeability to oil is then maximum (k_{rocw}). In the course of production life, the water saturation increases, thereby increasing the relative permeability to water and, hence, decreasing the relative permeability to oil. It is known that when the reservoir is tending to depletion, it produces more water than oil. When the oil saturation decreases to S_{orw}, the residual saturation, there will be no more oil flow. When oil production is being enhanced by CO_2 injection and disposal, CO_2 will play the role of water, and the risk of recovering the gas to be stored from the production wells must be taken into account. Assessment must thus be made as accurately as possible of the instant at which CO_2 injection has to be stopped.

2.2.2 Fluid Diffusivity Law

To assess the movement of fluids in deformable solids, the mass conservation principle expressed by the continuity equation must be used. This relates the rate of flow of fluid into a small volume to the rate of increase of the amount of fluid in this volume (Jaeger and Cook 1979). Combining the continuity equation with the transport law (Darcy's law in this case) will lead to a diffusivity equation.

When working in great depth conditions, the movement of fluids will be mainly driven by pressure, and the permeability will depend on the deformation of the solid skeleton (Charlez 1991; Coussy 1991). In such conditions, if the fluid is assumed to be non-compressible, the diffusion equation can be derived as:

$$k\nabla^2 p = -\frac{1}{\rho_{fl}}\frac{\delta m}{\delta t} \tag{6}$$

where m is the variation of the amount of fluid by unit volume, p the pore pressure, t the time and ρ_{fl} the density of the fluid.

The use of this equation, coupled to the mechanical behaviour of the medium, can lead to an assessment of the evolution of the reservoir and the fluid transfer (or leakage) over time.

The diffusivity equation cannot be solved on its own because it contains two unknowns. When working in geomaterials, it has to be combined with constitutive laws like the thermo-poro-elasticity law to provide more equations. An example of state equations when the effect of temperature is neglected is given below:

$$[\sigma]=[\sigma_0]+\frac{E}{1+\nu}[\varepsilon]+\frac{E\nu}{(1+\nu)(1-2\nu)}tr[\varepsilon]\cdot[1]+b(p-p_0)\cdot[1] \tag{7}$$

$$p=p_0+M\left(b\ \mathrm{tr}[\varepsilon]+\frac{m}{\rho_{fl0}}\right) \tag{8}$$

where $[\sigma]$ is the stress tensor (the subscript 0 is relative to the initial state of stresses); $[\varepsilon]$ is the strain tensor; $[1]$ is the unit tensor; $\mathrm{tr}[\varepsilon]=\varepsilon_1+\varepsilon_2+\varepsilon_3$ is the volumetric strain; m is the fluid mass increment; b and M are the Biot coefficient and modulus; p is the pore pressure (the subscript 0 relates to the initial pressure); E is the Young modulus; ν is the Poisson ratio; b, M, E and ν are known as the poro-elastic parameters of the porous medium.

The triaxial test with a pore pressure control is the most useful experimental system for determining the poro-elastic parameters. In fact, it is easy to measure the components of the strain tensor $[\varepsilon]$, which describes the deformation of the skeleton

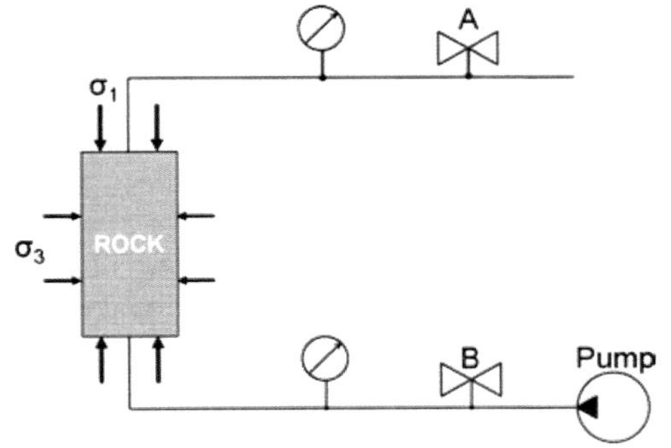

The rock sample is submitted to mechanical stresses σ_1 and σ_3; σ_1 is the major principal stress (*thick arrows*) whereas σ_3 is the minor principal or confining stress (*thin arrows*). The pore fluid is injected by means of a pump. Two valves (*A and B*) allow drained or undrained experiments. Pressure sensors give the inlet and outlet pore pressure.

Fig. 4 Triaxial test principle with pore pressure control of the rock sample, to drive the components of the stress tensor [σ], and to a lesser extent the variation of the pore pressure ($p - p_0$). The sketch in Fig. 4 shows the principle of a triaxial test that enables both drained ($p = p_0$) and undrained ($m = 0$) experiments.

2.2.3 Generalization of the Transport Theory

When dealing with diffusion mechanisms, the transport equation can be generalized as follows (Sherwood 1993; Tshibangu et al. 1996):

$$q_i^r = -\sum_s L_{ij}^{rs} \nabla_j C^s \tag{9}$$

where q_i^r is the mass flux of the r-th ionic species in direction i; C^s is the concentration of species s; L_{ij}^{rs} is the diffusion coefficient of the ionic species r in the presence of species s with a concentration C^s.

The mass conservation of ionic species r can be written as:

$$\nabla \cdot q^r = -\frac{\partial m^r}{\partial t} \tag{10}$$

Combining this continuity equation with the transport law (9) will give the classical diffusion equation. When considering a one-dimensional problem (direction x) in which only the own concentration gradient of species r is taken into account (Put and Henrion 1988), the following simplified diffusion equation can be derived:

$$\frac{\partial^2 C^r}{\partial x^2} = \frac{1}{L^r} \frac{\partial C^r}{\partial t} \tag{11}$$

L^r is the apparent diffusion coefficient for the ionic species considered. It depends on the specific conditions of the experiments. Put and Henrion (1988) define such a coefficient as being dependent on the diffusion coefficient in the liquid and the retardation factor to be applied to radionuclide diffusion mechanisms.

If a relationship can be established between the concentration and the mechanical behaviour of the solid skeleton, then the generalized poro-elasticity law can be written as:

$$d\varepsilon_{ij} = S_{ijkl} d\sigma_{kl} + \sum_r Q_{ij}^r \, d\mu^r \tag{12}$$

$$dm^r = Q_{ij}^r d\sigma_{ij} + \sum_s B^{rs} d\mu^s \tag{13}$$

where μ^r is the chemical potential of species r; m^r is the mass of species r per unit volume; ε_{ij} is the strain (or deformation) tensor; σ_{ij} is the stress tensor; S_{ijkl} is the matrix containing elastic properties (Young's modulus, Poisson's ratio, shearing modulus, etc.); Q^r_{ij} and B^{rs} are parameters to be determined by specific experiments in which the strains of the solid or mass of a given species can be measured with respect to variation of the chemical potential.

Prior to designing the geological disposal, a preliminary study of the transport mechanisms of solutes and gas in the porous media will be critical to the choice of potential sites. In fact, as stated in Sect. 3, collecting rock samples from the field and performing typical experiments will allow the identification of the most relevant transport mechanisms and assessment of the physical parameters needed. The RW repository study, on the one hand, will need a good knowledge of the natural barrier constituted by the host formation: porosity, permeability, and thermo-hydro-mechanical parameters (i.e. the poro-elastic parameters described in Eqs. 12 and 13). The repository study of CO_2, on the other hand, will address not only the issue of the barrier concept (caprock formation), but also deal with the injection capacity in the potential reservoir (mainly driven by the pressure gradient). When CO_2 is stored in coal formations, for example, Darcy's law can be used to assess the volume to be injected and the generalized poro-elastic equations to deal with matrix deformation (swelling) due to the adsorption phenomenon that is driving the volume to be stored.

When a sufficient amount of data is collected from field and laboratory, databases and 3-D geological models can be built to enable the design of suitable techniques for underground disposal.

3 Designing and Building Underground Openings

According to Bieniawski (1992), engineering design is the process of devising a system, component, or process to meet desired needs. It is a decision making process (often iterative), in which the basic sciences, mathematics, and engineering sciences are applied to convert resources optimally to meet a stated objective. Among the fundamental elements of the design process are the establishment of objectives and criteria, synthesis, analysis, construction, testing and evaluation. Central to the process are the essential and complementary roles of analysis and synthesis. In addition, sociological, economic, aesthetic, legal and ethical considerations need to be included in the design process.

The engineering work to design and build underground facilities can be summarized in the following main steps:

- Site investigation;
- Laboratory characterization;
- Rock mass characterization;
- In situ and field tests;
- Modelling the behaviour of the planned underground openings;

- Construction;
- Monitoring during construction and use;
- Operation;
- Closure and post-closure monitoring;
- Long-term safety analyses and performance assessment.

3.1 Site Investigation and Laboratory Characterization

Depending on the geological information available, this step can start with field visual observations. Geological maps must first be consulted. In a further approach, geophysical studies can be undertaken to ascertain the geometry of geological formations underground, and samples can be collected for laboratory tests (Brown 1981). The most common sampling method is to drill to collect cores (coring) or cuttings (destructive drill bits) of rocks.

The samples collected can be submitted to various tests depending on the intended use of the future underground opening: petrographic analysis, physical properties (porosity, permeability, density), mechanical properties such as, for instance, the poro-elastic properties described in Sect. 2 (Young's modulus, Poisson's ratio, shearing modulus, bulk compressibility modulus, Biot's coefficient and modulus, etc.), and the rock failure mechanisms with the associated parameters (cohesion, friction angle, pore collapse strength, etc.).

3.2 Rock Mass Characterization

The physical parameters measured in the laboratory should be scaled up so that they are applicable to a large volume of rock in accordance with the size of the structure to be developed. Structural analysis, assessment of the quality of the rock mass, and evaluation of the mechanical properties of the rock mass all need to be performed.

The structural analysis is intended to identify discontinuities in terms of type (fault, fracture, joints, bedding planes), orientation (dip and direction), frequency, quality of filling materials (rough surfaces in contact or joints filled with soft gouge materials), and presence of water.

To qualify the rock mass, different indices have been developed like the Rock Quality Designation (RQD) (Deere 1963), Rock Mass Rating (RMR) (Bieniawski 1984), the Geological Strength Index (GSI) (Hoek and Brown 1998) and Barton's Q-index (Barton et al. 1977). These indices use structural data collected from cores or outcrops and combine with some typical rock strength parameters like the unconfined compressive strength (UCS) to give a numerical value of the quality of the rock mass. The RMR method, for instance, uses five parameters (the UCS, the

RQD, the spacing of the joints, the nature of the joints, and the water inflows/seepage) to all of which a score is attributed. By adding the five scores a characterization of the quality of the rock mass can be reached; this amount can be corrected to take into account the direction of fractures with respect to the orientation of the future opening (i.e. a tunnel).

The quality indices can also be used to assess the mechanical properties of a rock mass: strength, deformability and risk of failure. For instance, the Hoek-Brown failure criterion given in Eq. 14 is intended to assess the strength of a rock mass based on the assessed GSI index:

$$\sigma_1 = \sigma_3 + \sigma_{ci}\left(m_b \frac{\sigma_3}{\sigma_{ci}} + s\right)^a \qquad (14)$$

where σ_1, σ_3 are the major and minor principal stresses; σ_{ci} is the unconfined strength of an intact rock sample; m_b is the Hoek-Brown constant for the rock mass; s and a are constants depending on the rock mass quality ($s=0$ for an aggregate and 1 for laboratory tests on intact rock samples).

The quality indices will be assessed more efficiently if databases are built that can be manipulated by numerical modelling software codes to allow 3-D geological models to be built. Modern mining or petroleum reservoir codes enable data from different sources to be used: cores, outcrops, faces of workings, results of mechanical tests, etc. Figure 5 gives an example of the description of a cored well and a geological model that can be built with data collected from many boreholes.

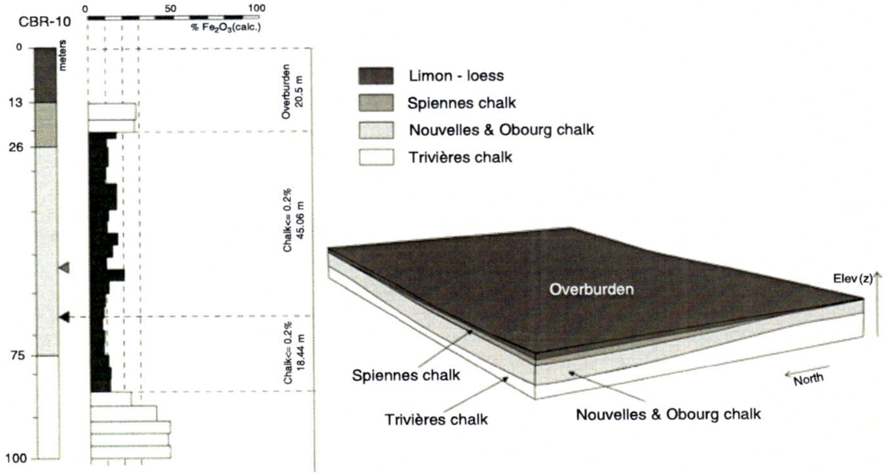

Fig. 5 A description of a cored well in terms of chemical composition and a simple geological model built with data from cores and essays.

3.3 From Modelling to Construction and Monitoring

After the collection of physical and geomechanical data and the building of geological models, the shape and size of specific underground structures or openings need to be designed, depending on what objectives are being pursued. If underground cavities are to be created for disposal purposes, then such cavities will be expected to remain open for the whole life of the operations. To predict the stability of underground openings, analytical and/or numerical modelling are used.

Underground openings can be of various shapes and types, and can be isolated or close to each other; thus a stress disturbance in a point of the rock mass situated in the neighbourhood of two openings can have effects that are superposed. Engineering practice distinguishes the following underground workings:

- Mining galleries and tunnels;
- Mining shafts;
- Mining working faces (areas in which the ore is being mined out);
- Wells for fluid extraction and/or injection;
- Large underground spaces (for example, space for a primary crusher in the mine, artificial cavities for storage of hydrocarbons, etc.).

The cross section of these openings can be circular, elliptical, rectangular, etc.

When dealing with the stability of underground openings, the equilibrium of a given opening has to be assessed over time. The equilibrium of solid bodies is governed by equilibrium equations obtained by balancing the forces acting on an infinitesimal element of the body (Jaeger and Cook 1979).

When considering a rectangular coordinate system *Oxyz*, an infinitesimal element can be represented as shown in Fig. 6. The six faces of the element are submitted to normal (σ_i) and tangential stresses (τ_{ij}), with subscripts being related to directions x, y and z. Figure 6 also describes the variation of stresses in the element for a given axis; this is expressed in terms of partial derivatives. Finally, the equilibrium equations are expressed as partial derivative equations, which are the most commonly used type in engineering problems:

$$\frac{\partial \sigma_x}{\partial x} + \frac{\partial \tau_{yx}}{\partial y} + \frac{\partial \tau_{zx}}{\partial z} + \rho X = 0 \tag{15}$$

$$\frac{\partial \tau_{xy}}{\partial x} + \frac{\partial \sigma_y}{\partial y} + \frac{\partial \tau_{zy}}{\partial z} + \rho Y = 0 \tag{16}$$

$$\frac{\partial \tau_{xz}}{\partial x} + \frac{\partial \tau_{yz}}{\partial y} + \frac{\partial \sigma_z}{\partial z} + \rho Z = 0 \tag{17}$$

where ρ is the density; X, Y and Z are the components of body forces per unit volume (in this case, we will consider only gravity forces); σ_x, σ_y and σ_z are the normal stresses acting on sides perpendicular to axis x, y and z respectively; τ_{xy}, τ_{xz} and τ_{yz} are the tangential stresses.

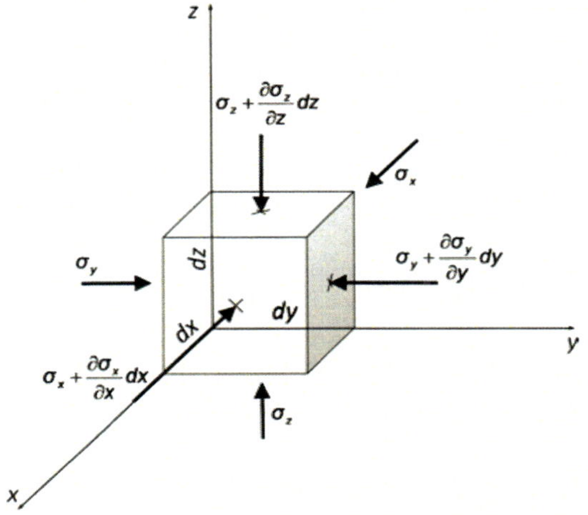

Only normal stresses are shown to ensure legibility. The arrows represent the normal stresses on each face of the cube. For instance, in the x-direction, the normal stress is σ_x for x=0.

Fig. 6 Equilibrium of an infinitesimal element (dx, dy, dz) in a Cartesian coordinate system

These are three equations with six unknowns (three components of normal stresses and three components of tangential stresses).

By assuming different behaviour laws for the material composing the rock mass, as described in Sect. 2, the stresses can be expressed in terms of the strains and the equilibrium equations then written in terms of strain or displacements.

Equilibrium equations have to be satisfied over the entire body under consideration, and also on the boundaries. In the latter case, the stresses have to balance the external forces applied to the body. This condition can be expressed in two dimensions by:

$$\overline{X} = l\sigma_x + m\tau_{xy} \tag{18}$$

$$\overline{Y} = m\sigma_y + l\tau_{xy} \tag{19}$$

where $\overline{X}$ and $\overline{Y}$ are applied surface forces in directions x and y respectively; l and m are the direction parameters of the boundary linear element.

As stated earlier, a knowledge of equilibrium equations is inadequate for assessing the fields of stresses, strains, displacements, pore pressure, temperature and state of failure over the entire body and for modelling them. It is therefore necessary to look for additional equations (or constitutive laws) by setting typical assumptions on the behaviour of the geomaterial: elastic isotropic, poro-elastic, visco-elastic, perfectly plastic failure, etc. By combining the constitutive laws with equilibrium equations and boundary conditions, enough relationships are derived to solve the problem and an analytical or numerical approach can be used to assess the field variation of interesting variables. The so-called physical methods, which consist of building reduced-size models, have been used intensively in the past but are now of less importance because of the development of computers and numerical software codes.

The *analytical method* is used in homogeneous and isotropic media for simple geometrical shapes such as, for example, circular tunnels or galleries (Fig. 7). To a lesser extent, elliptical and rectangular openings can also be evaluated in this way. The general working method is built by combining equilibrium equations with elastic constitutive laws and typical boundary conditions. The final partial derivative system of equations (generally expressed in terms of displacements) is integrated, and integration constants are assessed using the boundary conditions. The solution is given in terms of simple formulae describing the variation of state variables (for example, stress, strain and displacement components) with respect to spatial coordinates and, sometimes, to time. Figure 7 gives an illustration of the variation of principal stresses (σ_θ and σ_r) in a polar coordinate system for a circular cavity with an isotropic natural stress σ_0 at infinity (Bouvart et al. 1988).

For *numerical methods*, different and widespread approaches exist: the finite element method (FEM), the boundary element method (BEM), the finite difference method (FDM) using Lagrangian elements, and the distinct element method (DEM). These methods allow more complex shapes of underground cavities that can be dug in complex geology environments to be studied. They generally consist

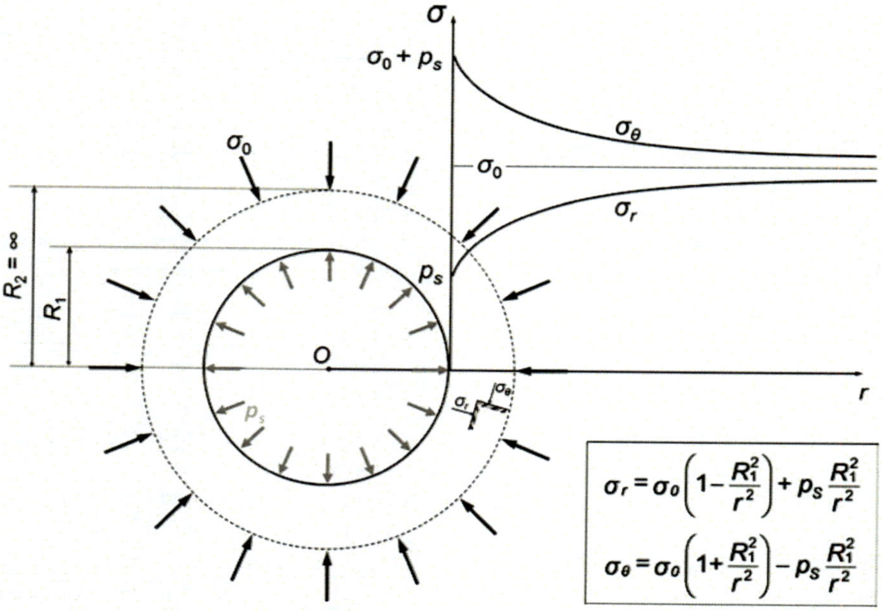

R_1 is the radius of the tunnel. σ_0 represents a lithostatic stress acting at an infinite radius R_2 whereas a supporting pressure p_s is applied on the wall. The evolution of the stresses versus the distance r from the tunnel centre shows a big difference between the radial stress σ_r and the tangential stress σ_0; for a so-called infinite distance the two stresses tend to reach the magnitude of the virgin rock stress $\sigma_{0.}$

Fig. 7 Example of a circular tunnel model in a homogeneous and isotropic medium of a subdivision of the studied model into different elements (meshing) with given shapes from which the unknown variables can be assessed. This is done by replacing the continuous function of the spatial variables (i.e. stresses, strains, displacements, etc.) by discrete approximations. This transforms continuous partial differential equations into discrete algebraic equations that can be solved by numerical computing methods. Figure 8 gives an example of a finite difference numerical model showing the distribution of pore pressure (a) and failed material or damaged (plastic) zone (b) for the bottom of the second shaft of the Mol research facility (Vereycken 2000). The technological development of computers during the last decades has brought a tremendous development in numerical computing methods. In fact, the software packages are designed so that big models can be run efficiently on personal computers.

Depending on the mechanical quality of the geological material, the excavation being designed can be self-supporting, or fail because of high induced stresses. To avoid the failure of underground openings, engineers need to design supporting structures and/or linings to ensure long-term stability. In the case of waste repositories, the structures also have to avoid or limit the leakage of the pollutants into aquifers. This means that supports need to be strong enough to balance the deformation of the rock mass and tight enough to limit the transfer of pollutants over the course of time. The mechanical characteristics of relevant supporting systems (concrete, steel arches, timber, etc.) can be used in numerical models to assess a new equilibrium of the excavation. To check the stability of the openings various failure criteria or strength envelopes that can be expressed in terms of stress functions then need to be assumed.

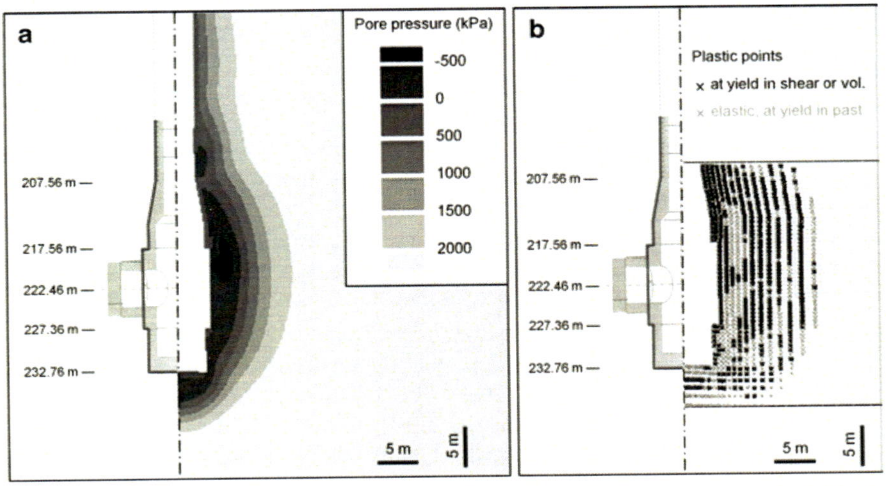

Fig. 8 Example of a finite difference numerical model showing the distribution of pore pressures (**a**) and failed material or damaged (plastic) zone (**b**) for the bottom of the second shaft of the Mol (Belgium) research facility (Vereycken 2000)

After construction, an underground opening has to be monitored, either visually or with the help of instruments. In engineering, structural monitoring may be carried out for different reasons, of which the two main ones are mentioned here (Brady and Brown 1999): (1) to ensure safety during construction and operation by giving warning of, for example, the development of excess ground deformations, groundwater pressures and loads in support elements; and (2) to check the validity of the assumptions, conceptual models and values of rock mass properties in design calculations. The monitoring measurements can be used to correct the mechanical parameters in the so-called back analysis.

Section 3 summarizes the working steps required to design, construct and use underground facilities. In the case of RW disposal, depending on the geomechanical properties of the targeted geological formation, the method described will be implemented with the objective of assessing the size, shape and support of cavities intended to receive the containers (canister or shroud). Combining geomechanical approaches with transport mechanisms (for instance, permeability will be modified in the plastic or damaged zone) will allow an evaluation of potential radionuclide migration in the rock mass and, hence, a sealing method to be designed accordingly. In the case of CO_2 disposal, the same working method can be applied to assess the stability of wells during both the drilling (calculation of the drilling fluid density) and the injection (calculation of casings and the production tubing) phases. The transport mechanisms will also be used to simulate the displacement of the injected gas in the reservoir and to check the sealing capacity of the caprock and the cemented well (see Branskill and Wilson 2011).

4 Disposing of Radioactive Waste in Underground Cavities

4.1 General Disposal Method

The waste material is deposited in a mine-like facility by moving the containers from the surface to underground. The disposal system must be based on the multi-barrier concept in which three subsystems can be considered (Marivoet et al. 1997): (1) the near field including the waste package, engineered barriers and the immediate part of the host rock that is significantly affected by the presence of the repository; (2) the far field (geosphere or natural barrier), including the host rock which surrounds the disposal system but which is not immediately affected by the presence of the cavity; and (3) the biosphere with the environment easily accessible by humans. This chapter focus mainly on the first two subsystems.

Storing RW, mainly high-level waste (HLW), will induce different phenomena in the near field, the physics of which has been invoked in Sects. 2 and 3 of this chapter: thermal processes, mechanical effects, chemical processes and radiological effects.

To fulfil the multi-barrier concept in both the near and far fields, selection of the geological host formation is critical. In Western Europe, three typical geological formations have been targeted.

- *Granite (France)*: found in massive rock formations, but always fractured, so that the issue of permeability and leakage must be very carefully addressed;
- *Clay (Belgium, France)*: generally impermeable and found in thick formations; but the thermal effects, especially in the near field must be addressed to avoid a thermo-hydro-mechanical coupling that could alter the isolation capability;
- *Salt (France, Germany, Netherlands)*: found in massive impermeable geological formations.

Figure 9 shows an example of the design of the multi-barrier concept for the repository of HLW in the Boom Clay formation (ONDRAF/NIRAS 2001, 2008). The engineered barrier system (EBS) must prevent the release of radionuclides for as long as possible. The period of time for which the EBS is designed depends, in fact, on the disposal concept: in the Belgian case, the EBS is intended to prevent the release of radionuclides during the thermal phase (i.e. only a few thousand years) but in other concepts (e.g. Sweden), the EBS plays a more important role and on a longer timescale. In the current design, it consists of a supercontainer placed in a gallery lined with wedge blocks that is sealed by a cementitious backfill. The supercontainer comprises a carbon steel overpack and a Portland cement concrete buffer, with or without an outer stainless steel envelope. The overpack encloses the canisters of HLW or the spent fuel assemblies and is designed to contain and prevent the release of RW during the thermal phase.

Figure 10 gives the layout of a schematic repository. This is composed of a network of galleries connected to at least two entrances (i.e. shafts or declines)

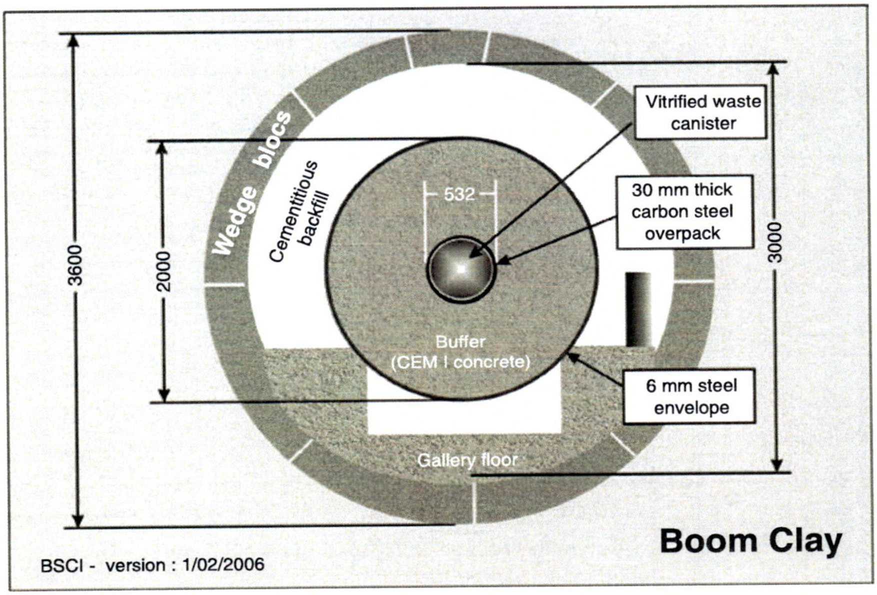

Fig. 9 The current reference concept for radioactive waste disposal in Belgium (© ONDRAF)

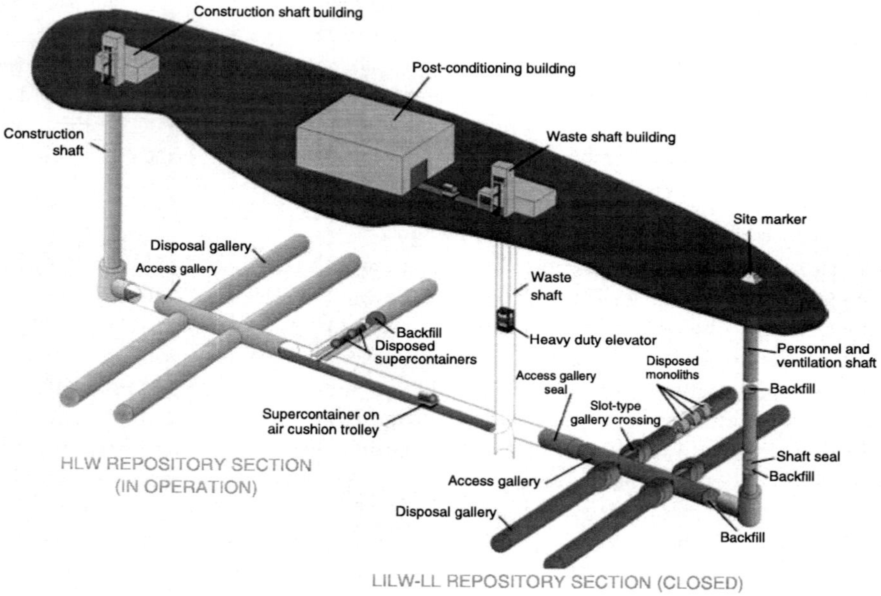

HLW: High-level waste
LILW-LL: Long-lived low- and intermediate-level waste

Fig. 10 Layout of the repository concept as it can be developed in the Boom Clay in Belgium (© ONDRAF)

to ensure ventilation of the underground facility and operational safety (personnel evacuation in case of emergency). HLW and intermediate-level waste (ILW) are stored in distinct areas. The dimension of the gallery will depend, among other things, on the diameter of the containers and the equipment used to handle them. All galleries and shafts will be backfilled in the closing phase of the repository.

4.2 Shaft Sinking and Gallery Digging

Accessing deep geological formations using mining methods can be performed in two ways:

- *By sinking a vertical or inclined shaft from the surface*: this structure needs to be a straight line and equipped with guides for the use of cages (elevators intended for men, equipment, and transport of broken rock material) or skips (buckets or containers used to handle broken rocks);
- *By digging a decline* (spirally inclined gallery) that can be used by road vehicles to access the deep galleries: conveyor belts can also be used to move broken rock material; the method is cheaper but is used mainly for shallow workings.

From the access structures (shaft or decline), near-horizontal galleries or rooms need to be developed to access the targeted areas. Pillars, whose dimensions can be assessed by mathematical modelling, are left between the galleries to ensure long-term stability. The digging method depends on the mechanical properties of the rock and the hydrological conditions:

- *Hard rocks–no water flow*: use of the classical drill and blast method or the mechanical method of tunnel boring machines (TBMs), but this latter method is cost-effective only for long tunnels;
- *Hard rocks–water flow*: use of cement grouting or other chemical to fill the cracks and faults before drilling and blasting; TBMs with compressed air or mud confinement;
- *Soft rocks–no water flow*: mechanical digging with use of open shield TBMs; the digging machine is a roadheader, back hoe, pneumatic or hydraulic hammer, etc.;
- *Soft rocks–water flow*: use of closed shield machines (TBMs generating confined space at the face with compressed air, mud pressure or mechanical support). The machine must be waterproof.

The drill and blast method in gallery digging is a cyclical method in which different operations follow each other in a repeating order. Each cycle should produce a certain length of excavated gallery or tunnel, a so-called round. The advance per round is usually 1–5 m depending on the characteristics of the rock mass. At a minimum, the following phases are included in a round:

- Drilling;
- Explosives charging;
- Blasting and ventilation;
- Scaling (removal of unstable rock pieces);
- Loading and hauling the blasted rock (mucking operations).

In addition to these, a supporting phase is normally needed, depending on the mechanical quality of the rock mass. This can comprise: rock bolting (use of steel rods), shotcreting (projection of concrete on the walls of the cavity), supporting arches and concrete lining.

A good knowledge of the ground conditions is required to estimate a schedule for a tunnelling project. Heavy immediate support, for instance, will lengthen the work cycle considerably. Figure 11 shows the typical operations included in a cycle of gallery digging.

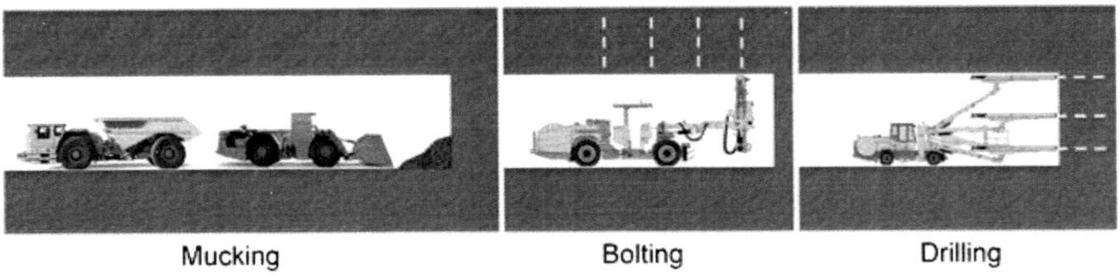

Fig. 11 The working cycle in the drill and blast method

Mechanized gallery digging can be performed either by continuous miners (road headers) or full face TBMs.

Continuous miners (also called road headers or point attack machines) are equipped with a rotating head with cutters or spikes to cut the rock. This type of machine is suitable for any cross sectional shape of the tunnel (circular, square, etc.). The technique is used only in soft to medium strength grounds. The abrasiveness of the rock is an important parameter in terms of addressing wear problems.

The *full face TBM* is composed of the boring machine itself followed by a trailer. The boring machine has a head (rotating or not) to cut circular tunnels by using cutting tools, the choice of which depends on the mechanical properties of the ground (strength, abrasiveness, water flow). The machine advances by use of gripping and pushing actuators. A mucking system is also included to remove the broken material from the face to the rear via a conveyor. The trailer carries all the technical equipment (compressor, support erecting systems, etc.).

Drilling and blasting are also cyclical operations in shaft sinking, as for horizontal openings. Drilling is performed by means of hand-operated pneumatic or hydraulic hammers. Sometimes the hammer can be secured on an upper platform and a mechanical pushing device can be used.

The mechanical shaft-sinking technique is used mainly for special working conditions like soft aquiferous ground. We indicate here two of various working methods: (1) the large- diameter boring machine system (up to 5–6 m) uses the drilling technique with the walls being supported during sinking by hydrostatic mud pressure; and (2) the pre-excavation ground freezing system. This second method uses a curtain of boreholes containing pipes in which brine refrigerated at –30°C is circulated.

5 Disposing of CO_2 by Injection from Deep Wellbores

5.1 *General Disposal Method*

Different mechanisms exist for storing CO_2 in geological formations (Benson et al. 2005): stratigraphic and structural physical trapping (below low permeability seals or caprocks), hydrodynamic physical trapping (fluids migrate very slowly over long distances, mainly in saline formations that do not have a closed trap), and geochemical trapping (solubility and mineralization).

One method for geological CO_2 storage is to drill wells and inject the gas in its supercritical state into permeable formations (reservoirs) situated at great depths (of at least 800–1,000 m to keep the CO_2 at the desired pressure). Different types of reservoirs can be used to meet the targeted conditions (Fig. 12): storing in depleted petroleum reservoirs, using the CO_2 pressure to improve the recovery of oil from producing fields (enhanced oil recovery or EOR); storing in deep saline aquifers; and storing in unmineable coal seams (with enhanced coalbed methane, or ECBM, production). Other possibilities like the use of abandoned mines or natural caverns

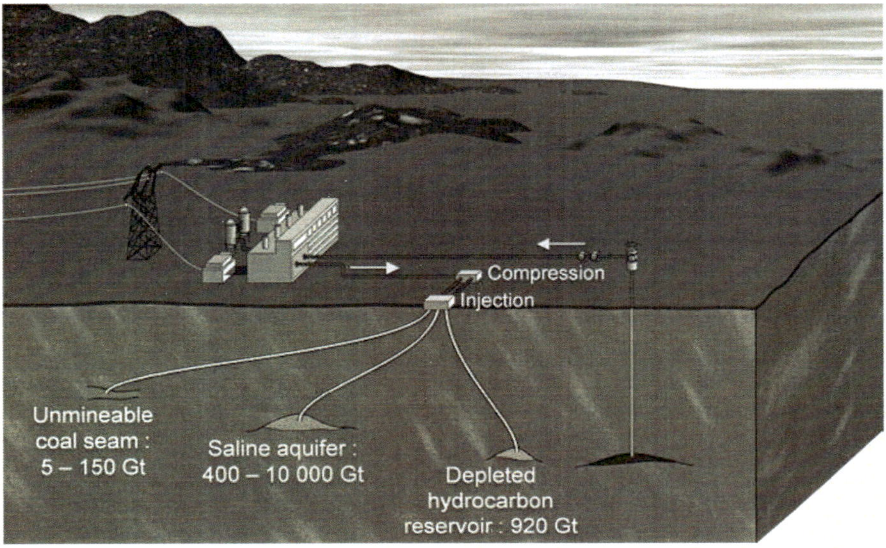

Fig. 12 General scheme of CO_2 injection from surface (© IFP)

have been put forward, but the capacity is limited and there is a high risk of leakage to surface.

During and after the injection phase, the well has to be sealed to prevent any migration of CO_2 to surface using preferential pathways. This will be achieved by placing cement and/or mechanical plugs in all parts of the well.

5.2 *Deep Drilling Technology to Access Reservoirs*

Deep drilling operations are generally performed by means of rotary drilling rigs, as shown in Fig. 13. In this technique, the hole is drilled by rotating a bit to which a downward force is applied. The hole can be initiated from the ground (onshore) or the surface of the sea (offshore), depending on the position of the targeted reservoir. Typically, the following operations are required to construct a production well: (1) put the drilling string in the hole and drill; (2) pull out the drill string and case the section; (3) perforate the casing to give access to geological formations from which formation fluids are to be collected; and (4) lower the production tubing in the hole and pump out the fluids. In the case of CO_2 disposal, the perforation technique will be used to allow injection. If the formation targeted for production or injection exhibits low permeability, the hydro-fracturing technique (performed by increasing the hydrostatic pressure in the well) can be used to create artificial fractures that will improve the fluid flow. This technique also enables the measurement of the in situ stresses.

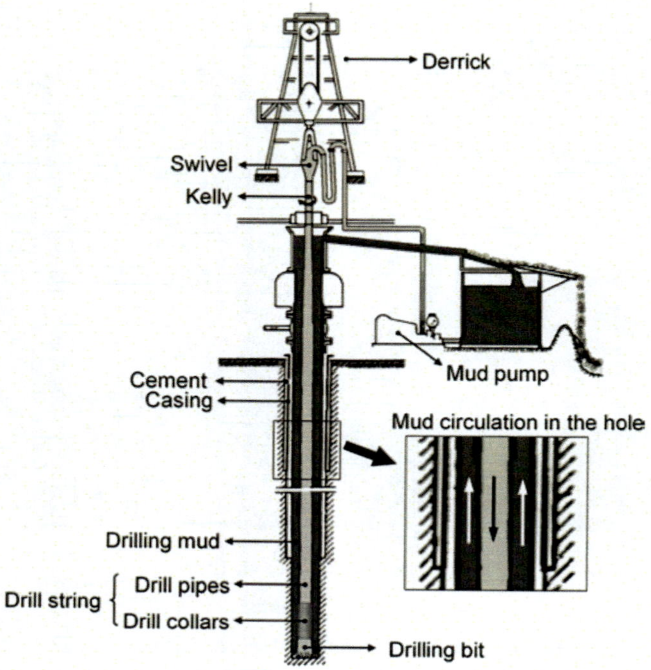

Fig. 13 Components of a typical rotary drilling rig

Nowadays it is possible to reach true vertical depths of 5,000 m and also to drill horizontally to distances of some 10,000 m from the vertical projection of the drill rig in the case of extended reach wells. Some experimental drilling projects such as the German Continental Deep Drilling Program (KTB) (Bram et al. 1995; Wohlgemuth et al. 1996), and the Kola Superdeep Borehole in Russia (Kozlovsky 1987) reached true vertical depths of about 10,000 m.

The main challenges for drilling operators can be summarized as follows:

- Choosing the suitable drilling bit in order to achieve the highest rate of penetration and the longest drilled distance or metrage;
- Ensuring the stability of the well during both the drilling phase by means of the drilling fluid and the production phase by use of cemented casings;
- Equipping the well to perform production operations.

5.2.1 Choosing the Drill Bit and Fluid

In rotary drilling, different cutting tools can be used depending on the mechanical properties of the geological formations (Bourgoyne et al. 1991; Moore 1981):

- *Rolling cutter bits* (or roller cones) are the most used bits and different technologies enable the drilling of soft to medium-hard rock formations;
- *Drag bits using polycrystalline diamond cutters* that can be used for soft to medium-hard rock with a reasonable abrasiveness;
- *Drag bits using surface set or the impregnated diamond technique* for hard to very hard rock formations.

Rolling cutter bits are at the lower end of the price range and impregnated bits at the higher end. Cost is thus an important factor in choosing the right bit to drill a given section of the well. To assess the theoretical performance of a bit, typical rock properties, including hardness, abrasiveness, mechanical behaviour (in terms of plasticity, for instance) must be measured.

Depending on the quantity of debris generated by the cutting tool, drilling fluid or mud should be used for cleaning, and the flow rate assessed accordingly. If the circulation is stopped, the fluid must be able to 'freeze' to avoid the settlement of cuttings and enable a good restart to circulation (thixotropic property). The usual drill fluids are water-based bentonite muds, oil-based muds and polymer-based muds. The bentonite-based muds are cost-effective, but they can cause clayey geological formations to swell. Oil-based muds are better for drilling such formations but they are forbidden in many countries for environmental reasons. In such cases, polymer muds can be a solution, even if they are expensive.

During drilling of each section, there is direct contact between the geological formations and the drilling fluid. Mechanical equilibrium is then provided by the hydrostatic pressure of the fluid. This is provided by the depth, on the one hand, and by the density of the mud, on the other. To ensure a suitable fluid pressure, drillers use additional materials like barium sulphate (barite) to increase density, but the fact that a very heavy mud can cause hydraulic fracturing of given formations must be taken into account. In such a case, the operation will be characterized by a loss of fluid in the formation, which could be very costly.

The drilling fluid has other functions: cooling the drill bit, avoiding ingress of formation fluids and guarding against mud loss by forming a *cake* (i.e. a thin layer of clay deposited on the wall of the well).

5.2.2 Ensuring Long-Term Stability and Sealing

Final support is achieved by means of a steel casing sealed to the walls by cement. First, the casing is introduced into the hole and then the cement is pumped into it to fill the annular space between the well wall and the casing. This sealing, when used with packers inside the casing, will avoid loss of fluids during the production phase (i.e. injection of CO_2). The casing will be designed to be strong enough to support the stresses from the ground induced by the well drilling. Deep wellbores are drilled in many stages or sections to ensure wall stability, depending on the geomechanical properties of the drilled formation. For instance, soft superficial formations should be drilled in large diameter and cased to avoid failure and enlargement when drilling deeper rocks. This obliges the driller to change to a smaller size of drill bit and casing when shifting from one section to the next (Fig. 13).

Different scenarios have been put forward to assess the CO_2 trapping mechanisms (see Sect. 5.1) and their evolution over time for as long as a million years (Benson et al. 2005). Because of the long time periods involved, well integrity in terms of mechanical behaviour (time-dependent stresses for creeping geomaterials) and corrosion (mainly of steel casing and cement) must be addressed carefully. In the abandonment procedure, special care then must be taken to use sealing plugs and cements that are resistant to degradation from CO_2. To tackle the problem of steel corrosion, the injection casing can be pulled out after operation and replaced by more resistant sealing materials (i.e. special cement).

5.3 Horizontal and Extended Reach Wells

From one onshore or offshore position, many targets can be reached using specific deviated well techniques: classical rotary drilling and use of deviating tools (i.e. whipstock), or steering motor with measuring while drilling (MWD) equipment (Bourgoyne et al. 1991). In the latter case, the drill string does not rotate and the downhole motor is operated by the pressure of the drilling fluid and is equipped with an electronic device, or MWD, to collect the data and transfer them to surface. This enables good real-time control of the trajectory of the well. The MWD can supply the following data:

- *Directional information*: taking real-time directional surveys using accelerometers and magnetometers to measure the inclination and azimuth of the wellbore, and then transmit the information to surface;
- *Drilling mechanics information*: provides information about the conditions at the drill bit, such as the rotational speed of the drill string, smoothness of the rotation, type and severity of any vibration, downhole temperature, torque and weight on bit measured near the drill bit, mud flow rate;
- *Formation properties*: when combined with logging while drilling tools, can take measurements of formation properties like density, porosity, resistivity, pseudo-caliper (measurement of the size of the hole), inclination at the drill bit, magnetic resonance and formation pressure.

Horizontal well technology is useful in storing CO_2 because of the high number of targets that can be accessed from one position of the drilling rig. The volume to be stored will also be increased if the trap (or reservoir) has a high horizontal extension.

5.4 The Particular Case of CO_2 Disposal in Coal Formations

CO_2 can be stored in combination with methane recovered from coal through ECBM. From a technical point of view, the method is similar to that used for petroleum reservoirs and deep aquifers. However, for economical reasons mining drilling techniques (smaller drilling rigs) can be tried.

Storing CO_2 in coal seams uses two physical mechanisms: filling the porous space constituted by fractures (cleats) and micropores, and adsorption (physical ability of fixing gas molecules) on the coal grain surface. This latter mechanism is typical of coal formations and other formations containing organic matter like shales.

The big challenge in terms of storing CO_2 in coal formations is the low permeability of this material and, hence, the difficulty of accessing large volumes. Some successful projects have been carried out in the world among which we can name the Alberta Research Council project in Canada (Gunter et al. 1997, 1998, 2005) and the Allison Unit CO_2-ECBM Pilot (Reeves et al. 2003). One recent trial, which was undertaken in Poland (the Recopol project in 2003–2005) (Pagnier et al. 2006) gave relatively poor results in terms of injected volumes and injectivity.

6 Comparing Engineering Issues Between Radioactive Waste and CO_2 Disposal

Table 1 summarizes some of the criteria we propose for use as a basis for comparison of the engineering issues involved in the geological disposal of CO_2 and RW. In both applications, some requirements are similar: the fluid propagation is (at least, partly) controlled by diffusion mechanisms and typical systematic studies (geophysical measurements, coring, laboratory testing, geomechanical and reservoir modelling) are necessary to assess the quality of the targeted medium and design the suitable disposal technique.

Regarding temperature, injected CO_2 can have a cooling action during the injection phase whereas RW will generate heat for a long time. In both cases, extensive study of thermo-hydro-mechanical coupling is necessary to take all possible effects into account: mechanical stability, water and gas migration, vapour formation, separation between openings.

However, some aspects are different. Because of the physical nature of the waste to be disposed of, CO_2 will be injected as a liquid (in its critical state) using deep drilling technologies as practised in the petroleum industry; RW, on the other hand, will be disposed of in mine-like facilities (using shafts and galleries). The depth of burial of RW will, in general, be shallower even if it is located deep enough to ensure isolation (ONDRAF/NIRAS 2001). However, if there are workers underground, the construction technique for RW disposal can be more hazardous, and the issue of handling hot materials must be addressed.

In both cases, an EBS, made of cement, steel and clay components, can be used. With RW disposal, the quantities involved are limited (from a few thousand to a million cubic metres) and the EBS is a complex multi-barrier system that isolates the waste from the host formation for a given span of time. For CO_2 disposal, the EBS is limited to the well, which is a small component of the big reservoir involved in the injection (quantities to be disposed of are several millions cubic metres).

The construction of the disposal site will create a damaged zone around the openings. This phenomenon may be of interest for CO_2 injection, as this can

Table 1 Comparison of radioactive waste and CO_2 disposals

Criteria	CO_2	Radioactive waste
1. Objective	Inject in existing porous medium	Create voids in host formations and secure them
2. Main transport mechanisms	Advection, diffusion, buoyancy	Diffusion
3. Retardation mechanisms	Sorption, mineralization	Sorption
4. Natural barrier	Caprock formation	Host formation mainly
5. Preliminary studies	Geophysics, coring, lab tests	Geophysics, coring, lab tests
6. Necessary properties of host and caprock formations	Porosity, permeability, thermo-hydro-mechanical (THM) parameters (elasticity, strength and plasticity, Biot's parameters, dilatation coefficients)	Porosity, permeability, THM parameters (elasticity, strength and plasticity, Biot's parameters, dilatation coefficients)
7. Construction method	Petroleum technology: deep drilling	Mining technology: shaft and galleries
8. Engineered barrier system (EBS)	Cement plugs with or no steel casing: limited control on the well	Multi-barrier concept: container, shroud(s), backfilling (cement, clay)
9. Geomechanical modelling	Stability of wells by mud or casing	Stability of created cavities and lining assessment
10. Reservoir modelling	Assess injected volume and sealing capacity	Migration of radionuclides in the host formation
11. Depth of burial	Generally from 800 to 1,000 m (injection at supercritical state)	Generally a few hundred metres or more
12. Temperature	Injected fluid cooler than host formation	Heat generation by the waste
13. Risk of hazard during construction and disposal	Limited, no workers underground	Higher because of the presence of workers
14. Relevance of disposal with respect to produced volumes	High capacities needed (millions of tonnes) Can be achieved by combining different reservoirs	Lower quantities (from few thousands to more than a million cubic metres)
15. Role of a damaged zone around the opening	Increases or decreases permeabilities Controlled by mud density	Generally increases permeabilities Controlled by the digging technique
16. Long-term sealing of the disposal	Favoured by trapping mechanisms Issues: leakage through the caprock, casing corrosion	Depending mainly on resistance of EBS Issues: corrosion of the shroud
17. Safety assessment	The methodology is the same Very long timescale phenomena (thousands of years to a million)	

increase the permeability of the host formation (fracturing or dilatancy). However, a decrease in permeability is also possible, particularly in coal seams (swelling). It is therefore important to control the mud density during the drilling phase. When an RW disposal site is being constructed, permeability of the host formation will generally increase due to the damaged zone, the size of which should be controlled through the digging technique (sequence of dig and support).

The long-term sealing of the reservoir is an important issue. This will be favoured by trapping mechanisms for CO_2 disposal whereas the EBS will play a major role when disposing of RW (especially when the host formation, for instance, granite, is not very tight). For RW, corrosion of the metallic shroud may occur and lead to migration of radionuclides into the host formation. For CO_2, the corrosion (or other chemical mechanisms) of the cemented steel casing is also an issue, but this can be avoided by pulling the casing out after injection and before cementing and plugging. Moreover, the leakage through the caprock (except for depleted reservoirs that exhibited sealing during geological times) must also be addressed.

For safety assessments, the same approaches are used, and studies have been performed for spans of time of up to 70,000 (RW) to a million (CO_2) years, taking into account different scenarios. Some criteria can be set for engineering approval and licensing: quality of the host formation (physical parameters, modelling of transport mechanisms), disposal and sealing technology (sequencing and security during operations and after abandonment), quality of the EBS components (resistance to corrosion, etc.). The monitoring system will play a major role regarding the long-term safety (Brunskill and Wilson 2011).

7 Conclusions

The chapter focuses on engineering issues related to the challenges of the geological disposal of both RW and CO_2. In both cases, engineering techniques exist to access deep geological formations with suitable characteristics, dispose of the waste and ensure long-term sealing. Engineering studies to design the disposal are similar: sample collection, laboratory testing, in-field qualification of geological formations, geomechanical and reservoir modelling. In fact, the displacement of potentially polluting fluids in the porous media needs to be assessed and the geological material both for resistance to excavation techniques (i.e. mechanical digging) and long-term stability to be characterized.

The RW repositories use mining techniques with some specificities like the handling of hot materials by workers underground during the disposal operations. The hosting geological formation is to be as impermeable as possible and the residual voids will be filled by a low permeability material (clay or cement) that will contribute to the EBS intended to limit the diffusion of radionuclides.

CO_2 disposal will use techniques developed in the petroleum field when targeting deep reservoirs; this means deep drilling with deviated trajectories. For long-term stability and sealing, there is a need to install cemented casings with sealing plugs in the well. One of the challenges for the future is the durability of this system in the presence of CO_2; special cements are being developed and it is possible to remove the casing before abandonment. The long-term leakage through the natural barrier or caprock must be addressed carefully.

In this chapter we have presented the main transport mechanisms of pollutants and the parameters needed to characterize the potential host geological formations. The methodology for designing and constructing the underground openings was then presented, with an emphasis on the techniques to be used for the disposal of RW and the CO_2. In Sect. 6, we have applied some criteria to compare the engineering issues related to the two approaches, and have accompanied this with a number of relevant comments.

Acknowledgements F. Descamps' thanks are due to the Belgian National Fund for Scientific Research (Fonds National de la Recherche Scientifique, FNRS) for supporting her post-doctoral research position.

References

Barton N, Lien R, Lunde J (1977) Estimation of support requirements for underground excavations. In: Fairhurst C, Crouch SL (eds) Proceedings of the 16th Symposium on Rock Mechanics. American Society of Civil Engineers, New York, pp 163–177

Benson S, Cook P et al (2005) Underground geological storage. In: Metz B, Davidson O, de Coninck HC, Loos M, Meyer LA (eds) IPCC Special Report on Carbon Dioxide Capture and Storage. Cambridge University Press, New York, pp 195–276

Bieniawski ZT (1984) Rock Mechanics Design in Mining and Tunneling. A. A. Balkema, Rotterdam

Bieniawski ZT (1992) Design Methodology in Rock Engineering. A. A Balkema, Rotterdam

Bourgoyne AT, Milheim KK, Chenervert ME, Young FS (1991) Applied Drilling Engineering, vol 2, SPE Textbook Series. Society of Petroleum Engineers, Richardson

Bouvart-Lecoanet A, Colombet G, Esteulle F (1988) Ouvrages souterrains, conception, réalisation, entretien. Presses de l'Ecole Nationale des Ponts et Chaussées, Paris

Brady BHG, Brown ET (1999) Rock Mechanics for Underground Mining. Chapman & Hall, Oxford

Bram K, Draxler J, Hirschmann G, Zoth G, Hiron S, Kühr M (1995) The KTB Borehole–Germany's superdeep telescope into the earth's crust. Oilfield Rev 7(1):4–22

Brown ET (1981) Rock Characterization Testing and Monitoring: ISRM Suggested Methods. Pergamon Press, Oxford

Brunskill B, Wilson M (2011) Monitoring methods used to identify the migration of carbon dioxide and radionuclides in the geosphere. In this volume

Charlez PA (1991) Rock Mechanics, vol 1, Theoretical Fundamentals. Technip, Paris

Coussy O (1991) Mécanique des milieux poreux. Technip, Paris

Dake LP (1978) Fundamentals of Reservoir Engineering. Elsevier, Amsterdam

Deere D (1963) Technical description of rock cores for engineering purposes. Felsmechanik Ingenieurgeologie 1(1):16–22

Gomit JM, Marivoet J, Raimbault P, Recreo F (1997) Evaluation of Elements Responsible for the Effective Engaged Dose Rates Associated with Final Storage of Radioactive Waste: EVEREST Project, vol 1, Common Aspects of the Study. Final Report EUR 17449/1. European Commission, Luxembourg

Gunter WD, Gentzis T, Rottenfusser BA, Richardson RJH (1997) Deep coalbed methane in Alberta, Canada: a fuel resource with the potential of zero greenhouse gas emissions. In Proceedings of the 3rd International Conference on Carbon Dioxide Removal. Energy Conserv Manage 38 (Suppl), pp S217–S222

Gunter WD, Wong S, Cheel DB, Sjostrom G (1998) Large CO_2 sinks: their role in the mitigation of greenhouse gases from an international, national (Canadian) and provincial (Alberta) perspective. Appl Energy 61(4):209–227

Gunter WD, Mavor MJ, Robinson JR (2005) CO_2 storage and enhanced methane production: field testing at Fenn-Big Valley, Alberta, Canada, with application. In: Rubin ES, Keith DW, Gilboy CF (eds) Greenhouse Gas Control Technologies: Proceedings of the 7th International Conference on Greenhouse Gas Control Technologies, vol 1. Elsevier, London, pp 413–422

Hoek E, Brown E (1998) Practical estimates of rock mass strength. Int J Rock Mech Min Sci 34(8):1165–1186

Horseman ST, Higgo JJW, Alexander J, Harrington JF (1996) Water, gas and solute movement through argillaceous media. Report CC-96/1. OECD Nuclear Energy Agency, Paris

Jaeger JC, Cook NGW (1979) Fundamentals of Rock Mechanics, 3rd edn. Chapman and Hall, New York

Kozlovsky YA (ed) (1987) The Superdeep Well of the Kola Peninsula. Springer, Berlin

Marivoet J, Volckaert G, Wermaere I, Wibin J (1997) Evaluation of Elements Responsible for the Effective Engaged Dose Rates Associated with the Final Storage of Radioactive Waste: EVEREST Project, Vol 2a: Clay Formation, Site in Belgium. Final Report EUR 17449/2a. European Commission, Luxembourg

Mody F, Hale A (1993) Borehole stability model to couple the mechanics and chemistry of drilling-fluid/shale interactions. J Petrol Technol 45(11):1093–1101

Moore WM (1981) Fundamentals of Rotary Drilling. Energy Publications, Dallas

ONDRAF/NIRAS (L'organisme national des déchets radioactifs et des matières fissiles enrichies–Belgian agency for radioactive waste and enriched fissile materials) (2001) SAFIR 2 Report: Safety Assessment and Feasibility Interim Report 2 Report NIROND 2001-06E. ONDRAF/NIRAS, Brussels

ONDRAF/NIRAS (L'organisme national des déchets radioactifs et des matières fissiles enrichies–Belgian Agency for Radioactive Waste and Enriched Fissile Materials) (2008) Evolution of the Near-Field of the ONDRAF/NIRAS Repository Concept for Category C Wastes: First Full Draft Report. NIROND-TR 2007-07E. ONDRAF/NIRAS, Brussels

Pagnier H, van Bergen FF, Krzystolik P, Skiba J, Jura B, Kobiela Z, Hadro J, Wentink P, De-Smedt G, Kretzschmar HJ, Fröbel J, Müller-Syring G, Krooss B, Busch A, Wolf KH, Mazumder S, Bossie-Codreanu D, Choi X, Grabowski D, Hurtevent D, Gale J, Manancourt A, Winthaegen P, van der Meer B, Bruining J, Reeves S, Stevens S (2006) RECOPOL (Reduction of CO_2 emission by means of CO_2 storage in coal seams in the Silesian Coal Basin of Poland). Final Report. TNO, Delft

Put MJ, Henrion PN (1988) An improved method to evaluate radionuclide migration model parameters from flow-through diffusion tests in reconsolidated clay plugs. Radiochim Acta 44(45):343–347

Reeves S, Taillefert A, Pekot L, Clarkson C (2003) The Allison Unit CO_2-ECBM Pilot: a reservoir modeling study. DOE Topical Report

Sherwood JD (1993) Biot poroelasticity of a chemically active shale. Proc R Soc London Ser A 440(1909):365–377

Tshibangu JP, Sarda JP, Audibert-Hayet A (1996) Etude des interactions mécaniques et physico-chimiques entre les argiles et les fluides de forage, application à l'argile de Boom. Revue de l'Institut Français du Pétrole (IFP) 51:497–526

Vereycken, M. (2000). Interprétation et simulations numériques de la convergence d'un revêtement de type cintres coulissants durant l'excavation d'un puits dans l'argile de Boom. Thesis in Mining Engineering, Faculty of Engineering, Mons

Wohlgemuth L, Tran Viet T, Engeser B (1996) Drilling-experiences at the ultradeep well KTB-HB. In Proceedings of the VIIIth International Symposium on the Observation of the Continental Crust Through Drilling. Tsukuba, Japan, 6 Feb–2 Mar, pp 39–45

The Costs of the Geological Disposal of Carbon Dioxide and Radioactive Waste

Ferenc L. Toth and Asami Miketa

Abstract Cost assessments of geological disposal of carbon dioxide and radioactive waste are presented. The scope of the cost assessments covers a range of activities from research, site identification, licensing and construction to operation, closure and post-closure monitoring of the disposal sites. The most meaningful indicator for comparison is the disposal cost per unit of electricity produced. The comparative assessment reveals important differences between the two waste products in the volume of material involved and the precautions to be taken that determine the cost per kWh indicator. The timing of investment to establish the disposal site is an important difference with significant cost implications: investments must be completed before starting CO_2 capture from fossil power plants whereas investments in radioactive waste repositories can be postponed for decades after the waste emerges from nuclear power reactors. The investment costs are significant and mid-course corrections are expensive; hence, both technologies need stable regulatory systems.

Keywords Geological disposal • Carbon dioxide • Radioactive waste • Disposal costs • Power generation costs

1 Introduction

The geological disposal of carbon dioxide (CO_2) and radioactive waste (RW) is undertaken as the final stage in the long fossil fuel electricity and nuclear power generation fuel chains, respectively. Although new clean coal technologies increasingly involve pre-combustion operations, the pathway of coal from mining to the

F.L. Toth (✉) and A. Miketa
International Atomic Energy Agency (IAEA), Wagramer Str. 5, P.O. Box 100, 1400 Vienna, Austria
e-mail: F.L.Toth@iaea.org; A.Miketa@iaea.org

F.L. Toth (ed.), *Geological Disposal of Carbon Dioxide and Radioactive Waste: A Comparative Assessment*, Advances in Global Change Research 44,
DOI 10.1007/978-90-481-8712-6_8,

power plant where it is converted into heat and/or electricity is somewhat simpler than the longer and more complex route for uranium from mines through enrichment and fuel fabrication to nuclear reactors. In the downstream phase (post-combustion of coal or burnup of nuclear fuel), the opposite is true. Depending on the carbon capture technology applied, it might take a series of chemical processes to separate and condition CO_2 into a supercritical state suitable for transport and disposal in geological formations. No such operations are needed for spent nuclear fuel (SNF). In the case of once-through fuel cycles, the most important factor between removing SNF from the power reactor and placing it in a geological disposal site is the time period required to reduce the amount of heat produced by the fuel rods before they can be packaged for disposal. If SNF is recycled, the resulting high-level RW also needs conditioning and packaging for transport and emplacement in the final repository. The costs associated with the various steps in these two fuel chains vary widely, depending on the geographical circumstances and the technologies chosen.

Quite clearly, the most meaningful approach to comparing the costs of electricity generation technologies should encompass not only the respective fuel cycles from cradle to grave but all other costs associated with building, operating and decommissioning the associated power plants. Such levelized cost estimates are regularly prepared by the OECD International Energy Agency (IEA), with the most recent update being published as IEA and NEA (2010). (This publication includes estimates of CO_2 capture costs, but not transport and disposal costs.) Complying with the mandate of this book, we limit the scope of our analysis largely to comparing the costs of the geological disposal of CO_2 and RW and to providing some insights into the related economic and financing aspects of fossil and nuclear power generation.

In all energy-economy models that include all the necessary technological and economic parameters of CO_2 capture and disposal (CCD) options in their technology dataset, the costs of reducing greenhouse gas emissions from fossil fuel electricity by capturing CO_2 and disposing of it in geological formations are implicitly compared with all other energy supply technologies. The resulting energy mix from these models represents the least-cost supply portfolio in which fossil energy sources, with and without CCD, and all other energy sources and technologies are utilized under a given set of external assumptions (about resource prices and availability, technological performance and costs, etc.) and constraints (e.g. CO_2 emission ceiling).

Some studies explicitly compare the costs of selected technologies to reduce CO_2 emissions from the energy system. Rogner et al. (2008) analyse the cost-effectiveness of two main baseload power technologies: fossil-based electricity with and without CCD and nuclear power in nine countries. They calculate life cycle electricity costs in present value terms from national energy studies comprising country-specific technological and economic data. The authors find that adding CCD results in a considerable increase in the cost of fossil fuel electricity (mainly coal). Relative to the reference cases, this would completely eliminate the cost advantage of fossil-based electricity without CCD over nuclear power in some of the countries analysed (Argentina, Bulgaria, China, India) and significantly increase the cost advantage of nuclear power in other countries included in the study (Korea, Pakistan, Poland, Romania, Russia).

Tavoni and van der Zwaan (2009) use the global integrated assessment model WITCH (Bosetti et al. 2006) to compare the competitiveness of coal-based electricity with CCD and nuclear power. They find that, using the standard formulation and assumptions of the WITCH model, but removing the expansion constraints usually adopted for nuclear power in most energy-economy models (to reflect perceived limits due to public acceptance, political and other non-economic reasons), global nuclear power capacity is projected to expand by 15–20 GW/year under a target of stabilizing global greenhouse gas emissions at 550 ppm CO_2-equivalent, while the expansion rate is projected to increase from about 20 GW/year in 2020 to almost 35 GW/year in 2050 under a 450 ppm stabilization scenario. The authors suggest that, provided public acceptance and politics do not prevent the expansion of nuclear capacities to their full economic potential, considerable improvements will be needed in CO_2 capture costs and technological efficiencies if CCD is to compete with nuclear power and provide a large share of CO_2 mitigation.

Framing a meaningful comparison of the disposal costs of CO_2 and RW is difficult for various reasons elaborated in preceding chapters: the volumes and physical and chemical properties (Bachu and McEwen 2011), the human health and environmental hazards posed (West et al. 2011), the transport (Gómez and Tyacke 2011) and site engineering (Tshibangu and Descamps 2011) technologies of CO_2 and RW heavily influence the various cost components and thus the total disposal costs. Relating these disposal costs to the amount of electricity generated, to the total fuel cycle costs and to the levelized electricity costs could provide information about the similarities and differences between the waste disposal cost profiles of the two technologies; however, the unavailability of consistent and comparable data makes the comparison largely unreliable.

An additional problem with comparing costs in both technologies over time is the general escalation of all technology costs (all constructions and equipment for CO_2 and RW disposal) over the past 8 years. The reviews presented in this chapter provide the original values from the cited studies but for the summary and comparison tables a common metric of 2010 US$ is adopted by applying the appropriate GDP deflators and converting other currencies at average 2010 exchange rates.

Section 2 of this chapter presents recent estimates of the costs of the geological disposal of CO_2, followed by a similar overview of RW disposal cost estimates in Sect. 3. Based on the issues raised in these assessments and on the resulting cost estimates, Sect. 4 provides a comparative assessment, followed by a concluding summary in Sect. 5.

2 Costs of the Geological Disposal of CO_2

This section provides a short overview of recent studies on the cost of CCD. First, the main cost elements are itemized. This is followed by cost estimates published over the past few years. The sample of studies involves rather different approaches ranging from specific disposal case studies (project costs) to aggregate supply curves of disposal potential. Care is needed when drawing general conclusions from such a diverse sample. Finally, the relative importance of the disposal costs in the fuel cycle and levelized electricity costs are analysed.

2.1 Cost Elements

The three main phases in the CCD chain include capture, transport and disposal. The cost estimates for all three steps are presented in the next section, but here the focus is on the disposal phase. This phase starts at the point of CO_2 delivery to the disposal site which, in most cases, is the end of a pipeline.

The first cost items include the preparatory steps in the site establishment process that starts with geological screening of potential sites, site exploration and characterization (geological, geophysical and engineering feasibility analyses) that lead to site selection. This is followed by reservoir characterization and evaluation. Licensing and stakeholder engagement also involve costs. The bulk of the disposal costs arise from the site construction: design and drilling the injections well, mounting in-field pipelines if needed, installation of the surface facilities: scrubber (to remove residual liquids), compressor and other equipment. Infrastructure costs may arise from road construction and energy supply connections. Operating costs cover labour, maintenance and energy.

Most cost estimates published to date include the above site characterization, capital and operating costs, but ignore other, potentially important cost items. Monitoring costs accrue throughout the whole process from establishing the baseline monitoring system through the post-closure monitoring of the disposal site, possibly over decades, but they are usually not included in the cost estimates. It is worth noting that monitoring costs are also interlinked with other cost items. A system consisting of many small reservoirs that are close to sources of CO_2 will have higher monitoring costs than a system with a few large-scale reservoirs that are connected by a well developed pipeline system. Longer pipelines add to the cost of the system, but the cost can be offset by lower disposal costs in large saline aquifers. The difference in monitoring costs (depending on how far out in time those costs are counted and the discount rate used) can tip the scale in favour of a pipeline network with a few major disposal sites.

Additional cost items might arise from remediation and liability. Remediation measures might be required when leakage occurs. Additional expenditures may occur to address long-term liability issues. These items are difficult to estimate and are excluded from most cost estimates published to date.

In the absence of industry-scale CCD experience, cost estimates draw on industrial analogues related to the relevant technological components. The main source is the oil and gas industry with its well drilling and injection technologies and associated knowledge base on monitoring and modelling of reservoirs. Enhanced oil recovery (EOR), acid gas and other liquid waste injection projects as well as experience with underground natural gas storage sites provide a good basis for cost assessments.

A comprehensive technology and cost analysis for geological CO_2 disposal is presented by the US Environmental Protection Agency (US EPA 2008). The report presents a general costing methodology together with the costs of specific technologies and operating practices that might be adopted for CO_2 disposal costing studies. Unit costs are estimated in terms of cost per site, per well, per unit of area or other relevant characteristic, depending on the nature of the cost item. The EPA report includes the following cost categories:

- Geological site characterization;
- Monitoring;
- Injection well construction;
- Area of review and corrective action;
- Well operation;
- Mechanical integrity tests;
- Post injection well plugging and site care;
- Financial responsibility;
- General and administrative.

The principal difficulty in all estimates is that disposal costs vary widely depending on the geological characteristics of the site. The depth, thickness and permeability of the host geological formation determine both the construction (number of wells, in-field pipelines) and the operation (injection) costs. Location is another key cost driver: offshore geological disposal is significantly more expensive than onshore disposal. (Offshore deep ocean disposal is not considered in this chapter.) Hence cost estimates need to be handled with care: extrapolation from one site to another could be rather imprecise or outright wrong.

2.2 Cost Estimates

The most expensive item in the CO_2 capture–disposal chain is capture. The cost of capture is calculated and presented in the literature in two ways: one is based on incremental cost of the capture equipment and its operation (a simple engineering cost) while the other takes net cost and emissions of a plant with capture equipment and compares this to a reference plant without capture (an economic cost). Accordingly, the costs quoted below include both types: incremental engineering as well as net avoided costs. Comparing these costs to each other is difficult because there is not sufficient information available to convert these estimates to a common metric, since the ratio of capture and avoided costs varies by 10–40% between different projects. Yet even with this caveat, ballpark comparisons are still meaningful.

The starting point for our cost review is the report by the Intergovernmental Panel on Climate Change (IPCC 2005) which summarized the then current state of knowledge about CCD. Chapters 3–7 of the IPCC report provide detailed assessments of the technologies involved, environmental, monitoring, risk and legal aspects, as well as the cost estimates for CCD phases, from capture and transport to disposal in underground geological formations, the ocean at great depth and via mineral carbonation. Chapter 8 on costs and economic potential (Herzog et al. 2005) draws on these chapters and summarizes a large number of the studies that were then available.

Not surprisingly, most of the attention of the studies assessed by Herzog et al. (2005) focuses on CO_2 capture costs, which are by far the most expensive component of the capture, transport and disposal chain. The authors find the following cost ranges for CO_2 capture expressed in US$/t CO_2 captured (that is simple engineering costs) for newly built plants (all in 2002 US$): 33–57 for new natural gas combined cycle (NGCC) plants, 23–35 for pulverized coal (PC) plants and 11–32 for integrated gasification combined cycle (IGCC) plants. The main factor determining the cost of CO_2 transport via pipelines is the volume transported (see also Gómez and Tyacke 2011). For a mass flow rate of more than 15 million tonnes (Mt) CO_2/year onshore on 'normal' terrain (low population density and no high mountains), the cost range is between US$1 and US$2/t CO_2 for a distance of 250 km.

In addition to site-specific features, the crucial factor determining the disposal costs is whether the CO_2 can serve a revenue generating activity through EOR, enhanced gas recovery (EGR) or enhanced coalbed methane (ECBM) recovery, or not. Onshore disposal with EOR can generate a revenue of US$10–16/t CO_2, depending on the prevailing oil prices. The cost range for disposal in saline formations and depleted oilfields is wide: US$0.5–8.0/t CO_2. Cost estimates for monitoring are in the range of US$0.1–0.3/t CO_2 (Herzog et al. 2005). The IPCC report (IPCC 2005) also estimates the costs of disposal in deep oceans, but this option remains controversial because of environmental concerns, and it is more expensive than the geological alternatives. At the high end of the spectrum, mineral carbonation cost estimates are in the range of US$50–100/t CO_2.

The IPCC report (IPCC 2005) identifies many factors that make the comparison of CCD cost estimates difficult. They include technology- and location-specific factors, the differing boundaries of the capture and disposal system and the different metrics adopted: the investment costs for the capture system, the incremental product costs (e.g. cost of electricity), the cost of CO_2 avoided and the cost of CO_2 captured and removed. Some studies also add a temporal dimension by assuming different learning rates resulting in declining unit costs over time. These difficulties are also involved in comparing the estimated or projected costs of the whole or of various components of the CCD system reviewed in this section.

The focus of the CCD studies published since the IPCC report remains on the capture phase. The studies cover a wide range of issues such as: retrofitting options for existing coal-fired power plants (e.g. Korkmaz et al. 2009), their operability with CO_2 capture equipment installed that is significantly different from their normal design conditions (see Alie et al. 2009), and the environmental impacts of the processes and materials involved, like the amines used in post-combustion capture of CO_2 (Eide-Haugmo et al. 2009). Large utilities report on their CCD pilot projects (Renzenbrink et al. 2009; Strömberg et al. 2009). As capture costs represent the largest share in the overall CCD costs, many studies are devoted to them (e.g. Ho et al. 2009). In addition, assessments of CO_2 disposal capacities in many countries and world regions attempt to provide more precise estimates (see the report by Vangkilde-Pedersen et al. (2009) about the European Union (EU) GeoCapacity project).

The number of studies reporting CCD cost estimates has been increasing since the publication of the 2005 IPCC report. Most of them focus on the capture component and omit transport and disposal costs altogether or add these items as lump sum figures from other sources. A selected set of cost studies are summarized in the paragraphs that follow. Table 1 summarizes the results of the cost estimates.

McCoy and Rubin (2005) review the CCD literature and find 'frequent inconsistencies and lack of clarity in defining the scope of the CO_2 capture, transport and [disposal] components'. They present engineering and economic models of CO_2 transport via pipelines and geological disposal of CO_2 in deep saline aquifers. The aim of their models is to provide first-order cost estimates that are sensitive to site-specific or project-specific technical and financial parameters. The authors analyse a case study in the Wabamun Lake area of Alberta, Canada. Their results indicate significant uncertainties in the cost of CO_2 transport and disposal, primarily due to the variability of the geological parameters of the reservoir, as well as to other factors such as transport distance and power plant capacity factor. McCoy and Rubin (2005) find that the combined cost of transport and disposal (on a cost per tonne CO_2 basis) could represent more than 32% of the total CCD cost, as opposed to other estimates of less than 15%. The case they analyse involves CO_2 disposal from a 500 MW coal-fired power plant. The median cost for CO_2 transport is US$0.44/t, US$1.44/t for disposal, and US$1.94/t CO_2 for the combined median cost of transport and disposal. The authors find that disposal costs are more variable than transport costs, and that the total cost varies between a fifth percentile of US$0.78/t and a 95th percentile of US$14.59/t. Based on these results, the median cost of transport and disposal seems to be a small part of the total cost of CO_2 disposal, but according to their study there will be cases in which the cost of transport and disposal are large.

Vosbeek and Warmenhoven (2007) provide a comprehensive assessment of the opportunities and prospects for CCD in the Netherlands. They conclude that retrofitting existing power plants with CO_2 capture equipment would reduce their efficiency to an unacceptably low level. They analyse three business concepts: stand-alone (one CO_2 source with a short pipeline to a dedicated disposal site), network (several sources with a pipeline network to several disposal sites) and a network with CO_2 utilization (for EOR or other industrial purposes). The report calculates the integral costs of capture and disposal from the costs per tonne of CO_2 emissions avoided in 2006 €. Based on an Ecofys report (Hendriks et al. 2007), the authors derive total CCD costs of €29 and €39/t CO_2 for the stand-alone and network cases, respectively. The transport costs are estimated at €1/t CO_2 for the stand-alone case (one plant located 40 km from a good disposal site) and €4/t CO_2 for network projects, while the disposal cost in both cases is taken to be €2/t CO_2. When CO_2 is used for EOR in the North Sea, transport costs are assumed at €5/t CO_2 and disposal costs range from –€3 to –€22/t CO_2, depending on the amount of oil yield and assuming an oil price of €20/barrel. Accordingly, the total CCD costs vary between €16 and €35/t CO_2.

Table 1 Overview of CCD costs estimated by selected studies

Source	Capture	Transport	Disposal	Total	Notes
IPCC (2005)	NGCC: 33–57 PC: 23–35 IGCC: 11–32	1–2 (15 Mt CO_2/year, 250 km)	−10 to −16 (EOR) 0.5–8 (saline formations) 50–100 (mineral carbonation)		All in 2002 US$/t CO_2 captured
McCoy and Rubin (2005)		0.44	1.44	1.94	US$/t CO_2; median transport and storage costs, total excludes capture costs
Vosbeek and Warmenhoven (2007)	26 for stand-alone 33 for network project	1 for stand-alone 4 for network project	2 −3 to −22 with EOR	29 (stand-alone) to 39 (network) 16–35 with EOR	€/t CO_2 captured
Ploumen et al. (2007)	IGCC: 25–30 PC: 32–42 PC retrofit: 40–52				€/t CO_2
Hendriks et al. (2007)		1.0–1.6			20–10 Mt CO_2/year
Pöyry Energy Consulting (2007)	21–27 for coal retrofit 31–35 for gas retrofit	1–9	Up to −16 with EOR, 1 in aquifers, 1–20 in oil- and gasfields	21–44 (for a range of 10–180 Mt CO_2/year)	£/t CO_2 abated
IEA (2008)		2–6 for 2 Mt/year to 100 km	10–25		US$/t CO_2 captured
Captured:					
NGCC	50–90	1–3 for10 Mt/year to 100 km			
PC and IGCC	40–55				

Avoided:						US$/t CO_2 avoided
NGCC	60–110 in 2010					
	55–90 in 2030					
PC and IGCC	60–75 in 2010					
	50–65 in 2030					
BCG (2008)	25	2–3		4–5	31–33	€/t CO_2
Hamilton et al. (2008)	52		10		62	US$/t CO_2 avoided
Dooley et al. (2008)	–	1–14		1–14	15	6 cases, US$/t CO_2
McKinsey & Company (2008)						€/t CO_2 abated in 2020 and 2030, respectively
Early (after 2020)	25–32	4–6		4–12	35–50	
Mature (after 2030)	18–25	6–8		6–14	30–45	
Hildebrand (2009)	30–40		5–15		35–55	US$/t CO_2 captured
Eccles et al. (2009)	–	–		2–7	–	2005 US$/t CO_2

Numbers indicate cost per tonne of CO_2 captured unless otherwise indicated

Transport costs are for stand-alone cases unless otherwise indicated

IGCC: integrated gasification combined cycle: NGCC natural gas combined cycle: PC pulverized coal

Ploumen et al. (2007) provide an in-depth analysis of the capture costs, but their analysis does not include any assessment of transport and disposal costs. They conclude that new IGCC power plants are the most likely candidates for CO_2 capture at costs in the range of €25–30/t CO_2. This is considerably lower than PC plants, where the capture cost range is €32–42/t CO_2. The authors note, however, that IGCC is a less mature and more expensive technology than PC and this may involve additional costs to utilities. The capture costs in power plants retrofitted with capture equipment are estimated to be in the range of €40–52/t CO_2.

Hendriks et al. (2007) assess transport costs in the complex terrain of the Netherlands (densely populated areas, waterways, freeways). The average cost for 100 km is estimated at €1.6/t CO_2 and €1.0/t CO_2 for transporting 10 and 20 Mt CO_2/year, respectively, in large networks.

Pöyry Energy Consulting has developed a model to examine how the economics of the entire CCD chain might evolve in the UK with the increasing deployment of this technology (Pöyry Energy Consulting 2007). The Pöyry Energy Consulting model estimates the abatement costs of the three main stages of the CCD process, these being:

- CO_2 capture at the emission sources (large fossil-based power plants and industrial sites in the UK);
- CO_2 transport to the disposal site (including optimization of the transportation network);
- CO_2 disposal in offshore oil- or gasfields and aquifers.

The report concludes that using CO_2 for EOR can generate revenue up to £16/t CO_2, which partly compensates for other CCD costs (before any taxation issues are considered); the cost of disposing of CO_2 in aquifers is close to £1/t; and the cost of CO_2 disposal in oil- and gasfields ranges from £1/t to £20/t. The cost for abatement of up to 100 Mt CO_2/year is estimated to be in the range of £21–28/t.

The Pöyry Energy Consulting report (Pöyry Energy Consulting 2007) also includes an example with detailed cost calculations for a 485 MW coal-fired unit at Aberthaw B in 2015, from where the captured CO_2 is transported by pipe (over a distance of 370 km) to a gas terminal and then to an offshore aquifer (an additional 85 km) with ideal features (shallow water, shallow depth of the disposal media, large disposal capacity to support 46 wells from two platforms). Transport costs of the captured CO_2 from all three units of the Aberthaw B power plant to the disposal site are estimated in terms of CO_2 abated and amount to £4.53/t (£3.65/t CO_2 captured) while disposal costs coming to £1.21/t (£1.01/t CO_2 captured).

An IEA report (IEA 2008) presents a comprehensive technology analysis for CCD. The study includes an overview of the prospects and costs, the legal and regulatory frameworks as well as financing issues, together with status reports of CCD activities in many countries. The chapter on capture technologies also includes cost assessments presented as unit costs of CO_2 captured and avoided for coal- and gas-fuelled electricity generation (see Table 1). The cost estimates of transport and disposal technologies are less detailed and arrive at a cost of US$1–6/t CO_2 transported. The disposal costs in Europe are estimated in the range of US$10–20/t CO_2 in saline aquifers and US\$10–25 US/t CO_2 in depleted oil- and gasfields. Cost estimates for North America are similar: US\$15–25/t CO_2 in similar geological formations.

The report by The Boston Consulting Group (BCG 2008) confirms that CCD is a technically feasible solution. It briefly discusses examples of ongoing CCD activities in several countries and concludes that 'Over the long term, the technology would pay for itself at a stable carbon price of €30/t' (p. 4). The main cost components are estimated in the order of €25 to capture, €2–3 to transport and €4–5 to dispose of a tonne of CO_2. These proportions are assessed to be stable across the main global regions and in accordance with the assumed declining costs between 2008 and 2030.

Hamilton et al. (2008) present a financial analysis for new supercritical PC plants with CCD for the nth plant (i.e. involving some cost decline due to accumulating experience). The authors estimate the costs of *avoided* CO_2 emissions (always larger than the capture costs by a factor of 1.1–1.4) and focus on the cost of CO_2 capture—transport and disposal are considered at an estimated US\$10/t CO_2 avoided. Their CCD estimate shows that post-combustion capture in a new supercritical coal plant would cost US\$52/t CO_2 avoided, amounting to US\$62/t CO_2 taking account of transportation and disposal costs of US\$10/t CO_2. This avoidance cost is significantly higher than carbon price estimates resulting from various CO_2 regulation proposals in the USA; slightly above even the highest carbon price of US\$61/t CO_2 estimated for the year 2030 under the Lieberman-Warner bill.

Looking beyond individual site- and project-specific conditions and costs, Dooley et al. (2008) estimate the long-term average price for CO_2 transport and disposal (including measurement, monitoring and verification) at US\$15/t CO_2 (2005 US\$). The authors present six actual modelled cost split cases across which both the transport and disposal costs vary between about US\$2–14/t CO_2 (depending on the CO_2 source and flow rate, the transport distance and terrain and the features of the disposal site) but the sum of these two components is around US\$15/t CO_2 in each case. Hence the authors argue that the US\$15/t CO_2 is a robust estimate for transport and disposal costs, likely to prevail in many CCD projects in the USA.

The study by McKinsey and Company (2008) provides an overview of CCD in Europe in three phases, with the primary focus on the economic aspects: the demonstration phase up to 2015, the early commercial phase around and shortly after 2020 and the mature commercial phase commencing after 2030, if by then at least 80–120 projects are implemented in Europe to foster the learning effect. Along this path, CCD costs are estimated in the order of €30–45/t CO_2 *abated* by 2030 (see Table 2). In the reference cases, transport and disposal costs remain relatively stable across the various phases at €4–8 and €4–14/t CO_2, respectively.

The McKinsey report identifies capital costs (cost of CCD equipment per kW plant capacity) and the average cost of capital as the main uncertainty factors affecting the total costs of CCD. Construction and material costs for CO_2 pipelines are highly proportional to their length; hence, distance is the most sensitive factor in the transport cost. Yet, because of the low share of transport in the total costs, this effect is limited: doubling the transport cost would lead to only about a 10% increase in the total CCD cost.

Table 2 Costs of CCD in the deployment pathway in €/t CO_2 abated

	Assumptions	Capture	Transport	Disposal	Total[a]
Demonstration phase (2012–2015)	300 MW hard coal or lignite, 25 years lifetime, 80% utilization rate, 100 km onshore, 200 km offshore transport	51–64	5–7	5–13	60–90
Early commercial phase (after 2020)	900 MW hard coal or lignite, 40 years lifetime, 86% utilization rate, 200 km onshore, 300 km offshore transport, 1,500 m injection depth, onshore DOGF—offshore saline aquifers	25–32	4–6	4–12	35–50
Mature commercial phase (after 2030)	900 MW hard coal or lignite, 40 years lifetime, 86% utilization rate, 300 km onshore, 400 km offshore transport	18–25	6–8	6–14	30–45

Source: McKinsey & Company 2008
DOGF: depleted oil- and gasfields
[a] Totals rounded to nearest 5

Disposal costs largely depend on the location (onshore versus offshore) and the nature of the disposal site. The McKinsey report estimates that disposal in saline aquifers might cost 10–15% more than in depleted oil- and gasfields because of the limited amount of information available about the former; hence the need for more exploration and site characterization. (This indicates the massive dependence of the disposal cost on local conditions: cf. the Pöyry Energy Consulting (2007) model, in which disposal in aquifers is estimated to cost less than in oil- and gasfields.) The importance of this cost component is the basis for economies of scale: unit disposal costs at a large disposal site serving two commercial power plants could be 30–35% lower than at sites serving only one plant, whereas they might be 60–70% higher if two small sites are needed for disposing of the captured CO_2 from a single plant. Finally, the McKinsey report notes that EOR and EGR could considerably reduce the overall cost of CCD. However, these options increase the possible range of CO_2 disposal costs even further, making them the most variable cost factor in relative terms. Nonetheless, given the relatively small fraction of global CO_2 generation that could be used for EOR/EGR and the low share of the disposal component in the total CCD costs, the ultimate effect of this large cost variation is modest in global CCD cost accounting.

Hildebrand (2009) presents an in-depth assessment of capture costs by exploring the options of partial capture as opposed to full capture (capturing nominally 90% of CO_2), which involves significant penalties on the technology, plant performance and capture costs. She presents spreadsheet models for both PC and IGCC plant technologies and investigates plant performance and economics as a function of capture percentage. The results show that partial capture can preserve efficiency and that, for a PC plant, the cost savings associated with partial capture are significant. The costs of captured emissions are estimated to be in the range of US$30–40/t CO_2. These numbers do not include the costs for transportation, disposal and monitoring, which can add US$5–15/t.

Focusing on the disposal part of the CCD chain, Eccles et al. (2009) present a general model that represents the maximum CO_2 disposal potential, the maximum injection rate and the cost of CO_2 disposal. By applying the model to deep saline aquifers in sandstone reservoirs in the USA, the authors observe essential linkages between injection rates and the amount of CO_2 that can be stored (increasing with depth, hence decreasing unit costs) and the cost of drilling and injection equipment (increasing with depth, hence increasing unit costs). Characteristics of the reservoirs vary significantly even within the same basin and the actual disposal costs will diverge accordingly, between US$0.01 and US$100/t CO_2. What is more important, however, is that the model estimates US$2–7/t CO_2 for the full range of depth and basin properties for formations not deeper than 3,000 m with what the authors consider base case thickness (65 m) and permeability (22 mD). The authors conclude that in the USA regions with extremely low injections costs exist in many reservoirs but the total capacity of low-cost regions is 'likely to be much lower than the thousands of gigatons often cited as the potential storage capacity of deep saline aquifers' (Eccles et al. 2009, p. 1967).

There are many other issues beyond the direct cost figures that will influence the diffusion of CCD. Narita (2009) maintains that the absence of secondary benefits and uncertainties associated with CCD would require thorough cost-benefit comparisons with other CO_2 mitigation options to be conducted. The author frames his CCD assessment as utilization of a non-renewable resource with a limited capacity. In this framework, scarcity of geological disposal capacity should involve a shadow price which could raise the effective price according to a Hotelling rent. By using a simple analytical dynamic optimization model, Narita examines the optimal paths of CCD use by comparing the operational price with the real price, including the shadow price. He concludes that the inclusion of the shadow price of CCD could make the technology more expensive and thus relatively less attractive compared to, for example, renewable energy sources.

A report by the National Energy Technology Laboratory estimates the costs of CCD in terms of both capital cost and the long-term cost of electricity (NETL 2007). The study includes a detailed breakdown of the likely cost of CCD for the major types of fossil fuel-fired power plants. This report also found that retrofitting CO_2 capture to today's power plants using existing technology is expensive.

For PC plants, the cost of CO_2 capture, transport and disposal in an underground formation could add 70–100% to the cost of electricity. IGCC power plants can achieve a lower cost for CCD because, in this case, CO_2 can be captured from the gas stream from gasifying coal. Yet NETL (2007) estimates that adding present-day CCD technology to IGCC plants would increase the cost of electricity by at least 30%.

Groenenberg and de Coninck (2007) investigate a series of policy instruments for reducing greenhouse gas emissions in the EU. They conclude that while the European Union Emission Trading Scheme (EU ETS) is the most cost-effective instrument, it is difficult to project the level of incentives that it provides in the future for CCD activities. The incentives may remain too weak if the allocation continues to be based on National Allocation Plans, grandfathering and limited harmonization by the European Commission. In this case, the EU ETS is not likely to provide sufficient deployment of CCD in the short term, or even in the longer term because of its short time horizons and because of a lack of commitment by the EU to deep emission reductions over the long term.

Celebi and Graves (2009) observe that CO_2 mitigation policies using cap and trade schemes with a drastic near-term mitigation limit are likely to lead to highly volatile CO_2 prices. This volatility will significantly increase investment risks in mitigation projects like CCD, raise the cost of capital, and thus discourage investments. The authors estimate that CO_2 price volatility could delay investment in CO_2 mitigation technologies by 10 years or more. Their proposed solution to this problem is a safety valve mechanism that involves both a floor and a ceiling on CO_2 prices.

The small sampler of recent cost estimates presented in this section indicates large variations in all phases of the CCD chain. (Some variation in capture costs is due to the timing of the cited report, since costs have been increasing significantly over recent years (see the note about cost escalation above)). This is despite the possibility of drawing on the actual costs of many components observed in related industries, especially oil and gas drilling and transport. In relative terms the smallest variation is in capture costs followed by a somewhat larger variation in transport costs. Disposal costs tend to have the widest ranges in relative terms because of the large range of possible disposal formations and the possibility of revenue generation. However, as capture costs dominate by far the total CCD price tag, even the larger variation in disposal costs causes only a small variation in downstream costs and an even smaller one in the total electricity cost.

2.3 *Relative Importance of the Disposal Costs*

Considering the diversity of the CO_2 disposal options and the resulting wide range of disposal cost estimates, any assessment of the level of disposal cost per unit of electricity or of the relative importance of the disposal costs in the total fuel cycle costs and in the levelized cost of electricity (LCOE) should be handled with care. This section presents some rather broad calculations based on the disposal costs found in recent literature and presented in the previous section.

The estimated disposal costs per unit of electricity generated vary across a very wide range (see Table 3). The main driver of this variation is the targeted geological formation (saline aquifers versus depleted oil- and gasfields), while the CO_2 intensity of the power generation technologies plays only a minor role.

Combining the estimated CO_2 disposal costs with the fuel costs (taken from recent cost estimates of capture-relevant technologies) allows us to calculate the share of the former in the total fuel cycle costs. Cheap disposal options represent only a low share (a few per cent) in the total fuel cycle costs, while expensive geological targets can amount to 35–40% of the fuel cycle costs. Not surprisingly, this pattern can also be observed when we calculate the share of the disposal cost in the extended LCOE (base plus disposal costs). This portion represents a very low share for low-cost disposal options, but can reach 15–20% for expensive geological options.

It is important to emphasize that the numbers in Table 3 result from, at best, indicative conceptual calculations. The basic insights concerning the dominance of the geological formations as the main driver of the disposal cost are robust, but the numbers should not by any means be considered as precise estimates.

3 Costs of Radioactive Waste Disposal

3.1 Overview of the Main Cost Items

The International Atomic Energy Agency's (IAEA) new General Safety Guide on Classification of Radioactive Waste (Safety Standards Series No. GSG-1) (IAEA 2009a) classifies RW primarily on the basis of long-term safety considerations and the associated disposal options. High-level waste (HLW) is waste with radioactivity levels high enough to generate significant quantities of heat through the radioactive decay process or with large amounts of long-lived activity. Disposal in deep stable geological formations with engineered barriers is an option considered appropriate for the disposal of HLW. Intermediate level waste (ILW) is waste which, because of its content, requires a higher level of containment and isolation than is provided by near-surface disposal, however, with no or only limited provision for heat dissipation during its storage and disposal. A repository for ILW is distinguished from a repository for HLW by the degree of integrity and stability of the geological formation, and not necessarily by the depth of the repository, although the repository for ILW is sometimes referred to as an intermediate-depth disposal as opposed to deep geological disposal for HLW. Deep disposal of ILW has also been discussed, but mainly for social and economic rather than safety reasons.

For the purposes of comparing the cost of RW and CO_2 disposal, we focus on HLW, including SNF, for which the deep geological repository concept is generally envisaged on the grounds of long-term safety considerations.

Cost studies for the following waste repositories are the main sources of the data discussed in this section: a final waste repository (which has now been suspended) at Yucca Mountain in the USA (OCRWM 2008a), a final waste repository facility at Olkiluoto in Finland (Kukkola and Saanio 2005), a final waste repository at

Table 3 The share of CO_2 disposal costs in the fuel cycle and levelized electricity costs

	Disposal cost (US$/t CO_2)	Emissions (t CO_2/MWh)	Disposal cost (US$/MWh)	Fuel costs (US$/MWh)	F+D costs (US$/MWh)	Disposal cost share in F+D (%)	LCOE (US$/MWh)	LCOE+ disposal cost (US$/MWh)	Disposal cost share in LCOE (%)
McCoy and Rubin (2005) – PC	1.63	1.07	1.74	19.78	21.52	8.08	88.64	90.38	1.92
McCoy and Rubin (2005) – IGCC	1.63	0.9	1.46	19.81	21.27	6.88	93.44	94.91	1.54
Pöyry Energy Consulting (2007) aquifers	1.68	0.81	1.36	29.06	30.42	4.46	100.90	102.26	1.33
Pöyry Energy Consulting (2007) oil/gas fields	33.53	0.81	27.16	29.06	56.22	48.31	100.90	128.06	21.21
Vosbeek and Warmen-hoven (2007)	3.81	0.77	2.94	29.01	31.94	9.19	91.88	94.82	3.10
Dooley et al. (2008) – LDC	1.03	1.02	1.05	20.99	22.04	4.77	91.11	92.16	1.14
Dooley et al. (2008) – HDC	14.44	1.02	14.73	20.99	35.71	41.23	91.11	105.84	13.91

McKinsey & Company (2008) LDC, LE	11.18	0.90	10.06	29.06	39.12	25.72	90.81	100.87	9.98
McKinsey & Company (2008) LDC, HE	11.18	1.00	11.18	29.06	40.24	27.79	90.81	101.99	10.96
McKinsey & Company (2008) HDC, LE	26.09	0.90	23.48	29.06	52.54	44.69	90.81	114.29	20.55
McKinsey and Company (2008) HDC, HE	26.09	1.00	26.09	29.06	55.15	47.31	90.81	116.90	22.32
Eccles et al. (2009) LDC	2.02	1.02	2.06	20.99	23.05	8.93	91.11	93.17	2.21
Eccles et al. (2009) HDC	7.06	1.02	7.20	20.99	28.19	25.55	91.11	98.32	7.33

Disposal costs: based on Table 2, converted to 2010 US$ by using currency-related GDP deflators and 2010 exchange rates
Specific emissions: taken from the source study if included or industry average from the IEA or US DOE online databases
Fuel costs: based on fuel cost estimates in IEA and NEA (International Energy Agency and Nuclear Energy Agency) (2010)
LCOE (levelized cost of electricity): based on IEA and NEA (2010) at 10% interest rate
F + D costs: fuel cost + disposal cost (without capture and transport): LDC low disposal cost: HDC high disposal cost: LE low CO_2 emissions/MWh: HE high CO_2 emissions/MWh

Forsmark in Sweden (SKB 2003), an unidentified final waste repository in Belgium (ONDRAF/NIRAS 2001a), an unidentified final waste repository in Japan (METI 2008a), an unidentified regional joint waste repository of 14 EU countries (Chapman et al. 2008) and an unidentified final waste repository in the UK (Nirex 2005). These studies vary in terms of their coverage, assumptions, level of detail and uncertainty, and transparency of the cost estimation methodology. The costs quoted are mainly overnight costs (i.e. interest to be accrued during construction and price escalation effects is ignored and future costs streams are not discounted, unless otherwise noted).

Generally speaking, the cost estimates for the radiological waste disposal consist of the following elements in the three main phases:

- *Pre-operation phase*: site investigation and characterization, and development and construction of a repository, transportation system to a repository, and encapsulation plants and other above-ground facilities;
- *Operation phase*: transportation, encapsulation and emplacement of wastes;
- *Post-operation phase*: decommissioning of the above-ground facilities and closure and monitoring of the repository.

Throughout all the phases, costs for programme administration are incurred. R&D may or may not be explicitly included. When SNFs are reprocessed, reprocessing costs and disposal costs of associated long-lived low and intermediate level wastes (LILW-LL) are not included. Table 4 summarizes what the total costs include in the above-mentioned studies. To the extent possible, the terminology used in the respective original studies is used when the cost components are being presented. Details of each study and definitions of some of the terminology are given in Sect. 3.2.1.

During the *pre-operation phase*, activities accounted for in the cost estimates for site investigation and characterization may include land acquisition costs (Japan and Belgium), costs related to site selection and conceptual designs for the development of the repository (USA). This process may take 10–30 years, including a period for preliminary siting studies and licence approval.

Construction of the underground facility (repository) and the above-ground facilities (encapsulation plants, on-site/off-site infrastructure, administration buildings, etc.) are also major cost items for the pre-operational phase. It may also include costs related to licensing, design, management, engineering and procurement. Construction of the underground facility may take 5–10 years. Typically, the operation and construction are planned to go partially in parallel, and for that reason not all construction costs occur during the pre-operational phase. For example, in the UK estimates, the construction costs after the first waste emplacement account for 32% of the total repository construction costs, mainly due to the construction and fit-out of the remaining disposal tunnels.

Some estimates include the costs for a waste transportation system. In the USA, the waste transport was planned to be handled mostly by rail, using dedicated trains. Costs for acquiring rail, truck cask systems and rolling stock for the national transportation system, as well as the costs for providing the interface between the

Table 4 Cost components and estimates for radioactive waste disposal in selected studies[a]

Country	Site investigation/ characterization	Underground facility	Above-ground facility	Post closure monitoring	Transportation	Programme administration	Other	Total
USA 2007 US$ (million)	Repository development and evaluation costs US$8,330 M	Licensing costs, engineering, procurement and construction cost, emplacement operation costs, pre-closure monitoring costs and closure costs US$56,400 M (4% for licensing, 28% for other engineering, procurement and construction, 47% for emplacement operations, 18% for monitoring, and 2% for closure)			Transportation costs US$20,250 M	Balance of programme costs US$11,200 M (21% for development and evaluation, 7% for quality assurance, 3% for waste management, 29% for programme management, 28% for benefit paid to the local entities, 12% for funding other agencies)	–	US$96,180 M
Finland December 2003 € (million)	Not included	Repository costs €714 M (investment 50%, operation 34%, closure 15%)	Above-ground facility (encapsulation plant, operating building, ventilation shaft building, and research building) €1,827 M (investment 8%, operation 92%, decommissioning less than 1%)	Not included	Accounted as a part of operational costs of above-ground facilities (€0.3 M)[b]	–	Contingency of 20% is included in each cost item	€2,542 M

(continued)

Table 4 (continued)

Country	Site investigation/ characterization	Underground facility	Above-ground facility	Post closure monitoring	Transportation	Programme administration	Other	Total
Sweden January 2003 SEK (million)[c]	Deep repository—siting and site investigation SEK 2,058 M	Deep repository—spent fuel SEK 8,150 M (investment 56%, operation and maintenance 14%, decommissioning and backfilling 30%)	Encapsulation plant, deep repository off-site facilities, deep repository above-ground facilities SEK 13,783 M (investment 32%, operation and maintenance 66%, decommissioning 2%)	Not included	Transport (for all waste types) SEK 3,024 M (investment 52%, operation and maintenance 48%)[d]	RD&D and administration costs SEK 9,692 M	Incurred costs are included under each cost categories	SEK 36,707 M
Japan 2008 ¥ (billion)	Site investigation and selection and land acquisition ¥174 B	Design and construction, operation, monitoring, and decommission and closure ¥1,823 B (design and construction 50%, operation 40%, monitoring 5%, decommissioning and closure 6%)		Included implicitly in the project administration costs		Project administration ¥553 B	Value-added tax ¥103 B Technology development ¥104 B	¥2,757 B
Belgium 2000 € (million)	Land acquisition, site characterization and site preparation	Building licence, off-site infrastructure, surface facilities, excavation, equipment for post-conditioning, handling equipment, backfilling 0of galleries, operating costs and personnel costs					Neither incurred R&D expenditure of €150 M nor additional R&D expenditure of €75–300 M is included in the total costs	

Reprocessing option	€23.4 M	€554.8 M (63% for investment phase, 22% for waste emplacement phase, 15% for closure phase)[e]		Not included	Not included	Not explicitly reported	50% contingency on average is included under each cost item	€578 M
Direct disposal option	€41.4 M	€1452.4 M (68% for investment phase, 10% for waste emplacement phase, 22% for closure phase)		Not included	Not included	Not explicitly reported	61% contingency on average is included under each cost item	€1,494 M
UK June 2004 £ (million)	Site characterization and rock characterization facility				Transport costs	Nirex internal and other programme works		
Non-retrievable option[f]	£755 M	£1,735 M (46% for construction, 47% for operation, 7% for sealing and closure)	£928 M (41% for construction, 57% for operation, 1% for sealing and closure)	Not included	£302 M[g]	£960 M		£5,226 M
Retrievable option		Common to above-ground and underground facilities £546 M (73% for construction, 27% for sealing and closure)		Maintenance and refurbishment for 250 years £1207 M[g]				£6,765 M
European Commission 2006 € (million)	Repository siting	Repository construction, operation and closure	Encapsulation facility		Transportation	Repository related R&D and administration		

(continued)

Table 4 (continued)

Country	Site investigation/ characterization	Underground facility	Above-ground facility	Post closure monitoring	Transportation	Programme administration	Other	Total
Swedish Cost Model	€231 M	€4,121 M (45% for construction, 34% for operation, 21% for closure)	€2,633 M	Not included	€1,065 M	€1,091 M		€9,141 M
Swiss Cost Model	€512 M	€4,698 M (45% for construction, 28% for operation, 27% for closure)	€2,433 M	Not included		€320 M (including compensation)		€9,029 M
Finnish Cost Model	Not explicitly reported	Repository facilities, operations and closure €2,514 M (54% for facilities, 40% for operations, less than 6% for decommissioning)	Above-ground facilities, operations and decommissioning €7,083 M (2% for facilities, 98% for operations, less than 1% for decommissioning)	Not included		Not included		€10,662 M

[a] The shares may not sum up to 100% due to rounding
[b] Already included in the cost of the above-ground facility
[c] The incurred costs through 2003 are given in current prices, and future costs estimates after January 2004 are given in the constant prices of January 2003. In this presentation, inconsistencies of the reported prices areignored and a simple sum of incurred costs and future costs is presented
[d] This share is applicable only to the future costs, as the share for the incurred costs are not reported
[e] The Belgium report presents the costs according to the phases
[f] Non-retrievable costs do not include maintenance and refurbishment costs
[g] This estimation is taken from a separate study (Nirex 2006), in which the price level of September 2003 was used. It has been adjusted to the price level of June 2004 using a conversion rate of 1.006

national transportation system and the repository, are included in their cost estimation. In Finland, the SNF will be transported by road. The transportation costs include transporting trailers and SNF transport casks. The SAPIERR II project (details below) also provides a rough estimate based on the weight of waste to be handled, without considering the transportation distances.

R&D costs may or may not be included explicitly in the total cost estimates. In the case of Japan, a cost item called 'technology development' is included. In the case of Finland and Belgium, R&D costs are explicitly excluded.

For the *operation phase* of a project, the costs largely depend on the amount of waste to be disposed of. Execution of waste transport, waste handling, purchase/ manufacturing of the casks/canisters (USA and UK), buffer material production (Japan, UK, Finland), encapsulation and emplacement of waste packages into a repository are activities often accounted for in the total cost estimates.

During the *post-operation* phase of RW disposal, the main activities include decommissioning of the above-ground facilities, restoration of the surface area and closure and monitoring of the underground facility. Monitoring may include pre-closure and post-closure monitoring. In the case of the USA, permanently installed sensors would monitor waste packages, emplacement drift and the surrounding rock, providing the data to confirm performance during the pre-closure monitoring period of 50 years. In the US estimate, fabrication of drip shields which would be emplaced during this period constitutes a major part of the pre-closure monitoring costs. Closure activities include backfilling of shafts and ramps, sealing, and protection of the repository from unauthorized intrusion. The period assumed in the cost estimates for closure cover 3–20 years. The UK estimates involve longer periods for post-closure monitoring, foreseeing up to 300 years.

Administration costs may include safeguards and security activities, regulatory, infrastructure and management support costs. Other miscellaneous costs included in the cost estimates are benefits paid to state and local entities (governments and tribes) (USA), contingency (Finland, Belgium) and value added tax (Japan).

3.2 Costs of Deep Geological Disposal of High-Level Waste

3.2.1 Cost Estimates from Various Countries

In the USA, Congress passed and the President signed a public law which approved Yucca Mountain as the site for a waste repository in 2002. The US Government announced suspension of all activities in 2009 (and the final decision is still pending), but the cost studies still provide valuable information. The Nuclear Waste Policy Act (NWPA) of 1982 put a limit on the emplacement of the SNF to 70,000 tonnes of heavy metal (tHM) in the first repository. Of the 70,000 tHM, 63,000 tHM is allocated to civilian waste. However, more than 58,000 tHM commercial SNF is already in storage, and the total inventory of commercial SNF is expected to grow at a rate of about 2,000 tHM/year. In 2008 the Secretary of Energy submitted, in accordance with the 1982 NWPA, a recommendation to the President and Congress that the current 70,000 tHM statutory limit should be removed, otherwise a second repository would be needed (OCRWM 2008b). Although the operation was expected to start no sooner than 2020, the programme has been at a standstill since February 2009 (WNN 2009).

Costs of disposal of the SNF and vitrified HLW are estimated by the Office of Civilian Radioactive Waste Management (OCRWM), providing a basis for assessing the adequacy of the Nuclear Waste Fund Fee as required by the NWPA (OCRWM 2008a). The latest estimates assume that all currently projected SNF and vitrified HLW from civilian and defence use will be disposed of at the Yucca Mountain repository. The projected amount referenced in the cost estimation is 122,100 tHM of SNF and vitrified HLW. This estimate is based on the discharge projections from all reactors operating until the end of licensed lifetimes, taking into account 47 reactors to which licence extensions were granted by the National Regulatory Commission (NRC) as of January 2007.

The total system life cycle costs span the period from 1983 to the assumed closure date of 2133, and total US$96.18 billion at a constant price of 2007. The total costs consist of costs for the repository, transportation and the balance of the programme, accounting for US$64.7 billion, US$20.3 billion, and US$11.2 billion, respectively. Of the repository costs, US$9.9 billion had been disbursed as of 2006, of which the 'development and evaluation' phase accounts for the major part (US$8.3 billion). The remaining repository costs are accounted for by the 'engineering, procurement, and construction' phase (32%, nearly half of which had already been disbursed), the 'emplacement operation' phase (the largest cost item, accounting for 47%, nearly half of which is due to the fabrication of waste packages), the 'monitoring' phase (18%, mostly due to fabrication of drip shields), and the 'closure' phase (2%). Part of the costs of the 'engineering, procurement, and construction' phase are accounted for by the licensing costs (4% points out of 35%). 'Monitoring' phase costs refer to costs related to pre-closure monitoring, which is assumed to last for 50 years after the emplacement operations.

Transportation costs consist of those related to the design of the transportation system, the National Transportation Project (transport from waste generating sources to the state of Nevada) and the Nevada Rail infrastructure project (providing an interface between the nationwide transportation system and the repository).

In Finland the Parliament ratified a decision-in-principle in 2000 for the construction of a final disposal facility for SNF at Olkiluoto (Finnish Ministry of Employment and the Economy 2001). The disposal facility will be constructed after the licence from Government is received during the period 2013–2019. The disposal facility will start operating in 2020, and continue its operation for over 100 years (Kukkola and Saanio 2005). Licences to construct and operate the final disposal facility are currently under development by Posiva, a company responsible for the final disposal of SNF. It is planned to dispose of the SNF generated from four existing and one new nuclear power plant (NPP) in the repository; the amount of SNF is estimated to be equivalent to approximately 6,500 tonnes of uranium (tU) (Finnish Ministry of Employment and the Economy 2002). Another application for a decision-in-principle on an extension to the final disposal facility for disposing of SNF from the second new NPP (Olkiluoto 4) was submitted to the Government in 2008 for approval (Posiva 2008a). If it is approved, the total amount of SNF to be disposed of increases to 9,000 tU.

The costs of the disposal facility at Olkiluoto were estimated by Posiva in 2005 (Kukkola and Saanio 2005). The estimates are based on the disposal of SNF corresponding to 5,643 tU. The total costs are estimated at €2,542 million, at a constant December 2003 price. They include transportation costs and contingency of 20% and do not include R&D costs and site selection costs. SNF will be transported to the encapsulation plant by road. The costs are divided into three periods, namely, pre-construction/construction, operation and closure, and account for 11%, 85% and 5% of the total costs, respectively. Across all these periods, investment-related costs account for 21%, whereas operation-related costs account for the rest. The main cost components are the operation costs of above-ground facilities (66% of total costs), of which 80% is accounted for by costs for encapsulation materials and personnel. Transportation costs are insignificant. These estimates are based on the reference design, in which the canisters with SNF are emplaced vertically in individual deposition holes. A separate study conducted by Posiva and SKB (Posiva 2008b) evaluated costs for an alternative design, KBS-3 H, in which the canisters are serially emplaced in long horizontal drifts. It was estimated that for the Olkiluoto site, KBS-3 H would realize savings of €96 million (at a constant price for an unspecified time period).

In Sweden site investigations began in 2002 at Forsmark and Laxemar. In June 2009 Forsmark was selected as a site for the final repository based on the results of these investigations. The repository is expected to have a capacity of 6,000 canisters (about 12,000 of SNF) and be located at a depth of about 500 m. The operation is expected to start by the beginning of the 2020s and to continue for about 40 years (SKB 2010). The costs for management and disposal of all kinds of RW were estimated in 2003 (SKB 2003). They cover costs for RD&D, transportation, a central interim storage facility for SNF, encapsulation of SNF, a deep repository for SNF, final repositories for LILW-LL, reactor waste, short-lived waste and waste produced during decommissioning, as well as costs of decommissioning the NPPs. The attribution of costs specific to a deep geological disposal of SNF is not provided in the study. However, it is fair to assume that the majority of the RD&D costs, some transportation costs, and all costs related to encapsulation and a deep repository for SNF are considered as costs associated with deep geological disposal of SNF.

The number of canisters referenced in the cost estimate is 4,500, corresponding to the existing and expected SNF of 9,493 tU. Of the total waste management costs for all waste categories, namely (Swedish krona) SEK 49,600 million, the subtotal related specifically to disposal of SNF was estimated to be SEK 29,870 million (approximately €3.2 billion) at a constant 2003 price. It includes SEK 4,860 million for RD&D and administration (including costs attributable to waste disposal other than SNF), SEK 2,230 million for investment, operation and maintenance of transport (including costs attributable to transportation of waste other than SNF),

SEK 7,920 million for investment, operation, maintenance and decommissioning of an encapsulation plant (including canister plants), and SEK 14,860 million for the deep repository. The costs for the deep repository include the following cost categories: siting, off-site facilities (investment and operation), above-ground facilities (investment, operation, maintenance and decommissioning) and underground facilities (investment, operation, maintenance, decommissioning and backfilling), accounting for 7%, 2%, 36% and 55%, respectively, of total deep repository costs.

The above does not include incurred costs through 2003, which, at current prices, are estimated to be SEK 4,832 million for RD&D and administration (including costs attributable to waste disposal other than SNF), SEK 794 million for transportation (including costs attributable to waste disposal other than SNF), SEK 192 million for an encapsulation plant, and SEK 1,018 million for siting and site investigations for the deep repository, totalling SEK 6,837 million. For presentation in Table 4, the incurred costs are added to the projected costs for each cost item, making the total costs SEK 36,707 million.

In Japan, under the Specified Radioactive Waste Final Disposal Act adopted in 2000, the Basic Policy on disposal of vitrified HLW was established in 2000 and revised in 2008. The policy sets out a timeline, starting from site selection (to be completed by around 2028), to construction of a final repository facility and operation of the repository (starting from around 2033–2037). The minimum capacity of the repository should be 40,000 canisters, which are estimated to be generated from reprocessing of SNF from nuclear electricity generation by 2021 (METI 2008b). The basis of the calculation is that 1 GW/year of NPP operation produces 30 units of vitrified HLW (METI 2008c; NUMO 2004). This indicates a reference energy production of 11,670 TWh.

The final disposal costs were estimated by the Advisory Committee for Natural Resources and Energy for two rock types (soft rock and hard rock) (METI 2008a). They are not substantially different (costs for the soft rock type are 5% higher than the hard). The average of the total costs for the two rock types is (Japanese yen) ¥2,757 billion (approximately US$27 billion) at a constant 2008 price. Costs are given according to the following cost categories: technology development (8.5%), site investigation and land acquisition (12.1%), design and construction (20.8%), operation (21.8%), monitoring (9.7%), decommissioning and closure (1.3%), project administration (22.0%) and value added tax (3.7%). The expenditure for decommissioning and closure is assumed to be due between 2075 and 2099. Project administration costs are assumed to be due between 2100 and 2395, presumably for post-closure monitoring purposes. Underground and above-ground facilities considered in the cost estimates include off-site infrastructure (harbour facilities and dedicated roads), an encapsulation plant and buffer material production facilities. Information on the way that the costs are attributed to each of these facilities is not provided. Costs related to two underground research laboratories do not appear to be included.

In Belgium deep disposal of HLW is considered as the reference solution. Research will continue for several years before a concrete decision is taken on the way the waste will actually be disposed of and where (ONDRAF/NIRAS 2009). The

Belgian Agency for Radioactive Waste and Enriched Fissile Materials (ONDRAF/NIRAS) published a report in 2001 (ONDRAF/NIRAS 2001a, b) that includes a cost estimate for a deep disposal facility. A reference site used for cost estimating purposes is the Boom Clay beneath the Mol-Dessel nuclear zone. The reference design for the vitrified HLW shows that there will be a total of 3,915 waste packages (corresponding to 4,860 tHM of conventional uranium fuel, all reprocessed) to be disposed of. The design also assumes that it will not only be used for the HLW but also for LILW-LL, although the cost assessments relate solely to disposal of vitrified HLW and SNF. According to the reference timetable, detailed design and safety studies take 10 years, and construction, operation and closure altogether take 30 years (including 20 years attributable to LILW-LL), assuming they are carried out partially in parallel. The total cost (attributable only to disposal of vitrified HLW and SNF), including construction, operation and closure of the repository, as well as the contingency margins for each of them, is estimated at €578 million (of which 50% is accounted for by the contingency margin) at a constant 2000 price. The costs are divided into three implementation stages: construction, operation and closure, each accounting for 64%, 21% and 15%, respectively, of total costs. The contingency margins for the construction and operation stages are 95%, and for the closure phase 138%. The cost estimates do not account for the R&D costs, which were approximately €150 million for the period 1974–2000 (at a constant 2000 price). ONDRAF/NIRAS estimated that additional R&D spending of €75–100 million should be enough to enter a pre-project phase, which is site specific, assuming that disposal and the Boom Clay are confirmed, respectively, as the long-term management option and the host formation. Should the authorities indicate their preference for another geological formation, then R&D spending to attain the same objective would be €250–300 million.

The report also estimates the costs of a direct disposal option, in which reprocessing is assumed to stop after the reprocessing of the 630 tHM SNF foreseen under the existing contract, the remaining SNF being disposed of without reprocessing. In this case, construction, operation and closure are assumed to take 40 years (of which 22 years are specifically for deposition of the vitrified wastes and SNF). The cost would then be estimated at €1,494 million (61% being accounted for by a contingency margin) at a constant 2000 price. The shares for three implementation stages are 70% for construction, 9% for operation and 21% for closure. The respective contingency margins for each step are 140%, 170% and 200%. The costs are much higher with this option, even discounting the fact that much higher contingency margins are assumed. It primarily reflects the fact that the galley space required for the *disposal* of vitrified HLW and SNF is much larger than is the case with a reprocessing option (a total length of 44 km instead of 6.5 km) and that the total length of the *main* gallery, to which the disposal galleries are connected, needs to be longer (4,245 m rather than 760 m) to allow the increased gallery spacing required for the SNF disposal.

In the UK, Nirex (2005) estimated costs for a repository for vitrified HLW and SNF, based on the Swedish repository concept, KBS-3. Cost estimates are based on the assumption that the repository would be a stand alone facility for HLW/SNF.

A reference timetable assumed in the cost estimation is: site characterization from 2007 to 2020, construction and underground research from 2020 to 2040, operation of the facility from 2040 to 2090, and closure from 2090 to 2100. The total number of canisters to be disposed of is 7,088 units, of which 3,700 units are for vitrified HLW, 572 units for SNF from pressurized water reactors (PWRs), and 2,816 units for SNF from advanced gas-cooled reactors (AGRs). To what extent these units correspond to weight is not stated. Although there is no direct reference to it, the UK Committee on Radioactive Waste Management (CoRWM) provides the national baseline inventory of RW in the UK (CoRWM 2005). Assuming conservatively that reprocessing of SNF will be discontinued (although reprocessing of all the existing and future SNF is planned), it consists of 54,500 tU of existing reprocessed SNF, which corresponds to 1,290 m^3 of HLW, 1,200 tU of PWR SNF, 3,500 tU of AGR SNF and 125 of plutonium and highly enriched uranium.

Total costs are estimated at £4.9 billion (approximately US$9 billion) at a constant June 2004 price. This total cost does not include transportation costs, post-closure costs and contingencies. The total costs of £4.9 billion are broken down into 'site characterisation', 'rock characterisation facility', 'repository construction to first waste emplacement', 'repository construction post first waste emplacement', 'repository operation', 'repository sealing and closure', and 'Nirex international and other programme works'. Repository construction is the major cost category (32%), followed by operation (27%) and programme works (19%). Above-ground facilities include a canister factory, an encapsulation plant, a bentonite/backfill plant, off-site infrastructure and other on-site infrastructure.

There is no explicit mention as to whether disposal of plutonium and enriched uranium is included in the cost estimates. In a similar study conducted by Nirex (2006), costs for disposal of plutonium (mainly from civilian sources) and enriched uranium (mainly from military sources) are estimated at £1.6 billion at a constant September 2003 price, on top of the £5 billion estimated for disposal of vitrified HLW and SNF. Furthermore, the operation period would be extended by 15 years. The transportation costs of canisters from waste generating sites to the repository were also estimated in this study but turned out to be minor (£0.3 billion). The total costs presented in Table 4 include the transportation costs. Should retrievability be retained as an option, the costs for maintenance and refurbishment before complete sealing of the repository, estimated at £1,207 million, should be added. It is assumed that maintenance and refurbishment will take place between the 50th year (at the end of the emplacement phase) and the 300th year from the first waste emplacement.

The SAPIERR II (Strategic Action Plan for Implementation of European Regional Repositories) project, supported by the European Commission and with the participation of 14 European countries (Austria, Belgium, Bulgaria, Croatia, Czech Republic, Hungary, Italy, Latvia, Lithuania, the Netherlands, Romania, Slovakia, Slovenia, Switzerland), published a report that includes cost estimates for a multi-national common repository programme (Chapman et al. 2008). Cost estimates were based on the waste inventory data and time schedule established in the predecessor project, SAPIERR (Support Action: Pilot Initiative for European

Regional Repositories) (Štefula 2006). The volume of non-processed SNF stored in SAPIERR countries by 2040 was estimated to amount to 25,637 tHM, based on the assumptions that (1) no new nuclear power reactors will be built, (2) the existing ones will operate until the end of their operational life time, and (3) there will be no plant life extension. The volume of vitrified HLW from SNF reprocessing by 2040 is estimated at 355 m^3, which roughly corresponds to 3,220 tHM of SNF, with the remaining volume of SNF being disposed of as SNF without reprocessing. (Note that for Bulgaria, Italy and the Netherlands, the volume of HLW was given only in terms of mass (150 m^3)). A conversion rate of 9 tHM/m^3, obtained based on Belgian and Swiss inventory reports (Štefula 2004), was used to derive the SNF equivalent of 1,350 tHM. The reference time schedule for a repository is start of repository operation in 2035, with the total length of operation being 50–60 years.

The cost estimates were prepared for six scenarios, four of which assume the joint disposal of HLW and ILW, and two of which assume repositories for disposing exclusively of SNF and vitrified HLW. Two scenarios correspond to the different rock types (hard rock and sediment rock). For the hard rock type, €8.1 billion (using a Swedish cost model) and €9.6 billion (using a Finnish cost model), both at a constant 2006 price, were estimated as costs for a repository and an encapsulation plant. For the sedimentary rock, €8 billion at a constant 2006 price was estimated (using a Swiss model). The costs for an encapsulation plant and for a repository were distinguished under the Swedish and Swiss cost models, and the costs for an encapsulation plant account for slightly over 40% in both cases whereas costs for the repository account for the rest. Three cost models were applied assuming that the disposed waste would be the half the reference volume. In this case the costs were estimated at €4.7–5.2 billion, indicating economies of scale effects. Indicative transportation costs were estimated at up to €1 billion, assuming unit costs for SNF transport were €40,000/t for the international transports that a European regional repository would require. The mode of transport is not specified and the estimate is based solely on mass. The total cost presented in Table 4 includes this transportation costs.

3.2.2 Amount of Radioactive Waste from Nuclear Power Generation and the Disposal Capacity

Most HLW arises as SNF from the operation of NPPs and as vitrified HLW from reprocessing of SNF. The amount of waste arising is determined mainly by the amount of electricity produced and the choice between direct disposal or reprocessing of SNF. The amount of waste generated is then used as a key parameter in estimating disposal costs. Assumptions regarding electricity production and the extent to which reprocessing of SNF is applied are used in the cost estimates discussed in the previous section, and are summarized in Table 5. Among the reports reviewed, only the Swedish report mentions explicitly the corresponding electricity generation. The value for Japan was calculated by the authors using the published ratio between the electricity production and the amount of vitrified HLW. Note that in the reviewed reports, the amounts of SNF are reported in different units. In subsequent sections, we present the amount of RW in terms of tHM (post-irradiation weight). In doing so, it was assumed that the unit quoted as tHM in various reports refers to the post-irradiation weight, rather than pre-irradiation weight, and that the unit quoted as tU refers to the initial weight of uranium in a UO_2 fuel assembly before irradiation. A ratio of conversion from an initial 1 tU of fresh fuel into fission products is applied to obtain the heavy metal weight of irradiated fuel. The conversion ratio is proportional to the burnup ratio (i.e. for a burnup ratio of 10 GWd/tU, the conversion ratio is 0.0105). In other words, for each 10 GWd/tU of burnup, the initial 1 tU becomes 0.9895 tHM, with the remaining 0.0105 having been converted into fission products.

There is a general relationship between the electricity generated and the weight of heavy metal in fresh fuel (the same as the weight of uranium in fresh fuel in the case of UOX fuel): the weight of SNF (tU) is approximated to be equal to electrical energy (GW/year) divided by the product of the efficiency (in per cent) and burnup ratio (GWd/tU). According to the authors' calculation using data from the IAEA's PRIS database (IAEA 2010a), the average net thermal efficiency and the burnup ratio of all power plants in the world including those shut down, weighted by the cumulative net electricity generated, are 32.9% and 35.7 GWd/tU, respectively. The average amount of SNF generated per GW/year of net electricity produced by all the reactors, weighted by the total cumulative net electricity production and converted into the weight of heavy metal using the above mentioned conversion procedure, is 39.9 tHM, while the averages for the PWR, boiling water reactor (BWR) and pressurized heavy water reactor (PHWR) subsets are 37.0, 41.8, and 165.1 tHM, respectively. In assessing how much SNF will be produced in the future, one should take into account that the amount of waste generated per unit of energy produced has been continuously reduced because of technological advances. For example, if we compute the average volume of SNF per GW/year excluding NPPs already shut down, then the amount of SNF per GW/year electricity is reduced to 38.5 tHM (with the PWR producing 29.2 tHM, the BWR 27.5 tHM and the PHWR 157.6 tHM). Note that in this calculation we assume that the weight of uranium in fresh fuel and the weight of heavy metal in fresh fuel is the same (i.e. UOX fuel is used). A fraction of this SNF is sent for reprocessing, producing HLW which is vitrified and stored for eventual final disposal. The remaining part of the SNF is also stored for eventual final disposal. As discussed in connection with the Belgium cost estimates, as the vitrified HLW requires about ten times less repository space than the equivalent amount of SNF, reprocessing reduces the overall space requirements of a repository for vitrified HLW and SNF, and thus decreases the costs. However, at the same time, reprocessing generates low-level waste (LLW) and ILW, which obviously increases the total costs by the amount required for their disposal. It is beyond the scope of this chapter to assess this trade-off.

Deep geological disposal of RW relies heavily on engineered barriers in addition to natural barriers. Construction of an underground facility requires massive underground engineering, which in turn implies some flexibility with respect to

Table 5 Assumptions on the amount of the wastes to be disposed of used in the respective cost estimates in Table 4

Study	Reference energy production	Spent fuel generated	Portion of spent fuel reprocessed	Expected number of waste packages to be disposed of
USA	Until the end of service life of respective NPP	122,100 tHM (of which 109,300 is commercial SNF)	10,300 tHM (belonging to US DOE with unknown share of commercial origin)	17,450 units of waste packaging
Finland	Until the end of service life of respective NPP	5,643 tU	None	2,899 units of canisters
Sweden	2,769 TWh	9,493 tU (all is due to commercial electricity production)	197 tU (producing no HLW)[a]	4,500 units of canisters
Japan	11,670 TWh until 2021	NA	All spent fuel is reprocessed	40,000 units of canisters
Belgium	Until the end of 40 years of service life	4,860 tU (UOX fuel) and 70 tHM (MOX fuel)		
Reprocessing option			4,860 tU of conventional uranium fuel	3,915 units of vitrified HLW, 144 units of MOX fuel
Direct disposal option			630 tU of conventional uranium fuel	420 units of vitrified HLW, 9,715 units of uranium spent fuel, 144 units of MOX fuel
UK	n.a.	n.a. Base line inventory: 59,200 tU	n.a. Base line inventory: 54,500 tU	3,700 units of HLW canisters, 3,388 units of SNF canisters
SAPIERR (EC)	Until the end of service life of respective NPP	25,637 tHM excluding those reprocessed	355 m^3 [b]	13,246 units of SNF canisters, 2,021 units of HLW canisters

NPP: nuclear power plant, *SNF*: spent nuclear fuel, t*HM*: tonnes of heavy metal, *tU*: tonnes of uranium, *US DOE*: United States Department of Energy

[a]This is because that reprocessing takes place abroad

[b]According to the author's calculation, this roughly corresponds to 3220 tHM of spent fuel

the capacity, compared to CCD, for which capacity is primarily defined by the availability of a suitable geological formation at a given site.

The density of HLW/SNF disposal in a disposal gallery is determined by thermal conditions, such as decay heat, and properties in the buffer and in the surrounding rock, as well as the requirement to ensure that the possibility of criticality will not be a concern. Greater thermal loads can be accommodated by extending the time that the repository is open and ventilated prior to repository closure. How the wastes are loaded in waste packages and whether the waste packages are stored to allow decay prior to emplacement are also key parameters determining the amount of waste that can be placed in a given volume of rock (OCRWM 2008b).

A repository for HLW is typically designed in such a way that all the expected waste to be dealt with in a given jurisdiction is disposed of at a selected site, and the capacity of a repository is thus determined primarily by the amount of expected waste in the foreseeable future. As discussed later, there is a strong economies of scale issue. Extending capacity at a later stage may be possible with a relatively small marginal increase in costs, as fixed cost components may account for a significant portion of the total costs, particularly for a smaller repository. When comparing the costs of RW and CO_2 disposal based simply on some sort of unit cost (i.e. costs per waste, or costs per electricity generated), this advantage might be difficult to capture. This is because, in principle, a few sites could receive all the globally generated RW, making it possible to fully realize economies of scale, whereas for the CO_2, a larger number of sites need to be explored, as the capacity of each disposal site for CO_2 is likely to be small in some geological formations compared with the amount of CO_2 generated from fossil fuel-based power plants.

3.2.3 Costs per Unit of Electricity Generation

The cost estimates from the reports reviewed above are converted into standardized units and are summarized in Table 6 to allow comparison. It has to be kept in mind that these cost estimates differ significantly in scope and coverage. Inclusion or exclusion of R&D, contingency and tax are sources of major differences. No attempt has been made here to harmonize the coverage of these cost estimates. This difficulty should be considered when comparing the numbers presented in Table 6.

All cost figures reviewed are given in overnight costs (i.e. without accounting for interest during construction and without cost escalation). Although it would be preferable to use net present values for such a comparison (e.g. discounted costs accounting for the time value of money), the published information is not detailed enough to allow the net present value to be calculated.

The cost data are first adjusted to a price level of 2000 and expressed in US dollars by applying market exchange rates. The costs are expressed in capacity units in terms of tonnes of heavy metal equivalent. Where the capacity is expressed in terms of tonnes of uranium in the original publication, conversion has been applied using the national average burnup rate and the 0.0105 coefficient explained above.

The cost data are also presented in relation to the unit of electricity generation corresponding to the amount of waste to be disposed of. The reference electricity production is not available from published studies, apart from the Japanese and Swedish studies. For all other studies, the reference electricity production corresponding to the SNF generated is estimated using the identity relating SNF generation and electricity production, as discussed in the previous section. The reference electricity production is estimated to correspond to the capacity of a repository; therefore the estimated reference electricity production may overestimate the actual electricity production, given that wastes of non-civilian origin may be included in the capacity estimates.

The amount of SNF generated per 1 GW/year of electricity production was estimated using gross thermal efficiency and the burnup ratio of each plant, and an average for a given country was calculated by weighting them with the lifetime generation as of December 2008. All data needed to estimate the amount of SNF are taken from the IAEA's PRIS database (IAEA 2010a).

For the above mentioned seven cost studies, unit costs for geological disposal of RWs range between US$113,000 and US$683,000/tHM when SNF is reprocessed and the waste comes mainly in the form of vitrified HLW, and between US$281,000 and US$650,000 when direct disposal of SNF is chosen. The costs of reprocessing and disposal of additional ILW/LLW are not included in the cost estimates for the reprocessing option. Costs for the interim storage of SNF are not taken into consideration in either case.

The Advanced Fuel Cycle Costs Basis study commissioned by the US Department of Energy (US DOE) assessed the costs of the SNF disposal at US$528,000 per tHM (in the range of US$381,000–900,000/tHM), and the costs of vitrified HLW disposal at US$211,000 (in the range of US$152,000–360,000/tHM) (at 2006 prices) (Shropshire et al. 2007).

In the SAPIERR project (Štefula 2006), international cost estimates for SNF disposal were compared. The unit cost of disposal of SNF ranges from €80,000–1,200,000/tU (at an undefined price level), with the most common values in the range of €300,000–600,000/tU (€264,000–529,000/tHM). Preliminary assessment of the data indicates the existence of the economies of scale—doubling the inventory will increase the costs (excluding the contingency and R&D expenditures) only by a factor of 1.5.

The SAPIERR II Project (Chapman et al. 2008) used linear cost models to estimate the costs for joint disposal of SNF by selected EU countries. The study is based on cost models developed by SKB (Sweden), Posiva (Finland) and Nagra (Switzerland). For each cost model, the portion of fixed costs and variable costs for several cost components was delineated. The fixed cost portions were identified as €770–1,973 million (constant December 2006 prices), and the variable costs per canister (which roughly corresponds to 2 tU) were about €650,000–880,000 (roughly €286,000–388,000/tHM).

Unit costs of RW disposal per kWh electricity generated are also computed and presented in Table 6. The unit costs are estimated to be in the range of 0.092–0.298 US cent per kWh in the case of direct disposal of SNF and of 0.036–0.221 US cent

Table 6 Summary of the unit costs of geological radioactive waste disposal

Study	Estimated total costs from Table 4	(A) Amount of SNF generated per 1 GW/year (tHM)[a]	(B) Capacity assumptions used for cost estimation (in parenthesis; estimated based on A and C) (tHM)	Estimated costs per tHM in constant 2000 US dollar prices (1,000 US$/tHM)	(C) Reference energy production (in parenthesis; estimated based on A and B) (TWh)	Estimated cost per kWh of electricity generation (in constant 2000 US cent prices)
USA	US$96.2 B (constant 2007 price)	40.2	122,100 – 111,800 (SNF) – 10,300 (HLW)[b]	650	(26,636)	0.298
Finland	€542 M (constant December 2003 price)	27.3	5,417 (SNF)[c]	427	(1,738)	0.133
Sweden	SEK 36,707 M (constant January 2003 price)	36.6	9,095 (SNF)	433	2,468	0.160
Japan	¥2,757 B (constant 2008 price)	28.3	(37,700 tHM)	683	11,670	0.221
Belgium *Reprocessing option*	€578 M (constant 2000 price)	27.5	4,697 – 4,627[c] (HLW) – 70 (MOX fuel)	113	(1,500)	0.036
Direct disposal option	€1,494 M (constant 2000 price)		4,697 – 4,097[c] (SNF) – 600[c] (HLW) – 70 (MOX fuel)	293		0.092

UK	£5,226–6,765 M (constant June 2004 price)	128.8	57,824[c] – 53,234 (HLW) – 4,590 (SNF)	133–172	(3,936)	0.195–0.252
SAPIERR (EC)	€9,029–10,662 M (constant December 2006 price)	39.5	28,800 – 25,600 (SNF) – approx. 3,200 (HLW)	281–332	(6,386)	0.127–0.150

HLW high-level waste, *SNF* spent nuclear fuel, *tHM* tonnes of heavy metal

[a] Authors' calculation

[b] The amount of commercial spent nuclear fuel stored by the end of 2007 was 57,700 tHM and the corresponding cumulative electricity production is 4,045 GW year

[c] In the respective original reports, this value was reported in tonnes of uranium. The conversion rate was calculated based on the average burnup rate for the respective country using the PRIS database

per kWh in the case of disposal of vitrified HLW. Note that as this calculation is based on non-discounted costs and non-discounted electricity production volume, the numbers are not comparable to the levelized electricity generation costs that reflect discounted costs and electricity production volumes.

The IAEA (1994) estimated the levelized unit costs of RW management and disposal for the reprocessing option (with the disposal of vitrified HLW) and for the once-through option (with disposal of SNF). Disposal costs include costs for storage and transportation, and are estimated at 0.121cent per kWh (31% of the total fuel cycle costs) and 0.192 US cent per kWh (51% of total fuel cycle costs), respectively. This was calculated using a conservative discount rate of 5% until the end of power plant life, with a zero discount rate thereafter. These costs may be compared with the cost of nuclear power electricity, which was given as 3–5 US cent per kWh at the time of publication of the 1994 IAEA report.

Finally, it is worth noting that the IEA and the OECD Nuclear Energy Agency (NEA) regularly publish levelized electricity cost estimates. In their report published in 2005 (IEA and NEA 2005) the nuclear fuel cycle cost estimates are presented for 13 OECD countries. The report distinguishes front-end and back-end fuel cycle costs, but the costs specific to the deep disposal of HLW and SNF are not provided. The back-end nuclear fuel cycle costs are in the range of US$0.07 (France) and 0.588 US cent per kWh (Japan) with the 5% discount rate, and between 0.05 and 0.479 US cent per kWh with the 10% discount rate. The prices are expressed at the 1 July 2003 level. When compared with the levelized costs of nuclear power electricity generation of the respective countries, the shares of the back-end fuel cycle costs are in the range of 2.6–12.3% with the 5% discount rate, and 1.3–7.5% with the 10% discount rate.

3.3 *Calculation of Financial Liability*

The IAEA Member States that signed the Joint Convention on the Safety of Spent Fuel Management and on the Safety of Radioactive Waste Management adopted basic financing principles aimed at avoiding burdens for future generations and ensuring that adequate funds are available for the proper discharge of all financial obligations for nuclear waste management (IAEA 2006).

According to the 'polluter pays principle', the responsibility for financing waste management lies primarily with the waste generator. In some countries, legislation mandates that the waste generator should post financial guarantees in the form of funds segregated from its normal operations. The legislative frameworks concerning financing RW management in selected OECD countries are reported to the NEA Radioactive Waste Management Committee (NEA 2003). The latest review of the status of the Joint Convention on the Safety of Spent Fuel Management and on the Safety of Radioactive Waste Management was conducted in May 2009, and it concluded that much would need to be done to meet the challenges of ensuring the availability of sufficient financial resources for effective and sustainable waste management (IAEA 2009b). Country reports are available on the IAEA's Joint Convention website (IAEA 2010b).

Many of the cost studies reviewed above were conducted to serve as the basis for deciding the contributions to waste management funds. Such cost calculations and their periodic updates are also prescribed by national legislation in some countries.

In the USA, 0.1 US cent per kWh fee is charged for civilian waste generators and deposited in the Nuclear Waste Fund. This fee does not include fees for disposal of waste from electricity generated and sold prior to 1983. The Government will share the defence part of the costs, which is, under the latest cost estimate assumptions, 19.6%. Fees collected through September 2007 totalled approximately US$21.9 billion (2007 US$) and through 2046 are expected to add a further US$19 billion (2007 US$) (OCRWM 2008c).

In Japan, NPP operators are required by law to pay contributions to the Nuclear Waste Management Organisation to cover the costs associated with the final disposal of vitrified HLW canisters. The total functional obligation was calculated as net present value of the total undiscounted cost. With a discount rate of 1.5%, the net present value of total cost was calculated as ¥1,446 billion (2008 prices), of which 39% was already paid by 2008. This total cost does not include expenditure incurred between 2000 and 2008. The contributions are paid per vitrified HLW canister, and the unit contribution is calculated and updated each year. In 2008 the contribution was set at ¥39.4 million per canister (note that the number of canisters is also discounted). This is the equivalent of ¥0.135 per kWh (approximately US$0.13 per kWh). In computing this contribution, all canisters generated in the past and in the future are taken into consideration, and the number of canisters is discounted. The fee computed in this way is more or less comparable to levelized electricity generation costs.

4 Comparing Disposal Costs of CO_2 and Radioactive Waste

This section first compares the results obtained in the previous two sections on the magnitudes and relative importance of the waste disposal costs in fossil and nuclear-based power generation. This is followed by a discussion of the main similarities and differences at the conceptual level.

4.1 Cost Comparison

The comparison is made in terms of one main indicator, the cost of disposal per unit of electricity produced. In the case of RW, this is computed by combining the cost of the minimum disposal capacity and the volume of waste to be disposed of per unit of electricity generated. There is no minimum capacity requirement for CO_2 disposal sites.

Several factors have to be taken into account when computing this indicator:

- Volume of waste to be disposed of by category: the various RW categories require different levels of safety measures that carry widely differing price tags, while CO_2 is a homogeneous waste product in this respect.
- Fuel cycle and waste management strategies: there is a choice between once-through and reprocessing cycles for RW, while no such choice exists for CO_2; fuel cycle strategies affect the volumes of HLW and SNF, whereas the choice of waste management strategy affects the volume of ILW significantly (90% volume reduction may be possible).
- Capacity of the disposal facility: for RW, the fixed cost component dominates the total disposal costs, and capacity expansion can be done at a relatively low cost; therefore the initially planned or licensed capacity of the repository is a good starting point for computing a unit investment cost for the construction of a repository. The fixed cost component is a much smaller fraction of the total disposal cost of CO_2 at any single site and, as many disposal sites are needed, the cost of CO_2 disposal is roughly proportional to the volume of CO_2 to be disposed of.
- Cost components of waste disposal are practically the same for both CO_2 and RW; they include costs related to handling the waste generated (such as pre-treatment, treatment, conditioning and transportation) and the life cycle costs of the waste disposal facility (including site exploration, engineering, operation, closure and post-closure expenditures).

A specific limitation for a meaningful cost comparison is that it is rather difficult to separate the costs strictly for disposal (establishing, operating and closing the disposal site) from the rest of the downstream fuel cycle costs of nuclear power. Yet the distortion is minor because the overwhelming share of the latter is in direct disposal costs. In contrast, by far the most expensive part of the CCD chain is capture, and transport costs are also significant.

With these caveats in mind, a comparison of the relevant columns in Tables 3 and 6 (disposal costs per unit of electricity generated for CO_2 and RW, respectively) reveals that the cost range is much smaller for RW, despite the diverging national circumstances (geological conditions, accounting rules, regulatory regimes, etc.). Skipping the reprocessing option, the costs span a range from US$0.92/MWh in Belgium to US$2.98 in the USA. There is a lot of variability in the CO_2 disposal costs, as shown in Table 3. Depending on the actual split of the US$15/t CO_2 combined transport and disposal cost in the Dooley et al. study (Dooley et al. 2008), the share of the disposal cost can vary between about 5–41% of the fuel plus disposal costs and between 1 and 14% of the LCOE.

The supply curve developed by Pöyry Energy Consulting (2007) for the UK covers a span from –€15 to €1 for the first 50 Mt CO_2 (EOR), then increases a little from €1 to €5 for the next 900 Mt CO_2 (saline aquifers), whereafter it jumps significantly to over €20 (depleted oil- and gasfields). The Eccles et al. study (Eccles et al. 2009) provides similar results for the USA.

Transport cost curves follow similar patterns. Obviously as a CCD system expands, it will first utilize low-cost combinations of transport and disposal. Over the medium and long term, the costs of other energy supply options, the prevailing CO_2 prices and other factors will determine to what extent more expensive disposal options will be used.

Because of the lack of comparable data for the different countries, it is not meaningful to attempt a quantitative comparison of the share of disposal costs in the total fuel cycle costs or in the LCOE. Yet the results in the previous two sections indicate that these shares are much larger for CCD than for RW disposal.

4.2 Selected Issues in Cost Comparison

The fundamental issue in waste disposal costs is that, to date, CO_2 emitters have been using a global public resource (the CO_2 abatement capacity of the biosphere and the global atmosphere) but will now need to shift to a similar arrangement for RW disposal that involves costs as well as using local, private or government-owned space, with some level of risk for the public. In economic terms, it was clear ever since the beginning of nuclear energy programmes that the costs of the safe management and final disposal of RW must be part of internal or private costs. This is because the health and environmental impacts of RW were never considered as a candidate for externality, and the possibility of compensation for damages was only raised in the case of unintended/accidental release of radioactive material from the RW management process. CO_2 has been vented from the burning of fossil fuels for centuries; its negative environmental impacts through the modification of the climate system have been understood only in the past few decades. This issue, together with the need to reduce emissions and compensate for climate change damage, has been raised only relatively recently in various international forums. This means that fossil fuel burners will need to internalize these external costs by paying for CCD themselves, purchasing CO_2 emission permits, paying the applicable carbon tax or not to operate at all. The first three cases imply a significant new cost element in fossil-based electricity costs, while disposal costs have always been included in nuclear power in one form or another.

An important difference is the related regulatory frameworks and the resulting decisions based on cost implications. Strict regulations for handling and disposing of RW have been in place for decades to minimize any inadvertent external effects from the release of radioactive material. New regulation will be required for internalizing the climate externality of CO_2. Investment decisions about CCD will depend on the nature of the regulation and the resulting carbon price. A command-and-control type technology standard (no new coal-fired power plants to be permitted without CCD) would bring some degree of certainty in terms of emissions, but the related costs might be high. A carbon tax would provide an input for deciding whether to build new fossil plants with CCD or just capture-ready (hedging against future carbon tax increases), while an emission permit trading scheme and the inherently large uncertainties about future carbon prices would make fossil power investment decisions even more difficult. Technology standards are the only regulatory option for RW. The associated costs may be high but there are no external sources of uncertainty influencing the decision as to whether or not they should be borne, whereas the market-based regulation of carbon prices heavily influences the decision about adding CCD to fossil plants.

A related issue with significant cost implications is whether leakage from the disposal site can be tolerated or not. Van der Zwaan and Gerlagh (2008) analyse the economic aspects of CCD in relation to the possibility of significant leakage of CO_2 from geological reservoirs. They review the economic and climatic implications of the large-scale use of CCD for reaching a stringent climate change control target when geological CO_2 leakage is accounted for. Their model includes three main CO_2 mitigation options: energy savings, transition to non-carbon energy sources and the use of CCD. The authors find CCD to be a valuable option, even with CO_2 leakage of a few per cent per year, well above the maximum seepage rates foreseen by geological assessments. However, this analysis focuses on the atmospheric and climate implications of CO_2 leakage and does not account for the potential environmental impacts, human health and economic damages at the local/regional scale where the leakage occurs. The possibility of leakage here is rather different from the case of RW, in which no leakage of radioactive material is tolerated over very long time horizons—practically until the level of radioactivity declines to that of natural uranium—except in rare cases in which sufficient dilution can be proven. The cost difference between imperfect and nearly perfect containment can be significant.

Both CO_2 and RW disposal involve a series of legal-economic issues that are linked to the ownership of the underground disposal space. Depending on the legal system, subterranean space may belong to the owner of the surface land area or to the community (government). For RW it is possible in principle to secure, at a relatively low cost, the ownership of the total surface area under which the disposal facility is constructed and operated because of the limited surface area required for even a large depository. As CO_2 disperses over large distances in the disposal media from the injection wells, this would be rather difficult for CO_2 technically and extremely expensive economically. A hitherto totally ignored aspect in CCD cost estimates is the price of using someone else's underground property in the first place and, more importantly, compensating the owner for making it unusable for any other purpose for a very long time (option value). This could be a particularly contentious issue in the case of disposal sites spanning national borders unless a joint operation is agreed between the states in question.

Irrespective of property rights, a related economic issue is the notion of underground space as a depletable resource and the associated scarcity rent. The ratio between the amount of SNF and HLW arising from even an extremely large-scale nuclear power expansion with a once-through fuel cycle (an unlikely scenario in itself) and the volume of geologically suitable space for their disposal is so low that the question of disposal space scarcity is irrelevant. In contrast, even optimistically assessed potential CO_2 disposal space would not be able to accept more than a few decades' (perhaps a century's) worth of CO_2 produced (although disposal space and thus the fill-up rate is highly dependent on the region) and as suitable disposal space becomes depleted, so the remaining space would have an increasing scarcity value. Yet currently this issue seems to be of conceptual interest only, as payment for using underground space is virtually absent from the existing literature.

Using up a finite resource raises the question of possible backstop technologies. This involves yet another difference, at least in the narrow sense. Irrespective of whether scarcity rents will or will not be reflected in disposal costs, with the depletion of suitable disposal space the only backstop technology for CO_2 is mineralization, which is a very energy-intensive and thus rather expensive process. If disposal space were ever raised as a limiting factor for RW, closing the fuel cycle with fast reactors burning minor actinides would be a technically feasible solution that, among other benefits, would reduce the volume and radiotoxicity of the ultimate waste products. In a broader sense, other power generation technologies or system solutions (e.g. smart grids with myriads of decentralized electricity storage capacities) might emerge as a backstop for both fossil and nuclear electricity if they can provide the same level and reliability of service at a lower price.

With a view to financing disposal costs, the most important difference in the economics of RW and CO_2 disposal concerns timing. In the case of RW it is possible to accumulate the funds necessary for all disposal-related costs as part of the operating costs from electricity sales because, in a given fuel cycle arrangement, the waste volume is proportional to the electricity generated. Moreover, it is safe and inexpensive to store SNF and HLW for decades until the disposal facility is established (acknowledging the undeniable ethical concerns and the existence of some risks). During this time the accumulating disposal fund can even earn interest. No fund accumulation option exists for CO_2. All capture facilities, transport lines and disposal sites must be put in place before the first molecules of CO_2 can be prevented from entering the atmosphere. This involves a need to finance all related investments and the corresponding costs of capital (interest during construction, etc.) before the CO_2 benefits can be harvested. For new coal power plants this means a significant increase in investment costs compared to the non-CCD alternative, increasing their share in total LCOE and thereby approaching the cost structure of nuclear electricity. Yet CCD also has a significant operating cost component related to the energy required for the capture and conditioning of CO_2.

The potential for using waste as a resource involves similarities and differences. Although the main objective of the geological disposal of both CO_2 and RW is to isolate these substances from the rest of the biosphere, at least a part of them could be used as a productive resource. CO_2 can be an indirect resource in EOR, EGR and in ECBM recovery for mobilizing economically valuable resources (oil, natural gas, methane) while RW itself is a potential resource as long as it contains material that can be separated and used in nuclear reactors. However, only a small portion of the capturable CO_2 can be used productively because of geological and economic considerations (how much CO_2 is needed and what the cost-benefit ratio is of transporting CO_2 to distant EOR or EGR sites), whereas most of the RW remains a potential resource for a long time. This leads to a major difference in the requirements for retrievability: it is sometimes a regulatory requirement to make the retrieval of RW possible for at least 100 years for possible reuse (or for improved disposal if better packaging material for encapsulation or advanced disposal technologies become available), while retrieval is at best an option for remediation in the case of CO_2 leakage from the disposal site (see Maul 2011).

The prevalence of uncertainty in the disposal cost assessments is a common feature for CO_2 and RW. Lacking any industry-scale full-chain CCD demonstration facility, cost estimates to date are derived from similar industrial processes (e.g. the oil and gas sector) and from experiments with separate components of the CCD chain, and remain rather speculative. The longer history of R&D, including the construction of underground research laboratories, provides some basis for estimating RW disposal costs. However, a considerable degree of uncertainty remains about the costs of all necessary materials, equipment, surface and underground facilities required for the full-scale operation of repositories.

Another common feature is that, in the midst of the prevailing uncertainty of cost estimates, cost will vary widely across countries and regions, mainly because of site-specific conditions and partly driven by the efficiency of the regulatory and implementing organizations.

Both fossil fuel and nuclear electricity generation involve a certain public good characteristic to the extent they enhance energy supply security. Having them in the electricity generation portfolio increases the diversity of supply. The use of coal from domestic or reliable foreign sources as well as availability of uranium from many politically stable regions and the competitive fuel industry make both technologies a secure source. The economic value of this public good is not reflected in the price of electricity but should be considered when public resources need to be made available for disposing of the related waste (e.g. providing underground space or direct financial support).

Another similarity is related to economic competitiveness. Fossil-based electricity and nuclear power compete with each other and with other electricity technologies. The costs of CCD and RW disposal are important factors in the competitive position of both technologies. However, the relationship between waste disposal costs and competitiveness is often blurred by various government interventions (special taxes or subsidies, explicit or hidden) with even greater impacts on competitiveness.

The stability of the regulatory system is crucial for both technologies. They involve expensive long-lived capital assets. Once these are in place, it would be a major economic loss not to use them at full capacity, let alone to retire them prematurely.

In summary, there are some similarities concerning the costs of CO_2 and RW disposal that can provide a basis for preparing in-depth analyses on specific issues, like the value of stable regulation and the extent to which it can foster relevant investments. Nonetheless, major differences dominate this comparative assessment: these range from the need to pay for waste disposal (an obvious element in the cost of nuclear power as opposed to a newly emerging cost item for fossil-based electricity) to the physical scarcity of the disposal space and the issue of accounting for scarcity rent in the cost calculations.

This section has also revealed the difficulties of framing meaningful comparisons of the disposal costs for CO_2 and RW. Despite all the caveats about accounting differences between CCD and RW disposal, and also within each domain, some might find it interesting to compare the disposal costs per unit of electricity generated. However, what really counts in public and private decision making is the LCOE that includes all investment, financing, fuel, operation and maintenance, waste disposal and decommissioning costs in the nation-specific geographic, natural resource, economic and political context.

5 Summary and Conclusions

In the final account: direct geological disposal costs are a small fraction of the LCOE for nuclear power and also for fossil fuel electricity. Accounting for the preceding steps (capture, conditioning and transport) and considering the total downstream fuel cycle costs, however, the full CCD component triggers a major increase in the fossil-based electricity cost, while remaining very small for nuclear power.

Insights from comparing the disposal costs per se are limited for the above reasons. The total downstream costs differ more significantly both in absolute terms (per MWh) and relative terms (as a share of the total electricity cost), with CCD being a much larger cost item in both instances. What ultimately counts in economic terms is LCOE together with all external costs (remaining CO_2 emissions, radiation risks, etc.). Yet the simple comparison exercise presented in this chapter might help find the most critical elements in the fuel cycles in which technological improvements could lead to cost reductions and thus enhance the competitiveness of the respective technologies.

The most profound difference in the costing of CO_2 and RW disposal is that the former represents a completely new cost item in fossil fuel electricity, whereas the latter has been an obvious item on the cost sheet since the 1960s, irrespective of whether the corresponding fee was collected and accumulated during the operation of NPPs or not. Accounting for CO_2 disposal costs and especially for the other related downstream costs (capture and transport) will trigger a significant increase in LCOE generated from the burning of coal or gas.

The other fundamental difference with severe implications for the disposal costs and thus for the LCOE stems from the timing of the investment into waste disposal relative to the time of the power generation. RW can be stored safely at low cost for decades before emplacement into the final repository, and this leaves ample time to accumulate the disposal costs by charging a small fee per unit of electricity generated. In contrast, CO_2 abatement requires immediate disposal after capture because temporary storage would be prohibitively expensive. Therefore, the investment portion of the disposal (as well as capture and transport) costs must be disbursed before CCD operation can commence.

The two waste management technologies share important regulatory concerns with cost implications. Sloppy or frequently changing rules, standards and other regulatory elements trigger significant increases in the disposal costs and increase the cost of capital because of uncertainties, as well as discouraging private investors. Therefore, clear and concise policies translated into stable and reliable regulation are crucial for both technologies.

Acknowledgements The authors thank Sarah Sollors and Romain Boniface for their assistance in the literature research and Kazumasa Hioki, Andriy Korinny, Paul Degnan and two anonymous referees for their valuable comments and suggestions for improvements. Any remaining deficiencies are the sole responsibility of the authors.

References

Alie C, Douglas PL, Davison J (2009) On the operability of power plants with CO_2 capture and storage. In: Gale J, Herzog H, Braitsch J (eds) Proceedings of the 9th International Conference on Greenhouse Gas Control Technologies. Energy Procedia 1:1521–1526

Bachu S, McEwen T (2011) Geological media and factors for the long-term emplacement and isolation of carbon dioxide and radioactive waste. In this volume

BCG (The Boston Consulting Group) (2008) Carbon capture and storage. A solution to the problem of carbon emissions. A Focus by The Boston Consulting Group. http://www.bcg.com/documents/file15263.pdf

Bosetti V, Carroro C, Galeotti M, Massetti E, Tanovi M (2006) WITCH: a world induced technical change hybrid model. Energy J 27(Special issue 2):13–38

Celebi M, Graves F (2009) Volatile CO_2 prices discourage CCS investment. The Brattle Group. http://www.hks.harvard.edu/hepg/Papers/2009/Celebi-Graves_CO2%20Long%20Volatility%20Paper_Final.pdf

Chapman NA, McCombie C, Richardson P (2008) Work package 3: economic aspects of regional repositories. SAPIERR II—strategic action plan for implementation of European Regional repositories: stage 2. European Commission. http://www.erdo-wg.eu/ERDO-WG_website/Documents_files/SAPIERR%20II%20WP-3%20web.pdf

CoRWM (Committee on Radioactive Waste Management) (2005) CoRWM's radioactive waste and materials inventory. CoRWM document 1279. Available at: http://www.corwm.org.uk

Dooley JJ, Dahowski RT, Davidson CL (2008) On the Long-term Average Cost of CO_2 Transport and storage. Joint Global Change Research Institute, Pacific Northwest National Laboratory, College Park, PNNL-17389

Eccles JK, Pratson L, Newell RG, Jackson RB (2009) Physical and economic potential of geological CO_2 storage in saline aquifers. Environ Sci Technol 43:1962–1969

Eide-Haugmo I, Brakstad OG, Hoff KA, Sørheim KR, da Silva EF, Svendsen HF (2009) Environmental impact of amines. In: Gale J, Herzog H, Braitsch J (eds) Proceedings of the 9th International Conference on Greenhouse Gas Control Technologies. Energy Procedia. 1:1297–1304

Finnish Ministry of Employment and the Economy (2001) The decision in principle by the Government on 21 December 2000 concerning Posiva Oy's application for the construction of a final disposal facility for spent nuclear fuel produced in Finland. Unofficial translation. http://www.tem.fi/files/19726/Decision_in_principle_2000.pdf

Finnish Ministry of Employment and the Economy (2002) A Government decision-in-principle issued on 17 January 2002 on Posiva Oy's application for the construction of a disposal facility for spent nuclear fuel generated in Finland regarding the unresolved part of the application, postponed by the Government to be considered in conjunction with the processing of Teollisuuden Voima Oy's application for a decision-in-principle on the construction of a new nuclear power plant unit, submitted to the Government on 15 November 2000. http://www.tem.fi/files/19727/Decision_in_principle_2002.pdf

Gómez DR, Tyacke M (2011) Transport of carbon dioxide and radioactive waste. In this volume

Groenenberg H, de Coninck H (2007) Technical support for an enabling policy framework for carbon dioxide capture and geological storage. Task 3: incentivising CO_2 capture and storage in the European Union. Energy Research Centre of the Netherlands (ECN). http://www.ecn.nl/docs/library/report/2007/o07007.pdf

Hamilton MR, Herzog HJ, Parsons JE (2008) Cost and U.S. public policy for new coal power plants with carbon capture and sequestration. In: Gale J, Herzog H, Braitsch J (eds) Proceedings of the 9th International Conference on Greenhouse Gas Control Technologies. Energy Procedia 1:4487–4494

Hendriks C, Hagedoorn S, Warmenhoven H (2007) Transportation of Carbon Dioxide and Organisational Issues of CCS in the Netherlands. Ecofys, Utrecht

Herzog H, Smekens K, Dadhich P, Dooley J, Fujii Y, Hohmeyer O, Riahi K (2005) Cost and economic potential. In: Metz B, Davidson O, de Coninck HC, Loos M, Meyer L (eds) IPCC Special Report on Carbon Dioxide Capture and Storage. Cambridge University Press, Cambridge/New York, pp 339–362

Hildebrand AN (2009) Strategies for demonstration and early deployment of carbon capture and storage: a technical and economic assessment of capture percentage. Masters thesis, Massachusetts Institute of Technology, Cambridge

Ho MT, Allinson GW, Wiley DE (2009) Factors affecting the cost of capture for Australian lignite coal fired power plants. In: Gale J, Herzog H, Braitsch J (eds) Proceedings of the 9th International Conference on Greenhouse Gas Control Technologies. Energy Procedia. 1:763–770

IAEA (International Atomic Energy Agency) (1994) Assessment and Comparison of Waste Management System Costs for Nuclear and Other Energy Sources, vol 366, Technical Reports. IAEA, Vienna

IAEA (International Atomic Energy Agency) (2006) Joint Convention on the Safety of Spent Fuel Management and on the Safety of Radioactive Waste Management, vol 1, IAEA International Law. IAEA, Vienna

IAEA (International Atomic Energy Agency) (2009) Classification of Radioactive Waste, vol GSG-1, IAEA Safety Standards. IAEA, Vienna

IAEA (International Atomic Energy Agency) (2009b) Joint Convention on the Safety of Spent Fuel Management and on the Safety of Radioactive waste Management: Summary Report. Third review meeting of the contracting parties, Vienna, 11–20 May. JC/RM3/02.Rev2.http://www-ns.iaea.org/downloads/rw/conventions/third-review-meeting/final-report-english.pdf

IAEA (International Atomic Energy Agency) (2010a) Power reactor information system. PRIS. http://www.iaea.org/programmes/a2/

IAEA (International Atomic Energy Agency) (2010b) Joint convention website. Joint convention on the safety of spent fuel management and on the safety of radioactive waste management. http://www-ns.iaea.org/conventions/waste-jointconvention.htm

IEA (International Energy Agency) (2008) CO_2 Capture and Storage: A Key Carbon Abatement Option. OECD/IEA, Paris

IEA and NEA (International Energy Agency and Nuclear Energy Agency) (2005) Projected costs of Generating Electricity: 2005 Update. OECD Publishing. http://www.iea.org/textbase/nppdf/free/2005/ElecCost.PDF

IEA and NEA (International Energy Agency and Nuclear Energy Agency) (2010) Projected Costs of Generating Electricity 2010 edition. OECD Publishing, Paris

IPCC (Intergovernmental Panel on Climate Change) (2005) IPCC Special Report on Carbon Dioxide Capture and Storage. Metz B, Davidson O, de Coninck HC, Loos M, Meyer LA (eds). Cambridge University Press, Cambridge/New York

Korkmaz Ö, Oeljeklaus G, Görner K (2009) Analysis of retrofitting coal-fired power plants with carbon dioxide capture. In: Gale J, Herzog H, Braitsch J (eds) Proceedings of the 9th International Conference on Greenhouse Gas Control Technologies. Energy Procedia. 1:1289–1295

Kukkola T, Saanio T (2005) Cost estimate of Olkiluoto Disposal Facility for Spent Nuclear Fuel. Working report 2005–10. Posiva Oy, Olkiluoto

Maul P (2011) Risk assessment, risk management and remediation for the geological disposal of radioactive waste and storage of carbon dioxide. In this volume

McCoy ST, Rubin ES (2005) Models of CO_2 transport and storage costs and their importance in CCS cost estimates. In: Proceedings of the Fourth Annual Conference on Carbon Capture & Sequestration. National Energy Technology Laboratory, US Department of Energy. http://www.netl.doe.gov/publications/proceedings/05/carbon-seq/Tech%20Session%20Paper%2092.pdf

McKinsey & Company (2008) Carbon capture & storage: assessing the economics. http://www.mckinsey.com/clientservice/sustainability/pdf/CCS_Assessing_the_Economics.pdf
METI (Ministry of Economy, Trade and Industry) (2008a) Revision of costs for the final disposal of specified radioactive waste and of contributions to the final disposal reserve fund. Press release, 5 Dec [in Japanese]. http://search.e-gov.go.jp/servlet/PcmFileDownload?seqNo=0000046113
METI (Ministry of Economy, Trade and Industry) (2008b) Revision of the basic policy on specified radioactive waste final disposal and the specified radioactive waste final disposal plan. Press release, 14 Mar. http://www.meti.go.jp/english/newtopics/data/nBackIssue20080314_01.html
METI (Ministry of Economy, Trade and Industry) (2008c) Methodology for estimating the amount of specified radioactive waste from the reprocessing of spent nuclear fuels resulting from the operation of nuclear power plants [in Japanese]. http://www.meti.go.jp/committee/materials/downloadfiles/g71218a05j.pdf
Narita D (2009) Economic optimality of CCS use: a resource-economic model. Kiel working paper No. 1508. Kiel Institute for the World Economy. http://www.ifw-members.ifw-kiel.de/publications/economic-optimality-of-ccs-use-a-resource-economic-model/KWP%201508.pdf
NEA (Nuclear Energy Agency) (2003) Topical Session on "Liabilities Identification and Long-term Management at National Level": Topical Session Held During the 36th Meeting of the RWMC. Proceedings. NEA/RWM(2003)/14. OECD Nuclear Energy Agency. http://www.nea.fr/rwm/docs/2003/rwm2003-14.pdf
NETL (National Energy Technology Laboratory) (2007) Cost and Performance Baseline for Fossil Energy Plants, vol 1, Bituminous Coal and Natural Gas to Electricity—Final Report. DOE/NETL-2007/1281, NETL.http://www.netl.doe.gov/energy-analyses/pubs/Bituminous%20Baseline_Final%20Report.pdf
Nirex (2005) Cost estimate for a reference repository concept for UK high-level waste/spent fuel. Technical Note. Nirex Document No. 484281. United Kingdom Nirex. http://www.nda.gov.uk/documents/upload/Cost-estimate-for-a-reference-repository-concept-for-a-UK-high-level-waste-spent-nuclear-fuel-A-Technical-Note-2005.pdf
Nirex (2006) Technical note on cost profiles for CoRWM option 7 (deep geological disposal) and option 9 (phased deep geological disposal). Nirex Document No. 506103. United Kingdom Nirex. http://www.nda.gov.uk/documents/upload/Technical-note-on-cost-profiles-for-CoRWM-Option-7-9depp-geolgical-disposal-and-Option-9-phased-deep-geological-disposal-June-2006.pdf
NUMO (Nuclear Waste Management Organization of Japan) (2004) Technology and Safety of Geological Disposal of High-level Radioactive Wastes. NUMO, Tokyo [in Japanese], NUMO-TR-04-01
OCRWM (Office of Civilian Radioactive Waste Management) (2008a) Analysis of the total system life cycle cost of the civilian radioactive waste management program, fiscal year 2007. DOE/RW-0591. US Department of Energy. http://www.energy.gov/media/Total-Life-Cycle.pdf
OCRWM (Office of Civilian Radioactive Waste Management) (2008b) The report to the president and the congress by the secretary of energy on the need for a second repository. DOE/RW-0595. US Department of Energy. http://www.energy.gov/media/Second_Repository_Rpt_120908.pdf
OCRWM (Office of Civilian Radioactive Waste Management) (2008c) Fiscal Year 2007: Civilian Radioactive Waste Management Fee Adequacy Assessment Report. DOE/RW-0593. US Department of Energy. http://www.energy.gov/media/2008_Fee_Adequacy.pdf
ONDRAF/NIRAS (L'organisme national des déchets radioactifs et des matières fissiles enrichies—Belgian Agency for Radioactive Waste and Enriched Fissile Materials) (2001a) SAFIR 2 Report: Safety Assessment and Feasibility Interim Report 2. Report NIROND 2001-06 E. ONDRAF/NIRAS. Available at: http://www.nirond.be/engels/Safir2_eng.php
ONDRAF/NIRAS (L'organisme national des déchets radioactifs et des matières fissiles enrichies—Belgian Agency for Radioactive Waste and Enriched Fissile Materials) (2001b) Technical overview of the SAFIR 2 Report: Safety Assessment and Feasibility Interim Report 2. NIROND 2001-05 E. http://www.nirond.be/engels/PDF/Safir2_apercutech_eng.pdf

ONDRAF/NIRAS (L'organisme national des déchets radioactifs et des matières fissiles enrichies—Belgian Agency for Radioactive Waste and Enriched Fissile Materials) (2009) Long-term management of category C waste. http://www.nirond.be/engels/7.5.3_CategoryC_eng.html

Ploumen PJ, Koetzier H, Turpin F, Smeets RD (2007) Investigations to CO_2 Storage: Strategy for CO_2 Capture. 30620089-Consulting 06-1064. KEMA Consulting, Arnhem

Posiva (2008a) Overall Description of An Extension to the Final Disposal Facility of Spent Nuclear Fuel to Accommodate for the Fuel From Olkiluoto 4 Plant Unit. Posiva Oy. http://www.posiva.fi/files/514/Posiva_YLPS_en.pdf

Posiva (2008b) Horizontal Deposition of Canisters for Spent Nuclear Fuel: Summary of the KBS-3 H Project 2004–2007. Report POSIVA 2008-03. Posiva Oy. http://www.posiva.fi/files/932/Posiva_2008-03web.pdf

Pöyry Energy Consulting (2007) Analysis of Carbon Capture and Storage Cost-supply Curves for the UK. Pöyry Energy Consulting, Oxford

Renzenbrink W, Ewers J, Keller D, Wolf KJ, Apel W (2009) RWE's 450 MW IGCC/CCS project – Status and outlook. In: Gale J, Herzog H, Braitsch J (eds) Proceedings of the 9th International Conference on Greenhouse Gas Control Technologies. Energy Procedia. 1:615–622

Rogner HH, Sharma D, Jalal AI (2008) Nuclear power versus fossil-fuel power with CO_2 capture and storage: a comparative analysis. Int J Energy Sect Manage 2(2):181–196

Štefula V (2004) Inventory of radioactive wastes. Deliverable D-1. SAPIERR—Support Action: Pilot Initiative for European Regional Repositories. European Commission, Brussels

Štefula V (2006) Final Report. Deliverable D-7. SAPIERR—Support Action: pilot initiative for European Regional Repositories. European Commission. http://www.arius-world.org/pages/pdf_2006_7/Report-SAPIERR%201-%20Final%20Report%20D7.pdf

Shropshire DE, Williams KA, Boore WB, Smith JD, Dixon BW, Dunzik-Gougar M, Adams RD, Gombert D (2007) Advanced Fuel Cycle Cost Basis, INL/EXT-07-12107. Idaho National Laboratory. http://www.inl.gov/technicalpublications/Documents/3667084.pdf

SKB (Svensk Kärnbränslehantering AB—Swedish Nuclear Fuel and Waste Management Company) (2003) Plan 2003: Costs for Management of the Radioactive Waste Products From Nuclear Power Production. SKB Technical Report TR-03-11. Svensk Kärnbränslehantering AB.http://www.skb.se/upload/publications/pdf/TR-03-11webb.pdf

SKB (Svensk Kärnbränslehantering AB—Swedish Nuclear Fuel and Waste Management Company) (2010) A repository for nuclear fuel in 1.9 billion year old bedrock. Svensk Kärnbränslehantering AB. http://www.skb.se/Templates/Standard____28848.aspx

Strömberg L, Lindgren G, Jacoby J, Giering R, Anheden M, Burchhardt U, Altmann H, Kluger F, Stamatelopoulos G-N (2009) Update on Vattenfall's 30 MWth oxyfuel pilot plant in Schwarze Pumpe. In: Gale J, Herzog H, Braitsch J (eds) Proceedings of the 9th International Conference on Greenhouse Gas Control Technologies. Energie Procedia 1:581–589

Tavoni M, van der Zwaan B (2009) Nuclear Versus Coal Plus CCS: A Comparison of Two Competitive Base-load Climate Control Options. Nota di lavoro 100.2009. FEEM working paper series, Fondazione Eni Enrico Mattei. http://www.feem.it/userfiles/attach/20091119123146410O-09.pdf

Tshibangu KJ-P, Descamps F (2011) Engineering challenges in the geological disposal of radioactive waste and carbon dioxide. In this volume

US EPA (United States Environmental Protection Agency) (2008) Geologic CO_2 sequestration technology and cost analysis. Technical Support Document. US EPA.http://www.epa.gov/ogwdw000/uic/pdfs/support_uic_co2_technologyandcostanalysis.pdf

van der Zwaan B, Gerlagh R (2008) The Economics of Geological CO2 Storage and Leakage. Nota di lavoro 10.2008. FEEM working paper series, Fondazione Eni Enrico Mattei. http://www.feem.it/userfiles/attach/Publication/NDL2008/NDL2008-010.pdf

Vangkilde-Pedersen T, Anthonsen KL, Smith N, Kirk K, Neele F, van der Meer B, Le Gallo Y, Bossie-Codreanu D, Wojcicki A, Le Nindre Y-M, Hendriks C, Dalhoff F, Christensen NP (2009) Assessing European capacity for geological storage of carbon dioxide—the EU GeoCapacity project. In: Gale J, Herzog H, Braitsch J (eds) Proceedings of the 9th International Conference on Greenhouse Gas Control Technologies. Energy Procedia. 1:2663–2670

Vosbeek M, Warmenhoven H (2007) Making Large-scale Carbon Capture and Storage CCS in the Netherlands Work: An Agenda for 2007–2020 Policy, Technology and Organization. Ecofys, Utrecht

West JM, Shaw RP, Pearce JM (2011) Environmental issues in the geological disposal of carbon dioxide and radioactive waste. In this volume

WNN (World Nuclear News) (2009) Obama dumps Yucca Mountain. 27 Feb. Available at: http://www.world-nuclear-news.org/

Managing Liability: Comparing Radioactive Waste Disposal and Carbon Dioxide Storage

Elizabeth J. Wilson and Sara Bergan

Abstract Liability issues are a major concern for final disposal of radioactive waste (RW) and for geological storage of carbon dioxide (CO_2). We develop a list of overarching questions that drive liability and present a discussion of where managing liability for geological CO_2 storage and RW disposal is fundamentally different and where it is similar. Governments have been trying to manage high-level RW from civilian reactors for over 40 years and there are ample lessons in the interplay between technology, policy, politics and society that are relevant for both future nuclear energy and geological CO_2 storage projects. We examine the history of managing liability for RW using case studies on Germany, France, Finland and the USA to better understand how liability for RW is currently structured. We compare this to potential liabilities for geological CO_2 storage and outline current proposals for managing liability in the US and European Union. From this, we develop 'lessons learned' from past management of RW that could help to both structure liability and ultimately deploy future RW and geological CO_2 storage projects. We conclude that while establishment of a legal framework is important for future development of nuclear energy and geological CO_2 storage, it is insufficient to guarantee deployment. Rather, legal liability is embedded within a larger socio-political context and addressing these broader concerns is vital for future RW disposal and geological CO_2 storage deployment.

Keywords Liability • Radioactive waste • Carbon capture and storage • Geological storage • Comparative analysis

E.J. Wilson (✉)
Humphrey Institute of Public Affairs, University of Minnesota, 301 19th Avenue South, Minneapolis, MN 55455, USA
e-mail: ewilson@umn.edu

S. Bergan
University of Minnesota Law School, 229 19th Avenue South, Minneapolis, MN 55455, USA
e-mail: berga011@umn.edu

F.L. Toth (ed.), *Geological Disposal of Carbon Dioxide and Radioactive Waste: A Comparative Assessment*, Advances in Global Change Research 44,
DOI 10.1007/978-90-481-8712-6_9, © International Atomic Energy Agency, 2011

1 Introduction

Both nuclear energy and carbon capture and storage (CCS) are important technologies that could–if widely deployed–help to significantly reduce greenhouse gas emissions from human activities. While both nuclear reactors and coal plants without CCS are already in wide use across the globe, new facilities must address cost, security, safety, and, importantly, how to manage radioactive waste (RW) and captured carbon dioxide (CO_2). Both coal and nuclear power have been deployed without solving the waste problem. Failing to do so in advance of technology deployment has created a socio-political quagmire for the nuclear industry as it searches for possible disposal sites. Failing to manage CO_2 in the use of coal and other fossil fuels has lead to our current climate crisis.

In some ways RW and captured CO_2 are the mirror image of one another: RW is the toxic by-product of a low-carbon energy technology, while captured and stored CO_2 is the pollution removed from a high-carbon energy technology. Nuclear reactors produce small amounts of high-level RW, while coal-fired power plants produce large quantities–billions of tons–of CO_2 annually. While civilian nuclear energy has existed for over 50 years without a final solution for high-level RW, it is impossible to temporarily store the large volumes of CO_2 produced at power plants. There is no CCS without the 'S'.

The challenges for future deployment of nuclear energy and CCS facilities are as much socio-political in nature as technical. One large problem facing both nuclear energy and CCS is managing legal liability throughout the entire technology life cycle–especially with regard to establishing a final repository for nuclear waste and geological storage sites for captured CO_2.

While most nations have adopted regulations to manage liability and ensure financial responsibility for nuclear facilities (NEA 2000), one area that remains unresolved is actually deploying high-level RW disposal repositories. Although legal responsibility for RW is clearly partitioned, no country–with the possible exceptions of Finland and Sweden–has successfully sited a final repository for RW. The absence of final disposal presents economic and security risks to both the operators and the public at large (Syrota 2008). It could also stymie future investment in nuclear energy and affect relicensing of existing reactors.

Meanwhile, the European Union (EU) and the USA are in the process of developing regulatory frameworks to manage CCS (EP 2009a; US EPA 2008). Many stakeholders have raised concerns about how CCS projects will be managed over their life cycle and especially how liability will be structured (CCSReg Project 2009; de Figueiredo 2007; Klass and Wilson 2008; Trabucchi and Patton 2008). Resolution of questions surrounding legal liability for CCS projects will affect both project developers and operators and have a strong influence on the ultimate industrial organization of the CCS industry.

In addition to environmental, health, and safety risks that occur over RW and CCS project lifetimes, both technologies have a 'long risk tail', which will require monitoring and management long after the RW is placed in the repository or the

CO_2 is injected underground. The site liability must be managed over long time frames, requiring regulatory and institutional arrangements that extend beyond the repository or injection site closure. RW remains toxic for hundreds of thousands of years, and injected CO_2 should remain underground for hundreds to thousands of years for a climate benefit (IPCC 2005; Macfarlane 2006).

This chapter focuses on managing the risks and liabilities of high-level RW from civilian RW repositories and geological storage of CO_2. While geological storage of CO_2 and final disposal of RW are fundamentally different technical problems and require different risk management strategies, many of the socio-political factors affecting technology deployment are similar. Additionally, the forty-plus years of experience in trying to site permanent RW repositories around the world provides ample lessons in the interplay between technology, policy, politics and society, relevant for both future nuclear energy and CCS endeavours. Section 2 presents a list of overarching questions that drive liability and a discussion of where managing liability for CCS and RW is fundamentally different and where it is similar. After presenting these overarching issues to establish the liability context, the remainder of this chapter will focus on the liability considerations for final disposal of RW and geological storage of CO_2. Section 3 describes how liability for RW management is structured in the USA, Germany, France and Finland and the implementation of RW waste policy. Section 4 briefly provides an overview of potential liabilities for CCS projects and outlines current proposals for managing liability in the USA and the EU. Section 5 distils 'lessons learned' from past management of RW that could help to both structure liability and ultimately deploy future nuclear energy and CCS projects.

2 Framing the Liability Issues for Radioactive Waste Disposal and Geological Storage of CO_2

Managing risk and legal liability associated with RW disposal or geological storage of CO_2 is central to guarding against harm to human health and the environment for current and future generations. It is also important for public acceptance and wide-scale deployment of the technologies. The risks of RW and CCS are described in West et al. (2011).

Liability–essentially a legal responsibility of one 'person' to another or to society (Garner 2004)–is fundamentally linked to risk. Liability can be civil or criminal and can be developed by common law, statutes or regulations. A party is 'liable' if they are legally obligated to another party (Garner 2004). Liability is established by demonstrating harm or injury and showing it resulted from a party's actions in breech of a duty of care (i.e. A's conduct caused B's harm). It can be joint or several if multiple parties are at fault (both A and B are responsible for C's harm). It can be limited through statutes of limitation or repose which specify a period of time a party is liable (A is only liable for 10 years). It can be limited, also often by statute, to particular types of incidents, damages or parties (B is only liable to C for X).

The clarity, certainty and extent of legal liability can affect new technology adoption and technology deployment. Companies considering adopting new technology are adverse to uncertain liability associated with commercial-scale deployment. Firms want stable and predictable liability terms to guide company investment decisions, as well as shareholder and financial community expectations. Legal liability is also important for government actors wishing to promote a technology because it helps ensure those with the most information about risks take appropriate care to avoid problems. Governments may wish to limit liability or take responsibility for liability over a certain amount of money or after a specified time as a tool to encourage technology deployment by the private sector. But ensuring that private investors in technology are at least partly liable limits government financial exposure and can make support of technology more financially and politically practical. Finally, transparent liability regimes help the public understand and have confidence that risks–to human health and the environment–will be actively managed and, in the event of an accident, effectively remediated and compensated. In these ways a transparent liability regime can help to deploy a new technology; but, as we will show, even clear liability regimes are insufficient on their own to ensure deployment.

When evaluating liability, the legal jurisdiction and the project stage are also crucial for valuing risk and the resulting harm. Liability in one location may be construed very differently from liability in another, as legal regimes across subnational states and countries vary significantly. Also, risk and the resulting liability will change across the technology life cycle. Risks associated with RW change as the waste decays, as the waste is transported and during disposal at the final repository. CCS risk will shift as CO_2 is captured, transported, injected underground, and stored and monitored post-closure. Each of these factors will affect the relative importance of liability and the tools used to managing it over the respective life cycles.

2.1 Overarching Questions for Managing Liability

For both RW and CCS, structuring liability requires careful balancing of private, government, and public interests and responsibility. It is embedded within a larger regulatory framework and changes over the technology life cycle. Discussions of liability are driven by the following questions:

1. What are the risks of the technology to humans? To the environment?
2. How severe or widespread are the risks?
3. How long will the risks be important?
4. Who is responsible if something goes wrong and for how long?
5. Can the liability be transferred?
6. Could any damages be remediated or compensated?
7. Are funds available to cover compensation and remediation?

2.2 *Similarities and Differences of Radioactive Waste and Carbon Capture and Storage That Affect Liability*

CCS and RW have several similarities that influence liability management: (1) both technologies depend on disposal systems that include natural geological features which increase uncertainty about risks; (2) both will need care long after operations at the site cease; and (3) both require that larger socio-political issues are resolved before the technology can be deployed. Aside from these three similarities, the risks posed by CCS and RW are quite different and these differences will be a significant driving factor in structuring respective liability regimes. RW and CCS liability will differ significantly in amount, time to recognize harm and the overall risk profile. The following list details some important differences–from a liability perspective–of RW and CCS management and maps them onto the overarching questions presented in Sect. 2.1.

Quality of waste (questions 1, 3, 6): RW is very toxic whereas CO_2 is harmless at low concentrations. High-level RW is lethal to humans for hundreds of thousands of years, toxic even at low doses or for short exposures (Macfarlane and Ewing 2006). Remediation of leaking RW is costly and difficult. Carbon dioxide is an inert gas and harmless at low concentrations. CO_2 is present in the atmosphere at 0.03% and is harmless to humans at concentrations of less than 1%. Concentrations of over a few per cent can cause health effects, and CO_2 is deadly within one minute at concentrations over 17–30% (OSHA 1989; US EPA 2000). Remediation technologies to reduce CO_2 concentrations are well understood and simple to deploy. By contrast, the severe toxicity of RW makes it much harder to remediate. As a result, an even greater premium is placed on careful RW management and strong liability provisions.

Quantity of waste (questions 1, 2, 6): Nuclear reactors produce small quantities of high-level RW; coal-fired power plants generally produce very large quantities–millions of tons–of CO_2. The annual production of high-level nuclear waste globally is roughly tens of thousands of tons, compared to the tens of billions of tons of carbon dioxide generated from coal-fired power plants and industrial facilities. Even individual geological storage projects will be large–injecting millions of tons of CO_2 annually, while entire countries are responsible for a cumulative total of hundreds to a few thousand tons of RW annually. Also, RW can be stored in interim facilities; given the volumes of CO_2, interim storage is impossible. Thus, the liability for RW centers around managing a small quantity of highly toxic waste, while geological storage projects must manage a large volume of relatively non-toxic waste. While nuclear energy facilities were built without agreement on final disposal of RW and nuclear energy facilities today wrestle with liability concerns associated with temporary storage, deploying CCS will require prior agreement on long-term storage.

Risks from transport to final disposal (questions 1, 2, 3, 4, 6): High-level RW will likely be transported to final disposal by train or truck (National Research Council 2006). CO_2, because of the large volumes, will likely be sent through pipelines constructed for this purpose. For RW transport, either the owner or operator of the nuclear reactor (most EU countries) or the federal government (USA) will be responsible and liable. For CCS, it is unclear if pipelines will be owned publicly or privately or who will control title to the CO_2 and ultimately who will be liable for damages incurred during transport. Liability during transport could remain with the producer of the CO_2–or be transferred to the party in charge of transport or final disposal.

Risk level over time (questions 2, 3): Where the risks of RW leakage to the environment *increase* over time as barrier stability likely decreases, risks of CCS leakage should *decrease* over time, as pressures decrease and geologic processes immobilize the CO_2. RW depends on waste isolation; CCS relies on natural attenuation. Current proposals call for the (relatively) small quantities of RW to be placed into secure geological repositories, within sealed canisters, and monitored closely. While the goal is to isolate the waste until it is no longer harmful, the risks of leakage become greater over the long time frames (hundreds of thousands of years) necessary for radioactive elements in the waste to decay. Canisters are generally designed for retrievability in case leakage occurs.

By contrast, millions of tons of carbon dioxide could be injected directly into saline formations or depleted hydrocarbon reservoirs. Natural mechanisms like dissolution into subsurface brines, and eventually, mineralization, should reduce risk over time (IPCC 2005). Projects should become more secure the longer the CO_2 stays in the formation. Once the CO_2 is injected, however, there will be no easy way to entirely retrieve it from the subsurface.

Whereas the liability risk for RW increases over time but is limited to highly monitored, discrete locations; CCS risks–and associated liability–could decrease over time, although they will spread over a much larger area.

Managing leakage (questions 3, 4, 5, 6, 7): Harm from leakage depends on where the leaking material goes, what it comes into contact with, and what, if any, damages occur. Although little experience exists managing RW repository leakage, as none have yet begun accepting waste, much experience exists with CO_2 or analogous gases (Beaubien et al. 2004). Analogues for underground injection for CCS include underground natural gas storage and enhanced oil recovery (EOR) projects, among others (Wilson and Pollak 2008). If a CCS project begins to leak CO_2 from the storage reservoir, it can be managed by plugging the leaking well or managing formation pressures to slow or stop leakage. Quick high-pressure leaks could happen during CO_2 transport or from a well blowout at the injection site. Because RW is very toxic and remediation is extremely costly, no leakage is permissible, while small leakage from CO_2 projects could be acceptable from a risk perspective but would constitute a 'climate liability'. How such leakage would be monitored and factored into different carbon accounting regimes is unclear but may amount to a relatively simple financial liability associated with carbon credit accounting. The importance of this liability is, of course, dependent on eventual climate policies, whereas RW has no climate liability.

Number of different sites (questions 2, 4, 5, 6): There will be a few RW repositories and many CCS injection projects. High-level RW management is coordinated at the national and international level, with countries trying to plan one or two carefully managed sites. If large-scale CCS deployment occurs, countries with suitable geology could have hundreds of different injection sites, located in different jurisdictions and subject to different liability regimes.

Institutional balance (questions 4, 5, 6, 7): Public and private institutions will play different roles in managing RW disposal and CCS. This balance will also differ significantly among nations because it is determined by differing regulation and legal precedent. While RW in the USA is the responsibility of the Federal Government, a consortium of private industries is in charge of repository construction and management in Sweden. For CCS, three of the world's four current projects are operated solely by private parties. No experience yet exists with long-term stewardship for closed CCS projects. How long the title–and associated liability–of the CO_2 rests with the site operator and when or if it could be transferred to the central or subnational government remains an open question.

Common or dread risk (questions 1, 2, 4, 6, 7): CO_2 is used regularly in soft drinks and industrial processes and, while the circumstances and exposures posed by geological storage are novel in many ways, CO_2 itself is common and familiar. It is uncertain how people will respond to the novel risk of geological storage, although initial analyses suggest possible caution (Palmgren et al. 2004; Parfomak 2008). RW is often associated with a dread risk which triggers a 'gut-level' fear in people (Slovic et al. 1990). Dread risk affects public perception of the danger of a technology and can influence siting, create intense resistance to a project from the potential host community, and place pressure on regulators to establish more stringent regimes for managing risk and associated liability. Indeed, the widespread failure to site a permanent RW repository is often attributed to dread risk.

2.3 Creating Liability Frameworks for Radioactive Waste Disposal and Geological CO_2 Storage

Integrating the overarching questions presented in Sect. 2.1 into RW disposal and geological CO_2 storage liability frameworks depends on a number of factors, and will be driven by the similarities and differences between RW and geological CO_2 storage. Overarching liability questions will be addressed through legislation, regulation and case law. Management of these questions will fundamentally shape technology deployment and the industrial structure for RW and geological CO_2 storage. As governments use liability provisions to support or thwart a particular technology and to ensure that risks are adequately managed within private financial calculations, the structure of liability will play an important role in future RW disposal and geological CO_2 storage projects. For RW, the liability framework is relatively well established and provides interesting institutional models for consideration while developing a framework for geological CO_2 storage.

3 Managing Liability of Radioactive Waste

Liability for RW is linked to waste escaping from the repository and adversely impacting human health or the environment. Leaking radioactive material could affect groundwater, flora or fauna. As RW remains toxic for hundreds of thousands of years, disposal regimes focus on absolute containment of waste. Given the time frame for disposal–tens to hundreds of thousands of years–discussing legal liability for potential damages that far into the future seems ridiculous. However, understanding how responsibility has been partitioned and what mechanisms are used to manage long-term risk is helpful for future discussions of both nuclear energy and CCS.

In both Europe and the US, industry and government have played a joint role in managing liability of RW. While existing liability regimes clearly delineate responsibility and establish funding for building and monitoring a final disposal site, this has proven insufficient to successfully complete a functioning repository. Public opposition and political controversy have thwarted attempts to establish a permanent disposal site in Europe and the USA (Macfarlane 2003).

The majority of efforts to structure liability for nuclear energy have focused on accidents at nuclear reactors. Three Mile Island in the USA, Chernobyl in the Ukraine, and even recent leaks from French nuclear facilities, shape much of the public and government debate on harm from nuclear power and test liability regimes, but these discussions have not advanced the construction of a permanent disposal repository (Mabe 2008).

Most third-party liability provisions for nuclear energy are focused upon potential accidents from transport or large-scale harm from reactor operation, and many are based–either directly or implicitly–upon the Paris Convention on Third Party Liability in the Field of Nuclear Energy and the Vienna Convention on Civil Liability for Nuclear Damage from the early 1960s. An explanatory text on the International Atomic Energy Agency's (IAEA) nuclear liability regime discusses the nuances of the different provisions (IAEA 2007).

Generally, the operator of the nuclear facility is exclusively liable for any nuclear damage (often referred to as a channelling provision) and strict (no fault) liability is imposed on the operator, in light of the special dangers from nuclear accidents. This includes liability for accidents or events at the nuclear installation as well as during transport of RW to or from the facility. The operator sending the RW remains liable until another party takes charge and assumes liability of the RW. Additionally, liability may be limited in both amount and in time (IAEA 2007).

Countries generally follow these provisions, with minor variations. All countries place strict liability on the operator, establish some type of liability channelling to the operator (the USA is an exception, as it has no channelling provision and guarantees insurance coverage), set time limits for claims resulting from a nuclear accident (many specify 10–30 years after an incident), require mandatory insurance and place caps on liability (these amounts vary; Austria is an exception in that it has no cap on liability) (NEA 2000). Some countries also have industry funded pools, and all rely on the national government in case damages exceed these amounts. Under the joint Paris Convention and 2004 Protocol to Amend the Brussels Supplementary Convention on Nuclear Third Party Liability, EUR 1.5 billion are available for liability coverage from nuclear-related accidents, and responsibility is divided into three tiers. The first tier of compensation comes from nuclear power plant operators, who are responsible for at least the first EUR 700 million; the EU country in which the liable operator is located is responsible for the second EUR 500 million; and all contracting parties to the Brussels Supplementary Convention are responsible for an additional EUR 300 million (NEA 2009). In most EU countries waste transport is the responsibility of the owners and operators of reactors and may include some liability for the waste transporter, while in the USA, the Federal Government is responsible.

RW is not explicitly mentioned in most countries' nuclear liability provisions, but it is covered implicitly in discussions of 'nuclear installations'. The OECD Nuclear Energy Agency (NEA) Steering Committee adopted a formal decision that permanent disposal was within the scope of the Paris Convention at least through the pre-closure phase, leaving open the question of liability post-closure. While the NEA Steering Committee decision is legally binding on contracting parties, a 2004 Protocol also amended the definition of 'nuclear installation' in the Paris Convention to cover final disposal. Separate waste management provisions deal with responsibility for waste– with most assigning financial responsibility to the operator, and most creating a government entity for long-term disposal. Many countries collect a small charge on electricity produced from nuclear reactors to fund development of a final repository. In the USA, the Federal Government is responsible for siting and managing the final disposal repository, but Sweden and Finland rely upon an industrial consortium and in France a public-private agency manages the process (NEA 2000).

3.1 Radioactive Waste and Liability in the European Union

European countries have been prominent advocates of international coordination on RW disposal. Densely populated regions, a history of controversy over close-to-border facilities, and memories of the Chernobyl accident remind EU countries that the scope of nuclear damage is likely to span borders (NEA 2008). Many members of the EU were also inaugural parties to the Joint Convention on the Safety of Spent Fuel Management and on the Safety of Radioactive Waste Management, which outlines guidelines for safety, communication and long-term regulation of RW disposal (IAEA 1997).

At the regional European level, the European Atomic Energy Community (Euratom) directs health and safety standards and encourages information exchange between the 12 member countries (Temple 1994). Initially Euratom was established to consider the safety of the reactors themselves, with little attention to the issue of RW management (Directorate-General of Energy and Transport 2008). But, in Section 2, Article 62, the Euratom Treaty states: 'such materials and any fertile wastes shall be left in the possession of the producer' (European Community 1957). It is important to note that at the time, nuclear reactors and the energy sector were largely state controlled. The business model of private ownership has evolved and necessitated a change in the liability regime.

As waste management became a more prominent concern, Euratom began to focus on information exchange, research, and eventually on the complicated issues of implementation. In late 2005 Euratom itself became a member of the previously mentioned Joint Convention on the Safety of Spent Fuel Management and on the Safety of Radioactive Waste Management (Directorate-General of Energy and Transport 2008).

Under Euratom, if an entity discharged RW in a fashion out of compliance with the set standards, the affected party had a right of action through the European Commission (EC) Court (Temple 1994). Under Article 37 of the Euratom Treaty, EU countries are also required to provide data to the Commission on planned nuclear or storage facilities. This has been interpreted by the Court of Justice as requiring that data be given to the Commission and the Commission's opinion on the project in turn given back to the EU country before authorization is issued (Cameron 2007).

Further regulation of nuclear installations, transport and long-term disposal of RW continues to be negotiated in EU directive proposals (see, for example, the packaged 'Amended proposal for a Council Directive (Euratom) laying down basic obligations and general principles on the safety of nuclear installations' and 'Amended proposal for a Council Directive (Euratom) on the safe management of the spent nuclear fuel and radioactive waste' (EC 2004), but progress toward agreement has been slow (Cameron 2007)). In April 2009, the European Parliament adopted a newly negotiated resolution on a proposal for a Council Directive (Euratom) setting up a 'Community framework for nuclear safety' (EP 2009b). Meanwhile the EC established a research project aimed at feasibility studies for regional long-term geological disposal of RW in Europe (SAPIERR I) which concluded that regional international waste management facilities would be particularly economically beneficial for countries with small nuclear programmes or no suitable geological formations (Dietze 2004; SAPIERR Project 2008). A second phase of the project, SAPIERR II, currently under way, is aimed at creating the framework for a European Development Corporation which would work with national waste agencies on regional waste facilities (SAPIERR Project 2008). While any country would retain the right to dispose of RW domestically (and some currently require it by law), Article 4, Number 6 of the Draft Directive on the Management of Spent Nuclear Fuel and Radioactive Waste acknowledges the need for shared repositories in the EU (Dietze 2004). How liability for RW could be partitioned within a shared repository is still unclear.

While some argue that liability issues are already among the most complex in nuclear law, shared disposal, increased transboundary movement of the waste and harmonization of various liability regimes only increase the complexity (Dietze 2004). Although the EU continues to tackle the issues of shared disposal and the complex liability issues associated with movement of waste, this paper focuses on existing liability regimes and processes for final disposal only.

More generally the traditional legal principles employed in EU countries affect implementation of the RW liability created in the EU (Louka 1993). Ultimately, managing liability for RW in Europe rests with the individual country. Details will vary by country with owners and operators of nuclear reactors and central governments deeply involved in the creation of final RW repositories. Among EU countries, success in establishing a permanent repository is dramatically different (Dawson and Darst 2006). Only Sweden and Finland have moved forward with final plans for waste disposal, while other countries continue to struggle to reach agreement.

3.1.1 Country Case Studies

The case study countries (Germany, France and Finland) are all party to the 1960 Paris Convention, established under the NEA (2008). While all case study countries are party to the same international treaty, their respective experiences in attempting to establish a permanent RW repository differ considerably. Germany is the largest European economy and has adopted an aggressive plan to phase out nuclear energy: research on long-term storage facilities has ceased and is facing what some call 'irreconcilable political differences.' France is the most widely recognized proponent of nuclear energy with the largest share of electricity coming from nuclear installations and a national commitment to reduce waste through reprocessing, but still no agreement on final disposal. By contrast, Finland is a relatively small European economy with few nuclear installations, but has reached agreement on final repository specifications and sites (although recent concerns over increased costs and radioactivity have proven significant obstacles to the project's success). In addition, the structure of liability differs somewhat among the three. In Germany the government is ultimately responsible for interim storage and final disposal; in France RW generators–who are predominantly owned by the government–are responsible for final disposal; and in Finland private waste generators are liable through final disposal, with responsibility passed to the government after closure of the repository.

Germany

Country Context: RW liability in Germany involves the nuclear power plant operators as well as the Länder (German states) and the central government. Germany currently has 17 reactors in operation, making up between 25% and 30% of the country's electricity production (WNA 2008a). Germany is the biggest electricity market in Europe and is the fourth largest producer of nuclear power in the world despite official plans to phase out nuclear power over time (EIA 2008). The phase-out plans were incorporated into the 2002 amendment of the Atomic Energy Act and remain in place today (Khune 2007). While the Red-Green coalition government initially intended to completely phase out nuclear energy, including nuclear reprocessing and transport, industry backlash over the lack of temporary RW storage stymied the legislation. Eventually parties agreed on the Phasing-out Act of 2002, which included the reauthorizing of RW transport, banning of nuclear reprocessing by 2005, disposal of waste by waste generators, and a nominal 32-year phase-out plan (NEA 2006; WNA 2008a). In 2003 the first reactor was closed under this plan (beyond those closed in Eastern Germany after reunification).

During the 1980s and 1990s, Germany required its spent fuel to be transported to and reprocessed in France or the United Kingdom (Rudig 2000). The significant amounts of reprocessed waste from past nuclear activities being stored outside the country (e.g. by AREVA) are expected to be returned by contract no later than 2022 (WNA 2008a).

Legal, Regulatory and Liability Frameworks: Owners and operators of nuclear reactors are liable for any human health, environmental or property damage caused by RW. Chapter 4, § 25–40, of the German Atomic Energy Act outline the liability structure for nuclear installations including those for RW storage and disposal. The Act stipulates that while a final repository is not available in Germany, plant operators are required to build storage on or close to the reactor site (NEA 2006). The operating companies have formed joint ventures to build and manage temporary surface facilities at Ahaus and Gorleben and are liable for any damages. Although the owner of the nuclear facility is liable for the RW no matter where an accident may occur, a transporter of waste can accept liability during transport by contract and thereby be deemed liable (BMU 2009, § 25). Owners of RW are subject to unlimited financial liability following the Paris Convention but the period of liability is limited to 30 years after the incident (or 3 years after the claimant discovered the damage) (BMU 2009, § 32).

The states (Länder) are responsible for establishing collection sites for the interim storage of RW produced within their borders. For domestic accidents, the state may share some of the liability with the Federal Government (BMU 2009, §§ 34 and 36). Federal authorities are responsible for final disposal and are authorized to levy contributions from any party that stands to benefit from the controlled disposal of RW (BMU 2009, § 21(b)(1)). The Federal Office of Radiation Management (Bundesamt für Strahlenschutz (BfS)) is responsible for final disposal of all RW but it has commissioned the German Company for the Construction and Operation of Waste Repositories to design, construct and operate the federal repositories. Although a third party has been commissioned to build and operate the facilities, responsibility for final disposal must remain solely with the Federal Government by law (BfS 2009a). By statute, the Federal Government must indemnify any third party used in building or operating the final disposal plans up to EUR 2.5 billion (BMU 2009, § 9a).

Siting History: Despite the country's official policy to phase out nuclear power, the Government has continued to affirm a commitment to establish a permanent RW repository. A site near Gorleben, currently home to an interim storage site, was selected as early as 1974 (BfS 2009b). Originally the site was planned as an integrated and centralized waste facility including a reprocessing plant that was later dropped from the plans (Vandenbosch and Vandenbosch 2007). Although a report commissioned in 1999 determined that the Gorleben salt dome was suitable for long-term disposal of spent fuel, the previous Red-Green coalition government halted exploration activities. Plans for the facility were stalled as a result of the 2001 moratorium on nuclear facilities and the site was criticized, as it was considerably different from other planned repository sites and did not allow for waste retrieval. (Khune 2007; Vandenbosch and Vandenbosch 2007). Public opposition to the facility has been strong and focused on RW transportation to and from Gorleben. In fact, opposition to Gorleben served as the centre of the German anti-nuclear movement in the 1990s. Khune writes that future decisions about lifting the moratorium for the site are 'one of the thorniest controversies the present government faces in the field of nuclear policy' (Khune 2007). While the moratorium remains in effect, the German Government is reassessing the suitability of salt as host rock compared with other rocks such as clay and granite and no statement has yet been made about the suitability of the Gorleben site under the new more comprehensive standards (BfS 2009b).

France

Country Context: Liability for RW in France is managed by a public-private partnership. France is home to roughly 59 reactors which make 75–80% of France's electricity (Lidskog and Andersson 2002; Lidskog and Sundqvist 2004). France has the largest share of electricity coming from nuclear power and is the world's largest net electricity exporter (WNA 2008b). After the oil shock of the 1970s, France invested heavily in nuclear energy. The reactors are operated by Electricité de France (EDF), which had been a government owned company until late 2004 when it was restructured and floated a small portion of shares on the Paris Stock Exchange. Despite a small move toward privatization, the Government retains the vast majority ownership–roughly 85% (Bennhold 2005).

France reprocesses its spent fuel to recover the uranium and plutonium for reuse and to reduce the long-term disposal volume. Spent fuel is not considered nor treated as a waste for final disposal and the French Environmental Code stipulates that it can only be considered waste when it can no longer be reused or recycled (NEA 2007). France also reprocesses spent fuel for other countries and this material must be returned to the originating country after the 'necessary storage period', as directed by the Environment Code (NEA 2007). The Nuclear Materials and Waste Management Programme Act of 2006 stipulated that French reprocessing facilities would not store foreign sources of nuclear waste for extended periods of time.

Reducing the total volume of waste to be stored through recycling was originally outlined as a priority RW management strategy in 1975 (Article L.541 of the Environmental Code) and was affirmed again in 2006 as part of the country's comprehensive RW management system (Birraux 2006). Waste generators include: EDF as the sole nuclear electricity provider; the Commissariat à l'Énergie Atomique

(CEA), the French regulatory body designed to oversee civilian and defence related nuclear activities; and AREVA, the private reprocessing company whose primary shareholder is CEA (NEA 2007). All highly tied to the French Government, each of these waste generators oversees its own fund for the associated waste management.

Legal, Regulatory and Liability Frameworks: The liability and RW management regime in France must consider damages from the associated fuel reprocessing, which decreases the volume but not the toxicity of material to be managed. Although reprocessing is a large part of France's waste management strategy and all the associated management and transport substantially complicates liability regimes, this chapter focuses on the liability associated with permanent waste disposal.

France passed its Waste Management Act in 1991 and updated it in 2006 (WNA 2008b). The Act established the Agence nationale pour la gestion des déchets radioactifs (ANDRA) as the national RW management agency and set in motion the research at Bure (Haute-Marne) underground rock laboratory in eastern France. The amendments in 2006 transformed ANDRA into an independent state-owned organization responsible for long-term RW management, recognizing it as the only type of entity suitable to provide short and long-term safety, stability and independence in the management of RW (Birraux 2006). ANDRA also fills inventory, industrial and research roles with regard to France's RW management (Parliament of France 2009).

A related 2006 Act (2006-686), concerning nuclear transparency and safety, set up the Nuclear Safety Authority (ASN) as the independent administrative authority tasked, on behalf of the State, with regulating nuclear safety and radiation protection (ASN 2009). ASN regulates civil nuclear activities, is charged with informing the public on safety matters and inspects installations and grants licenses. It also has the power to issue sanctions where problems are observed (ASN 2009).

Like other countries party to the Paris Convention, operators of nuclear installations are exclusively liable for damage resulting from the facilities or the transport of RW to and from the facility (NEA 1960). The Programme Act of 2006 (2006-739) also made clear that producers of spent fuel and RW are liable 'without prejudice to the liability of their holders as people responsible for nuclear activities' (NEA 2007). It also recognizes ANDRA's public service duty to collect, transport and take charge of radioactive material when requested by and at the expense of those responsible for it. This includes requisition when those responsible for the waste have defaulted. A dedicated fund fed by a tax paid by RW producers supports ANDRA's ongoing research activities and another dedicated fund fed by agreements between RW producers and ANDRA supports the agency's storage and management activities (Birraux 2006). ANDRA ultimately assumes legal responsibility for the long-term RW storage pursuant to Article L542-12 of the French Environmental Code (Parliament of France 2009). Although ANDRA is permitted by law to work with third parties in building and operating long-term storage facilities, liability stays with ANDRA.

Siting History: Although it had long been assumed that France would dispose of its waste in geological repositories, the first attempts to characterize a few sites were suspended in 1990 due to public opposition (Vandenbosch and Vandenbosch 2007). Afterwards a report commissioned by the Parliamentary Office for Scientific and Technological Decisions emphasized the need for transparency and accountability in future waste disposal decisions. As part of the obligations under the Aarhus Convention, France set up a National Commission on Public Debate in 1995 to gather public input on large industrial projects, including RW. Although the Bure facility managed by ANDRA continues to research the feasibility of long-term waste disposal, all other previously identified sites have been abandoned either for geological complications or local opposition. To try and entice local interest, the French Government had offered communities up to the equivalent of roughly US$5 million per year to accept the research laboratories in their area (Vandenbosch and Vandenbosch 2007). The 2006 Programme Act on the Sustainable Management of Radioactive Materials and Waste (French National Assembly 2006) more recently and formally prioritized deep geological disposal as the preferred solution for high-level RW. It sets a 2015 goal for licensing a repository and a 2025 goal for opening it, but did not specify a final disposal site (WNA 2008b). In accordance with the Act, ANDRA has also planned a corresponding schedule of public debate and input on the project between 2009 and 2015 (Dupuis 2009).

Finland

Country Context: Liability for RW in Finland rests primarily with private industry. By 2020 Finland will have an estimated 2,600 t of spent fuel to dispose of (Lidskog and Andersson 2002). It has only two nuclear power plants: one in Loviisa and the other in Olkiluoto, with four reactors which provide over one quarter of Finland's electricity (STUK 2008; WNA 2008c). In May of 2002, the Finnish parliament voted to build a fifth reactor to be operational in 2012, the first new nuclear power plant to be authorized in Western Europe for more than a decade (WNA 2008c). Just a year earlier, the Finnish Parliament had given final approval for siting the permanent waste repository at Olkiluoto. Given how controversial RW siting has proven in other countries, the low- and intermediate-level RW disposal facilities that have been developed in Sweden and Finland are among the very few successful implementation models (Lidskog and Sundqvist 2004).

Legal, Regulatory and Liability Framework: The management of RW and RW disposal is governed primarily by the 1987 Nuclear Energy Act (WNA 2008c). The Ministry of Trade and Industry (KTM) supervises operation of nuclear power and waste disposal, while the Radiation and Nuclear Safety Authority (STUK) regulates and inspects the industry, including confirmation of proper waste disposal.

The producers of the waste are responsible for the preparation, funding and safe implementation of nuclear waste disposal (Lidskog and Sundqvist 2004). A 1994 amendment to this act stipulated that waste from these plants must be managed and permanently disposed of in Finland (STUK 2008). Prior to this, the Loviisa plant, which uses Russian technology, returned the waste to Russia for reprocessing. After the decision to require domestic long-term disposal, the joint venture company Posiva Oy was set up to handle the waste (Nuclear Association 2008c). Operator licensees submit a waste management plan and, assuming the licensee implements the agreed upon measures and pays the required lump sum to the State, are certified by STUK as having fulfilled all obligations. Ownership, control and responsibility for the waste is transferred to the State once the final disposal facility is closed pursuant to sections 32 and 34 of the Act (Parliament of Finland 1987). If KTM considers the licensee's measures to be unsatisfactory or the licensee is unable to fulfil its obligation, the State has secondary responsibility and may assume ownership for the RW according to section 31 of the Act (Parliament of Finland 1987).

The liability regime for nuclear energy and RW management in Finland was modified when the Finnish Parliament passed the 2005 Nuclear Liability Act (NEA 2005). The Act raised the insurance coverage required of operators to EUR 700 million and ensured the ultimate liability of the operator may be unlimited in the case where the Brussels Supplementary Convention, which supplies up to EUR 1.5 billion, has been exhausted without compensating all damage. The Act also authorized the Finnish Council of State to decide on a lower amount of liability for transporters of RW, but created a floor of EUR 80 million. The definition of nuclear liability was expanded to include economic loss and the costs of repairing damaged environment, consistent with Article 1 of the revised Paris Convention. Interestingly, while other countries have expressly excluded acts of terrorism from such liability, the Act seems to include terrorist acts within the liability scheme (NEA 2005).

Siting History: Posiva Oy has been developing plans for a deep geological repository at Eurajoki near Olkiluoto (WNA 2008c). The Government set guidelines for long-term waste management in 1983 which included a local right of absolute veto on the siting process. Posiva Oy investigated several locations all of which were technically suitable. The Eurajoki site and plans were ratified by Parliament in May 2001 (by a vote of 159 to 3), recognizing the construction of the facility as a public good. The proposal had been carefully vetted with the public and had strong local support with the Eurajoki council voting heavily in favour of the proposal (20 in favour, 7 against) (Vandenbosch and Vandenbosch 2007). The project has recently encountered significant obstacles, however, after Posiva's environmental impact assessment suggested the RW from the new nuclear power plant (European Pressurized Reactor technology) would be much more radioactive and disposal far more costly than had been expected (Kanter 2009).

3.2 Radioactive Waste and Liability in the USA

Country Context: In the USA, the liability for RW has been central to the debates surrounding nuclear energy and long-term disposal, and it remains a difficult and intractable issue. Under the 1982 Nuclear Waste Policy Act, the US Federal Government is responsible for disposing of spent nuclear fuel, but the operator holds liability for RW still at the reactor site. As the US Government has failed to site and open a permanent RW repository, it faces significant liabilities–essentially a breach of contract with the nuclear reactor owners and operators (Cawley 2007). The long history of politically charged siting battles continues. In short, while responsibility for RW has been clearly delineated, this has not spurred the construction of a final repository. The debates surrounding RW have centred on the Government's ultimate responsibility to take title and liability of RW and its breach of contract. Spent nuclear fuel is currently kept at reactor sites, and the reactor owner bears liability in the event of an accident–although the amount of liability is capped under the Price-Anderson Act (see below).

The current fleet of nuclear reactors will produce an estimated 2,000 t of spent fuel per year–105,000 metric tons of spent fuel in total over the estimated lifetimes of nuclear energy facilities and potentially more if reactor lifetimes are extended (Cawley 2007; Deutch et al. 2003). Yet for the past 40 years, a comprehensive plan to manage RW has proven elusive. Even if there were agreement to build the proposed US RW repository at Yucca Mountain tomorrow, it is already full, as it is designed to accept only 70,000 t of RW.

3.2.1 Legal, Regulatory and Liability Framework: Liability While the Radioactive Waste is at the Reactor

While RW is stored at the reactor, in spent fuel ponds or dry storage casks, the owner of the nuclear reactor is liable for damage, up to the limits specified under the Price-Anderson Act. The Price-Anderson Act was developed to stimulate investment in civilian nuclear power by blending different risk management instruments into a coordinated framework of coverage (NRC 1957). First passed by Congress in 1957 (and recently renewed in 2005), the Price-Anderson Act was originally envisioned as a temporary 10-year provision to stimulate and support the development of civilian nuclear energy by creating funding for accident remediation while at the same time limiting tort liability for nuclear accidents (Anderson 1978). The Act's original purpose was to limit financial uncertainty arising from nuclear accidents by placing a cap on liability and guaranteeing that citizens could be compensated for damages to person and property. Criticized by opponents as a subsidy to the nuclear industry, Price-Anderson limits liability from potential 'extraordinary nuclear occurrences' and creates a tiered structure of financial responsibility combining private insurance, an industry pooled fund and a cap on total liability (NRC 2006a). Each nuclear reactor over 10 MW was required to have US$300 million per plant in insurance (NRC 2006b). Any additional claims are paid from an industry funded pool–the Price-Anderson fund–with each company contributing up to US$95.8 million if an accident occurs (US GAO 2004).

In the event of an accident, companies are required to pay US$15 million annually until the claim is met or the maximum reached. With 103 operating nuclear power plants in the USA, the fund contains approximately US$10 billion (US GAO 2004). Any claims beyond this amount would be covered by funds raised by the US Nuclear Regulatory Commission (NRC) from Congress using public monies. The Act indemnifies licensees from any amount over the liability cap and, since amendments in 1988, any nuclear incident–including those associated with interim RW storage–would fall under the jurisdiction of the federal district courts (United States Congress 1988). The 1988 Amendments divested the state courts of jurisdiction, specifically barred state law claims for punitive damages and pre-empted any state law inconsistent with the Act (United States Congress 1988). Subsequent appellate courts have barred other state law claims, reasoning that they are inconsistent with the federal claims standards set forth in the 1988 Amendments.

To date, the Price-Anderson fund has paid out a total of US$202 million (with US$70 million associated with the 1979 Three Mile Island incident) (Deutch et al. 2003). For proponents, the Act has aided nuclear industry development and has obligated the nuclear plant operators and the industry to hold a higher level of liability insurance coverage than might otherwise be the case. They argue that in the event of a large-scale accident the Fund may be cost effective for both the industry and the government (Deutch et al. 2003). For critics, the Act serves as a public subsidy to the nuclear industry and ends up limiting the ability of affected parties to recover adequate damages.

3.2.2 Legal and Regulatory Framework: Liability Once Waste Is Removed from the Reactor

Once RW is removed from the site of a nuclear reactor, the title and associated liability pass to the US Federal Government. However, in the 25 years since the Federal Government accepted this liability, no permanent disposal site has been created. The Nuclear Waste Policy Act of 1982 established a plan to explore five sites as permanent nuclear repositories, accepting a maximum of 70,000 metric tons of heavy metal waste. In this Act, the US Congress also established the Nuclear Waste Fund, charging US$0.001 per kWh produced from nuclear facilities and raising approximately US$29.9 billion since 1983 (Nuclear Energy Institute 2008). The Fund resources are used to plan and site final RW repositories. Because it charges a flat rate per kWh generated, it has been criticized for providing no economic incentive to reduce RW (Deutch et al. 2003). The Act also specified that a second future waste repository be located in the Eastern USA, where most of the nuclear power plants are located.

Siting History: By 1983, the Department of Energy had identified nine potential repository sites and three of these were recommended for further characterization (Vandenbosch and Vandenbosch 2007). Lawsuits were filed by all three target states. After the politically contentious 1987 amendments to the Nuclear Waste Policy Act the standards shifted. Only Yucca Mountain remained as a candidate site and the requirement for a second Eastern repository was dropped. While Yucca Mountain was officially approved in 2002, waste could not be accepted until 2017 at the earliest, and the state of Nevada remains opposed to siting the facility (Vandenbosch and Vandenbosch 2007).

As development of a permanent repository for RW remains stalled and centralized interim storage non-existent, the owners of nuclear reactors continue to face RW related liability, in spite of the monies paid and federal commitment to establish a permanent repository. Yucca Mountain was scheduled to begin receiving waste in 1998, but after the Federal Government did not meet its contractual deadline, several successful lawsuits awarded electric power companies damages for costs incurred from the delay (Cawley 2007). In 2002 the 11th Circuit US Court of Appeals held that the damages could not be paid from the Nuclear Waste Fund. The Government eventually ended up paying US$290 million from the Treasury Department's Judgment Fund to four utilities for costs of providing additional on-site RW storage. This tension between industry and government continues today, with government appeals over an additional US$337 million ongoing and 44 additional cases yet to be tried (Cawley 2007). Indeed the ownership and subsequent liability issue were prominent when President Clinton vetoed the Nuclear Waste Policy Act of 2000 (S. 1287):

> Finally, the bill passed by the Congress does little to minimize the potential for continued claims against the Federal Government for damages as a result of the delay in accepting spent fuel from utilities. In particular, the bill does not include authority to take title to spent fuel at reactor sites, which my Administration believes would have offered a practical near-term solution to address the contractual obligation to utilities and minimize the potential for lengthy and costly proceedings against the Federal Government. Instead, the bill would impose substantial new requirements on the Department of Energy without establishing sufficient funding mechanisms to meet those obligations (Clinton 2000).

These issues continue to play out in the US Courts and are no closer to being resolved. Indeed, the Energy Policy Act (United States Congress 2005) provided funding for nuclear energy R&D, construction, a production tax credit and decommissioning funds, but no resolution to the repository issue. More recently President Obama cut funding for the Yucca Mountain in his proposed budget and is directing his administration to consider new alternatives for RW disposal. Moreover, US Energy Secretary Steven Chu recently told the Senate that the Yucca Mountain site is no longer viewed as an option for storing RW (Hebert 2009), though whether this is final remains to be seen. With thirty-plus proposed new nuclear plants and the continuing controversy over RW repository siting, liability associated with RW will continue to influence the future of nuclear energy in the USA.

4 Managing Liability for Geological Storage of CO_2

Liability for geological storage of CO_2 is linked to risks of CO_2 escaping from the storage reservoir, displacing native brines, or inducing geological hazards, as outlined in Bachu and McEwen (2011). Risks, and associated liability, will shift over the geological storage project life cycle, and differ if the CO_2 is injected on- or offshore, or into a saline formation or hydrocarbon recovery project. The relative importance of liability will also vary by legal jurisdiction.

Liability regimes for geological storage are still evolving, but current regulatory proposals in Europe and the USA suggest a joint role for industry and government in managing liability for stored CO_2. The emerging model is that the operator would be liable during siting and injection and for a post-injection period sufficient to demonstrate that the stored CO_2 is stable and poses no risk, at which point the site could be closed. At closure, liability may be transferred to a public entity. How ownership of injected CO_2 would be transferred after site closure is currently unclear. This model is not yet fully implemented in either Europe or the USA, as discussed further in Sects. 4.1 and 4.2. Many first-generation CCS projects may be linked with enhanced oil and gas recovery operations because such operations already have CO_2 injection infrastructure in place and high oil prices make tertiary recovery increasingly cost-effective. Liability for these projects may be handled under existing oil and gas law, with unresolved questions about how the transition from EOR to pure storage would affect liability issues.

For a geological CO_2 storage site, risk and associated liability will evolve as the project moves from siting, injection, post-injection monitoring, to closure and long-term care. Regulators must ensure that geological storage projects have responsibility for harm and liability clearly assigned and delineated across the technology life cycle. This will help provide incentives for responsible behaviour and will help developers and operators to invest in projects with a better idea of future costs and liability. Private insurance companies have already begun to develop liability and financial assurance products over the design, operational, closure and post-closure periods of a geological storage life cycle (Business Wire 2009).

Liability associated with the siting phase includes acquiring the rights to characterize the site, then obtaining surface access and subsurface property rights for site development. Large-scale storage reservoirs to sequester millions of tons of CO_2 could potentially cover hundreds of square kilometres and require clearly defined property rights and liability arrangements. Such rights will intersect with mineral and subsurface resource management regimes and interact with environmental and natural resource legislation.

During the injection phase, there are four distinct risk areas that may yield potential liability to the developer or operator. These include: (1) CO_2 leakage to the surface; (2) groundwater contamination; (3) hydrocarbon damage; and (4) geological hazards (see Bachu and McEwen 2011). By better understanding the probability that a specific risk will occur, and the potential harm that the risk could cause, CCS project stakeholders will be able to better evaluate, manage and remediate any potential harm. This knowledge will also help to bound the cost of any damage and establish necessary mechanisms for financial responsibility.

Liabilities associated with CO_2 leakage to the surface could stem from damages to human health or the environment, or from obligations under climate regimes. Potential for harm from surface leakage is greater for onshore sites (prevalent in the USA), and reduced for offshore sites (prevalent in Europe). Recovery of damages to person or property resulting from CO_2 surface leakage likely will rely on established theories of liability (Klass and Wilson 2008). Precedents exist in the oil recovery and underground natural gas storage context (de Figueiredo 2007). Similarly, for environmental risks, countries in the EU and USA have long established regulations and protocols to value damage to crops or forests. Liability and required financial responsibility will depend on the perceived extent and permanence of damages as well as the ability to establish a causal link between damage and the injected CO_2 (Wilson et al. 2007a). Climate liabilities of CO_2 leaking to the surface result from the future cost of carbon at the moment of leakage. If CO_2 was injected when the cost was US$40/t, leakage when the CO_2 price is US$100/t could present a significant financial liability. How such leakage will be monitored and integrated into a larger carbon management and accounting scheme remains to be seen.

Damages to groundwater from geological storage of CO_2 could occur due to pressure-induced displacement of saline water into drinking water aquifers, or to CO_2 leaking into drinking water aquifers and unfavourably altering the groundwater chemistry. Establishing liability for damages to groundwater would require a causal chain proving that a particular geological storage project caused specific groundwater damage or other environmental harm, which could be difficult (Wilson et al. 2007b).

Hydrocarbon resources could also be affected if CO_2 were to leak into oil or natural gas bearing strata. While CO_2 is injected for EOR, if the CO_2 were to leak into a formation that was not covered, extra costs of stripping the CO_2 from the final product might be incurred.

Geological hazards, such as induced seismicity or ground heave, present another source of liability. In general, the risks from geological hazards seem to be addressed by existing regulations in the USA and the EU which manage injection pressures. However, the large volumes of CO_2 and large scale of CCS deployment might call for a re-evaluation of these risks.

For a climate benefit, injected CO_2 is expected to remain in the subsurface for hundreds (if not thousands) of years. Liability and responsibility for long-term monitoring can be divided into two phases: (1) a post-injection phase where an operator is liable, and (2) a post-closure phase where liability may be transferred to the government. Long-term risks to subsurface or surface resources are likely to be similar to those that exist in the operational phase, but decreasing over time as pressures decrease and the injected CO_2 is immobilized.

Post-closure liability differs in several fundamental ways from operational liability, and therefore may be of significant concern. For example, in the event of an accident or damage after a well has been plugged and abandoned, there may be difficulty in identifying responsible parties, delegating responsibilities for remediation, and apportioning damages associated with legacy liabilities (Wilson et al. 2007a). Clarification of which parties are financially responsible for long-term management of the site is crucial for CCS deployment.

4.1 Geological Storage of CO_2 and Liability in the EU

The relative importance of CCS will vary significantly across Europe. Some countries are very dependent on coal–Germany and new EU members–and others much less dependent (Wilson and Gibbons 2007). Offshore reservoirs could play a particularly important role in the EU, indeed two of the four existing geological storage projects are located in the North Sea and the Barents Sea. The fact that Europe has a climate policy in place provides an important framework for evaluating potential geological storage liabilities. The EU Environmental Directorate is developing an enabling framework and examining how to integrate geological storage into the Emissions Trading Scheme (Directorate-General of Environment 2008).

The EU has taken important steps toward establishing a legal and regulatory framework for geological storage of CO_2. The 1972 London Convention, which aims to prevent the dumping of wastes at sea (United Nations 1972), and the OSPAR Convention, focused on protecting the North Atlantic marine environment (OSPAR 2007), could both have prevented off-shore geological CO_2 storage, but amendments in 2006 have paved the way for future CCS projects by allowing for injection of CO_2 from CCS projects into the sub-seabed (OSPAR 2007). Additionally, several other EU directives are important for shaping liability from geologic storage: the Environmental Impact Directive, the Integrated Pollution Prevention and Control Directive, the Seveso Directive, the Environment Liability Directive and the EU Emission Trading Scheme, in addition to the Water Framework Directive and Landfill Waste Directive (EC 2007).

The European Directive on geological storage of CO_2 was adopted by the European Parliament in March 2009 (EP 2009a), and liability for damage to human health, natural resources or the environment, as well as climate damage, feature prominently within the text. The text states:

> Provisions are required concerning liability for damage to the local environment and the climate, resulting from any failure of permanent containment of CO2. Liability for environmental damage (damage to protected species and natural habitats, water and land) is regulated by Directive 2004/35/EC of the European Parliament and of the Council of 21 April 2004 on environmental liability with regard to the prevention and remedying of environmental damage, which should be applied to the operation of storage sites pursuant to this Directive. Liability for climate damage as a result of leakages is covered by the inclusion of storage sites in Directive 2003/87/EC, which requires surrender of emissions trading allowances for any leaked emissions (p. 14).

Project developers are required to ensure that sufficient financial resources are available for monitoring and any remediation activities. The directive states that

> The responsibility for the storage site, including specific legal obligations, should be transferred to the competent authority, if and when all available evidence indicates that the stored CO_2 will be completely and permanently contained (p. 15).

In summary, this directive assigns liability for damages during the siting, injection and post-injection periods to the project operator, and suggests that EU national governments assume liability after site closure. It is worth noting that liability associated with site characterization and development could be relatively straightforward in the EU because subsurface mineral resources and pore space are owned by the central governments and leased for mineral exploration. Member states must adopt this directive into their own legal frameworks within 2 years. Provisions are purposely vague, as countries must shape it within their own legal and regulatory contexts.

4.2 *Geological Storage of CO_2 and Liability in the USA*

The current US legal framework for geological storage of CO_2 is evolving rapidly, with recent rules proposed by the US Environmental Protection Agency (EPA), and regulatory initiatives under way in several states. In general, liability issues associated with geological storage are less well defined in the USA than in the EU. In the USA, liability associated with geological storage will be closely linked to state jurisdiction and property ownership regimes–as the USA has a strong tradition of private property rights including subsurface mineral and uncertain rights in subsurface pore space ownership. The USA has many onshore reservoirs suitable for geological storage of CO_2, but onshore storage increases concerns over liabilities due to health or environmental damages and complicates liabilities associated with siting. Finally, the lack of a US climate policy makes assessment of potential climate liabilities difficult.

Liability during the siting phase will largely be a function of state property rights law. Private ownership of mineral rights and pore space will make acquiring the rights to create large storage sites difficult. Also, the legal relationship between surface and subsurface property rights varies between states, further complicating liability issues if storage sites cross state boundaries. Private property rights could also create liabilities during the injection phase of a project if CO_2 migration gives rise to claims of subsurface trespass (Klass and Wilson 2010; Wilson and de Figueiredo 2005).

US regulations governing injection of CO_2 are based on the protection of underground sources of drinking water, under the authority of the 1974 Safe Drinking Water Act. This law allows for regulation of all underground injection activities in the USA and is managed by regulations promulgated by the EPA Underground Injection Control (UIC) Program. While the federal regulatory framework underlies all underground injection activities in the USA, states are allowed to implement the programme themselves, provided they meet minimum standards established by the Federal Government.

The EPA UIC Program recently released a final rule for regulating wells that inject CO_2 for geological storage (US EPA 2010). This proposal, which creates a new well class (Class VI) for CO_2 injection wells, establishes the operator's responsibilities during siting and injection and proposes a post-injection monitoring period of 50 years, during which the site owner/operator would be responsible for monitoring and remediation. Due to statutory limitations, this proposal does not cover: liabilities associated with damages other than to groundwater; property rights issues; the transfer of liability of the final storage site from one party to another; and climate liability. EPA plans to continue to regulate CO_2 injection projects associated with oil or gas recovery under Class II category for EOR and enhanced gas recovery (EGR). EOR and EGR projects that transition to long-term commercial storage projects may cause a change in well classification and associated regulatory requirements. Under UIC regulations, the geological storage site operator would be liable in perpetuity.

Table 1 Existing US state CO_2 storage policies (Based on Pollak and Wilson 2009)

State	Policy	Year	Description
Kansas	HB 2419	2007	Instructs the Kansas Corporation Commission to establish rules for CO_2 storage. State accepts long-term liability, fund created to pay for regulatory costs, remediation, long-term stewardship
	KAR 82-3-1100-120	2010	Sets requirements for CO_2 storage facility operating permits
Louisiana	HB 1117	2008	Gives the State Mineral Board and the Commissioner of Conservation authority over CO_2 storage. Declares injected CO_2 the property of the operator, authorizes eminent domain for CO_2 storage, and provides for leasing state land for CO_2 storage
Massachusetts	SB 2768	2008	Instructs the Department of Energy Resources to set sequestration definitions and standards
Montana	SB 498	2009	Instructs the Board of Oil and Gas to establish rules for CO_2 storage. Declares pore space the property of the surface owner, assigns ownership of injected CO_2 to the operator, gives mineral rights primacy and provides for unitization for CO_2 storage sites. State accepts long-term liability, fund created to pay for regulatory costs, remediation, long-term stewardship
New Mexico	EO 2006-69	2006	Requires the Department of Energy, Minerals and Natural Resources to study statutory and regulatory requirements for CO_2 storage
North Dakota	SB 2095	2009	Instructs the Industrial Commission to set rules for CO_2 storage. Assigns ownership of injected CO_2 to the operator, gives mineral rights primacy and provides for unitization for CO_2 storage sites. State accepts long-term liability, fund created to pay for regulatory costs, remediation, long-term stewardship
	SB 2139	2009	Declares pore space the property of the surface owner
Oklahoma	SB 1765	2008	Declares CO_2 a commodity. Creates a task force to make recommendations on CCS. Also declares that 'the capture, recovery and geologic storage of carbon dioxide will benefit the citizens of this state'
	SB 610	2009	Instructs the Corporation Commission and the Department of Environmental Quality to set rules for CO_2 storage. Assigns ownership of injected CO_2 to the operator and gives mineral rights primacy
Utah	SB 202	2008	Task force to recommend rules for CO_2 storage by 1 January 2011, interim report by 1 July 2009
Washington	ESSB 6001	2007	Directs the Department of Ecology to set rules for CO_2. Specifies that CO_2 storage can be used to meet greenhouse gas emission reduction goals.

(continued)

Table 1 (continued)

State	Policy	Year	Description
WAC	173-218-115	2008	Revises Washington UIC rules for CO_2 storage
WAC	173-407-110	2008	Sets performance standard for CO_2 storage
West Virginia	SB 2860	2008	Study group to recommend rules for CO_2 storage by 2011
Wyoming	HB 89	2008	Declares pore space the property of surface owner
	HB 90	2008	Department of Environmental Quality to propose rules for CO_2 storage permitting; no set date. Working group to recommend financial assurance and post-closure care requirements by 30 September 2009
	HB 57	2009	Amends HB 89 by establishing mineral rights primacy and indicating mineral owners may prevent CO_2 projects that interfere with their rights
	HB 58	2009	States that the injector owns and is legally liable for the CO_2 during operation

States are beginning to pass their own legislation, some of which covers the gaps in the federal regulations (Table 1). For example, Kansas, Montana and North Dakota have passed legislation authorizing the state to accept long-term liability for geological CO_2 storage sites. The state of Washington has adopted rules for CO_2 injection wells under its state UIC programme, but unlike the EPA UIC rules, the Washington state rules cover liabilities for any harm to human health, natural resources, or the environment, including climate liabilities, because they are complemented by state level legislation setting mandatory greenhouse gas reduction goals. Statutes in Louisiana, Montana, North Dakota, Oklahoma and Wyoming establish that the geological storage site operator is liable for the effects of injected CO_2 (as opposed to the surface owner or the pore space owner).

Ultimately, in the USA, both federal and state law will affect liability associated with geological storage of CO_2. The evolution of this legal framework depends strongly on future climate and energy policy in the USA (Pollak and Wilson 2009).

5 Lessons Learned

Liability is important, as early failures in managing legal issues can potentially forestall technology deployment and impact the technology for decades to come. Given the need to reduce greenhouse gases to avoid the most serious consequences of climate change, both nuclear energy and CCS could play important roles. However, the nature of the risks from RW and CCS projects is fundamentally different, except in three key areas: the importance of geological storage, the need for long-term stewardship and the need to resolve socio-political concerns surrounding technology deployment.

The past 40 years provide ample experience with the difficulties in establishing a legal framework to manage liability for RW and siting a final nuclear waste repository. They offer a cautionary tale for both new nuclear energy and future CCS projects. In over 40 years, only one country–Finland–has begun constructing a final repository and one other nation–Sweden–has chosen a final site for a RW repository. Given the relatively small quantities of high-level civilian waste generated, this slow time frame may be acceptable for existing RW facilities (Macfarlane 2006). However, the inability to resolve RW repository siting may affect continued and future use of nuclear power. If future CCS projects face the public opposition that has plagued RW disposal, the technology will fail. As hundreds of geological storage projects must be sited for the technology to significantly reduce CO_2, and there is no option for interim storage, establishing final geological storage sites is a must.

While the institutional framework for managing RW liability has been established in nations with civilian nuclear reactors, some socio-political lessons for structuring legal liability and for implementing the legal agreements can be drawn that are relevant to both future RW repositories and CCS facilities:

- *Legal framework alone is not enough*: In addition to resolving questions of liability, projects need to be successfully sited. While the legal framework is an important step towards deployment, it alone is insufficient.
- *Socio-political context*: Siting and managing both RW and CCS projects will occur within a charged socio-political context with different actors, rules and negotiation of interests. Recognizing the inherent pluralistic nature of technology deployment and identifying the interests of relevant parties is crucial for navigation. For future energy operators, geopolitical and local concerns meld together to influence both technology selection and eventual deployment. How the liability regime for both RW and CCS is structured will influence the socio-political context.
- *Clear and transparent criteria for site selection, performance, closure and monitoring*: The two countries that used clear technical site selection criteria and developed several sites for comparison were successful in final siting. The data was available for all interested stakeholders to examine and compare. Macfarlane (2006) suggests that technical judgments, possibly supplemented with comparative analysis, should underlie any policy decision. This approach allows policy makers and the public to better frame the crucial issues that underlie site selection. Understanding what is required, when it is required and how the technical evidence supports site selection and management will be important for geological CO_2 storage projects and future RW repositories. Delineating clear responsibility and liability across RW and geological CO_2 storage life cycle is a fundamental piece of this.
- *Clear decision points and deadlines are important for all parties*: The US RW repository siting experience highlights how shifting criteria–for the Yucca Mountain repository–and political maneuvering have affected both the technical and institutional credibility of the process and created a quagmire played out in multiple institutions: owners and operators of nuclear facilities, federal and state agencies, the courts, Congress and the Executive Branch. For geological CO_2

storage projects, understanding the factors necessary for liability transfer will be important. While the proposed directive in the EU specifies that national governments will accept ultimate responsibility, no conditions for transfer of liability from the project owner/operator to a public entity were developed. In the USA, the current regulatory framework governing underground injection does not allow for transfer of liability, although individual states are proposing solutions. Any liability framework will shape and be shaped by the political environment.

- *Local community engagement and benefits*: Both Sweden and Finland worked with the local communities from the initial stages of the siting process. They were able to highlight employment benefits and respond to community concerns throughout the process. In Finland, the legislation even specified that the local community could veto the repository if it was unacceptable to them (Macfarlane 2006; Vandenbosch and Vandenbosch 2007). Communities that were familiar with nuclear energy were also more willing to potentially host a repository. In Finland, the repository is located in a town with a nuclear reactor and an existing disposal site for low- and intermediate-level RW. It will be interesting to see if the same pattern emerges with CCS–if communities which are used to oil and gas recovery or underground injection are more accepting of geological CO_2 storage projects. Initial investigation on public perceptions of CCS at potential pilot project host sites have uncovered different community perceptions based on past community experience (Parfomak 2008).
- *Clarification of roles and responsibilities*: In every country with high-level civilian RW, regulation clarifies who is responsible and for what duration. Breaches of this contractual relationship have led to lawsuits (Cawley 2007). While future nuclear energy facilities will benefit from a pre-existing institutional relationship, the institutional and regulatory roles a CCS project must fulfil are still being developed. It is nearly inevitable that they will vary by jurisdiction, as many geological storage risks intersect traditional regulations surrounding land and water use, hydrocarbon recovery and environmental protection. This highlights the need for inter-jurisdictional coordination for both RW disposal and geological storage of CO_2. As many prospective basins for CO_2 injections cross subnational state or country boundaries, coordination could be necessary if multiple projects were active.

  The role of private parties in managing RW varies by country, with some only responsible for storage at the reactor site and others responsible for this plus transporting the waste, siting, building and operating the final repository. Importantly, the roles are clear and well defined. For CCS, the model of industrial organization, responsibility and ultimate liability is still emerging. Questions of long-term liability have not yet been resolved. Future regulation and perceived risk could largely determine the available capital and industrial organization

- *Temporary policies often become permanent*: The US Price-Anderson Act was established as a *temporary* aid to the nuclear energy industry. Interim RW storage has persisted for decades. Whether proposed subsidies for CCS or future nuclear energy could also become long-lasting remains to be seen.

Of the seven areas discussed above, there are many lessons for both RW disposal and geological CO_2 storage projects. While RW management and geological CO_2 storage are very different technologies with different risk profiles and liability frameworks, the socio-political forces that will shape and govern any future deployment share several common threads. Within the context of future deployment, liability frameworks will also play an important role in shaping the industrial organization. While the RW liability structure is well defined, that for geological CO_2 storage is still under development and the past experience with siting a high-level RW repository offers important cautionary tales and lessons mentioned above. For both future nuclear energy and CCS projects, while establishment of a legal framework is important for future development, it is insufficient to guarantee deployment. Rather, legal liability is embedded within a larger socio-political context and addressing these broader concerns is vital for future RW and geological CO_2 storage deployment.

Acknowledgements Thanks to the two anonymous reviewers, Ferenc Toth of the IAEA and to Melisa Pollak of the University of Minnesota for help with this article.

References

Anderson DR (1978) Limits on liability: the Price-Anderson Act versus other laws. J Risk Insur 45:651–674

ASN (Autorité de sûreté nucléaire–Nuclear Safety Authority) (2009) The French Nuclear Safety Authority. http://www.asn.fr/french-nuclear-safety-authority/about-asn. Accessed 10 Apr 2009

Bachu S, McEwen T (2011) Geological media and factors for the long-term emplacement and isolation of carbon dioxide and radioactive waste. In this volume

Beaubien SE, Lombardi S, Ciotoli G, Annunziatellis A, Hatziyannis G, Metaxas A, Pearce JM (2004) Potential hazards of CO_2 leakage in storage systems–learning from natural systems. In: Rubin ES, Keith DW, Gilboy CF (eds) Greenhouse Gas Control Technologies: Proceedings of the 7th International Conference on Greenhouse Gas Control Technologies, vol 1. Elsevier, Oxford, pp 551–560

Bennhold K (2005) EDF shares fail to light up market: no price change in first day of trading. International Herald Tribune. http://www.iht.com/articles/2005/11/21/business/edf.php

BfS (Bundesamt für Strahlenschutz–Federal Office for Radiation Protection) (2009a) English description of who is responsible for operation of federal radioactive waste repositories. Atomic Energy Act (AtG) § 9a para. 3. http://www.bfs.de/en/endlager/publika/verant_betrieb_endl.html

BfS (Bundesamt für Strahlenschutz–Federal Office for Radiation Protection) (2009b) Introduction to the Gorleben site. http://www.bfs.de/en/endlager/gorleben/einfuehrung.html. Accessed 10 Apr 2009

Birraux C (2006) The French 2006 Act on the sustainable management of radioactive materials and wastes: a major breakthrough. Parliamentary Office for Science and Technology Assessment. http://www.icgr2007.org/downloads/background/Birraux_2006-Act-Presentation.pdf

BMU (Bundesministerium für Umwelt, Naturschutz und Reaktorsicherheit–Federal Ministry for the Environment, Nature Conservation and Nuclear Safety) (2009) Act on the peaceful utilization of atomic energy and the protection against its hazards (Atomic Energy Act). http://www.bmu.de/files/pdfs/allgemein/application/pdf/atg_english.pdf

Business Wire (2009) Zurich creates two new insurance policies to support green house gas mitigation technologies, addressing the unique needs of carbon capture and sequestration. Business Wire, 19 Jan. http://www.businesswire.com/news/home/20090119005535/en

Cameron PD (2007) The revival of nuclear power: an analysis of the legal implications. J Environ Law 19:71–87

Cawley K (2007) CBO Testimony: statement of Kim Cawley, Chief, Natural and physical resources cost estimates unit, on the federal government's liabilities under the nuclear waste policy act before the committee on the budget, US House of Representatives. Congressional Budget Office. http://www.cbo.gov/doc.cfm?index=8675&type=0

CCSReg Project (2009) Carbon capture and sequestration: framing the issues for regulation. http://www.ccsreg.org/pdf/CCSReg_3_9.pdf

Clinton WJ (2000) Message to the Senate returning without approval the nuclear waste policy amendments act of 2000. United States Government Printing Office. http://frwebgate.access.gpo.gov/cgi-bin/getdoc.cgi?dbname=106_cong_documents&docid=f:sd019.106.pdf

Dawson JI, Darst RG (2006) Meeting the challenge of permanent nuclear waste disposal in an expanding Europe: transparency, trust and democracy. Environ Polit 15:610–627

de Figueiredo M (2007) The liability of carbon dioxide storage. Ph.D dissertation,. Massachusetts Institute of Technology, Cambridge

Deutch J, Moniz EJ, Ansloabehere S, Driscoll M, Gray P, Holdren J, Joskow P, Lester R, Todreas N (2003) The future of nuclear power. Massachusetts Institute of Technology, Cambridge. http://web.mit.edu/nuclearpower/

Dietze W (2004) Overview on the legal issues involved in the international disposal of high-level radioactive waste–with special consideration to the regional disposal in the European Union. SAPIERR Working Group Meeting. Piestany, 19–20 Feb 2004. www.sapierr.net/files1/Documents/Dietze.doc

Directorate-General of Energy and Transport (2008) Radioactive waste. http://ec.europa.eu/energy/nuclear/waste/index_en.htm. Accessed 13 Aug. 2008

Directorate-General of Environment (2008) CCS in Europe. http://ec.europa.eu/environment/climat/ccs/work_en.htm. Accessed 13 Aug 2008

Dupuis MC (2009) The implementation process for a deep geological repository for radioactive waste in France. In a common objective, a variety of paths. Third International Conference on Geological Repositories. Berne, Switzerland, 15–17 Oct 2007. http://www.nea.fr/html/rwm/reports/2009/6875-icgr2007.pdf

EC (European Commission) (2004) Amended proposal for a Council Directive (Euratom) laying down basic obligations and general principles on the safety of nuclear installations. Amended Proposal for a Council Directive (Euratom) on the Safe Management of the Spent Nuclear Fuel and Radioactive Waste. COM(2004)526. http://eur-lex.europa.eu/LexUriServ/LexUriServ.do?uri=COM:2004:0526:FIN:EN:PDF

EC (European Commission) (2007) Draft report of the meeting of the ECCP working group on carbon dioxide capture and geological storage (CCS). The impact assessment of the enabling legal framework for CCS. http://ec.europa.eu/environment/climat/ccs/pdf/070626_ccs_eccp_meeting_report.pdf

EIA (Energy Information Administration) (2008) Germany energy profile. http://tonto.eia.doe.gov/country/country_energy_data.cfm?fips=GM. Accessed 13 Aug 2008

EP (European Parliament) (2009a) Directive of the European Parliament and of the Council on the geological storage of carbon dioxide. http://ec.europa.eu/environment/climat/ccs/pdf/st03739_en08.pdf

EP (European Parliament) (2009b) European Parliament Legislative Resolution of 22 Apr 2009 on the Proposal for a Council Directive (Euratom) Setting Up a Community Framework for Nuclear Safety (COM(2008)0790 – C6-0026/2009 – 2008/0231(CNS)). http://www.europarl.europa.eu/sides/getDoc.do?type=TA&reference=P6-TA-2009-0254&language=EN

European Community (1957) Treaty establishing the European Atomic Energy Community (Euratom). http://eur-lex.europa.eu/en/treaties/dat/12006A/12006A.htm

French National Assembly (2006) The 2006 programme act on the sustainable management of radioactive materials and waste. http://www.asn.fr/sites/default/files/files/radioactive-waste-management-act-280606.pdf

Garner BA (2004) Black 's Law Dictionary. West Publishing Company, St. Paul

Hebert HJ (2009) Nuclear waste won't be going to Nevada's Yucca Mountain, Obama official says. Chicago Tribune. http://www.chicagotribune.com/news/nationworld/chi-nuke-yucca_frimar06,0,2557502.story

IAEA (International Atomic Energy Agency) (1997) Joint convention on the safety of spent fuel management and on the safety of radioactive waste management. INFCIRC/546. Article 20. http://www.iaea.org/Publications/Documents/Infcircs/1997/infcirc546.pdf

IAEA (International Atomic Energy Agency) (2007) The 1997 Vienna Convention on Civil Liability for Nuclear Damage and the 1997 Convention on Supplementary Compensation for Nuclear Damage–Explanatory Texts, vol 3, IAEA International Law Series. IAEA, Vienna

IPCC (Intergovernmental Panel on Climate Change) (2005) IPCC Special Report on Carbon Capture and Storage. Metz B, Davidson O, de Coninck HC, Loos M, Meyer LA (eds). Cambridge University Press, Cambridge and New York

Kanter J (2009) Debating next-generation nuclear waste. New York Times. http://greeninc.blogs.nytimes.com/2009/02/02/debating-next-generation-nuclear-waste/?hp

Khune G (2007) Judicial progress in Germany's nuclear waste disposal policy: the Konrad repository decisions of 26 Mar 2007. Nucl Law Bull 80:9–19

Klass AB, Wilson EJ (2008) Climate change and carbon sequestration: assessing a liability regime for the long-term storage of carbon dioxide. Emory Law J 58:103–179

Klass AB, Wilson EJ (2010) Climate change, carbon sequestration, and property rights. University of Illinois Law Review. Spring 2010. 363–428

Lidskog R, Andersson A-C (2002) The management of radioactive waste. a description of ten countries. Svensk Kärnbränselhantering AB, Stockholm, SKB Report

Lidskog R, Sundqvist G (2004) On the right track? Technology, geology and society in Swedish nuclear waste management. J Risk Res 7:251–268

Louka E (1993) Bringing polluters before transnational courts: why industry should demand strict and unlimited liability for transnational movement of hazardous and radioactive waste. Denver J Int Law Pol 22:79

Mabe M (2008) Will French leaks harm nuclear's revival? Business Week. http://www.businessweek.com/globalbiz/content/jul2008/gb20080728_585698.htm?chan=globalbiz_europe+index+page_top+stories

Macfarlane A (2003) Underlying Yucca Mountain: the interplay of geology and policy in nuclear waste disposal. Soc Stud Sci 33:783–807

Macfarlane AM (2006) Uncertainty, models, and the way forward in nuclear waste. In: Macfarlane AM, Ewing RC (eds) Uncertainty Underground: Yucca Mountain and the Nation's High-Level Nuclear Waste. MIT Press, Cambridge, pp 393–410

Macfarlane AM, Ewing RC (2006) Uncertainty Underground: Yucca Mountain and the Nation's High-Level Nuclear Waste. MIT Press, Cambridge

National Research Council (2006) Going the Distance: The Safe Transport of Spent Nuclear Fuel and High-Level Radioactive Waste in the United States. The National Academies Press, Washington

NEA (Nuclear Energy Agency) (1960) Convention on Nuclear Third Party Liability, as Amended by the Additional Protocol of 28 Jan 1964 and by the Protocol of 16 Nov 1982. Article 4(a). www.nea.fr/html/law/nlparis_conv.html

NEA (Nuclear Energy Agency) (2000) Nuclear Legislation: Analytical Study–Regulatory and Institutional Framework for Nuclear Activities. OECD Publishing, Paris

NEA (Nuclear Energy Agency) (2005) Legislative updates. Nuclear Liability Bill (2005). OECD Nuclear Energy Agency, Finland. http://www.nea.fr/html/pub/newsletter/2005/23-2-legislative-updates.pdf. Accessed 14 Aug 2008

NEA (Nuclear Energy Agency) (2006) Nuclear legislation in OECD countries: regulatory and institutional framework for nuclear activities. OECD Nuclear Energy Agency, Germany. http://www.nea.fr/html/law/legislation/germany.html

NEA (Nuclear Energy Agency) (2007) Radioactive waste management programmes in OECD/NEA member countries. Country profile France. OECD Nuclear Energy Agency. http://www.oecdnea.org/html/rwm/profiles/France_profile_web.pdf

NEA (Nuclear Energy Agency) (2008) International nuclear third party liability. OECD Nuclear Energy Agency. http://www.nea.fr/html/general/press/press-kits/nuclear-law.html
NEA (Nuclear Energy Agency) (2009) 2004 Protocol to amend the Brussels supplementary convention on nuclear third party liability. OECD Nuclear Energy Agency. http://www.nea.fr/html/law/brussels-supplementary-convention-protocol.html. Accessed 6 Apr 2009
NRC (Nuclear Regulatory Commission) (1957) Financial protection requirements and indemnity agreements. 10 CFR 140. United States Government Printing Office, Washington, DC. http://www.nrc.gov/reading-rm/doc-collections/cfr/part140/
NRC (Nuclear Regulatory Commission) (2006a) Amounts of financial protection for certain reactors. 10 CFR 140.11. United States Government Printing Office, Washington, DC. http://edocket.access.gpo.gov/cfr_2006/janqtr/pdf/10cfr140.12.pdf
NRC (Nuclear Regulatory Commission) (2006b) Determination of extraordinary nuclear occurrence. 10 CFR 140.83. United States Government Printing Office, Washington, DC. http://edocket.access.gpo.gov/cfr_2006/janqtr/pdf/10cfr140.85.pdf
Nuclear Energy Institute (2008) Nuclear waste fund payment information by state. Nuclear Energy Institute, Washington DC. http://www.nei.org/resourcesandstats/documentlibrary/nuclearwastedisposal/graphicsandcharts/nuclearwastefundpaymentinformationbystate/
OSHA (Occupational Safety and Health Administration) (1989). Carbon Dioxide, Industrial Exposure and Control Technologies for OSHA Regulated Hazardous Substances, vol. I of II: substance A – I. Occupational Safety and Health Administration, US Department of Labor, Washington, DC
OSPAR (Convention for the Protection of the Marine Environment of the North-East Atlantic) (2007) Northeast Atlantic OSPAR amendment Annex II, Article 3 (2), Annex III 3(3). OSPAR Commission, London. http://www.ospar.org
Palmgren CR, Morgan MG, de Bruin WB, Keith DW (2004) Initial public perceptions of disposal of CO_2. Environ Sci Technol 38:6441–6450
Parfomak PW (2008) Community acceptance of carbon capture and sequestration infrastructure: siting challenges. Congressional Research Service. http://assets.opencrs.com/rpts/RL34601_20080729.pdf
Parliament of Finland (1987) Nuclear Energy Act. 11.12.1987/990. http://www.edilex.fi/stuklex/en/lainsaadanto/19870990
Parliament of France (2009) French environmental code, Book V, Title IV, Chapter II: specific provisions relating to radioactive waste. Article L542-12. Legifrance: le Service Public de la Diffusion du Droit. http://195.83.177.9/code/liste.phtml?lang=uk&c=40&r=5069
Pollak M, Wilson E (2009) Regulating geologic sequestration in the United States: early rules take divergent approaches. Environ Sci Technol 43(9):3035–3041
Rudig W (2000) Phasing out nuclear energy in Germany. Ger Polit 9:43–80
SAPIERR Project (2008) SAPIERR Project information website. European Commission. http://www.sapierr.net/index_01.htm. Accessed 13 Aug 2008
Slovic P, Fischoff B, Lichtenstein S (1990) Rating the risks. In: Glickman TS, Gough M (eds) Readings in Risk. RFF Press, Washington, DC, pp 61–74
STUK (Finnish Radiation and Nuclear Safety Authority) (2008) Nuclear power plants. http://www.stuk.fi/ydinturvallisuus/ydinvoimalaitokset/en_GB/ydinvoimalaitokset/
Syrota J (2008) L'avenir du nucléaire civil. Politique étrangère 50:161–171
Temple RK (1994) Regulation of nuclear waste and reactor safety within the commonwealth of independent states: toward a workable model. Chicago-Kent Law Rev 69:1071–1120
Trabucchi C, Patton L (2008) Storing carbon: options for liability, risk management, financial responsibility. BNA Daily Environ 170:1–22
United Nations (1972) Convention on the prevention of marine pollution by dumping of wastes and other matter. http://www.imo.org/Conventions/contents.asp?topic_id=258&doc_id=681
United States Congress (1988) 1988 Amendment to Price-Anderson Act. 42 U.S.C. 2210. US Government Printing Office, Washington, DC
United States Congress (2005) Energy Policy Act of 2005. 42 U.S.C. 15801. US Government Printing Office, Washington, DC

US EPA (United States Environmental Protection Agency) (2000) Carbon dioxide as a fire suppressant: examining the risks. Report EPA430-R-00-002. http://www.epa.gov/Ozone/snap/fire/co2/co2report.pdf

US EPA (United States Environmental Protection Agency) (2008) Federal requirements under the underground injection control program for carbon dioxide geologic sequestration wells. Federal Register. Office of the Federal Register, vol. 73. US National Archives and Records Administration, College Park, pp. 43491–43541

US EPA (United States Environmental Protection Agency) (2010) Federal Requirements Under the Underground Injection Control (UIC) Program for Carbon Dioxide (CO_2) Geologic Sequestration (GS) Wells Final Rule, November, 22, Docket No. EPA-HQ-OW-2008-03902, pp 282

US GAO (United States General Accounting Office) (2004) Nuclear regulation: NRC's liability insurance requirements for nuclear power plants owned by limited liability companies. GAO-04-654. http://www.gao.gov/new.items/d04654.pdf

Vandenbosch R, Vandenbosch SE (2007) Nuclear Waste Stalemate. The University of Utah Press, Salt Lake City

West JM, Shaw RP, Pearce JM (2011) Environmental issues in the geological disposal of carbon dioxide and radioactive waste. In this volume

Wilson EJ, de Figueiredo MA (2005) Geologic carbon dioxide sequestration: an analysis of subsurface property law. Environ Law Rep 36:10114–10124

Wilson EJ, Gibbons A (2007) Deploying carbon capture and storage in Europe and the United States: a comparative analysis. J Eur Environ Plann Law 5:343–352

Wilson EJ, Pollak M (2008) Regulation of Carbon Capture and Storage, Policy Brief. International Risk Governance Council, Geneva

Wilson EJ, de Figueiredo MA, Trabuchi C (2007a) Liability and Financial Responsibility FrameWorks for Carbon Capture and Sequestration, WRI Issue Brief. World Resources Institute, Washington, DC

Wilson EJ, Friedmann SJ, Pollak MF (2007b) Research for deployment: incorporating risk, regulation and liability for carbon capture and sequestration. Environ Sci Technol 41:5945–5952

WNA (World Nuclear Association) (2008a) Nuclear power in Germany. http://www.world-nuclear.org/info/inf43.html. Accessed 13 Aug 2008

WNA (World Nuclear Association) (2008b) Nuclear power in France. http://www.world-nuclear.org/info/inf40.html. Accessed 13 Aug 2008

WNA (World Nuclear Association) (2008c) Nuclear power in Finland. http://www.world-nuclear.org/info/inf76.html. Accessed 13 Aug 2008

Public Acceptance of Geological Disposal of Carbon Dioxide and Radioactive Waste: Similarities and Differences

David M. Reiner and William J. Nuttall

Abstract Public acceptance of geological disposal of carbon dioxide (CO_2) and that of radioactive waste (RW) are fundamentally different problems because of the history, scale and nature of the two issues. CO_2 capture and storage (CCS) is a technology in its infancy with no full-scale commercial application and there are only a handful of full-scale storage projects globally. CO_2 storage is almost completely unknown whereas RW disposal has been the subject of highly charged (often unresolved) political debates for decades and all matters nuclear are viewed as both the subject of fear and fascination in the broader cultural and political context. Nevertheless, there are some notable similarities, including: the difficulty of extricating not-in-my-backyard (NIMBY) considerations from other concerns; the inability to divorce the politics of waste streams from the underlying electricity generating technologies; the challenge of communicating the highly technical nature of both issues; and the role that both CO_2 storage and RW play in the larger debate over energy policy, particularly as a proxy issue for non-governmental organizations. A key question identified is whether CCS will continue to be portrayed as the saviour of fossil fuels or whether it becomes an Achilles' heel, much as resolving RW has become a necessary condition for further expansion of nuclear power. It is too early to draw any firm conclusions regarding the acceptability of CO_2 storage because of the current low levels of awareness. Nevertheless, the nature of the CO_2 storage problem tends to support the view that it will be less controversial than RW because of the large number of storage sites needed, public familiarity with CO_2 and the need to resolve storage at the very beginning before CCS can proceed on large point source facilities.

Keywords Public attitudes • Social acceptance • Geological disposal • CO_2 storage • NIMBY

D.M. Reiner (✉) and W.J. Nuttall
Judge Business School, University of Cambridge, Trumpington Street, Cambridge, CB2 1AG, United Kingdom
e-mail: dmr40@cam.ac.uk; wjn21@cam.ac.uk

F.L. Toth (ed.), *Geological Disposal of Carbon Dioxide and Radioactive Waste: A Comparative Assessment*, Advances in Global Change Research 44,
DOI 10.1007/978-90-481-8712-6_10, © International Atomic Energy Agency, 2011

1 Introduction

The differences between the geological disposal of carbon dioxide (CO_2) and radioactive waste (RW) would seem, on its face, to be enormous, both technically and with respect to the attitudes of both local communities and the wider public. It is a reflection of difference in public acceptability, or at least the image that proponents wish to generate, that supporters of CO_2 capture and storage (CCS) assiduously avoid use of the term 'disposal', in spite of the fact that, in its most straightforward sense, there is no interest in ever retrieving the 'stored' CO_2. In the case of RW, however, 'storage' is used to describe an interim measure, often above ground, where the wastes are subject to human oversight and monitoring. 'Disposal' of RWs refers primarily to the waste being placed in a deep geological repository, where the need for monitoring is expected to last for perhaps 100–300 years, and where the ultimate goal is for a passive facility where the waste will be permanently sealed. By contrast, the term 'disposal' is rarely used in the case of CO_2 (Palmgren et al. (2004) being a notable exception). Instead, in virtually all cases, 'storage' of CO_2 refers to a similar situation to 'disposal' of RWs, whereby the CO_2 is stored in a deep geological reservoir and monitored for an extended period during the injection and post-injection phases.

The nature of the interaction with the geological formation is also very different. Unlike RW management, which uses a multi-barrier approach to waste containment, the large volume of CO_2 pumped underground means that from the outset the CO_2 will be contained only by the geological reservoir itself. By comparison, regulatory and scientific analysis of CO_2 storage is in its relative infancy and the duration of monitoring needed and the responsibility for monitoring post-closure are the subjects of active ongoing debate.

The scale differences and levels of experience are striking. A 1,000 MW light water reactor will generate some 800 t of low- and intermediate-level waste and 30 t of spent nuclear fuel per year IAEA (1997). Although debates over final storage are ongoing in many countries, in the meantime wastes have been managed worldwide for five decades. By contrast, a new coal-fired plant of similar size will produce perhaps 6 million tonnes of CO_2 (Mt CO_2) per year. To date, the largest CO_2 injection sites of roughly 1 million tonnes per year each are Sleipner off the coast of Norway (1996), Weyburn in Canada (2000) and In Salah in Algeria (2005). Total monitored CO_2 storage worldwide is thus still less than would be needed for a single power plant.

If CCS were to become a major climate mitigation option, the scale of CO_2 storage activities would be comparable to the current operations of the oil and gas industry. One gigatonne (Gt) of carbon (Gt C) (~3.6 Gt CO_2) is equivalent to capture from 600 1-GW plants and would require the equivalent of 3,600 injection projects at the scale of Statoil's Sleipner project (MIT 2007). The storage sites would require injection of roughly 60 million barrels of supercritical CO_2 each day, or two thirds the current global petroleum production volume (Friedmann 2006). Nuclear power, by contrast, is already operating on a scale of two thirds of a Gt C; as of April 2008, there were 372 GW of nuclear power in operation (IAEA 2008).

Albeit difficult to compare easily, CO_2 is non-toxic at lower concentrations, although at high concentrations it acts as a simple asphyxiant. Air normally contains 0.03% CO_2; at concentrations of 2.5–5% headaches and upper respiratory problems may result, at 10% unconsciousness within 1 min and at 20% respiratory arrest. The threshold limit value is set at 0.5% or 5,000 ppm (Kent 1998). By contrast, even stipulating the existence of any threshold effects for RWs has proven extremely controversial and on a precautionary basis it has become conventional to extrapolate linearly from known high radiation dose effects down to lower doses with no assumed safe dose threshold. Assumptions are required because statistically reliable robust data is very difficult or impossible to obtain for low radiation exposures. For a critique and review of the no-threshold-linear-dose–response assumption, see Prasad et al. (2004).

The political and public context is also vastly different. High-level RW disposal, in particular, has been the subject of intense debate, usually at the national level and has often continued unresolved for decades; for example, the US National Academy of Sciences first proposed deep geological disposal of spent nuclear fuel in 1957 (NAS 1957). By contrast, CO_2 storage is a recent subject that is still largely unknown to the vast majority of the public (EC 2007; Reiner et al. 2006). CCS is playing an increasingly important role in the larger debate over climate change, the future of coal and decentralized generation, but that awareness is largely restricted to policy elites rather than to the general public.

RW is an inescapable problem, in the sense that even if no additional nuclear power plants are built, there will still be a need to deal with the legacy waste that has accumulated. By contrast, concerns over CO_2 are currently only hypothetical, based on the expectation of first large-scale demonstration, then commercialization and widespread expansion of CCS technologies over the next few decades. The converse is that, given the volumes from even a single plant, it will be essential to resolve the storage question upfront for CO_2 whereas RW, in the absence of agreed long-term solutions, can be, and has been, dealt with on a temporary or ad hoc basis for many years. The physical characteristics of CO_2 would seem to lead to far more local (and far more frequent) debates over siting than the national debates over RW siting that usually focus on very few (often only one) site. Nevertheless, we will also explore the similarities in terms of the way in which controversies over storage impact on the wider debates over energy and climate policy, the engagement and attitudes of non-governmental organizations (NGOs), and the basis for local opposition or support.

We divide our analysis into four parts: (1) a brief review of the history of each subject and a discussion of the role that both CO_2 storage and nuclear waste play in the larger debate over energy policy, particularly as a proxy issue for NGOs; (2) general public opinion on the subjects; (3) the role of NIMBY (Not In My Back Yard) and compensation to local communities in facilitating the siting of storage facilities; and finally (4) the extent to which culture, fear and iconography influence public perceptions and political debate (on which, see also de Groot and Steg 2011).

2 History, Energy Choices and the Views of NGOs

2.1 Radioactive Waste

During the phase of rapid nuclear development of the 1950s and 1960s, the speed with which the first nuclear power plants were designed and sited was breathtaking in the context of the infrastructure siting and energy policy debates of the past 30 years. Consider the case of, arguably, the world's first commercial nuclear power station at Calder Hall: that four-reactor station went from concept to power generation in only 42 months (Jay 1956, cited in NDA 2007).

These developments in the years after the Second World War (WWII) led RW to become a new problematic topic for science and technology public policy. Of course, awareness of radiation as a cause of biological harm was already known scientifically before WWII, but that itself had not been sufficient to generate widespread fears. In fact the genesis of societal fear of radiation and nuclear technologies is a complex and fascinating story explored extensively by Weart (1988), whose thesis is that nuclear science and technology manifested numerous sources of fear that had long existed in society: nuclear power just happens to be intrinsically scary.

Over time, nuclear power became increasingly politically controversial, especially following the 1979 accident at Three Mile Island, Pennsylvania, USA, and the disaster in 1986 at Chernobyl in Ukraine. However, even before these events the seeds of later policy difficulties had already been sown. For instance, in the UK in 1976 the report of the Royal Commission on Environmental Pollution (RCEP), known informally as the 'Flowers Report', famously proposed that no commitment should be made to a 'large programme' of nuclear power until a 'method exists to ensure the safe containment' of RWs 'for the indefinite future' (RCEP 1976, p. 202). The Flowers Report provided those with a firm resolve to oppose all nuclear energy developments with the opportunity to block future nuclear power developments merely by rendering the RW question unanswerable. Similar seeds of such difficulties linking resolution of the waste disposal question to nuclear power development were also being sown in Germany and elsewhere (Darst and Dawson 2008). Indeed, the pre-eminence of disposal (in Germany and elsewhere) is inextricably linked to the decisions over reprocessing. Until 1994, German utilities were obliged to reprocess spent fuel to recover the usable portion and recycle it. From 1994 to 1998 reprocessing and direct disposal were equally acceptable to the federal government, but the policy of the coalition government from 1998 is for direct geological disposal of spent fuel and no reprocessing after mid-2005 (WNA 2008).

In this way RW became the Achilles' heel of nuclear power. In such a paradigm RW takes on an importance far beyond the narrow issues of waste and the associated hazards. Arguably waste becomes a proxy battle for much wider questions about nuclear energy, the nature of electricity systems and associated infrastructures and, in extremis, the very nature of industrial and post-industrial society.

2.2 Carbon Capture and Storage

CCS is often put forward as the saviour of fossil-fired generation, and especially in preserving coal as an element in the fuel mix of a carbon constrained world. One might consider, therefore, a situation where CCS might take on the status of Achilles' heel for the fossil fuel industry. To some extent the recent insistence that no new coal plants be built without CCS requires the same resolution. Reflecting the large scale of the problem, the main barrier to penetration of CCS is, however, costs, and, in particular, the costs of capture (IPCC 2005). Resolving the ongoing debate over long-term liability is viewed by many investment firms as essential to the financing of CCS (de Figueiredo 2007). Experience from the RW debate might imply that success for some might be achieved by merely preventing any resolution of questions concerning CCS deployment.

The political debates over both RW and CCS have been shaped by many leading environmental NGOs, almost all of which are strongly anti-nuclear. Nuclear issues catalyzed many of the major environmental groups that were founded in the late 1960s and early 1970s. Greenpeace's original concern was opposition to French nuclear weapons tests in the Pacific, and Friends of the Earth was founded by David Brower, in part out of frustration at the unwillingness of the Sierra Club to oppose nuclear power in general and the Diablo Canyon nuclear plant in California in particular (Shabecoff 1993). Opposition to nuclear power was also central to the creation of many Green Parties (Richardson and Rootes 1995).

This anti-nuclear disposition on the part of most NGOs has remained steadfast in the face of growing concerns over climate change. Indeed, opposition to nuclear power in part explains the willingness of NGOs to remain neutral or even to be slightly favourably disposed towards CCS. Some, such as the Natural Resources Defence Council and Environmental Defence, adopt a pro-CCS position in the hopes of pushing a more aggressive CO_2 concentration target and bringing countries such as China into an emissions control regime (Wong-Parodi et al. 2008). In the US, support among NGOs is also combined with the drive for greater use of coal gasification technology, which would also reduce emissions of traditional air pollutants. By contrast, other NGOs, such as World Wide Fund for Nature (WWF), express support for CCS as a 'necessary evil', in the hopes that the success of CCS will signal the demise of any efforts to revive nuclear power. Stefan Singer, its European Policy Office director, has described WWF's support for CCS as contingent on a move away from nuclear (Singer 2007).

Other NGOs, such as Greenpeace, are concerned at the possibility of increased focus on CCS diverting public resources away from renewables and increasing support for the fossil fuel economy (Rochon 2008). In a survey of over 500 European stakeholders, NGO respondents were also far more likely to take many of the associated risks of deployment quite seriously and, in particular, to worry about the potential for investment in CCS to divert resources away from favoured technologies such as renewables (Shackley et al. 2007).

CCS, although largely unfamiliar to the vast majority of the public, has come to play an increasingly central role in the debates over energy policy and climate change policy in many countries. Perhaps the country where the greatest attention has been paid to CCS is Norway, where a coalition government fell in 2000 over proposals to include CCS in Norway's first ever natural gas-fired power plant (Quiviger 2001). In its so-called 'Soria Moria Declaration' of October 2005, the three coalition parties agreed that all new licenses for gas-fired power plants require CCS. The Bellona Foundation, a major Norwegian NGO, has taken a lead in promoting CCS as an environmentally friendly energy source not just in Norway, but in Europe and beyond. Nevertheless, cost considerations forced the plant at Mongstad to scale back to capture 100,000 t CO_2 in its first years of operation rather than full-scale capture (which would be roughly 1.3 Mt CO_2) and the project has decided to simply release the CO_2 to the atmosphere (Berglund 2007).

Other countries where CCS has played an increasingly important role in national energy and climate policy include Australia, the Netherlands, the USA, the UK and Germany. In all cases, the debate over CCS is tied in closely to ongoing debates over energy security and intra-fuel competition. In Europe, concerns over increased reliance on Russia for natural gas has increased the appeal of domestic coal as well as imports from countries considered more stable (Williams 2008). In the USA and Australia, the two largest coal producers in the developed world, CCS is intimately tied to the continuation of coal-fired electricity generation.

Opposition to continued use of unabated coal-fired generation has increased dramatically in the past few years. In the USA, Texas Utilities was sold in 2007, in large part because of opposition to unabated coal plants; Germany has recently seen proposals for large new coal plants defeated in local referenda (Deggerich 2008), and plans for a 1.6 GW coal-fired plant at Kingsnorth in the UK has come under fire from the Royal Society as well as from over 200 Members of Parliament and activists in the Camp for Climate Action (Adam and Macalister 2008).

2.3 *Similarities and Differences*

One important distinction between RW and CO_2 is that RW is not a single well-characterized entity. Even before WWII, industrial activities involving radioactive materials had already generated significant volumes of materials equivalent to RW. Examples of harmful materials that pre-date capture by RW policy include materials associated with: pre-war radium therapies, luminous paints used in WWII aircraft and pre-war clocks, uranium used in the glassware and lamp mantle industries. To this day such materials (i.e. those created before 1946) are still not officially regarded as RWs in the UK, despite the equivalence of content and hazard that they have with later official wastes (Nuttall 2005). Historical context and administrative classification can be important in defining RWs in addition to the various science-based issues and hazard-related considerations that necessarily affect such processes.

There are numerous classifications of RW and numerous conditions in which it can be found. The main UK classifications of waste are therefore high-level waste (HLW), intermediate-level waste (ILW) and low-level waste (LLW). LLW is relatively unproblematic, as evidenced by the many countries with LLW disposal facilities. Of these British formal classifications of RW, HLW and ILW are defined so as to suit the output streams of aqueous nuclear fuel reprocessing.

Regardless of the fuel cycle, the dominant paradigm is geological disposal. This is near universally agreed as being either current policy, or an eventual policy goal. The slow pace of progress towards these goals has, however, in many cases motivated significant work into surface and near-surface managed storage options, albeit usually framed as an interim measure. Such measures have, however, lasted for decades in many countries.

Interestingly, the UK Committee on Radioactive Waste Management (CoRWM) endorsed the concept of geological disposal in 2006 and rejected formal moves towards monitored 'retrievability'. As such, the committee aligned itself with orthodox scientific approaches to the problem and away from moves that had started to take root that were trading small amounts of notional safety off against popular preferences of inexpert groups of the public (CoRWM 2006). By contrast, in France, it is only when waste cannot be reused or recycled under current technical and economic conditions that it may be disposed of (Warin 2007).

The paradigm of deep geological disposal bears superficial similarity to issues of CO_2 storage and hence it is this approach that we shall focus on in this chapter.

Threats to an RW repository fall into two classes. Those which are more amenable to scientific analysis relate to natural geological and hydrological processes, together with the materials science of immediate waste encapsulation. These natural processes can be analysed for the potential for harmful radionuclides to be released and for the pathways by which they might be transported so as to bring them into contact with the biosphere and human populations. Timescales of such risks are typically measured in tens or hundreds of thousands of years or more. The second class of threat is more difficult to analyse and involves human intrusion into a geological repository either accidentally or deliberately. Key to appreciating these latter risks is the need to reflect upon the timescales involved. Even at 10,000 years old an RW repository would still be young compared to its design life. Human society, however, if it still exists, could by then have gone through two or more cataclysmic collapses and rebuildings. There are few artefacts left from the Mesolithic era 10,000 years ago, when humans first cultivated grains and domesticated animals. Who knows what the future will hold, but it is not unimaginable that millennia from now citizens of a semi-industrialized world might intrude on an RW repository by boring a deep well or that they might seek to excavate, in a primitive fashion, a long sealed repository poorly understanding its contents. In other imaginable futures, spent nuclear fuel might be viewed as a resource that could be extracted and used. The timescales and the risks of deliberate and accidental intrusion into sequestered RW or CO_2 differ from one another, and in each case are difficult to assess or quantify.

Arguably all considerations of environmental sustainability can usefully be expressed in terms of the interests and needs of our great-grandchildren 100 years from now. Commentators (including eminent economists) have pointed out that conventional economic tools of discounting undervalue the needs and interests of future generations (Weitzman 1998). By implication, much smaller, or perhaps even negative, discount rates should be considered. By contrast, most public surveys have supported the view of Charles Galton Darwin that 'most human beings do not care in the least about the distant future. Most care about the conditions that will affect their children and grandchildren, but beyond that the situation seems too unreal, and... uncertainties are too great' (Darwin 1952).

Even more so than for RW, storage of CO_2 underground is *nominally* a matter involving lifetimes of thousands of years, but is primarily a question of the next century, during which the adequacy of the global response to climate change will be revealed (Herzog et al. 2003). Aside from localized effects, such as migration to someone's basement, leakage is of concern because it will add to the atmospheric burden of CO_2 and thereby reduce the effectiveness of CCS. Some have argued that the only acceptable leakage rate when viewed from the perspective of public explanation is zero (Ha-Duong and Loisel 2009), but there have been no studies on how the issue will be framed and what counterfactuals will be assumed. The British Geological Survey, for example, has argued that currently 'leakage' from fossil generation is effectively 100%, so even accounting for the energy penalty and the occasional leak, CCS is a far more climate friendly option (HCSTC 2006).

3 Demographics and Opinion

3.1 Radioactive Waste

Data from Eurobarometer surveys reveals quite stable patterns in public attitudes to RW (EC 2005, 2008a). The dominant opinion of Europeans polled is that roughly three quarters consider themselves to be 'not well informed' on these matters. Generally, northern Europeans report higher levels of understanding than those in southern Europe. Of respondents reporting that they are inclined to support nuclear energy, 65% claim to be well informed about RW, whereas for those averse to nuclear energy 79% report being poorly informed on RW. Even though a large majority (71%) of Eurobarometer respondents correctly understood that there are several types of RW but, tellingly, 78% incorrectly believed that all types of RW are very dangerous, which is roughly the same level as surveys conducted in 2001 (EC 2002) and 2005 (EC 2005).

Although almost all Europeans (93%) believe that there is an urgent need to finding a solution to RW now, rather than leaving it unsolved for later generations, over 70% do not believe there is any safe way of getting rid of HLW (EC 2008a). Deep underground disposal is seen as the single most appropriate solution for managing high-level RW over the long term, but support is only moderate (43% vs. 36% opposed). Although the overall view of nuclear power improved between 2005 and 2008, there was relatively little change in the views towards waste disposal. In spite of decades-long public debate over nuclear power, the public remains divided when asked whether nuclear power was a major contributor to global warming (EC 2003; Reiner et al. 2006).

Information does not necessarily bring support. The 2008 survey found that those who felt well informed were *more* likely to agree with the statement: 'There is no safe way of getting rid of highly radioactive waste' (EC 2008a). There is also keen interest for affected individuals to be directly involved in decisions. Few amongst the public (15%) would defer to the authorities in the siting of an underground storage facility or would even want local NGOs to be consulted on their behalf (22%); instead, the majority (56%) wanted to participate directly in the process.

It is sometimes assumed that knowledge, interest and enthusiasm in nuclear matters are correlated, but it is important to stress that there are many people firmly opposed to nuclear energy who are expert in its intricacies, which further calls into question the 'deficit model' view of science which argues that support is linked to knowledge and that opposition can be overcome via education (Sturgis and Allum 2004). Such anecdotal observations prompt us to question whether the observed correlations are causal. Women are more nervous about nuclear power and RW and also know less about it, but is the hostility to all matters nuclear related to the lack of knowledge and if so, how? Furthermore, do they know less about nuclear issues because they are less likely to have studied physics and maths in school? Is the 'gender' aspect of public attitudes to RW merely a reflection of more fundamental sociological or perhaps sociobiological issues relating to teenage girls and boys and their interests in school or with regard to the technologies in question? These issues will be addressed immediately below and in the next section.

Public attitudes to RW differ according to the sex of the respondent reporting. Women tend to hold much more negative opinions—46% of men favoured nuclear power compared to 29% of women in the 2005 Eurobarometer poll. A 2008 ABC News/Stanford University poll in the USA found that 60% of men supported expansion of nuclear power versus only 29% of women (Langer 2008). Women are also less likely to favour deep underground storage (37% vs. 49% for men) and less likely to believe that nuclear power allows for diversification of the energy supply (57% vs. 72% for men). Aside from such negative views, women are generally less well informed about the issues—in the 2005 Eurobarometer report, men outperformed women on a range of knowledge questions.

While it is true that women are less likely to have training in the sciences and are more sceptical of technology, Barke et al. (1997) found that even female physical scientists judged the risks from nuclear technologies to be higher than their male counterparts. Flynn et al. (1994) found that white males, in particular some 30% of white males, judged risks to be lower for every hazard described. Slovic (1999) described this subgroup as 'characterized by trust in institutions and authorities and by anti-egalitarian attitudes'. In particular, the subgroup were far less likely to agree that local residents should be able to close a nuclear power plant if they feel it is not run properly and that the public should vote on issues such as nuclear power, but were far more likely to trust the experts who build, operate and regulate nuclear power stations and to believe that government and industry can be trusted to make the right decision when managing technological risks.

3.2 Carbon Capture and Storage

By contrast, at a basic level, the lay public has a quite good familiarity with CO_2. Studies of US, British, Japanese and Swedish publics find a clear understanding that automobiles, coal-fired power plants and steel mills produce CO_2 and that trees absorb CO_2 (Reiner et al. 2006). CO_2 storage is less familiar than RW storage and studies in various countries find that there is very little awareness of CCS or even clear recognition that CCS addresses climate change as opposed to other air pollutants or even other environmental problems such as toxic waste or water pollution (Reiner et al. 2006). Similar results have been found in opinion surveys in Spain and in Australia.

The major concern voiced in focus groups (Shackley et al. 2005) was concern over leakage of CO_2 into the atmosphere followed by ecosystem and human health effects. Surveys of stakeholder groups (government, industry, academia and NGOs) have found that both CO_2 storage and CCS generally are considered to be relatively low risk (Shackley et al. 2007). Nevertheless, NGOs tend to view both CCS and storage in particular as somewhat riskier than other stakeholders. The major concern expressed is not over the risks of deployment of CCS per se, but over the additional fossil fuel use necessary because of the energy penalty in the capture process. Other concerns include human health and safety from onshore CO_2 storage and environmental damage from both onshore and offshore CO_2 storage.

Unlike in the case of nuclear power and RW, most studies have not found any appreciable gender gap. The Australian study by Miller et al. (2007) is the most prominent to find that women were more sceptical than men about CCS (as opposed to CO_2 storage specifically), but the survey was non-representative and fully 79% of respondents were female, making any extrapolation of their findings, even to the Australian population, questionable.

Whether more information increases acceptance of CCS is also difficult to study because of the novelty of the issue. Itaoka et al. (2009) have extended their studies of information effects and find that although greater knowledge is associated with stronger support, after information is provided support drops, which the authors explain as being related to lack of awareness of the risks. More generally, Dutch social psychologists working in this area have conducted a number of studies on the stability of individual preferences when faced with information on a novel and complicated technology (see, for example, de Best-Waldhober et al. 2009). They find that many respondents provide 'pseudo-opinions', or 'non-attitudes', whereby respondents are willing to provide an opinion even on topics they know nothing about. These pseudo-opinions are found to be unstable and easily changed according to the specific information provided. This instability of public opinion should provide a caution when drawing conclusions from any study of attitudes towards CCS no matter how carefully designed. Finally, even more problematic is that the current status of risk communications on CCS has been judged to fall far short of best practice and in many cases is extremely weak, so that what information that is out there for the interested layperson is actually not up to the task of providing a clear exposition of the basic facts (Reiner 2008).

4 Location, NIMBY and Compensation

Siting RW facilities has proven exceedingly difficult around the world. As Gerrard (1996) notes in the context of the USA, 'Despite scores of siting attempts and the expenditure of several billion dollars since the mid-1970s... there is only one small radioactive waste disposal facility; only one hazardous waste landfill... and a small handful of hazardous waste treatment and incineration facilities' (Gerrard 1996).

The Facility Siting Credo (Kunreuther et al. 1993) offers a series of suggestions on how to successfully site a major infrastructure project: (1) instituting a broad-based participatory process; (2) seeking acceptable sites through either a volunteer or a competitive siting process; (3) keeping multiple options open at all times; (4) guaranteeing stringent safety standards; (5) ensuring geographic equity; and (6) making the host community better off. Most national-level processes aimed at choosing an RW site have been unwilling or unable to comply with many of these recommendations (e.g. competitive siting, geographic equity, keeping many options open). Although there are few existing examples of siting CO_2 storage facilities near a concerned community, the scale of CO_2 storage means that there will inevitably be many sites at a national level, which means that it will be easier to meet some of the elements of the credo than would be the case for a single national RW repository.

One area that has drawn considerable attention is the possibility of making the host community better off. Compensation combined with other incentives has been used successfully to gain public acceptance of locally contested infrastructure projects in settings as diverse as Japan, France, Australia and the USA (Lesbirel and Shaw 2005). For example, in France public utilities offer reduced electricity prices to host communities and in Japan compensation is provided to both the host community and surrounding communities. By contrast, other studies have found that compensation may prove counterproductive (Frey et al. 1996). Singleton's study (Singleton 2007) of the potential for compensation in the case of CCS is largely sceptical of the potential role that might be played.

If the problem is purely one of NIMBY, then one would expect that compensating for losses in property values or other negative impacts should be relatively simple. If, however, the issue is fear of a technology or waste product or distrust in those, then straightforward compensation will be made more difficult or perhaps impossible.

NIMBY or NUMBY (Not Under My Backyard) as coined by Huijts (2003) poses a serious challenge to the siting of CO_2 storage. Jaeger (2007) argues that the necessary public trust can be gained: 'If the businesses involved in CCS would accept collective liability for the safety of CCS, they could establish the kind of credibility the nuclear industry is lacking.' Huijts et al. (2007) offer one of the few case studies of the attitudes of local residents ($n = 103$) in the vicinity of a potential storage site for CO_2. They found that public attitudes towards CCS in general were slightly positive, but attitudes towards storage nearby were slightly negative. In spite of having little knowledge about CO_2 storage, the lay public showed little desire to learn more. Therefore it is not surprising that trust in those providing information was seen as particularly important. NGOs were found to be trusted most, and industry least, by the general public. Trust in different actors appeared to depend on perceived competence and intentions. Moreover, previous experience with the organizations or actors involved, concerns over accountability, and openness can also play important roles in shaping trust (see generally, Cvetkovich and Löfstedt 1999).

Wong-Parodi et al. (2007) conducted focus groups in two communities in California's Central Valley and found that compensation is critical for technology acceptance and that community involvement was essential for the success of the project, but that past experience was critical for defining a community's willingness to believe they would receive compensation. Rio Vista's experience with royalties from natural gas and mineral rights which accrued to the long-time landowners left them more favourably disposed to siting of CCS facilities whereas in Thornton, experience with water treatment left residents distrustful of further projects.

In a survey of 1,001 Nevada residents, Kunreuther et al. (1990) found that perceived risk (e.g. risk to future generations) depends in part on the trust placed in the US Department of Energy to manage the repository safely. Opposition did not decrease significantly if compensation of US$1,000–5,000 in rebates per year for 20 years was offered to residents. Rather, the public needs to be convinced before compensation is considered that the repository will possess minimal risks to themselves as well as to future generations, and that the site currently targeted is suitable.

Of course, success is not simply a function of compensation. In the cases where a high-level RW facility has been successfully sited, such as in Sweden and Finland, a key element in the success has been public engagement (Litmanen 1999). One of the more successful examples of consensus building was the CoRWM process in the UK, which differed from all previous (unsuccessful) approaches to policy for the management of RW in that from the outset it was not constructed to be simply a scientific and technical problem. CoRWM recognized from the outset that it was as much a sociological and political problem. In addition to issues considered by previous policymaking bodies, CoRWM devoted much energy to what the committee termed 'ethics', and in particular 'intra-' and 'intergenerational' ethics (CoRWM 2006). CoRWM suggests that intergenerational equity must balance the needs and interests of future generations with the needs and interests of those living today. As such, it is not appropriate to discount the future in ways that are commonplace in modern economics. Intra-generational equity should consider the question of where to locate a waste disposal facility and, in so doing, seek to properly handle the needs and interests of spatially separated communities living at the same time as one another. Such thinking led CoRWM to recommend 'community packages' of compensation to communities willing to accept an RW facility but subject to negative externalities such as property blight and disturbance.

As the Nuclear Decommissioning Authority in the UK seeks to implement policy recommendations emerging from the Government in response to CoRWM it seems possible that communities might actually compete to host an RW repository, if the 'compensation' on offer is sufficiently attractive. As such NIMBYism might even be replaced with PIMBYism (Please In My Back Yard or YIMBY (Yes, In My

Back Yard)). Polls have found, for example, stronger support for nuclear power in the vicinity of operating nuclear power plants (e.g. Wikdahl 1991, for the case of Sweden).

It is not unimaginable that, at least for many of the first projects, CCS might relate more to PIMBYism than to more conventional notions of NIMBYism. Such a response seems especially likely where the reservoir in question is a depleted oil and gas reservoir and where the community has hosted oil and gas operations and benefited from employment and built trust in the companies involved. This situation is true of enhanced oil recovery in the Permian Basin in Texas or of acid gas injection in Alberta (Heinrich et al. 2004) as well as the Lacq project in France.

Locations for RW repositories are usually in isolated and economically distressed regions. The former criterion might have a rational basis in the event that the proposed facilities are not as safe as is stated by their proponents. The latter argument is perhaps more compelling, that poor isolated communities lack political influence and hence make it easier for proponents of controversial installations to win the day.

5 Culture, Fear and Iconography

Fundamental to attitudes to RW are attitudes to nuclear technologies generally, including especially nuclear weapons. The interrelationship between the Cold War and the Bomb are culturally resonant, attracting the attention of Stanley Kubrick (Dr Strangelove, 1964), Andy Warhol (Atomic Bomb, 1965), and Salvador Dali (Atomic and Uranian Melancholic Idyll, 1945) among many others (Jones 2002).

The interrelationship between matters nuclear and pop culture extended in time beyond nuclear weapons to include aspects of civil nuclear power such as RW. The timing was such that opposition to nuclear energy followed on directly from previous protest movements, which had followed trajectory in the USA from Civil Rights through opposition to the Vietnam War.

One observer of the 1960s describes the close link between environmentalism and opposition to nuclear power as follows (Morgan 1991, p. 244):

> 'One of the primary early targets of ecological activism was the nuclear power industry. In fact, of all forms of environmental politics, the antinuclear movement was the most directly reminiscent of Sixties activism. With citizens' referenda, lobbying, litigation, and administrative intervention; civil disobedience and other forms of direct action; and mass rallies aglow with countercultural trappings, the antinuclear movement recalled the antiwar movement that had just ended. In its early days, it was largely populated by former peace activists as well as feminists, assorted environmentalists, and counterculture communards.'

It is far from clear whether opposition to CCS would fall naturally into line with a continuous tradition of countercultural protest, although opposition to coal without CCS would seem to have increasingly fallen into that category. For example, as mentioned in Sect. 2, coal-fired power stations have increasingly become the focus for direct action. The Camp for Climate Action, a grassroots movement which originated in the UK, but which has spread across Europe since 2007, has set up camp at UK coal-fired power stations for two of the past 3 years and has sought to engage in various forms of direct action including efforts to shut the plants down or block coal trains (Joyce 2008). In 2007, two major new proposed German coal-fired stations were defeated in local referenda on sites that had previously been occupied by coal-fired generation units.

Ocean storage had already been effectively ruled out as a viable option as a result of the major international experiment being torpedoed by opposition. The project was planned first for Hawaii, where opponents delayed the project and then, when it relocated to Norway, Greenpeace sailed the Rainbow Warrior to meet with the Norwegian environment minister who withdrew permission for the experiment (de Figueiredo 2002).

The primary advocates of CCS—national governments and the energy industry—are precisely those least trusted by the public, especially when compared to high levels of trust in NGOs and independent scientists (EC 2008b). For RW, the reality is that it has been there before, with large-scale protests in, for instance, the 1980s. It seems likely that, by extension, plans for geological disposal of RW will be disrupted by protest, but it is far from certain that they will be. If the counterculturalists of yesteryear are now too old to stand up and protest and they failed to pass their politics to the next generation, then RW developments might progress relatively unimpeded by protest.

One aspect of 1960s protest may continue to echo in today's attitudes to RW and this relates to the attitude of women to nuclear technologies. As noted above, polling reveals that a significantly larger number of women than men oppose nuclear energy. Perhaps the greater tendency for women to have negative attitudes to nuclear technologies is something more intrinsic to these technologies themselves. If so, then this would expose a key difference between RW perceptions and those relating to CCS. Given the low overall levels of awareness regarding CCS it is too early to determine whether there will be any significant gender split.

The thesis that says that the aversion of some women to nuclear technologies is more intrinsic points to observations such as:

- The relationship between radiation and genetic damage tapping into, and arguably subverting, a woman's ability to control her own fertility. Such issues became resonant in the 1960s given the then growing interrelationship between feminism and fertility after the introduction of the contraceptive pill in 1957.
- The emergence of the notion of deep ecology, which posits that mankind is merely a component of a broader living and evolving environment within which it has no special status. This philosophy draws much upon the concept of Gaia developed and popularized by James Lovelock.

It is with the growth of Gaia as a popular construct that the interplay between environmentalism and nuclear energy arguably comes full circle. In The Revenge of Gaia, Lovelock (2006) argues that anthropogenic climate change is a threat to the entire biosphere. In comparison, the risks associated with nuclear energy and RWs are small and manageable.

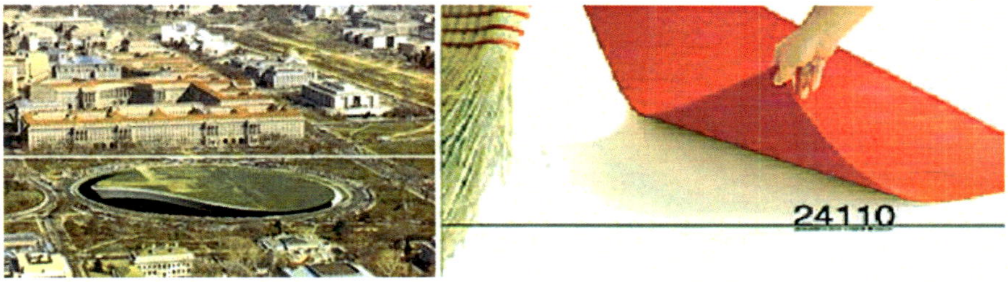

Fig. 1 Michael Simonian's Plutonium Memorial concept '24110'. (Images copyright Simonian; see: http://www.designboom.com/eng/cool/simonian.html) (*see* Colour Plates). The artist imagines a central Washington DC location for a plutonium store just under the Ellipse, a field 1 km in circumference, near the White House, which takes to an extreme the notion that plutonium storage should not be *out of sight and out of mind*

There is another link between culture and RW that has few, if any parallels, in CCS policy, namely the notion of possible warning signs on RW repositories to protect against the risk of accidental intrusion referred to earlier. The Bulletin of the Atomic Scientists has supported creative responses to this problem with, respectively, the Universal Warning Sign competition (ECYMIO 2003) and the Plutonium Memorial Design Contest, won in 2002 by Michael Simonian with his concept '24110', which takes its name from the half-life in years of the main plutonium isotope Pu-239 (Bulletin of the Atomic Scientists 2002), and which is shown in Fig. 1.

Although often thought benign, at high enough concentration CO_2 may lead to asphyxiation caused by oxygen displacement. Being heavier than air, leakage may lead to accumulation in low lying areas or basements and may therefore pose a minimal threat to local populations in the vicinity of storage sites or CO_2 pipelines. There are a number of natural analogues: CO_2 seeps at Poggio dell'Ulivo in Central Italy discharge 200 t CO_2/day via soil degassing and at least ten people have been reported to have died from CO_2 releases in the Lazio region in the past 20 years (IPCC 2005); in April 2006, at Mammoth Mountain in California, three ski patrollers died while trying to fence off a volcanic vent (USGS 2001; Doyle 2006).

Far more dramatically, in 1986, 1,700 people died after a massive CO_2 explosion at Lake Nyos in Cameroon (Kling et al. 1987). In 1984, a smaller explosion in Lake Monoun, also in Cameroon, killed 37 people. A third lake, Lake Kivu, on the Congo–Rwanda border, is also known to be a reservoir of CO_2 and methane. Accumulation of CO_2 begins when CO_2-rich gas of volcanic origin comes into contact with groundwater, which is then discharged into the bottom of the lake. Before the gas events, these lakes were strongly stratified, such that surface and bottom waters did not mix, thus allowing the gas that was being input in CO_2-charged springs to build up in the bottom waters of the lakes.

The trigger mechanism responsible for the gas release from the lake has been the subject of much speculation. Although there were some claims that there was a volcanic event, it now seems likely that a large landslide entered the lake causing the lake stratification to break down enough to initiate the gas release. Although there is no physical analogue to CCS, to the consternation of CCS advocates, Lake Nyos is often cited as a reason to fear large-scale storage of CO_2 (Brown 2007). Given the low level of public awareness of CCS in the first place, this fear-mongering is unlikely to have had much influence on the public to date, but it remains one element of the arguments put forward by groups opposed to CCS (Rochon 2008).

6 Conclusions

So how does public acceptability of RW disposal compare to that for CO_2 storage? In some respects, it is nonsensical to provide an answer at this stage when less than 10% of the public in most countries have even heard of the concept of CCS and when there are no full-scale operating CCS projects. Nevertheless, there are some reasons to believe that CO_2 storage is fundamentally less controversial. The need for many storage sites avoids the painful debates over equity associated with choosing a single national storage site. The sheer volume of CO_2 from a single large coal-fired plant requires resolution of any local (or national) concerns long before a project starts whereas the tiny volume of RW means that final decisions on ultimate disposal can be, and have usually been, deferred for not just years, but many decades. The inextricable association of RW with nuclear power and the potential for meltdown and with nuclear weapons and security concerns such as proliferation imbues the subject with dread fear of a nature that is rare for any subject. Although there are some such fear-inducing associations with CO_2 storage such as the disaster at Lake Nyos, most people commonly relate CO_2 to exhaling and to the uptake of CO_2 by trees and other vegetation. CO_2 is something familiar and, as such, largely uninteresting even if the phase concerned is liquefied and at high pressure. Conversely, all matters nuclear, including RW, are unfamiliar, so perhaps it is ironic, but in part as a result of its exotic nature, nuclear issues have become part of our popular culture and are often regarded with great interest.

Table 1 summarizes some of the main themes that have emerged in this chapter. Controversies over local siting, the links to the associated energy technology and trust in the actors providing information and communicating with the public are the issues that bear the greatest similarities across the two areas but, with the exception of the actors involved, even these broad similarities are quite different in practice because of differences in attitudes towards coal and nuclear power and the likelihood of a single onshore repository for RWs versus multiple CO_2 storage sites that might be onshore or offshore. As noted above, it is the nature and history of the two subjects that result in the greatest differences. Attitudes towards RW are well monitored and unlikely to change other than marginally bar a major event, whereas opinions on CCS and CO_2 storage have only been studied for a few years and the public remains largely ignorant. Even those opinions voiced in current surveys are unlikely to be stable and will depend on the framing and evolution of CCS as the first demonstration plants are funded and launched.

Table 1 Comparison of key attributes associated with public acceptance of CO_2 storage and geological disposal of radioactive wastes

Subject	Radioactive waste disposal	CO_2 storage
Public awareness	Broad public awareness	Minimal public awareness of any aspect of CCS
Public understanding	Generally weak in spite of high awareness	Basic understanding of carbon cycle but minimal to none on CO_2 storage itself
Public acceptability of solution	Acceptability poor and greater acceptance not necessarily linked to greater understanding	Linked to climate change and perceived adequacy of other solutions, but still too early to determine
Demographics	Strong female opposition across time and region	Little evidence of major differences visible at this stage
Timing	Not necessary to address immediately; in most cases deferred for decades	Essential to resolve storage before operation begins because of volume of waste stream
Risk communications	Extensively studied but practice remains weak	Few examples of good practice, poorly studied
Trust in actors	Involves energy industry and government, some of least trusted actors in society Eroded by image of 'nuclear priesthood'	Involves energy industry and government, some of least trusted actors in society
Views of grassroots and environmental NGOs	Generally hostile although there has been successful engagement on narrow question of repository siting	Main environmental groups are neutral to moderately positive Some resistance from grassroots groups less concerned with climate change alone
Support for associated energy technology	Support for nuclear power remains divided and this division has continued for decades	Unabated coal is becoming increasingly unpopular, although there remains support for coal miners in many countries

CCS Carbon capture and storage, *NGO* Non-governmental organization

It is interesting to consider why the RW problem has been so difficult. One compelling idea is that the RW problem is an example of a *wicked problem* (Conklin 2006). Such problems are characterized by an odd circular property that the question is shaped by the solution. As each solution is proposed it exposes new aspects of the problem. Wicked problems are not amenable to the conventional linear approaches to solving complex problems. Linear approaches go from gathering the necessary data, through analysing the data and formulating a solution, towards implementation of a final agreed solution. By contrast, wicked problems can at one moment appear to be on the verge of solution, yet the next moment the problem has to be taken back to its complete fundamentals for further progress to be made. As such, any opinion that the problem is almost solved is no indication that it actually is. Wicked problems can persist for decades and, for a true wicked problem, no solution will ever be possible. Wicked problems typically combine technical factors and social factors in complex multi-attribute trade-offs. A problem that is not wicked is said to be 'tame'. A key question for consideration by the CCS community is whether they too have found themselves in a similar situation. The key difference, as noted earlier, is that if there is no resolution to concerns over CO_2 storage there will be no possibility for large-scale implementation of CCS to proceed.

According to MacKerron (2004), nuclear power has not merited the same government support as renewables because of the associated non-climate change externalities. The economic and, especially, political risks of nuclear power are perceived as balancing its climate advantages from being a low-carbon source of electricity. MacKerron then lists a series of ways in which nuclear power might become 'ordinary' and hence more attractive to private investors, chief among these being 'resolution, to the satisfaction of the wider public, most stakeholders and any affected local communities, of the radioactive waste management problem.' A similar question for CCS is whether it might command subsidies needed to allow for construction of the first tranche of large-scale projects. The future of fossil-fired generation is therefore wrapped up in questions both of the fuels themselves but also of the ultimate fate of CO_2 underground. As described above, nuclear power and RW have never been perceived as 'ordinary'. Although CO_2 storage is still unfamiliar to the vast majority of the public, the familiarity with CO_2 itself and its comparatively benign nature may allow CO_2 storage to proceed even though individual CO_2 storage projects may well be halted for a variety of NIMBY or other local considerations much as would be the case for many other types of waste facilities.

Acknowledgements This work is supported through a contract with the International Atomic Energy Agency. We would like to thank Ferenc Toth for his support and his helpful comments as well as for those of two anonymous referees.

References

Adam D, Macalister T (2008) Energy: coal-fired power stations will lock UK into a high-emissions future, say MPs. The Guardian, 22 July. http://www.guardian.co.uk/environment/2008/jul/22/carbonemissions.carboncapturestorage

Barke R, Jenkins-Smith H, Slovic P (1997) Risk perceptions of men and women scientists. Soc Sci Quart 78:167–176

Berglund N (2007) Government blasted for dropping carbon deal. Aftenposten, 18 December

Brown VJ (2007) Of two minds: groups square off on carbon mitigation. Environ Health Perspect 115:A546–A549

Bulletin of the Atomic Scientists (2002) Plutonium memorial: design contest. B Atom Sci 58(3):38–39

Conklin J (2006) Dialogue Mapping: Building Shared Understanding of Wicked Problems. Wiley, Chichester

CoRWM (Committee on Radioactive Waste Management) (2006) Managing Our Radioactive Waste Safely. Final Report, CoRWM Doc 700. Committee on Radioactive Waste Management, London
Cvetkovich G, Löfstedt RE (eds) (1999) Social Trust and the Management of Risk. Earthscan, London
Darst RG, Dawson JI (2008) "Baptists and bootleggers, once removed": the politics of radioactive waste internalization in the European Union. Global Environ Polit 8:17–38
Darwin CG (1952) The Next Million Years. Rupert Hart-Davis, London
de Best-Waldhober M, Daamen D, Faaij A (2009) Informed and uninformed public opinions on CO_2 capture and storage technologies in the Netherlands. Int J Greenhouse Gas Control 3:322–332
de Figueiredo MA (2002) The Hawaii carbon dioxide ocean sequestration field experiment: a case study in public perceptions and institutional effectiveness. Masters thesis, Massachusetts Institute of Technology, Cambridge
de Figueiredo MA (2007) The liability of carbon dioxide storage. PhD dissertation, Massachusetts Institute of Technology, Cambridge
de Groot JIM, Steg L (2011) Psychological perspectives on the geological disposal of radioactive waste and carbon dioxide. In this volume
Deggerich M (2008) Germans discover direct democracy. Spiegel Online, 25 April
Doyle A (2006) Concerns rise about CO_2 burial. Reuters, 11 July
EC (European Commission) (2002) Europeans and radioactive waste. Special Eurobarometer EB 165. http://ec.europa.eu/public_opinion/archives/ebs/ebs_165_en.pdf
EC (European Commission) (2003) Energy: issues, options and technologies science and society. Special Eurobarometer EB 169. http://ec.europa.eu/public_opinion/archives/ebs/ebs_169.pdf
EC (European Commission) (2005) Radioactive waste. Special Eurobarometer EB 227. http://ec.europa.eu/public_opinion/archives/ebs/ebs_227_en.pdf
EC (European Commission) (2007) Energy technologies: knowledge, perception, measures. Special Eurobarometer EB 262. http://ec.europa.eu/public_opinion/archives/ebs/ebs_262_en.pdf
EC (European Commission) (2008a) Attitudes towards radioactive waste. Special Eurobarometer EB 297. http://ec.europa.eu/public_opinion/archives/ebs/ebs_297_en.pdf
EC (European Commission) (2008b) Attitudes of European citizens towards the environment. Special Eurobarometer EB 295. http://ec.europa.eu/public_opinion/archives/ebs/ebs_295_en.pdf
ECYMIO (Eureka County Yucca Mountain Information Office) (2003) Art exhibition in Eureka: universal warning sign. Nucl Waste Update 9:1–2, http://www.yuccamountain.org/pdf/summer_2003_blue.pdf
Flynn J, Slovic P, Mertz CK (1994) Gender, race, and perception of environmental health risks. Risk Anal 14:1101–1108
Frey BS, Oberholzer-Gee F, Eichenberger R (1996) The old lady visits your backyard: a tale of morals and markets. J Polit Econ 104:1297–1313
Friedmann SJ (2006) Defining an end state for CO_2 sequestration and EOR in North America. In: Røkke N, Bolland O, Gale J (eds) Proceedings of the 8th International Conference on Greenhouse Gas Technologies. Elsevier, London, CD-ROM
Gerrard MB (1996) Whose Backyard, Whose Risk: Fear and Fairness in Toxic and Nuclear Waste Siting. MIT Press, Cambridge
Ha-Duong M, Loisel R (2009) Zero is the only acceptable leakage rate for geologically stored CO_2: an editorial comment. Clim Change 93:311–317
HCSTC (House of Commons Science and Technology Committee) (2006) Meeting UK energy and climate needs: the role of carbon capture and storage. First Report of Session 2005–2006. Oral and Written Evidence, vol II. Report HC 578-II. Her Majesty's Stationery Office, London
Heinrich JJ, Herzog HJ, Reiner DM (2004) Environmental assessment of geologic storage of CO_2. Report MIT LFEE 2003-002 RP. MIT Laboratory for Energy and Environment, Cambridge
Herzog H, Caldeira K, Reilly J (2003) An issue of permanence: assessing the effectiveness of temporary carbon storage. Clim Change 59:293–310
Huijts N (2003) Public perception of carbon dioxide storage: the role of trust and affect in attitude formation. Masters thesis, Department of Technology Management, Eindhoven University of Technology, Eindhoven

Huijts NMA, Midden CJH, Meijnders AL (2007) Social acceptance of carbon dioxide storage. Energ Policy 35:2780–2789

IAEA (International Atomic Energy Agency) (1997) Sustainable Development and Nuclear Power. IAEA, Vienna

IAEA (International Atomic Energy Agency) (2008) Energy, electricity and nuclear power estimates for the period up to 2030. Reference Data Series No. 1. http://www-pub.iaea.org/MTCD/publications/PDF/RDS1-28_web.pdf

IPCC (Intergovernmental Panel on Climate Change) (2005) IPCC Special Report on Carbon Dioxide Capture and Storage. Metz B, Davidson O, de Coninck HC, Loos M, Meyer LA (eds). Cambridge University Press, Cambridge/New York

Itaoka K, Okuda Y, Saito A, Akai M (2009) Influential information and factors for social acceptance of CCS: the 2nd round survey of public opinion in Japan. In: Gale J, Herzog H, Braitsch J (eds) Proceedings of the 9th International Conference on Greenhouse Gas Control Technologies. Energy Procedia 1:4803–4810

Jaeger CC (2007) Carbon capture and storage: risk governance and rent seeking. Paper presented at the conference on regulating and financing carbon capture and storage, Swiss Re Centre for Global Dialogue, Ruschlikon, 8 November. http://www.irgc.org/IMG/pdf/IRGC_CCS_Jaeger07.pdf

Jay K (1956) Calder Hall—The Story of Britain's First Atomic Power Station. United Kingdom Atomic Energy Authority, Didcot

Jones J (2002) Magic mushrooms. The guardian, 6 August. http://www.guardian.co.uk/artanddesign/2002/aug/06/art.artsfeatures

Joyce J (2008) Showdown looms for climate camp. BBC News, 6 August. http://news.bbc.co.uk/1/hi/uk/7544475.stm

Kent C (1998) Basics of Toxicology: Preserving the Legacy. Wiley, New York

Kling GW, Clark M, Compton HR, Devine JD, Evans WC, Humphrey AM, Lockwood JP, Tuttle ML (1987) The 1986 Lake Nyos gas disaster, Cameroon, West Africa. Science 236:169–175

Kunreuther H, Easterling D, Desvousges W, Slovic P (1990) Public attitudes toward siting a high-level nuclear waste repository in Nevada. Risk Anal 10:469–484

Kunreuther H, Fitzgerald K, Aarts TD (1993) Siting noxious facilities: a test of the facility siting credo. Risk Anal 13(3):301–318

Langer G (2008) Fuel costs boost conservation efforts; 7 in 10 reducing 'carbon footprint'. ABC News, 9 August. http://abcnews.go.com/print?id=5525064

Lesbirel SH, Shaw D (2005) Managing Conflict in Facility Siting: An International Comparison. Edward Elgar, Cheltenham

Litmanen T (1999) The social shaping of radwaste management: the cases of Sweden and Finland. Current Sociol 45:59–79

Lovelock J (2006) The Revenge of Gaia: Why the Earth is Fighting Back—And How We Can Still Save Humanity. Penguin, London

MacKerron G (2004) Nuclear power and the characteristics of ordinariness—the case of UK energy policy. Energ Policy 32:1957–1965

Miller E, Bell L, Buys L (2007) Public understanding of carbon sequestration in Australia: socio-demographic predictors of knowledge, engagement and trust. Aust J Emer Technol Soc 5:15–33

MIT (Massachusetts Institute of Technology) (2007) The Future of Coal: Options for a Carbon Constrained World. MIT, Cambridge

Morgan EP (1991) The 60s Experience: Hard Lessons About Modern America. Temple University Press, Philadelphia

NAS (National Academy of Sciences), (1957) The Disposal of Radioactive Waste on Land. National Research Council, Washington

NDA (Nuclear Decommissioning Agency) (2007) Calder Hall nuclear power station feasibility study. Atkins Engineering. http://www.nda.gov.uk/documents/loader.cfm?url=/commonspot/security/getfile.cfm&pageid=18053

Nuttall WJ (2005) Nuclear Renaissance: Technologies and Policies for the Future of Nuclear Power. IOP/CRC Press, Bristol

Palmgren CR, Morgan MG, Bruine de Bruin W, Keith DW (2004) Initial public perceptions of deep geological and oceanic disposal of CO_2. Environ Sci Technol 38:6441–6450

Prasad KN, Cole WC, Hasse GM (2004) Health risks of low dose ionizing radiation in humans: a review. Exp Biol Med 229:378–382

Quiviger G (2001) Building new power plants in a CO_2 constrained world: a case study from Norway on gas-fired power plants, carbon sequestration, and politics. Masters thesis, Massachusetts Institute of Technology, Cambridge

RCEP (Royal Commission on Environmental Protection) (1976) Nuclear Power and the Environment. Sixth report. Her Majesty's Stationery Office, London

Reiner DM (2008) A Looming Rhetorical Gap: A Survey of Public Communications Activities for Carbon dioxide Capture and Storage Technologies. Electricity Policy Research Group Working Paper 0801, University of Cambridge, Cambridge

Reiner DM, Curry TE, de Figueiredo MA, Herzog HJ, Ansolabehere SD, Itaoka K, Johnsson F, Odenberger M (2006) American exceptionalism? Similarities and differences in national attitudes towards energy policy and global warming. Environ Sci Technol 40:2093–2098

Richardson D, Rootes C (1995) The Green Challenge: The Development of Green Parties in Europe. Routledge, London

Rochon E (2008) False Hope: Why Carbon Capture and Storage Won't Save the Climate. Greenpeace International, Amsterdam

Shabecoff P (1993) A Fierce Green Fire: The American Environmental Movement. Hill & Wang, New York

Shackley S, McLachlan C, Gough C (2005) The public perception of carbon dioxide capture and storage in the UK: results from focus groups and a survey. Clim Policy 4:377–398

Shackley S, Waterman H, Godfroij P, Reiner DM, Anderson J, Draxlbauer K, Flach T (2007) Stakeholder perceptions of CO_2 capture and storage in Europe: results from a survey. Energ Policy 35:5091–5108

Singer S (2007) Presentation to meeting of the ECCP working group on carbon dioxide capture and geological storage (CCS): first consultation on impact assessment of the enabling legal framework for CCS, Brussels, 8 May

Singleton GR (2007) Geologic storage of carbon dioxide: risk analyses and implications for public acceptance. Masters thesis. Massachusetts Institute of Technology, Cambridge

Slovic P (1999) Trust, emotion, sex, politics, and science: surveying the risk-assessment battlefield. Risk Anal 19:689–701

Sturgis P, Allum N (2004) Science in society: re-evaluating the deficit model of public attitudes. Public Underst Sci 13:55–74

USGS (US Geological Survey) (2001) Invisible CO_2 gas killing trees at Mammoth mountain, California. US Geological Survey Fact Sheet 172-96. Online Version 2.0. http://pubs.usgs.gov/fs/fs172-96/

Warin D (2007) Status of the French research program on partitioning and transmutation. J Nucl Sci Technol 44:410–414

Weart SR (1988) Nuclear Fear: A History of Images. Harvard University Press, Cambridge

Weitzman M (1998) Why the far-distant future should be discounted at its lowest possible rate. J Environ Econ Manag 36:201–208

Wikdahl C-E (1991) Sweden: nuclear power policy and public opinion. IAEA Bull 1(1991): 29–33

Williams S (2008) Clean coal could revive Wales mining heritage. Western Mail, 31 March

WNA (World Nuclear Association) (2008) Nuclear power in Germany. http://www.world-nuclear.org/info/inf43.html

Wong-Parodi G, Ray I, Farrell A (2007) Perceptions of geologic sequestration in California's Central Valley. Paper Presented at Conference on Regulating and Financing Carbon Capture and storage, Swiss Re Centre for Global Dialogue, Ruschlikon, Switzerland, 8 November. www.irgc.org/IMG/pdf/14.30_14.45_Gabrielle_Wong-Parodi_IRGC.pdf

Wong-Parodi G, Ray I, Farrell AE (2008) Environmental non-government organizations' perceptions of geologic sequestration. Environ Res Lett 3. doi:10.1088/1748-9326/3/2/024007

Comparative Ethical Issues Entailed in the Geological Disposal of Radioactive Waste and Carbon Dioxide in the Light of Climate Change

Donald A. Brown

Abstract Many governments are at various stages of planning to dispose of radioactive waste (RW) in geological formations. Many governments expressly expect to use geological formations to dispose of the carbon dioxide (CO_2) produced in fossil fuel combustion. This chapter compares, in the light of climate change, the ethical issues involved in disposing of RW and CO_2 in geological formations, given their potential to cause harm and given the risks involved in the deployment of these disposal technologies. It highlights the ethical issues triggered by the need for a high level of scientific certainty regarding the ability of potential disposal sites to counteract the migration of substances of concern away from disposal areas. However, the ethical issues entailed in geological disposal of RW and CO_2 may need to be re-evaluated in light of the fact that disposal contributes to climate change mitigation. It is concluded, however, that as long as alternative methods for mitigating climate change are available that do not involve geological disposal of CO_2 and as long as scientific uncertainty remains about the efficacy of disposal sites to contain RW and CO_2, such alternative methods are ethically preferable. The chapter identifies another ethical issue, namely that the development and deployment of alternative technologies to reduce greenhouse gas emissions could be delayed by reliance on geological disposal of CO_2.

Keywords Ethics • Geological disposal of radioactive waste • Geological disposal of CO_2 • Climate change mitigation • Geological site characterization • Nuclear waste disposal

---

D.A. Brown (✉)
Program on Science, Technology, and Society, Penn State University,
201 Willard Building, University Park, PA 16802, USA
e-mail: dab57@psu.edu

F.L. Toth (ed.), *Geological Disposal of Carbon Dioxide and Radioactive Waste: A Comparative Assessment*, Advances in Global Change Research 44,
DOI 10.1007/978-90-481-8712-6_11, © International Atomic Energy Agency, 2011

1 Introduction

This chapter compares the main ethical issues entailed in two approaches to climate change mitigation: (1) the geological disposal of radioactive waste (RW), and (2) the geological disposal of carbon dioxide (CO_2). As all approaches to climate change mitigation can cause harm, and climate change itself is a threat to human health and the environment, the ethical issues raised by potential climate change mitigation options, such as the two considered in this chapter, must be evaluated in the light of the risks from human-induced climate change. Thus, more specifically, this chapter compares (1) the ethical issues entailed in geological disposal of RW and CO_2 with (2) the ethical issues created by climate change.

The use of nuclear technology to produce electricity raises ethical questions because of the potential risk to human health and the environment arising from, say, a nuclear power plant meltdown or the release of radiation into the environment. So too does the expansion of nuclear power expertise around the world, which could facilitate production of fissionable material by individuals or nations for use in nuclear weapons. This chapter considers neither of these ethical questions. The only ethical question discussed here is that related to the need to dispose of RW that will remain hazardous for millennia. Nor does the chapter discuss all the ethical issues that could arise from greater and more widespread use of coal due to the increased feasibility of geological CO_2 disposal.

A proper ethical analysis of the geological disposal of nuclear waste and of CO_2 must begin with a description of the known environmental, economic and social impacts and risks of the technologies involved, based on the current scientific and economic understanding of their potential harmfulness, risks and benefits. If this scientific understanding changes, then the ethical conclusions reached in this chapter may also change, for which reason they must be considered as provisional.

The term 'ethics' in this chapter means the domain of inquiry that examines claims about what is right or wrong, obligatory or non-obligatory, or the circumstances under which responsibility attaches to human actions (Brown et al. 2006). An ethical analysis of the geological disposal of nuclear waste examines claims made about whether and under what circumstances geological disposal of nuclear waste should be pursued. In a similar way, an ethical analysis of geological disposal of CO_2 is concerned with claims made about whether CO_2 disposal is ethically justified.

The ethical issues entailed in any potential environmental problem, including those discussed here, are often dependent on the nature of the harmfulness of related human activities. Identifying ethical issues arising from potential technological harm does not necessarily lead to agreement about what ethics requires—ethical theories themselves often differ in this respect. To guide ethical conclusions, one may, for instance, look to utilitarian, rights-based, biocentric, or ecocentric theories, or theories of ethical relationships, to name just a few. (For a general discussion of environmental ethics, see Hargrove 1996). However, these theories may reach different conclusions about what is ethical under the same facts. Therefore, ethical issue spotting does not necessarily lead to ethical consensus.

For some projects or problems there is an overlapping consensus among ethical theories about what ethics requires, even though ethical theories might differ (Brown et al. 2006). For other matters, although there is no ethical consensus about what ethics requires, most ethical theories would agree that certain proposals are ethically problematic. In other words, ethical criticism of proposed projects is possible, even if there is no consensus on what ethics requires under those circumstances. Thus, identification of ethical issues may lead to: (1) conflict about what ethics requires; (2) consensus about what ethics requires; and (3) consensus that a proposed activity is ethically problematic despite there being no consensus as to what ethics requires. It is the goal of this chapter to identify ethical issues entailed in the geological disposal of RW and CO_2 rather than to draw absolute ethical conclusions about all the issues identified here.

It will nonetheless be assumed in this chapter that ethics requires those who are the proponents of a project to protect others from serious harm caused by the project, particularly those who have not consented to be put at risk, and especially in cases where the harm is potentially significant and irreversible (Shrader-Frechette 2002). This ethical duty is a matter about which there is large overlapping consensus among ethical theories, particularly if the harm affecting others involves death or serious adverse health effects. However, some ethical theories, including some forms of utilitarianism, would allow a balancing of harm and benefits. Even so, most utilitarians would require those who are harmed by the actions of others to, at a minimum, be compensated for the harm done to them. In addition, most philosophers believe that those who are potentially at harm from the actions of others have a right to exercise fully informed consent to being put at risk of harm (Shrader-Frechette 2002). Thus, even utilitarians often recognize the right to exercise fully informed consent to be harmed by others. (See Shrader-Frechette (2002) for an example.) The duty to protect others from harm is usually considered to be in proportion to the harm that may be inflicted upon them. For a discussion of how ethical duties increase in relationship to the magnitude of harm, see Jonas (1984).

This chapter covers: (1) ethical issues entailed in the geological disposal of RW; (2) ethical issues raised by the geological disposal of CO_2; and (3) a comparative ethical analysis of these technologies in light of climate change. The findings are then summarized in the conclusions.

2 Ethical Issues Entailed in the Geological Disposal of Nuclear Waste

2.1 Potential Harm from High-Level Nuclear Waste Disposal

Geological disposal of RW is one of several disposal methods that has been considered and pursued at the global level, including, for example, sub-seabed disposal and reprocessing. Geological disposal has been recommended by most scientists as a preferred way of disposing of nuclear waste, although some advocate the reprocessing of fuel for use in breeder reactors (US DOE 2008a). Any complete ethical analysis of RW disposal should take into account the disposal methods that potentially create the least harm.

Most countries that generate electricity in nuclear power plants support geological disposal of nuclear waste. Belgium, Canada, China, Finland, France, Germany, Japan, Russia, Spain, Sweden, Switzerland, the UK and the USA all support deep geological disposal as the best method of isolating highly radioactive, long-lived waste (US DOE 2008a). Many of these countries have performed detailed geological studies, or characterizations, by drilling numerous boreholes and exploratory shafts and ramps in underground research laboratories. These data, it is believed, will be useful in determining the predicted safety performance of future nuclear waste repository sites (US DOE 2008a). There are currently no final disposal facilities in any country for high-level and long-lived RW produced during the generation of nuclear energy. That not a single final disposal site has been established in the nuclear industry's more than 50 years of existence and that RW is currently provisionally held in interim storage facilities (EU 2008) would imply lack of public acceptance, and thus possible ethical significance, regarding the siting of future RW disposal facilities.

Other chapters in this book discuss the potential risks and adverse effects of RW disposal. Of particular significance from an ethical perspective is the very long length of time that RW remains hazardous to human health. The risk of highest concern regarding long-term disposal of RW in geological formations is migration of radionuclides from the disposal site into the environment over the millennia during which waste remains hazardous.

As RW will remain hazardous for longer than the history of modern human civilisation (over 10,000 years), its safe disposal raises a number of ethical issues, which are discussed below. Driving these ethical concerns is the need to assume responsibility for such hazardous substances for such a long period of time to ensure that existing and future populations will not be exposed to them while they remain hazardous. This is the core ethical imperative.

2.2 *Preventing Release of Radioactive Substances from Disposal Sites*

The purpose of a deep geological repository is to provide future generations, especially those in the far future, with passive protection against any harmful release of radioactive material. This concept must prevail, even in the event of the repository's existence being forgotten and irrespective of what the technical knowledge of the future generations may or may not be (Allègre 1999).

Common elements of repository systems include RW, the containers enclosing the waste, the tunnels housing the containers, and the geological make-up, including rock types, of the surrounding area (US DOE 2008b). All elements of site design must do their part in preventing RW from escaping into the environment. Critical to the design of the facility is the ability of the geological structure to isolate the waste during the entire period for which the radioactive material is hazardous. The most likely route of leakage from a disposal facility is through water that penetrates the disposal facility and transports radionuclides to the environment, with subsequent possible exposure of people and water resources to ionizing radiation.

The 'ideal repository' would be located in a stable area and would be deep enough to be protected against surface erosion, major climatic changes (such as a new ice age), earthquakes (much less severe at depth) and human intrusion. It would be located in an impermeable formation, with sedimentary salt or clay layers being the most suitable. A continuing challenge to repository design is to find geological formations that are not vulnerable to water intrusion over the life of the facility and/or rock fracturing caused by tectonic events. Ethics, which requires protection of others from exposure to hazardous wastes, also requires proponents of geological disposal sites for RW to select repositories that will counteract any threat to others posed by the waste.

The basic idea behind this is the need to find stable geological environments that have retained their integrity for millions of years and are therefore likely to provide a suitable isolation capacity for a long time to come. Yet, it must be said that even geological sites that have been stable for long periods of time may not be stable over future periods of concern extending to thousands of years. The key to the acceptability of any geological disposal site is to find stable geological formations that are impermeable to water, impermeable in this case meaning that no water, or only a very small volume of water, can circulate in the geological formation.

A key element of disposal site design is the creation of barriers that further isolate the waste from the environment. Possible barriers, frequently considered in repository design, are glass, copper, ceramic, additional zirconium, stainless steel, nickel and titanium. Most RW disposal sites are designed to incorporate barrier methods that supplement the environmental isolation that can be expected from the site's geology. Although in some sites under certain conditions RW can be successfully isolated from the environment, at other locations, it may create unacceptable risks. Thus a key element in terms of understanding the risk of exposure to radioactive materials is the level of confidence that can be achieved about the geological properties and long-term stability of any disposal site under consideration.

For a variety of reasons, one of the important questions regarding repository design is whether the waste should be retrievable after being placed in the disposal site, so that new information about the waste or disposal site can be accommodated in the future (NEA 2001). In the design of waste repositories, many organizations with a need to dispose of nuclear waste have considered the concept of reversibility and retrievability.

The geological structures investigated to date throughout the world for possible nuclear waste disposal have included salt, granite, volcanic tuff and basalt (Warf and Plotkin 1996). Every site has been selected after much consideration of its geological and scientific suitability, and all have proven to be flawed in some way, making irretrievable burial problematic. In some instances fractures in the structure have occurred or have been discovered that would have allowed the nuclear waste to eventually escape its confines. Other problems have included the build-up and then outflow of water (Warf and Plotkin 1996). Earthquake susceptibility is always of concern and automatically precludes the use of some sites.

Reasons for incorporating reversibility and retrievability into disposal site design include the ability of such sites to:

- Deal with safety concerns that are recognized only after waste placement or when new safety standards have been introduced;
- Recover resources from the repository (e.g. where the waste has been discovered to contain a new and valuable component or where the site has been found to have an unforeseen amenity value);
- Use alternative waste treatment technology that is developed in the future;
- Respond to changes in social perceptions of risk or changes in policy requirements (NEA 2001).

As long as there is uncertainty about the long-term stability of a site, and thus its ability to isolate the waste from the biosphere for the entire time for which it is hazardous, the building of that site for potentially retrievable waste is of ethical significance. Ethics would require whoever builds a site to protect others from exposure to potential hazardous substances from that site. Thus a site that allows RW to be removed would be ethically preferable, if experience with that site demonstrates that assumptions of long-term stability were unfounded.

Reasons against incorporating reversibility and retrievability into disposal site design include:

- Uncertainty about negative effects, including conventional radiological exposure of workers, radiological exposure of workers engaged in extended operations or monitoring, and marginal increases in worker exposure;
- The difficulties involved in sealing a repository properly because of the implementation of longer-lasting or more complex operational plans designed to assure retrievability;
- Irresponsible attempts to retrieve or interfere with waste during times of political and/or social turmoil when safeguards or monitoring procedures are no longer in place;
- A possible need for enhanced nuclear safeguards (NEA 2001).

In addition, the decision to incorporate reversibility and retrievability into the design of disposal sites may greatly increase the costs of disposal.

If the risks from building a disposal site with retrievability of RW would put people at significant risk, then ethics might require that repositories should not include retrievability as part of their design.

The high costs of including reversibility and retrievability in a design could be ethically relevant if the costs of adding this feature to a disposal site were so great as to make the site unfeasible and if there were no other ways of safely disposing of the RW. However, as there is little evidence that the cost of building retrievability into storage design is prohibitive, the additional costs of including retrievability in a disposal facility design are not likely to be ethically justifiable as a basis for not including it. However, some utilitarians would modify the duty to build retrievability into site design if the costs were excessively disproportionate to the harm avoided.

It is possible to build retrievability into the early stages of the life of a repository while planning a decision for permanent waste isolation that would virtually eliminate irretrievability options at some time in the future. Even after the closure, mining techniques could allow containers to be retrieved as long as their integrity was maintained. Every decision about retrievability of waste has costs and benefits that need to be considered during the design of the disposal site along with the ethical duty to protect others from exposure to hazardous waste.

Fulfilling a campaign promise, US President Obama has proposed a budget for 2009 that cuts off most money for the Yucca Mountain nuclear waste project in Nevada, which has been under consideration by the USA as its first high-level RW disposal site. The decision could cost the Federal Government additional billions in payments to the utility industry, and if the cuts go ahead, it would mean that most of the US$10.4 billion spent since 1983 to find a place to put nuclear waste has been wasted. The Yucca Mountain project has been hotly debated and widely opposed, partly because of concerns about the adequacy of the site geology to retard water migration through the site during the life of the repository, and partly because of potential tectonic activity. Whether these issues should be a basis for denying operating permits for the Yucca Mountain site are matters of considerable controversy in the USA. Assuming the Yucca Mountain site is abandoned, new ethical concerns arise about the safety of the growing amounts of high-level nuclear waste currently being stored at nuclear power plants around the USA.

As a matter of ethics, it can be argued that the Obama decision is the right one, if one assumes that proponents of Yucca Mountain had not demonstrated that the site was sufficiently geologically stable to isolate the waste for 10,000 years, as required by US regulations.

Proponents of geological disposal of RW often argue that although risks of harm from the release of ionizing radiation into the environment exist, the benefits of nuclear energy outweigh the potential harm. At the centre of this debate are the difficulties in reaching a high level of confidence about any site's adequacy in terms of nuclear waste disposal.

The purpose of disposal of RW in geological formations is to isolate the waste from the environment during the period for which the waste is dangerous. Given that this period is millennial, the ability to know, with a high level of confidence, whether the site may leak ionizing radiation in the long term is extremely challenging scientifically. One author has identified numerous irreducible scientific uncertainties that limit the ability to predict with a high level of confidence the future performance of geological disposal sites. These uncertainties are related, among other things, to geological conditions very far in the future and to climate conditions (Shrader-Frechette 1993).

To perform adequate site characterization for geological disposal of RW, the geological structure of the site needs to be determined with a high level of precision. This includes all potential pathways of water movement throughout the site for thousands of years into the future. Predictions of groundwater flow rates and direction for some sites where nuclear waste has been stored have often proved to be wrong in the past in just a few decades. In the USA, for instance, predictions about groundwater transmission rates at Department of Energy facilities at Maxey Flats, Fernald, Savannah River, Hanford, Idaho Falls and Rocky Flats have led to problems (Shrader-Frechette 1993). Given the enormous time that RW repositories must isolate the waste, risk assessments about site safety must necessarily make unprovable assumptions—and therefore methodological value judgments—about future geological variability and climate conditions (Shrader-Frechette 1993). If these assumptions are wrong, the interests of future generations will be most affected, while the benefits of nuclear-produced electricity will be highest for present generations.

Decision making in the face of scientific uncertainty about the environmental impacts of human activities raises three types of ethical question:

1. On whom should the burden of proof rest regarding a project's safety or danger: the proponent(s) of the project, the government, or those who might be harmed by the project?
2. What quantity of proof should satisfy the burden of proof? Should a 95% confidence level—a norm followed by many scientific disciplines, the balance of the evidence, or other level of proof determine when a project is deemed to be safe or dangerous? (For a discussion of ethics and uncertainty, see Jonas 1984). A strong ethical argument can be made that the quantity of proof needed to satisfy the burden of proof should be in proportion to how dangerous or potentially catastrophic the project is. Given (1) the difficulty of reaching a high level of certainty about the safety of a proposed RW disposal site, and (2) the potentially great harm if people are exposed to waste that migrates from a disposal facility, there is a strong ethical argument that the burden of proof should rest with the proponents of the use of geological sites to show, with a high level of proof, that the site will isolate the RW for the entire period of concern.
3. Decision making in the face of scientific uncertainty raises questions of procedural justice about the rights of those who are put at risk by decisions under uncertainty and/or the rights of potentially affected parties to participate in the decisions.

2.3 *Ethical Duties to Existing Local Populations*

Risks from geological disposal of RW include serious threats to local populations near the disposal site if radioactive materials migrate either from the disposal site or are released during transportation of the waste to the site into the air or into water resources, thereby exposing populations. As discussed in West et al. (2011), such releases could be lethal or cause a variety of diseases to exposed populations.

Risks of harm to local populations from doses of ionizing substances can be minimized by facility design, engineering controls, monitoring of site performance, adaptive management techniques and careful site selection.

A number of ethical issues are particularly important as far as local populations are concerned because of the scientific challenges of site characterization.

- As methodological assumptions made in the course of on-site characterization projects cannot predict future long-term geological site conditions, they should err on the side of preventing harm because of the duty to protect local populations and future generations.
- To the maximum extent that is economically feasible, and for reasons stated above, site design and operational procedures should include the potential for reversibility and retrievability if very high levels of scientific certainty cannot be achieved about a site's ability to fully contain the radioactive materials.
- Among the practices related to the disposal of nuclear waste in geological repositories, there should be obligatory financial insurance which would be used to compensate local populations and future generations in the event of a disposal site failure.
- Proponents of the disposal of RW in geological formations (hereafter, proponents) and governments with authority to approve such projects (hereafter, governments) have an ethical duty to protect local populations from exposure to ionizing radiation through adequate site selection, site design, operational procedures, and engineering and monitoring controls.
- Proponents and governments must assure adequate representation of local populations in site approval, site design and site operations, before giving final approval to RW disposal projects.
- Proponents and governments must assure adequate education of local populations about potential risks to local populations.
- Approval procedures for geological disposal sites for RW must place the burden of proof on the proponents to demonstrate that a proposed site does not create risks to local populations from exposure to ionizing radiation.

2.4 Ethical Duties to Future Generations

Because of the extraordinarily serious long-term risks from ionizing radiation, and given the millennial time spans over which nuclear waste remains hazardous, the responsibility to protect future generations from radioactive exposure is an important ethical consideration that needs to be evaluated in the course of decision making. (For a discussion of ethical duties to future generations, see Partridge 1981.) Those ethical considerations include the following:

- It is the duty of proponents to demonstrate that the site will not incur exposure to ionizing radiation during the millennia for which the waste will be hazardous.
- Disposal of RW in geological sites should be limited to those sites for which it can be demonstrated, with a high level of confidence, that either the site will isolate the ionizing radiation at millennial scales or that the waste can be removed if a site's stability becomes questionable. If these conditions cannot be met, other technologies for generating the energy needed are ethically preferable. However, as we will see, other energy generation technologies create serious risks to human health and the environment in that they intensify climate change. Thus all energy-related technologies must be compared in terms of the harm they might cause.

- The generation producing the waste is responsible for the safety of the waste disposal site on behalf of future generations. This is a particular ethical challenge because of the propensity of existing generations to consider only their ethical obligations to contemporaries. The duty to protect future generations from exposure to high-level nuclear waste is not covered by normal economic methods for calculating cost-benefit analyses, which rely on discounting future benefits. (For a discussion of ethical limits of discounting future benefits in cost-benefit analysis, see Brown et al. 2006).

3 Ethical Issues Raised by the Geological Disposal of CO_2

3.1 Potential Harm from the Geological Disposal of CO_2

As described in other chapters of this book, there are a number of CO_2 capture and storage technologies for removing (capturing) CO_2 from fuel combustion emissions, and then injecting it into geological formations for long-term storage, instead of releasing it to the atmosphere. This section focuses on ethical issues arising from CO_2 storage in geological reservoirs and does not consider the risks of capturing CO_2 from combustion processes.

Risks of harm from geological disposal of CO_2 have been described in other chapters. These risks can be grouped into the following categories: (a) risks to local populations living near the site, and (b) risks of long-term leakage (Brown 2008).

3.1.1 Risks to Local Populations

Risks from geological disposal of CO_2, described in other chapters of this book, include serious risks to local populations near the injection site or along feeder pipelines, should CO_2 above certain concentrations leak from injection wells, pipelines and other elements of a storage system. Particularly vulnerable to such releases are people who live in the proximity of injection wells. However, this risk can be virtually eliminated by locating injection wells in unpopulated areas. In addition, if sites selected for geological disposal of CO_2 do not provide adequate caprock isolation of injected gases from groundwater systems, contamination of groundwater in the site vicinity is also a potential risk.

3.1.2 Risks from Long-Term Leakage of CO_2 into the Environment

As the purpose of geological disposal of CO_2 is to keep CO_2 out of the atmosphere in order to mitigate climate change, long-term leakage from storage sites could constitute failure of this technique. In addition, if CO_2 eventually leaks into the atmosphere from geological storage sites in sufficient quantities, it could make climate change impacts worse than they would have been if CO_2 emissions had been reduced by other methods.

The amount of potential leakage from CO_2 storage sites will determine the magnitude of the risks. Very small amounts of long-term leakage may have trivial impacts on climate change, while large leakage rates could exacerbate the adverse impacts. For this reason, it is important to be able to predict with sufficient accuracy the long-term fate of the CO_2 at each proposed geological storage site. Yet, as was the case in characterizing nuclear waste disposal sites, describing all potential leakage pathways over the area that might be the zone of impact for CO_2 storage is scientifically challenging for many potential sites. However, the time period of concern regarding site integrity is shorter in the case of geological CO_2 disposal (a few hundred years) than in the case of nuclear waste disposal (tens of millennia).

There is considerable experience of CO_2 injection over several decades in petroleum and gas recovery operations and considerable experience of, and understanding about, the natural storage of CO_2 and natural gas. However, there is little experience of long-term leakage from sites expressly chosen for the purpose of long-term CO_2 storage in places where petroleum or gas has not been naturally stored. Leakage of CO_2 from gas and petroleum production sites where CO_2 has been injected to enhance fossil fuel recovery is believed to be almost zero in the short term, but adequate long-term performance in terms of CO_2 leakage has not been demonstrated. Experience with leakage of CO_2 injected as part of petroleum and gas recovery operations may not be applicable to sites that do not have geological confining layers to the extent of those present in petroleum and gas fields.

To prevent threats from long-term leakage of CO_2, it is critically important that any potential site be properly characterized to determine not only the presence of an adequate caprock that will trap injected CO_2 but also the absence of other pathways through which CO_2 could leak into the environment. Whereas in the case of RW disposal there is a need to assure that the disposal site is stable for tens of thousands of years, for geological carbon storage it is usually assumed that after a hundred years or so from final CO_2 injection, the CO_2 will no longer pose a threat of leaking into the environment. This difference has ethical significance because it is technically more challenging to predict geological structure performance over very long periods.

To perform adequate site characterization, it is necessary to determine the geological structure of the site and all potential pathways of leakage from it, including leakage that could come from caprock dissolution. For this reason, most of the scientific challenges entailed in characterizing a site for RW disposal are also relevant to CO_2 disposal, with geological CO_2 disposal areas probably having the added problem of being much larger than RW storage spaces. As is the case in geological disposal of RW, significant ethical questions arise in the characterization of geological sites for CO_2 disposal about who the burden of proof should rest with, what quantity of proof is satisfactory, and the role in decision making of those who may be susceptible to harm from leaking CO_2.

The injection of CO_2 captured from the emissions of large coal-fired power plants will have a large area of impacts. Thus it may be particularly challenging to determine the variability of the geology over the entire impact area of a proposed site, particularly in parts of the world that have a highly variable geological structure. A site's zone of impact will increase with time as CO_2 injection continues, and it is thus necessary to understand the site geology over the entire zone of impact throughout the life of the project.

Risks of long-term leakage can be minimized by adaptive management techniques that are based on adequate monitoring of injection pressures and storage rates, which will limit further injections of CO_2 if leakage potential is identified. For this reason, regulatory controls of storage operations are necessary to assure adequate performance of the site in storage terms.

It may be necessary to install institutional controls over the site to prevent the creation of new leakage pathways in the course of time. For this reason, it could be necessary to restrict some aspects of future land use over the entire zone of impact.

3.1.3 Earthquake Risks

Underground injection of CO_2 or other fluids into porous rocks at pressures substantially higher than formation pressures can induce fracturing and movements along faults (IPCC 2005). Induced fracturing and fault movement activation can both increase pathways of leakage and induce earthquakes large enough to cause damage. Reduction of the risk of earthquakes can be accomplished by keeping injection pressures below pressures that will induce seismic activity and by not locating storage sites in seismically active zones. It is believed that risks of earthquake induction can be greatly minimized by regulatory controls over injection pressures and site selection (IPCC 2005).

3.2 *Ethical Issues Entailed in the Geological Disposal of CO_2*

For the reasons given above, proponents of geological storage of CO_2 have an ethical duty to demonstrate that a proposed disposal site will not harm others. As discussed in the chapter on environmental issues (West et al. 2011), in addition to adverse potential impacts on human life and health from geological disposal of CO_2, there are potential impacts on plants, animals and ecological systems. While there is a duty to protect people from harmful levels of exposure to CO_2 that is recognized by most ethical theories, different ethical theories would reach different conclusions about duties to plants, animals and ecological systems. Some utilitarians would find no absolute duty to protect plants, animals or ecological systems that could not be modified based on cost-benefit considerations. Other ethical theories such as biocentric and ecocentric ethical theories would demand their protection. Thus, one of the ethical issues raised by geological disposal of CO_2 is what are the ethical duties to prevent harm to plants, animals and ecosystems in cases where there are no risks to human life or health.

3.2.1 Ethical Issues Concerning Risks to Local Populations and Ecosystems from Geological Carbon Storage

Ethical issues raised by risks to local populations and ecological systems from geological CO_2 storage include:

- Proponents of geological CO_2 disposal projects (hereafter, proponents) have ethical duties to protect local populations from toxic doses of CO_2 through adequate site selection, design, engineering, and monitoring controls. For this reason, approval procedures for geological disposal of CO_2 must place the burden of proof on proponents to demonstrate that a proposed CO_2 storage project does not create unacceptable toxic risks to local populations or ecological system through leakage of CO_2.
- Proponents must ensure adequate representation of local populations in site approval and design.
- Proponents should ensure adequate education of local populations about the potential risks they face to assure fully informed consent about being put at risk.
- Governments responsible for approval of CO_2 disposal sites must insist upon adequate regulatory controls over project design and site selection criteria.
- Proponents must acknowledge their ethical duties to compensate local populations or insure them from harm caused by leakage of CO_2, should this occur.

3.2.2 Ethical Issues Entailed by Potential Earthquake Triggering

Ethical issues raised by risks of potential earthquake triggering by injected substances include the following:

- Proponents must demonstrate that injection of CO_2 and associated liquids or gases will not trigger earthquakes. This will involve demonstrating that injection rates will not be sufficient to induce seismic movement and that the site is not located in a seismically active zone for as long as earthquakes threaten human health or the environment.
- Governments responsible for approval of CO_2 disposal sites must make site approval conditional upon compliance with regulatory controls to guard against seismic movement.
- To assure the absence of potential triggering of earthquake by CO_2 injection, the burden of proof must rest with proponents to show that they have adequately characterized the geology of a proposed injection site to determine potential seismic response from CO_2 injection.

4 The Comparative Ethical Issues Involved in Geological Disposal of Nuclear Waste and CO_2 and Human-Induced Climate Change

Enormous and unprecedented challenges and threats to the human race are raised by climate change (Brown et al. 2006). Among the challenges are numerous profound ethical questions that emerge on at least four grounds:

1. The nations and people who are the main contributors to climate change are often not those who are most vulnerable to its impacts.
2. The impacts of climate change are potentially catastrophic. That is, climate change threatens people and ecosystems around the world with, inter alia, droughts and floods, rising seas, vector-borne disease, killer heat waves, and reductions in agricultural productivity (see Brown et al. 2006).
3. To address the threats posed by climate change, those who cause the problem need to consider the adverse impacts of climate change on people and their environment separated from them in time and space.
4. Most of the options for mitigating climate change carry potential harm and risks that must be considered through an ethical lens. That is, although most approaches to climate change mitigation, including geological disposal of RW and CO_2, raise ethical issues that need to be considered before the technologies are deployed, the ethical issues raised by these solutions must always be evaluated in the light of ethical issues raised by the threats posed by climate change itself. In particular, the ethical dimensions of each approach must be compared both with ethical issues entailed in harm arising from business-as-usual use of fossil fuels and with ethical issues raised by specific efforts to mitigate climate change.

Climate change raises many different civilisation-challenging ethical issues, some of which are relevant to the choice of climate change mitigation options, including nuclear power and the geological disposal of CO_2 (Brown et al. 2006). Among other ethical issues, climate change creates an immediate duty for nations to reduce their share of global greenhouse gas emissions to a level that is fair. This duty in turn creates a responsibility on the part of those nations currently exceeding their fair share of safe global emissions to consider approaches to climate change mitigation, including nuclear power and the geological disposal of CO_2. Climate change also raises many other ethical issues such as what atmospheric stabilization level should be the goal of all nations and who should pay for damage caused by climate change (Brown et al. 2006).

There follows an evaluation of ethical issues arising from human-induced climate change compared with ethical issues arising from the two potential approaches to mitigating climate change considered in this chapter, namely nuclear power and the geological disposal of CO_2.

4.1 *The Ethics of Climate Change and of High-Level Nuclear Waste Disposal*

As seen above, high-level nuclear waste disposal creates potential risks to local populations and future generations because of the millennia during which the RW must be isolated from the biosphere. If geological disposal facilities for nuclear waste fail, resulting in contamination of the surrounding environment, those who live in the vicinity of the disposal site are at highest risk, with the contamination likely to be local or regional, rather than global. There are numerous ethical issues arising from these risks.

The potential harm from RW and associated ethical questions, as serious as they are, must be reconsidered in the light of the potential harm of climate change and related ethical issues. From this, a number of things follow:

- Climate change creates an ethical duty for nations exceeding their fair share of greenhouse gas emissions to reduce their emissions. These nations must seriously consider alternatives to conventional fossil fuel technologies, including nuclear power and the geological disposal of CO_2. Ethically speaking, proponents of climate change mitigation options have a duty to deploy technologies that minimize potential adverse effects on human health and the environment. For this reason, technologies such as nuclear power or geological disposal of CO_2 may be ethically superior to fossil fuel technologies, although not necessarily superior to other technologies that can meet energy demand while reducing the threat of climate change.
- Those with greatest interest in the potential harm resulting from RW disposal are the current and future generations that could be adversely impacted by release of ionizing radiation. Given the nature and danger of these risks, a strong case can be made that as long as there are other ways of generating energy, then, as a matter of ethics, methods that are less risky should be chosen for generating energy until technologies are invented that allow radioactive geological waste disposal without generating contamination threats for future generations at a millennial scale.
- As fossil fuel technologies like carbon capture and storage also create potentially catastrophic impacts for existing and future generations, as well as for plants and animals and ecosystems around the world, there is a strong ethical imperative to move away from current methods of generating energy from fossil fuels.
- Given the ability of nuclear energy to produce power with minimum greenhouse gas emissions, there is a need to re-evaluate ethical issues associated with RW disposal.
- To the extent that alternatives to fossil fuel and nuclear energy can meet energy needs, a strong case can be made that these energy generation technologies should be given priority over both nuclear energy and fossil fuel-derived energy that releases large amounts of greenhouse gases. In fact, until problems associated with long-term isolation of ionizing radiation can be resolved, other approaches to climate change need to be considered before the use of nuclear power is extended.

4.2 *The Ethics of Climate Change and of Geological Disposal of CO_2*

Geological disposal of CO_2 raises ethical issues that need to be considered before the technology is deployed. However, this disposal technology also represents a way of mitigating climate change by not releasing into the atmosphere greenhouse gases currently being generated by coal combustion. Given that climate change-related threats are enormous, the ethical issues raised by the geological disposal of CO_2 must be evaluated in the context of climate change-related ethical problems that it could help to mitigate if the technology were deployed. Because of climate change concerns, the ethics of geological disposal of CO_2 must consider: (a) short-term risks from geological disposal of CO_2, (b) ethical issues entailed in the risk of long-term leakage of CO_2 back into the environment, and (c) ethical issues created by inappropriate reliance on geological disposal of CO_2.

4.2.1 Ethical Issues Entailed in Short-Term Risks from Geological CO_2 Disposal in the Context of Climate Change

As seen, geological disposal of CO_2 creates ethical issues regarding the need to protect people living near a CO_2 disposal site from leakage and the need to prevent the triggering of earthquakes. Yet these risks, comparatively speaking, are less serious and more local than the potentially catastrophic risks of climate change. Therefore, although there are continuing duties and responsibilities related to the geological disposal of CO_2, an ethical argument can be made regarding the short-term risks from geological disposal of CO_2, namely that this technology should be deployed for as long as there is no viable alternative to fossil fuel energy production and for as long as there is a reasonable prospect that the technology will meet design criteria and is economically feasible.

4.2.2 Ethical Issues Entailed in Risks of Long-Term Leakage of CO_2 Back into the Environment

If geological sites for CO_2 disposal leak CO_2 into the atmosphere in sufficient quantities, the requisite reductions in greenhouse gases cannot be provided and financial resources that could have been used for more effective greenhouse gas reduction technologies will have been wasted. Therefore, ethical duties with regard to long-term leakage of greenhouse gases from geological CO_2 disposal sites include the following:

- Proponents of geological CO_2 disposal sites have an ethical duty to demonstrate that there will be no long-term leakage of CO_2. This will be a difficult duty to meet because of the scientific challenges involved in characterizing the geology of a large site in parts of the world where the geology may be highly variable or fractured.
- As long-term leakage of CO_2 from a geological CO_2 disposal site could harm people at great distances from the disposal site, there is a global interest in assuring that geological CO_2 disposal sites do not contribute to climate change. Given the potential international climate change impacts from geological CO_2 disposal sites, those responsible have the ethical duty to ensure that siting criteria satisfy international standards.

4.2.3 Ethical Issues Entailed in Delays in Reducing CO_2 Emissions Because of the Potential for Geological Carbon Storage

Most observers agree that significant research is needed before geological storage of CO_2 can be widely deployed at coal-fired power plants around the world. There needs to be a high level of confidence that they will not leak CO_2 into the atmosphere; they should not pose short-term risks to persons and ecosystems near the site; and they must be economically feasible. A recent report of the Intergovernmental Panel on Climate Change (IPCC 2005) on geological storage identified the following knowledge gaps that need to be filled before this technology, although very promising, can be extensively used:

- There are major gaps regarding storage capacity at global, national and regional scales.
- There are significant knowledge gaps regarding storage capacity in those parts of the world that are likely to experience the greatest energy growth, such as China, South-East Asia, India, Russia, other successor states of the former Soviet Union, Eastern Europe, parts of South America, and southern Africa.
- There is a need for greater knowledge about some storage mechanisms including: (a) the kinetics of geochemical trapping and the long-term impact of CO_2 on reservoir fluids and rocks; and (b) the fundamental processes of CO_2 adsorption and methane (CH_4) desorption on coal during storage operations.
- There is some need to improve knowledge about: (a) risks of leakage from abandoned wells caused by material and cement degradation; (b) temporal variability and spatial distribution of leaks that might arise from inadequate storage sites; (c) microbial impacts in the deep subsurface; (d) environmental impact of CO_2 on the marine sea floor; and (e) methods to conduct end-to-end quantitative assessment of the risks to human health and the environment.
- There is a need to improve knowledge about the quantification of potential leakage rates from a greater number of storage sites.
- There is a need to improve reliable coupled hydrological–geochemical–geomechanical simulation models to predict long-term storage performance;
- There is a need for better monitoring technology at the surface and subsurface for: (a) location of CO_2 in the subsurface; (b) detection of sub-aquatic CO_2 see page; (c) leak detection at the surface; (d) fracture detection and characterization of leakage potential; and (e) long-term monitoring techniques (Benson et al. 2005).
- This research will be quite expensive. The highest-profile cancellation of geological sequestration research involved a project known as FutureGen, which President Bush announced in 2003. The project had been funded by a utility consortium with subsidies from the US Government, which intended to build a plant in Mattoon, Illinois, that tested the most advanced techniques for coal gasification, capturing pollutants, and burning the gas for power (Wald 2008). The project design called for CO_2 from coal combustion to be compressed and pumped underground, with monitoring devices determining whether gases would escape into the atmosphere. According to a New York Times article (Wald 2008),

about US$50 million had been spent on FutureGen before the Government pulled out of the project in January 2008 when the projected costs more than doubled to US$1.8 billion accompanied by fears that costs would go even higher. In addition, electricity utilities have also been cancelling their commitments to coal gasification plants that would make geological sequestration more affordable because they would produce less CO_2 per kWh of electricity generated.

Because of these research needs, it has been predicted that CO_2 from geological CO_2 disposal technology may not be technically feasible at the scale for which it is needed until 2030 (WBCSD 2006). Thus, the efficacy and magnitude of geological CO_2 disposal as an effective method of mitigating the effects of climate change may not be ascertained for perhaps decades.

Some observers of the development of geological CO_2 disposal technology are concerned about the potential for scarce research finance to be consumed on costly geological carbon storage research that could be more effectively used to research other climate change mitigation options, such as wind, solar power and advanced biofuels. For these reasons, waiting perhaps several decades for preliminary geological carbon storage research to be concluded may delay the introduction of other technologies that could reduce the threats from climate change.

Moreover, the additional costs of geological CO_2 disposal may make this technology economically undesirable compared to other climate change solutions. Thus, even if geological disposal of CO_2 is proven to be an effective means of keeping greenhouse gases from entering the atmosphere, it may not be economically viable.

Given the urgency of reducing greenhouse gas emissions in large quantities over the next several decades, using the potential of geological carbon storage as an excuse to delay deployment of other greenhouse gas emission technologies and strategies could exacerbate the impacts of climate change still further. Geological disposal of CO_2 therefore raises the following ethical issues.

- Given that greenhouse gas emissions are already causing harm to populations and ecosystems at a global scale, no nation that is already exceeding its fair share of safe global emissions may delay taking steps to reduce its emissions on the basis that new less costly technologies, including geological disposal of CO_2, may be invented in the future (Brown et al. 2006).
- For these reasons, nations that are already exceeding their fair share of safe global emissions need to use all currently available means to reduce greenhouse gas emissions, for example by using renewable energy and energy demand side management to reduce emissions to their fair share of safe global emissions while other technologies such as geological CO_2 disposal are being tested and developed.
- A nation that delays deploying available technologies to reduce greenhouse gas emissions on the basis that new less costly technologies such as geological CO_2 disposal may be available in the future should be liable for any damage caused by the delay.
- The promise of geological CO_2 disposal should not be used as an excuse for not implementing other available greenhouse gas emissions reduction strategies in those countries that are already exceeding their fair share of safe global emissions.

5 Conclusions

As we have seen from the above, the geological disposal of both nuclear waste and CO_2 creates several different types of potential risk which trigger ethical issues and concerns. Common to both types of disposal is the need to achieve an appropriately high level of confidence that the geological structures will contain the substances of concern during the period for which the substances could cause harm.

Ethically speaking, the higher the degree of seriousness of the potential adverse impacts, the higher the level of care that needs to be provided. Making a comparison of the ethical obligations triggered by the use of these two technologies is useful for contrasting the nature of the potential harm that could be caused if the geological structures fail to contain the substance disposed of. As geological disposal of RW and that of CO_2 entail different time periods of potential concern and different threats to human health and the environment if, once disposed of, they leak into the environment, ethics requires that different levels of scientific scrutiny be achieved about the suitability of the geological structures that are relied upon to contain each.

Without doubt, safe geological disposal of RW requires the geological structure in which the RW is disposed of to isolate the nuclear waste for the millennia during which the nuclear waste is hazardous, unless the disposal facility allows for retrieval of the waste where there may be evidence that the site is not performing as designed. The period of concern for geological sites for CO_2 disposal is considerably less (several hundred years) than that for nuclear waste disposal.

As the period of concern is so long for RW, and as ethics requires care in proportion to the degree of potential harm, a strong case can be made that as a matter of ethics, extreme care about the long-term suitability of any given nuclear waste disposal site is required. Because of the scientific challenges involved in reaching a high level of confidence that any geological structure will prevent radioactive substances from migrating into the environment during such a long period of concern, the duty of care regarding the suitability of geological disposal sites for RW is a particular technical challenge. Although the period of concern regarding CO_2 containment is significantly shorter than that for RW, there are nevertheless significant technical challenges in meeting relevant ethical obligations to prevent CO_2 leakage, given the large areas that will be needed to dispose of the CO_2 generated by large coal-fired power plants.

As we have seen, ethics requires care in proportion to the degree of potential harm from the proposed activity under consideration. As the radioactive substances that will be disposed of in geological sites will be extremely toxic, high levels of care are warranted regarding the ability of such sites to prevent exposure. Exposure to CO_2 is not, under most circumstances, likely to be a threat to human health except in cases where very high concentrations of CO_2 could leak from injection wells or other large leakage pathways, potentially harming local populations or reversing climate change mitigation efforts.

If it can be shown that the risks from the geological disposal of RW and CO_2 are less problematic than the threat of climate change and that there were no reasonable alternatives to the geological disposal of RW and CO_2 as a way of mitigating the threat of climate change, an ethical justification for the use of these technologies, despite their risks, can be made; however, ethical duties remain to deploy these technologies in a consistent way with other ethical obligations to the maximum degree feasible. If, however, there are reasonable alternatives to nuclear power and the use of fossil fuels with geological disposal of CO_2 as methods of mitigating the threat of climate change, and an appropriate level of confidence cannot be attained that the geological structures can prevent radioactive substances or CO_2 from harming human health or the environment, then an ethical argument can be made that other alternatives should be preferred.

References

Allègre M (1999) The deep geological repository: an unavoidable and ethically correct solution. In: Proceedings of the 24th Annual International Symposium, The Uranium Institute. http://www.world-nuclear.org/sym/1999/allegre.htm. Accessed 17 Nov 2008

Benson S, Cook P et al (2005) Underground geological storage. In: Metz B, Davidson O, de Coninck HC, Loos M, Meyer LA (eds) IPCC Special Report on Carbon Dioxide Capture and Storage. Cambridge University Press, Cambridge, pp 195–276, http://www.ipcc.ch/pdf/special-reports/srccs/srccs_summaryforpolicymakers.pdf. Accessed 17 Nov 2008

Brown D (2008) The ethics of allocating public research funds for carbon capture and storage. Rock Ethics Institute, Penn State University. http://climateethics.org/?p=48. Accessed 24 Nov 2008

Brown D, Tuana N, Averill M, Bear P, Born R, Lessa Brandão CE et al (2006) White paper on the ethical dimensions of climate change. In: The Collaborative Program on the Ethical Dimensions of Climate Change, Rock Ethics Institute, Penn State University. http://php.scripts.psu.edu/dept/rockethics/climate/whitepaper/whitepaper-intro.shtml. Accessed 17 Nov 2008

EU (European Union) (2008) Management of spent nuclear fuel and radioactive waste. http://europa.eu/legislation_summaries/energy/nuclear_energy/l27048_en.htm. Accessed 17 Nov 2008

Hargrove EC (1996) Foundations of Environmental Ethics. Environmental Ethics Books, Denton

IPCC (Intergovernmental Panel on Climate Change) (2005) IPCC Special Report on Carbon Dioxide Capture and Storage. Metz B, Davidson O, de Coninck HC, Loos M, Meyer LA (eds). Cambridge University Press, Cambridge/New York, Available in full at www.ipcc.ch. Accessed 12 July 2009

Jonas H (1984) The Imperative of Responsibility: In Search of an Ethics for a Technological Age. University of Chicago Press, Chicago/London

NEA (Nuclear Energy Agency) (2001) Reversibility and retrievability in geologic disposal of radioactive waste. Reflections at the international level. NEA Report 3140. http://www.nea.fr/html/rwm/reports/2001/nea3140.pdf. Accessed 17 Nov 2008

Partridge E (ed) (1981) Responsibilities to Future Generations. Prometheus Books, Amherst

Shrader-Frechette KS (1993) Burying Uncertainty: Risk and the Case Against Geological Disposal of Nuclear Waste. University of California Press, Berkeley

Shrader-Frechette KS (2002) Environmental Justice: Creating Equality, Reclaiming Democracy. Oxford University Press, New York

US DOE (US Department of Energy) (2008a) Radioactive waste: an international concern. http://www.ocrwm.doe.gov/factsheets/doeymp0405.shtml. Accessed 17 Nov 2008

US DOE (US Department of Energy) (2008b) Managing nuclear waste: options considered. http://www.ocrwm.doe.gov/factsheets/doeymp0017.shtml. Accessed 17 Nov 2008

Wald ML (2008) Mounting costs slow the push for clean coal. The New York Times. http://www.nytimes.com/2008/05/30/business/30coal.html. 30 May 2008
Warf JC, Plotkin SC (1996) Disposal of high-level nuclear waste. Nuclear Age Peace Foundation. http://www.wagingpeace.org/articles/1996/09/00_warf_disposal.htm. Accessed 17 Nov 2008
WBCSD (World Business Council for Sustainable Development) (2006) Facts and trends: carbon capture and storage. http://www.wbcsd.org/web/publications/facts&trends-ccs.pdf. Accessed 17 Nov 2008
West JM, Shaw RP, Pearce JM (2011) Environmental issues in the geological disposal of carbon dioxide and radioactive waste. In this volume

Psychological Perspectives on the Geological Disposal of Radioactive Waste and Carbon Dioxide

Judith I.M. de Groot and Linda Steg

Abstract Public acceptability of risky technologies is not only related to the objective risks involved, but to a number of subjective factors as well. Therefore, various studies examined psychological factors related to acceptability judgements. In this chapter we demonstrate the relevance of psychological factors that contribute to the explanation of the acceptability of radioactive waste disposal and carbon dioxide (CO_2) disposal technologies. The acceptability of CO_2 disposal has received far less attention in psychological studies than the acceptability of radioactive waste disposal, and therefore we have made an assessment of possible psychological determinants based on research on the acceptability of the latter. We conclude that the acceptability of CO_2 disposal may be explained by similar factors to those influencing the acceptability of radioactive waste disposal, i.e. risk characteristics (dread and the unknown), affect, values and worldviews, fairness and trust. We argue that these psychological factors are directly related to the acceptability of CO_2 disposal as well as indirectly, via the perceived risks and benefits of CO_2 disposal. Furthermore, we discuss group differences (i.e. lay *versus* experts, and cross-cultural differences) in acceptability of radioactive waste disposal and, again, translate these results for possible consequences in psychological research in the area of the acceptability of geological disposal of CO_2. Finally, we integrate the psychological factors into a conceptual model and discuss the limitations of current research, future research directions and policy implications for the acceptability of both types of technologies.

J.I.M. de Groot (✉)
School of Design, Engineering and Computing, Bournemouth University, Psychology, Poole House, Fern Barrow, Poole, BH12 5BB, United Kingdom
e-mail: jdgroot@bournemouth.ac.uk

L. Steg
Faculty of Behavioural and Social Sciences Department of Psychology, University of Groningen, Grote Kruisstraat 1/2, 9712 TS, Groningen, The Netherlands
e-mail: E.M.Steg@rug.nl

F.L. Toth (ed.), *Geological Disposal of Carbon Dioxide and Radioactive Waste: A Comparative Assessment*, Advances in Global Change Research 44,
DOI 10.1007/978-90-481-8712-6_12,

Keywords Acceptability • Risk perception • Affect • Values • Fairness • Trust • Radioactive waste • CO_2 disposal

1 Introduction

Responses to new and large-scale technologies are related to individual perceptions of the risks they pose. For this reason, research on risk perception is of great importance to industries and governments trying to assess and implement new technologies. Two relevant examples of current risky technologies involve the geological disposal of radioactive waste (RW) and carbon dioxide (CO_2). Their fortunes ultimately depend on social acceptability of the technology rather than its technological advantages or disadvantages. Therefore, we focus on these technologies in this chapter.

We will demonstrate the relevance of various psychological factors to explain the acceptability of technologies associated with RW and CO_2 disposal. Acceptability is affected by perceived risks caused by uncertainty and lack of control, and therefore risks are interpreted as the perceived negative consequences or costs related to these technologies. The expected benefits are seen as the positive consequences of the risky technologies (Hisschemöller and Midden 1989). Thus, the perception of risks and acceptability of technologies are not the same. However, a lot of studies focus on 'perception of risks', 'acceptability of risks' and/or the 'acceptability of a technology' and use these concepts interchangeably. Throughout this chapter, we refer to 'acceptability' as the acceptability of RW or CO_2 disposal technologies. In Sect. 4, we will explain how we assume the perception of risks and benefits are related to acceptability.

There is a growing consensus that public involvement is essential for the success of virtually any risky technological facility (Short and Rosa 2004). Therefore social scientists have worked to understand why the public is highly concerned about new technologies such as the geological disposal of RW for many years. As a result, there is now considerable understanding of which factors determine public support or opposition to RW disposal. The geological disposal of CO_2 has received far less attention, mainly because this technology has been developed more recently. In this chapter we review psychological factors that contribute to the acceptability of RW disposal (Sect. 2). We also explain to what extent these factors may be relevant for the acceptability of CO_2 disposal (Sect. 3). Finally, we summarize our findings (Sect. 4), we discuss the theoretical implications of current psychological research in research on the acceptability of RW and CO_2 disposal (Sect. 5), and suggest future research directions and policy implications for the acceptability of both types of technologies (Sect. 6).

2 Psychological Factors of the Acceptability of Radiation Waste Disposal

Individuals are inclined to overestimate the probability that a serious accident may happen with risky, large-scale technologies such as RW disposal (e.g. Daamen et al. 1986; Fischhoff et al. 1978). For example, results show that the average yearly fatality rates related to RW and the subjective judgements of the seriousness of the risk posed by RW are barely correlated, indicating that judgements of the general public are based on other factors than the objective probability of the risks of death involved with the disposal of RW (Gowda and Owsley-Long 1998). Thus, acceptability of large-scale technologies is not based on purely quantitative aspects such as expected morbidity. Subjective or qualitative aspects of risks, such as values and affect, play an important role in supporting or opposing these technologies.

Because public acceptability is not only related to the objective perception of risks involved with RW disposal, a lot of research has focused on psychological factors related to acceptability. Especially (general and specific) attitudes towards nuclear energy and RW disposal are seen as main predictors of the acceptability of RW disposal (van der Pligt and Midden 1990). Attitudes are assumed to be based on expectancy-valence models as proposed by Fishbein and Ajzen (1975). In these models, the acceptability of risky technologies is seen as a trade-off between risks and benefits: the more a person believes that the attitude object (e.g. RW disposal) has beneficial rather than negative consequences, the more favourable the attitude towards the object tends to be and the more acceptable the technology is judged.

Which factors are important in the formation of attitudes towards RW disposal? Psychological research has focused on different factors. These factors include the extent to which the technology is perceived as 'dreaded' and 'unknown', affect, moral aspects (i.e. values and worldviews), fairness and trust (Sect. 3.1–3.5). There is also some psychological research concerning group differences in the acceptability of RW disposal (Sect. 3.6). In Sect. 3, we explain how these aspects are related to the acceptability of the geological disposal of RW and we also indicate to what extent these factors may be relevant in the acceptability of the geological disposal of CO_2.

3 Psychological Factors Related to the Acceptability of Radiation Waste and CO_2 Disposal

3.1 Dread and Unknown Risk Factors

Most studies on nuclear energy show that individual attitudes are based upon perceptions of a limited number of potential negative and positive aspects of nuclear energy in general and RW disposal in particular (van der Pligt 1989). These aspects can be characterized along two dimensions, namely the *dread* and the *unknown* risk factor (e.g. Peters et al. 2004; Slovic 1987). Dread risk refers to the extent to which individuals experience: (1) a lack of control; (2) feelings of dread; (3) a catastrophic potential of the technology; and (4) an unfair distribution of risks and benefits involved with the technological risks. Unknown risk is characterized by the extent to which a hazard is perceived as unobservable, unknown, new, and delayed in producing harmful impacts. RW disposal tends to be judged highly on both dread and unknown risk factors (Peters et al. 2004; Fischhoff et al. 1978; Vlek and Stallen 1981; Verplanken 1989). The radiation risk is, for example, described as unknown, invisible and dangerous to health and for the environment, in the short as well as in the long run (MacGregor et al. 1994). Therefore acceptability of RW disposal is generally low.

The geological disposal of CO_2 may probably also be judged as highly dreaded and unknown because CO_2 disposal carries potential risks, such as leaks from underground disposal or CO_2 seepage. These risks are uncertain and potentially dangerous to health and for the environment. It could be argued that the possible consequences of CO_2 disposal are even more unknown than the consequences of RW disposal, because experts as well as lay people have less experience with CO_2 disposal compared to RW disposal, and consequently the acceptability of CO_2 disposal may be lower. However, some explorative research on the role of psychological factors in explaining the acceptability of CO_2 disposal does not fully confirm this view (Bradbury et al. 2009; Tokushige et al. 2007). In contrast with the acceptability of RW disposal, the perceived benefits of CO_2 disposal seem to be the most predictive factors when explaining the acceptability of this technology. These initial results could imply that there is a difference in perception of the costs (i.e. risks) and benefits of CO_2 disposal and RW disposal. Consequently, the risks of CO_2 disposal may be perceived as less dreaded. Future research should examine why the perceived benefits seem more important when explaining the acceptability of CO_2 disposal than when explaining the acceptability of RW disposal.

The perception of a risk as 'unknown' may also affect the stability of acceptability judgements. New information shapes attitudes towards a technology more when people know little about the subject than when people are familiar with it. However, various studies show that many people have a strong and relatively stable opinion on the pros and cons of the acceptability of RW, while RW technologies are perceived as highly unknown. The 'affect heuristic' may explain why people oppose RW disposal without knowing exactly why (see Sect. 3.2).

We think that the affect heuristic does not play an essential role yet in explaining the acceptability of CO_2 disposal. Because the consequences of CO_2 disposal are more unknown to the general public, and less emotionally charged discussions are known regarding these than on the consequences of RW disposal, the acceptability toward this type of technology may be based more on 'non-attitudes': 'Apparent attitudes that have little meaning in the world outside the interview' (Rosema et al. 2008: p. 353). For example, de Best-Waldhober et al. (2009) examined the differences in attitudes towards CO_2 disposal technologies among Dutch inhabitants with well informed and uninformed opinions. They concluded that the Dutch public is largely uninformed about CO_2 disposal, while they are still inclined to express their opinion about this technology. The few studies on public perceptions of CO_2 disposal confirm that this technology is largely unknown to the general public (de Best-Waldhober et al. 2009; van Alphen et al. 2007). Obviously, non-attitudes are less stable and relatively easy to change via communication (de Best-Waldhober et al. 2009; Sjöberg 2003). In comparison to RW disposal, more people will be 'indecisive' about CO_2 disposal; therefore discussions with the general public about this type of technology will be less controversial than for RW disposal.

3.2 *Affect*

Recently it has been suggested that affect may be an important factor in risk perception of large-scale technologies (Peters and Slovic 1996; Peters et al. 2004; Siegrist et al. 2006; Slovic et al. 1991a, 2004; Summers and Hine 1997). Along this line of reasoning, scholars argue that humans base their acceptability towards RW disposal not only on how they think about it (i.e. cognitive aspects), but also, or even more strongly, on how they feel about it (i.e. affective aspects).

Various studies on (the geological disposal of) RW have found that affect is indeed related to the acceptability of RW (Peters and Slovic 1996; Peters et al. 2004; Slovic et al. 1991a, 2004; Siegrist et al. 2006; Summers and Hine 1997). One of the findings was that perception of risks and society's responses to these risks were strongly related to the extent to which a hazard evoked feelings of dread (e.g. Slovic 1987; Slovic et al. 2007). Activities associated with the disposal of RW are seen as riskier, less acceptable, and more in need of regulation than activities with less dreaded forms of energy generation, such as energy produced by windmills. The amount of dread of a certain risk, as mentioned in Sect. 3.1, and someone's affect therefore seem closely related.

Another result of research on affect and RW disposal was that individuals who had positive feelings about nuclear energy and RW evaluated the negative consequences associated with this technology low and its potential benefits high, indicating that they found the risks associated with RW disposal more acceptable (Alhakami and Slovic 1994; Slovic et al. 2007). Furthermore, Slovic et al. (1991a) found that affect provoked by images of an RW repository was related to voting in favour or against an RW repository, and to how risky people judged activities related to RW disposal to be. When the image was judged negatively and thus provoked negative feelings, participants were more inclined to vote against an RW repository and believed that the risks involved with RW disposal were higher than when the image was judged less negatively. This finding was replicated by a study of Peters and Slovic (1996). Slovic and colleagues (Slovic et al. 2004, 2007) refer to the *affect heuristic* to explain these results. The affect heuristic implies that representations of objects and events in people's minds are marked with positive and negative feelings to varying degrees. Individuals consider these positive and negative feelings about the object or event to make a decision on the acceptability of RW disposal. If individuals only experience negative feelings, this will result in less acceptability of RW disposal without them rationally considering the costs and benefits of this technology. It is assumed that affective reactions may serve as a quicker, easier and more efficient way to make decisions in a complex and uncertain world than cognitive reasoning, and therefore serve as a 'heuristic'.

Like RW disposal, and as reasoned in Sect. 3.1, CO_2 disposal is probably perceived as highly dreaded and unknown. Consequently, it will evoke negative affect in a similar way as RW disposal does. However, people will probably be less able to use an affect heuristic with CO_2 disposal; because CO_2 disposal is very new and unknown, no stigma is associated with the object yet. Peters et al. (2004) show that

RW from nuclear power plants, radiation from nuclear weapons, and nuclear power plants in general, are strongly stigmatized subjects. Consequently, they provoke negative feelings such as fear and anger, which in turn results in a higher risk perception and less acceptance of the stigmatized object. The disposal of RW is stigmatized and it can be used as a heuristic partly because people have received a lot of negative information associated with nuclear energy and RW. For example, most people are aware of major or minor accidents that have occurred with nuclear energy, so they know something bad can happen (e.g. Slovic et al. 1991b). Research showed that attitudes towards nuclear energy became more negative when people were faced with an accident (e.g. Hohenemser and Renn 1988; Verplanken 1989). CO_2 disposal is not as yet associated with unfortunate events such as accidents, evidence of mismanagement or discoveries of CO_2 releases. Therefore CO_2 disposal is probably less stigmatized and consequently the affect heuristic will be less important than for RW disposal.

The influence of affect and the affect heuristic is one of the most thoroughly examined aspects in explaining the acceptability of nuclear energy and the disposal of RW. In contrast to most studies that explain the acceptability of RW disposal, correlational as well as experimental designs were used and a clear theoretical paradigm was followed for which support outside the domain of RW disposal was found as well. Future studies should reveal to what extent and under what specific conditions affect will influence the acceptability of CO_2 disposal.

3.3 Values and Worldviews

Various studies focused on relationships between moral aspects and the acceptability of RW disposal (van der Pligt 1989). It is argued that public reactions to the disposal of RW are not only based on perceptions of health and environmental risks, but are based on values and worldviews as well (e.g. Gowda and Easterling 2000; Peters and Slovic 1996; Short and Rosa 2004; Sjöberg and Drottz-Sjöberg 2001). The acceptability of RW disposal and CO_2 disposal may be viewed as moral issues, that is, a function of general beliefs on what is the right or wrong thing to do. Acting on the basis of moral considerations generally implies choosing behavioural options that will result in public (or environmental) benefits (de Groot and Steg 2009; Thøgersen 1994, 1996). We will review to what extent two types of general beliefs are related to the acceptability of RW disposal and CO_2 disposal: values and worldviews.

3.3.1 Values

Several scholars suggest that the importance of various risks and benefits of a new technology depends on the values someone upholds (Short and Rosa 2004; van der Pligt 1989). Values are defined as 'desirable transsituational goals, varying in importance, that serve as guiding principles in the life of a person or other social entity' (Schwartz 1994: p. 21). Two value orientations are particularly relevant in explaining the acceptability of RW disposal: self-transcendence or altruistic *versus* self-enhancement or egoistic value orientations. Some scholars have proposed that a third value orientation is important in the environmental domain. This 'biospheric' value orientation emphasizes the intrinsic value of nature (e.g. de Groot and Steg 2007, 2008; Stern 2000; Stern et al. 1993). These three value orientations seem to be important in explaining the acceptability of RW, since most studies support the belief that economic (i.e. egoistic), community health and safety (i.e. self-transcendent, 'prosocial' or altruistic), and environmental (i.e. biospheric) considerations are important for understanding the acceptability of RW, as we will explain next.

Most studies on the acceptability of the disposal of RW assume a conflict between benefits and risks of RW. For example, nuclear energy is relatively cheap; however, this conflicts with the perceived risks involved with RW disposal, which threaten other people and the environment. Or, nuclear energy produces less CO_2 emissions, which helps to reduce global warming. However, the geological disposal of RW is a problem in the long term and it is hard to estimate the risks for future generations.

An often mentioned concept within RW research which emphasizes these conflicts is NIMBY (Not In My Back Yard) (Gervers 1987; Sjöberg and Drottz-Sjöberg 2001). People who are driven by NIMBY motives are supposed to profit from the benefits of nuclear power, but at the same time they refuse to accept the associated risk involved such as the siting of an RW repository within a nearby area.

NIMBY assumes that a low acceptability of RW disposal is rooted in egoistic or self-interest. However, is this assumption a realistic perspective? People may have profound prosocial or environmental reasons why they oppose RW disposal, for example concerns about the risks involved for the health or safety of people in the community. Therefore, is it really selfish to oppose RW disposal? Results of studies on RW suggest the opposite. Krannich and colleagues (Krannich et al. 1993) showed that the opposition to the siting of an RW facility does not reflect a NIMBY response on the part of area residents. They emphasized that risk perceptions are mainly influenced by concerns about future generations (i.e. prosocial considerations), and that these concerns are especially important in determining responses to an RW repository. However, they did not correct for the fact that residents may use future and environmental concern arguments as an excuse for not wanting a waste facility anyway. Other studies also showed that prosocial and environmental consequences, such as consequences for health, community, safety and environment, are more predictive of acceptability of nuclear power in general and RW disposal in particular, than are personal consequences (Sjöberg and Drottz-Sjöberg 2001).

Results of these studies indicate that, next to personal or egoistic concerns, the acceptability of RW disposal, and presumably of CO_2 disposal, depends on the conflict between considerations that are 'non-selfish' in origin. For example, is it morally more correct *to oppose* RW disposal or CO_2 disposal because people are concerned about the health of future generations and a decrease in environmental quality than *to support* RW or CO_2 disposal because of concerns about global warming for future generations? Both concerns are real and based on unselfish considerations, but which choice do people regard as morally most correct when altruistic and/or biospheric considerations conflict? As yet, no studies have focused on the effect of these conflicts between various altruistic and/or environmental values on the acceptability of RW or CO_2 disposal.

3.3.2 Worldviews

Worldviews can be important in explaining the acceptability of RW disposal (e.g. Peters and Slovic 1996; Peters et al. 2004; Sjöberg and Drottz-Sjöberg 2001). Worldviews are defined as generalized attitudes toward the world and its social organization and function as orienting dispositions that guide people's responses in complex situations (Dake 1991, 1992). In this way, the definition of worldviews is highly compatible with the definition of values as proposed by Schwartz (1992).

The cultural theory of risk perception (Douglas and Wildavsky 1982) is the most common theory used when explaining risk perception with worldviews. According to this theory, people decide upon the riskiness of a technology on the basis of their cultural orientation. Dake (1991, 1992) proposes four basic worldviews that differ on two dimensions. The first dimension distinguishes people who are more group-oriented from those who are more individually oriented. The second dimension focuses on the extent to which someone believes that socially stratified rules are needed to control behaviour. Based on these dimensions, four basic worldviews emerge that determine a person's risk perception: hierarchical, fatalistic, individualistic and egalitarian. In a hierarchical worldview, people are believed to be group-oriented and prefer a high level of stratified prescriptions. A fatalistic worldview suggests that someone is focused on individuals instead of groups, but believes that socially stratified rules are necessary. The individualist is individually oriented, but believes that only few rules are needed to guide behaviour. Finally, people with an egalitarian worldview are group-oriented, but believe in low levels of stratified rules.

Support for the cultural theory in relation to the perception of nuclear energy and RW disposal is mixed. Peters and Slovic (1996) found some support for the relationship between worldviews and support for nuclear energy. Especially fatalistic, hierarchical and individualistic worldviews were associated with a stronger support towards nuclear energy. An egalitarian worldview was negatively related to support for nuclear energy. However, correlations were moderate indicating that other factors may be more important when explaining the acceptability of RW disposal. Indeed, some scholars argue that cultural theory hardly adds any additional variance when more powerful determinants such as lack of fairness and risk for future generations are entered into the same model (Sjöberg and Drottz-Sjöberg 2001; see also Sjöberg 1997).

In conclusion, values and, to a lesser extent, worldviews may influence the acceptability of RW disposal. People will evaluate the acceptability of RW disposal and CO_2 disposal largely on the basis of the extent to which important values are perceived to be affected by the consequences of these technologies. We believe that it is important to study to what extent values and worldviews are related to acceptability of RW and CO_2 disposal because specific attitudes towards new objects must be built on something more stable and relatively enduring, and general antecedents (i.e. values and worldviews) may provide such a basis (Stern et al. 1995; Stern 2000). This is especially relevant in the domain of relatively new technologies, such as RW disposal, but even more for CO_2 disposal, because this technology is even more unknown.

3.4 *Fairness*

Fairness is another factor that is relevant for explaining the acceptability of RW disposal. Scholars argue that the acceptability of policies, including policies to implement repository sites, is strongly related to their perceived fairness, that is, policies are more acceptable when they are perceived to be fair (e.g. Cvetkovich and Earle 1994; Tyler 2000). Some studies on the acceptability of RW disposal measure fairness in general, or do not distinguish between various types of fairness (e.g. Summers and Hine 1997; Sjöberg and Drottz-Sjöberg 2001), which makes it difficult to draw conclusions about which specific type of fairness contributes to acceptability. Others distinguish between different types of fairness (e.g. Ahearne 2000; Gowda and Easterling 2000; Hisschemöller and Midden 1989; Short and Rosa 2004; Shrader-Frechette 2000), but mostly use non-theory-based data to provide support for their distinction. Therefore, we will use literature from both within and outside the RW and CO_2 disposal domain to provide possible frameworks for explaining the relationship between fairness and acceptability of RW and CO_2 disposal.

It is important to distinguish distributional from procedural fairness (e.g. Gowda and Owsley-Long 1998; Schuitema 2010). Distributional fairness concerns how risks and benefits that are associated with policies, such as implementing a repository, are distributed across various groups in society (Deutsch 1975, 1985). Some people may be disproportionally affected by a decision to implement a waste repository in their neighbourhood, because they are exposed to risks without receiving any compensation for the potential risks. Therefore distributional fairness seems to be crucial for implementing hazardous waste facilities such as those for RW and CO_2 disposal (see van der Pligt 1989).

Various principles may be followed when deciding whether a particular distribution of outcomes is fair, such as the equity principle, the equality principle, social justice, and environmental justice (Schuitema 2010). As yet, most studies do not differentiate between these different principles, and it is not known which principle is most influential in acceptability judgements. The equity principle implies that risks and benefits should be distributed in proportion to an individual's contribution (Adams 1965). Those who benefit most should carry the most risk. Policies to implement a repository site would be acceptable if people believed that the risks of implementing a repository (e.g. potential risks) did not exceed the benefits of the repository (e.g. financial compensation, possibilities for work). The equality principle suggests that everyone should be affected to the same extent by the policy (Deutsch 1985), that is, no groups may be affected disproportionally. This principle implies that implementing a repository site would be fair and acceptable if the risks were the same for everyone. When equality is the most dominant fairness principle, people are likely to oppose RW and CO_2 disposal unless the disproportionally affected group is highly compensated. Social justice refers to striving for a greater degree of equality in general, that is, outside the domain of RW and CO_2 disposal as well. Finally, environmental justice refers to the protection of nature, environment and future generations (e.g. Clayton 2000; Montada and Kals 1995; Opotow and Clayton 1994). This principle overlaps with intergenerational equity (Ahearne 2000; Gowda and Easterling 2000; Shrader-Frechette 2000), which refers to concerns about how future generations and the environment may be affected by the current generation's choices. In contrast to Gowda and Easterling (2000) and Clayton (2000), we interpret this principle as a specific type of distributional fairness because it concerns a distribution of risks and benefits among the present and the future generations. All four principles seem important in relation to the acceptability of RW and CO_2 disposal. Future research should examine which fairness principle is prevalent in acceptability judgements.

Distributional fairness is closely related to values, that is, which distribution of risks and benefits is considered to be fair depends on one's value orientation (cf. Deutsch 1975). Nuclear energy and implementing a waste repository may have egoistic, altruistic and biospheric benefits and risks. People who value egoistic aspects most will judge the implementing of an RW or CO_2 repository to be fair when the egoistic benefits (e.g. employment) outweigh the risks associated with it (i.e. they prefer the equity principle). For people who value altruistic aspects, a policy to implement a repository would be considered fair when the altruistic benefits (e.g. cheap energy for everyone) of the repository outweigh its risks (i.e. they prefer the social justice principle). And, when people have a strong biospheric value orientation, they perceive implementing a repository as fair when the outcomes of this policy would benefit nature and the environment (e.g. no CO_2 emissions, i.e. they prefer the environmental justice principle).

It is difficult to decide how the risks of an RW facility can be distributed in a fair way. When the potential host community perceives itself as bearing an unjust burden (i.e. unequal distribution), most people will oppose the siting even though they are compensated by financial or economic benefits to increase distributional fairness. In this case, the question is how to translate subjective risks to health, safety and the environment into financial or economic compensation to make the distribution fair again. The few studies on monetary compensation and acceptability of RW disposal show that such measures have mixed success only (Sjöberg and Drottz-Sjöberg 2001; Summers and Hine 1997; van der Pligt 1989). Compensations do often not result in higher acceptability levels. People may view the financial compensations as a 'bribe', which may intensify concerns about unequal distributions and increases suspicion and distrust of relevant authorities (van der Pligt 1989). Other ethical considerations are at play as well: A relatively poor community may be more in need of monetary payments than a rich community. Consequently, residents of a poor community may also be more inclined to accept a repository because they benefit more from it, although they still believe that the risks and benefits are distributed unequally.

The second type of fairness is procedural fairness, which involves the use of fair procedures (e.g. Lind and Tyler 1988), for example, to come to a decision over an RW repository. These procedures should be perceived as fair and consistent towards all parties involved. When a potential host community perceives the decision making process as unfair or inconsistent, opposition is more likely to occur (e.g. Gowda and Owsley-Long 1998; Sjöberg and Drottz-Sjöberg 2001; Summers and Hine 1997). Procedural fairness can be promoted via communication and public involvement. For example, people were slightly more willing to accept an underground RW repository when they were involved in the planning process (Summers and Hine 1997). Sjöberg (2004) showed how extensive information programmes in four Swedish municipalities have positively changed the extent to which people accepted a local RW repository. However, no (field) experiments have been conducted in the area of perceived procedural fairness and the acceptability of RW disposal, thus conclusions about changes in acceptability judgements remain tentative.

The extent to which and how (distributional and procedural) fairness considerations influence acceptability of RW and CO_2 disposal may vary across situations. In the case of a large physical distance between the host community of a siting and the repository site, the relevance of distribution and procedural fairness can decrease because people experience less direct individual risks of the repository. At the moment, different possibilities of CO_2 disposal locations are being explored. For example, in the Netherlands experts propose the possibility of offshore CO_2 repositories. When the exact policies hardly affect people directly (e.g. large physical distance of repository), people are less committed and less likely to experience benefits and risks involved with CO_2 disposal directly. In such cases, aspects related to distributional and procedural fairness may play a less prominent role. Therefore we expect that structural factors, such as the location of a repository, will influence concerns about distributional and procedural fairness and this will affect acceptability of CO_2 disposal.

Both procedural and distributive fairness are important for public support of policies (e.g. Clayton and Opotow 2003; Cohen 1987; Cook and Hegtvedt 1983; Deutsch 1975, 1985; Rawls 1999; Tyler 2000) and often the two types of fairness interact. Unfair distributions may result in perceived procedural unfairness and *vice versa*. For example, the acceptability of a siting for RW or CO_2 disposal could be increased by monetary payments or by emphasizing economic benefits of the repository (i.e. distributional fairness), but only when relevant stakeholders are involved in the planning process (i.e. procedural fairness). Reasonably, both types of fairness are necessary for explaining the acceptability of RW and CO_2 disposal, and future studies should examine possible interaction effects.

3.5 Trust

Trust is seen as a crucial factor for the acceptability of RW disposal (e.g. Binney et al. 1996; Earle and Cvetkovich 1995; Flynn et al. 1992; Gowda and Owsley-Long 1998; Katsuya 2002; Kasperson et al. 1992; Slovic et al. 1991c; Summers and Hine 1997). Among other things, it is reasoned that trust may enhance feelings of general and personal control, and therefore people experience less dread. They will perceive the threat as less risky and consequently they are more willing to accept the technology (Slovic 1993).

Although the assumption of a strong relationship between trust and risk perception seems plausible, empirical data shows mixed support (Sjöberg 2001). Some studies reveal that trust in government and risk management agencies explains risk perception and acceptability of RW disposal to a large extent (e.g. Biel and Dahlstrand 1995; Katsuya 2002; Summers and Hine 1997), but results of other studies have shown moderate to weak relationships (e.g. Bord and O'Connor 1990, 1992; Mushkatel et al. 1993; Pijawka and Mushkatel 1991/1992). A possible explanation for the weak relationship between trust and risk perception is that people believe that science and the experts themselves also do not fully understand the effects of the technology of RW disposal yet (Sjöberg 2001). Thus, even though experts, governments and corporations promoting nuclear energy are perceived to be trustworthy, the general public can still disagree with the conclusion that the risks associated with RW disposal are negligible. In this case, the public does trust that authorities are honest, but they do not believe that authorities can control the technology. For example, Sjöberg (2001) showed that a lack of scientific knowledge of RW technologies tends to be a more important predictor of risk perception than trust in authorities that communicate this knowledge to the public. In the case of the acceptability of CO_2 disposal, a similar line of reasoning may be followed: the public may be uncertain about whether experts have sufficient knowledge of the risks of CO_2 disposal; consequently, the public may perceive that authorities in this domain have little control over the situation, which will decrease the acceptability of this technology. The limited research in the area of the acceptability of CO_2 disposal and trust indicates that trust is indeed an important factor that affects the acceptability of CO_2 disposal (Huijts et al. 2007; Tokushige et al. 2007). More specifically, research among Japanese university students showed that trust had an indirect impact on the acceptability of CO_2 disposal via the perceived risks and benefits of this technology. Higher trust was associated with perceiving more benefits and slightly less risks from CO_2 disposal.

Trust has not been integrated into a general theoretical model, and studies that did include trust have hardly been replicated. Typically, trust is assessed in different ways and within specific samples and specific areas of risk. Therefore the concept is still relatively unexplored empirically, and even more so in the field of the acceptability of RW and CO_2 disposal.

3.6 Group Differences in the Acceptability of Radiation Waste and CO_2 Disposal

Several group differences exist in the acceptability of RW and CO_2 disposal. Below we discuss the two most common group differences that are studied in the domain of RW disposal, namely lay *versus* experts and cross-cultural differences.

3.6.1 Lay People *Versus* Experts

Studies on risk perception of RW disposal show that lay people assess risks very differently from experts (Sjöberg 1998). A common finding is that lay people exhibit higher perceptions of hazardous risks involved with RW compared to experts (e.g. Flynn et al. 1993; Purvis-Roberts et al. 2007; Sjöberg 1998). For example, experts assessed risks associated with high-level RW as lower, they showed more trust in programme managers, they perceived more positive consequences of a repository project, and they had more positive images of an RW repository than the general public (Flynn et al. 1993). In another study, Purvis-Roberts and colleagues (Purvis-Roberts et al. 2007) found that lay people were the most risk-averse group, followed by physicians and scientists.

There are several reasons for the difference in risk perception between experts and the public. Experts tend to evaluate the acceptability of a technology on quantitative aspects, while the public focuses on qualitative characteristics (Drottz-Sjöberg and Sjöberg 1991; Gardner and Stern 2002). Experts tend to assess the risks in terms of the probable number of human deaths and the costs of building and operating a power plant, and tend to overlook damage caused to ecosystems and non-human forms of life. Experts judge whether the technology is acceptable overall to society based on whether the technology's quantitative benefits outweigh its quantitative costs. If the benefits exceed the costs, the technology should be acceptable to society. As described in the sections above, various qualitative aspects are important for the acceptability of a technology to the public, such as the extent to which the risk affects future generations or the environment or whether the benefits are equitably distributed among those who bear the risk. Thus, the public uses a broader and more complex definition of risks and acceptability than do experts (Gardner and Stern 2002).

Furthermore, scholars argue that experts perceive hazardous technologies such as RW or CO_2 disposal as more acceptable than the public because they perceive higher levels of personal control and are more familiar with the risky activity than the public (Sjöberg 1998). Either way, the public perceives the risk of dreaded and unknown technologies such as RW disposal as more severe than experts and therefore judge it as less acceptable. As this is a general phenomenon, we have no reason to expect this to be different for the perception of risks and the acceptability of CO_2 disposal.

3.6.2 Cross-Cultural Differences

There are only few psychological studies that have focused on cross-cultural differences in the acceptability of nuclear energy. And, to the authors' knowledge, there are no cross-national studies that have specifically examined cross-cultural differences of attitudes towards RW disposal. From the studies that have been conducted, we can draw some general conclusions on the relationships between cross-cultural aspects on the acceptability of RW and CO_2 disposal.

Wiegman et al. (1995) made a cross-national comparison of risk perception of nuclear energy between France and the Netherlands. In contrast to their expectations, they found that the French had a higher risk perception and a more negative attitude toward nuclear energy than the Dutch. They provided two possible explanations for these results. First, they indicated that French citizens have less power in the decision making process. The government seems to mobilize aversive reactions against nuclear power, because the general public can participate less in decisions on, for example, whether to build a nuclear power plant. This probably results in decreasing perceived procedural fairness. Second, Wiegman and colleagues argue that nuclear technology is more developed in France and therefore the French are probably more exposed to these technologies and the risks they entail. However, other explanations are possible as well, and future research should more specifically examine which explanation is most plausible.

Hohenemser and Renn (1988) showed that attitude stability differs cross-culturally. They assumed that in countries with well formed nuclear attitudes (i.e. where respondents score less on 'don't know' categories), such as the USA, Finland and the UK, acceptability towards RW disposal may be more stable than in countries in which individuals have less formed attitudes, such as Greece or former Yugoslavia. They provided some empirical data on the acceptability of RW before and directly after the Chernobyl accident and a year after the accident. Results showed that in countries in which respondents were more indecisive about supporting or opposing nuclear energy, acceptability changed to a larger extent in a negative direction after the accident than for other respondents. Furthermore, results indicated that countries with well formed attitudes returned faster to their pre-accident level of acceptability towards nuclear power than countries with less formed attitudes. Therefore, for countries with citizens with well formed attitudes, acceptability judgements might be more stable over time even after a negative event than in countries with a large proportion of undecided citizens.

In conclusion, countries vary in the degree to which they oppose or support nuclear energy and RW disposal, and the extent to which these attitudes towards RW are stable. Institutional and structural factors, such as the political system, technological advances and knowledge, have been proposed as possible determinants to explain these differences. However, results of studies on cross-cultural differences have not tested and validated these assumptions. Furthermore, they have not focused specifically on the acceptability of RW and CO_2 disposal. Therefore conclusions remain tentative and future research should reveal which of these factors explain cross-cultural differences in relation to the acceptability of RW and CO_2 disposal most.

4 Summary of Psychological Factors

As explained in Sect. 1, the perception of risks and acceptability of the technology are different constructs. However, some studies used these terms interchangeably. We think both concepts need clear conceptualizations. In our view, the acceptability of

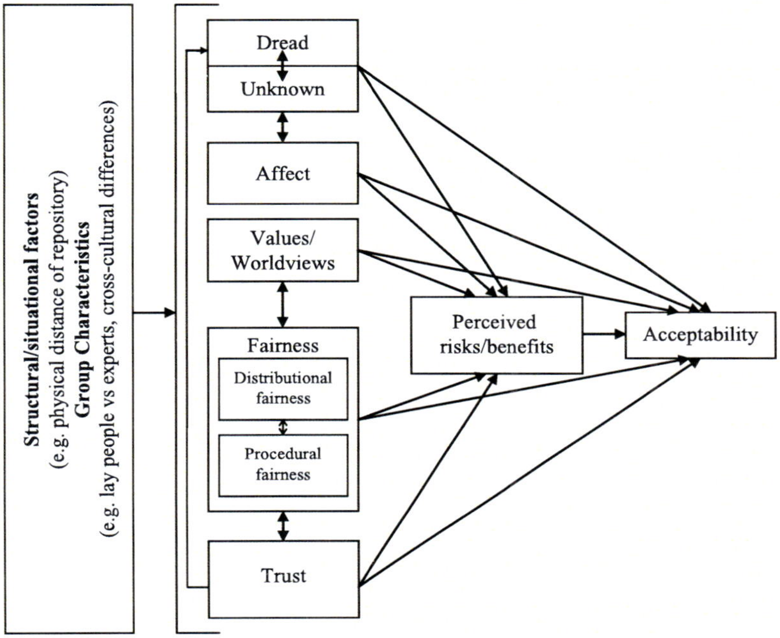

Fig. 1 Psychological factors related to the acceptability of CO_2 disposal

RW and CO_2 disposal is based on the trade-off between perceived costs (the negative consequences) and benefits (the positive consequences), in which the perceived costs are mostly interpreted as the perceived risks involved with RW or CO_2 disposal, as many costs are uncertain. Consequently, acceptability is not solely an evaluation of risks, but is based on weighing risks and benefits (Hisschemöller and Midden 1989): the more a person believes that RW and CO_2 disposal have positive rather than negative consequences, the more RW and CO_2 disposal are evaluated as acceptable.

Figure 1 summarizes the assumed relationships between the psychological factors, perceptions of risks/benefits and acceptability of CO_2 disposal. All these relationships are based on our current knowledge of research in the area of the acceptability of the disposal of RW. The acceptability of CO_2 disposal has received far less attention in psychological studies than the acceptability of RW disposal; therefore we have only made some assumptions on possible psychological determinants of the acceptability of CO_2 disposal based on research on the acceptability of RW disposal. As shown in Fig. 1, the acceptability of CO_2 disposal depends on risk characteristics (dread and unknown), affect, values and worldviews, fairness and trust. We believe that these factors are directly related to the acceptability of CO_2 disposal as well as indirectly, via the perceived risks and benefits of CO_2 disposal. For example, people with strong egoistic values will especially consider risks and benefits of technologies for them personally: when the perceived personal benefits exceed the perceived personal risks they will more likely accept the technology and *vice versa*. Another example: If people show more trust in the information that the national government provides about the risks and benefits of the disposal of CO_2, they will more likely perceive the risks and benefits in accordance with this information than people who do not trust this information. These differences in perceived risks and benefits will result in a difference in the acceptability of CO_2 disposal.

Figure 1 also includes some interrelationships between the different psychological factors that we have described in Sect. 3. For example, dread factors are related to affect: previous research showed that the perception of risks and the affective responses to these risks were strongly related to the extent to which a hazard evoked feelings of dread (e.g. Slovic 1987; Slovic et al. 2007). The potential of a high perception of dread for risks associated with CO_2 disposal, such as CO_2 seepage, provide people with negative feelings, and this will consequently result in an acceptability judgement of CO_2 disposal that is not only based on how they think about it, but also on how they feel about it. Another example is the relationship between fairness and values: what people perceive as 'fair' depends on what they value in life. For people who mostly value altruistic aspects in life, a policy to implement a CO_2 disposal would be considered fair when the perceived altruistic benefits (e.g. cheap energy for everyone) of implementing a disposal site would outweigh its perceived costs or risks (i.e. social justice). We also expect that trust may be related to the perceived fairness: if people lack trust in the authorities' judgement about the distribution of risks and benefits when planning to implement a CO_2 disposal site, they will perceive these distributions probably as more unfair than if people do trust the choices that governments or other relevant stakeholders make. Also, it is reasoned that trust may enhance feelings of general and personal control, and therefore people experience less dread. Figure 1 summarizes all the potential relationships between these psychological factors.

Finally, Fig. 1 also shows how structural, situational and group characteristics directly affect the psychological factors of dread and unknown risk characteristics, affect, values and worldviews, fairness and trust, while they affect the perceived risks and benefits of CO_2 disposal and the acceptability of this technology indirectly. For example, in some countries the discussion about CO_2 disposal sites may already be more advanced than in other countries. A difference of this kind may influence to what extent people perceive the risks and benefits based on cognitive aspects or affective aspects (e.g. stigmatizing CO_2 disposal) and, consequently, the extent to which they perceive it as an acceptable technology. We also described how structural and situational factors could change perceived fairness. In the case of a large physical distance between the host community of a CO_2 disposal site and the actual disposal site, the significance of fairness for the perception of risks and benefits and acceptability can decrease because people experience less direct individual risks from the disposal site. So structural factors, such as the location of a CO_2 disposal site, will result in concerns about distributional and procedural fairness, and this will affect acceptability of CO_2 disposal.

Figure 1 summarizes *how* some potential psychological factors may be related to the acceptability of CO_2 disposal. Table 1 summarizes *to what extent* these factors could contribute to the explanation of the perceived risks and benefits and

Table 1 Summary of factors that are regarded as most important determinants of acceptability of radioactive waste and CO_2 disposal

	Acceptability of	
	RW disposal	CO_2 disposal[a]
Risk characteristics:		
– Dread	+	+
– Unknown	+	+
Affect	+	+/–
Values	+	+
Worldviews	+/–	+/–
Fairness:		
– Distributional	+	+[b]
– Procedural	+	+[2]
Trust	+	+

+ = important
+/– = sometimes important, sometimes not important
– = not important
[a] These comparisons are based on assumptions by the authors, as only limited empirical research on CO_2 disposal is available yet
[b] Depends on structural aspects such as the location of site

the acceptability of CO_2 disposal. Table 1 is based on theories and empirical research on the acceptability of RW disposal; we assume that relationships are similar for the acceptability of CO_2 disposal. We emphasize once more that these comparisons are based on our assumptions, because hardly any empirical research on the acceptability of CO_2 disposal is available yet.

The first factor in Table 1 is the amount of dread and the extent to which the risks are unknown. As discussed previously, research has shown that the risks of RW disposal are generally perceived as highly dreaded and unknown and we argued that CO_2 disposal may also be judged as highly dreaded and unknown. Therefore we assume that these risk characteristics are also important factors for explaining the acceptability of CO_2 disposal.

Although research shows that affect is an important factor for the acceptability of RW disposal, we think that the affect heuristic does not play an essential role yet for explaining the acceptability of CO_2 disposal. The consequences of CO_2 disposal are more unknown, and people generally know little or nothing about this type of technology (e.g. de Best-Waldhober et al. 2009). Also, no stigma is associated with this technology yet. Therefore the acceptability towards CO_2 disposal may be mostly based on 'non-attitudes' rather than on affect.

People evaluate the acceptability of RW disposal on the basis of the extent to which important values and worldviews are perceived to be affected by the consequences of this technology. We argued that it is important to study to what extent values are related to acceptability of CO_2 disposal as well, because specific attitudes towards new objects, such as CO_2 disposal, must build on something more stable and relatively enduring in life, and values may provide such a basis (Stern 2000). Worldviews are also perceived as relatively stable; however, empirical research on the acceptability of

RW disposal shows mixed support regarding the contribution of worldviews to explain acceptability judgements. Therefore, we assume that worldviews and especially values will be relevant when explaining the acceptability of CO_2 disposal.

We also argued that policies, such as decisions on RW or CO_2 disposal, are more acceptable when they are perceived to be fair. There is some support for the assumption that fairness is an important predictor for the acceptability of RW disposal (e.g. Ahearne 2000; Summers and Hine 1997; Sjöberg and Drottz-Sjöberg 2001). We believe that both procedural and distributional fairness may be important to explain acceptability of RW and CO_2 disposal. Future research should examine which distributional fairness principle is most prevalent in acceptability judgements.

The final psychological factor that we list in Table 1 is trust. It is assumed that when people have higher levels of trust in decision making authorities and experts, they will perceive the threat as less risky and consequently they evaluate the risky technology as more acceptable. Although empirical data of the relationship between trust and the acceptability of RW shows mixed support (Sjöberg 2001), we do assume that trust is important to explain the acceptability of RW and CO_2 disposal. The limited research in the area of the acceptability of CO_2 disposal and trust shows that trust is indeed an important factor that affects the acceptability of CO_2 disposal (Tokushige et al. 2007). Whether trust is a strong contributor mainly depends on the way it is measured because research measures trust in different ways (i.e. Trust in whom? Trust in what?).

5 Theoretical Implications

Various studies have focused on the acceptability of nuclear energy and, more specifically, on the acceptability of the disposal of RW. Most of these studies were descriptive and explorative in nature. Although of great importance, they have provided less information about which and to what extent psychological factors uniquely contribute to the explanation of acceptability. Moreover, a clear theoretical framework or model on factors influencing acceptability is generally lacking. This makes it hard to compare and relate results from different studies. In this section, we provide frameworks from other domains which might be relevant to understand the acceptability of RW and CO_2 disposal.

A relevant theory that may be used to explain and change acceptability of RW and CO_2 disposal is the 'protection-motivation' theory (Rogers 1983) or its modified version (Gardner and Stern 2002). The theory assumes that acceptability depends on two aspects, namely the perceived costs and benefits of the risks, and the perceived efficacy or amount of control one experiences. We assume that the cost-benefit assessment of the risky activity depends on the risk characteristics of the technology (i.e. dread and unknown), affect, values/worldviews and distributional fairness. In the case of RW and CO_2 disposal, the perceived efficacy and control depends on (possible) responses of relevant authorities to the risky technology, and therefore procedural fairness and trust may be the most relevant factors in this respect.

Another model that may be relevant in explaining the acceptability of RW and CO_2 disposal is the norm activation model (NAM) (Schwartz and Howard 1981). This model focuses on the role of moral obligations to act in favour of the common good, and some extended versions of this model (see e.g. Stern 2000) also explain how egoistic, altruistic and biospheric values may be related to acceptability (e.g. de Groot et al. 2008; Stern 2000). According to the NAM, personal norms, i.e. 'feelings of moral obligation to perform or refrain from specific actions' (Schwartz and Howard 1981, p. 191), influence the acceptability of policies related to RW and CO_2 disposal. When personal obligations towards accepting nuclear energy and CO_2 disposal are strong, there will be more support for policies promoting RW and CO_2 disposal and *vice versa*. Personal norms are activated when someone acknowledges that not accepting RW/CO_2 disposal will lead to negative consequences for self, others or the environment (awareness of consequences), and when someone feels responsible for these negative consequences (ascription of responsibility). If actors fail to activate personal norms, no actions will be recognized as appropriate and no change in acceptability of RW or CO_2 disposal will follow.

The NAM has successfully been applied to explain moral acceptability judgements, such as the acceptability of policies to reduce household energy consumption (Steg et al. 2005) and the acceptability of policies aimed at reducing car use (de Groot et al. 2008). Various scholars have indicated that moral considerations are of primary importance for explaining the acceptability of high risk technologies such as RW and CO_2 disposal (e.g. Gowda and Easterling 2000). De Groot and Steg (2010) provided some first support that the NAM is indeed useful to explain risky technologies such as opposing or supporting nuclear energy. Therefore, the NAM may function as a relevant framework for explaining the acceptability of RW and CO_2 disposal.

Knowing which and how factors are related to acceptability of RW and CO_2 disposal can assist decision makers in choosing which antecedents can best be targeted in programmes to change acceptability. In order to do so, we should systematically study the acceptability of RW and CO_2 disposal via questionnaire as well as (field) experimental studies from a clear theoretical perspective. Such 'diagnostic' studies give specific insight into which factors are most important for changing acceptability. Based on this information, decision makers can select a strategy to change acceptability and monitor how determinants and acceptability are affected by such strategies. The protection-motivation theory and the NAM could function as a point of departure for such studies.

6 Conclusions

Based on this review, we expect that the acceptability of RW and CO_2 disposal has some important commonalities but also some differences that policymakers should take into account when translating psychological research from the acceptability of RW disposal to the acceptability of CO_2 disposal. In this final section we will summarize our findings and discuss some practical implications.

First, CO_2 disposal tends to be evaluated as a highly dreadful and unknown risk because, like RW disposal, CO_2 disposal carries potential short- and long-term risks that are uncertain and potentially dangerous to human health and for the environment. The possible consequences of CO_2 disposal are even more unknown than the consequences of RW disposal because the technology is relatively new. Consequently, the stability of acceptability judgements of RW disposal is probably higher than for CO_2 disposal. A policy implication is that acceptability judgements of CO_2 disposal can be changed more easily than acceptability judgements of RW disposal. Attitudes that are generally stable, such as for the acceptability of RW disposal, are more difficult to change by, for example, media and communication strategies. The 'social judgment theory' (Sherif and Hovland 1961) suggests that the more extreme one's attitude (i.e. towards RW disposal), the greater the amount of rejection of new information, and thus the more difficult it is to persuade someone, no matter what kind of strong or weak arguments you use. This is especially the case when the new information deviates strongly from one's attitude.

We argued that CO_2 disposal is relatively more unknown. The technology suffers less from stigmatization. Yet, and consequently, affect is expected to play a less dominant role in explaining acceptability compared to RW. But still, for both domains, affect should be considered when explaining and changing acceptability judgements. A study by Meijnders et al. (2001) showed that high levels of fear of global warming resulted in more positive attitudes towards the fear reducing object (i.e. using energy saving bulbs), no matter whether arguments were weak or strong, while moderate levels of fear had only a positive effect on attitudes when strong arguments were used (Meijnders et al. 2001). These results indicate that, also in the domain of RW and CO_2 disposal, decision makers should take affect (e.g. fear) into account in their communication about new technologies to the public because communication strategies have to be adjusted based on the amount of affect people experience.

Another important conclusion is that for both the acceptability of RW and CO_2 disposal, conflicts between egoistic, altruistic and biospheric considerations. Therefore moral aspects, such as values, are important when considering acceptability of both technologies. For implementing CO_2 disposal, values and worldviews are even more relevant, as research indicates that acceptability towards new objects is mostly built on stable and relatively enduring antecedents of behaviour, such as values and worldviews (Stern 2000; Stern et al. 1995). Thus, the acceptability of RW and especially CO_2 disposal strongly depends on the extent to which important values are perceived to be affected by these technologies.

Future research should focus on how and to what extent policies related to RW and CO_2 disposal threaten or support values and worldviews. Decision makers can adjust their policies based on this information. For example, when altruistic considerations contribute most to the explanation of acceptability of CO_2 disposal, acceptability should increase when policies focus on benefits for other people (e.g. everybody should have equal access to energy sources; better for the health of people in the community because of less CO_2 emissions). Another advantage of knowing which values and worldviews are threatened by certain policies is that relevant authorities can provide tailored information based on this knowledge.

Tailored information refers to highly personalized and specific information (Abrahamse et al. 2005). For example, for people who are egoistically oriented, information about individual (dis)advantages will be more effective, while for someone who is biospherically oriented, environmental (dis)advantages should be emphasized and environmental risks minimized in interventions to change acceptability. Therefore it is important to study which value orientation (i.e. egoistic, altruistic or biospheric) or worldview is most relevant in explaining the acceptability of RW and CO_2 disposal in more detail.

Distributional and procedural fairness are also important to consider in policies related to RW and CO_2 disposal. Authorities should examine to what extent and why the general public evaluates policies on RW and CO_2 disposal as fair or not, because this will affect acceptability. Different fairness principles may play a role in this respect. However, which distributional fairness principle influences acceptability most? Studies that explicitly studied the role of fairness principles in the transport domain (Schuitema 2010) revealed that environmental justice plays an important role. This suggests that respondents judge policies as fair when these policies are believed to protect nature, environment and future generations. The contribution of various fairness principles in explaining acceptability should also be examined when trying to change the acceptability of policies in the domain of RW and CO_2 disposal.

Policy and decision makers should also consider procedural fairness when implementing policies related to both RW and CO_2 disposal. When the general public does not believe that fair procedures have been used, trust in relevant authorities and acceptability decreases. Therefore communication and public involvement seem pivotal for increasing procedural fairness and, simultaneously, trust and acceptability.

Finally, trust in authorities involved with RW and CO_2 disposal is relevant for explaining acceptability judgements. Trust enhances feelings of general and personal control, which affects the acceptability of these technologies (see Fig. 1). Again, communication and public involvement is of major importance for decision makers to decrease the perceived uncertainty and lack of control, which in turn may increase acceptability as well.

In this chapter we have described psychological factors that have been most relevant in studies on the acceptability of the geological disposal of RW. We have also discussed how these factors may explain the acceptability of the geological disposal of CO_2. On the basis of these findings, we described how acceptability of RW and CO_2 disposal can be changed. Policymakers may adjust or design policies for changing acceptability in connection with RW and CO_2 disposal. We hope that this chapter will help researchers and decision makers to better address acceptability issues in their work and to develop plans that will change acceptability in the intended way.

References

Abrahamse W, Steg L, Vlek C, Rothengatter T (2005) A review of intervention studies aimed at household energy conservation. J Environ Psychol 25:273–291

Adams JS (1965) Inequity in social exchange. In: Berkowitz L (ed) Advances in Experimental Social Psychology. Academic, New York, pp 267–299

Ahearne JF (2000) Intergenerational issues regarding nuclear power, nuclear waste, and nuclear weapons. Risk Anal 20:763–770

Alhakami AS, Slovic P (1994) A psychological study of the inverse relationship between perceived risk and perceived benefit. Risk Anal 14:1085–1096

Biel A, Dahlstrand U (1995) Risk perception and the location for a repository of spent nuclear fuel. Scand J Psychol 36:25–36

Binney SE, Mason R, Martsolf SW, Detweiler JH (1996) Credibility, public trust, and the transport of radioactive waste through local communities. Environ Behav 28:283–301

Bord RJ, O'Connor RE (1990) Risk communication, knowledge, and attitudes: explaining reactions to a technology perceived as risky. Risk Anal 10:499–506

Bord RJ, O'Connor RE (1992) Determinants of risk perceptions of a hazardous waste site. Risk Anal 12:411–416

Bradbury J, Ray I, Peterson T, Wade S, Wong-Parodi G, Feldpausch A (2009) The role of social factors in shaping public perceptions of CCS: results of multi-state focus group interviews in the U.S. In: Gale J, Herzog H, Braitsch J (eds) Proceedings of the 9th International Conference on Greenhouse Gas Control Technologies; Energy Procedia 1:4665–4672

Clayton S (2000) Models of justice in the environmental debate. J Soc Issues 56:459–474

Clayton S, Opotow S (2003) Justice and identity: changing perspectives on what is fair. Pers Soc Psychol Rev 7:298–310

Cohen RL (1987) Distributive justice: theory and research. Soc Just Res 1:19–40

Cook KS, Hegtvedt KA (1983) Distributive justice, equity, and equality. Annu Rev Sociol 9:217–241

Cvetkovich G, Earle TC (1994) The construction of justice: a case study of public participation in land management. J Soc Issues 50:161–178

Daamen DDL, Verplanken B, Midden CJH (1986) Accuracy and consistency of lay estimates of annual fatality rates. In: Brehmer B, Jungermann H, Lourens P, Sevón G (eds) New Directions in Research on Decision Making. North Holland, Amsterdam, pp 231–243

Dake K (1991) Orienting dispositions in the perception of risk: an analysis of contemporary worldviews and cultural biases. J Cross-Cult Psychol 22:61–82

Dake K (1992) Myths of nature: culture and the social construction of risk. J Soc Issues 48:21–37

de Best-Waldhober M, Daamen D, Faaij A (2009) Informed and uninformed public opinions on CO_2 capture and storage technologies in the Netherlands. Int J Greenhouse Gas Control 3:322–332

de Groot JIM, Steg L (2007) Values, beliefs and environmental behavior: validation of an instrument to measure egoistic, altruistic and biospheric value orientations in five countries. J Cross-Cult Psychol 38:318–332

de Groot JIM, Steg L (2008) Value orientations to explain environmental attitudes and beliefs: how to measure egoistic, altruistic and biospheric value orientations. Environ Behav 40(3):330–354

de Groot JIM, Steg L (2009) Mean or green? Values, morality and environmental significant behavior. Conserv Lett 2:61–66

de Groot JIM, Steg L, Dicke M (2008) Transportation trends from a moral perspective: value orientations, norms and reducing car use. In: Gustavsson FN (ed) New Transportation Research Progress. Nova Science Publishers, New York, pp 67–91

de Groot JIM, Steg L (2010) Morality and Nuclear Energy: Perceptions of Risks and Benefits, Personal Norms and Willingness to Take Action Related to Nuclear Energy. Risk Anal 30 (9):1363–1373

Deutsch M (1975) Equity, equality, and need: what determines which value will be used as the bases of distributive justice? J Soc Issues 31:137–149

Deutsch M (1985) Distributive Justice. Yale University Press, New Haven

Douglas M, Wildavsky A (1982) Risk and Culture: An Essay on Selection of Technological and Environmental Dangers. California University Press, Berkeley

Drottz-Sjöberg BM, Sjöberg L (1991) Adolescents' attitudes to nuclear power and radioactive wastes. J Appl Soc Psychol 21:2007–2036

Earle TC, Cvetkovich GT (1995) Social Trust: Toward a Cosmopolitan Society. Praeger Publishers, Westport
Fischhoff B, Slovic P, Lichtenstein S, Read S, Combs B (1978) How safe is safe enough? A psychometric study of attitudes towards technological risks and benefits. Policy Sci 9:127–152
Fishbein M, Ajzen I (1975) Belief, Attitude, Intention and Behavior: An Introduction to Theory and Research. Addison-Wesley, Reading
Flynn J, Burns W, Mertz CK, Slovic P (1992) Trust as determinant of opposition to a high-level radioactive waste repository: analysis of a structural model. Risk Anal 12:417–429
Flynn J, Slovic P, Mertz CK (1993) Decidedly different: expert and public views of risks from a radioactive waste repository. Risk Anal 13:643–648
Gardner GT, Stern PC (2002) Environmental Problems and Human Behavior. Pearson, Boston
Gervers JH (1987) The NIMBY syndrome: is it inevitable? Environment 29:18–29
Gowda MVR, Easterling D (2000) Voluntary siting and equity: the MRS facility experience in native America. Risk Anal 20:917–929
Gowda MVR, Owsley-Long P (1998) The relevance of social science to societal technology choices. In: Neuby BL (ed) Relevancy of the Social Sciences in the New Millennium. State University of West Georgia, Carrollton, pp 104–113
Hisschemöller M, Midden CJH (1989) Technological risk, policy theories and public perception in connection with the siting of hazardous facilities. In: Vlek C, Cvetkovich G (eds) Social Decision Methodology for Technological Projects. Kluwer Academic Press, Dordrecht, pp 173–194
Hohenemser C, Renn O (1988) Chernobyl's other legacy. Shifting public perceptions of nuclear risk. Environment 30:4–45
Huijts NMA, Midden CJH, Meijnders AL (2007) Social acceptance of carbon dioxide storage. Energ Policy 35:2780–2789
Kasperson RE, Golding D, Tuler S (1992) Social distrust as a factor in siting hazardous facilities and communicating risks. J Soc Issues 48:161–87
Katsuya T (2002) Difference in the formation of attitude toward nuclear power. Polit Psychol 23:191–203
Krannich RS, Little RL, Cramer LA (1993) Rural community residents' views of nuclear waste repository siting in Nevada. In: Dunlap RE, Kraft ME, Rosa EA (eds) Public Reactions to Nuclear Waste: Citizens' Views of Repository Siting. Duke University Press, Durham, pp 263–287
Lind EA, Tyler TR (1988) The Social Psychology of Procedural Justice. Plenum Press, New York
MacGregor D, Slovic P, Mason RG, Detweiler J, Binney SE, Dodd B (1994) Perceived risks of radioactive waste transport through Oregon: results of a statewide survey. Risk Anal 14:5–14
Meijnders AL, Midden CJH, Wilke HAM (2001) Communications about environmental risks and risk-reducing behavior: the impact of fear on information processing. J Appl Soc Psychol 31:754–777
Montada L, Kals E (1995) Perceived justice of ecological policy and proenvironmental commitments. Soc Just Res 8:305–327
Mushkatel AH, Nigg JM, Pijawka KD (1993) Nevada urban residents' attitudes toward a nuclear waste repository. In: Dunlap RE, Kraft ME, Rosa EA (eds) Public Reactions to Nuclear Waste: Citizens' Views of Repository Siting. Duke University Press, Durham, pp 239–262
Opotow S, Clayton S (1994) Green justice: conceptions of fairness and the natural world. J Soc Issues 50:1–11
Peters E, Slovic P (1996) The role of affect and worldviews as orienting dispositions in the perception and acceptance of nuclear power. J Appl Soc Psychol 26:1427–1453
Peters E, Burraston B, Mertz CK (2004) An emotion-based model of risk perception and stigma susceptibility: cognitive appraisals of emotion, affective reactivity, worldviews, and risk perceptions in the generation of technological stigma. Risk Anal 24:1349–1367

Pijawka KD, Mushkatel AH (1991/1992) Public opposition to the siting of the high-level nuclear waste repository: the importance of trust. Policy Stud Rev 10:180–194
Purvis-Roberts KL, Werner CA, Frank I (2007) Perceived risks from radiation and nuclear testing near Semipalatinsk, Kazakhstan: a comparison between physicians, scientists, and the public. Risk Anal 27:291–302
Rawls J (1999) A Theory of Justice. Belknap Press of Harvard University Press, Cambridge
Rogers R (1983) Cognitive and physiological processes in fear appeals and attitude change: a revised theory of protection motivation. In: Cacioppo J, Petty R (eds) Social Psychology: A Sourcebook. Guildford Press, New York, pp 153–176
Rosema M, Jost JT, Stapel DA (2008) Social psychology and the study of politics. In: Steg L, Buunk AP, Rothengatter T (eds) Applied Social Psychology. Understanding and Managing Social Problems. Cambridge University Press, Cambridge, pp 291–315
Schuitema G (2010) Priceless Policies. Factors Influencing the Acceptability of Transport Pricing Policies. University of Groningen, Groningen, the Netherlands
Schwartz SH (1992) Universals in the content and structure of values: theoretical advances and empirical tests in 20 countries. In: Zanna M (ed) Advances in Experimental Social Psychology. Academic, Orlando, pp 1–65
Schwartz SH (1994) Are there universal aspects in the structure and contents of human values? J Soc Issues 50:19–45
Schwartz SH, Howard JA (1981) A normative decision-making model of altruism. In: Rushton JP, Sorrentino RM (eds) Altruism and Helping Behavior. Lawrence Erlbaum, Hillsdale, pp 89–211
Sherif M, Hovland CI (1961) Social Judgment: Assimilation and Contrast Effects in Communication and Attitude Change. Yale University Press, New Haven
Short JF, Rosa EA (2004) Some principles for siting controversy decisions: lessons from the US experience with high level nuclear waste. J Risk Res 7:135–152
Shrader-Frechette K (2000) Duties to future generations, proxy concent, intra- and intergenerational equity: the case of nuclear waste. Risk Anal 20:771–777
Siegrist M, Keller C, Cousin ME (2006) Implicit attitudes toward nuclear power and mobile phone base stations: support for the affect heuristic. Risk Anal 26:1021–1029
Sjöberg L (1997) Explaining risk perception: an empirical and quantitative evaluation of cultural theory. Risk Decis Policy 2:113–130
Sjöberg L (1998) Risk perception: experts and the public. Eur Psychol 3:1–12
Sjöberg L (2001) Limits of knowledge and the limited importance of trust. Risk Anal 21:189–198
Sjöberg L (2003) Attitudes and risk perception of stakeholders in a nuclear waste siting issue. Risk Anal 23:739–749
Sjöberg L (2004) Local acceptance of a high-level nuclear waste repository. Risk Anal 24:737–749
Sjöberg L, Drottz-Sjöberg BM (2001) Fairness, risk, and risk tolerance in the siting of a nuclear waste repository. J Risk Res 4:75–101
Slovic P (1987) Perception of risk. Science 236:280–286
Slovic P (1993) Perceived risk, trust, and democracy. Risk Anal 13:675–682
Slovic P, Flynn JH, Layman M (1991a) Perceived risk, trust, and the politics of nuclear waste. Science 254:1603–1607
Slovic P, Layman M, Kraus N, Flynn J, Chalmers J, Gesell G (1991b) Perceived risk, stigma, and potential economic impacts of a high-level nuclear waste repository in Nevada. Risk Anal 11:683–696
Slovic P, Layman M, Flynn JH (1991c) Lessons from Yucca Mountain. Environment 33:7–30
Slovic P, Finucane ML, Peters E, MacGregor DG (2004) Risk as analysis and risk as feelings: some thoughts about affect, reason, risk, and rationality. Risk Anal 24:311–322
Slovic P, Finucane ML, Peters E, MacGregor DG (2007) The affect heuristic. Eur J Oper Res 177:1333–1352
Steg L, Dreijerink L, Abrahamse W (2005) Factors influencing the acceptability of energy policies: testing VBN theory. J Environ Psychol 25:415–425

Stern PC (2000) Toward a coherent theory of environmentally significant behavior. J Soc Issues 56:407–424
Stern PC, Dietz T, Kalof L (1993) Value orientations, gender, and environmental concern. Environ Behav 25:322–348
Stern PC, Dietz T, Kalof L, Guagnano GA (1995) Values, beliefs, and proenvironmental action: attitude formation toward emergent attitude objects. J Appl Soc Psychol 25:1611–1636
Summers C, Hine DW (1997) Nuclear waste goes on the road: risk perceptions and compensatory tradeoffs in single-industry communities. Can J Behav Sci 29:210–222
Thøgersen J (1994) Monetary incentives and environmental concern: effects of a differentiated garbage fee. J Consum Policy 17:1–36
Thøgersen J (1996) Recycling and morality: a critical review of the literature. Environ Behav 28(4):536–558
Tokushige K, Akimoto K, Tomoda T (2007) Public perceptions on the acceptance of geological storage of carbon dioxide and information influencing the acceptance. In: Gale J, Bolland O (eds) 8th International Conference on Greenhouse Gas Control Technologies – GHGT-8. Int J Greenhouse Gas Control 1(1):101–112
Tyler TR (2000) Social justice: outcome and procedure. Int J Psychol 35:117–125
van Alphen K, van Voorst tot Voorst Q, Hekkert MP, Smits REHM (2007) Societal acceptance of carbon capture and storage technologies. Energ Policy 35:4368–4380
van der Pligt J (1989) Nuclear waste: public perception and siting policy. In: Vlek C, Cvetkovich G (eds) Social Decision Methodology for Technological Projects. Kluwer Academic Press, Dordrecht, pp 235–252
van der Pligt J, Midden CJH (1990) Chernobyl: four years later: attitudes, risk management and communication. J Environ Psychol 10:91–99
Verplanken B (1989) Beliefs, attitudes, and intentions toward nuclear energy before and after Chernobyl in a longitudinal within-subjects design. Environ Behav 21:371–392
Vlek C, Stallen PJ (1981) Judging risks and benefits in the small and in the large. Organ Behav Hum Perf 28:235–271
Wiegman O, Gutteling JM, Cadet B (1995) Perceptions of nuclear energy and coal in France and the Netherlands. Risk Anal 15:513–521

Part II
Regional Assessments

Comparative Assessment of Status and Opportunities for Carbon Dioxide Capture and Storage and Radioactive Waste Disposal in North America

Curtis M. Oldenburg and Jens T. Birkholzer

Abstract Aside from the target storage regions being underground, geologic carbon sequestration (GCS) and radioactive waste disposal (RWD) share little in common in North America. The large volume of carbon dioxide (CO_2) needed to be sequestered along with its relatively benign health effects present a sharp contrast to the limited volumes and hazardous nature of high-level radioactive waste (RW). There is well documented capacity in North America for 100 years or more of sequestration of CO_2 from coal-fired power plants. Aside from economics, the challenges of GCS include lack of fully established legal and regulatory framework for ownership of injected CO_2, the need for an expanded pipeline infrastructure, and public acceptance of the technology. As for RW, the USA had proposed the unsaturated tuffs of Yucca Mountain, Nevada, as the region's first high-level RWD site before removing it from consideration in early 2009. The Canadian RW programme is currently evolving with options that range from geologic disposal to both decentralized and centralized permanent storage in surface facilities. Both the USA and Canada have established legal and regulatory frameworks for RWD. The most challenging technical issue for RWD is the need to predict repository performance on extremely long timescales (10^4–10^6 years). While attitudes toward nuclear power are rapidly changing as fossil fuel costs soar and changes in climate occur, public perception remains the most serious challenge to opening RW repositories. Because of the many significant differences between RWD and GCS, there is little that can be shared between them from regulatory, legal, transportation or economic perspectives. As for public perception, there is currently an opportunity to engage the public on the benefits and risks of both GCS and RWD as they learn more about the urgent energy–climate crisis created by greenhouse gas emissions from current fossil fuel combustion practices.

C.M. Oldenburg (✉) and J.T. Birkholzer
Earth Sciences Division 90-1116, Lawrence Berkeley National Laboratory, Berkeley, CA, 94720, USA
e-mail: cmoldenburg@lbl.gov; jtbirkholzer@lbl.gov

F.L. Toth (ed.), *Geological Disposal of Carbon Dioxide and Radioactive Waste: A Comparative Assessment*, Advances in Global Change Research 44,
DOI 10.1007/978-90-481-8712-6_13, © Lawrence Berkeley National Laboratory, 2011

Keywords Carbon dioxide capture and storage • Geologic carbon sequestration • Geologic CO_2 storage • Geosequestration • Carbon sequestration • Radioactive waste disposal • Radioactive waste repository

1 Introduction

Accelerating emissions of carbon dioxide (CO_2) from fossil fuel combustion (Raupach et al. 2007) and associated threats to global climate are motivating an urgent search for low-carbon energy sources. With renewable energy sources such as hydroelectric, solar and wind projected to supply less than 10% of the world's energy needs by 2030 (EIA 2007), two low-carbon sources of electricity, namely nuclear (fission) power and coal with pre- or post-combustion carbon dioxide capture and storage (CCS), have been proposed as key components of a multifaceted approach to meeting the global energy–climate challenge (e.g. Pacala and Socolow 2004). While producing a majority of electricity from nuclear fission and from coal with CCS in the future will drastically reduce atmospheric CO_2 emissions relative to today, nuclear power produces radioactive waste (RW) that must be isolated from the environment, and the CO_2 from coal combustion must be sequestered. For nuclear power, the waste stream is radioactive, highly toxic, and, depending on the nuclear fuel cycle (i.e. after reprocessing), may present a security risk owing to its capacity for use in producing nuclear weapons (IAEA 2005). In the case of electricity production using coal, the concern about CO_2 is its role as the main greenhouse gas responsible for climate change.

Research into radioactive waste disposal (RWD) has been going on for decades in North America, and geologic disposal of RWs is planned in many countries worldwide (see e.g. Witherspoon and Bodvarsson 2006). Of the countries often considered geographically as part of North America, only the USA and Canada have substantial amounts of RW. We therefore restrict our comparisons in this chapter to US and Canadian RWD activities. In North America, as well as elsewhere, the volumes of high-level RW expected to accumulate over the foreseeable future are typically small enough to allow for geologic disposal in engineered underground facilities, usually excavated horizontal galleries and emplacement tunnels. One centralized repository for the USA and one for Canada are likely sufficient to store the expected volumes of domestic high-level RW over the next several decades. A geologic repository for transuranic RWs has been operating in the USA since 1999. The so-called WIPP (Waste Isolation Pilot Plant) facility, located approximately 42 km (26 mi) east of Carlsbad, New Mexico, stores wastes about 600–700 m underground within a 1,000 m (3,000 ft) thick salt formation.

It has only been in the last 10–20 years that geologic carbon sequestration (GCS), the geologic storage part of CCS, has been considered seriously as an emissions reduction strategy, but because of CO_2's natural abundance and utility in enhancing hydrocarbon recovery, CO_2 injection has been carried out continuously in the USA for enhanced oil recovery (EOR) for approximately 35 years (Bondor 1992). The experience and knowledge gained from CO_2 EOR has provided a solid foundation for GCS. In the last 10 years or so, the pace of investigations into GCS has grown rapidly. For example, studies of capacity (e.g. Bradshaw et al. 2007), cost (e.g. Friedmann et al. 2006), effectiveness (e.g. Hepple and Benson 2005), potential impacts (Oldenburg 2007), regulatory and legal aspects (Wilson et al. 2003) and pilot projects (Litynski et al. 2006), among many others, have been published in the last few years.

The purpose of this chapter is to compare and contrast GCS and RWD, and evaluate the opportunities and discuss the challenges of implementation in North America, specifically the USA and Canada. Space does not allow us to thoroughly review the science, history and current activity of RWD, and we refer interested readers to the high-level RW worldwide review(s) (e.g. Witherspoon and Bodvarsson 2006) for a comprehensive summary of RWD efforts around the world. Similarly, for GCS the Intergovernmental Panel on Climate Change (IPCC) Special Report on Carbon Dioxide Capture and Storage (IPCC 2005) provides a comprehensive explanation of the technical basis for GCS that is beyond the scope of this chapter. With the ultimate purpose of the collected chapters in this volume aimed at comparing options for electricity generation, we focus on sequestration of CO_2 from coal-fired power plants (the largest stationary point sources of CO_2 in North America) rather than on industrial sources (e.g. refineries and cement plants). As to the recent removal of Yucca Mountain from consideration as a RWD site as of March 2009, the concepts for storage and related challenges, especially as they relate to similarities and differences with GCS, will likely be shared with any future repository site.

2 Geologic Carbon Sequestration in North America: Status, Opportunities, and Challenges

2.1 Current Status of Geologic Carbon Sequestration

Activities involving the injection of large volumes of CO_2 into deep geologic formations in North America are associated with EOR (see e.g. Bondor 1992), mostly in West Texas, but also in New Mexico, Oklahoma, Kansas, Arkansas, Louisiana, Mississippi, Wyoming, Colorado, Utah, Montana, Alaska, Pennsylvania, and the Canadian provinces of Alberta and Saskatchewan. The majority of the US CO_2 EOR operations utilize CO_2 produced from natural CO_2 reservoirs, such as those at St Johns-Springerville (Arizona), Bravo Dome (New Mexico), McElmo and Sheep Mountain Domes (Colorado), and Jackson Dome (Mississippi), which are connected by pipelines to the oil fields. Other sources include gas processing plants (e.g. LaBarge in Wyoming) and an ammonia plant in Oklahoma (Moritis 2008). Early implementations of CO_2 EOR started in the mid-1970s before there was widespread interest in reducing CO_2 emissions. Notwithstanding the past and ongoing sourcing of CO_2 for EOR from natural CO_2 reservoirs, co-optimization of CO_2 EOR and GCS are areas of current research interest (e.g. Kovscek and Cakici 2005) anticipating a future in which anthropogenic CO_2 is abundantly available and carbon credits are awarded for CO_2 retained in the formation (sequestered). The most well known CO_2 EOR projects that use anthropogenic CO_2 are being carried out in Canada (in Saskatchewan) by EnCana at the Weyburn oilfield and by Apache at the nearby Midale oilfield (White et al. 2004). The source of CO_2 for these projects is a coal gasification plant in Beulah, North Dakota, from which CO_2 is sent by a 330 km-(200 mi-) long pipeline to southern Saskatchewan. In these projects, around one half of injected CO_2 is produced with oil and recycled, while the other half is sequestered in the reservoir. In Alberta, PennWest has been using CO_2 from a nearby petrochemical plant since 1984 for CO_2 EOR in its Joffre field. Another significant activity in North America involves injecting hydrogen sulphide and CO_2 that are stripped from produced natural gas (mostly methane) and injected for disposal. Although the injected gas stream is often over 95% CO_2 (Bachu and Gunter 2004), the primary purpose is to dispose of hydrogen sulphide, hence the name acid gas injection.

The other major class of GCS activity in North America involves pilot CO_2 injection projects. The first was the Frio Brine Pilot in Dayton, Texas, that began in 2002 with a small 1,700 t injection into a brine formation in a mature oilfield and finished recently following a second phase of injection (Hovorka et al. 2006). Shortly after the first Frio Pilot injection, the US Department of Energy (DOE) launched its Regional Carbon Sequestration Partnership (RCSP) Program (Litynski et al. 2006). In this programme, seven RCSP groups representing different regions of the USA and Canada embarked in 2003 on a long-term effort to: (1) characterize their individual regions in terms of opportunities for GCS; (2) validate the opportunities; and (3) deploy GCS to demonstrate the feasibility of storing hundreds of years' worth of industrial-scale CO_2 emissions. All seven of the RCSPs have advanced to Phase III to develop sites for the deployment of industrial-scale pilot projects involving the injection of at least 10^6 t CO_2.

2.2 *Opportunities for Geologic Carbon Sequestration*

Opportunities for large-scale GCS in North America are abundant and well documented. The primary reference for North American GCS opportunity is the NATCARB database, an ongoing development of the US DOE. To allow real-time queries on sources of CO_2, pipeline transport, potential sequestration sites (sinks), and more, NATCARB can be accessed online at http://drysdale.kgs.ku.edu/natcarb/eps/natcarb_alpha_content.cfm. Focusing here on electricity generation, which represents approximately 86% of CO_2 point-source emissions in North America (NATCARB 2006), we present in Fig. 1 the locations of large power plant sources of CO_2 in North America as compiled by NATCARB in its 2006 database. The sources shown emit an estimated 3.3 gigatonnes of CO_2 per year (Gt CO_2/year) (NATCARB 2006). The locations of CO_2 sources from power generation in the USA reflect the distribution of population, with most sources concentrated in the more densely populated and industrial eastern part of the country.

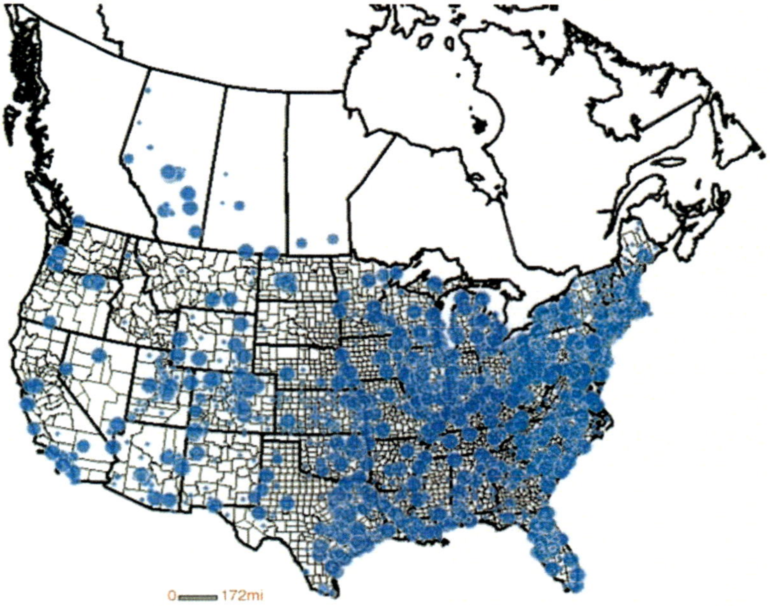

Fig. 1 North American CO_2 sources from electricity generation (Source: NATCARB: http://drysdale.kgs.ku.edu/natcarb/eps/natcarb_alpha_content.cfm) (*see* Colour Plates)

Shown in Fig. 2 is a map from the NATCARB online database showing deep saline formations (blue) and oil- and gasfields (red) that provide North America with enormous potential sequestration capacity for CO_2. For brevity, we have omitted from the maps unmineable coal, organic-rich shale, and basalt formations that, while potentially significant, are thought to be less important to the North American region than saline formations and oil- and gasfields. Comparing Figs. 1 and 2, we observe that many suitable sedimentary basins with saline formations and oil and gas resources underlie the large stationary sources of CO_2, particularly in the industrial Ohio Valley, Midwest and Texas Gulf Coast areas, potentially minimizing the need for long pipelines for CO_2 transport. Because of early industrialization in the USA, oil- and gasfields are mostly mature, meaning production over the last 50–100 years has left them depleted or nearing depletion. These mature reservoirs offer the advantages of demonstrated sealing against upward migration of buoyant fluid, potential to inject CO_2 to make up for net extraction of oil and gas, detailed knowledge of the local subsurface from decades of oil and gas production, and a history of land use similar to that associated with GCS.

GCS capacity for North America has been studied by US and Canadian researchers with data compiled by NATCARB. While not all potential sinks have been evaluated, Table 1 shows the documented capacity as determined by NATCARB using a consistent capacity estimation methodology for the various sequestration targets. Assuming total CO_2 emissions from fossil fuel power plants in North America

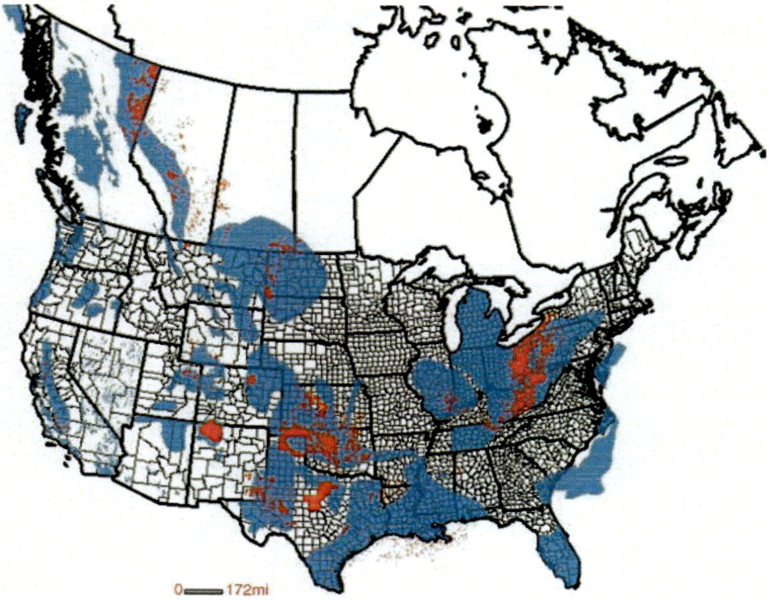

Fig. 2 North American deep saline reservoirs (*blue*) and oil and gas reservoirs (*red*) potentially available for GCS (Source: NATCARB: http://drysdale.kgs.ku.edu/natcarb/eps/natcarb_alpha_content.cfm) (*see* Colour Plates)

Table 1 Capacity estimates for North America as compiled by NATCARB

Geological carbon sequestration sink	Gt CO_2	
	Low	High
Oil and gas reservoirs	82	–
Unmineable coal seams	156	183
Saline formations	920	3400

amount to 4 Gt CO_2/year, and using the low estimates of capacity, the saline formations would provide capacity for over 200 years of CO_2 injection, depleted oil and gas reservoirs for 20 years, and unmineable coal seams for 40 years. Putting aside the uncertainty in emissions growth, capacity estimates (e.g. Bradshaw et al. 2007), practical and economic barriers to capturing CO_2 from all power plants, CO_2 transportation considerations, etc., research to date strongly suggests that very significant amounts of CO_2 can be sequestered safely and effectively in the deep subsurface in North America.

In addition, GCS represents a business opportunity currently being pursued by a variety of companies and sectors. For example, oil and gas companies are viewing depleted reservoirs and associated brine formations as potential revenue generating sinks for CO_2. Pipeline companies see the potential need for pipeline infrastructure to transport CO_2 from power plant sources to GCS sites. And oil services companies see a large opportunity for characterization and monitoring of field operations of GCS. These nascent industrial efforts are anticipating widespread implementation of GCS as a revenue generating enterprise.

2.3 Challenges Associated with Geologic Carbon Sequestration

Although opportunities of GCS are many in North America, so are the challenges to implementation. The basic barrier delaying industrial GCS in the region and worldwide is economic; simply put, CCS does not pay at present. Without either a cap and trade policy or carbon tax in place in North America, the economic incentive for CCS is missing. The capture part of CCS in particular is expensive and technologically challenging, especially for existing coal-fired power plants. Nevertheless, forward-looking scientific, government, industrial and power utility leaders are looking beyond today's economics to develop the science, technology, policy and regulations that will be needed to implement large-scale CCS. Assuming there was today an economic advantage or policy imperative for CCS, implementation of CCS in North America faces secondary challenges such as the lack of: (1) transportation infrastructure; (2) established GCS injection regulations; and (3) applicable laws regarding ownership and liability of CO_2 in the subsurface. In addition, there are challenges for GCS surrounding public perception and environmental justice that need to be addressed. Finally, there are technical challenges being addressed by scientists and experts, but these do not appear to present insurmountable barriers to implementation. Below we elaborate on each of these challenges.

As shown by comparison of Figs. 1 and 2, many of the region's CO_2 sources are located on top of or in close proximity to large saline formation sinks for GCS. However, the scale of the maps belies the details of the extent of need for new networks of pipelines to transport CO_2 from power plants to permitted CO_2 injection wells. In addition, there are large areas of the region with power generation sources of CO_2 without local onshore GCS capacity, e.g. the southern Atlantic coast area. Simply put, it will be challenging to establish new pipeline transport corridors through areas of existing and growing populations where there is an ever-increasing concern for the local environment – the so-called NIMBY (Not In My Back Yard) syndrome. Because CO_2 transportation represents another business opportunity for GCS, and pipeline infrastructure is well established in North America for hydrocarbon transport, utilities such as power, water and sewer, as well as fibre-optic data transmission purposes, building such an infrastructure is not expected to be an insurmountable hurdle.

At the time of this writing, the regulatory environment for GCS is rapidly evolving in North America. The existing framework for deep underground injection in the USA is covered by the Environmental Protection Agency's (EPA) Underground Injection Control (UIC) Program, designed to protect underground sources of drinking water (i.e. potable water, defined as water with less than

10,000 mg/L total dissolved solids) (e.g. USEPA 2001; Benson et al. 2002). This programme is currently being extended in the USA to cover deep injection of CO_2 through the addition of Class VI injection well (a CO_2 injection well) to augment the existing Class I-V wells. Over half of US states retain primary enforcement responsibility, or primacy, and enforce UIC regulations on their own using regulations that are equivalent or more restrictive than EPA's, while seven states jointly enforce UIC regulations with EPA (e.g. California), and the rest (e.g. Arizona) rely on EPA alone for regulation. In Canada, fluid injection is entirely a matter of provincial rather than federal jurisdiction. The Canadian province of Alberta has a regulatory framework potentially applicable for GCS through its CO_2 EOR and acid gas injection programmes (Bachu 2008). The Interstate Oil and Gas Compact Commission recently published guidelines intended as a reference for states and provinces in their development of GCS regulations (IOGCC 2007), but this does not carry the same force of law as the EPA Class VI proposed regulation. Layered on top of the ongoing and rapid developments in establishing regulations for widespread implementation of GCS is the prospect of future changes in the definition of potable water as changes in climate, water demand and desalination technology promote use of waters that are considered today non-potable.

If the regulatory framework for GCS can be said to be developing, a GCS-specific legal framework must be said to be non-existent. In short, mineral rights and surface property rights are kept distinct, a system that has worked in practice for the awarding of royalties for mineral and hydrocarbon extraction for over 100 years. However, when it comes to storing fluids in the pore space of rock below the surface owner's property, there are few laws in place (Wilson et al. 2007). The state of Wyoming decided by statute in 2008 to place pore space ownership in the hands of the surface owner. While it seems likely that other jurisdictions will follow this same model, other statutes and laws are lacking at the time of this writing. Details and analysis of liability issues associated with GCS and RWD are presented in Wilson and Bergan (2011). The working hypothesis of leaders in the field is that the long sequestration time frames (several hundred to thousands of years) will necessitate the eventual assumption of liability by a governmental entity that can persist longer than any corporation has been known to exist. However, this desire to hand off the long-term liability for the injected CO_2 is creating suspicion among sceptics that GCS is risky or unsafe and thereby causing a public perception problem.

Public perception of GCS is a growing challenge, but also an opportunity. In North America, the problem of climate change and its relation to greenhouse gas emissions is fairly well known thanks to Al Gore's popular movie, An Inconvenient Truth, and widespread and growing media coverage of climate change and future emission projections involving the developing world, primarily China and India. When it comes to tackling the energy–climate problem, renewable energy sources, particularly solar and wind, are familiar to people, while the inclusion of CCS as part of a portfolio of approaches is much less well known. In our personal experience talking about GCS in public meetings and university classes, there is great scepticism of any solution promoted by the same entities (power utilities, coal companies, oil companies, governments) perceived as having caused the greenhouse gas emission problem to start with. However, because many people are aware of the problem of climate change and so few people have heard of GCS, there is the opportunity to carefully present the case for GCS to an audience receptive to technologies that will address climate change. For a comparative analysis of public perception issues see Reiner and Nuttall (2011).

Summaries of GCS physical and chemical processes, technical challenges and research areas being addressed by scientists worldwide can be found in the literature (e.g. IPCC 2005; Wilson et al. 2007; Bachu and McEwen 2011). None of these technical challenges prevents GCS from being initiated today on a single-source scale as the existing worldwide industrial projects (e.g. Sleipner and Snovhit in Norway, In Salah in Algeria, Weyburn-Midale in Canada) demonstrate. However, it must be mentioned that while capacity in North America is demonstrably large, so too are national point-source CO_2 emissions. Large-scale implementation of GCS involving multiple power plant sources with injection into the same saline formations can lead to widespread pressure rise (Nicot 2008), which raises the risk of induced seismicity and brine displacement into shallower groundwater resources (Birkholzer et al. 2009) that are highly valued especially in the arid parts of North America. The brine displacement hazard arises mainly in areas of North America with depleted hydrocarbon reservoirs where there are many abandoned wells that are considered the most likely conduits for leakage of CO_2 out of GCS sites (Gasda et al. 2004; Nordbotten et al. 2004) and which may also serve as brine migration conduits, depending on the conditions of the well. The US EPA Class VI draft rule on CO_2 injection addresses the hazard of abandoned wells by requiring their identification and plugging prior to CO_2 injection. It is interesting that many of the same wells that cause concern for leakage (typically the wells that are very old and also very deep) have been the sources of information for indicating the presence of high-quality seals against upward migration, and favourable injectivity and capacity in the storage interval. Although the identified wells of concern will be properly plugged prior to injection, monitoring of these wells may still be needed along with pressure management to reduce pressure resulting from large-scale injection.

3 Radioactive Waste Disposal in North America: Status, Opportunities, and Challenges

In North America, both the USA and Canada have RWD programmes that have been working for decades on providing a permanent and reliable method of isolating high-level RW from the biosphere. Because of public acceptance concerns, Canada has been in a phase of re-evaluating storage options that range from geologic disposal in mined underground facilities to both decentralized and centralized permanent storage in surface facilities. Similarly in the USA, the Yucca Mountain site in Nevada had been designated and studied for decades as a geologic disposal site, and, as a major step forward, the licence application for this site was submitted to the regulatory authorities on 3 June 2008. However, on March 5, 2009, the incoming Obama administration stated in a Senate hearing that the Yucca Mountain site was no longer considered an option for storing nuclear spent fuel and other radioactive waste. On March 3, 2010, DOE filed a motion with the regulatory authorities to withdraw its licence application for Yucca Mountain, in contradiction to the Nuclear Waste Policy Act. Since then, multiple lawsuits to stop this action have been filed by states, counties, and individuals across the country, many of these still ongoing at the time of finalizing this paper. Meanwhile, DOE initiated a long-term R&D program with focus on advanced fuel cycle solutions and alternative repository options. With this state of uncertainty, we focus below on the United States geologic disposal program as it existed prior to 2009, while elaborating in lesser detail on the Canadian programme.

3.1 Current Status of Radioactive Waste Disposal in the USA

3.1.1 Selection of Yucca Mountain

In 1982 the US Congress passed the Nuclear Waste Policy Act (NWPA), a federal law that established the US policy for the permanent disposal of high-level RW. While the site screening and selection process, including various alternative sites, was still ongoing, and before a clear technical ranking had been established (e.g. in volcanic rocks, in basalts, and in bedded and dome salt sites), Congress amended the act in 1987, directing the US DOE to study only Yucca Mountain, Nevada, as the permanent geologic repository. Although the Yucca Mountain site has many technical advantages over other sites, the congressional decision for Yucca Mountain has since been criticized by some as political and arbitrary. The state of Nevada in particular viewed the decision as singularly unfair, made at the expense of a state with less political clout than other states (Carter 2006; Macfarlane 2006).

Since then, Yucca Mountain has been characterized and evaluated in numerous scientific studies to determine its suitability. The site characterization phase ended in 2002, when the Secretary of Energy recommended the Yucca Mountain site as suitable for further development. Following this event, DOE began the process of preparing a licence application for authority to construct a geologic repository at Yucca Mountain, which was submitted to the regulating authority on 3 June 2008. The following subsections introduce the Yucca Mountain site, discuss the barriers critical to geologic RW isolation and provide some details on the regulatory standards. Much of this discussion is based on the summary description given in Arthur and Voegele (2006).

3.1.2 Yucca Mountain Site Description

The proposed Yucca Mountain site is located on federal land in southern Nevada, approximately 90 miles north-west of Las Vegas. The location is remote, far away from population centres. Yucca Mountain is one of a series of north-south trending ridges, consisting of successive layers of volcanic tuffs, millions of years old. The alternating layers of welded and non-welded volcanic tuffs have differing hydrogeologic properties that significantly influence the manner in which downward percolating water moves through the mountain. The proposed repository horizon would be in the welded tuff, which is highly fractured and thus relatively permeable. However, the climate is arid, infiltration into and percolation through the mountain is very small, and the water table at Yucca Mountain is deep, about 600 m below the mountain crest. The repository horizon is located in this thick unsaturated zone, more than 200 m above the water table, such that the repository tunnels would remain relatively dry, accessible by ramps from outside the mountain, and amenable to monitoring and inspection for centuries (Carter 2006). This specific repository setting, i.e. waste emplacement in a thick unsaturated zone with small rates of water movement but rather permeable rocks, is fairly unique worldwide. Other proposed disposal concepts typically involve repositories in low permeability rocks such as sparsely fractured granite, claystone and/or salt situated below the groundwater table in saturated conditions. While Yucca Mountain offers major advantages over sites beneath the water table for reasons listed above, it was also suggested that the number, complexity, and interaction of relevant processes makes prediction of repository behaviour more difficult and possibly more uncertain (Macfarlane 2006).

Figure 3 shows a schematic of the proposed repository design. An existing U-shaped tunnel, named the Exploratory Studies Facility (ESF), allows access into the mountain from two entry points, the North Portal and the South Portal. The ESF was built in the mid-nineties and has been used as an underground rock laboratory, where processes and properties have been studied in multiple in situ experiments (see e.g. Bodvarsson et al. 1999). Access to the repository would be made possible via an additional ramp as well as via ventilation shafts. The waste, contained in cylindrical canisters made from corrosion resistant material, was to be emplaced into circular drifts of about 5 m diameter. More than 11,000 waste packages may be stored in more than 40 miles of emplacement drifts. Titanium drip shields were planned to protect packages from dripping water and rock fall. Waste canisters would be transported to the site and into the repository primarily by railroad.

The operational period at Yucca Mountain was anticipated to last for decades, during which waste packages would have been received at the site, transported underground, and emplaced. Regulations for Yucca Mountain required that the wastes be retrievable from the repository beginning at any time up to 50 years after emplacement had begun and before final closure. An ongoing performance confirmation programme would ensure that further site characterization activities will be conducted in the to-be-drilled emplacement drifts and that monitoring of the drift and near-field environment will take place, with the objective of creating confidence in performance predictions. Because of the extensive engineered barrier systems for the RWs, it was considered highly unlikely that radionuclides would be released from the repository during or soon after closure, as discussed in DOE's licence application for authorization to construct a repository at Yucca Mountain. Thus, monitoring was expected to confirm the predicted system behaviour with respect to various heat-related coupled processes, but would not be able to provide insight into the barrier capability of the natural system to prevent or delay radionuclide transport.

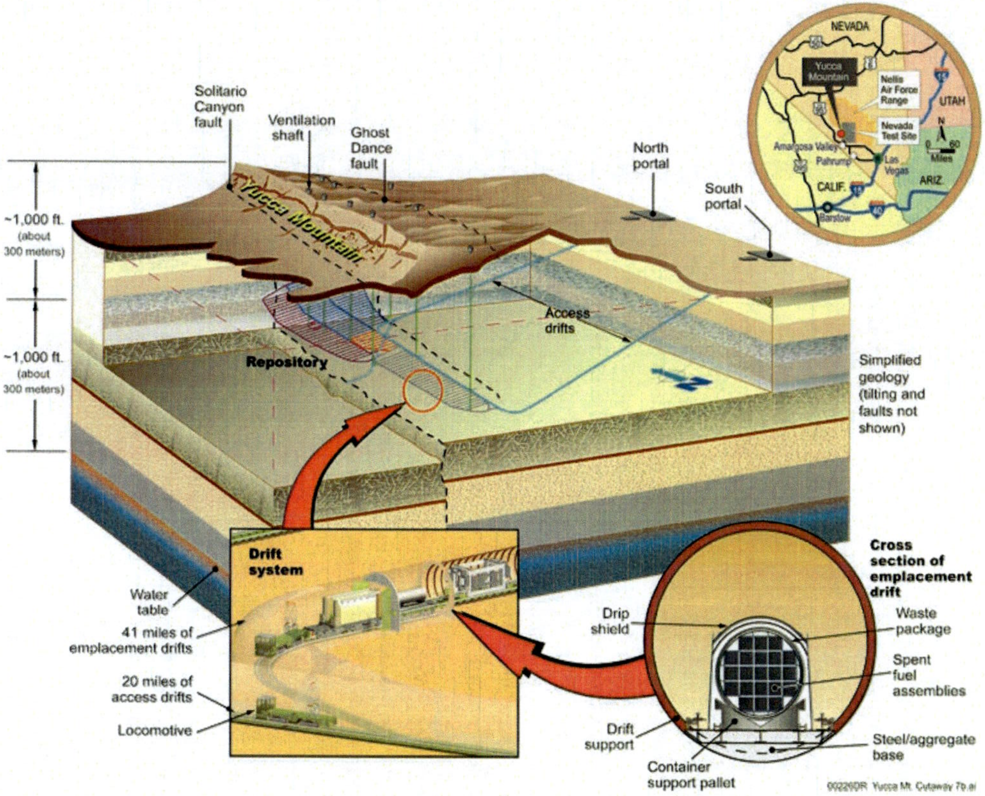

Fig. 3 Location of Yucca Mountain and underground development design (Source: Arthur and Voegele 2006)

3.1.3 Long-Term Safety and Performance Criteria

The long-term safety of a repository for RW depends on the performance of natural and engineered barriers, which together prevent or delay the transport of radionuclides to where the public could eventually be exposed. Three major barriers were relied upon at Yucca Mountain: the upper natural barrier, the engineered barrier system and the lower natural barrier. The characteristics of these barriers, and their overall barrier capabilities, are described in detail in DOE's licence application, and shall only be briefly reviewed below.

The upper natural barrier, comprising the topography, surficial soils, bedrock, as well as the unsaturated zone above the repository horizon, would prevent or limit water from entering the emplacement drifts. According to DOE's licence application, important factors contributing to barrier capability would be the small net infiltration into Yucca Mountain, on the order of a few millimetres per year, and the specific hydrogeological conditions in the volcanic tuffs above the repository that function to divert the infiltrating water, dampen episodic pulses of infiltration, and limit seepage of water into the emplacement drifts. Future changes in climate may increase net infiltration at the site and were therefore considered in the performance assessment. However, the assumed climatic changes reflect natural cycles, not the possible impacts from global warming.

The engineered barrier system was planned to have two functions: (1) to prevent or limit water contact with the RW; and (2) to limit the release of radionuclides from the waste into the lower natural barrier. The first function would be achieved by corrosion resistant drip shields and waste packages; both of which prevent seepage water from contacting the waste as long as they remain intact. Over long periods of time, corrosion (generalized corrosion or stress corrosion resulting from mechanical damage) is expected to deteriorate some drip shields and may also create local breaches in waste packages. It is important to note that the unsaturated zone at Yucca Mountain is a naturally oxidizing environment in which metals can corrode if they become wet or damp, as from high humidity or water seepage (Carter 2006). If breaches occur, the release of radionuclides from waste packages will be limited by the slow rate of waste-form degradation, by sorption on iron corrosion products within the waste package, as well as by sorption onto the granular material on the floor of the drifts.

Radionuclides released from the emplacement drifts would enter the lower natural barrier, which comprises the unsaturated zone and the groundwater zone (saturated zone). Radionuclides would then migrate downward with the flow in the unsaturated zone to the water table and will then be transported by groundwater flow towards the accessible environment. Both the unsaturated zone and the saturated zone would contribute to barrier capability, delaying the migration of radionculides with slow advective flow. In addition, several processes would cause the movement of radionuclides to be retarded compared to the rate of movement of the water. These processes include diffusion of radionculides from the advective fracture flow into the matrix pores, sorption onto mineral surfaces, and colloid filtration.

Radiation protection standards for a Yucca Mountain repository have been developed by the US EPA, initially issued in 2001. These standards address all potential pathways of radiation exposure and limit an individual's annual exposure to a maximum value of 15 mrem (0.15 millisievert). For comparison: the radiation exposure for an average person is about 350 mrem/year, most of which comes from natural sources of radiation (i.e. cosmic radiation and terrestrial sources such from soils and rocks), the remainder from exposure to artificial radiation sources, such as medical X-rays. The time period initially defined to evaluate radiation exposure was 10,000 years after closure. However, in 2001, a federal court ruled that a 10,000-year performance period is not sufficient. EPA has since then proposed a rule that defines performance standards for a significantly longer period, up to 1 million years after closure, which includes the expected time of peak dose and is within a defined period of geologic stability at Yucca Mountain. Because predictive uncertainties increase as the time period for which the predictions are made increases, the maximum annual dose for the time period after 10,000 years is 350 mrem.

The US Nuclear Regulatory Commission (NRC) is the independent federal entity assigned with the responsibility of regulating the nation's civilian use of radioactive materials, which includes regulating their safe disposal in a geologic repository. As mentioned above, DOE had submitted the licence application for Yucca Mountain to the NRC in summer 2008, only to withdraw it shortly after as the political circumstances changed. NRC was then in the midst of conducting a detailed review in a process that was expected to last a minimum of 3 years. The review included consideration of conformity to regulations based on the proposed EPA standard. Performance of the repository was to be measured based on the expected dose received by the so-called reasonably maximally exposed individual (RMEI), a person assumed to live 18 km downstream of the Yucca Mountain site drinking 2 L/day from one or more wells. The regulations comprised specific criteria beyond evaluating maximum radiation exposure, such as the requirement of multiple barriers acting in concert and the requirement to incorporate all significant aspects of uncertainty and variability in a probabilistic assessment of the repository's performance. As guidance to NRC review staff, specific review acceptance criteria were provided in the Yucca Mountain Review Plan (NRC 2003).

3.2 *Current Status of Radioactive Waste Disposal in Canada*

3.2.1 Evolving Programme

Canada is one of the countries where public reactions to RW isolation plans have recently led to a change in the programme approach. For decades, Atomic Energy of Canada Limited (AECL) had been developing the concept for the emplacement of nuclear fuel wastes in a geologic repository excavated in the plutonic crystalline rock of the Canadian Shield. An underground research laboratory was established in 1989 near Lac du Bonnet, Manitoba, in a large, previously undisturbed, nearly fracture-free granitic pluton, allowing for a comprehensive research programme for developing characterization methods, developing in situ stress measurement techniques, demonstrating full-scale canister emplacement, evaluating rock stability, modelling groundwater flow and transport, and conducting grouting and tunnel-sealing experiments (Russell et al. 2001). In 1994, AECL submitted the Environmental Impact Statement on the concept of a repository excavated in the plutonic rock of the Canadian Shield to a federal Environmental Assessment Panel (EAP) for review. Public hearing associated with the review took place during 1996 and 1997. In 1998, the Federal Government completed its review of the concept and found it to be technically safe, and in compliance with regulatory requirements. However, the review also concluded that there was not sufficient public support at that time to implement a repository siting programme (Witherspoon and Bodvarsson 2006).

The Government of Canada responded to the recommendations of the EAP and issued the Nuclear Fuel Waste Act (NFWA), which in 2002 initiated the establishment of a new agency, the Nuclear Waste Management Organization (NWMO), to study different disposal options and develop collaboratively with Canadians an approach for long-term management of used RW that is technically sound, socially acceptable, environmentally responsible and economically feasible (Russell and Facella 2006). Three options were then studied in a comparative evaluation of strengths and limitations, as detailed below.

3.2.2 Three Options for Long-Term Management

The first option would involve geologic disposal in an engineered repository in the stable crystalline rock of the Canadian Shield, a concept similar to the one initially developed by AECL. The underground facility would be constructed at a depth of 500–1,000 m (CTECH 2002), consisting of a network of horizontal access galleries and emplacement rooms or boreholes. Used fuel would be placed into long-lived containers, such as steel-lined copper containers, and surrounded by clay-based sealing materials. Approximately 11,000 containers of used fuel would be emplaced and filled-up rooms would be backfilled with bentonite and then progressively sealed (McMurry et al. 2003). Possible designs include in-room placement, in-floor borehole placement and horizontal borehole placement. Initial performance assessments demonstrate that the Canadian Shield has favourable geologic and hydrologic features for waste isolation.

The second option involves permanent or indefinite storage at the nuclear reactor sites. Existing interim dry storage facilities would have to be expanded or new long-term dry storage facilities would have to be established. The key disadvantage as expressed by the Canadian authorities is the need for continuing control and operation, including the necessary funding, for the thousands of years that the used RW remains hazardous (Russell and Facella 2006). Compared to geologic repositories, surface facilities are also more readily accessible to malevolent intervention. The third option would involve storage of all wastes in a new long-term dry storage facility at one site in Canada. Designs of a new central facility have been prepared for both above-ground and shallow below-ground storage (in rock caverns excavated to a depth of 50 m in competent rock). Similar to the second option above, the key disadvantage is the need for long-term control and operation of such a facility.

3.2.3 A Fourth Option and Path Forward

In a period of dialogue with the public, many Canadians suggested that an additional option should be considered, an option that would attempt to capitalize on the advantages of the other three approaches. This led NWMO to develop the Adaptive Phased Management (APM) approach, and to launch a dialogue with Canadians about its appropriateness. Overall, the majority of those engaged in these discussions considered APM to be a reasonable approach for Canada, and the final report that was submitted by NWMO to the Government in November 2005 recommended this path forward (Russell and Facella 2006). The key attributes of the approach are:

- Centralized containment and isolation of waste in a geologic repository within a suitable rock formation, either in the Canadian Shield or in deep sediments such as the Ordovician Shale;
- Flexibility in the pace and manner of implementation through a phased decision making process, supported by an R&D programme;
- Provision of an optional step in the implementation process in the form of shallow underground storage at a central site, prior to final placement in a geologic repository;
- Continuous monitoring of the spent fuel to support confirmation of the safety and performance of the storage facility or repository;
- Potential for retrievability of the waste for an extended time period, until a future society makes a determination on the final closure and the appropriate form and duration of post-closure monitoring.

The staged approach in Canada would start with an approximately 30-year phase that focuses on: (1) technology development for fuel management; and (2) selection of a site that has rock formations suitable for shallow underground storage, an underground characterization facility, and a geologic repository at greater depth. The second phase, which would take an additional 30 years, would concentrate on central storage implementation. Only the third and last phase, which would start after the two 30-year phases, would involve the implementation of long-term containment in a geologic repository (Russell and Facella 2006).

3.3 Opportunities for Radioactive Waste Disposal

Both the USA and Canada have established paths forward to providing permanent and reliable methods of isolating RW in a geologic repository. The legal and regulatory framework for RWD has been defined, state-owned organizations have been commissioned that develop and pursue disposal plans, and regulating authorities have been identified. Canada pursues a staged implementation approach that may include temporary storage in shallow underground storage prior to long-term containment and isolation in a geologic repository. While no specific site has been selected, Canada has various options with several areas hosting rock formations deemed suitable for RWD, such as the crystalline rock of the Canadian Shield.

In the USA, the Yucca Mountain site in Nevada had been designated for high-level RWD, and the licence application for this site was submitted in 2008. While there was some controversy among scientists about the scientific basis for the licence application and related uncertainties (e.g. Long and Ewing 2004; Macfarlane and Ewing 2006), DOE maintained until early 2009 that the plan for disposing of RW at Yucca Mountain was technically sound and that the repository safely meets the proposed performance standards set by the EPA. If the DOE had continued with the Yucca Mountain site and regulators had shared the DOE's assessment of safe disposal at Yucca Mountain during the 3- to 4-year review process and finally granted construction authorization, high-level RW could have been emplaced at the site within the next 10–20 years. With a change in political leadership, however, DOE withdrew the licence application for Yucca Mountain in the Spring of 2010 and started an R&D program to develop alternative disposal options. At the time of finalizing this paper, the USA has no proposed site for high-level RW.

A solution to the RW problem becomes particularly important in light of the worldwide renaissance of nuclear power. In the USA, the further expansion of nuclear power is seen as a promising avenue to meet the substantial demand growth for energy, while addressing national energy security and greenhouse gas emission concerns. This change in attitude towards nuclear power, also evident in many countries worldwide, is fuelled by the rising costs of fossil fuels and the increased attention to the environmental threats associated with burning fossil fuels.

While economic challenges are a primary barrier for industrial-scale GCS, the implementation of a national RWD facility is typically not governed by market forces and thus dependent instead on political will and/or a legal framework to guarantee sufficient federal funding. In the USA, for example, the NWPA requires utilities which generate electricity using nuclear power to pay a fee of one tenth of one cent (US$0.001) per kWh into the so-called Nuclear Waste Fund. As of 31 December 2008, payments and interest credited to the fund totalled US$29.6 billion, out of which approximately US$9.5 billion have been spent so far by the RWD programme in the USA. In other words, RWD programmes are typically more affected by political/social challenges than by economics. While the RWD programme in the USA has received substantial financial support over the past decades, it has also experienced significant funding fluctuations, depending on the annual budget negotiations between the president and the US Congress.

While capacity constraints and sink/source matching are important issues for GCS, the amounts of RW expected to accumulate over the foreseeable future are typically small enough to allow for storage in one national underground storage facility. In other words, once a suitable rock formation has been identified and characterized, a large enough system of tunnels and galleries can usually be built to accommodate the expected waste volumes. In the USA, the situation was a bit different, as the capacity of Yucca Mountain was limited – by law – to 70,000 t of high-level RW. While this is more than the current waste volume in the USA (mostly spent fuel from power plants, but also defence wastes), this capacity would not have accommodated the future quantities to be produced from the currently 103 US commercial reactors (Peterson 2003). The USA is engaged in advanced fuel cycle initiatives, with the goal of substantially accelerating efforts to develop new reactor and reprocessing technology. The goals of the programme are: (1) to guarantee that nuclear materials from the fuel cycle are protected from proliferation and misuse for non-peaceful purposes; (2) to recover the substantial energy value in used nuclear fuel to make sure that sufficient fuel remains available for centuries in the face of depleting fossil fuels; and (3) to significantly reduce the burden related to the geologic disposal of nuclear fuel in terms of waste volume, heat load and radiotoxicity, thereby avoiding capacity problems.

3.4 *Challenges for Radioactive Waste Disposal*

Public acceptance is arguably the most serious challenge to RWD. High-level RW is known to be extremely dangerous, and some members of the public are very sceptical about the usefulness and reliability of long-term performance predictions. Some are also concerned about the irretrievability of the RW once it is emplaced and the repository has been closed off. If radionuclides were to escape because of unexpected early failure, so goes the argument, there is no simple way of mitigating the consequences. While in many countries the reaction of the public to the development of RW isolation has resulted in site selection delays, re-evaluations of disposal approaches, and the development of new ways of engaging voluntarily local communities (Witherspoon and Bodvarsson 2006), the RWD programme in the USA and the site selection of Yucca Mountain has, over the years, been relatively unaffected by public acceptance problems, at least on a national level. The State of Nevada, however, has been strongly opposed to the Yucca Mountain repository, and raised technical issues and legal and political roadblocks. Indeed, recent developments demonstrate that political challenges can impact nuclear waste programmes even as far advanced as those in the USA. The incoming Obama administration questioned the suitability of Yucca Mountain, and in January 2010 installed a blue-ribbon commission to devise a new strategy toward fuel cycle options and RW disposal. The expected new strategy will include: (1) developing alternatives to Yucca Mountain as the nation's permanent repository; and (2) starting aggressive R&D programmes for reprocessing of spent fuel. All the above would require legislative action to revise the NFWA.

In Canada, the RWD programme changed its organization and repository approach completely in response to the insufficient public support for implementing a siting programme, despite the fact that the geologic storage concept was found to be technically safe. In both countries, further acceptance problems may be expected from the public perception of risks related to transporting RWs (e.g. by rail) from various interim storage locations to a permanent central repository site.

A technical challenge most difficult to address is the necessity to consider the evolution of a very complex geologic environment over extremely long timescales ranging anywhere from 10^4 to 10^6 years. As discussed in length in Macfarlane and Ewing (2006) for the Yucca Mountain site, there are many factors that make it difficult to predict repository behaviour over geologic times, including climate, fractured rock flow and transport, saturated and unsaturated zone behaviour, volcanism, seismicity, thermal processes and geochemical interactions, all of which contribute to significant uncertainty in natural and engineered barriers. The question of the timescale over which a quantitative performance assessment should be undertaken has been widely debated in the RW community (Maul et al. 2007), and regulations differ from country to country. Prediction uncertainty increases with time, and thus less reliance should be placed on calculation far into the future. As mentioned earlier, the question about the appropriate performance period for Yucca Mountain was settled eventually by a court decision, ruling for a performance period of up to 1 million years. This decision is in line with a recent review by the NEA (2004) emphasizing the need for such long timescales because: (1) good sites with well performing barriers imply release of contaminants only very far into the future; and (2) ethical considerations would expect the same level of environmental protection in the far future as in the short term. The same review points out that even the most stable materials and geologic environments, over long enough timescales, are subject to perturbing events and long-term changes, which makes quantitative predictions more and more uncertain. Several suggestions are made how a performance assessment should take into account the changes and uncertainties associated with long timescales (NEA 2004).

4 Comparison of Geologic Carbon Sequestration to Radioactive Waste Disposal in North America

4.1 General Characteristics

We present in Table 2 a comparison of GCS and RWD processes with emphasis on their general characteristics in North America. Although some of the same fundamental physical and chemical processes apply to both GCS and RWD, and methods of characterizing the subsurface and predicting these fundamental processes with simulation models are often similar (e.g. DOE 2007), GCS and RWD have little in common with respect to their general characteristics in North America, as shown in the table. For a comparative analysis of the physical differences between GCS and RWD see Bachu and McEwen (2011). Of the many profound differences between GCS and RWD, one difference that stands out is that the amounts of materials (high-level RW in one case, CO_2 in the other) are orders of magnitude different. The total mass of high-level RW in the USA that was intended for disposal at Yucca Mountain is restricted by law to 70,000 t. In contrast, the annual production of CO_2 from a single 1,000 MW coal-fired power plant is approximately 9 million tonnes, and the total North American power generation production of CO_2 is over 3 Gt per year (NATCARB 2006). The much larger mass of CO_2 produced (over 40,000 times more CO_2 than RW) presents a great challenge compared to RW.

Table 2 Comparison between geologic carbon sequestration and radioactive waste disposal in North America[a]

Characteristic	Geological carbon sequestration	Radioactive waste disposal
Target geologic formations	Sedimentary basins with brine formations, sometimes also containing depleted hydrocarbon reservoirs known to have trapped oil and gas over geologic time	Unsaturated volcanic tuffs (US), crystalline rock or clay (Canada)
Volume/mass	Very large volume/mass from power generation; large coal-fired power plants will require multiple injection wells	Volumes of high-level waste small enough that storage in one national underground facility is generally possible
Transportation	Pipeline as liquid CO_2, injection through wells	Waste in containers likely to be transported by railroad
Form	Liquid or supercritical CO_2 at injection; supercritical in storage formation	Solid waste typically encapsulated in corrosion resistant containers; only container failure allows for waste form dissolution and radionuclide migration
Trapping or storage mechanism	Stratigraphic, residual phase, dissolution, mineral trapping	Multiple barrier concept, with engineered (waste form, container, bentonite) and natural barriers (low permeability rock)
Timescale for isolation	Hundreds to thousands of years	Up to 1 million years
Possible migration mechanisms	Buoyant upward flow of CO_2; displacement of brine; seepage out of the ground	Radionuclide transport in groundwater; dispersal by future volcanic eruption
Retrievability	Injection and observation wells can be used as production wells to bring CO_2 and other fluids out of the formation if desired	USA and Canada require that waste can be retrieved during the first decades to centuries after emplacement

[a] See Bachu and McEwen (2011) for comparison of geologic carbon sequestration and radioactive waste disposal in general

Another significant difference between GCS and RWD is the timescale required for containment. Regulations in the USA for RW originally required performance demonstration for at least 10,000 years, but that was extended up to 1 million years to make them consistent with recommendations by the National Academy of Sciences (National Research Council 1995). Clearly, the possibility of major yet infrequent disruptive events (such as volcanism, earthquakes) and the effects of long-term changes in the hydrology or geologic environment need to be accounted for over such timescales. For CO_2 storage, the necessary time period of containment can be related to the purpose of reducing atmospheric emissions, namely the mitigation of climate change effects. Timescales of a few centuries to a few thousand years are likely sufficient for sequestered CO_2 to remain out of the atmosphere globally to address climate change over the next few hundred years while fossil fuel resources remain abundant, although to date, no timescale has been established by regulation for CO_2 retention in GCS systems in North America. Differences between GCS and RWD can also be seen in the existence of an acceptable global leakage rate for CO_2 calculated at between 0.01% and 0.1% per year of sequestered CO_2 (Hepple and Benson 2005). For RWD, an acceptable rate of radionuclide migration is not defined a priori; it is typically constrained indirectly by regulatory requirements limiting radiation exposure to individuals. Although long by human standards, the timescale for isolation for GCS is short relative to geologic time and permits a large degree of confidence in GCS site performance prediction.

RWD and GCS differ with respect to the time period when failure of containment is most likely to occur. Radionuclide releases from the waste canisters into the subsurface environment are most likely only very long after repository closure, when the engineered barriers may have (partially) degraded, and performance may then depend mostly on the barrier capabilities of the natural system. In contrast, escape of CO_2 from deep storage is most likely in early operational project stages, when injection-induced pressures are high, thus providing an additional leakage driving force, while trapping mechanisms such as solubility trapping or mineral trapping have not yet fully developed.

Careful site characterization is required in the evaluation of the expected performance of both RWD and GCS sites (Maul et al. 2007). Both GCS and RWD require understanding of the basic geologic system and system behaviour on a large scale (i.e. kilometre scale), and use similar characterization methods to achieve such understanding (e.g. borehole data, geophysics). However, the level of effort expected to characterize an RW repository is arguably not necessary in CO_2 storage projects, for which the timescale of containment is shorter and limited leakage is tolerable. The level of effort needed for an RWD site would also not be economically feasible for GCS given the large number of sites required to accommodate the enormous volumes of CO_2 that need to be sequestered to mitigate climate change. RWD requires a detailed understanding of the near field (i.e. the immediate surroundings of the repository tunnels) and how its properties may change with tunnel excavation and waste emplacement. Such understanding has been achieved in the USA by characterization studies and in situ testing carried out in the ESF at Yucca Mountain, Nevada. For GCS sites, the existence of mitigation options, the relatively benign nature of CO_2, and the ability to monitor injection operations provide the opportunity to continue to characterize a GCS site and monitor its performance throughout the process of injection (see e.g. Doughty et al. 2008).

While GCS and RWD share little in common, the contrast in two aspects, namely: (1) the relatively small amount of RW compared to the large amount of CO_2 that needs to be isolated; and (2) the health hazard (RW is highly toxic, CO_2 is relatively benign) stand out as the most significant differences between the processes from the perspectives of health, safety and environmental risk, as well as economic and practical feasibility. These two aspects – volume and consequences of storage failure – are at the root of the challenges facing North America, as discussed in the following sections.

4.2 Shared Challenges and Opportunities

Both GCS and RWD are widely believed to be technically feasible in North America. A summary of the main points of comparison between GCS and RWD is presented in Table 3, and elaborated upon here. GCS and RWD require available lands and suitable subsurface geologic characteristics to contain CO_2 and RW, respectively, for long periods of time. The large area and varied subsurface geology and physical geography of North America, including areas of sparse population, provide excellent potential opportunities for both GCS and RWD over the next 100 years or more. Both GCS and RWD will require transportation infrastructure. In the case of RWD, rail and road transport infrastructure that could be used for transport exists in many parts of North America, but its use for the purpose of conveying RW is expected to be challenged by some in the communities through which it passes.

Table 3 Comparison of challenges/opportunities for geologic carbon sequestration and radioactive waste disposal in North America

Challenge/opportunity	Geological carbon sequestration	Radioactive waste disposal
Availability of sites to contain required volume	Good opportunities for the next 100 years or more	Volume is sufficiently small that capacity is not an issue
Transportation from source to site	Pipeline infrastructure needs to be built	Transport by rail and road is technically feasible but will be subject to protest and security concerns in practice
Public perception	Large opportunity to educate the public on the benefits and risks of geologic carbon sequestration	Negative but potentially evolving as impacts of climate change become more well known
Suburbanization/land use changes	Large volumes of CO_2 that need to be injected may end up underneath the lands owned by neighbours	Not an issue for much smaller-volume and government-controlled radioactive waste disposal sites
Legal and liability issues	Uncertain and evolving	Well established that government will take long-term responsibility
Evolving drinking water standards	The classification of potable water may change to disallow injections into what are considered today non-potable water resources	Not an issue as the nation's water resources are not affected by the possible contamination of groundwater near one repository

This could be a challenge in the USA because the proposed repository is located in the western USA (Nevada), while the bulk of nuclear spent fuel is produced in the eastern part of the country. For GCS, additional pipeline infrastructure for transporting CO_2 from point sources throughout the country to GCS sites needs to be built. GCS has the challenge of very large volumes of CO_2 required to be injected to be effective against climate change.

As for legal, policy, and economic considerations, GCS and RWD differ considerably. First, implementation of GCS currently faces legal complications because the USA lacks laws regarding ownership of pore space and its contents. No such ambiguity exists with RW, for which government ownership and control are a given. By current law, RWD has the challenge of long timescales for isolation (10^4–10^6 years), making predictions of geologic and hydrologic stability and isolation of waste difficult. Even the somewhat shorter sequestration timescales relevant to GCS create complications for assigning long-term liability, and suggest the need for eventual government ownership and control, the policy and funding for which has yet to be established. On the economic side, GCS is widely viewed as a business opportunity once a carbon trading market is developed or a carbon tax is imposed. The oil industry in particular is positioned well for this opportunity with its experience, resources, and capital to transport, inject and monitor fluids in the subsurface. RWD is by and large a government run and operated enterprise from which no significant new industries are expected to arise.

RWD and GCS share the technical challenge of characterizing and interpreting geologic systems using relatively sparse data, making long-term predictions about flow and transport processes in the underground over long time periods, and providing quantitative analysis of the future performance of a site in a regulatory and legal environment. Emplacement of hot RW as well as injection of large volumes of CO_2 will both perturb the natural systems and induce complex hydrologic, mechanical, geochemical and thermal processes, the similarities and differences of which are laid out in Bachu and McEwen (2011). These technical issues make it inevitable that performance predictions have uncertainties, some of which may be critically important for evaluating whether a site can be suitable, others may be of little consequence. Of the many possible RWD sites worldwide, the Yucca Mountain site in Nevada may have involved some of the most daunting technical challenges, because of its unique setting in an unsaturated fractured tuff and the strong heat-induced flow perturbations expected from emplacement of waste. Such technical challenges, while they may have been met by sound science or may not even be relevant for performance, can make a safety case for a site very complex and hard to convey to regulators and the public. A lesson learned from RWD for GCS may thus be to choose sites where performance can be demonstrated without having to rely on features and processes that are very difficult to quantify. As an example: a GCS site with a proven seal for trapping of CO_2 or other gases would likely be viewed very favourably.

Public perception is both a challenge and an opportunity for GCS and RWD. The challenge comes from the public's legitimate concerns about safety and environmental impacts of both technologies. The status of RWD in North America may offer a valuable lesson for the less mature GCS development, in that the screening and selection of any geologic repository site needs to be conducted in an environment of open communication with all stakeholders, focusing on sound technical standards and, ideally, by comparing possible alternatives. The early phase of the RWD programme in the USA is a negative example for such a process, where the 1987 congressional decision to stop the ongoing screening of several alternative sites in favour of the Yucca Mountain site was viewed by many as political and arbitrary (Carter 2006). The opportunity comes from the sea-change in perception that we anticipate will accompany the growing concern about greenhouse gas impacts on climate. Once the public and elected officials understand the magnitude of climate-related impacts predicted to occur due to burning fossil fuels as currently carried out, i.e. with emission of CO_2 directly into the atmosphere, there may be a large shift in thinking about benefits and risks associated with transportation and storage of RW and CO_2 underground.

5 Conclusions About Geologic Carbon Sequestration and Radioactive Waste Disposal in North America

GCS and RWD in North America share little in common. Vast differences exist in the volumes of material that need to be stored, the means of transportation and emplacement, the geologic environments suitable for the two processes, the depths of emplacement, and the consequences of leakage, among others. These differences mean the legal and regulatory frameworks in place for RWD are not entirely appropriate for GCS. The general need to predict future performance of geologic systems is one commonality, and already there is crossover as researchers in the USA who were formerly focused on RWD are now applying their expertise in site characterization and performance assessment to GCS. The main technical challenge for GCS is to learn how large-scale CO_2 injection involving many wells will perturb and impact the hydrologic, geochemical and geomechanical systems that provide secure storage while avoiding significant impacts (e.g. on groundwater quality) arising from CO_2 or brine migration. Given the urgent energy–climate challenge, the best way to achieve this understanding is to begin GCS as soon as possible and to 'learn while doing' as early projects are implemented with strong and thorough monitoring and verification programmes. As for RWD, the licence application submitted in 2008 makes the case for the technical feasibility of Yucca Mountain for safely storing RW; the subsequent removal of Yucca Mountain from consideration does not refute this case. As for public perception, we see an opportunity to engage the public on both GCS and RWD as the public becomes better informed about the urgent need to reduce anthropogenic greenhouse gas emissions to minimize climate change.

Because of the severe potential consequences of anthropogenic climate change to global environments and economies, the increased use of nuclear fission and deployment of CCS for fossil fuel-derived power are promising options for North

America. Both approaches present challenges but also opportunities. To the extent that some of these challenges may not be overcome, or that unforeseen events or circumstances may impair one or the other technology, we suggest that policies be enacted to pursue both approaches, with processes in place to ensure health, safety and minimal environmental impacts, to help reduce North American CO_2 emissions as soon as possible.

Acknowledgements We thank our LBNL colleagues Larry R. Myer, Karsten Pruess and Patrick F. Dobson, along with three external anonymous reviewers, for critical reviews and constructive comments which have allowed us to make significant improvements in the manuscript. This manuscript has been authored by a contractor of the US Government under Contract No. DE-AC02-05CH11231 with the US Department of Energy. The views and opinions of authors expressed in this article do not necessarily state or reflect those of the United States Government or any agency thereof or The Regents of the University of California.

References

Arthur WJ, Voegele MD (2006) Preparing to submit a license application for Yucca Mountain in the United States. In: Witherspoon PA, Bodvarsson GS (eds) Geological Challenges in Radioactive Waste Isolation: Fourth Worldwide Review. Technical Report LBNL-59808. Lawrence Berkeley National Laboratory, Berkeley, pp 257–266

Bachu S (2008) Legal and regulatory challenges in the implementation of CO_2 geological storage: an Alberta and Canadian perspective. Int J Greenhouse Gas Control 2(2):259–273

Bachu S, Gunter WD (2004) Acid-gas injection in the Alberta Basin, Canada: a CO_2-storage experience. In: Baines SJ, Worden RH (eds) Geological Storage of Carbon Dioxide, vol 233, Geological Society Special Publication. Geological Society of London, London, pp 225–234

Bachu S, McEwen T (2011) Geological media and factors for the long-term emplacement and isolation of carbon dioxide and radioactive waste. In this volume

Benson SM, Hepple R, Apps J, Tsang C-F, Lippmann M (2002) Lessons learned from natural and industrial analogues for storage of carbon dioxide in deep geological formations. Report LBNL-51170. Lawrence Berkeley National Laboratory, Berkeley

Birkholzer JT, Zhou Q, Tsang C-F (2009) Large-scale impact of CO_2 storage in deep saline aquifers: a sensitivity study on pressure response in stratified systems. Int J Greenhouse Gas Control 3(2):181–194

Bodvarsson GS, Boyle W, Patterson R, Williams D (1999) Overview of scientific investigations at Yucca Mountain – the potential repository for high-level nuclear waste. J Contam Hydrol 38(1–3):3–24

Bondor PL (1992) Applications of carbon dioxide in enhanced oil recovery. Energy Convers Manage 33(5–8):579–586

Bradshaw J, Bachu S, Bonijoly D, Burruss R, Holloway S, Christensen NP, Mathiassen OM (2007) CO_2 storage capacity estimation: issues and development of standards. In: Gale J, Bolland O (eds) 8th International Conference on Greenhouse Gas Control Technologies – GHGT-8: Int J Greenhouse Gas Control 1(1): 62–68

Carter LJ (2006) The path to Yucca Mountain and beyond. In: Macfarlane A, Ewing RC (eds) Uncertainty Underground: Yucca Mountain and the Nation's High-Level Nuclear Waste. MIT Press, Cambridge, pp 381–392

CTECH (2002) Conceptual design for a deep geologic repository for used nuclear fuel. Report 1106/MD18085/REP/01. CTECH, Toronto

DOE (Department of Energy) (2007) Basic research needs in geosciences: facilitating 21st century energy systems. http://www.sc.doe.gov/bes/reports/abstracts.html#GEO. Accessed 08 December 2010

Doughty CA, Freifeld BM, Trautz RC (2008) Site characterization for CO_2 geologic storage and vice versa: the Frio brine pilot, Texas, USA as a case study. Environ Geol 54(8):1635–1656

EIA (Energy Information Agency) (2007) International Energy Outlook 2007. http://www.eia.doe.gov/oiaf/ieo/index.html. Accessed 20 June 2008

Friedmann SJ, Dooley JJ, Held H, Edenhofer O (2006) The low cost of geological assessment for underground CO_2 storage: policy and economic implications. Energy Convers Manage 47:1894–1901

Gasda SE, Bachu S, Celia MA (2004) Spatial characterization of the location of potentially leaky wells penetrating a deep saline aquifer in a mature sedimentary basin. Environ Geol 46:707–720

Hepple RP, Benson SM (2005) Geologic storage of carbon dioxide as a climate change mitigation strategy: performance requirements and the implications of surface seepage. Environ Geol 47(4):576–585

Hovorka SD, Benson SM, Doughty C, Freifeld BM, Sakurai S, Daley TM, Kharaka YK, Holtz MH, Trautz RC, Nance HS, Myer LR, Knauss KG (2006) Measuring permanence of CO_2 storage in saline formations: the Frio experiment. Environ Geosci 13(2):105–121

IAEA (International Atomic Energy Agency) (2005) Non-proliferation of nuclear weapons & nuclear security. https://www.iaea.org/Publications/Booklets/nuke.pdf. Accessed 08 December 2010

IOGCC (Interstate Oil and Gas Compact Commission) (2007) CO_2 storage: a legal and regulatory guide for states. http://iogcc.myshopify.com/collections/frontpage/products/co2-storage-a-legal-and-regulatory-guide-for-states-2008. Accessed 08 December 2010

IPCC (Intergovernmental Panel on Climate Change) (2005) IPCC Special Report on Carbon Dioxide Capture and Storage. Metz B, Davidson OR, de Coninck HC, Loos M, Meyer LA (eds). Cambridge University Press, New York/Combridge, UK http://www.ipcc.ch/publications_and_data/publications_and_data_reports.shtml#2. Accessed 08 December 2010

Kovscek AR, Cakici MD (2005) Geologic storage of carbon dioxide and enhance oil recovery II. Co-optimization of storage and recovery. Energy Convers Manage 46:1941–1956

Litynski JT, Klara SM, McIlvried HG, Srivastava RD (2006) The United States Department of Energy's regional carbon sequestration partnerships program: a collaborative approach to carbon management. Environ Int 32(1):128–144

Long JCS, Ewing RC (2004) Yucca Mountain: earth-science issues at a geologic repository for high-level nuclear waste. Annu Rev Earth Planet Sci 32:363–401

Macfarlane A (2006) Uncertainty, models, and the way forward in nuclear waste disposal. In: Macfarlane A, Ewing RC (eds) Uncertainty Underground: Yucca Mountain and the Nation's High-Level Nuclear Waste. MIT Press, Cambridge, pp 393–408

Macfarlane A, Ewing RC (eds) (2006) Uncertainty Underground: Yucca Mountain and the Nation's High-Level Nuclear Waste. MIT Press, Cambridge

Maul PR, Metcalfe R, Pearce J, Savage D, West JM (2007) Performance assessments for the geological storage of carbon dioxide: learning from the radioactive waste disposal experience. Int J Greenhouse Gas Control 1(4):444–455

McMurry J, Dixon D, Garroni J, Ikeda B, Stroes-Gascoyne S, Baumgartner P, Melnyk T (2003) Evolution of a Canadian deep geologic repository: base scenario. Nuclear Waste Management Division Report 06819-REP-01200-10127-R00. Ontario Power Generation, Toronto

Moritis G (2008) Special Report: More US EOR Projects Start but EOR Production Continues Decline. Oil Gas J 106(15):41–46

NATCARB (2006) Carbon sequestration atlas of the United States and Canada. National Energy Technology Laboratory, Office of Fossil Energy, US Department of Energy. http://www.natcarb.org/Atlas/ims_map.html. Accessed 08 July 2008

National Research Council (1995) Technical Basis for Yucca Mountain Standards. The National Academies Press, Washington, DC

NEA (Nuclear Energy Agency) (2004) The handling of timescales in assessing post-closure safety. Report NEA No. 4435. OECD Nuclear Energy Agency, Paris

Nicot J-P (2008) Evaluation of large-scale CO_2 storage on fresh-water sections of aquifers: an example from the Texas Gulf Coast Basin. Int J Greenhouse Gas Control 2(4):582–593

Nordbotten JM, Celia MA, Bachu S (2004) Analytical solutions for leakage rates through abandoned wells. Water Resour Res 40:4204–4213

NRC (Nuclear Regulatory Commission) (2003) Yucca Mountain review plan. Report NUREG-1804. US Nuclear Regulatory Commission, Washington, DC

Oldenburg CM (2007) Migration mechanisms and potential impacts of CO_2 leakage and seepage. In: Wilson EJ, Gerard D (eds) Carbon Capture and Sequestration: Integrating Technology, Monitoring, and Regulation. Blackwell, Ames, pp 127–146

Pacala S, Socolow R (2004) Stabilization wedges: solving the climate problem for the next 50 years with current technologies. Science 305(5686):968–972

Peterson PF (2003) Will the United States need a second repository? Bridge 33(3):26–32

Raupach MR, Marland G, Ciais P, Le Quéré C, Canadell JG, Klepper G, Field CB (2007) Global and regional drivers of accelerating CO_2 emissions. Proc Natl Acad Sci 104(24):10288–10293

Reiner DM, Nuttall WJ (2011) Public acceptance of geological disposal of carbon dioxide and radioactive waste: similarities and differences. In this volume

Russell S, Facella J (2006) Progress towards long-term management of used nuclear fuel in Canada. In: Witherspoon PA, Bodvarsson GS (eds) Geological Challenges in Radioactive Waste Isolation: Third Worldwide Review. Technical Report LBNL-59808. Lawrence Berkeley National Laboratory, Berkeley, pp 39–54

Russell S, Gierszewski PJ, Jensen MR, Kempe TF (2001) Long-term management of nuclear waste fuel in Canada: Technical developments in the concept for a deep geologic repository. In: Witherspoon PA, Bodvarsson GS (eds) Geological Challenges in Radioactive Waste Isolation: Fourth Worldwide Review. Technical Report LBNL-49767. Lawrence Berkeley National Laboratory, Berkeley, pp 67–76

USEPA (United States Environmental Protection Agency) (2001) Technical program overview: underground injection control regulations. Office of Water 4606, EPA 816-R-02-025. http://www.epa.gov/safewater/uic/index.html. Accessed 08 December 2010

White DJ, Burrowes G, Davis T, Hajnal Z, Hirsche K, Hutcheon I, Majer E, Rostron B, Whittaker S (2004) Greenhouse gas sequestration in abandoned oil reservoirs: the International Energy Agency Weyburn pilot project. GSA Today 14(7):4–10

Wilson EJ, Bergan S (2011) Managing liability: comparing radioactive waste disposal and carbon dioxide storage. In this volume

Wilson EJ, Johnson TL, Keith DW (2003) Regulating the ultimate sink: managing the risks of geologic CO_2 storage. Environ Sci Technol 37(16):3476–3483

Wilson EJ, Friedman SJ, Pollak MF (2007) Research for deployment: incorporating risk, regulation, and liability for carbon capture and sequestration. Environ Sci Technol 41(17):5945–5952

Witherspoon PA, Bodvarsson GS (2006) Introduction to geological challenges in radioactive waste isolation: fourth worldwide review. In: Witherspoon PA, Bodvarsson GS (eds) Geological Challenges in Radioactive Waste Isolation: Fourth Worldwide Review. Technical Report LBNL-59808. Lawrence Berkeley National Laboratory, Berkeley, pp 1–14

Comparing the Geological Disposal of Carbon Dioxide and Radioactive Waste in Western Europe

Ferenc L. Toth, Richard A. Roehrl, Asami Miketa, and Nadira Barkatullah

Abstract The current status of and prospects for the geological disposal of carbon dioxide (CO_2) and radioactive waste (RW) are assessed for Western Europe by focusing on three large countries: Germany, France and the UK. The relative importance of the associated electricity generation technologies (coal-based and nuclear generation) varies across countries but extensive efforts are under way to explore the feasibility of and available capacities for disposing of the resulting waste. Suitable geological formations seem to be available for both CO_2 and RW disposal in all three of these countries. The main thrust of the disposal for both waste products is on national solutions despite many research projects coordinated by the European Union, and economic and energy collaboration. Research into RW disposal has a much longer history than CO_2 disposal. Yet there are learning opportunities in many areas, ranging from geology and risk assessment to regulation and liability, as well as in public information and participation in decision making, particularly with regard to site selection. Despite well-established (RW) and emerging (CO_2) European Union and international standards and regulatory principles, there are marked differences in the disposal strategies for CO_2 and RW in the three countries.

Keywords Carbon dioxide • Radioactive waste • Geological disposal • Geological formations • Disposal capacity • Western Europe • Germany • France • UK

1 Introduction

Despite declining energy intensities of economies and decreasing carbon intensities of energy systems, greenhouse gas (GHG) emissions have been steadily increasing in most countries of Western Europe. These trends seem to be difficult to reverse, despite the region's aspiration to become the global leader in climate protection and the

F.L. Toth (✉), R.A. Roehrl, A. Miketa, and N. Barkatullah
International Atomic Energy Agency, Wagramer Str. 5, P.O. Box 100, 1400 Vienna, Austria
e-mail: F.L.Toth@iaea.org; aroehrl@gmail.com; A.Miketa@iaea.org; N.Barkatullah@iaea.org

F.L. Toth (ed.), *Geological Disposal of Carbon Dioxide and Radioactive Waste: A Comparative Assessment*, Advances in Global Change Research 44,
DOI 10.1007/978-90-481-8712-6_14,

policies and instruments to foster compliance with the commitments under the Kyoto Protocol to the United Nations Framework Convention on Climate Change (UNFCCC). Additional and more drastic measures will be required to reduce GHG emissions according to the pledges made in preparation for the 15th Conference of the Parties to the UNFCCC (EC 2009). This chapter explores in a comparative framework the implications of two key strategies currently being considered in many Western European countries to reduce energy-related carbon dioxide (CO_2) emissions: the increased use of nuclear power and CO_2 capture from fossil fuel combustion, both resulting in the need to dispose of the waste products in suitable geological formations.

For the purposes of this chapter, Western Europe is delineated so as to include all member states of the European Union (EU) as of 1993 (EU15) plus Cyprus, Iceland, Malta, Norway, Switzerland, and other small countries of the region. Western Europe has a large population, GDP and energy densities per unit of area. Although it is geographically small in comparison with other main world regions, the relative importance of different energy sources varies widely across countries. This results in substantial disparities in energy-related environmental problems, particularly in GHG emissions. Given the mandate of this chapter, we focus here on CO_2 emissions from the power sector.

Figure 1 presents the CO_2 intensities and the shares of non-fossil sources in power generation for selected countries of the region. Countries well endowed with hydropower sources (e.g. Austria, Norway, Sweden, Switzerland) can secure significant shares of their electricity requirements from this low-carbon source. Some of them (Sweden and Switzerland) complement hydropower with nuclear electricity while others (Belgium, France) generate large shares of their power from nuclear. Several countries still use huge quantities of coal and will need to find ways to reduce CO_2 emissions under the increasingly stringent EU restrictions on GHG emissions.

As a result, research on various aspects of geological disposal for the two main waste products, CO_2 and radioactive waste (RW), and the search for suitable disposal sites are pursued intently in most countries. This chapter presents lessons from comparative assessments for three large countries in the Western European region: Germany (Sect. 2), France (Sect. 3) and the UK (Sect. 4). Each section presents an overview of the current status and the main issues of the geological disposal of CO_2, followed by a similar status review for RW. This material is then used as the foundation for the three national comparative assessments. The closing section summarizes the main lessons learned.

2 Germany

2.1 CO_2 Sources and Geological Disposal in Germany: Status and Issues

This section provides a brief overview of the most salient large-point CO_2 sources and geological disposal options that are being explored in Germany.

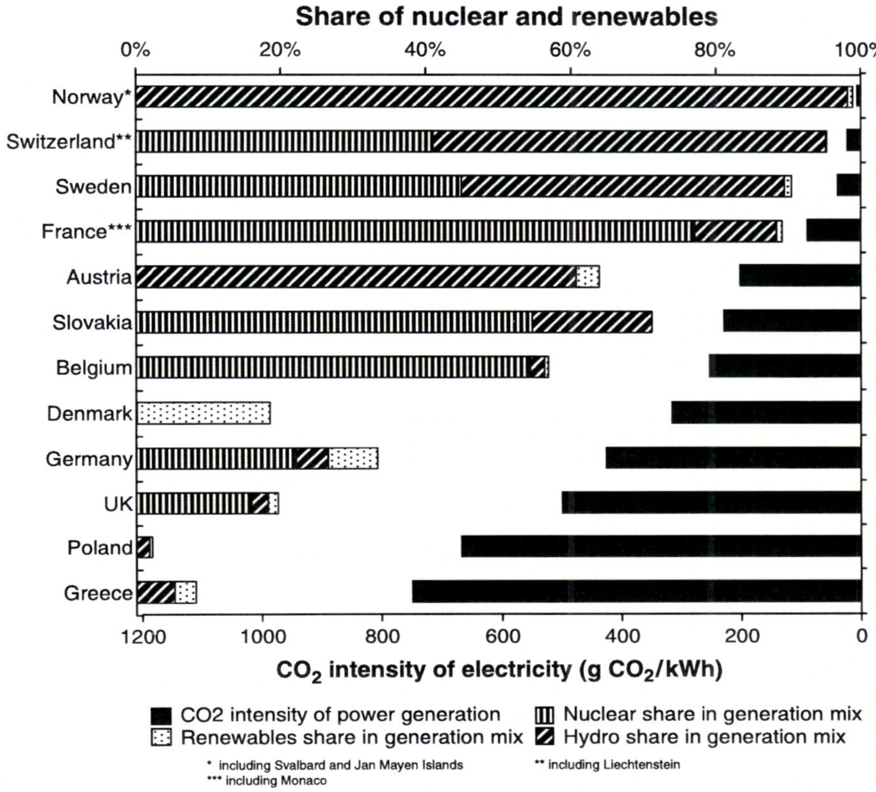

Fig. 1 CO_2 intensity and the shares of non-fossil sources in the electricity sector of selected countries (Source: IAEA calculations based on IEA (2008a) data)

2.1.1 Fossil-Based Electricity and CO_2 Emissions

In 2007, CO_2 emissions in Germany were 824.2 million tonnes (Mt) or roughly one-fifth of EU27 emissions (UBA 2009a; EEA 2008). Total energy-related CO_2 emissions were estimated at 755.3 Mt, of which 51% originated from the energy industries (385.5 Mt). Public electricity and heat production accounted for 345.7 Mt CO_2 of which 291.1 Mt originated from solid fuels (coal, lignite), 40.7 Mt from gaseous fuels (natural gas and other gases), 3.6 Mt from liquid fuels, 10.4 Mt from biomass and 10.3 Mt from other fuels (UBA 2009a). In other words, 84% of emissions from public electricity and heat production originated from coal and lignite power plants and 12% from gas-based power plants. These large stationary sources are particularly suitable for CO_2 capture and disposal (CCD).

By 2007, net CO_2 emissions in Germany had decreased by 18.2% from their 1990 level, according to the Federal Environment Agency (UBA) (UBA 2009a), and per capita emission levels are now similar to the 1950s levels (Marland et al. 2009).

Germany's GHG emissions in 2008 were down by 23.3% relative to 1990; thus it appears that Germany has already reached its 21% reduction target under the Kyoto Protocol and the EU burden-sharing agreement (UBA 2009b).

In 2007 the Government announced an eight-point plan to reduce GHG emissions by 40% from 1990 to 2020, corresponding to an additional reduction of 270 Mt CO_2-equivalent from the 2007 level. While the plan does not include CCD, the Government has recognized the need to explore this option, as evidenced by the number of ongoing pilot studies and applications. The Government also announced the need to reach GHG emission reductions of 80% by 2050, an ambitious target that will require the consideration of all possible options, including CCD. It should be noted that in 2006 the UBA issued a position paper that examined the disposal potential and the environmental impacts of CCD and concluded that CCD was only an interim solution and would not be available for large-scale power plants in Germany before 2020 (UBA 2006).

There are considerable coal reserves and resources in Germany. The estimated lignite reserves (40,818 Mt) and resources (36,760 Mt) are some of the largest in the world and could serve current German consumption levels for another 430 years. Hard coal resources (82,947 Mt) would not run out for centuries. However, in 2007 hard coal reserves were estimated at 118 Mt which was equivalent to only 5 years of production. Estimated natural gas reserves and resources were relatively small (418 giga m^3 (Gm^3)) compared to consumption (96 Gm^3), and oil reserves and resources were relatively negligible (57 Mt), as reported by the Federal Institute for Geosciences and Natural Resources (BGR) (BGR 2008). Thus Germany has become increasingly dependent on the import of fossil fuels. Almost all the oil (97%), most of the natural gas (82%) and two thirds of the hard coal consumed in Germany was imported in 2007. Ten years earlier, only one third of hard coal was imported. In contrast, almost all lignite was produced domestically in 2008 (BGR 2008).

Coal resources are concentrated in the Rhineland in West Germany. The Tagebau Garzweiler mine near Düsseldorf is the largest lignite surface mine in the world and produces more than one quarter of the fuel for Germany's electricity. Other large coal mines are located at Heimbach and Inden close to the border with the Netherlands. Natural gas fields located in the north-western German Basin, the Upper Rhine Graben and the Molasse Basin spread over 41% of German territory. Currently, Germany's natural gas refining and production occurs mostly in the north-western state of Niedersachsen, but the country also has sizeable natural gas reserves in the North Sea. The country's largest oil producing field, Mittelplate, is located off the western coast of the North German state of Schleswig-Holstein.

In 2007 electricity use in Germany was more than 1,525 petawatt-hour, with a mix of hard coal (24.5%), lignite (27.0%), gas (12.6%), hydro and wind (4.2%), nuclear (27.9%), and oil and other solids (2.4%) (AGEB 2009). While overall coal use has decreased in recent years, more than half of electricity is still derived from coal and lignite. In view of the need for baseload power, together with the nuclear power phase-out decision and the high oil and natural gas prices over the last years, two dozen coal plants are currently in the planning or construction stage in Germany. In fact, the Federal Ministry for the Environment, Nature Conservation and Nuclear

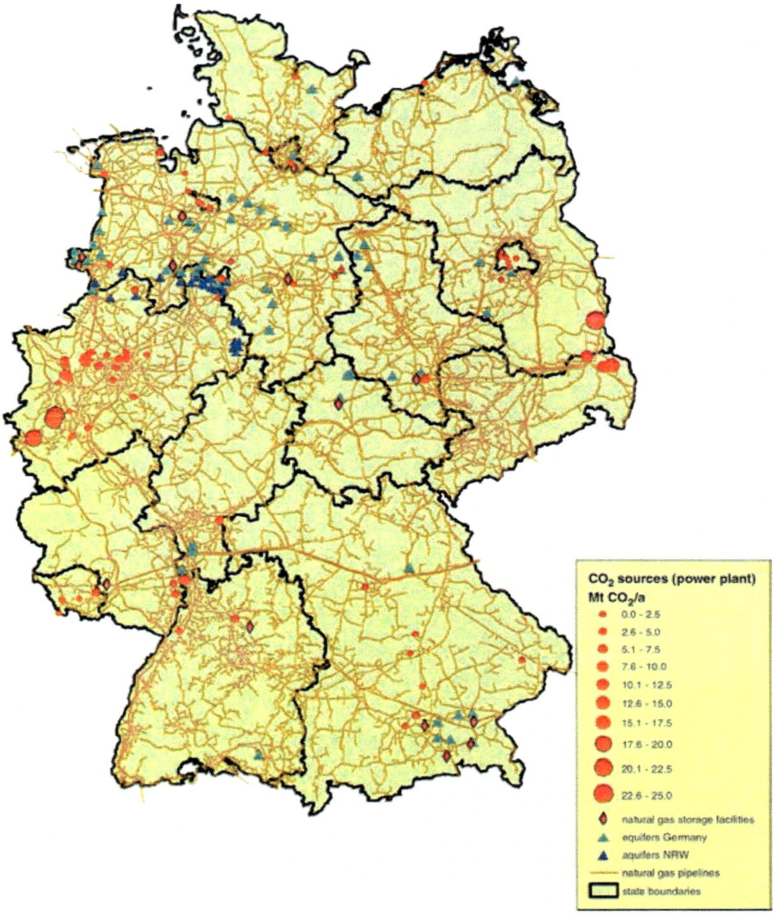

Fig. 2 Major stationary sources of CO_2 (*power plants*), potential disposal in saline aquifers and natural gas storage facilities, and the existing gas pipeline network in Germany (Source: Fischedick et al. 2007) (*see* Colour Plates)

Safety (BMU) projects a continued reliance on coal for electricity production over the next decades. Therefore CCD is expected to play an important role in CO_2 mitigation strategies in the future.

Figure 2 shows the locations of major stationary sources of CO_2 emissions (red circles) which are mainly coal power plants and some gas-based power plants. Such plants are located close to major coal mines and/or consumption centres (cities). For example, the Schwarze Pumpe power plant in the Ruhr is the largest lignite power plant in Germany, with a capacity of 2 GW and more than 10 Mt CO_2 emissions per year (Kreft et al. 2007). Figure 2 also indicates the locations of suitable geological formations for CO_2 disposal, as well as the existing gas pipelines.

The GeoCapacity project developed a geographical information system (GIS) mapping tool for the analysis of sources, potential sinks and CO_2 transport scenarios (Fischedick et al. 2007).

2.1.2 Geological Formations for CO_2 Disposal

In Germany a wide range of geological formations are being explored for CO_2 disposal (Stroink 2006). Most disposal options are based on the permeability of high-porosity geological formations. In particular, deep saline aquifers, depleted oil- and gasfields and deep (presently unexploitable) coal seams are considered to be promising options. Based on Herzog et al. (1997), Gerling (2004), Ziesing (2006) and May et al. (2003), Fischedick et al. (2007) carried out an assessment of the pros and cons of the options and found closed coal mines and salt caverns (which had also been considered) to be unsuitable. In addition to the full range of CO_2 disposal options in deep geological formations, disposal options in the sea and in biomass have also been explored. However, the marine options (in the German seabed) were considered too risky and are no longer being pursued (Stroink 2006). An increasing number of R&D activities on CO_2 disposal have been carried out in Germany, especially since the start of the EU Emissions Trading Scheme in 2005 (Krooss and May 2006). Many of these R&D activities involve partnerships between academia, government and the private sector. The main research findings are summarized below for each geological formation under consideration in Germany.

Deep saline aquifers have been identified as the option with the largest disposal capacity in Germany, with estimates ranging from 12 to 28 gigatonnes of CO_2 (Gt CO_2) (May et al. 2005; Fischedick et al. 2007). Earlier estimates were even higher, of the order of 23–43 Gt (see Bentham and Kirby (2005), based on results of the project on Geological Storage of CO_2 from Combustion of Fossil Fuel (GESTCO) (Christensen and Holloway 2004)), and 33 Gt (Kuckshinrichs et al. 2004). In fact, saline aquifers in Germany have the largest capacities in Europe, according to the Federal Ministry of Economics and Technology (BMWi 2009). However, significant research efforts will be needed to refine estimates and better assess the storage quality and potential of this option in Germany. Moreover, the risks of leakage from the geological formation (and from pipelines) require more research. Suitable saline aquifers are being explored at depths of roughly 1 km. While the option is, in principle, technologically feasible, the costs of this disposal option are expected to be relatively high (Fischedick et al. 2007). The possibility of long-term fixation of the CO_2 in the form of solid carbonate is being explored, but more research will be needed into the corresponding chemical reaction rates and the optimal mineral composition of the aquifers to support the formation of carbonates (Fischedick et al. 2007). Potential future conflicts with the use of geothermal energy (hydrothermal/hot-dry-rock approaches) and with the use of deep aquifers for seasonal energy storage have also been noted.

Depleted gasfields are the most promising option for CO_2 disposal in Germany in terms of economics and technical feasibility. These are mainly located in the North and Middle German Sedimentary Basin in Permian and Triassic sandstones (Stroink 2004).

CO_2 is stored in liquid supercritical phase. Depleted gasfields appear to be the cheapest options for geological disposal of CO_2. This is because of the use of CO_2 injection for enhanced gas recovery (EGR), and because existing gas infrastructure and technology can be used with relatively few modifications. Key technical challenges relate to the development of new materials (different types of cement and steel), and simulation and monitoring. However, conflict of use may arise in the future because of CO_2 contamination of the remaining natural gas (Fischedick et al. 2007). The estimated CO_2 storage capacity in depleted gasfields in Germany is 1.77–2.56 Gt CO_2, which is small compared to the annual emissions of almost 0.4 Gt CO_2 from large (>0.1 Mt) stationary sources (Fischedick et al. 2007). In fact, there are only 66 gasfields of adequate size in Germany to store CO_2 (Stroink 2006). An average German gasfield would be large enough to hold roughly 3–5 years of the CO_2 emissions from a typical German large lignite power plant, which emits roughly 8–10 Mt CO_2/year (BMWi 2009).

The CO_2 disposal capacity in *depleted oilfields* in North and East Germany is very limited, being estimated at less than 0.11 Gt CO_2. Proven technologies exist, and it is expected to be a low-cost option in view of the extensive experience with enhanced oil recovery (EOR). Similar to the disposal option in depleted gasfields, leakage and materials issues need to be addressed (Fischedick et al. 2007).

Deep and presently unexploitable coal seams appear to be a promising CO_2 disposal option for Germany because of the large coal resources in close proximity to coal power plants, the economic benefits of enhanced coalbed methane (ECBM) recovery. However, much greater R&D efforts will be needed, especially into the physico-chemical properties of coal under in situ conditions. The adsorption potential for CO_2 depends on the type of coal and depth. The adsorption method requires depths of roughly 1.5 km. CO_2 disposal in coal seams may make future recovery of such coal resources difficult or impossible. It will also be important to fully capture all the resulting coalbed methane, which is also a GHG (Fischedick et al. 2007). While the estimated technical potential for CO_2 disposal in deep coal seams in Germany is up to 3.7–16.7 Gt CO_2 in the regions of Münsterland and the Saar-Nahe Basin, the economic potential is probably much lower. Industrial pilot projects already exist, for example, with German participation in a project in Katowice, Poland.

CO_2 disposal in *closed coal mines* appears to be an attractive option, as these mines are located in close proximity to major CO_2 sources. However, very high safety risks have been noted, due to connections between mines that are closed and those in use, and because some mines, especially in the densely populated Ruhr, are only a few metres below the surface. There is also a conflict of use with mine gas. The estimated storage capacity is 0.7 Gt CO_2, or 15% of the mined coal seams, most of which are located in the Ruhr und Saar region (Fischedick et al. 2007).

Salt caverns suitable for CO_2 disposal exist mainly in the states of Sachsen-Anhalt and Thüringen. The estimated storage potential is only 0.03 Gt CO_2 in Germany, even smaller than in oilfields. The disposal technology exists. Safety is a major issue because of flooding with water, as well as negative experience of explosive leakage of natural gas stored in salt caverns. Salt caverns are preferred geological formations for the disposal of highly toxic waste and RW, and even the storage of documents in salt caverns for the purposes of data security is being explored in

Germany. In view of the small volumes of suitable salt caverns, such conflict of use is being taken seriously (Fischedick et al. 2007).

2.1.3 Locations and Capacity Estimates

German participation in R&D on geological disposal of CO_2 has been carried out mainly through EU research projects together with foreign partners. These projects include Joule II, GESTCO, GeoCapacity, NASCENT, RECOPOL, CASTOR, CO_2SINK, CO_2STORE, CO_2GeoNet, ICBM and Dynamis. Noteworthy national research projects on CO_2 disposal include the programme GEOTECHNOLOGIEN with ten projects (14 research institutions and 15 companies), as well as CSEGR and the Speicherkataster. It should be noted that government support for R&D on CO_2 subsurface disposal has been relatively small, particularly given the German Government's focus on renewable technologies (Krooss and May 2006).

A sandstone aquifer near the town of Ketzin (west of Berlin) is the location of a field trial of CO_2 injection and disposal. The disposal site is situated at the flank of an anticline above a salt pillow at a depth of 1,500–2,000 m. The saline aquifer formation for CO_2 injection is a Stuttgart Formation of Triassic age at a depth of 650 m. It has a thickness of up to 80 m and a Triassic Weser Formation as top seal (Förster et al. 2008). The overburden of the storage formation contains several aquifers and aquitards, including an abandoned gas storage facility. Since April 2004 preparations and measurements have been performed within the framework of the EU project CO_2SINK, including flow experiments with water and CO_2 in various sandstone types. In 2004 a seismic survey provided 3-D information of the formation. The research showed the caprocks at the Ketzin site to have good sealing properties. The CO_2 injection started in June 2008, and by April 2009, 13,077 t CO_2 had been injected (CO_2SINK 2009). It is planned to inject at least 60,000 t CO_2 over a period of 2 years (Förster et al. 2008).

Another noteworthy field trial of CO_2 injection and storage is taking place in the Altmark natural gas field, which is Europe's second largest natural gas field. The field is located in the Altmark region in the state of Sachsen-Anhalt in north-eastern Germany, roughly 120 km south–east of Hamburg, Germany's second largest city. In geological terms, the Altmark is part of the North German Basin and part of the Mid-European Basin. It contains several sub-reservoirs (Rebscher et al. 2006). The reservoir rocks are located at a depth of 3.5 km and are formed of red sandstone and siltstone with shale layers, with a wide range of porosity and permeability. Above the reservoir, there is a several hundred metre thick Zechstein salt bedrock with very low permeability which forms an effective caprock. The CO_2 injection and storage project is part of an EGR project. CO_2 has been injected in the depleted Altmark natural gas reservoirs to test their technical feasibility for EGR. The storage capacity is estimated at up to 508 Mt or roughly one fifth of the total storage potential of German gasfields. It is the only depleted gasfield available in Germany that can store the entire lifetime CO_2 emissions of a large coal power plant. Carbon capture plants are being built at the nearby Schwarze Pumpe power plant, and small 250–350 MW units are planned by Vattenfall (Vattenfall 2009).

In the CO_2STORE project, a field trial is being carried out in a saline aquifer below the village of Schweinrich, roughly 100 km north–west of Berlin and 250 km north–west of the Schwarze Pumpe power plant. The Schweinrich structure follows an elongated anticline which covers almost 100 km^2. The reservoir formations are within the Lower Jurassic and Uppermost Triassic, and are located between two large salt diapirs at a depth of roughly 1,500–1,600 m. The reservoir is about 150 m thick and consists of several layers of fine-grained, highly porous sandstones overlaid with thick Jurassic clay formations. The storage capacity is estimated to be at least 400 Mt CO_2 (Kreft et al. 2007; CO2STORE 2009).

In the NASCENT project, the BGR and its partners carried out a series of geological studies and soil gas surveys at Oechsen in the Vorderrhön region of Central Germany. In that region, natural CO_2 occurs below and within Permian Zechstein salts and was previously produced commercially (Krooss and May 2006). As part of the GESTCO project, two case studies were selected for numerical simulations of CO_2 injection, the Buntsandstein aquifer near a planned power plant at Lubmin and the abandoned natural gas field Alfeld-Elze. Another site at Kalle was also analysed (Krooss and May 2006).

The total geological disposal potential for CO_2 in Germany is estimated at 19–48 Gt CO_2, which is of the order of 30–60 years of CO_2 emissions from all large stationary CO_2 sources in the country (based on 2007 emissions) (Fischedick et al. 2007). The alternative estimate of the BGR is similar: 20 ± 8 Gt CO_2 (BGR 2009). These are estimates of the technical potentials, only a fraction of which may become economically feasible. Generally, the CO_2 disposal potential is relatively large in the north of Germany and relatively small in the middle and south of the country, compared to current German CO_2 emissions.

A study commissioned by the BMU and carried out by several research organizations assessed the storage potentials for Germany against ecological and techno-economic criteria. The study concluded that only deep saline aquifers, depleted gasfields and deep coal seams were of practical relevance for CO_2 disposal. Table 1 provides an overview of key results in terms of capacity, long-term stability, costs, state of respective technology, utilization conflicts and general risks (Fischedick et al. 2007).

Fischedick et al. (2007) combined their German capacity estimates (based on, for example, Hendriks et al. 2004) with cost estimates for Western Europe from the earlier GESTCO report to create cumulative capacity–cost curves for deep saline aquifers, depleted gasfields and deep coal seams in Germany. The authors find that 2.56 Gt CO_2 could be stored for roughly 6.5 €/t in depleted gas fields, 12–28 Gt CO_2 for roughly 8 €/t in saline aquifers, and 3.7–16.7 Gt CO_2 for roughly 13 €/t in deep coal seams. However, large uncertainties remain regarding both costs and capacities. Cost estimates were derived from German case studies and range widely, especially for CO_2 transport and disposal in saline aquifers.

2.1.4 Implementation Issues

In addition to the techno-economic issues mentioned above, a range of political, social and institutional issues will determine the overall feasibility of carbon

Table 1 Assessment of geological disposal options

Type of disposal	CO_2 storage capacity (Gt)	Long-term stability	Costs	State of technology	Utilization conflicts	General risks
Depleted gasfields	2.3–2.5	Good	Good	Very good	May be resolvable	Good
Deep saline aquifers	12–28	Good	Very problematic	Good	May be resolvable	Good
Deep coal seam	3.7–16.7	Good	Very problematic	May be resolvable	May be resolvable	May be resolvable
Depleted oilfields	0.11	Good	Very good	Very good	May be resolvable	Good
Salt cavern	0.04	Very problematic	n.a.	Good	Very problematic	Very problematic
Disused coal mine	0.78	Very problematic	Very problematic	Very problematic	Very problematic	May be resolvable

Sources: Fischedick et al. 2007; May et al. 2005

disposal options in Germany. A 2006 survey conducted by the University of Marburg showed that 93% of Germans considered climate change an important issue and that most people living close to power plants welcomed CCD. Rostock (2008) also reviewed public acceptance of CCD in Germany. While identifying public resistance as the biggest argument against this technology, especially with regard to the perceived risks during transport and disposal, he noted that the public was not yet debating the pros and cons of CCD. Whereas experts and industry representatives were generally optimistic, environmental organizations were either generally uneasy with or outright opposed to CCD. Hansson and Bryngelsson (2009) recently carried out interviews with CCD experts and reported a discrepancy between the uncertainties and the experts' optimism.

German environmental organizations have increasingly warned about the risks of CCD and its negative implications in terms of energy demand and coal lock-in. In particular, the concern has been voiced that CCD may delay efforts to move towards renewable, low-emission technologies. For example, the World Wide Fund for Nature (WWF) has generally welcomed the development of CCD, but warns against fossil lock-in, lack of transparency and potential environmental consequences (WWF 2009). At a more extreme end of the spectrum, Greenpeace Germany strictly opposes CCD and uses language identical to that in its battle against nuclear waste transport and disposal, namely 'CO_2 repository time bomb' (*Zeitbombe CO_2 Endlager*) (Greenpeace 2009).

Prominent German research institutions have focused on techno-economic assessments of CCD, typically without reference to the potential socio-political limits in Germany. For example, the Potsdam Institute for Climate Impact Research (PIK) calls for as many as 12 CCD demonstration projects to be carried out before 2015, which should demonstrate all the steps: from CO_2 capture, through transport, to sequestration, and which should also demonstrate leakage of below 0.01% per year. PIK also suggests mandating operators to buy CCD bonds for each unit of CO_2 sequestered that would be held by a state authority and handed back only after 30 years (Helda and Edenhofer 2009). According to the Öko-Institut e.V. (Matthes et al. 2009), CCD, together with renewable energy, energy efficiency and combined heat and power, can play an important role in addressing anthropogenic climate change.

The Sustainability Council (Nachhaltigkeitsrat) of the Federal Government views CCD as a necessary technology to support a transition to renewable energy, while the German Advisory Council on the Environment (Der Sachverständigenrat für Umweltfragen) has expressed its concern that CCD may become available too late and turn out to be too expensive. The German Advisory Council on Global Change (Wissenschaftliche Beirat der Bundesregierung Globale Umweltveränderung) has advised against CO_2 disposal in the sea and argued that safe disposal would need to be provided for more than 1,000 years. The UBA itself considers CO_2 capture and disposal as an interim solution at best (UBA 2006). In contrast, German industry associations are very optimistic about CCD. The German Lignite Association (DEBRIV) recommends the use of CCD, while the Hard Coal Association (GVSt) categorizes CCD as a long-term option and focuses on the further increase in power plant efficiency.

The legal basis for CO_2 disposal remains unclear. In fact, elements of the mining act (Bundesberggesetz BBergG), the recycling and waste act (Kreislaufwirtschafts- und Abfallgesetz KrW/AbfG), and the federal water act (Wasserhaushaltsgesetz) apply. On 1 April 2009 the Federal Government adopted a CCD act (Gesetz zur Regelung von Abscheidung, Transport und dauerhafter Speicherung von Kohlendioxid) that sets basic parameters and limits the liability of private operators to 30 years after the CO_2 disposal site is closed, after which the state takes over responsibility (BMU 2009). However, in June 2009 the act failed to be passed by the national parliament. It should be noted that the German CCD act is rather general and leaves a number of key questions open. The Government plans to carry out an evaluation and impact report in the year 2015 based on experience gained with CO_2 disposal from the three German pilot plants in Hürth (Nordrhein-Westfalen), Jänschwalde (Brandenburg) and Wilhelmshaven (Niedersachsen).

The BGR is developing standards and criteria for CO_2 disposal sites. To date, two relevant DIN (standing for Deutsches Institut für Normung (German Institute for Standardization)) standards exist. DIN EN 1918-1 (Untertagespeicherung von Gas in Aquiferen) provides functional and safety recommendations for design, construction, commissioning, operation, maintenance and surveillance of underground gas storage in aquifers, and DIN EN 1918-2 (Untertagespeicherung von Gas in Öl-/Gasfeldern) describes procedures and practices which are safe and environmentally acceptable, covering the subsurface aspects of design, construction, testing, commissioning, operation and maintenance of underground storage facilities in oil- and gasfields.

The German CCD act has been heavily criticized by environmental organizations, such as Greenpeace and WWF. Among other things, they have criticized the characterization of CO_2 as an economic good rather than as waste, which has important legal implications. For example, there are legal restrictions on the transport of waste, especially across national borders.

2.2 *Sources of Radioactive Waste and Geological Disposal in Germany: Status and Issues*

This section provides a brief overview of the generation and geological disposal options for RW in Germany.

2.2.1 Nuclear Installations and Waste Generation

Nuclear power has been an important source of baseload electricity in Germany since the 1970s. Thirty-one per cent of electricity had been generated by 19 nuclear reactors by the end of the 1990s. However, the Government took a nuclear phase-out decision in 2000. An agreement between the Government and nuclear power plant (NPP) operators mandated early decommissioning of reactors (after 32 years of operation). The two oldest reactors were shut down in 2003 and 2005, but the

Fig. 3 Nuclear power plants and storage facilities in Germany (Source: Sailer 2008) (*see* Colour Plates)

phase-out law was revised in 2010 and now the phase-out is expected to be completed by 2036. In 2008, 17 nuclear reactors were being operated at 12 different sites (see Fig. 3) in Germany (Sailer 2008) with a capacity of 21.5 GWe (BMWi 2009). In 2007, 27.9% of electricity in Germany was produced by NPPs (AGEB 2009), which provided 45% of the national baseload. Nuclear power is the second cheapest method of electricity generation in Germany after lignite, and much cheaper than hard coal, hydro or renewables (BMWi 2009).

Uranium is supplied primarily from Canada, Australia and the Russian Federation, and imports amount to 3,800 t/U/year. The construction of a nuclear fuel reprocessing facility at Wackersdorf was stopped amidst widespread public protests in 1989, after which German nuclear fuel was reprocessed mainly in France at the La Hague facility (86%) and to a lesser extent at Sellafield, UK, and at other locations. A smaller reprocessing facility was operated in Karlsruhe until 1990. Since 2005, transport of fuel from German NPPs to reprocessing facilities is prohibited by law (according to a revision of the 1959 Atomic Energy Act) and transport from these facilities is limited. Thus, interim storage and eventual final geological disposal have been the only remaining options since 2005.

In addition to the international classification into high-, medium- and low-level RW, Germany distinguishes between heat-generating waste (HGW) and negligible heat-generating waste (NHGW) (Sailer 2008). NHGW is basically defined as waste that will be disposed of in the Konrad repository, i.e. according to the Konrad waste acceptance requirements (Brennecke 1995; Bund 1989). In practice, HGW is more or less the same as high-level waste (HLW) according to the respective international classification (IAEA 1994). Most of Germany's HLW is kept in reactor pools and at dry interim storage facilities. By the end of 2005, 11,810 tonnes of heavy metal (tHM) of HGW in terms of spent fuel (SF) had been produced by nuclear reactors, of which 5,140 tHM had been stored in Germany and 6,670 tHM had been shipped for reprocessing (Alter et al. 2006) (see Table 2). This corresponds roughly to a volume of 14,000 m^3. Another 1,859 m^3 of HGW were produced from other sources.

In Germany, the utility companies are responsible for interim storage of SF, and they have formed joint companies to build and operate off-site surface facilities. By the end of 2007, 118,124 m^3 of NHGW had been stored at the 12 NPP sites, according to the Federal Office for Radiation Protection (BfS) (BfS 2008), at interim storage facilities in Greifswald, Jülich, Karlsruhe, Mitterteich and Gorleben (see Fig. 3), as well as at state facilities for RW from nuclear applications in research and the health, food and industrial sectors (Table 2). Sailer (2008) estimates the amount of low-level waste (LLW) and intermediate-level waste (ILW) at 100,000 m^3. Another 36,753 m^3 of NHGW had been disposed of in the Morsleben repository and 47,000 m^3 in the Asse research mine (Table 2).

Seventeen experimental and commercial reactors have been shut down and are being decommissioned, including all the reactors in former East Germany after reunification in 1990, producing roughly 10,000 m^3 of RW (WNA 2009a). Decommissioning of all reactors that are currently operating in Germany may produce an estimated 115,000 m^3 of RW (WNA 2009a).

The cumulative amount of NHGW is expected to increase to 277,000 m^3 by 2040 (see Table 3), according to the year 2000 phase-out law, i.e. a maximum life-time of NPPs of 32 years (BfS 2008). This is based on an average 60 m^3 of NHGW produced per reactor per year. More recently BfS (2008) quotes lower estimates of 45 m^3 per reactor and year and 5,000 m^3 per reactor for decommissioning. Witherspoon and Bodvarsson (2006) report a somewhat higher estimate of 297,000 m^3 of NHGW by 2040 (see Table 3). Roughly two thirds of this amount is expected to originate from the public sector, one third from electricity utilities and the nuclear industry (NEA 2006). Energiewerke Nord (EWN) explored waste minimization strategies for electric utilities which would lead to significantly lower NHGW amounts of 192,000 m^3 by 2040, also assuming mandated 32 year maximum licences (Table 3). Waste optimization for public institutions (e.g. research, medicine and the reprocessing facility in Karlsruhe) may prove more difficult. Thus, in this scenario only 45% of NHGW would originate from electric utilities and the nuclear industry.

In 2005 the 17 operating NPPs in Germany produced 417 tHM of SF, leading to cumulative total production of 11,810 tHM by the end of 2005 (Alter et al. 2006).

Table 2 Inventory of radioactive waste in Germany, as of 31 December 2007

		Untreated waste	Interim products	Conditioned waste	Total radioactive waste	Conditioned waste added in 2007	Waste disposed of in geological repositories
Negligible heat generation waste		18,506	8,541	91,077	118,124	2,383	36,753 (Morsleben) 47,000 (Asse)
Heat-generating waste	Without spent fuel[a]	63	1,252	544	1,859	0	0
	Spent fuel from reactors	n.a.	n.a.	n.a.	(~14,000 m^3) 11,810 tHM produced[b]	n.a.	0

Sources: BfS 2008; Alter et al. 2006; and authors' estimates

[a] But including spent fuel from a thorium high-temperature reactor

[b] By the end of 2005

The cumulative amount of SF is expected to increase to 17,200 tHM or roughly 29,000 m^3 by 2040 (BfS 2008). This amount includes 20,600 m^3 of SF elements in pollux containers; 3,400 m^3 of waste conditioning facility components; 660 m^3 of vitrified HLW; 1,340 m^3 of medium-active vitrified waste from reprocessing plants; 130 m^3 from research reactors; and 2,000 m^3 from an experimental reactor and a thorium high-temperature reactor (BfS 2008). Sailer (2008) reports lower estimates of 22,000 m^3 of HGW in 2040.

The cumulative amount of HGW from all sources was expected to reach 22,000 (Sailer 2008) to 29,000 m^3 (BfS 2008) by 2040 under the year 2000 phase-out law. The BfS estimate corresponds to 17,200 tHM by 2040, 46% higher than today (Table 3). In 2010 the implementation of the nuclear phase-out was postponed until 2036. Assuming licences would not be limited to 32 years but extended to 60 years, similar to what has been common practice in the USA, this would imply an additional 21,400 m^3 of NHGW by 2040 (Table 3), or an increase of roughly 8% (BfS 2008). In this scenario, cumulative amounts of HGW by 2040 would increase by 11,700 tHM or 68% (BfS 2008). The difference in relative change is due to the large share of NHGW in decommissioning.

2.2.2 Geological Formations for Radioactive Waste Disposal

Deep geological disposal of RW has been the only legal option for final disposal of both NHGW and HGW in Germany since the amendment of the Atomic Energy Act in 1975. Disposal is considered a national responsibility and therefore disposal abroad is illegal. An extensive knowledge base has been built in Germany on suitable geological formations, especially salt domes, which have been thoroughly surveyed, researched and field tested since the 1960s. The focus has been on salt formations, but crystalline rock formations and, more recently, argillaceous rock formations have been explored in detail (BGR 2007). Results of this work have been summarized in a series of reports by the BGR, commissioned by the German Government, in particular on HGW disposal in salt formations (so-called *Salzstudie*) (Kockel and Krull 1995), HGW disposal in crystalline formations (so-called *Kristallinstudie*) (Bräuer et al. 1994), and NHGW disposal in claystone (Hoth et al. 2005 and 2007). In addition to these technical reports, the BMU has commissioned a comprehensive review study of RW disposal that also includes socio-political issues (Brasser et al. 2008).

A long series of lists of minimum requirements and criteria for repository sites have been suggested and used over the past 40 years. While the more recent lists also include socio-political elements that were not part of the earlier lists, there are hardly any differences in terms of the geological criteria considered (Appel 2008). The geological criteria recommended by the German Government's task force AkEnd (2002) are summarized in Table 4. The criteria contained in the first evaluation step imply that salt formations and argillaceous rock formations are the only suitable formations satisfying the criterion of very low permeability, as crystalline rock formations may be permeable because of fractures.

Table 3 Projections for cumulative radioactive waste in Germany 2000–2040 (in m^3)

			2040	2040	2040	2040
Year (end)	2000	2007	BfS estimate (32-year licences, no reprocessing)	WRB estimate (32-year licences, higher unit waste assumptions)	EWN estimate (assuming waste minimization strategy)	BfS estimate (licence extensions to 60 years)
Negligible heat generating waste	76,000	118,124	277,000	297,000	192,000	298,400
Heat generating waste	8,400	14,900	22,000[a]–29,000 (17,200 tHM)	24,000	n.a.	48,000 (28,900 tHM)

Sources: BfS 2008; Witherspoon and Bodvarsson 2006; and authors' estimates
BfS Bundesamt für Strahlenschutz, *EWN* Energiewerke Nord, *WRB* World Reference Base
[a] Sailer 2008

Table 4 Main requirements and criteria for repository sites suggested by AkEnd (2002)

Criterion	First evaluation step	Second evaluation step
Seismic activity	Must not exceed Earthquake Zone 1 (DIN 4149)	–
Volcanic activity	No quaternary or expected future volcanism	–
Thickness of the isolating rock zone	>100 m; rock types with field hydraulic conductivity of $<10^{-10}$ m/s	>500 m for rock salt deposits in salt domes (Kockel and Krull 1995)
Depth of the top of the isolating rock zone	>300 m	Salt roof above repository zone > 300 m; cover rock over salt dome > 200 m and impermeable to water
Underground depth of the repository	<1,500 m	<1,000 m for argillaceous rock formations
Minimum area of the isolating rock zone	>10 km^2 in claystone	>3 km^2 (AkEnd 2002) and >9 km^2 (Kockel and Krull 1995) for salt dome
Research findings	No findings that raise doubt that field hydraulic conductivity, thickness and extent of the isolating rock zone can be fulfilled for 1 million years	–
Other	–	Rock salt not affected by any other mining or drilling

Note: Second evaluation step supplemented with recommendations by Kockel and Krull (1995), as reported in BGR (2007)

There is an extensive body of knowledge on rock salt formations in Germany, which have been thoroughly researched for the past 60 years; several hundred years of salt mining experience in Germany can also be drawn upon. For example, the BGR draws on data sources from more than 25,000 boreholes across Germany at depths of more than 300 m (Bräuer 2008). Disposal of RW is planned in drifts and deep boreholes at a maximum depth of roughly 900 m, using crushed salt as backfill. Rock salt has a number of favourable properties for RW disposal. In particular, it is almost impermeable to liquids and gas, has a very high heat conductivity and heat resistance, and shows visco-plastic deformation behaviour. The design temperature is 200°C, and no drift reinforcement structures are necessary, which makes rock salt suitable for disposal of both NHGW and HGW.

Rock salt formations in northern Germany (and to a lesser extent southern Germany) occur in the form of salt domes and stratiform rock salt deposits. BGR's *Salzstudie* (Kockel and Krull 1995) assessed more than 200 salt formations in Germany for their suitability as repositories for RW. BGR (2007) considers the Hauptsalz of the Staßfurt Formation in North Germany to be the only formation which 'is known to have uniformly good host rock properties throughout, and to form very thick deposits'. The stratiform salt deposits in the Zechstein Basin are considered as a backup option. While the Rotliegend rock salt in north-west Germany is very thick in some places, it occurs 'in salt domes with very complicated internal structures' (BGR 2007). The Zechstein salts of the Aller to Mölln Formation, as well as the Upper Bunter, Muschelkalk and Tertiary rock salts are too thin. The Keuper salts, the Upper Jurassic rock salts and the stratiform salt deposits of the Werra district are considered unsuitable. In addition to the Gorleben salt dome, in 1995 the BGR reassessed the salt domes in northern Germany and identified a range of salt formations worth investigating at Wahn, Zwischenahn, Gülze-Sumte and Waddekath (Bräuer 2008).

Comparatively less knowledge exists on argillaceous rock formations and their suitability as repositories. Disposal of RW is planned in drifts or shallow boreholes at depths of roughly 500 m, using bentonite as backfill. Among the advantages of argillaceous rock formations are their low permeability and low dissolution behaviour. However, their low heat conductivity and low heat resistance is considered a problem and limits design temperatures to less than 100°C. There is also a need for man-made drift reinforcement structures, which would be a particular problem at great depths (BGR 2007).

While argillaceous rock formations at desired depths and thickness are found in the Tertiary, Cretaceous and Jurassic in both northern and southern Germany (BGR 2007), a wide range of such formations have been considered unsuitable by BGR (2007), including the argillaceous rock formations in the Upper Rhine Graben (earthquake zone), Tertiary clays in northern Germany (low level of consolidation), Tertiary clays and claystones of the Alpine Foreland Basin (minor consolidation only), Opalinus Clay formation (proximity to exploited karst aquifer, partly in an earthquake zone) and areas with extremely steep bedding near salt structures. The investigation focus is thus on thick argillaceous rock formations in the Northern Cretaceous sequence and the North and South German Jurassic sequences (BGR 2007; Hoth et al. 2005, 2007).

Crystalline rock formations are geologically well mapped in Germany, and it is possible to draw on significant mining experience. Disposal of the nuclear waste is planned in drifts or boreholes at a depth of 500–1,200 m, using bentonite as backfill (BGR 2007). The advantages of crystalline rock are its high strength and cavity stability, its low heat sensitivity and very low dissolution properties. However, its brittle deformation behaviour and anisotropic in situ stress behaviour is considered problematic. Most importantly, crystalline rocks when fractured show unsuitably high permeability. Man-made drift reinforcement would be necessary in fractured zones, limiting design temperatures to less than 100°C (because of the bentonite backfill). In 1995 the BGR identified ten crystalline formations for further investigation, including formations at Saldenburg, Nördlicher Oberpfälzer Wald, Fichtelgebirge, Graugneis, Granulitgebirge, Pretzsch, Prettin, Pulsnitz, Radeberg-Löbau and Zawidow (Bräuer 2008). In 2007 the BGR concluded that it is 'unlikely that Germany has zones of homogenous and unfractured crystalline rocks large enough for the construction of a nuclear repository mine' (BGR 2007).

2.2.3 Locations and Capacity Estimates

Since the 1960s West Germany has stored a total of 47,000 m^3 of NHGW at a 'test disposal facility' in the Asse salt mine (Sailer 2008). Former East Germany operated the Morsleben salt mine where 36,753 m^3 of NHGW were disposed of between 1971 and 1998. After a quarter of a century of legal battles, a final court decision (*unanfechtbarer Planfestellungsbeschluss*) awarded an operating licence for the Konrad iron ore mine, and it is expected to open for NHGW disposal in 2013 (Sailer 2008). The Gorleben salt dome was selected as a disposal site for HGW some 30 years ago; however, its development has been constrained by strongly opposing political views.

The *Konrad Mine* is a former iron ore mine near the town of Salzgitter in the state of Lower Saxony in northern Germany. The target layer for disposal in the Konrad mine is the iron ore layer at depths of 800–1,300 m. The ore deposit is quite unique in that it is very dry and fairly deep and was deposited in the Upper Jurassic 150 million years ago (Biurrun and Hartje 2003). The iron ore is overlain by highly impermeable Cretaceous claystone and marlstone (Sailer 2008). From 1960 to 1976, iron ore was mined at Konrad at great depths of 900–1,300 m. The mine extends over a 1.4 by 3.0 km area. Only 6.7 Mt of iron was mined, accounting for 0.5% of the resources. Extensive geoscientific exploration and investigations assessed the site's suitability to host a final repository for RW, concluding that the mine was very suitable for the disposal of both HGW and NHGW (Biurrun and Hartje 2003). From 1976 to 1982 the German Government commissioned the Gesellschaft für Strahlen- und Umweltforschung mbH (GSF) to conduct a geological, seismic and geotechnical study which showed that the site was ideal for the final disposal of NHGW. From 1983 to 1990, the site was further investigated, and a safety report, the Konrad Plan, was issued in 1991. While the Konrad mine could accommodate an estimated 650,000 m^3 of waste, the approved licence is only for 303,000 m^3, which would be more than enough for all NHGW from German reactors, including decommissioning and all other sources.

The *Gorleben salt dome* is one of many salt domes in the North German Basin. The suitability of Gorleben as a final repository for disposal of all types of RW has been under investigation since 1979. An extensive number of seismological surveys and geophysical measurements were carried out until the government moratorium on exploration in 2000. The salt dome consists of massive formations of Zechstein salt. Large homogeneous salt areas were found in the Staßfurt sequence of the Zechstein, which are particularly suitable for RW disposal (Brasser et al. 2008). It should also be noted that an almost complete sequence of principally clayey-silty marine sediments from the Upper Paleocene onwards is preserved. The salt dome covers an area of about 14 by 4 km. The top of the salt dome is 250 m below the surface and the salt base at depths of 3,200–3,400 m. In 1986 two shafts (Gorleben 1 at 933 m and Gorleben 2 at 840 m) were constructed with the main gallery at a depth of 840 m. In total, about 7 km of drifts and galleries with a volume of 234,000 m^3 have been excavated, and geological and geotechnical boreholes with a total length of 16 km have been drilled (Brasser et al. 2008). In order for Gorleben to become operational, political agreement would need to be reached and a site plan approval procedure completed.

The former *salt mine Asse*, close to the town of Remlingen in the district of Wolfenbüttel, was explored and used as a repository for R&D from 1965 to 1995 (Brasser et al. 2008). From 1967 to 1978, LLW and ILW was stored at Asse in 13 chambers at depths of 511, 725 and 750 m. In contrast to Gorleben, extensive salt mining took place at Asse from 1909 to 1964, which has led to mechanical instabilities that make the site unsuitable for long-term disposal.

The Former German Democratic Republic (East Germany) licensed the *Morsleben Repository for Radioactive Waste* (Endlager für radioaktive Abfälle Morsleben (ERAM)) in 1981. It was operated for NHGW until 1998. There was storage of LLW and ILW in the twin salt mines of Bartensleben and Marie in the state of Sachsen-Anhalt near the villages of Morsleben and Beendorf. The twin mine is 5.6 km long and 1.7 km wide, whereas the overall salt deposit covers an area of 50 by 2 km. Mining took place for 70 years until 1969 (Brasser et al. 2008). Two shafts connect to a system of drifts, cavities and blind shafts at depths of 320–630 m below the surface, amounting to a volume of roughly 6 million m^3. (Another 2 million m^3 were backfilled with crushed salt.) The drifts for the final disposal are located in the mine's periphery. The centre appears to be stressed (Kreienmeyer et al. 2004). The ERAM was constructed in Zechstein salt strata, with Staßfurt, Leine and Aller Formations being exposed in the repository mine (Behlau and Mingerzahn 2001). ERAM is located in the Allertalzone structure, which is a fault structure separating the Lappwald block and the Weferlinger Triassic block. Permian evaporate strata intruded into the fault zone and accumulated in a plug, forming the present salt structure. The Zechstein salt deposit has a thickness of 380–500 m and the salt leaching surface is about 140 m (maximum 175 m) below mean sea level. The salt body includes a high amount of anhydrite layers of the Leine sequence which stabilize the salt structure and lead to low convergence of mine excavations (Kreienmeyer et al. 2004). It also includes potash seams, mainly carnallitite and kiseritic hard salt. The caprock has a very low hydraulic conductivity and isolates the salt structure from the aquifers in the overlying upper Cretaceous formations. Above the aquifers there are unconsolidated or semi-consolidated glacial sediments and the surface cover consists of Quaternary sediments (Kreienmeyer et al. 2004).

We have not been able to find any published overall national estimates of geological RW disposal capacity for Germany. Quoted capacities are 650,000 m^3 for the Konrad mine, several million cubic metre for the Morsleben mines. Assuming conservatively that at least 10 of the 140 salt domes previously investigated in northern Germany would prove suitable for geological disposal of nuclear waste, national capacity will be at least 10 million m^3. This exceeds the country's cumulative expected nuclear waste volume from all sources for 1970–2040 by one to two orders of magnitude; this implies that the geological storage capacity is large enough for hundreds of years of large-scale nuclear power generation, assuming no waste minimization strategy.

2.2.4 Implementation Issues

Compared to fossil-fired power plants, the use of nuclear power in Germany means that 100–150 Mt CO_2 emissions are avoided every year, which is similar to the annual national emissions from vehicular traffic (BMWi 2009). This has been a convincing argument against the nuclear phase-out, as most Germans are increasingly concerned about anthropogenic climate change. In fact, a public survey carried out in June 2007 showed that 63% of Germans did not believe in the feasibility of the phase-out and that there was a stable majority of Germans in favour of nuclear power in the long run (Koecher 2007). In other words, a great deal of uncertainty remains about the future of nuclear power in Germany.

In Germany, geological disposal of RW is governed by the Atomic Energy Act of 1959 and its subsequent revisions, as well as the mining law (Bergbaugesetz). Disposal of RW is the sovereign task of the Federal Government. The BMU is responsible for nuclear safety and radiation protection. Operational tasks are managed by the BfS which is supervised by the ministry. The BMWi supervises the BGR, which advises the German Government in all geological and geotechnical matters.

The issues of nuclear power in general and RW disposal in particular have been highly politicized both in the national public debate as well as at government level. While a nuclear phase-out decision was taken in 2000 by the then ruling government, somewhat contradictory views are expressed within the main political parties and also within the current federal government that postponed the phase-out by revising the law in 2010. In fact, while the BMU has taken a rather anti-nuclear stance, the BMWi has highlighted the importance of the continued use of nuclear power and the need to make geological repositories for RW disposal operational. The anti-nuclear side succeeded in imposing an investigation moratorium on the Gorleben site and in setting up the government task force, AKEnd, in 1999, which suggested that a new selection process for repository sites be started with a 'white map of Germany' (Sailer 2008). Another point of disagreement in the Government is the issue of whether to pursue the development of a single national geological repository or several. A recent study carried out by the BfS and the Gesellschaft für Reaktor- und Anlagensicherheit (GRS) mbH showed that the single-repository concept would cause additional costs of several billion euros which would be more than the total cost of the construction, operation and decommissioning of the Konrad repository. While the additional costs for the single-repository concept would have to be fully financed from public funds, two thirds of the cost of constructing and operating the Konrad repository had to be borne by the industry (Pfeiffer 2007).

Public opinion was also polarized on the issues. While in the late 1980s and early 1990s the majority of the public had concerns about nuclear power use, this now seems to have changed. In fact, in 2007, 63% of Germans believed that the country would not abstain from the use of nuclear power in the long run, compared to only 18% who believed that the year 2000 phase-out agreement would be completed (Koecher 2007). Eighty per cent of German businesses are in favour of extending the operating lifetime of the country's nuclear power plants beyond the phase-out dates of the year 2000 law, according to a survey of the German Association of Chambers of Industry and Commerce (WNN 2008).

Because of such polarized views in Germany, the history of developing geological repositories for RW has been characterized by decades of legal challenges and socio-political conflicts. In 1982 the predecessor of BfS applied for a construction and operating licence for a NHGW final disposal site at the Konrad mine. Following the Konrad plan in 1991, extensive public consultations were held in which 289,387 persons formally raised issues that were summarized into more than 1,000 themes. In view of the political and legal opposition, the German state of Niedersachsen approved the licence only in 2002, and it took until 2007 for the highest administrative court to rule in favour of the site. All legal means have been exhausted (*unanfechtbarer Planfestellungsbeschluss*), but political opposition continues. The technology for storage and backfilling of the cavities is available and was tested by DBE. Planning for the facility is under way and it is expected to open in 2013 (Sailer 2008).

In 1976 the state government of Niedersachsen preselected 4 out of 140 salt domes that had been investigated as potential sites for NHGW/HGW repositories (Gorleben, Lichtenhorst, Mariaglueck and Wahn). Using geological and socio-political criteria, and also in view of the fact that it is one of the largest unmined salt domes in Germany, the state government selected Gorleben. In 1977, the German Federal Government confirmed the choice (Brasser et al. 2008). As part of the nuclear phase-out policy decision in 2000, the Government imposed a moratorium (of 3–10 years) on further exploration and preparation of the Gorleben site. To start implementation, a site plan approval procedure needs to be completed and all legal challenges considered. This process took 25 years in the case of the Konrad site.

The former German Democratic Republic (East Germany) carried out safety and techno-economic assessments and explored the disposal of LLW and ILW at Morsleben from the 1960s onward. The site was selected as a geological repository in 1972, and between 1981 and 1998 some 36,800 m^3 of RW were stored there. A few years after German reunification, the German Government decided to stop waste disposal at the site in 1998 and to prohibit it in 2002. Since 2005 the site has been under licensing for closure. In the next 10–15 years backfilling and sealing is planned. Nevertheless, some geologists continue to believe that the potash and rock salt cavities would have been promising properties for a long-term repository (Preuss et al. 2002).

There is a wide range of cost estimates for geological disposal of RW in Germany, many of which appear politically motivated. The most objective estimates are available for the Konrad repository. These data are the most reliable as they relate to the real financial liabilities of the private and public sectors. Aggregate costs for exploratory and planning activities for the Konrad repository amounted to €945 million by 2007. Costs for converting the mine will amount to approximately €900 million. Annual costs for keeping the Konrad mine open are €18.5 million. Overall life cycle cost estimates are around €10,000–25,000/m^3 (BfS 2009). The low estimate is based on low waste volumes (200,000 m^3) and a long life cycle until 2080, and the high estimate assumes higher waste volume (290,000 m^3) and a short life cycle until 2040.

2.3 Comparison of Geological Disposal of CO_2 and Radioactive Waste in Germany

This section compares the geological disposal of CO_2 and RW in Germany. It identifies the major differences and similarities in terms of geological environment, rock type and characteristics, safety potential, mode and purpose of disposal, volume (disposal capacity), disposal depth, containment mode, site selection and public acceptance, and implementation issues. Table 5 summarizes the key results.

Disposal of CO_2 and RW is pursued in rather different geological environments. All promising CO_2 disposal options are based on the permeability of high-porosity geological target formations below low-permeability caprock cover. Examples are deep saline aquifers, deep-lying depleted oil- and gasfields, and deep coal seams. Exceptions are closed coal mines and salt caverns that had once been investigated, but are not longer considered suitable. In contrast, RW disposal is pursued in low-permeability rocks with geological stability and low groundwater fluxes. These include rock salt formations in the form of salt domes and stratiform rock salt deposits (e.g. the Gorleben, Morsleben and Asse repositories), argillaceous rock formations, and the unique case of the deep and very dry Konrad iron ore mine. Crystalline rock formations are considered unsuitable because of fractures.

The most promising target rock types for disposal differ greatly. It is interesting to note, however, that the preferred caprocks for potential CO_2 storage reservoirs include Zechstein salt and Jurassic clay formations, both of which are also preferred rock types for RW repositories. For example, CO_2 disposal in the gasfield at Altmark occurs in red sandstone and siltstone with shale layers, overlain by several hundred metres of Zechstein salt bedrock. CO_2 disposal in the sandstone aquifer at Ketzin occurs in a Stuttgart Formation of Triassic age with a Triassic Weser Formation as top seal. CO_2 disposal in the saline aquifer at Schweinrich occurs in layers of sandstones (Lower Jurassic and Uppermost Triassic) overlain with thick Jurassic clay formations. In contrast, RW disposal is preferred in rock salt formations, as they are almost impermeable to liquids and gas, show very high heat conductivity and heat resistance, visco-plastic deformation behaviour, and achieve design temperatures of 200°C with no drift reinforcement structures necessary. The preferred rock type is the Hauptsalz of the Staßfurt Formation (e.g. the Gorleben repository). Stratiform salt deposits at the Zechstein Basin are considered a backup option. Disposal in argillaceous rocks is also explored because of their low permeability and low dissolution behaviour, despite the low heat conductivity and heat resistance with lower design temperatures of 100°C. In this context, investigation focuses on thick argillaceous rock formations in the Northern Cretaceous sequence and the North and South German Jurassic sequences.

CO_2 and RW disposal both have a high safety potential. However, whereas the technology for RW disposal is mature and safe, this is only the case for CO_2 disposal in depleted oil-/gasfields. General risks of CO_2 disposal are considered manageable for depleted oil-/gasfields, whereas important challenges and concerns (e.g. usage conflicts) remain in the case of saline aquifers and coal seams, even though long-term stability is considered good in these two cases.

The technology for RW disposal in salt formations has been developed for several decades and is considered safe by experts. It also takes into account extreme risks such as earthquakes, tectonic movements and the potential impact of a new ice age. Furthermore, safety regulations limit radioactive exposure close to the repository to

Table 5 Comparison of geological disposal of CO_2 and radioactive waste in Germany

Criteria	Carbon dioxide	Radioactive waste
Geological environment	Primarily: deep, permeable, high-porosity geological formations with low-permeability caprock cover	Low-permeability rock with geological stability and low groundwater fluxes
Rock type and characteristics	*Sandstone and saline aquifers*: Stuttgart Formation of Triassic age with a Triassic Weser Formation as top seal; or Lower Jurassic and Uppermost Triassic sandstone layers with Jurassic clay formations on top	*Rock salt formations*: Hauptsalz of the Staßfurt Formation, or stratiform salt deposits at the Zechstein Basin
	Gasfield: red sandstone and siltstone overlayed by Zechstein salt bedrock	*Thick argillaceous rock formations* in the Northern Cretaceous sequence and the North and South German Jurassic sequences
Mode and purpose of disposal	*Mode*: Injection of liquid supercritical CO_2 through well and boreholes, or controlled heating of liquid CO_2 at high pressures	*Mode*: Emplacement in gallery via shafts and boreholes
	Purpose: EGR, EOR, ECBM or just disposal in aquifer	*Purpose*: Safe and secure, final disposal
Volume (disposal capacity)	*National technical disposal capacity:* 19–48 Gt CO_2 or 30–60 years of CO_2 emissions from all large stationary sources in Germany	*National technical disposal capacity*: >10 million m^3 or hundreds of years of expanded nuclear power generation, not taking into account waste minimization
	By type: Saline aquifers (12–28 Gt CO_2), depleted gasfields (2.56 Gt), oilfields (0.110 Gt), coal seams (3.7–16.7 Gt)	*Konrad site alone*: 650,000 m^3, but licensed for 303,000 m^3 (i.e. more than the country's cumulative radioactive waste from all sources 1970–2040)
Depth	650 m–3,500 m	320–1,300 m
Containment mode	Natural barriers with very low permeability	*Natural barriers* of highly impermeable formations
		Man-made barriers: (a) Backfill/sealing with crushed salt or betonite; (b) drift reinforcement structures in clay and crystalline formations

(continued)

Table 5 (continued)

Criteria	Carbon dioxide	Radioactive waste
Site selection and public acceptance	Researchers and private sector select sites	Government-organized selection process among over 200 salt formations. Forty years of official site selection criteria. Licensing of the Konrad site took 25 years
	No public debate due to limited knowledge. Experts and industry representatives are optimistic, environmental NGOs increasingly uneasy or opposed to CCD	Radioactive waste issue highly politicized. Polarized views on Government's nuclear phase-out decision. Majority of Germans do not believe in the phase-out
Implementation issues	German CCS Act passed in 2009, but strongly criticized by environmental NGOs. *Standards* and criteria for CO_2 disposal sites	Sovereign task of the government (German Atomic Energy Act of 1959 and revisions, Mining Law). Konrad site (operational by 2013) the only geological repository with a valid licence
	Estimated costs: 2.56 Gt CO_2 at €6.5/t in depleted gas fields, 12–28 Gt/CO_2 at €8/t in saline aquifers, 3.7–16.7 Gt/CO_2 at €13/t in deep coal seams	*Estimated costs*: €10,000–25,000 per m^3 of RW (life cycle basis, Konrad mine)

CCD carbon capture and disposal, *ECBM* enhanced coalbed methane (recovery), *EGR* enhanced gas recovery, *EOR* enhanced oil recovery, *NGOs* non-governmental organizations

levels within the natural range between different regions (less than 0.8 mSv/year at the Konrad site). CO_2 is injected whereas RW is emplaced. In contrast to RW disposal, some CO_2 disposal options serve additional purposes besides final disposal. More specifically, CO_2 is injected in liquid supercritical state through wells and boreholes or, alternatively, liquid CO_2 at high pressures is heated in a controlled way. In the case of storage in an aquifer, the only purpose is final disposal, whereas CO_2 injection can also be used for EGR, EOR and ECBM. In contrast, the only purpose of RW emplacement in galleries (via shafts and boreholes) is its safe and secure final disposal.

In absolute terms, the technical potential for CO_2 disposal is large and about two orders of magnitude larger than the technical potential for geological disposal of RW in Germany. Yet, relative to the waste volumes to be disposed of, the potential for RW disposal is at least one order of magnitude larger than for CO_2. The national technical disposal potential is estimated at 19–48 Gt CO_2, which is equivalent to 30–60 years of CO_2 emissions from all large stationary sources in Germany (although the BGR estimate is more conservative, namely 20±8 Gt CO_2). More specifically, capacity estimates are in the range of 12–28 Gt CO_2 in saline aquifers, 2.56 Gt in depleted gasfields, 0.11 Gt in oilfields and 3.7–16.7 Gt in coal seams.

The total national technical geological RW disposal capacity is more than 10 million m^3 (about 200 Mt), which is large enough for hundreds of years of expanded nuclear power generation, not taking into account any waste minimization strategy. The technical storage potential is about 650,000 m^3 for the Konrad site alone and several million cubic metre for the Morsleben site. The Konrad site is licensed for only 303,000 m^3, which is still more than the country's cumulative expected RW from all sources from 1970 to 2040.

While CO_2 disposal is explored mainly at depths of more than 1,000 m, RW disposal is pursued primarily at depths of less than 1,000 m. Examples of CO_2 disposal depths include 650 m (Ketzin aquifer), 1,500–1,600 m (Schweinrich aquifer) and 3.5 km (Altmark gasfield). Examples of RW disposal depths include 800–1,300 m (Konrad iron ore mine), 840–933 m (Gorleben salt dome) and 320–630 m (Morsleben salt mine).

CO_2 disposal is based on natural barriers with very low permeability while RW disposal includes both natural and man-made barriers. Examples of natural barriers with very low permeability include 100 m thick Zechstein salt bedrock in the case of the Altmark gasfield and thick Jurassic clay formations in the case of the Schweinrich saline aquifer. Examples of natural barriers in the case of RW disposal include several hundred metres of highly impermeable Cretaceous claystone and marlstone in the case of the Konrad site, and several hundred metres of unmined salt dome in the case of the Gorleben site. Engineered barriers around the waste packages include backfill and sealing for which crushed salt is used in salt formations and bentonite in clay and crystalline formations. Man-made drift reinforcement structures are needed for potential repositories in clay and crystalline formations.

Whereas the site selection for CO_2 disposal is carried out by researchers and the private sector with hardly any government involvement, site selection is a government-driven process in the case of RW disposal. The German public does not yet debate the pros and cons of CCD because of limited knowledge. While experts and industry representatives are generally optimistic about CCD, environmental organizations have expressed their uneasiness or outright opposition. In the case of RW disposal, site selection criteria have been officially adopted and have barely changed in the past 40 years, except for the increasing prominence of socio-political aspects. Despite an exhaustive selection process covering more than 200 salt formations organized by the Government, a government task force in 2002 suggested that the site selection process be restarted from scratch. The licensing of the Konrad site took 25 years and included public consultations in which 289,387 persons formally raised issues on over 1,000 themes. The RW disposal issue has been highly politicized and polarized both in government and among the public. The majority of Germans do not believe in the feasibility of the nuclear phase-out in the long run, and the overwhelming majority of German businesses favour an extension of the operating lifetimes of Germany's NPPs.

While the legal basis for RW disposal has been in place for 50 years, that for CO_2 disposal has emerged only recently. Estimated disposal costs are about two orders of magnitude greater per tonne of RW compared to CO_2. The German CCD act was passed in early 2009. It has been criticized by non-governmental organizations. The BGR is also developing standards and criteria for CO_2 disposal sites. For example, DIN standards exist, such as DIN EN 1918-1 on gas storage in aquifers and DIN EN 1918-2 on gas storage in oil-/gasfields. Geological disposal of RW has been governed by the German Atomic Energy Act of 1959, its revisions, and the Mining Law. The RW disposal is a sovereign task of the government. To date, only the Konrad site has a valid licence that is no longer subject to legal challenges. The site will be operational by 2013.

While large uncertainties remain in terms of costs and capacities, an estimated 2.56 Gt CO_2 could be stored for about 6.5 €/t in depleted gasfields, 12–28 Gt CO2 for 8 €/t in saline aquifers, and 3.7–16.7 Gt CO_2 for 13 €/t in deep coal seams. In contrast, the costs of storing the cumulative RW of Germany from 1970 to 2040 in the Konrad mine are about €10,000–€25,000/m^3 on a life cycle basis.

In conclusion, while CO_2 disposal differs greatly form RW disposal in Germany in technical terms, important lessons can be learned for CO_2 disposal from the RW experience. In particular, similar public acceptance issues are likely to surface in the future requiring a similarly large-scale need for public consultation and very long time frames. A big difference is the much larger amounts of CO_2 needing to be disposed of compared to RW, which has important implications for their management. It may very well be that experts greatly overestimate the socio-political potential for CO_2 disposal in Germany.

3 France

3.1 *CO_2 Sources and Geological Disposal in France: Status and Issues*

3.1.1 Fossil-Based Electricity and CO_2 Emissions

In France an estimated 390 Mt CO_2 was emitted in 2005 from fossil fuel combustion, of which electricity and heat production accounted for 14.6% (56.6 Mt CO_2) (IEA 2008a, b). In 2006, 78% of electricity was produced by nuclear power in France (IEA 2008c), which, together with hydropower, supplies most of the baseload power. Fossil fuel-based plants, accounting for 9% of gross electricity production, are mainly operated to meet peak demands, which generally occur under extreme weather conditions. France's dependence on nuclear power is partly due to its lack of domestic fossil energy resources. Fifty-two per cent of total primary energy supply is accounted for by fossil fuels, of which only 1.5% is produced domestically.

Consequently, France has relatively low CO_2 emissions per unit of electricity generated (91 g CO_2/kWh in comparison to the world average of 502 g CO_2/kWh and the OECD average of 442 g CO_2/kWh in 2005). Total CO_2 emissions per capita are much lower than the OECD average, and the CO_2 reduction commitment of France under the Kyoto Protocol is modest. However, the Government has highlighted the need for further CO_2 emission reductions. Thus, the technological challenges to further reduce GHG emissions are a high-priority R&D issue in France (Brosse 2005).

France's principal CO_2 emission sources are concentrated in five main areas, as presented in Fig. 4 (Bonijoly et al. 2003): Nord-Lorraine (Lorraine region), Basse-Seine (Haute-Normandie region), Golfe de Fos (Provence-Alpes-Côte d'Azur region), Dunkerque (Nord-Pas de Calais region) and the Loire estuary (Pays de la Loire region). The regions of the Paris Basin alone account for 61% of CO_2 emissions of the industrial and energy sectors in France. The search for CO_2 disposal sites has been limited primarily to the immediate proximity (not more than tens of kilometre) of the major emission sources in view of concerns about accidents and the high costs related to CO_2 transportation by pipeline (Bonijoly et al. 2003).

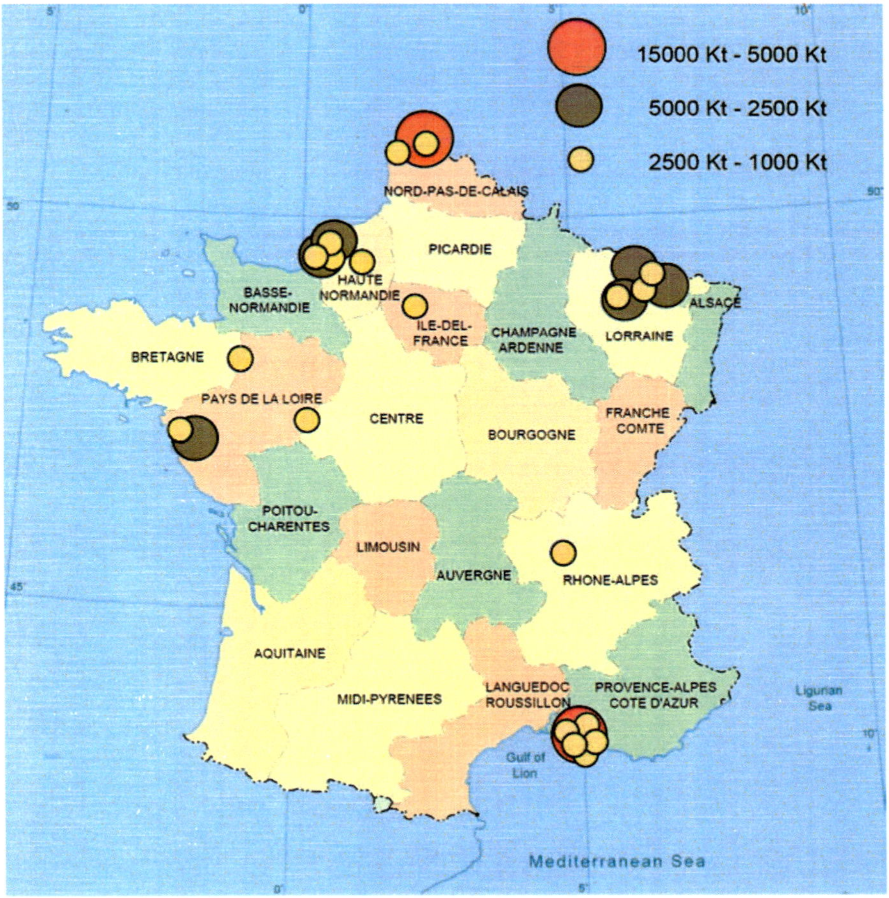

Fig. 4 Major sources of CO_2 emissions (Data taken from the Registre français des emissions polluantes 2009) (see *Colour Plates*)

3.1.2 Geological Formations for CO_2 Disposal

In France, four types of geological formations are under consideration for the disposal of CO_2: aquifer reservoirs, hydrocarbon deposits, coalbeds and basic and ultrabasic formations (such as basalts, periodotites or serpentinites), with decreasing expected disposal potential in this order. Aquifer reservoirs are found in sedimentary basins. There are three major basins: the Paris Basin, the Aquitaine Basin and the South-East Basin (see Fig. 5). Many of the assessments to date have been conducted for the Paris Basin, in view of its proximity to the largest sources of emissions. The locations and capacities of hydrocarbon deposits are well known to major oil and gas companies. Details are not necessarily disclosed to the public. Coalbeds have been evaluated in terms of their potential CO_2 storage capacity in the area of Marseille. The principal form of disposal in basic and ultrabasic formations is mineral sequestration. In Europe, 11.7% of French territory has these formations, especially the Massif Central. New Caledonia, Reunion and Corsica also have such formations (Bonijoly et al. 2009; BRGM 2008).

Within the framework of the European GESTCO project, aquifers in the Paris Basin were identified and assessed (Bonijoly et al. 2003). The most favourable geological conditions were defined as: (1) permeable rock more than 1,000 m deep; (2) an impermeable cover to ensure storage security by preventing gas return to the biosphere; and (3) a suitable structure (i.e. trap) to limit lateral transfers of CO_2.

Fig. 5 Location of the Paris Basin (*upper marked area*), the Aquitaine Basin (*lower left marked area*) and the South-East Basin (*lower right marked area*) (Adapted from Bonijoly et al. 2006)

Injected CO_2 is expected to rise buoyantly to the top of the reservoir structure and accumulate beneath the caprock, a porous material of low permeability saturated with brine. Efficient caprocks are usually composed of salt or clay formations. Such low-permeability rocks are well known in France, because they are considered to be good candidates for RW disposal.

Environmental issues also play an important role in the search for CO_2 disposal sites. Under the PICOREF (Piégeage du CO_2 dans les réservoirs géologiques en France (CO_2 trapping in reservoirs in France)) project, environmental reviews of potential CO_2 disposal sites in the Paris Basin were carried out (Blanchard 2006). The project included R&D on CO_2 disposal with a focus on site identification and evaluation in France (Brosse 2005). The main environmental issues considered were the protection of water resources and biodiversity. The project created maps to support decision making on the question of siting.

3.1.3 Locations and Capacity Estimates

One third of the land area of France is underlain by sedimentary basins that could contain aquifers suitable for CO_2 disposal. The EU project JOULEII provided estimates of the national CO_2 disposal capacities. In particular, the capacity of the trapped fraction of all aquifers in France is estimated at 1.5 Gt CO_2, with 0.3 Gt for the Paris Basin and the rest for the Aquitaine Basin (Barbier 1996).

The feasibility of CO_2 disposal and estimates of capacities in the Paris Basin were evaluated under the GESTCO project (Bonijoly et al. 2003). The Paris Basin occupies about half of northern France and is composed primarily of Mesozoic rock. The main reservoir beds are shown in Fig. 6. Among these reservoirs, only Triassic sandstone-conglomerate layers of the Bundsandstein (upper part of Triassic sandstone), the Keuper (lower part of Triassic sandstone), and the Dogger oolitic limestone were identified as having the desirable geological properties for CO_2 disposal.

The Bundsandstein reservoirs are found mainly in the Lorraine region and the lower part of Champagne-Ardennes region, covering an area of about 21,000 km^2. The depth of the top of the Bundsandstein sandstone increases westwards from the edge of the exposure, reaching 1,800 m. The average thickness is 200 m, with some areas exceeding 400 m. The Keuper sandstone is found mainly in the Île-de-France region and the western part of the Centre region, also stretching into neighbouring regions and covering an area of about 27,500 km^2. The average thickness is 25 m with some areas exceeding 300–400 m, and the maximum depth of the top of the layer is about 2,800 m. These two reservoir beds in Triassic formations are among the largest aquifer reservoirs in the Paris Basin.

The Dogger reservoir covers a large area including the regions of Haute Normandie, Picardie, Île-de-France, a large part of Champagne-Ardennes and the northern part of Bourgogne, covering a total area of 15,000 km^2. In the central and the western sector, the thickness of the reservoir is more than 150 and 175 m, respectively. The depth of the top of the layer is in the range of 1,100–1,800 m.

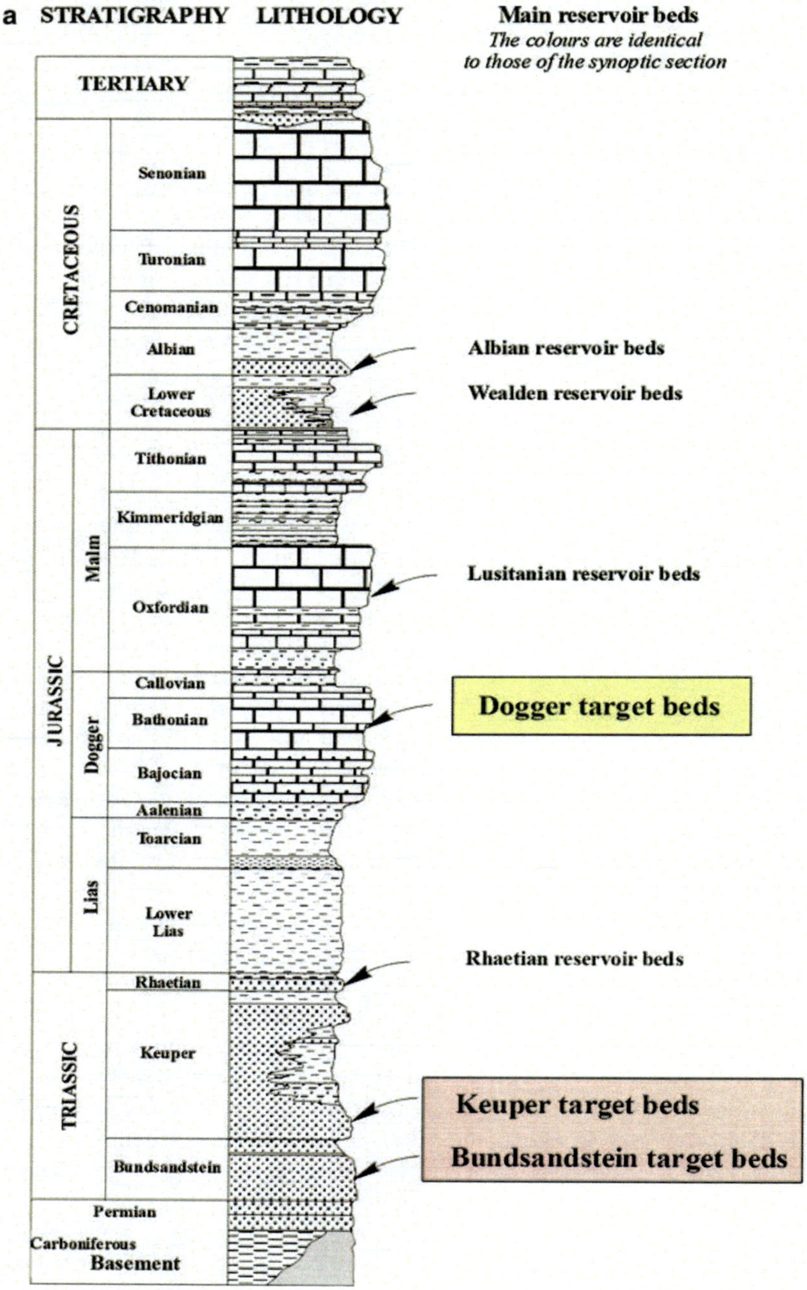

Fig. 6 Geological formations and main CO_2 reservoirs in the Paris Basin (Source: Bonijoly et al. 2003) (*see* Colour Plates) Panel **a**. Synoptic log of sedimentary formations in the Paris Basin

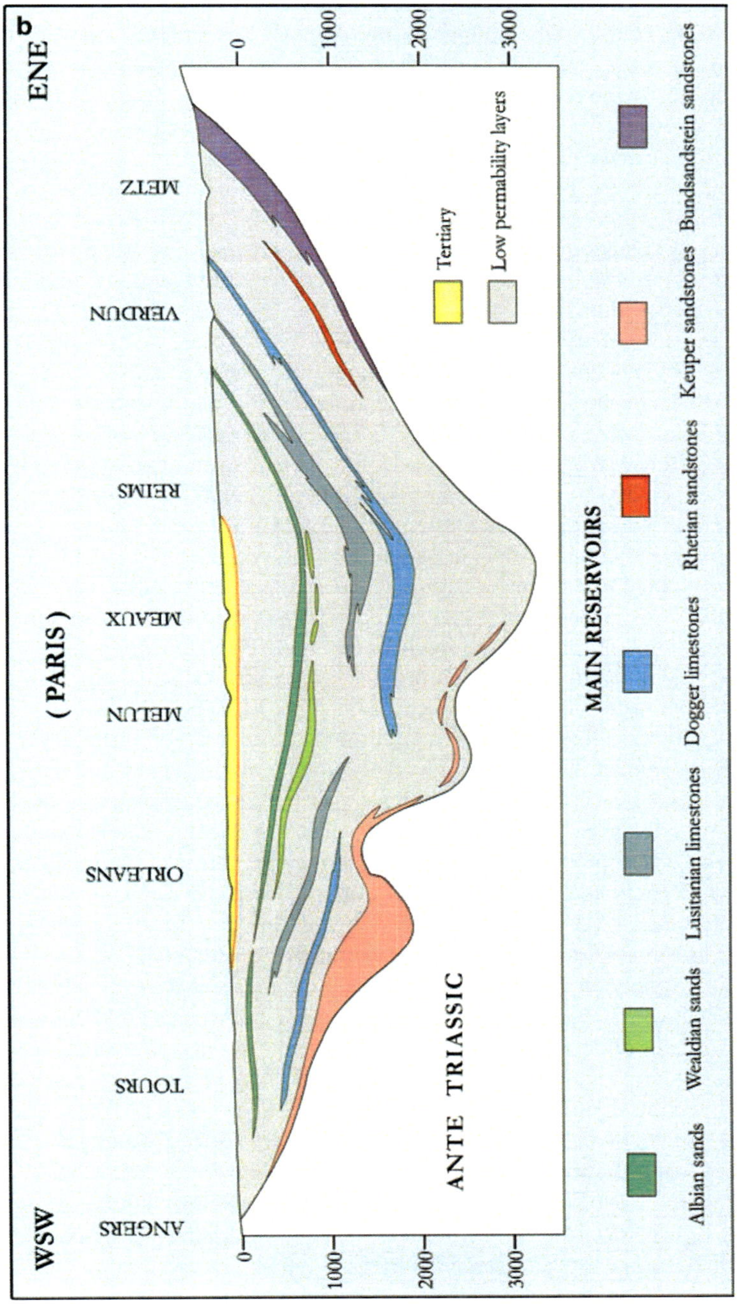

Fig. 6 (continued) Panel **b**. Main reservoirs identified in the Paris Basin

The METSTOR (Méthodologie de présélection des sites de stockage du CO_2 dans les réservoirs souterrains en France (Site preselection methodology for CO_2 storage in subsurface reservoirs in France)) project involves most of the institutions that participated in the GESTCO project. The project estimated the total CO_2 disposal capacities of the entire aquifer at 15.5 Gt for the Trias reservoir (Bundsandstein and Keuper reservoirs) and 13.6 Gt for the Dogger reservoir (Bonijoly et al. 2009). These estimates correspond to 'effective' or 'realistic' capacities that assume realistic reservoir behaviour, as opposed to 'theoretical' capacities that would comprise the entire porous volume accessible to CO_2, fluid saturation and maximum adsorption available in coal.

The earlier capacity estimates of the GESTCO project (22 Gt for Trias and 4.3 Gt for Dogger) corresponded to the theoretical capacities. They also provided the effective capacities, by applying a coefficient that represents the ratio of the disposal capacity of the aquifer confined in traps to the capacity of the entire aquifer: 3% was assumed for the Trias reservoir, and 0.2% for the Dogger reservoir (Bonijoly et al. 2003). A confined structure facilitates the monitoring of injected CO_2, as it is retained in defined areas and reservoir models can be constructed with a higher degree of certainty than in unconfined aquifers (Bentham and Kirby 2005). It should be noted that there is a significant difference between the estimates made by the GESTCO project and those of the METSTOR project. However, no new discussion on this point has been made under the METSTOR project.

In 2008 Veolia launched a CCD project. Claye-Souilly near Paris was selected as the site for a pilot plant. The plant will handle 200,000 t CO_2/year. The gas will be injected into a saline aquifer at a depth of more than 1,500 m for several years (Veolia Environment 2008). Building on previous preliminary evaluation studies of the Paris Basin for CO_2 disposal in depleted hydrocarbon fields and deep saline aquifers, the PICOREF project narrowed the list of potential sites for a pilot injection project to areas about 120 km south-east of Paris, where the roof of the Dogger reservoir is located at a depth of about 1,500 m (Durst and Kervevan 2007).

CO_2 disposal capacity in hydrocarbon fields in the Paris Basin (oilfields) and the Aquitaine Basin (oil- and gasfields) was roughly estimated using static-equilibrium assumptions, implying that the estimates may be conservative. For oilfields in the Paris Basin, it was estimated at 100 Mt CO_2 (with a minimum and maximum of 83 Mt and 117 Mt). For oil- and gasfields in the Aquitaine Basin, it was estimated at 283 Mt (with a minimum and maximum of 140 Mt and 327 Mt), and at 277 Mt (with a minimum and maximum of 170 Mt and 383 Mt), respectively (Brosse 2009). The south-eastern part of the Paris Basin has been thoroughly explored by oil and gas companies. In this area, several oilfields are located either in the uppermost limestone formation of the Dogger Group or in the sand-rich units of the Keuper Group. The data for carbonate reservoirs of the Saint-Martin-de-Bossenay oilfield were made available to the PICOREF project by an operating company (Brosse et al. 2006).

In the Lacq basin (part of the Aquitaine Basin) in south-western France, the company Total launched a CO_2 capture and disposal project (Total 2007). The injection site is the depleted gasfield at Rousse near Chapelle de Rousse. The reservoir lies 4,500 m below the surface and is about 2 km long. It is part of the Adour-Arzaq sub-basin, which is one of four sub-basins of the Aquitaine Basin (Gapillou et al. 2009). The plan is to inject 150,000 t CO_2 during the first 2 years of the project. The selection of the site was made as a result of preliminary studies on all depleted fields operated by Total in the region (the studies were not published).

The METSTOR project provided the first estimates of the theoretical capacities for CO_2 disposal in coalbeds in southern France (Bonijoly et al. 2009). The assessment was limited to 100 km^2 of an unexploited area of the Gardanne coal deposit (near Marseille), which is located at a depth of 500–1,500 m. The result was an estimated theoretical capacity of CO_2 storage of 70 Mt.

3.1.4 Implementation Issues

At present, there is no comprehensive regulatory framework for the geological disposal of CO_2 in France. The PICOREF project included a review of the current regulatory environment (Blanchard 2006), including the mining code (incorporating the waste act, the water act, environmental protection and liability for damage resulting from mining), the environmental code (legislation on industrial facilities, environmental impact assessment, waste management, protection of groundwater and surface water) and the regulation for underground gas storage.

The GESTCO project discussed the potential for simultaneous processes of geothermal operation and injection of CO_2, either in dissolved or supercritical form (Bonijoly et al. 2003). The technically and economically most acceptable scenario for the injection of CO_2 in dissolved form is based on an injection rate of 36 t/day for an average geothermal injection flow rate of 150 m^3/h. The injection cost for this operation is estimated at €100/t CO_2 injected. The injection of CO_2 in supercritical form has the advantages of larger quantities of CO_2 (up to 500 t/day per well) and lower injection costs (€15.6/t CO_2). However, the latter requires preliminary processing and transport of CO_2 to the injection sites, while the risk of a vertical leakage of supercritical CO_2 through the caprock is not negligible. The estimated investment cost for the dissolved form injection is €4 million and for the supercritical form injection about €4.3 million per site.

The first CO_2 injection in France will most likely be the above-mentioned CCD project by Total. The authorization for the injection project for a maximum of 120,000 t CO_2 was granted in May 2009 (Préfecture des Pyrénées-Atlantiques 2009). The injection was planned to commence in June 2009 (Carbon Capture Journal 2009). The total cost of the project, including construction of a unit to extract oxygen from the air and a compression plant for the CO_2, provision of new boiler burners, the modifications to the boiler to enable combustion in the presence of pure oxygen (at the capture site), the work-over of the injection well, the installation of a new unit to compress the CO_2 before injection (at the storage site) and the operating expenses for 2 years are estimated to be about €60 million (equivalent to a total system of cost of €500/t CO_2). Although CO_2will be transported for 27 km from the capture site to the injection site, no extra investment is needed for transportation facilities, as an existing gas pipeline will be utilized as a dedicated CO_2 pipeline. The capture and transport phases of the project will be carried out in accordance with the existing regulatory framework for Environmental Protection at Industrial Sites and Pipeline Transport of Mineral Resources. The injection phase is covered by existing petroleum regulations, as injection will take place on an existing gas production permit. The results of Total's project are expected to provide the authorities with data to help draft appropriate legislation tailored for larger-scale future CCD projects (Total 2007).

The company Total has made outreach and information efforts, notably through the Local Commission for Information and Monitoring (la Commission locale d'information et de surveillance (CLIS)) of the Pyrénées-Atlantiques prefecture. A public opinion survey conducted in 2007 among 1,076 respondents showed that the French public was not strictly opposed to CCD, but was more suspicious than supportive. CCD is simply not known to the public. Only 6% of the respondents were able to define it (Ha-Duong et al. 2009).

3.2 Sources of Radioactive Waste and Geological Disposal in France: Status and Issues

3.2.1 Nuclear Installations and Waste Generation

In 2008, 59 nuclear power plants were operating in France which generated 418 TWh of electricity, 76% of the total electricity generated (IAEA 2009). All SF from reactor operation is being reprocessed at a plant at La Hague in the Basse-Normandie region. The reprocessing plant includes waste processing facilities for treatment and conditioning, and storage areas. An earlier reprocessing plant at Marcoule in the Languedoc-Roussillon region is currently under decommissioning (IAEA 2008).

The National Radioactive Waste Management Agency (Agence nationale pour la gestion des déchets radioactifs (ANDRA)) is mandated by the Planning Act of 2006 (see Sect. 3.2.2) to publish an RW inventory every 3 years. According to the latest report (ANDRA 2009), the reprocessing plants had produced 2,208 m^3 of HLW by the end of 2007, all of which were in storage (1,650 m^3 at la Hague and 558 m^3 at Marcoule). A small fraction (74 m^3) of these volumes consists of HLW from various research activities carried out by the French Atomic Energy Commission (CEA). In addition, 11 m^3 of vitrified HLW packages produced in the PIVER (standing in English for 'first industrial pilot plant for the vitrification of solutions of fission products') pilot plant before 1980 are stored at Marcoule. Some 54.5 m^3 are stored in Cadarache in the Provence-Alpes-Côte d'Azur region and 19.5 m^3 are stored in Saclay in the Île-de-France region.

ANDRA (2009) also provides estimates of the expected volumes of RW for 2020, 2030 and after 2030 (Table 6). The volumes are based on the following assumptions: the existing 58 NPPs (one plant was closed during 2009) and one new European Pressurized Reactor (starting from 2013) operate until each NPP reaches

Table 6 Expected volume of high-level waste in France

	2007	2020	2030	2030–2055
Total HLW (m^3)	2,293	3,679	5,060	7,910
of which: spent fuel	74	74	74	74
of which: PIVER	11	11	11	11

Source: ANDRA 2009
PIVER: Vitrification pilot plant (premier pilote industriel de vitrification de solutions de produits de fission)

the end of its plant life of 40 years; annual power output is assumed to be 430 TWh/year (plus 13 TWh from 2013 onward); and all SF are reprocessed, with reprocessing of MOX fuel starting in 2031.

3.2.2 Geological Formations for Radioactive Waste Disposal

In 1991 the Act on Research on Radioactive Waste Management (the so-called Bataille Act) was adopted. Article 4 stipulates the directions for research on geological disposal of HLW. Article 4 specifies that: (1) the Government shall submit to Parliament a report on the progress of research on HLW management, in which, among other things, the possibilities of reversible or irreversible disposal in geological formations shall be explored through the implementation of underground laboratories; and (2) within 15 years the Government will submit to Parliament a comprehensive report evaluating the establishment of an HLW disposal facility, together with a bill on the establishment of an HLW storage centre. The Act prohibits the storage or disposal of RW in these laboratories.

Among 30 sites nominated as potential locations for a laboratory, a few sites were identified as a result of a mediation mission (mediation mission on the establishment of underground research laboratories) (Bataille 1994), in which geological feasibility criteria and expressions of interests from local communities were taken into account. The geological criteria for the implementation of underground laboratories are: (1) rock with very weak permeability with sufficient volume and at sufficient depth; (2) geological stability at a depth over 200–300 m; (3) a depth of under 1,000 m for safe operation of facilities; and (d) non-occurrence of natural resources at the site. The first two are considered particularly important.

The area straddling the Haute-Marne and the Meuse sites (later referred to as Bure) is characterized by a layer of clay of 130 m thickness at 400 m below the surface. The site of Gard near Marcoule is characterized by a layer of clay over 300 m in thickness. The site is close to a fault zone, and therefore seismic risks are present. The site of Vienne (later referred to as la Chapelle-Bâton) is characterized by a granite massif (Bataille 1996). The review by the Nuclear Installation Safety Directorate (DSIN) prioritized the sites in the order of Bure, Gard and le Chapelle-Baton, while technical reservations against la Chapelle-Bâton were noted (Bataille and Galley 1998). Bure was then selected as the location for the laboratory. An in situ experimental chamber became operational at the end of November 2004. It is located in a layer of Callovo-Oxfordian clay, with a thickness between 100 m in the south-west and 160 m in the north-west, at an average depth of about 450 m and with a surface area of around 100 km^2 (Bataille and Birraux 2005).

In 2005, 15 years after the 1991 Act, the reports and the bill stipulated by the Act were submitted to the Parliament. ANDRA submitted two reports on two types of geological formation, one on clay (ANDRA 2005a) and the other on granite (ANDRA 2005b), for deep geological disposal of HLW. Both types of geological formation were assessed positively.

According to the above-mentioned study on clay by ANDRA (2005a), the clay layer of Callovo-Oxfordian is argillite (i.e. the formation is made up of 40–45% clay minerals, with the rest being other minerals, mainly quartz and carbonates). It is a sedimentary rock with very little permeability, and elements dissolved in water move only very slowly because their migration results mainly from their own movement rather than from being driven by water circulation. It has a chemical environment that enables absorption of chemical disturbances. Furthermore, the argillite has good mechanical strength while being sufficiently deformable to adapt to long-term movements that occur very slowly over time. When the actual site is being selected, the geological environment must be very stable over a long period, without exposure to earthquakes and erosion. The rock must be homogeneous in terms of its structure and mineral composition, and it should have stable chemical properties. It should also be drillable.

The ANDRA study (ANDRA 2005b) on granite referred to above indicates that granite also presents some favourable properties for HLW disposal: it is hard, strong, slightly porous, and shows very low permeability and good thermal conductivity. Most of the massive granite in France reaches significant depths, offering great flexibility for disposal design. Any changes in the composition of the rock from one point to another of the mass do not significantly alter its properties. However, up to a few tens of metres, small fractures can affect the local permeability of the rock. Faults that can reach several kilometres are far less numerous. In the actual implementation of the disposal facility, the identification of granite blocks without fault is a major issue. Nonetheless, priority is given to clay for further development in France.

In June 2006, based on the reports, the Parliament adopted the Planning Act on the sustainable management of radioactive materials and waste, which stipulates that studies on reversible disposal in deep geological formations are to be pursued, so that an application for authorization can be filed by 2015, with operation of the disposal facility from 2025 (OECD 2009).

3.2.3 Locations and Capacity Estimates

The research by ANDRA on the clay formations confirmed favourable site-specific conditions at the Meuse/Haute-Marne area, whereas for the granite the main uncertainty concerns the existence of sites without 'too many faults' in the granite massifs, as they would be exceedingly dependent on engineered barriers. The area with clay formation north-west of the Meuse/Haute-Marne laboratory with a size of

200 km^2 was defined as a transposition zone, which has equivalent geological properties to the laboratory site. The exact location of the disposal site could be decided by 2013. A basic design for the architecture of the disposal facility is proposed by the same study. It adopts the modular approach, which allows gradual construction, operation and closure within each zone.

The overall capacity of potential geological disposal sites in France is clearly much larger than any existing and foreseeable amounts of RW generated in the country. In other words, capacity constraint for geological disposal of RW is not an issue. Thus, the capacity of the geological repository to be developed will be determined by need (i.e. the cumulative amount of RW produced in France).

3.2.4 Implementation Issues

Public consultation and dialogue with the local population is an important issue in France. There was a 1 year moratorium for the site selection process for underground research laboratories in 1990. This was in response to strong local opposition to the research initiated by ANDRA on HLW that aimed to study the possibility of implementing laboratory research in four *départements* between 1988 and 1989. The opposition was due to the proceedings having insufficient prior information and no legal guarantees (Bataille 1994). In response, more importance was attached to local consultations thereafter. In 1991 the Bataille Act set out a procedure for public consultations in the search for the underground laboratory, and mandated dialogue with the local population before undertaking any preliminary exploration work for a site.

The Planning Act in 2006 defined procedures for implementation of a deep geological disposal facility. It stipulated that application for a repository licence be reviewed in 2015 and that (subject to granting of the licence) the repository be commissioned by 2025. The application must relate only to a geological formation that has been investigated through an underground laboratory, and the facility must guarantee the reversibility for at least 100 years. The Act further defined a public consultation process, including an obligation for public debate at specified milestones (Article 12), the formation of public interest group (Article 13), and the establishment of a local information and oversight committee for monitoring research activities at the underground laboratory (Article 18). The Act also established a fund to finance the construction, operation, termination, maintenance and monitoring of the facility, together with a committee to oversee its financing.

In 2004–2005, the French Government, ANDRA and waste producers (Eléctricité de France (EDF), AREVA and CEA) conducted a joint study to estimate the cost of deep geological disposal of HLW in clay formations (DGEMP 2005). In the baseline case (industry scenario) the total costs are estimated in the range of €13.5–16.5 billion. These cost estimates are given jointly for HLW and long-lived ILW, and their volumes correspond to those generated throughout the lifetime (assumed to be 40 years) of the current 58 NPPs. The latest cost estimates by

ANDRA are cited in the same report, showing that costs estimated for long-lived ILW alone are about 10% of total costs. The estimate is based on a scenario in which reprocessing of all SF is assumed.

3.3 Comparison of Geological Disposal of CO_2 and Radioactive Waste in France

This section provides a concise comparison of the geological disposal of CO_2 and RW in France. The main points are summarized in Table 7.

Research on CO_2 disposal in France has reached a stage where three major pilot projects are presently under preparation. The research has been advanced mainly through the participation of French research institutions in EU projects on CCD. The assessment of geological formations in France has been focused on the Paris Basin because of its close proximity to the largest emission sources. To date, no comprehensive regulatory framework exists for CO_2 disposal in France.

In contrast to the CCD activities, the research on the disposal of HLW has been strongly guided by laws. A candidate site for a repository was narrowed down to a 200 km^2 transposition zone. Site selection is primarily guided by the interest expressed by local governments in hosting a repository, as the law mandates consultation with local authorities prior to preliminary studies. Proximity to waste generation sources is not an important factor in the site selection process, presumably because there are only three sites in France where HLW is being generated.

France produced approximately 0.4 Gt CO_2 in 2005. Geological CO_2 disposal capacity in France is estimated at about 30 Gt, a technical estimate that does not consider socio-economic and regulatory constraints on disposal potential or trapping efficiency. The potential for geological disposal of RW is much larger than the cumulative amounts of RW generated to date and projected over the next few decades. Moreover, the law mandates commissioning of a single repository. Thus, its capacity is basically determined by the amount of RW generated. The volume of HLW is expected to amount to approximately 5,000 m^3 by 2030.

Favourable geological conditions for CO_2 disposal include permeable rocks covered by impermeable rocks. Impermeable rock is a favourable condition for RW disposal, and geological assessments aimed at selecting possible CO_2 sites benefit from geological knowledge obtained through the search for RW disposal sites. Aquifers in the Paris Basin, in particular, the Bundsandstein, Keuper and Dogger layers, are assessed to have favourable geological conditions and sufficient capacities for CO_2 disposal, whereas the argillite formation of the Callovo-Oxfordian layer is a target formation for RW disposal. As far as the depth of the disposal is concerned, geological formations deeper than 1,000 m are targeted for CO_2 disposal, whereas formations of less than 1,000 m are targeted for the disposal of HLW.

During the search for a potential site for an underground research laboratory for RW in the late 1980s, local opposition led to the termination of research at several

Table 7 Comparison of geological disposal of CO_2 and nuclear waste in France

Criteria	Carbon dioxide	Radioactive waste
Geological environment considered for disposal	Permeable rock more than 1,000 m deep	Little permeability
	An impermeable cover to ensure storage security	Geological stability at a depth of at least 200–300 m, but less than 1,000 m for safe operation
	A stable structure (i.e. trap) to limit lateral transfers	
Rock type and characteristics	Aquifer reservoir (sandstone, conglomerate, oolitic limestone etc.)	Argillite (sedimentary rock)
	Hydrocarbon deposits	Granite
	Coalbeds	
	Basic and ultrabasic formations	
Mode and purpose of disposal	Injection of CO_2 in dissolved form into the geothermal water, or injection into the supercritical form	Emplacement in engineered galleries
Volume (disposal capacity)	29.1 Gt in the Paris Basin aquifer, unconfined;	The amount of radioactive waste is very small compared to the potential geological disposal capacity
	393–827 Mt in hydrocarbon deposits;	
	70 Mt in coalbeds	
Depth	Maximum depth of 1,800 m (Bundsandstein aquifer), 2,800 m (Keuper aquifers), 1,100–1,800 m (Dogger aquifer), >1,500 m (Veolia project)	Average depth of 450 m (Callovo-Oxfordian formation at Bure)
Containment mode	Natural barriers with very low permeability	Combination of natural and engineered barriers with very low permeability
Site selection and public acceptance	An area for a government-led CO_2 injection pilot project has been defined, and its geological characteristics studied.	The underground laboratory at Bure opened in 2004
	Two CCD projects led by multinational companies (Total and Veolia) initiated	Public opposition was strong in some candidate areas for the underground laboratory
	CCD is not yet well known by the public	The area and target geological formation for the repository have been selected and studied
Implementation issues	No regulatory framework exists, but it is expected soon following the recent EU directive on CCD	The Planning Act was adopted in 2006. It defines the implementation schedule. Some costing studies exist
	Injection costs for the current pilot projects range between €15 and 100. Total CCD system costs including transport for the first pilot project are as high as €500/t CO_2	

sites. This was because procedures did not allow for a sufficient level of local consultations. In 1991 a law was passed mandating local consultations when researching sites for underground research laboratories. A 2006 law likewise stipulated the procedure for public consultation in selecting the site for a final repository. Research into CO_2 disposal is much more recent than into RW disposal. A recent public opinion survey shows that CCD is not widely known about by the public.

Detailed costing studies for CO_2 disposal in France are not available. There are rough cost estimates of about €60 million provided by the company Total for its 120,000 t CCD project at Lacq. Estimates of CO_2 injection costs consisting only of investment and operation of an injection well in the case of simultaneous operation of CO_2 injection and geothermal energy production are available. When the CO_2 is injected in a dissolved form into the geothermal water, the investment and operation costs are estimated at €4 million or €100/t CO_2 (with disposal rates of up to 36 t/day). If CO_2 is injected in a supercritical form, the cost estimates are €4.3 million total or €15.6/t CO_2 (with a disposal rate of up to 500 t/day). However, these estimates do not include the costs of the necessary preliminary processing and transport of CO_2. For RW, ANDRA and other companies have published cost estimates which are in the range of €13.5–16.5 billion for handling the cumulative amounts of HLW and long-lived ILW over the complete lifetime of all previously existing and present NPPs.

4 United Kingdom

4.1 CO_2 Sources and Geological Disposal in the UK: Status and Issues

4.1.1 Fossil-Based Electricity and CO_2 Emissions

The UK emits more than 500 Mt CO_2 every year. GHG emissions have increased, and reached an estimated 640 Mt CO_2-eq. in 2007. The most important GHG is CO_2, which accounts for 85% or 544 Mt (Defra 2008). Fossil fuel-based power plants are the main sources of CO_2, but steel plants, refineries and the petrochemicals sector also contribute significantly to GHG emissions. Most of the 50 largest CO_2 sources are concentrated in the southern part of the UK (see Fig. 7). These comprise 37 combined heat and power plants, 8 refineries, 3 integrated steel plants, a chemical plant and a cement plant (Holloway et al. 2006).

In 2004, 61% of total CO_2 emissions in the UK originated from fossil fuel power plants. Fitting CCD equipment to the 20 largest power plants in the UK would reduce total CO_2 emissions by approximately 20% (Holloway et al. 2006). CCD can reduce the emissions of a typical fossil-fired power plant by roughly 90% (DECC 2009a). The Government has taken steps to promote this technology and has announced the target of making CCD commercially viable by 2020 (DECC 2009b). In April 2009, the UK Government took new measures to encourage CCD

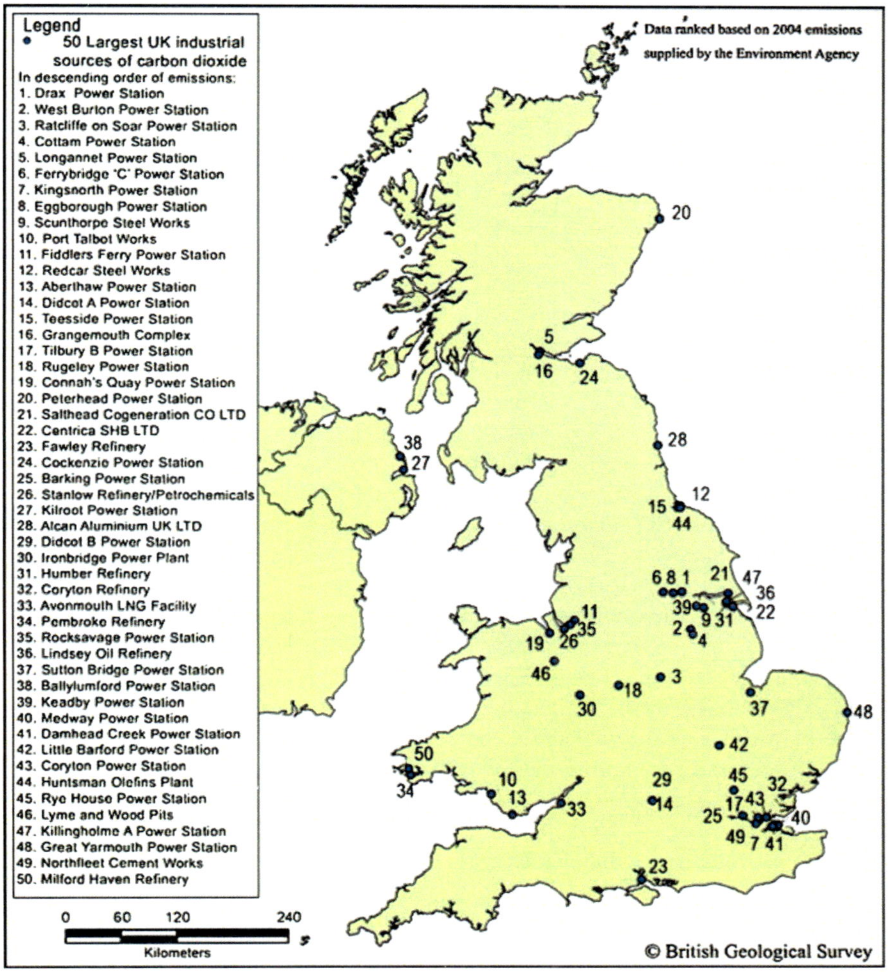

Fig. 7 The largest industrial sources of CO_2 in the UK (Source: Holloway et al. 2006)

development, and confirmed a 'no new coal without CCD' policy. Any new combustion power plant in excess of 300 MW (net output), regardless of whether it is running on gas, coal, oil or biomass, would have to be built with carbon capture ready technology. Five years after the technology is proven to be commercially ready, a full-scale retrofit of CCD will be required (DECC 2009a).

The UK has committed to national and European CO_2 reduction targets: the EU targets to reduce GHG emissions by 20% from 1990 to 2020 (DECC 2009c) and by 80% from 1990 to 2050, as well as the legally binding targets of the UK Climate Change Act 2008 that require UK CO_2 emissions to be reduced by 26% from 1990 to 2020 (UK Parliament 2008). In 2007 the Government launched a competition for construction of the world's first commercial-scale CCD power plant in the UK (capturing CO_2 from a coal-fired power plant of 300 MW net capacity and with offshore CO_2 disposal). In June 2009, the Government proposed a new financial and regulatory framework to assist with the development and delivery by establishing an Office of Carbon Capture and Storage within the Department of Energy and Climate Change (DECC 2009c). The UK Low Carbon Transition Plan released in July 2009 (DECC 2009c) and the UK Low Carbon Industrial Strategy (BIS and DECC 2009) aim to promote CCD in the power sector.

4.1.2 Geological Formations for CO_2 Disposal

In the case of the UK, geological formations considered suitable for long-term geological disposal of CO_2 are oil- and gasfields, as well as saline aquifers (i.e. saline water-bearing reservoir rocks). EGR and EOR technologies are expected to bring additional economic benefits to CO_2 disposal projects, given the long experience with such technologies and the large amount of data available.

The quantifiable CO_2 disposal potential in coal seams in the UK is considered small because of low permeability, which makes unmineable coal seams a less viable option. There are significant coal resources in the UK at depths greater than 1,500 m, but their permeability is expected to be even lower than the seams located at shallow depths (Jones et al. 2004). Conflict of use between CO_2 disposal and future coal extraction has been emphasized. Moreover, knowledge about CO_2 disposal in deep coal seams is limited, especially in view of uncertainties regarding the diffusion of CO_2 into the coal above the critical temperature of 31.1°C. This makes coal seams a less likely option for CO_2 disposal in the foreseeable future.

4.1.3 Locations and Capacity Estimates

Following Bradshaw et al. (2007)——and as illustrated in Figure 8——the total CO_2 disposal capacity in the UK can be categorized as: (a) theoretical disposal capacity that consists of a large but speculative capacity or potential, is poorly known or poorly constrained, and includes uneconomic opportunities; (b) realistic disposal capacity that meets both geological (permeability, porosity, heterogeneity) and engineering criteria and is estimated using existing basin data; and (c) viable capacity, which is built upon realistic estimates and considers various additional economic, legal or regulatory issues regarding CO_2 disposal. If not otherwise stated, capacity estimates in this section refer to the theoretical capacity.

Disposal of CO_2 in the offshore sedimentary basins that contain most of the UK oil- and gasfields is considered the most relevant option (Holloway et al. 2006). The capacity of onshore oil- and gasfields in the UK is considered too small, and major aquifers are widely used for potable water extraction. Formations that trap gas and oil are quite extensive and many of them are considered suitable for CO_2 disposal. Generally, major basins have been identified for potential CO_2 disposal, including the southern North Sea Basin (gas), the central and northern North Sea Basins (oil and

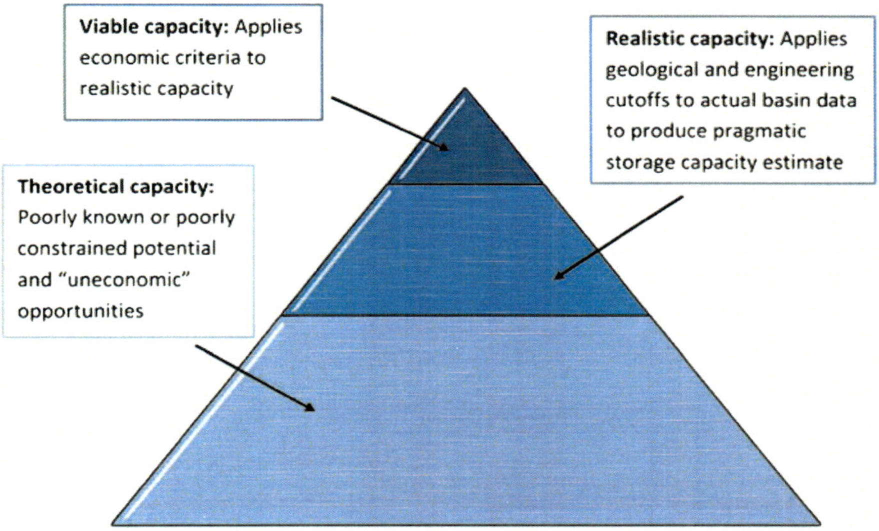

Fig. 8 Techno-economic resource pyramid for geological CO_2 storage space (Adapted from Bradshaw et al. 2006)

gas) and the Irish Sea (gas). In the case of saline aquifers, the potential disposal sites are the southern North Sea gasfields. Figure. 9 shows the locations of the offshore hydrocarbon fields and the major oil-bearing and gas-bearing sedimentary basins.

A recent study by the Scottish Centre for Carbon Storage (SCCS 2009) includes a comprehensive assessment to identify the potential disposal sites for CO_2 in Scotland and north-eastern England. Most of the potential CO_2 disposal sites lie in offshore saline aquifers, as well as in a few depleted hydrocarbon fields. The study identified 29 potential hydrocarbon fields for CO_2 disposal. Amongst these fields, the most promising disposal sites are four gas condensate fields (the Brae North, Brae East, Britannia and Bruce fields), a gasfield (the Frigg Field, UK) and an oilfield (the Brent Field), with an estimated total CO_2 disposal capacity of between 300 and 1,000 Mt.

Unlike hydrocarbon reservoirs, detailed information about saline aquifers beneath the North Sea is not readily available. Therefore, a generic figure of disposal efficiency was estimated (SCCS 2009) based on other regional studies and numerical models, using a disposal efficiency between 0.2 and 2% of pore volume, which implies a total CO_2 disposal capacity of 4,603–46,012 Mt. The study (SCCS 2009) also identified ten saline aquifers that meet the geological and disposal requirements. The analysis showed that the oil- and gasfields pose a low risk and lowest cost options and are thus more promising than saline aquifers. Without EOR, oilfields offer only limited capacity, mainly because of the past replacement of extracted oil with water for pressure support. Thus, the depleted gas and gas condensate fields show the best prospects for CO_2 disposal.

The UK, in a collaborative effort with the Government of Norway, also participates in the monitoring programme of Statoil Hydro in the Sleipner field, the

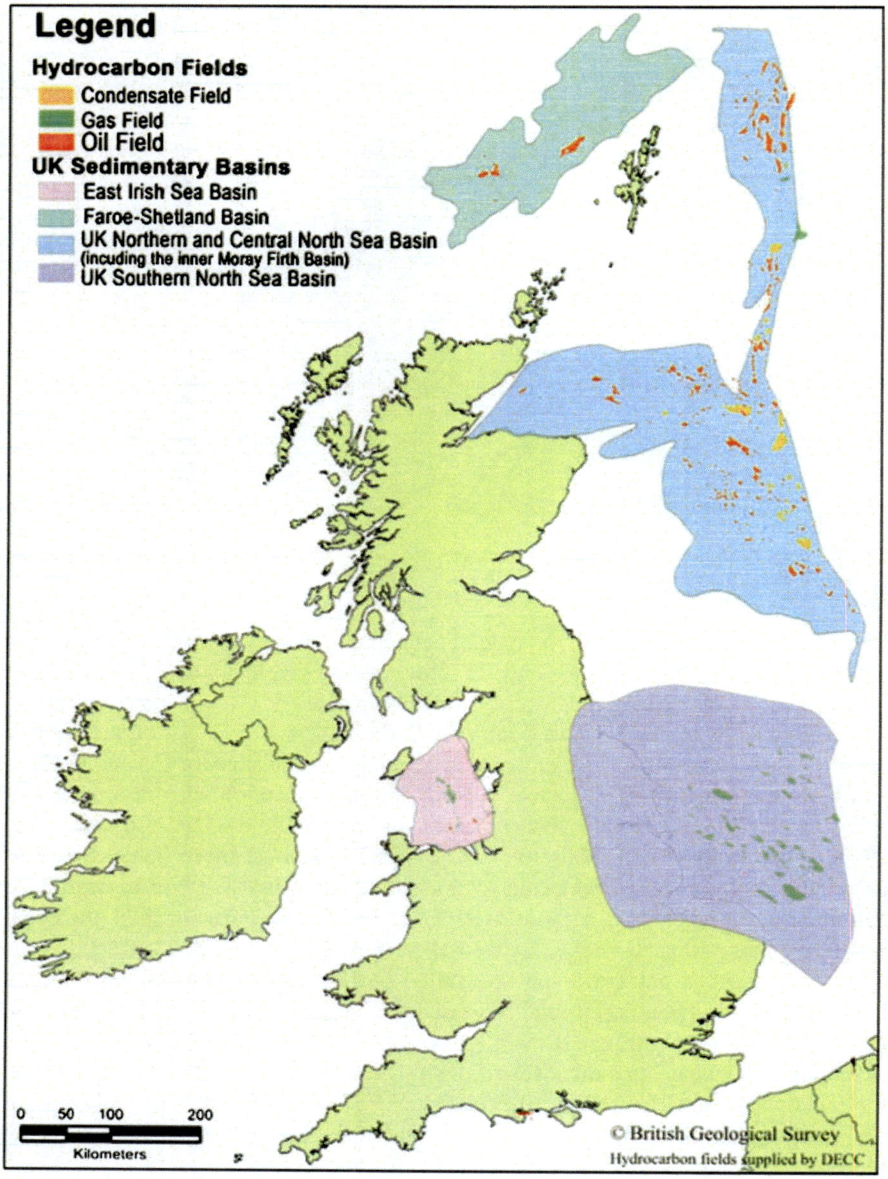

Fig. 9 Offshore hydrocarbon fields and the major oil- and gas-bearing sedimentary basins (Source: Holloway et al. 2006) (*see* Colour Plates)

world's first commercial CO_2 disposal project. Statoil Hydro also plans to establish a full-scale CCD project at the Mongstad refinery in the future. In 2008 a second CCD project at Snohvit was initiated by Statoil Hydro. The UK and Norway are also working to draft regulations for transport of CO_2 in the North Sea (DECC 2009c).

Table 8 Theoretical estimates of the gross CO_2 disposal capacity in the UK

Type of disposal	Potential CO_2 capacity
Gas and condensate fields	5,982 Mt (75 fields)
Oilfields	4,225 Mt (74 fields)
Saline aquifers	14,446 Mt (32 sites)

Source: ACCAT 2009

In the UK the total theoretical gross capacity for CO_2 disposal is estimated at 24.7 Gt (BERR 2007). Table 8 shows the breakdown of the total gross capacity, although the estimate is speculative and theoretical. Such potential is likely to be much smaller when socio-economic factors have been taken into account (ACCAT 2009).

Nevertheless, it is believed that these numbers are initial estimates with large uncertainties which require further testing against empirical data. Further validation and verification are required, especially as 60% of the capacity is associated with saline formations for which data quality is considerably poorer than for oil and gas reservoirs and coal seams. Disposal in such geological formations requires a combination of a porous and permeable reservoir rock that will act as the disposal reservoir and an aquitard or aquiclude in a configuration that will isolate the CO_2 from the atmosphere. Only a few studies are available in the public domain that aim to estimate the disposal capacity.

4.1.4 Implementation Issues

Public response to the use of CCD in the UK has been generally favourable because it is seen as allowing increased energy production without an increase in CO_2 emissions. However, this may be because there has not yet been a real public debate about the subject and because all suggestions for CCD have only included areas in the UK sector of the North Sea, that is, there is a limited NIMBY (Not In My Backyard) effect. The CO_2 disposal projects in the Sleipner and Snohvit fields in the North Sea have received broad support from the main environmental organizations which may have had a positive effect on the general public's acceptance. Surveys of primary and secondary stakeholder opinion of CCD have been conducted at the EU level by the ACCSEPT project (Shackley et al. 2007) and at the UK level, by the UK Carbon Capture and Storage Consortium (UKCCSC) survey in 2006 (Gough 2008). The ACCSEPT project survey reveals that British respondents were enthusiastic about the role of CCD in reducing carbon emissions, but the UKCCSC survey cited some challenges to CCD, including the lack of long-term policy support, the costs and the requirement for an international regulatory framework. The results from the Fossil Energy Coalition (FENCO) project, which is a comparative study funded by six European governments to study the effectiveness of CCD communication by comparing focus groups and Information-Choice Questionnaire (2009–2010), will be published in 2010 and will shed further light on public perception regarding CCD technology.

As mentioned in Sect. 4.1.1, the UK Government has taken firm measures to implement and develop the CDD technology. These comprise, for example, both the inclusion in the Draft Legislative Programme 2009/10 (OLHC 2009) of the pertinent part of the Energy Bill which proposes financial support for four CDD demonstration plants, as well as the establishment by DECC of an office responsible for CCD-related matters to assist with the implementation process.

The UK Government is also working with other organizations to develop a long-term stable regulatory strategy. For example, it works with the OSPAR Commission (OSPAR Commission for the Protection of the Marine Environment of the North East Atlantic) to provide a legal basis for CCD that requires an amendment to the London Protocol (1996 Protocol to the Convention on the Prevention of Marine Pollution by Dumping of Wastes and Other Matter 1972) to allow for sub-seabed CO_2 disposal. The Government proposed amendments to the EU Emissions Trading System (ETS) Directive regarding CCD (UK Parliament 2008), and it is working with EU partners on a potential agreement to use allowances from the EU ETS to support 12 CCD demonstration projects by 2012. The Government agreed with G8 leaders in July 2008 to support 20 large-scale CCD demonstration projects by 2020. It was involved in the development of the EU–China Near Zero Emissions Coal Initiative for a commercial-scale CCD demonstration project in China; it also co-hosted (with Norway) the Carbon Sequestration Leadership Forum Ministerial Meeting in October 2009 (DECC 2009c).

The potential economic benefits of CCD due to EOR and EGR depend on the oil price and, to some extent, on the price of CO_2 in the European market. Some initial estimates by the Scottish study (SCCS 2009) carried out recently showed that CO_2 EOR may be economical in North Sea oilfields at an oil price of US$80–110 per barrel, depending on whether the cost of CO_2 (US $28–56 per tonne) is included in the project cost. If risk premiums are included, then it is unlikely that CO_2 EOR will be commercially viable in North Sea fields at an oil price of less than US$100 per barrel. As offshore CO_2 EOR has not been applied in the early projects, it implies significant financial risks, as detailed engineering design and economic appraisals will require full risk assessments. Other important findings of the study are: the financial cost of initiating CCD will be high but comparable with costs of commercial renewable energy sources; the levelized costs of CCD gas and CCD coal are similar; and the carbon prices have to be high and stable over the long term for the financial viability of large-scale CCD.

4.2 *Sources of Radioactive Waste and Geological Disposal in the UK: Status and Issues*

4.2.1 Nuclear Installations and Waste Generation

In 2006, 19% of the UK's electricity was generated by NPPs. This share dropped to 15% or 57.5 TWh in 2007 and further declined to 13.5% or 52.5 TWh in 2008 (WNA 2009b). At present, the UK has 19 operating reactors (IAEA 2009), 18 of which are expected to be retired by 2023. The NPPs are spread over ten different sites around the country with 14 advanced gas-cooled reactors (AGRs), four magnesium non-oxidizing (Magnox) reactors, and one pressurized water reactor (PWR). The UK expects to bring online a new generation of NPPs, at the very earliest by 2017. Against the background of energy security and the Government's ambitious target (announced in 2008) to reduce GHG emissions by 80% by the year 2050, the UK Government's position has recently become favourable to nuclear power (Summers and Carrington 2008).

The main sources of RW in the UK are NPPs and the activities related to the fuel cycle (Figure 10). Other sources are industry, medical applications and research. To review options for long-term storage and disposal of HLW, the Government established a representative committee in 2003: the Committee on Radioactive Waste Management (CoRWM). In 2006, after 3 years of research, the CoRWM recommended the solution of deep geological disposal for long-lived HLW and ILW and 'robust interim storage' (Defra et al. 2008). In October 2007, a new CoRWM was announced which was given the task of reporting on progress in the geological disposal of RW.

The Nuclear Decommissioning Authority (NDA) has the task of managing this long-lived waste and of developing a suitable geological disposal facility (GDF). The UK Government has mandated the NDA with planning and delivering the GDF, which is to be 'a safe, environmentally sound, publicly acceptable geological disposal solution' for this waste (NDA 2007). As part of the process, the NDA will reach out to and engage the regulators, stakeholders and relevant communities (Defra et al. 2008). Eventually, it is expected that the Radioactive Waste Management Division (RWMD) of the NDA will develop into a Site Licence Company that will be responsible for construction and operation of the GDF and will be known as a 'delivery organisation'. The NDA will also develop a Disposal System Specification that will support the GDF implementation programme (NDA 2009a).

4.2.2 Geological Formations for Radioactive Waste Disposal

Geological Formations

Suitable and stable rock formations for hosting a GDF for long-lived waste are present in the UK (Defra et al. 2008) and about one third of the said area might be suitable for geological disposal (NDA and Defra 2008). A broad range of generic disposal concepts can be applied to the UK. The White Paper for the NDA (Baldwin et al. 2008) reviewed five geological environments and their applicability to typical rock formations found in the UK (see Table 9).

The geological environments across the UK are highly variable, providing various options for the manner in which a geological disposal facility can be implemented at a suitable site. The study by Baldwin et al. (2008) evaluated a wide range of concepts, with the focus on HLW and SF. For example, disposal in boreholes in evaporate formations with no overpack might be a less expensive option for HLW;

Fig. 10 Locations of major UK radioactive waste producers (Source: Defra 2008) (*see* Colour Plates)

Table 9 Rock formations in the UK that could be considered potentially suitable for hosting a geological disposal facility

Host rock	Overlying rock formation	Relevant geological environment in the study
Crystalline rock	Low-permeability sedimentary rock formations	G1 or G2
Crystalline rock	High-permeability sedimentary rock formations	G1 or G2
Crystalline rock	Crystalline rock to surface	G2
Indurated low-permeability sedimentary rock formation	High-permeability sedimentary rock formations	G3 or G4
Plastic low-permeability sedimentary rock formation	Sedimentary rock formations (permeability unspecified)	G3 or G4
Evaporites——salt dome and bedded salt	Sedimentary rock formations (permeability unspecified)	G3
Carbonate	Sedimentary rock formations(permeability unspecified)	G5

Source: Baldwin et al. 2008

G1: Stronger rocks with very low flow of likely saline waters

G2: Stronger rocks with higher water flow; probably relatively fresh water

G3: Weaker rocks with no effective flow and relatively saline waters in pores

G4: Weaker rocks with very low water flow and relatively saline waters in pores

G5: Evaporite formations: plastic, with no water flow and little accessible water (brine) content

however, SF would require an overpack. Disposal in very deep boreholes seems more suitable for HLW than for SF. Baldwin et al. (2008) suggest that the NDA would need to focus on a subset of more appropriate concepts and develop for one or more site-specific conditions in collaboration with stakeholders. CoRWM (2009a) have also expressed the need to assess a wide range of options.

Geological Disposal Facility

As it will take many years before a GDF is ready to receive waste, the UK Government accepted CoRWM's (2006) recommendation of robust interim storage. The Government issued a White Paper stating: 'The Government considers that waste can and should be stored in safe and secure interim storage facilities until a geological facility becomes available' (BERR 2008). Figure 11 displays an interim storage facility able to prevent hazardous release to the outside environment. The four layers of engineered barriers include: (1) a waste form, which is the primary barrier; (2) the waste container; (3) the control of the store environment, which is the tertiary barrier; and (4) the external store structure, as the final layer of protection. The existing stores for waste packages usually have a service life of 50–100 years. The facility will provide interim storage until the GDF programme is developed. To develop a robust programme for the disposal facility, the NDA is reviewing the existing UK waste storage arrangements, including the Sellafield storage, currently the only storage facility for HLW.

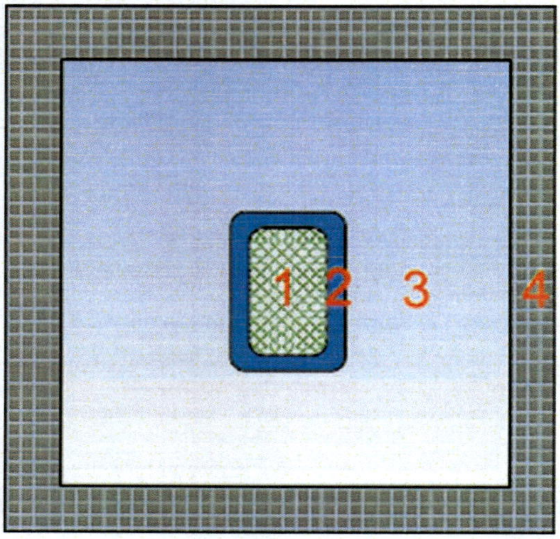

Fig. 11 Interim storage of radioactive waste (Source: Defra et al. 2008) 1: A waste form 2: The waste container 3: Control of the store environment 4: External store structure

Countries like France, Finland, Sweden and the USA have made good progress towards geological disposal. Although no decision has been made in the UK regarding the disposal concept to use, the methodology used in Finland and Sweden is potentially applicable to the HLW and SF in the UK. This involves waste being sealed in copper canisters and put into individual deposition holes that are drilled in the floor of the deposition tunnels. As copper under suitable conditions can be extremely resistant to corrosion, it is expected that in a suitable geochemical environment such canisters could last for a long time and maintain their integrity for hundreds of thousands of years (Defra et al. 2008).

The potential range of depth of the underground areas of a disposal facility for ILW or LLW and HLW/SF would be of the order of 200 m–1 km. However, the exact geological site environment and the design of the disposal facility will depend on the baseline inventory (Defra et al. 2008). Over the coming decades, exchanging experiences and international benchmarking will constitute a key part of the GDF development process in the UK.

4.2.3 Estimates of Waste Volumes and Site Selection for a Geological Disposal Facility

There were no formal plans for geological disposal in the UK between 1997 and 2007. The recent process was initiated following the CoRWM recommendations in July 2006, which proposed geological disposal as a long-term solution for managing HLW. The current target date for an operational GDF for HLW is 2040.

The UK Radioactive Waste Inventory includes three levels of waste: HLW, ILW and LLW. HLW is defined as: 'wastes in which the temperature may rise significantly as a result of their radioactivity, so that this factor has to be taken into account in designing storage or disposal facilities' (Wilson 1996). It is expected that by 2015, most of the HLW in the UK will have been made 'passively safe' by converting it from a liquid to a solid form using the vitrification process. The treated HLW is poured into stainless steel containers (each with 150 litre capacity) in which the waste will solidify. To significantly reduce its radioactivity through the natural decay process, the vitrified HLW is planned to be stored for at least 50–100 years before final disposal (Defra et al. 2008). At present, all HLW is stored at Sellafield in stainless steel canisters in silos (WNA 2009b).

As of 1 April 2007, the volume of RW in the UK was about 290,000 m^3 (NDA and Defra 2008). The inventory data is updated every 3 years. Table 10 shows the volumes of HLW, ILW and LLW in the UK. The 1,730 m^3 of HLW represent less than 1% of the total volume of RW. On the other hand, 196,000 m^3 of LLW account for 60% of the total volume, but less than 0.1% of the overall radioactivity. The volume of RW is expected to increase in the coming decades and will depend on the quantity and the type of the next generation of NPPs.

It should be noted that reprocessing of SF will not take place for any new reactors, so there is likely to be SF (an estimated volume of 8,150 m^3 based on a variety of assumptions regarding the number of new reactors) as well as HLW in a GDF. However, the bulk of the LLW will not go to a GDF but to a surface-based LLW facility (the estimate for this long-lived LLW is 37,200 m^3) (see CoRWM 2006). It is only some of the longer-lived LLW that will go to a GDF.

Location and Site Selection

The location of the GDF is still not known. The CoRWM report released in March 2009 (CoRWM 2009b) addressed the issue of an interim storage facility, which is the

Table 10 Inventory of high-, intermediate- and low-level waste in the UK

Waste Type	Volume (m^3)	Radioactivity
HLW	1,730	Very high
ILW	92,500	Medium
LLW	196,000	Very low (0.1%)

Source: NDA and Defra 2008

HLW high-level waste, *ILW* intermediate-level waste, *LLW* low-level waste

Note: *Intermediate-level waste (ILW)* in the UK is defined as waste 'with radioactivity levels exceeding the upper boundaries for low-level wastes, but which do not require heating to be taken into account in the design of storage or disposal facilities' (HMSO 1995). ILW is generated mainly from spent nuclear fuel resulting from operations and maintenance at nuclear sites. Typically, ILW is packaged for disposal by encapsulation in cement in highly engineered 500 litre stainless steel *Low-level waste (LLW)* is defined as waste having a content not exceeding 4 gigabecquerels per tonne of alpha activity. The majority of the LLW will go to the LLW disposal facility at Drigg. Only a small volume of the LLW——that containing radionuclides with long half-lives——will go a to a geological disposal facility. In 2008 the estimate for this long-lived LLW was 37,200 m^3. In addition, there is the possibility of civil plutonium and civil uranium being declared as waste. Estimates for these are plutonium: 3,720 m^3 and uranics: 74,950 m^3

first step towards the development of a GDF. This was followed by a report on R&D for interim storage and geological disposal (CoRWM 2009c). This report also highlighted the recent review by the NDA (NDA 2009b) of the UK-wide waste storage options for higher-level waste, including 19 ILW (e.g. at Sellafield, Dounreay, Harwell, Winfrith, Trawsfynydd, Hunterston, Sizewell B, Aldermaston, Amersham and Cardiff) and one HLW store at Sellafield. The NDA review also detailed the plans of the UK nuclear industry for some new storage facilities, such as the plans to construct five new ILW stores at Sellafield, the construction of one new store at Dounreay and British Energy's plan to have one new ILW store at each AGR site. The NDA review process indicated that some of these stores can be 'made fit', after appropriate refurbishment and replacement, to provide safe and secure storage until a GDF is available.

The NDA has developed a Geological Disposal Facility Provisional Implementation Plan (GDF-PIP) and is developing a generic Disposal System Safety Case (CoRWM 2009a). The GDF-PIP assumes that perhaps two potential sites for geological disposal will have been identified by the Government by mid-2012. A GDF is expected to be available from 2040 for ILW and from 2075 for HLW/SF (NDA 2009b), although the NDA recognizes the possibility, highlighted by CoRWM, that a GDF may be delayed beyond this point. Given that the high-activity waste in the interim storage facilities would need to be transported to a GDF, the transport process has to be planned and scheduled very carefully; it is expected that it might take many decades to move all such high-activity waste to a GDF.

At this stage it is not known whether there will be one or perhaps two GDFs. However, the Government has indicated a preference for a single site for all HLW/SF (Defra et al. 2008) and for the concept of a single GDF with two separate parts (one for ILW and long-lived LLW and the other for HLW and SF), also known as a combined or co-located GDF.

Currently no site has been selected but the Government is engaged in a site selection process based on the principles of voluntarism and partnership of local communities. As of autumn 2009 Copeland and Allerdale Borough Councils and Cumbria County Council had submitted expressions of interest in opening discussions with the Government (CoRWM 2009a). A flexible approach is preferred to facilitate and promote confidence among the stakeholders in the project. An important aspect of this approach is the right of withdrawal, which would allow any community to withdraw its involvement in the process (CoRWM 2006). As discussed above, the first step towards a GDF is to define an interim storage facility for a storage period of up to 100 years (CoRWM 2006).

4.2.4 Implementation Issues

Public opinion in the UK has become increasingly favourable towards nuclear power. For example, in a survey carried out in November 2008, 65% agreed that nuclear is needed as part of the UK's energy mix, 44% were of the view that old NPPs should be replaced with new ones, and 40% expected an increased role for nuclear power (WNA 2009b). Among Members of Parliament, support for nuclear power was 72%

in 2008, up from 66% in 2006. In October 2008, Defra initiated a one-day open meeting on the geological disposal of RW to discuss developments in the characterization of deep geological and hydrogeological environments and the potential for geological disposal facilities in the UK (GS 2008). Both the UK Government and the NDA have been involved in public and stakeholder engagements. The UK Government issued a White Paper (Defra et al. 2008) and set up a dedicated website for public information on the topic. The NDA issued consultation documents and organized workshops. However, the Government recognizes that additional efforts will be needed to better inform the public and local authorities (CoRWM 2009a).

The UK has a regulatory regime for the management and storage of RW. Planning and delivering the GDF is a collaborative effort between the NDA and the Government, with the NDA as the implementing organization. In April 2007, the NDA established a department for the implementation of geological disposal, which is planned to evolve into a 'delivery organisation' in the future. It is recognized that the NDA will need to reach out to relevant communities and stakeholders, including regulators, for the development of a coordinated strategy for the planning permission and regulatory approvals.

Based on the CoRWM recommendations of September 2008, a Joint Regulatory Office will be established (CoRWM 2009ba) to ensure more 'coherence and coordination' among the current regulators, the Health and Safety Executive (HSE) (Nuclear Installations Inspectorate, Office of Civil Nuclear Security and the UK Safeguards Office), the Environment Agency (EA), the Department for Transport (DfT) and the planning authorities. Legislative modifications are envisaged, for example changes to the provisions of the Radioactive Substances Act 1993 (RSA 93) to permit the authorization of GDFs in several stages, and changes to the Nuclear Installation Regulations 1971, such that disposal becomes a 'prescribed activity under the Nuclear Installations Act [1965]', thus enabling a GDF to be licensed 'as such' instead of purely as a storage facility (CoRWM 2009a).

The construction and operation of a GDF will be a long-term engineering project. The NDA's estimate of the undiscounted lifetime costs of a GDF is £12.2 billion (at 2008 prices), including research, design, construction, operation and closure (although this assumes that only one GDF will be required). The NDA's share of this amount is £10.1 billion, which is then discounted at 2.2% to give a discounted cost of £3.4 billion, the balance being payable by other users. Various factors will influence the actual cost, including the inventory of waste, the timing of waste production, the geology of the site in question and the design of the GDF (NDA 2009c).

4.3 Comparison of Geological Disposal of CO_2 and Radioactive Waste in the UK

A comparison of geological disposal of CO_2 and RW in the UK is provided in Table 11, which highlights both the similarities and the differences. The evaluation of the geological environment shows that offshore gas- and oilfields, as well as saline

Table 11 Comparative analysis of geological disposal of CO_2 and radioactive waste in the UK

Criteria	CO_2 disposal	Radioactive waste
Geological environment	Promising disposal options are offshore depleted oil- and gasfields; offshore and onshore saline aquifer formations. Unmineable coal seams are a less likely option because of their low permeability.	One third of the UK territory has geological environments that are in principle considered potentially suitable for the geological disposal of RW
Rock type and characteristics	Hydrocarbon fields Saline water-bearing reservoir rocks	Crystalline rock, indurated low-permeability sedimentary formations, plastic low-permeability sedimentary formations, evaporites——salt dome and bedded salt and some carbonates
Mode and purpose of disposal	The use of EGR and EOR for depleted oil- and gasfields is an advantage Injection of liquid supercritical CO_2 through wells for saline aquifers.	No specific disposal concept has been decided but the methodology employed in Sweden and Finland could be potentially applicable for emplacing HLW and spent fuel in tunnels
Volume (disposal capacity)	*Gasfields and condensate fields:* 5,982 Mt (75 fields); *Oilfields:* 4,225 Mt (74 fields); *Saline aquifers:* 14,446 Mt 32 sites)	HLW: 1,730 m^3 ILW: 92, 500 m^3 LLW: 196,000 m^3; the majority of LLW will not go to a GDF, but to a surface-based disposal facility; potential disposal capacity far exceeds waste volumes
Depth	Not above 800 m on account of the low density of CO_2	An engineered facility is likely to be located in the depth range of 300–1,000 m. If deep borehole disposal is used for some waste forms (HLW and spent fuel only) then depths as great as 5,000 m might be considered
Containment mode	Natural barriers with low permeability	Combination of natural barriers with engineered barrier systems
Site selection and public acceptance	Offshore oil- and gasfields, offshore and onshore saline aquifers identified as potential CO_2 disposal sites To date there is no significant public opposition to CCD, but no specific sites have yet been proposed	No site has been selected Public consultation is in progress

(continued)

Table 11 (continued)

Criteria	CO_2 disposal	Radioactive waste
Implementation issues	*Regulation:* The regulatory arrangements are under development. New coal plants to be built in a design ready for later CCD fitting. The Energy Bill 2009–10 proposes financial support for four CCD demonstration plants. Office of CCS is to be set up to assist with the development and delivery of these.	*Legal and regulatory:* The legislation is in place, as the geological storage of RW is governed by the Nuclear Installation Act of 1965, but additional legislative changes have been recommended. A Joint Regulatory Office will also be set up by the current regulators (HSE, EA, DfT and the planning authorities) for greater coordination. NDA is the implementing organization, and the RWMD of the NDA is the delivery organization.
	Economics: CO_2 EOR may be economical in North Sea oilfields at an oil price of US $70–110 per barrel.	*Economics:* The NDA's current estimate of the undiscounted lifetime costs of a geological disposal facility is £12.2 billion (at 2008 values).
	Public acceptance: Favourable public support, although no specific sites mentioned; however, there is a need for long-term policy support in collaboration with international partners, as well as a reduction of the costs.	*Public acceptance:* Public consultation is in progress and both the UK Government and NDA are involved in public and stakeholder engagements but additional efforts are necessary to inform the public and local communities.

CCD carbon capture and disposal, *DfT* department for transport, *EA* environment agency, *EGR* enhanced gas recovery, *EOR* enhanced oil recovery, *GDF* geological disposal facility, *HLW* high-level waste, *HSE* health and safety executive, *ILW* intermediate-level waste, *LLW* low-level waste, *NDA* nuclear decommissioning authority, *RW* radioactive waste, *RWMD* radioactive waste management division

aquifers, are likely options for future CO_2 disposal. It is thought that approximately one third of the UK has geological environments which are, at least in principle, suitable for the geological disposal of RW. In the UK, hydrocarbon fields and saline water-bearing reservoir rocks are considered most suitable for CO_2 disposal. For RW disposal, a range of rock formations are considered as being potentially suitable. These include crystalline rocks, indurated low-permeability sedimentary formations, plastic low-permeability sedimentary formations, evaporates——salt dome and bedded salt——and some types of carbonates.

EOR or EGR provide potential advantages for CCD. Another option is using injection wells for saline aquifers, but the actual saline formations are not known and need further verification and testing to explore the viability of this option. In the case of RW, no decision has been made regarding the disposal concept. However, it is estimated that the disposal facility is likely to be located in the depth range of 300–1,000 m. If deep borehole disposal is used for some waste forms (this would be limited to HLW and SF), then depths as great as 5,000 m might be considered, while for CO_2 a depth of at least 800 m is required because of the low density of CO_2.

Major basins, offshore hydrocarbon fields and the major oil- and gas-bearing sedimentary basins have been identified for CO_2 disposal, including the southern North Sea Basin (gas), the central and northern North Sea basins (oil and gas) and the Irish Sea (gas). In the case of saline aquifers, the potential disposal sites are the southern North Sea gasfields. In the case of RW, no GDF site has been selected but there are several interim storage sites and more are planned.

The regulation of CO_2 disposal is still in progress, with some regulatory and legislative arrangements in place, for example: the UK Government announcement in April 2009 that all new coal plants are to be built with carbon capture ready technology; the Energy Bill, as part of the Draft Legislative Programme 2009/10, proposing financial support for four CCD demonstration plants; and the establishment of an Office of Carbon Capture and Storage to assist with the development and delivery of CCD. Compared with CCD, the regulations associated with the management and disposal of RW are mature. RSA 93 provides the legal framework for controlling the management of RW in a way that protects the public and the environment. It imposes requirements for registering the use of radioactive materials and for authorizing the accumulation or disposal of RW. Subject to the outcome of a UK Government review, RSA 93 may be replaced in England and Wales, possibly by 2010, by new regulations. New guidance on requirements for authorizing the geological disposal of RW was published in 2009 (EA and NIEA 2009), which supersedes the 1997 guidance, and allows for phased authorization, as the disposal programme proceeds. For more efficient regulatory mechanism a Joint Regulatory Office will be established among the current regulators, HSE, EA, DfT, and the planning authorities. On the implementation front, the NDA is the implementing organization, the RWMD is the delivery organization.

Some recent figures from the Scottish study (SCCS 2009) show that the cost for CO_2 EOR may be economical in North Sea oilfields at an oil price of US$70–110 per barrel, but no gross estimates are available for CO_2 disposal. Regarding RW, the NDA reported a figure of £12.2 billion (at 2008 prices) for the GDF, based on the undiscounted lifetime costs of a GDF, including costs related to research, design, construction, operation and closure.

With regard to the possibility of CCD, in general the public response has been favourable, as the technique is seen as a possible method for increased energy production without a concomitant increase in CO_2 emissions. The EU ACCSEPT survey results (Shackley et al. 2007) showed that British respondents were enthusiastic about the role of CCD in reducing carbon emissions. The UKCCSC survey cited some challenges to CCD, including the lack of long-term policy support, the cost and a requirement for an international regulatory framework. Public support for nuclear power has increased over the last few years. Consultations are currently in progress with interested communities on the possibility of locating a GDF, and both the UK Government and the NDA are involved in public outreach work. However, it is recognized that additional effort is necessary to inform the public and local authorities, especially those in the areas that have no previous experience of nuclear activities (CoRWM 2009a).

5 Summary and Conclusions

The broader socio-economic context and the many general energy and environmental regulations are similar in the three large EU countries analysed in this chapter. The EU-level energy and climate policies (particularly GHG and CO_2 mitigation targets) and the international conventions on RW management also provide a common framework for the national disposal strategies for CO_2 and RW. Moreover, the three countries cooperate in EU projects in both areas. Nonetheless, they seem to follow somewhat different strategies in their respective R&D and implementation.

Germany has considerable technical potential for both CO_2 and RW disposal. The optimistic estimate of the CO_2 disposal potential is in the range of 19–48 Gt CO_2, which is equivalent to 30–60 years of CO_2 emissions from all large stationary sources. The conservative estimate of 20 ± 8 Gt CO_2 is considerably lower. The total RW disposal capacity is assessed at more than 10 million m^3 (about 200 Mt), which could accommodate RW for hundreds of years of expanded nuclear power generation, even without any waste minimization strategy.

While the legal basis for RW disposal has been in place for 50 years, that for CO_2 disposal has emerged only recently. The German CCD act was passed in early 2009, and the BGR is developing standards and criteria for CO_2 disposal sites. Geological disposal of RW is governed by the German Atomic Energy Act of 1959, the revisions thereof, and the Mining Law, and is the exclusive responsibility of the government. As a result, an interesting dichotomy can be observed in the management process in Germany. The site selection for CO_2 disposal is carried out by researchers and the private sector with very little government involvement; site selection for RW disposal is entirely a government-driven process.

So far there has not been much public discussion about the benefits and drawbacks of CCD owing to limited knowledge about this technology. Experts and industry representatives tend to be optimistic about CCD, whereas environmental organizations have declared serious reservations or outright opposition. As far as RW disposal is concerned, site selection criteria have been officially adopted and have barely changed over the past 40 years, but public discussion and socio-political issues have become increasingly important. The political debate culminated in the decision by the Federal Government in 2000 to suspend all exploration at the Gorleben site, which had been selected in a long and thorough assessment process about 20 years before. A government task force in 2002 suggested that a completely new site selection process be started. While the Gorleben moratorium remained in place through mid-2010, no significant effort has been made to start a new site selection process.

In France, CO_2 disposal capacities in the Paris Basin have been partially estimated for two targeted types of geological formations that host aquifer and hydrocarbon fields. Other basins may have a bigger capacity, but given the proximity to the major emission sources, which is one of the key issues in the search for the disposal site, the Paris Basin has been studied the most extensively. The capacities in the Paris Basin have been estimated to lie within the range of 0.3–29.1 Gt CO_2 for the aquifer and 83–117 Mt for hydrocarbon fields. In comparison to France's annual emissions of 390 Mt of CO_2 for 2005, the estimated capacity is viewed as limited in this region.

RW that will have been produced by 2055, including that already produced and stored for final disposal, is estimated to have a volume of 7,912 m^3. The overall capacity of potentially suitable sites in France is much larger than this and a single site, such as the one currently being investigated for its suitability at Meuse/Haute Marne, is expected to host all the existing and foreseen HLW and SF. This is in contrast to the situation for CCD, which would likely require multiple sites for disposing of the greater part of the CO_2 expected to be generated in France over the foreseeable future.

Implementation efforts in the area of RW disposal in the 1990s faced difficulties, as the lack of a public consultation procedure led to strong local opposition against underground research laboratories, which halted the site selection process for a year. Learning from this experience, research on the disposal of HLW has since been strongly regulated by laws, and steps and procedures for public consultations have now been established. For the geological disposal of CO_2, there is at present no comprehensive regulatory framework. Therefore experience from the RW management process might provide useful lessons for the management of CO_2 disposal.

Compared to some other EU countries, the UK has proposed tougher targets to mitigate climate change. It aims to reduce GHG emissions by 20% from 1990 to 2020 and by 80% from 1990 to 2050. With respect to CO_2, legally binding targets have been set in the Climate Change Act of 2008 that require UK CO_2 emissions to be reduced by 26% from 1990 to 2020.

The most significant option for the disposal of CO_2 is offshore sedimentary basins that contain most of the UK's oil- and gasfields. About one-third of the UK might be appropriate for geological disposal of RW due to the availability of suitable and stable rock formations for hosting a geological disposal facility.

The UK Government has taken firm measures to implement and develop CCD technologies and has proposed financial support for four CCD demonstration plants. It has also initiated steps towards the implementation of a geological RW disposal facility and has tasked the Nuclear Decommissioning Authority with managing HLW.

Considerable R&D and implementation-related activities to foster the geological disposal of CO_2 and RW are underway in many other West European countries. In-depth comparative assessments in the national context may well lead to interesting insights, similar to those emerging from the analyses presented in the preceding sections of this chapter.

Acknowledgements The authors wish to thank Romain Boniface and Karine Langlois for the initial literature survey, and Aurora Badulescu for her assistance in preparing this chapter. They are also indebted to Etienne Brosse, Bernard Neerdael, Jürgen Kupitz, Tim McEwen, Julia West and two anonymous referees for their comments and suggestions for improving the various drafts.

References

ACCAT (Advisory Committee on Carbon Abatement Technologies) (2009) Accelerating the deployment of carbon abatement technologies — with special focus on carbon capture and storage. Advisory document from ACCAT. URN 09/Z108. http://www.berr.gov.uk/files/file50470.pdf

AGEB (Arbeitsgemeinschaft Energiebilanzen e.V.) (2009) Auswertungstabellen. http://www.ag-energiebilanzen.de/viewpage.php?idpage=139 [in German]

AkEnd (Arbeitskreis Auswahlverfahren Endlagerstandort) (2002) Auswahlverfahren für Endlagerstandorte. Empfehlungen des AkEnd — Arbeitskreis Auswahlverfahren Endlagerstandorte [in German]. http://www.bfs.de/de/endlager/faq/langfassung_abschlussbericht_akend.pdf

Alter U, Götz C, Koch W, Mester W, Mohr S (2006) Management of radioactive waste and spent fuel, balance of reprocessing. Presentation at the Second Review Meeting of the Joint Convention on the Safety of Spent Fuel Management and on the Safety of Radioactive Waste Management, IAEA, Vienna, 17 May. http://www.bmu.de/files/pdfs/allgemein/application/pdf/jc_waste_storage_reprocessing.pdf

ANDRA (Agence nationale pour la gestion des déchets radioactifs — National Radioactive Waste Management Agency) (2005a) Dossier 2005 argile. [in French]. http://www.andra.fr/download/site-principal/document/editions/266.pdf

ANDRA (Agence nationale pour la gestion des déchets radioactifs — National Radioactive Waste Management Agency) (2005b) Dossier 2005 granite. [in French]. http://www.andra.fr/download/site-principal/document/editions/267.pdf

ANDRA (Agence nationale pour la gestion des déchets radioactifs — National Radioactive Waste Management Agency) (2009) Inventaire national des matières et déchets radioactifs: rapport de synthèse [in French]. http://www.andra.fr/download/site-principal/document/inventaire/351.pdf

Appel D (2008) Das Gorleben-Moratorium und Argumente für ein Standortauswahlverfahren. Paper presented at Internationales Endlagersymposium Berlin 2008, Berlin, 30 October – 1 November [in German]. www.bundesumweltministerium.de/files/pdfs/allgemein/application/pdf/endlagersymp_vortrag_appel.pdf

Baldwin T, Chapman T, Neall F (2008) Geological disposal options for high-level waste and spent fuel. Report for the UK Nuclear Decommissioning Authority. http://www.nda.gov.uk/loader.cfm?csModule=security/getfile&pageid=20941

Barbier J (1996) The CO2 storage capacity of France. In: Holloway S (ed) The Underground Disposal of Carbon Dioxide: Final Report of the Joule II Project Number CT92-0031. British Geological Survey, Keyworth, pp 41–46

Bataille C (1994) Mission de médiation sur l'implantation de laboratoires de recherche souterrains: rapport au premier ministre. La Documentation française, Paris, [in French]

Bataille C (1996) L'évolution de la recherché sur la gestion des déchets nucléaires à haute activité. Tome I: Les déchets civils. Sénat, rapport no. 299 (1995–1996). Office parlementaire d'évaluation des choix scientifiques et technologiques [in French]. http://www.senat.fr/rap/r95-299/r95-2991.pdf

Bataille C, Birraux C (2005) Report on progress and prospects of research on the management of radioactive wastes. Parliamentary Office for Scientific and Technological Assessment. http://www.senat.fr/opecst/eng/rapport/rapport_dechets_anglais.pdf

Bataille C, Galley R (1998) Rapport sur l'aval du cycle nucléaires. Tome I: Etude générale. Sénat, rapport no. 492 (1997–1998). Office parlementaire d'évaluation des choix scientifiques et technologiques [in French]. http://www.assemblee-nationale.fr/11/rap-off/i0978.asp

Behlau J, Mingerzahn G (2001) Geological and tectonic investigations in the former Morsleben salt mine (Germany) as a basis for the safety assessment of a radioactive waste repository. Eng Geol 61(2–3):83–97

Bentham M, Kirby G (2005) CO_2 storage in saline aquifers. Oil & gas science and technology – Rev. IFP. 60 (3):559–567. Available at: http://ogst.ifp.fr/index.php?option=article&access=doi&doi=10.2516/ogst:2005038

BERR (Department for Business, Enterprise and Regulatory Reform) (2008) Meeting the energy challenge: a white paper on nuclear power. http://www.berr.gov.uk/files/file43006.pdf

BERR (Department for Business, Enterprise and Regulatory Reform) (2007) Development of a CO_2 transport and storage network in the North Sea. Report to the North Sea Basin Task Force. http://www.berr.gov.uk/files/file42476.pdf

BfS (Bundesamt für Strahlenschutz — Federal Office for Radiation Protection) (2008) Abfallmengen//Prognosen. Status of 18 december [in German]. http://www.bfs.de/de/transport/endlager/abfall_prognosen.html

BfS (Bundesamt für Strahlenschutz — Federal Office for Radiation Protection) (2009) Kosten und Kostenverteilung des Endlagerprojekts Schacht Konrad [in German]. http://www.endlager-konrad.de/cln_189/nn_1072978/DE/Themen/Kosten/__node.html?__nnn=true

BGR (Bundesanstalt für Geowissenschaften und Rohstoffe — Federal Institute for Geosciences and Natural Resources) (2007) Nuclear waste disposal in Germany: investigation and evaluation of regions with potentially suitable host rock formations for a geologic nuclear repository. http://www.bgr.bund.de/cln_144/nn_335086/EN/Themen/Geotechnik/Downloads/WasteDisposal__HostRockFormations__en,templateId=raw,property=publicationFile.pdf/WasteDisposal_HostRockFormations_en.pdf

BGR (Bundesanstalt für Geowissenschaften und Rohstoffe — Federal Institute for Geosciences and Natural Resources) (2008) Reserves, resources and availability of energy resources 2007: annual report. http://www.bgr.bund.de/cln_101/nn_335066/EN/Themen/Energie/Downloads/energiestudie__kurzf__2007__en,templateId=raw,property=publicationFile.pdf/energiestudie_kurzf_2007_en.pdf

BGR (Bundesanstalt für Geowissenschaften und Rohstoffe — Federal Institute for Geosciences and Natural Resources) (2009) CO_2 storage: storage options. http://www.bgr.bund.de/cln_101/nn_334990/EN/Themen/Geotechnik/CO2-Speicherung/Speichermoeglichkeiten/speichermoeglichkeiten__inhalt__en.html

BIS and DECC (Department for Business Innovations and Skills and Department of Energy and Climate Change) (2009) The UK low carbon industrial strategy. www.berr.gov.uk/files/file52002.pdf

Biurrun E, Hartje B (2003) License for the Konrad deep geological repository. In: Proceedings of the Waste Management 2003 Symposium. WM symposia Inc. http://www.wmsym.org/archives/2003/pdfs/607.pdf

Blanchard F (2006) Piégeage du CO_2 dans les réservoirs géologiques en France: projet PICOREF. Cadrage réglementaire et enjeux environnementaux. Rapport final BRGM/RP-54638-FR. Bureau de recherche géologiques et minières [in French]. www.brgm.fr/Rapport?code=RP-54638-FR

BMU (Bundesministerium für Umwelt, Naturschutz und Reaktorsicherheit — Federal Ministry for the Environment, Nature Conservation and Nuclear Safety) (2009) Entwurf: Gesetz zur Regelung von Abscheidung, Transport und dauerhafter Speicherung von Kohlendioxid [in German]. http://www.bmu.de/files/pdfs/allgemein/application/pdf/gesetzentwurf_ccs.pdf

BMWi (Bundesministerium für Wirtschaft und Technologie — Federal Ministry of Economics and Technology) (2009) Kernenergie in Deutschland [in German]. http://www.bmwi.de/BMWi/Navigation/Energie/kernenergie.html

Bonijoly D, Barbier J, Robelin C, Kervevan C, Thiery D, Menjoz A, Matray JM, Cotiche C, Herbrich B (2003) Feasibility of CO_2 storage in geothermal reservoirs: example of the Paris Basin — France. Rapport final BRGM/RP-52349-FR. Bureau de recherche géologiques et minières. http://www.brgm.fr/publication/pubDetailRapportSP.jsp?id=RSP-BRGM/RP-52349-FR

Bonijoly D, Brosse E, Rigollet C, Manai T (2006) Site selection for CO_2 storage in deep aquifers of the Paris Basin — France. Presentation made at the International Symposium on Site Characterization for CO_2 Geological Storage (CO_2SC 2006), Berkeley, CA, 20–22 March. http://esd.lbl.gov/CO2SC/co2sc_presentations/Regional_Scale_Site_Select/bonijoly.pdf

Bonijoly D, Ha-Duong M, Leynet A, Bonneville A, Broseta D, Fradet A, Le Gallo Y, Munier G, Lagny C, Lagneau V et al (2009) METSTOR: a GIS to look for potential CO_2 storage zones in France. In: Gale J, Herzog H, Braitsch J (eds) Proceedings of the 9th International Conference on Greenhouse Gas Control Technologies; Energy Procedia, 1:2809–2816

Bradshaw J, Bachu S, Bonijoly D, Burruss R, Holloway S, Christensen NP, Mathiasen OM (2007) CO_2 Storage capacity estimation: issues and development of standards. Int J Greenhouse Gas Control 1:62–68

Brasser T, Droste J, Müller-Lyda I, Neles J, Sailer M, Schmidt G, Steinhoff M (2008) Endlagerung wärmeentwickelnder radioaktiver Abfälle in Deutschland. Institut für angewandte Ökologie e.V. and Gesellschaft für Anlagen- und Reaktorsicherheit (GRS) mbH [in German]. http://endlagerung.oeko.info/dokumente/Hauptband.pdf

Bräuer V (2008) Ergebnisse der BGR-Studien zur Endlagerung hochradioaktiver Abfälle in Tongesteinen, Steinsalz und kristallinen Gesteinen. Paper presented at Internationales Endlagersymposium 2008, Berlin, 30 October – 1 November [in German]. http://www.bundesumweltministerium.de/files/pdfs/allgemein/application/pdf/endlagersymp_vortrag_braeuer.pdf

Bräuer V, Reh M, Schulz P, Schuster P, Sprado, K-H (1994) Endlagerung stark wärmeentwickelnder radioaktiver Abfälle in tiefen geologischen Formationen Deutschlands: Untersuchung und Bewertung von Regionen in nichtsalinaren Formationen. Report of the Federal Institute for Geosciences and Natural Resources (BGR) [in German]. http://www.bgr.bund.de/cln_145/nn_322852/DE/Themen/Geotechnik/Downloads/BGR__kristallinstudie,templateId=raw,property=publicationFile.pdf/BGR_kristallinstudie.pdf

Brennecke P (1995) Anforderungen an endzulagernde radioaktive Abfälle (Endlagerungsbedingungen, Stand: Dezember 1995) — Schachtanlage Konrad. Interner BfS-Bericht ET-IB-79. Bundesamt für Strahlenschutz, Salzgitter [in German]

BRGM (Bureau de recherche géologiques et minières — Bureau of Geological and Mining Research) (2008) METSTOR website [in French]. http://www.metstor.fr/

Brosse E (2005) PICOREF: a French R&D programme on CO_2 geological storage. Greenhouse Issues, 77. http://www.ieaghg.org/index.php?/2009120864/greenhouse-issues-march-2005-number-77.html#article3

Brosse E (2009) CO_2 capacity in the oil fields in the Paris Basin Personal communication.

Brosse E, de Smedt G, Bonijoly D, Garcia D, Saysset S, Manaï T, Thoraval A, Crepin S (2006) PICOREF: towards an experimental site for CO_2 geological storage in the Paris Basin? In: Gale JJ, Røkke N, Zweigel P, Svenson H (eds) Proceedings of the 8th International Conference on Greenhouse Gas Control Technologies. Elsevier, Oxford, CD-ROM

Bund (1989) Bekanntmachung der Richtlinie zur kontrolle radioaktiver Abfälle mit vernachlässigbarer Wärmeentwicklung, die nicht an eine Landessammelstelle abgeliefert werden vom 16. Januar 1989. Bundesanzeiger 41(63a):1–12, [in German]

Carbon Capture Journal (2009) TOTAL's Lacq pilot — Project update. Carbon Capture Journal, 8. http://www.carboncapturejournal.com/issues/CCJ8web.pdf

Christensen NP, Holloway S (2004) Geological storage of CO_2 from combustion of fossil fuel: the GESTCO project — Summary report. European Union Fifth Framework Programme for

Research & Development. Project no. ENK6-CT-1999-00010. European Union. http://www.geus.dk/program-areas/energy/denmark/co2/GESTCO_summary_report_2ed.pdf

CO_2SINK (2009). www.co2sink.de

CO_2STORE (2009). http://www.co2store.org/

CoRWM (Committee on Radioactive Waste Management) (2006) Managing our radioactive waste safely. CoRWM's recommendations to Government. CoRWM document 700. Available at: www.corwm.org.uk

CoRWM (Committee on Radioactive Waste Management) (2009a) Geological disposal of higher activity radioactive wastes. CoRWM report to Government. CoRWM document 2550. http://www.corwm.org.uk/Pages/Current%20Publications/2550%20CoRWM%20Report%20on%20Geological%20Disposal%20Final%2031%20July%2009.pdf

CoRWM (Committee on Radioactive Waste Management) (2009b) Interim storage of higher activity wastes and the management of spent fuels, plutonium and uranium. CoRWM report to Government. CoRWM document 2500. http://www.corwm.org.uk/Pages/Involving%20People/2500%20-%20CoRWM%20Interim%20Storage%20Report%20Final%2031%20March%2009.pdf

CoRWM (Committee on Radioactive Waste Management) (2009c) Report on national research and development for interim storage and geological disposal of higher activity radioactive wastes, and management of nuclear materials. CoRWM report to Government. CoRWM document 2543ES. http://www.corwm.org.uk/Pages/Current%20Publications/2543ES%20Executive%20Summary%20of%20CoRWM%20Report%20on%20RD%20Final%2030%20October%202009.pdf

DECC (Department of Energy and Climate Change) (2009a) No new coal without CCS — Miliband. Press release, 23 April. www.decc.gov.uk/en/content/cms/news/pn050/pn050.aspx

DECC (Department of Energy and Climate Change) (2009b) Strategic environmental assessment for a framework for the development of clean coal: post adoption statement. Available at: http://www.decc.gov.uk/en/content/cms/consultations/clean_coal/clean_coal.aspx

DECC (Department of Energy and Climate Change) (2009c) The UK low carbon transition plan: national strategy for climate & energy. Available at: http://www.decc.gov.uk/en/content/cms/publications/lc_trans_plan/lc_trans_plan.aspx

Defra (Department for Environment, Food and Rural Affairs) (2008) Sustainable development: climate change and energy — Indicators. Greenhouse gas emissions. http://www.defra.gov.uk/sustainable/government/progress/national/climate-change.htm

Defra (Department for Environment, Food and Rural Affairs), BERR (Department for Business, Enterprise and Regulatory Reform) and the devolved administrations for Wales and Northern Ireland. (2008) Managing radioactive waste safely: a framework for implementing geological disposal. A white paper. Cm 7386. http://www.official-documents.gov.uk/document/cm73/7386/7386.pdf

DGEMP (Direction générale de l'énergie et des matières premières) (2005) Rapport du groupe de travail relatif au coût d'un stockage souterrain de déchets radioactifs de haute activité et à vie longue [in French]. http://www.developpement-durable.gouv.fr/energie/nucleair/rapport-gt-cout-stockage.htm

Durst P, Kervevan C (2007) Modélisation 3D couplée chimie–transport de l'injection de CO_2 à l'échelle d'un site pilote en aquifère salin profond (Dogger, bassin de Paris). Rapport d'avancement. Rapport BRGM/RP-55667-FR. Bureau de recherche géologique et minière, Orléans [in French]

EA and NIEA (Environment Agency and the Northern Ireland Environment Agency) (2009) Geological disposal facilities on land for solid radioactive wastes: guidance on requirements for authorisation. http://publications.environment-agency.gov.uk/pdf/GEHO0209BPJM-e-e.pdf

EC (European Commission) (2009) EU Action Against Climate Change: Leading Global Action to 2020 and Beyond. European Communities, Luxembourg

EEA (European Environment Agency) (2008) Greenhouse gas emission trends and projections in Europe 2008: tracking progress towards Kyoto targets. Report no. 5/2008. http://www.eea.europa.eu/publications/eea_report_2008_5

Fischedick M, Esken A, Pastowski A, Schüwer D, Supersberger N (2007) RECCS Strukturell-ökonomisch-ökologischer Vergleich regenerativer Energietechnologien (RE) mit Carbon Capture and Storage (CCS). Wuppertal Institut für Klima, Umwelt, Energie GmbH, Wuppertal [in German]

Förster A, Giese R, Juhlin C, Norden B, Springer N, CO_2SINK Group (2008) The Geology of the CO_2SINK Site: from regional scale to laboratory scale. In: Gale J, Herzog H, Braitsch J (eds) Proceedings of the 9th International Conference on Greenhouse Gas Control Technologies; Energy Procedia, 1:2911–2918

Gapillou C, Thibeau S, Mouronval G, Lescanne M (2009) Building a geocellular model of the sedimentary column at Rousse CO_2 geological storage site (Aquitaine, France) as a tool to evaluate a theoretical maximum injection pressure. In: Gale J, Herzog, H, Braitsch J(eds) Proceedings of the 9th International Conference on Greenhouse Gas Control Technologies; Energy Procedia, 1:2937–2944

Gerling P (2004) COORETEC — Optionen zur CO_2-Speicherung in Deutschland. Paper presented at Fachkongress: innovative Technologien zur Stromerzeugung — auf dem Weg zu CO_2-freien Kohle- und Gaskraftwerken. Berlin, 10–12 May [in German]

Gough C (2008) State of the art in carbon dioxide capture and storage in the UK: an experts' review. Int J Greenhouse Gas Control 2:155–168

Greenpeace (2009) CO2-Endlager sind geologische Zeitbomben [in German]. http://www.greenpeace.de/themen/klima/nachrichten/artikel/co2_endlager_sind_geologische_zeitbomben-1/

GS (Geological Society) (2008) The geological disposal of radioactive waste: How geologists understand the earth beneath our feet. Conference organized by The Geological Society, London, 24 October

Ha-Duong M, Nadaï A, Compos AS (2009) A survey on the public perception of CCS in France. In: Gale J, Herzog H, Braitsch J (eds) Proceedings of the 9th International Conference on Greenhouse Gas Control Technologies; Energy Procedia, 1:4757–4764

Hansson A, Bryngelsson M (2009) Expert opinions on carbon dioxide capture and storage—A framing of uncertainties and possibilities. Energy Policy 37(6):2273–2282

Helda H, Edenhofer O (2009) CCS-Bonds as a superior instrument to incentivize secure carbon sequestration. In: Gale, J Herzog H, Braitsch J (eds) Proceedings of the 9th International Conference on Greenhouse Gas Control Technologies; Energy procedia, 1(1):4559–4566

Hendriks C, Graus W, van Bergen F (2004) Global carbon dioxide storage potential and costs. Report EEP-02001. Ecofys. http://www.ecofys.com/com/publications/documents/GlobalCarbonDioxideStorage.pdf

Herzog H, Drake E, Adams E (1997) CO_2 Capture, reuse and storage technologies for mitigating global climate change. A white paper. DOE Order No. DE-AF22-96PC1257. http://sequestration.mit.edu/pdf/WhitePaper.pdf

HMSO (Her Majesty's Stationery Office) (1995) Review of radioactive waste management policy: final conclusions. Cm 2919

Holloway S, Vincent CJ, Kirk KL (2006) Industrial carbon dioxide emissions and carbon dioxide storage potential in the UK. Study by the British geological survey. DTI report no. COAL R308, DTI/PubURN 06/2027. Department of trade and industry. http://www.berr.gov.uk/files/file35684.pdf

Hoth P, Wirth H, Krull P, Olea R, Feldrappe H, Reinhold K (2005) Tonstein-Formationen—eine mögliche Alternative für die Endlagerung radioaktiver Abfälle in Deutschland? Z Geol Wiss 33(4/5):209–241 [in German]

Hoth P, Wirth H, Reinhold K, Bräuer V, Krull P, Feldrappe H (2007) Endlagerung radioaktiver Abfälle in tiefen geologischen Formationen Deutschlands: Untersuchung und Bewertung von Tongesteinsformationen. Bundesanstalt für Geowissenchaften und Rohstoffe, Hannover, [in German]

IAEA (International Atomic Energy Agency) (1994) Classification of radioactive waste, IAEA Safety Series No. 111-G-1.1. IAEA, Vienna

IAEA (International Atomic Energy Agency) (2008) Country waste profile report for France. Reporting year: 2006. http://newmdb.iaea.org/getrepfile.aspx?ByCountry=FR&ByYear=6&RPart=0

IAEA (International Atomic Energy Agency) (2009) Power reactor information system: PRIS. http://www.iaea.org/programmes/a2/

IEA (International Energy Agency) (2008a) CO_2 Emissions from Fuel Combustion 2008, IEA Statistics. OECD Publishing, Paris

IEA (International Energy Agency) (2008b) Energy Balances of OECD Countries 2008, IEA Statistics. OECD Publishing, Paris

IEA (International Energy Agency) (2008c) Energy Statistics of Non-OECD countries 2008, IEA Statistics. OECD Publishing, Paris

Jones NS, Holloway S, Smith NJP, Browne MAE, Creedy DP, Garner K, Durucan S (2004) UK coal resource for new exploitation technologies. DTI report no. COAL R271, DTI/Pub URN 04/1879. DTI cleaner coal technology transfer programme. http://www.berr.gov.uk/files/file19153.pdf

Kockel F, Krull P (1995) Endlagerung stark Wärmeentwickelnder radioaktiver Abfälle in tiefen geologischen Formationen Deutschlands: Untersuchung und Bewertung von Salzformationen. Report of the Bundesanstalt für Geowissenschaften and Rohstoffe [in German]. http://www.bgr.bund.de/cln_092/nn_329630/DE/Themen/Geotechnik/Downloads/BGR__salzstudie.html

Koecher R (2007) Keine Angst vor Klimawandel und Kernenergie. Frankfurter Allgemeine Zeitung, FAZ.NET, 13 June. http://www.faz.net/s/RubFC06D389EE76479E9E76425072B196C3/Doc~E3A185421D9524B5ABA0A182D70FCBAFA~ATpl~Ecommon~Scontent.html

Kreft E, Bernstone C, Meyer R, May F, Arts R, Obdam A, Svensson R, Eriksson S, Durst P, Gaus I, van der Meer B, Geel C (2007) "The Schweinrich structure", a potential site for industrial scale CO_2 storage and a test case for safety assessment in Germany. Int J Greenhouse Gas Control 1:69–74

Kreienmeyer M, Schmidt H, Gläß F, Mauke R (2004) Central part of the Morsleben Repository — Geomechanical situation before stabilization. Paper presented at DisTec 2004: International Conference on Radioactive Waste Disposal, Berlin, 26–28 April. http://www.bfs.de/de/endlager/publika/distec2004lect_eram03.pdf

Krooss BM, May F (2006) State of CO_2 capture and subsurface storage activities in Germany. In: Lombardi S, Altunina LK, Beaubien SE (eds) Advances in the Geological Storage of Carbon Dioxide (289–301), vol 65, NATO Science Series: IV: Earth and Environmental Sciences. Springer, Dordrecht

Kuckshinrichs W, Markewitz P, Hake J-F (2004) CO_2 capture and sequestration in Germany: status and perspectives. In: Proceedings of the Third Annual Conference on Carbon Capture & Sequestration. National Energy Technology Laboratory. http://www.netl.doe.gov/publications/proceedings/04/carbon-seq/143.pdf

Marland G, Boden TA, Andres RJ (2009) Global, regional, and national fossil-fuel CO_2 emissions. Carbon Dioxide Information Analysis Center, Oak Ridge National Laboratory, US Department of Energy, Oak Ridge, Tenn. doi: 10.3334/CDIAC/00001

Matthes FC, Barth R, Schulze F, Hermann A, Acker H (2009) Regulierung von CO2-Abscheidung und Ablagerung (CCS): Der Entwurf für das Kohlendioxid-Speicherungsgesetz (KSpG) — Stellungnahme zum Entwurf für das Kohlendioxid-Speicherungsgesetz (KSpG) zur Anhörung des Ausschusses für Umwelt, Naturschutz und Reaktorsicherheit des 16. Deutschen Bundestages, Öko-Institut e.V. [in German]. http://www.oeko.de/oekodoc/901/2009-019-de.pdf

May F, Brune S, Gerling P, Krull P (2003) Möglichkeiten zur CO_2-Speicherung in Deutschland — eine Bestandsaufnahme im Jahre 2003. geotechnik 26(3):162–172, [in German]

May F, Müller C, Bernstone C (2005) How much CO_2 can be stored in deep saline aquifers in Germany? VGB PowerTech 6:32–37

NDA (Nuclear Decommissioning Authority) (2007) Nuclear Decommissioning Authority (NDA) launches new radioactive waste management directorate. http://www.nda.gov.uk/news/nuclear-decommissioning-authority-launches-new-radioactive-waste-management-directorate.cfm

NDA (Nuclear Decommissioning Authority) (2009a) NDA business plan 2009/2012. http://www.nda.gov.uk/loader.cfm?csModule=security/getfile&pageid=28874

NDA (Nuclear Decommissioning Authority) (2009b) UK radioactive higher activity waste storage review. Issue 1. http://www.nda.gov.uk/documents/upload/UK-Radioactive-Higher-Activity-Waste-Storage-Review-March-2009.pdf

NDA (Nuclear Decommissioning Authority) (2009c) Annual report & accounts 2007/08. http://www.nda.gov.uk/loader.cfm?csModule=security/getfile&pageid=21824

NDA and Defra (Nuclear Decommissioning Authority and Department for Environment, Food and Rural Affairs) (2008) The 2007 UK radioactive waste inventory: a summary of information for international reporting. Report prepared by Pöyry Energy Limited. Defra/RAS/08.002, NDA/RWMD/004. http://www.nda.gov.uk/ukinventory/documents/Reports/loader.cfm?url=/commonspot/security/getfile.cfm&pageid=10353

NEA (Nuclear Energy Agency) (2006) Disposal of radioactive waste: forming a new approach in Germany — FSC Workshop Proceedings, Hitzacker and Hamburg, Germany, 5–8 October 2004, NEA No. 6117. OECD Publishing, Paris

OECD (Organisation for Economic Co-operation and Development) (2009) Radioactive waste management and decommissioning in France. Report. OECD, Paris. http://www.nea.fr/rwm/profiles/France_report_web_2009.pdf

OLHC (Office of the Leader of the House of Commons) (2009) Energy Bill 2009–10, Draft Legislative Programme (DLP) for the fifth parliamentary session 2009–10 (Cm 7739). www.commonsleader.gov.uk/output/Page2830.asp

Pfeiffer F (2007) Neubewertung von wirtschaftlichen Auswirkungen aus der Verfolgung und Realisierung des Ein-Endlagerkonzeptes: Abschlussbericht zum Vorhaben SR 2542. Auftrags-Nr.: 857774. Gesellschaft für Anlagen und Reaktorsicherheit (GRS) mbH. http://www.bfs.de/de/endlager/publika/abschlussbericht_ein_endlagerkonzept.pdf [in German]

Preuss J, Eilers G, Mauke R, Müller-Hoeppe N, Engelhardt H-J, Kreienmeyer M, Lerch C, Schrimpf C (2002) Post closure safety of the Morsleben repository. In: Proceedings of the Waste Management 2002 Symposium. WM symposia, Inc. http://www.wmsym.org/archives/2002/Proceedings/17/486.pdf

Préfecture des Pyrénées-Atlantiques (2009) Arrêté no. 09/IC/122 autorisant la société TOTAL E&P France à exploiter un pilote de production, de captage, de transport et de stockage géologique de CO_2. www.pyrenees-atlantiques.pref.gouv.fr/sections/actions_de_l_etat/environnement_et_dev/actualites/arrete_prefectoral_d/downloadFile/file/AP_pilote_CO2signe.pdf?nocache=1243439258.8 [in French]

Rebscher D, May F, Oldenburg CM (2006) Sequestration of CO2 in the Altmark natural gas field, Germany: mobility control to extend enhanced gas recovery. Paper LBNL 60157. Lawrence Berkeley National Laboratory Berkeley, CA. http://repositories.cdlib.org/lbnl/LBNL-60157

Rostock S (2008) Status of CCS in Germany. Paper presented at the seminar on Carbon Storage Below the North Sea. E3G and Zero Emission Resource Organisation, London, 1 September

Sailer M (2008) Nuclear waste repository case studies: Germany. Bulletin of Atomic Scientists, 29 August. http://www.thebulletin.org/node/4227

SCCS (Scottish Centre for Carbon Storage) (2009) Opportunities for CO_2 storage around Scotland — An Integrated strategic research study. http://www.scotland.gov.uk/Resource/Doc/270737/0080597.pdf

Shackley S, Waterman H, Godfroij P, Reiner D, Anderson J, Draxlbauer K, Flach T (2007) Stakeholder perceptions of CO_2 capture and storage in Europe: results from a survey. Energy Policy 35:5091–5108

Stroink L (2004) The R&D programme GEOTECHNOLOGIEN — A national platform for CO_2 storage initiatives in Germany. In: Proceedings of the Third Annual Conference on Carbon Capture & Sequestration. National Energy Technology Laboratory. http://www.netl.doe.gov/publications/proceedings/04/carbon-seq/Stroink.pdf

Stroink L (2006) Underground Storage of CO_2 in Germany: a new research programme as part of the national R&D-programme geotechnologien. Greenhouse Issues, 81

Summers, D, Carrington D (2008) Government pledges to cut carbon emissions by 80% by 2050. guardian.co.uk, 16 October. http://www.guardian.co.uk/politics/2008/oct/16/greenpolitics-edmiliband

Total (2007) Lacq CO_2 capture and geological storage pilot project. Project information dossier. http://www.total.com/MEDIAS/MEDIAS_INFOS/1872/EN/CO2-Lacq-Total-Project-Information-dossier.pdf

UBA (Umweltbundesamt — Federal Environment Agency) (2006) CO_2 capture and storage — Only an interim solution: possible impacts, potential and requirements. Summary of a position paper. http://www.umweltdaten.de/publikationen/fpdf-k/k3074.pdf

UBA (Umweltbundesamt — Federal Environment Agency) (2009a) National inventory report for the German greenhouse gas inventory 1990–2007. Submission under the United Nations Framework Convention on Climate Change. http://unfccc.int/national_reports/annex_i_ghg_inventories/national_inventories _submissions/items/4771.php

UBA (Umweltbundesamt — Federal Environment Agency) (2009b) Climate protection: greenhouse gas emissions in 2008 at their lowest since 1990 — Germany reaches its Kyoto target. Press Release No. 16/2009. Joint press release by the Federal Environment Ministry and the Federal Environment Agency (UBA). http://www.umweltbundesamt.de/uba-info-presse-e/2009/pdf/pe09-016_climate_protection_greenhouse_gas_emissions_in_2008_at_their_lowest_since_1990.pdf

UK Parliament (2008) Climate Change Act 2008. http://www.opsi.gov.uk/acts/acts2008/pdf/ukpga_20080027_en.pdf

Vattenfall (2009) Vattenfall's project on CCS. http://www2.vattenfall.com/www/co2_en/co2_en/index.jsp

Veolia Environment (2008) CO_2 Storage: an important research and development challenge, http://www.sustainable-development.veolia.com/en/Articles/20080708,co2-storage.aspx

Wilson PD (ed) (1996) The Nuclear Fuel Cycle: From Ore to Wastes. Oxford Science Publications, Oxford University Press, New York, NY

Witherspoon PA, Bodvarsson GS (eds) (2006) Geological challenges in radioactive waste isolation: fourth worldwide review, Report LBNL-59808. Lawrence Berkeley National Laboratory, Berkeley, CA

WNA (World Nuclear Association) (2009a) Nuclear power in Germany. http://www.world-nuclear.org/info/inf43.html

WNA (World Nuclear Association) (2009b) Nuclear power in the UK. http://www.world-nuclear.org/info/inf84.html

WNN (World Nuclear News) (2008) Most German businesses support nuclear extension. 3 September. http://www.world-nuclear-news.org/newsarticle.aspx?id=22130&terms=german+80%25

WWF (World Wildlife Fund) (2009) Stellungnahme zum Gesetz zur Regelung von Abscheidung, Transport und dauerhafter Speicherung von Kohlendioxid (CCS). Positionspapier. WWF Deutschland. http://www.wwf.de/downloads/publikationsdatenbank/ddd/30517/ [in German]

Ziesing H-J (2006) Trotz Klimaschutzabkommen: Weltweit steigende CO_2-Emissionen. DIW Wochenbericht 73(35):485–499

Carbon Dioxide and Radioactive Waste in Central and Eastern Europe: A Regional Overview of Geological Storage and Disposal Potential

Zsuzsanna Hódossyné Hauszmann, Péter Scholtz, and György Falus

Abstract Given that carbon dioxide (CO_2) capture and storage is a promising option for reducing greenhouse gas emission levels and that nuclear power-based energy production is a proven carbon-free technology, we give a regional overview of the geological storage potential of CO_2 and disposal potential of radioactive waste in eight countries of Central and Eastern Europe (CEE): Bulgaria, Croatia, the Czech Republic, Hungary, Poland, Romania, Slovakia and Slovenia (referred to here as the CEE-8). A region-specific summary of CO_2 emission point sources and emission figures is given as well as of nuclear power plant facilities, their waste types and respective volumes. In addition, we provide a description of the geological storage types available for CO_2 in the CEE-8, namely depleted hydrocarbon fields, saline aquifers and coal reserves. We give insights into the determining factors for site selection for geological storage of CO_2. We review the countries in the region that are considering and/or working on a radioactive waste disposal facility. An assessment is provided on the status of the site selection programme, if any, in each country. Potential geological features are summarized in terms of possible disposal sites. We compare the identifiable similarities and differences in geological storage of CO_2 and disposal of radioactive waste among the countries studied and between the two types of substances to be disposed of.

Keywords CO_2 • Radioactive waste • Central and Eastern Europe • Geological storage • Radioactive waste disposal

Z.H. Hauszmann
Department of Strategy, Hungarian Atomic Energy Authority, Fényes Adolf utca 4, 1539 Budapest, Hungary
e-mail: hodossyne@haea.gov.hu

P. Scholtz (✉) and G. Falus
Eötvös Loránd Geophysical Institute, Kolumbusz utca 17–23, 1145 Budapest, Hungary
e-mail: scholtz@elgi.hu; falus@elgi.hu

F.L. Toth (ed.), *Geological Disposal of Carbon Dioxide and Radioactive Waste: A Comparative Assessment*, Advances in Global Change Research 44,
DOI 10.1007/978-90-481-8712-6_15,

1 Introduction

In view of rising global energy demand and the absence of a breakthrough in carbon-free technology, a portfolio of options is needed to manage the risks of global climate change, with as many sustainable options as possible being used and developed. Carbon capture and storage (CCS), that is, the capture of carbon dioxide (CO_2) produced from chemical and combustion processes and its storage in geological formations, is a relatively new option that is rapidly gaining support. In a study released in December 2004, the OECD International Energy Agency (IEA) states: ‘CCS is a promising emission reduction option with potentially important environmental, economic and energy supply security benefits.’ (IEA 2004) On a lifecycle basis, nuclear power is a low-carbon technology and has the potential to supply a substantial part of the world’s electricity needs. Unfortunately, the high initial capital needed to build a nuclear power plant (NPP), along with environmental and security issues, limit the capacity of countries to establish or extend their reliance on nuclear energy. A major obstacle in this process is the safe disposal of radioactive waste (RW), for which geological disposal is an established and accepted solution.

The official announcement by the European Union (EU) of Europe’s commitment to reduce greenhouse gas emissions by 20% by 2020 has put clear pressure on governments and industry in Europe to seriously address emission reduction options. This is particularly so in the case of the countries involved in this study: Bulgaria, Croatia, the Czech Republic, Hungary, Poland, Romania, Slovakia and Slovenia (referred to collectively here as the eight Central and Eastern European countries, or CEE-8). The threats, damage and the overall negative effect of climate change are starting to be understood by a wide cross section of the general public, and also by research and industry in the CEE-8. Most stakeholders in the region now readily agree that action must be taken. However, there is no consensus among the stakeholders as to what form this action should take; this, though, is probably due to a lack of knowledge regarding possible climate change mitigation options.

2 CO_2 Emissions and Radioactive Waste Generation

The structure of energy production is similar in each of the target countries, being largely based on fossil fuel combustion, in many cases with very low efficiency (for instance, a lignite power plant has only 30% conversion efficiency). The role of natural gas in energy production has risen in the last few decades and is expected to increase continuously in the short to long term.

According to the national greenhouse gas inventories of the CEE-8 for 2005, CO_2 emissions from point sources account for a total of approximately 510 million tonnes (Mt)/year. These point source emissions are responsible for more than 70% of the total emissions of the CEE-8 countries.

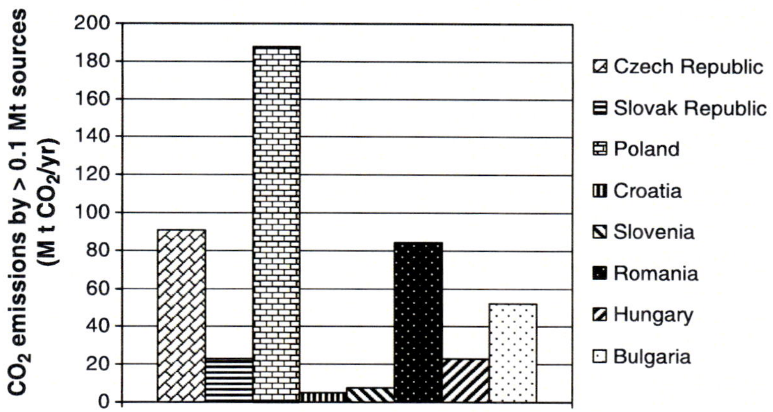

Fig. 1 Annual CO_2 emissions by larger point sources in the CEE-8 countries

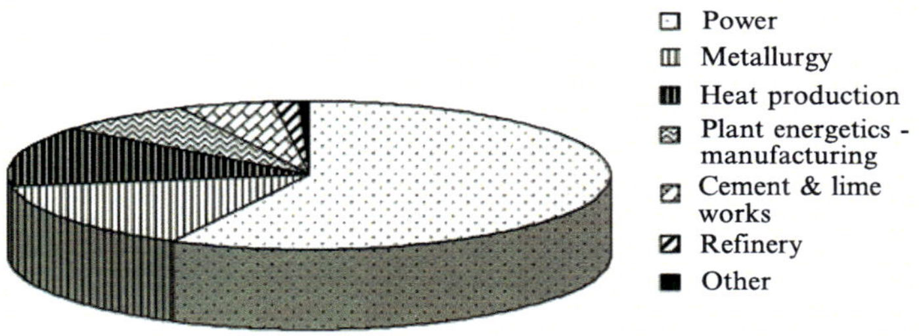

Fig. 2 Relative distribution of CO_2 emissions among industries in the Czech Republic

To obtain information about the geographical distribution of CO_2 emissions, an inventory of point sources greater than 100,000 t CO_2/year was created. The inventory output is presented in Fig. 1.

Power plants have the highest emission rates, followed by heating facilities and energy producers for manufacturing industry. The typical distribution of emissions across different industry segments in the Czech Republic is shown in Fig. 2. Metallurgical plants are also major emitters, particularly in Slovakia; cement and lime works and chemical factories are among other major contributors.

The CO_2 concentration in industrial exhaust gases is usually lower than 20% on account of the combustion processes used. The exception is chemical factories, where CO_2 concentration in the flue gas is much higher.

The generally accepted pathway to decreasing CO_2 emissions in line with EU reduction prescriptions is to increase energy production and consumption efficiency. Nevertheless, CCS is also beginning to be accepted as a viable way of mitigating the effects of climate change. While the research and industrial sectors in the

CEE-8 are preparing to start deployment of CCS in the near future, governments are lagging behind. Nevertheless, the regulations, financing strategy and overall understanding of the role of CCS in climate change mitigation and its context in energy and energy safety are improving at the political level.

Nuclear energy accounts for 16% of global electricity production. NPP operations produce most of the RW, making appropriate nuclear waste management and safe disposal indispensable. However, waste is also derived from the use of radioactive substances for medical, agricultural, research-related, industrial and educational purposes. RW can also originate from certain specific sources such as, for example, naturally occurring radioactive materials (NORM) and technologically enhanced naturally occurring radioactive materials (TENORM). As the overall radiotoxicity of the waste from these sources is low, we deal here only with RW from the nuclear industry.

Geological disposal of high-level RW is now the accepted disposal solution around the world. A range of host geological formations have been considered for deep repositories, including hard crystalline rocks (granite, gneiss and volcanic tuff), argillaceous rocks (clays, mudrocks, shales) and evaporate formations (dome and bedded salts).

The requirements for a deep geological repository are not just technical (although these are perhaps the main issue), but go above and beyond a straightforward technical feasibility study and a demonstration of safety to meet regulatory standards. They also include ethics, security, environmental acceptability, public acceptance and economic viability.

In the CEE-8 there are 22 nuclear power units at eight locations, as shown in Fig. 3. Two countries, Croatia and Poland, have no NPPs. Most of Poland's energy requirement is met by coal-based power plants. The rest comes from wind and hydroelectric plants. Krško NPP in Slovenia meets some of Croatia's energy requirements.

All CEE-8 countries, apart from Croatia and Poland, have RW disposal facilities (see Fig. 3). The existing facilities allow disposal of short-lived low- and intermediate-level waste (LILW-SL) and, in certain cases, long-lived low- and intermediate-level waste (LILW-LL). The repositories used are surface repositories, except in Romania, which has a geological facility. The current host rocks for these repositories are clay loess, sandy clay, granite, sedimentary and crystalline rocks.

In every CEE-8 country that has RW disposal facilities, the volume of high-level waste (HLW) is very low compared to other types of RW. The HLW produced through the operation of NPPs is usually represented by the spent fuel kept in interim storage facilities at these sites. The statistics that follow below regarding spent fuel and HLW on a country-by-country basis are taken from the International Atomic Energy Agency (IAEA 1997).

In Romania the total amount of fuel that will have accumulated by 2020 is estimated to be 4,170 t of heavy metal (tHM). In the Czech Republic HLW from decommissioning is estimated at about 20,000 m^3 and the volume of spent fuel at around 3,000 tHM. Bulgaria has sent some 1,168 spent fuel assemblies to Russia, which will all eventually be returned as vitrified HLW. This amount could rise to 7,331 in the future, effectively covering all the spent fuel currently in storage at NPPs, namely around 100 m^3 of vitrified HLW. In Hungary there could be 200 m^3

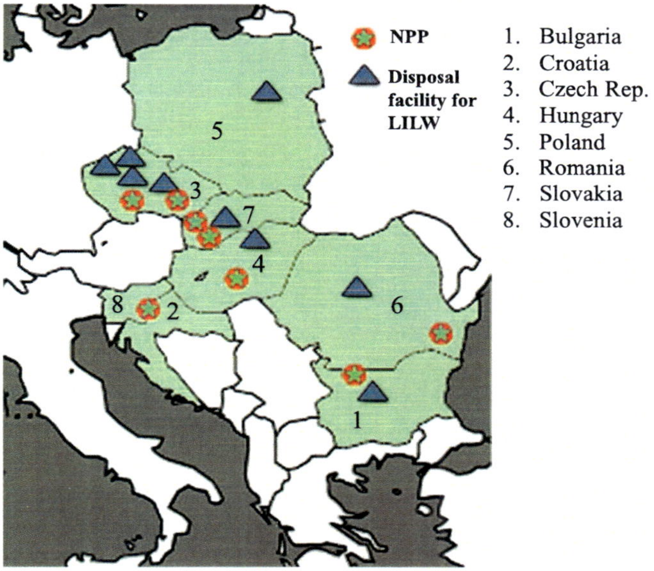

Fig. 3 Location of existing disposal facilities and nuclear power plants in the Central and Eastern European region (*see* Colour Plates)

of HLW resulting from NPP operation by 2032. The average amount of HLW generated is 5 m^3/year, and storage capacity at the Hungarian site is 220 m^3. Spent nuclear fuel (SNF) is kept in an interim spent fuel storage facility. The projected total number of SNF assemblies is 11,000 (1,286 tHM) in 2017. In Slovenia the total storage capacity of the spent fuel pools is 1,694 fuel positions. The decommissioning programme will result in a total of 620 tHM spent fuel assemblies and 16 m^3 of HLW. In Slovakia the total inventory of SNF in 2039 (after the decommissioning of all six nuclear power units) is estimated at 2,374 tHM (18,654 fuel assemblies). Slovakia also has another 2,600 m^3 of long-lived RW.

3 Current Status and Issues of Geological Storage of CO_2 in the Region

The main aims of geological storage of CO_2 are to prevent a large amount of anthropogenic CO_2 being emitted to the environment and to keep it isolated from the atmosphere in secure long-term storage. There is a strong tendency in Europe to standardize site selection criteria in order to reduce storage-related risks. The partners in the EU GeoCapacity project under the Sixth Framework Programme are assessing storage potential and have developed and adopted a series of criteria for use in their site selection procedures to ensure quality and consistency (see, e.g. Saftic et al. 2007, 2008; Hatziyannis et al. 2007, 2008; Sliaupa et al. 2007; Wojcicki 2008; Tarkowski et al. 2008; Martinez et al. 2008; Georgiev 2008; Kuharic 2008). The criteria are intended to serve as the basis for a standardized procedure in future site selection processes (see EC 2009), and focus primarily on the adequate storage potential and appropriate seal integrity of the potential storage complexes (storage formation and surrounding geological environment).

3.1 Main Selection Criteria Concepts

One of the main selection criteria (see Table 1) is that the reservoir should be deep enough to store CO_2 safely, as this ensures that CO_2 stays in its dense phase, which results in low positive buoyancy and makes storage more economically viable. As the storage depth is usually compromised with decreasing porosity and permeability, storage depth is recommended to be not greater than 2,500 m unless there are authoritative data to validate acceptable porosity and permeability values at greater depth (Chadwick et al. 2007).

Seal integrity is considered the most important criterion for site selection. In the geological storage of CO_2, only the geological barrier itself prevents CO_2 from escaping to the environment. The presence of an adequate seal, capable of restraining CO_2 in the reservoir at a given pressure, temperature and chemical composition, is thus essential. Seals are low permeability rocks, typically shales and mudstones with a minimum thickness of 20 m.

Any reservoir used for CO_2 storage should also possess effective petrophysical reservoir properties. Satisfactory reservoir properties (i.e. high permeability and high porosity) are essential to ensure sufficiently high injectivity to make the process economically viable and to guarantee sufficient storage volume.

As investment costs are expected to be the most important among the non-geology-related limiting factors in CO_2 storage, an important aspect of the ideal storage site is sufficient storage volume. The site selected must be capable of storing CO_2 released from the emission source.

3.2 Storage Types

Natural examples and a number of ongoing projects clearly demonstrate that CO_2 can be safely stored in appropriate geological formations such as depleted oil and gas reservoirs, deep saline aquifers and unmineable coal reserves (Fig. 4). Globally, geological formations represent a large storage capacity. Although there are wide differences in storage capacity depending on local geology, it can nevertheless be concluded that the capacity is sufficient to store worldwide anthropogenic CO_2 emissions for decades, and possibly centuries.

Table 1 Key geological indicators for CO_2 storage site suitability (Based on Chadwick et al. 2007)

Basic geology-related criteria	Influential geological and physical parameters	Criteria to be investigated in the screening process	
		Positive indicators	Cautionary indicators
Sufficient depth of reservoir	Pressure Temperature	Depth of crest of reservoir >1,000 m	Depth of crest of reservoir <800 m
		Depth of base of reservoir <2,500 m	Depth of base of reservoir >2,500 m
Petrophysical reservoir properties	Porosity	>20%	<10%
	Permeability	>300 mD	<200 mD
Integrity of seal	Lithology	Low permeable lithologies such as clay	
	Porosity		
	Permeability		
	Thickness	>100 m	<20 m
	Faults	Unfaulted	Faulted
	Heterogeneity	Homogenous	Heterogeneous
	Tectonic activity	No tectonic activity	Tectonic activity
Storage capacity	Reservoir	Total capacity of reservoir estimated to be much larger than the total amount produced from the CO_2 source	Total capacity of reservoir estimated to be similar to or less than the total amount produced from the CO_2 source
	Thickness	>50 m	<20 m
	Area	Well defined	Not well defined
	Heterogeneity		
	Faults	Unfaulted	Faulted
	Trap type	Well defined structures	Not well defined
	Petrophysical properties	Values given above	Values given above

mD millidarcy

3.2.1 Hydrocarbon Fields

Hydrocarbon fields could be the first geological sites to be used for CO_2 storage in many of the CEE-8 countries. This is because:

1. Their geological, structural and dynamic characteristics are well known and have been studied for a long time, in some cases for several decades;
2. It is possible to combine CO_2 storage with enhanced oil (and gas) recovery, which could offset the costs of CCS. Croatia, Hungary and Romania now have significant experience in enhanced oil recovery techniques, especially using CO_2, dating back to the early 1970s;
3. The regulatory framework in the CEE-8 (and overall in Europe) permits CO_2 storage in hydrocarbon reservoirs.

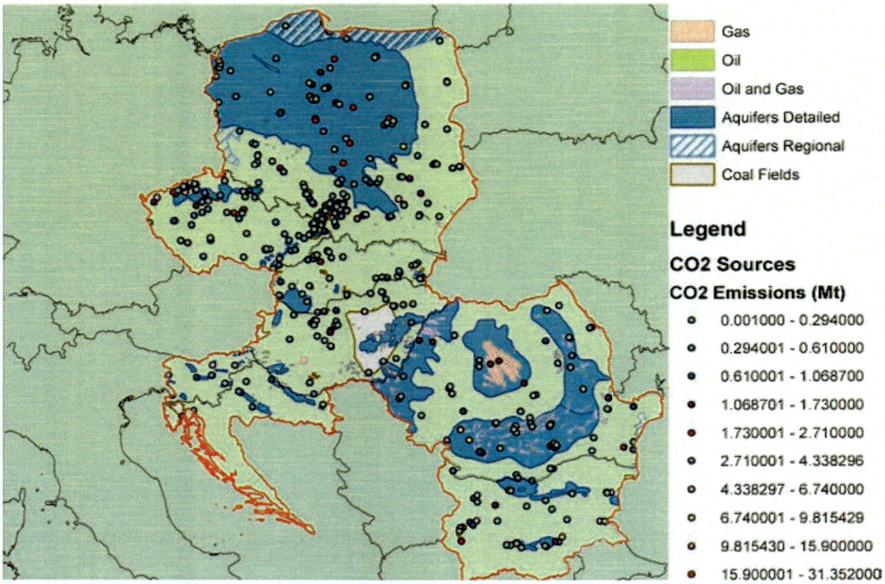

Fig. 4 The study area with larger emission point sources (*circles*), depleted hydrocarbon fields, saline aquifer and unmineable coal reserve sites (*see* Colour Plates)

Existing and depleted fields are mainly onshore, but Bulgaria, Croatia, Poland and Romania also have some offshore production. Hydrocarbon fields and associated storage capacities, although unevenly distributed, are concentrated in the Pannonian Basin and the Carpathian foredeep basins in the CEE-8 countries (see Fig. 4). The overall potential storage capacity is over 2,000 Mt; however, some fields will not be available for many years and this, together with strong competition between CO_2 and natural gas for storage, implies that there will not be much volume available for CO_2.

3.2.2 Unmineable Coal Reserves

There is far less geological and engineering information available on unmineable coal seams than on hydrocarbon fields. Nevertheless, there is a reasonable amount of knowledge about their geological and structural characteristics and their physical properties, especially in Poland where coal is the main source of energy production.

The methodology of CO_2 storage in unmineable coal seams is far less developed than that of storage in depleted hydrocarbon reservoirs. There are obstacles (i.e. low permeability, heterogeneity, timeliness of adsorption–desorption mechanisms) which need to be dealt with before unmineable coal seam storage can be used commercially.

Nonetheless, there are two important aspects related to CO_2 storage in coal seams in the CEE-8: (1) large coal-based power plants are usually very close to the potential coal storage sites; and (2) CO_2 storage could be linked with methane exploitation, which would compensate for CCS costs.

According to studies carried out in the course of the EU GeoCapacity project, the highest potential for storage with associated methane production is in the Czech Republic, Hungary and Poland (Fig. 4). There is less potential in Bulgaria, Romania and Slovenia, and in Croatia and Slovakia the potential is negligible. The overall storage capacity in the Czech Republic, Hungary and Poland is around 700 Mt associated with the production of about 180 billion m^3 of methane.

3.2.3 Saline Aquifers

The most promising areas for CO_2 storage are thought to be saline aquifers, which are present in all the CEE-8 countries and have a potentially much higher storage capacity than hydrocarbon fields and unmineable coal seams. Ideal aquifers have vertically closed structures with adequate sealing and significant pore volume capacity. Although saline aquifers have the greatest potential storage capacity, a lack of economic interest in this kind of storage option in the past means that available public data are not detailed enough for an accurate estimate and comprehensive characterization. The quality and the availability of data for reliable calculations vary from country to country. The potential storage areas in the CEE-8 region are shown in blue in Fig. 4. Details are as follows:

- In Poland, the approximate locations of 19 structures ranging from 100 to 625 km^2 were determined based on Mesozoic Formation maps by Dadlez (1998).
- In Croatia, five regional aquifers were identified, four of which lie in the Pannonian Basin and one offshore in the Adriatic Sea.
- Calculations are to be carried out regarding the storage capacity of two large basins in Slovakia: the Eastern Slovakian Basin and the Danube Basin. The most important strata with known aquifers are the Sarmatian, Pannonian and Pontian sediments.
- Altogether, 22 potentially suitable structures were identified in the Czech Republic, 17 of them in the Carpathians (the eastern part of the country) and five in sedimentary basins of the Bohemian Massif.
- In Slovenia, 35 potential locations were identified for CO_2 storage in aquifers.
- The first estimation of CO_2 storage potential for Bulgaria based on well logs and seismic investigations shows that there are several potential aquifers, related to Devonian, Lower Triassic, Middle Jurassic, Valanginian and Eocene reservoirs.
- In Romania, four large basins filled with clastic sediments represent enormous potential for CO_2 storage. Storage formations with potentially adequate capacity are mostly young Miocene–Pliocene clastic basin fills.
- There are six large regions in Hungary associated with basement highs and structural and lithological closures that are potential aquifer storage sites. Like the Romanian, Slovakian and Croatian examples, these aquifers are mainly related to young Miocene–Pliocene sediments.

The overall potential storage capacity in the CEE-8 is very high, representing several tens of gigatonnes. However, the capacities mentioned should be treated as first rough estimates of real storage capacity. Further research concerning porosity, permeability and structural closure parameters are essential if aquifer storage is to become realistic. It is expected that the actual storage potential will be considerably lower than stated here.

4 Current Status and Issues of Geological Disposal of Radioactive Waste in the Region

A rock formation is the most likely solution for HLW disposal. Geological disposal is the disposal of solid RW in a facility located underground in a stable geological formation (usually several hundred metres or more below the surface) that provides long-term isolation from the biosphere of the radionuclides in the waste. Based on a Safety Guide published by the IAEA (1994), we summarize here the different factors to be taken into account in the countries' site selection programmes.

The geological environment is expected to contribute towards ensuring safe disposal in three ways, namely by:

- Providing physical isolation of the waste from the near-surface environment and the potentially disruptive processes that occur there;
- Maintaining a geochemical, hydrogeological and geomechanical environment favourable to the preservation and performance of an engineered barrier system;
- Acting as a natural barrier to restrict the access of water to the waste and the migration of active radionuclides.

Siting of such a storage facility is a multistage process. Several factors need to be considered when a site is being selected. These are:

- Geological setting;
- Possible future natural changes;
- Hydrogeology;
- Geochemistry;
- Events resulting from human activities;
- Construction and engineering conditions;
- Transportation of waste;
- Protection of the environment;
- Land use;
- Social impacts.

Several CEE-8 countries have screened their territories for suitable geological sites and are considering construction of a geological HLW disposal facility (see Fig. 5). Certain selection criteria have been determined, leading to a more detailed

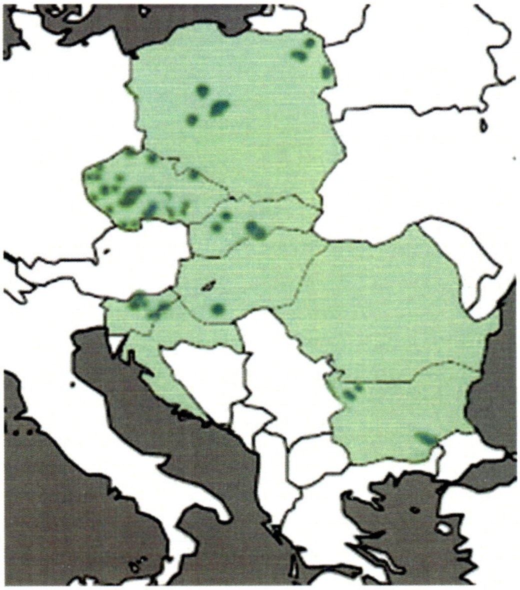

Fig. 5 Location of candidate areas for deep radioactive waste geological repositories in the Central and Eastern European region

analysis of prospective areas and the choice of a site. The countries' site selection programmes consist of four different stages:

1. Conceptual and planning;
2. Area survey;
3. Site characterization;
4. Site confirmation.

The works of Witherspoon and Bodvarsson (2001, 2006), Witherspoon (1991, 1996), Chapman (2006) and the OECD Nuclear Energy Agency (NEA 2004) have been used in research relating to deep geological repositories.

4.1 Bulgaria

The Bulgarian Government approved a national strategy for safe management of SNF and RW in 1999. This strategy includes deep geological disposal of HLW and near-surface disposal of conditioned LILW. After the preliminary screening of the national territory, 30 sites were selected using exclusionary criteria (28 mainly geological criteria), from which four sites were chosen. Two sites are composed of Lower Cretaceous clayey marls in north-west Bulgaria and two sites are in Sakar granite pluton in the south-east of the country.

The potential sites in the Lower Cretaceous marls are about 50–55 km south of the Kozloduy NPP (see Fig. 5). These two sites have similar conditions: they are about 750 m thick at one site and 1,000 m thick at the other; the formations consist mainly of clayey marls and rare thin-layered sandstones. The composition of the marls guarantees good sorption properties. The unconfined compressive strength values vary between 11 and 29 MPa. The Lower Cretaceous marls are known to be water-impervious layers. The sites are located in a region with a seismic intensity of VII on the MSK-64 scale. Seismicity has no connection with the fault structures of the area. Both marl sites are suitable not only for a deep HLW disposal but also for a near-surface LILW repository. This means that both types of repository could be constructed at the same site.

Two sites in the Paleozoic medium-grained granites of the Sakar pluton have been studied in detail, and these can also be discussed jointly. Both sites are situated about 300 km from the Kozloduy NPP. Their mineral composition is mainly plagioclase, orthoclase and quartz. The bulk density of granite is 2.62 g/cm^3, the density of solid particles 2.7 g/cm^3, the absorbed water content 0.35% and the unconfined compressive strength about 120–140 MPa. Analysis of the topographic and tectonic features of the sites suggests that the isolation capability of the deep disposal system will not be disturbed by erosion processes in the next million years. The sites are located in a region with a seismic intensity of VIII on the MSK-64 scale. The investigations indicate that the granite host rock at both sites is a suitable host medium for deep RW disposal.

In the early 2000s, analysis and explorations were carried out in a 25–30 km zone around the Kozloduy NPP to evaluate the geological conditions for RW disposal. The available data show that sediments with small discontinuities are represented in the geological profile. These sediments include the Middle and Upper Paleogene and Neogene formations. The Paleogene sediments consist of three formations: lower, marl–sandstone–limestone; middle, marl; and upper, silty clay. The middle and upper formations could be considered potential host media for geological disposal. They have a total thickness of 300–400 m. The Neogene, exceeding a depth of 1,000 m in the region of the NPP, is represented by sediments of the Miocene and the Pliocene.

Table 2 summarizes the features of the Neogene formations.

From the seismic point of view, the Kozloduy region appears to be one of the most geologically calm areas of Bulgaria. To summarize the recent research results, the main conclusion is that there is a possibility of developing a site in the vicinity of the Kozloduy NPP for deep HLW disposal and also for a near-surface LILW repository. Any decision will be based on further investigations and other important considerations (e.g. safety, waste transport, infrastructure, support of the local population, etc.).

4.2 Poland

Poland has no NPPs, but the country has its own nuclear power programme. A strategic government programme entitled Radioactive Waste and Spent Nuclear Fuel Management in Poland was conducted from 1997 to 1999. The aim of the

Table 2 The potential sites of a deep geological repository in Neogene formations in Bulgaria

Formations	Composition	Thickness (m)	Age
Miocene			
Delein	Greyish-blue clays with clayey limestones, silty clays and sandstones	200–440	Badenian
Krivodol	Grey and greyish-blue clay, stratified silty and calcareous clay with marls, dense clayey limestones and sandstones	120–240	Sarmatian
Pliocene			
Smirnenski	Grey and grayish-green low-calcareous, silty clays with clayey limestone and marl	200–250	Meotian–Lower Pontian
Archar	Sand	80–100	Upper Pontian
Brusarci	Clays with sand intercalations	50–200	Dacian–Romanian

programme was to investigate the legislative, institutional and technical issues relevant to RW and SNF, as well as public information issues, which were an essential element of the programme. Likewise, under this programme, a feasibility study of future repositories for SNF and HLW, as well as a study of all unmined deposits and rock formations in the existing deep excavations were performed. The study eliminated from consideration all deep mines currently under exploitation because of potential water threats, static distortion of formations or fissures caused by mining activities, the vicinity of current underground works and the seismicity of the area.

After a review of the geology of Poland, 44 rock formations were selected for further investigation. These included 17 sites in igneous (mainly granitic) and metamorphic rocks, 7 sites in shale and 20 sites in salt deposits. During the second stage of evaluation of potential sites, four geological structures were chosen as promising. These are Triassic clay rocks in south-west Poland and three salt domes in central Poland. The candidate sites were selected on the basis of preliminary geophysical investigations and the study of archive geological and hydrogeological data.

The general criteria for RW disposal in shale are:

- Shale beds to be at least 200 m thick;
- Overlying rocks to be at least 300 m thick.

Based on the general criteria with respect to shale, the candidate site was selected in the Triassic shale (near Jarocin), also known as the Upper Gypsum Beds.

The following initial criteria were considered for RW disposal in salt domes:

- Rock salt to occur at a maximum depth of 600 m below the ground surface;
- Overlying rocks to have a minimum thickness of 400 m;
- Homogeneous rock salt to have a minimum thickness of 250 m;
- Disposal zone thickness to range between 20 and 200 m;
- Maximum depth of repository to be <1,200 m below the ground surface.

Regarding disposal in salt domes, there are three candidate sites: Damaslawek, Klodawa and Lanieta. Lanieta has been explored in very great detail because of its economic importance for salt mining. In Damaslawek, two potential waste repository sites have been suggested in the central part of the salt dome, based on geophysical data. In the Klodawa salt dome, a future repository can be located some 2 km away from tunnels in the Klodawa salt mine.

4.3 Hungary

Because of the country's geology there are only a limited number of potentially suitable disposal sites for HLW in Hungary, which is why selection was carried out without preliminary national screening. The research regarding a suitable geological host site began with the Boda Claystone Formation (BCF) near the city of Pécs in south-western Hungary. Close to part of the BCF is a Permian sandstone formation. Information about the lithology, structure of the overlying sandstone and groundwater flow conditions of the sandstone was collected during operations at the Mecsek uranium mine (now closed) over the past 40 years. A specific study programme was started in 1993 to conduct a further examination of the BCF.

In 1994 the exploration tunnel excavated in the Mecsek uranium mine reached the claystone formation, and on-site underground data acquisition began in this area at a depth of 1,000 m (accessible from the former uranium mine). The possibility of implementing in situ examinations at this depth is very rare. Between 1995 and 1998, a short-term programme was launched to characterize the rock mass. The results are summarized below.

The recent 700–1,000 m thick layers of the BCF were settled in an alkaline basin under extreme climatic inflow and geochemical conditions and later buried at a depth of at least 3.5–4.5 km. The bulk porosity and hydraulic conductivity of the intact rock matrix are 0.6–1.4% and 10–15 m/s. The typical interval for the Young modulus is between 30 and 40 GPa, and the average unconfirmed strength exceeds 100 MPa. The dominant clay mineral in unweathered rock types of the BCF is illite (25–40%).

In 2000 the uranium mine was closed after plans for an underground research laboratory at the BCF site were rejected by the Government. A new policy initiative was launched, with the Public Agency for Radioactive Waste Management (PURAM) contracting the Empresa Nacional de Residuos Radiactivos (ENRESA) of Spain as a consultant organization to develop a strategy for disposal of high-level and/or long-lived radionuclide waste and SNF management. The long-term strategy is based on the ENRESA study. To ensure the safe disposal of HLW, the construction of a repository in a deep geological formation within Hungary is vital. Such a repository could also be used for direct disposal of SNF and, even more importantly, for the disposal of waste from the reprocessing of SNF assemblies.

Also in 2000, nationwide screening was carried out using desk studies to evaluate the potential rock formations in detail. This investigation confirmed the primacy of the BSF among the potentially suitable sites for an HLW repository. In 2004, in parallel with policy development, an exploration programme for an HLW repository restarted at Boda. The investigation had to be carried out from the surface because the uranium mine was no longer accessible. The aim was to pinpoint a location for an underground research laboratory (URL) from which rock investigation could be conducted.

The time schedule for the disposal of HLW and the management of SNF is presented below:

Time period	Tasks
2005–2008	Start of R&D work
	• Surface exploration of the BCF region for the construction of a URL
	• Preparation of a Preliminary Environmental Impact Study Report
	• Finalization and approval of the HLW management strategy
2009–2012	• Start of construction of the URL
	• Elaboration of a research/exploration programme
2013–2032	• Construction of the URL
	• Implementation of research/exploration
	• Completion of safety assessments
2033–2046	• Construction of the repository
2047–2069	• First phase operation of the HLW repository
	• Transfer of spent fuel assemblies from the Interim Spent Fuel Storage Facility to the repository
2070–2094	• Operation of the repository
2093–2094	• Extension of the capacity of the repository to accommodate the decommissioned HLW
2095–2104	• Second phase operation of the HLW repository
	• Transfer and loading of the decommissioned waste (HLW) from Paks NPP
2105–2108	• Sealing of the repository

4.4 Slovenia

In Slovenia a strategy on SNF and HLW was adopted by the Government in 1996, but it was recommended that any decisions on SNF disposal should be postponed until 2020 and that no significant action should be taken until then.

In 2004 the disposal strategy was reinvestigated, after the dual ownership of the Krško NPP had finally been clarified and agreed upon between Croatia and Slovenia. According to the agreement, the decommissioning and disposal of SNF and LILW from the NPP is the responsibility of both parties. To address this problem, a joint programme was instituted by the Croatian and Slovenian waste management organizations in 2004. The preliminary aim of the joint programme was to provide an accurate estimate of the future liabilities of the NPP.

As there are small quantities of HLW and SNF and limited financial resources at Krško NPP, a very rational approach was required. An example of the best practice available was followed: this was the Swedish KBS-3 concept of disposal in hard rock, and its cost analysis method. Many adjustments were required before this approach fitted the needs of Krško, and some additional control measures were also introduced.

In developing the disposal concept, the following additional assumptions were taken into account:

- Only direct disposal of SNF would be considered (no reprocessing).
- The repository would be developed for a hard rock environment at a depth of 500 m.
- The capacity of the repository would be such as to accommodate the estimated 620 tHM that would be generated over the plant's lifetime and the small quantities of HLW generated during decommissioning.

Regarding the timing, two alternatives were analysed:

1. The repository would be available and would start operation shortly after the plant shutdown in 2030. As SNF could be stored in the SNF pool on the NPP premises, no interim storage of SNF would be needed.
2. The repository would become available a few decades after the plant shutdown and would enter into operation in 2050. Until then, a longer interim dry storage period for SNF would be applied before final disposal.

A reference scenario was developed that:

- Covered only SNF and HLW management at the Krško NPP, which would cease operating in 2023;
- Assumed that all SNF would be disposed of in a single deep geological repository;
- Covered a generic location in hard rock media, given that no site investigations for a deep geological repository have been carried out in Slovenia and that no specific data for geological disposal are available;
- Was limited to those elements that were directly connected to disposal activities (packaging and disposal of SNF);
- Included an encapsulation plant for SNF at the site, as SNF would be sealed in a massive copper canister.

A comparison between the two alternatives for repository development reveals strong technological and economic preferences for the second one. In this alternative, the plan is for operation of the repository to begin almost 30 years after plant shutdown, allowing sufficient time for the site selection process. Heat release from SNF is low enough for the canisters to be filled optimally, thus almost halving the number required and consequently shortening the operation of the encapsulation plant and repository.

The agreement requires revision and updating of the joint decommissioning and the SNF and LILW disposal programme every 3–5 years. Revisions will focus on possible optimizations of the disposal system.

As the disposal activities are planned for the fairly distant future, there is time for other possibilities to be investigated. Disposal concepts in other geological environments will also be studied. Multinational shared repositories are another option that may be interesting for Slovenia with its small quantities of SNF and HLW.

4.5 Croatia

For Croatia, only information concerning LILW disposal is available (see also the details on Slovenia).

4.6 Czech Republic

The Czech programme for a deep geological repository began in 1993 with the project 'A selection of prospective HLW disposal sites in the Bohemian Massif'.

During the first stage of the site selection programme, 27 areas were identified from the geological, hydrogeological and geophysical viewpoints. Most of the national territory is crystalline rocks (more than 60%). These rocks exhibit favourable characteristics for hosting an HLW repository. These 27 localities were reviewed and the 13 most suitable sites were recommended for critical assessment based on archive data. The area of the recommended sites ranges from 20 to 120 km^2. As a result of this assessment, eight localities, all in granitic rocks, were recommended for further detailed geological survey. The Melechov Massif in the Central Moldanubium Pluton was chosen as a test site and for the first stage of research (an evaluation and study of its geological, hydrogeological, geophysical, tectonic and structural properties have already been completed). This test site represents an area that is analogous with the host geological environment for future HLW and spent fuel disposal in the Czech Republic. It is important to note that the deep repository will not be built at this site, although it is suitable for research targeting the sampling and collection of descriptive data using the most advanced scientific methods.

Next, four polygons were selected to represent all types of the Melechov Massif on which detailed geological, geophysical, hydrogeological, structural, geochemical etc., research were carried out. This work covered all non-destructive geoscience methods and prepared suitable data for the siting of boreholes for conducting:

- Well logging measurements;
- Geophysical, hydrogeological tests;
- Physical property estimation of different rock types;

- Petrographical and petrochemical study of samples;
- Mathematical modelling of fluid migration and micro and macro structures.

In 1997 the RW management system changed significantly, when a new law on the peaceful utilization of nuclear energy and ionizing radiation was passed. A key document, The Concept of Radioactive Waste and Spent Nuclear Fuel Management in the Czech Republic, was published in 2001. The Concept sets out the basic aims and direction for the development of the RW and SNF management system.

A number of studies aimed at locating a site for a future deep geological repository were carried out. Their main objective was to collect and evaluate existing geological information relevant to the selection of promising sites. On the basis of this knowledge, eight sites were recommended for further consideration. In 2001 a survey project was started on the entire geographical area of the nation. This project was divided into five steps. In steps 1–3, 11 sites were identified as suitable. In step 4 the number of sites was reduced to eight on the basis of accessibility, transport infrastructure, population density, land ownership and public acceptance. The national evaluation was made on the basis of existing information only. No new data were obtained.

In 2003 the preliminary site characterization stage began at six sites, all of which are located in granitoid bodies, in order to reduce the area of existing sites to ~40 km^2 each, and to recommend the optimal area for detailed site characterization. In the Czech Republic, a deep geological repository is expected to be built in granitic rock. Currently all siting activities have been postponed until 2009 by decree of the Government. These six sites will be evaluated, as scheduled, by 2015, and it is assumed that the repository will start operating in 2065.

4.7 Slovakia

In Slovakia three possible alternatives for the back end of the fuel cycle were taken into consideration:

1. SNF could be placed in interim storage for 40–50 years then disposed of directly; HLW would be disposed of in a deep geological repository constructed on Slovak territory.
2. SNF would be shipped and undergo final disposal outside the country.
3. HLW would be reprocessed and stored abroad, then disposed of on Slovak territory.

From the economic point of view, the first alternative, direct disposal after 40–50 years of interim storage, seems to be the most advantageous. The second and third alternatives have not, as yet, been considered but may be given further consideration.

Research and development for a deep geological repository in Slovakia began in 1996. The site selection programme began with a critical review of information (no field investigations) and included a survey of published and archive data on regional geology, hydrogeology, engineering and geophysics. The results identified 15 areas potentially suitable for a deep geological repository in granitic (7), metamorphic (3) and flyschoid (1) formations.

The next 4 years focused on screening via limited field verification and some technical measures. Taking into account the important geological, hydrogeological and mineralogical data, three areas in five localities were determined as suitable sites for the construction of a deep geological repository.

Three localities are situated in granitic rocks:

1. The central part of the Tribec Mountains, 46 km^2;
2. The southern part of the Veporske vrchy Mountains, 78 km^2;
3. The south-western part of Stolicke vrchy Mountains, 24 km^2.

Two are in argillaceous and pelitic formations:

1. The eastern part of the Cerova vrchovina Upland, 87 km^2;
2. The western part of the Rimavska kotlina Basin, 85 km^2.

The Central Tribec Mountains site is an area of granitic rock in the southern Tribec–Zobor block in the Tribec Mountains. The Zobor Massif is one of the largest crystalline complexes in the Western Carpathians. Tectonic deterioration of the site is generally low and thus hydrogeological conditions for a repository seem favourable. The southern part of the Veporske vrchy Mountains and south-western part of the Stolicke vrchy Mountains are adjacent to one other, but belong to two different geomorphologic units, which are divided by a Muran–Divin tectonic line. The Vepor granitic pluton is the largest in the Western Carpathians (~60 km in length) and is a complex pluton, consisting of several granitic rocks. Because of the pluton's size, it has been recommended for further investigations.

The eastern part of the Cerova vrchovina Upland and western part of the Rimavska kotlina Basin belong to different geomorphologic units. From a lithological, structural and spatial perspective, the most suitable host rocks appear to be two lithostratigraphic units: the Szecseny schlier of the Lucenec Formation and the Lenartovce beds of the Ciz formations. These units form the principal mass of the basin filling. The predominant lithology in both formations is a mixture of silstones and claystones. The maximum thickness of the Ciz Formations in the territory of Slovakia is 400–500 m, while that of the Lucenec Formation is 1,300 m.

The project activities are limited, but research and development work is expected to start in the near future. This should lead to a candidate site that is publicly acceptable and a demonstration of the feasibility of the proposed construction, operation and closure of the deep geological repository.

4.8 Romania

In Romania, spent fuel is classified as waste, and government policy aims for direct disposal of SNF around 2050, when the technology becomes commercially available. Romania has Canadian-type reactors, which use natural uranium as nuclear fuel.

It is planned to dispose of these spent fuel elements either in a salt or a hard-rock formation. A long-term safety assessment of a repository has been performed for spent CANDU (CANada Deuterium Uranium) fuel elements in a deep repository located in salt. Results from this report are compared to those of a hypothetical direct disposal of spent light water reactor (LWR) fuel elements in salt.

In the long-term safety assessment of a repository three scenarios have been considered:

1. A subrosion scenario, which represents normal evolution developing over a long period (millions of years). This scenario assumes that, over time, the salt dome is dissolved by groundwater in the caprock region.
2. A human intrusion scenario, which assumes that parts of a 1,000-year-old repository containing RW would be laid bare during the solution mining of a storage cavern.
3. A combined accident scenario, comprising a short overview of the general modelling procedure, which assumes a combination of brine intrusion from the overburden and undetected brine pockets.

The differences between modelling for CANDU and LWR fuel are related to the inventories and temperatures of the waste emplacement fields. Results have been discussed mainly in terms of effective dose. The calculated radiation exposure for the human intrusion scenario are between those for the subrosion and the combined accident scenarios and are of comparable magnitude for CANDU and LWR fuel.

5 Comparison of the Geological Storage of CO_2 and Disposal of Radioactive Waste

There are obvious interconnections between RW disposal and CO_2 storage, and between disposal/storage in the region studied and disposal/storage in general. In a broader sense, the list of interconnections can start with the fact that these materials are produced mainly during power generation. Any kind of technology used to replace them would reduce the need for disposal/storage; moreover, any shift in the balance between nuclear- and fossil fuel-based power generation would alter the type of disposal/storage needed. Here we consider the present-day situation where the replacement of common power-generation technologies with alternatives like solar or wind is still a slow and expensive process. As a result, RW disposal and CO_2 capture and storage have the same importance in terms of keeping all the existing options open.

In the countries of the region, legislation on many different aspects of interim storage and final disposal of RW is well accepted; however, CO_2 storage is a new concept, and it is only because of political decisions (based on long-term climate change issues) and economic influence (CO_2 emission quotas, emission trading) that it has become a possible solution for greenhouse gas reduction in the past few years.

In the region studied, the nuclear industry formerly relied on Russia taking back the spent fuel for reprocessing. Those times have gone, and now most countries are taking steps to establish their own storage and disposal facilities. Each country is in a different phase.

If one looks at the geological storage options considered in the previous chapters, one must conclude that there are different requirements depending on the different geological environments in which CO_2 would be stored and RW would be disposed of. Although both environments should have some kind of seal, the host rocks need to be quite different. In the case of CO_2, they should be permeable, porous or, for example, in the case of aquifers, the chemical solution should hold the CO_2 (for the long term). In the case of RW, the host rock would be part of the seal and should be impermeable; in other words, the opposite. As a result, there is no competition between CO_2 and RW for storage/disposal sites. For CO_2 storage, however, there are economically competitive uses of suitable sites, for example geothermal energy, natural gas storage or coal mining.

The amount of space needed for disposal/storage is also very different. In the case of RW, usually the volume is less than or around 100 t but in the case of CO_2 a few tens or hundreds of million tonnes are generated during 1 year for storage in a single country. In the context of the CEE-8 region this difference has consequences. If one compares the volume produced and the disposal/storage capacities needed for it, there is no obstacle to local solutions, either for RW or for CO_2 (although uncertainty exists with respect to saline aquifers).

Based on the detailed country-by-country discussion of RW disposal and geological storage of CO_2, we summarize some of the key features in Table 3.

In the row 'expected volume' we give comparable figures for RW, where possible. The data refer to the given date or to the end of the lifecycle of the NPPs. We know of no additional benefit resulting from the nuclear waste disposal process, but CO_2 storage can be combined with enhanced oil recovery (EOR) or enhanced coalbed methane (ECBM) production to achieve a more economical solution. Potential CO_2 storage sites should be close to the sources (several hundred in the region); hence, each country must find and exploit its own potential. From the technical and economic point of view one common site for RW at the best location would be the ideal solution. Some countries may not have the resources or the full range of expertise to build their own HLW repository. Countries with small amounts of HLW or with no national solutions in place need to face the problem that deep geological repositories are expensive. These nations also need safe and secure long-term waste management options. There is thus an increasing interest in the concept of shared deep geological repositories in Europe, with a number of countries agreeing to cooperate in implementing a regional facility. From the security point of view (availability, transport, etc.) and public acceptance, this solution can be hard to implement, but it is starting to be discussed. Currently all EU countries, even those with very small nuclear programmes, are under pressure to try to follow purely national programmes, even though the EU and the European Parliament support the concept of regional facilities.

In the period 2003–2005, the European Commission funded a project devoted to pilot studies on the feasibility of shared regional storage facilities and geological repositories for use by European countries. The goal of the second period (2006–2008) was to develop possible practical implementation strategies and organizational structures.

Table 3 CEE-8 region-specific features of radioactive waste disposal and CO_2 geological storage

Key features		Bulgaria	Czech Republic	Hungary	Poland	Romania	Slovakia	Slovenia	Croatia
Expected volume	CO2	510 Mt/yr from point sources (2005)							
	RW	7331 fuel assemblies	3000 tHM	1286 tHM (2017)	No NPP	4170 tHM (2020)	2374 tHM (2039)	620 tHM	No NPP
Estimated storage capacity	CO2	2000 Mt in *depleted hydrocarbon fields*							
		Several tens of gigatonnes in *saline aquifers*							
		No data	700 Mt in *unmineable coal*			No data			
	RW	No practical limitations							
Status of planning	CO2	Pilot projects considered mainly; industrial scale storage demonstration projects are also foreseen in Poland							
	RW	National Strategy (1999)	The Concept of Radioactive Waste and Spent Nuclear Fuel Management in the Czech Republic (2001)	Elaboration of a national policy for HLW management (2000)	Strategic Government Programme: Radioactive Waste and Spent Nuclear Fuel Management in Poland (1997 to 1999)	Generic studies	R&D for deep geological repositories (1996)	Joint Programme (2004)	
Additional benefits	CO2	EOR ECBM	EOR ECBM (180 billion m3)			EOR ECBM	negligible	EOR ECBM	EOR
	RW	No							
Source locations	CO2	Several 100s larger (>100,000 t/yr) point sources							
	RW	22 NPPs in 8 locations							
Competition with other use	CO2	Natural gas storage, geothermal energy, coal mining							
	RW	No							

Host rock types	CO2	Devonian, Lower Triassic, Middle Jurassic, Valanginian and Eocene	Sediments of various age and lithology	Miocene–Pliocene psammitic sediments	Mesozoic Formations	Miocene–Pliocene psammitic sediments			
	RW	Lower Cretaceous marls, Paleozoic grained granites	Granitic rocks	Boda Claystone Formation	Triassic shale	Generic salt and hard-rock formation	Granitic rocks argillaceous and pelitic formations	Generic location in hard rock	No inf.

ECBM enhanced coalbed methane,
EOR enhanced oil recovery,
HLW high-level waste,
NPP nuclear power plant,
RW radioactive waste,
tHM tonnes of heavy metal

As has been pointed out, geological storage of CO_2 is an emerging technique; however, RW disposal is an existing solution (but not in the region). At the moment none of the countries in the region have an HLW disposal site or a storage site for CO_2, but some countries have a facility for low-level RW or a storage site for natural gas or naturally occurring CO_2. Although the nuclear industry has the support of the public in the region, waste disposal is less welcome near the actual site. Obtaining public support for several tens or hundreds of CO_2 storage sites could be as difficult as solving the technical issues. As there are a number of natural gas geological storage facilities in the region, public acceptance of geological storage of another gas could be easier to obtain.

6 Conclusions

In this study we discussed geological CO_2 storage and RW disposal potential in Bulgaria, Croatia, the Czech Republic, Hungary, Poland, Romania, Slovakia and Slovenia. Data of CO_2 point sources were highlighted and possible geological storage locations (aquifers, oil- and gasfields and coalfields) were shown. Based on these reference data, the geological storage capacity is estimated at several decades, if not centuries, for all CO_2 emissions (510 Mt/year) from the larger point sources. These data were compared with the information on RW geological disposal. The total amount of HLW produced during the lifecycle of the existing 22 NPPs is not much more than 10,000 tHM; hence, there are no practical limitations on the quantity that can be disposed of, should a disposal site be built.

We also established that some of the geological sites suitable for CO_2 can be used for other purposes such as geothermal energy production, natural gas storage and coal mining, but these sites are not in competition with suitable RW disposal locations. In the region examined, the majority of the countries are dealing with the problem of HLW disposal. Investigations in all these countries are at an advanced phase, but it will be necessary to wait for several decades before construction of the first deep geological repository in the Central and Eastern European region can be started. As far as CO_2 is concerned, to date only pilot projects are under consideration, although less strict legislation and improved economic benefits, when combined with ECBM production or EOR, could speed up the implementation process.

References

IAEA (International Atomic Energy Agency) (1997) Joint Convention on the Safety of Spent Fuel Management and on the Safety of Radioactive Waste Management. IAEA INFCIRC/546. IAEA, Vienna

Chadwick A, Arts R, Bernstone C, May F, Thibeau S, Zweigel P (eds.) (2007) Best Practice for the Storage of CO_2 in Saline Aquifers: Observations and Guidelines from the SACS and CO_2STORE Projects. British Geological Survey, Nottingham

Chapman NA (2006) Geological disposal of radioactive wastes—concept, status and trends. J Iberian Geol 32(1):7–14

Dadlez R (ed) (1998) Tectonic Map of the Zechstein-Mesozoic Complex in the Polish Lowlands, 1:500 000. Polish Geological Institute, Warsaw

EC (European Commission) (2009) Directive 2009/31/EC of the European Parliament and of the Council of 23 April 2009 on the Geological Storage of Carbon Dioxide

Georgiev G (2008) CO_2 emissions and geological storage options in Bulgaria. Slovak Geol Mag 2008:43–52

Hatziyannis G, Falus G, Georgiev G, Sava CS (2007) Geological storage potential in the central-eastern group of countries (Hungary, Romania, Bulgaria, Greece, incl. Albania and FYROM). CO2NET Annual Seminar, Lisbon, 6–7 Nov

Hatziyannis G, Falus G, Georgiev G, Sava CS (2008) Assessing capacity for geological storage of carbon dioxide in central–east group of countries (EU GeoCapacity project). In: Gale J, Herzog H, Braitsch J (eds) Proceedings of the 9th International Conference on Greenhouse Gas Control Technologies. Energy Procedia 1:3691–3697

IAEA (International Atomic Energy Agency) (1994) Siting of Geological Disposal Facilities, vol 111-G-4.1, IAEA Safety Series. IAEA, Vienna

IEA (International Energy Agency) (2004) Prospects for CO_2 Capture and Storage. OECD/IEA, Paris

Kuharic LL (2008) Case for CO_2 geological storage-site Bzovik, Central Slovakia volcanic area. Slovak Geol Mag 2008:73–80

Martinez R, Suarez I, Zapatero MA, Saftic B, Kolenkovic I, Car M et al. (2008) The EU GeoCapacity project—saline aquifers storage capacity in group south countries. In: Gale J, Herzog H, Braitsch J (eds) Proceedings of the 9th International Conference on Greenhouse Gas Control Technologies. Energy Procedia 1:2733–2740

NEA (Nuclear Energy Agency) (2004) Stability and Buffering Capacity of the Geosphere for Long-term Isolation of Radioactive Waste: Application to Argillaceous Media. *"Clay Club"* Workshop Proceedings, Braunschweig, Germany, 9–11 Dec 2003. OECD Publishing, Paris

Saftic B, Martinez R, Donda F, Car M (2007) South group. CO2NET Annual Seminar, Lisbon, 6–7 Nov

Saftic B, Martinez R, Donda F, Car M, Zapatero MA, Suarez I, Vellico M, Persoglia S, Kolenkovic I, Vulin D (2008) Geological storage options in peri-Mediterranean countries—plans for case studies. 1st EAGE CO_2 Geological Storage Workshop, Budapest, 29–30 Sept

Sliaupa S, Lojka R, Tasaryova Z, Kolejka V, Hladik V, Kotulova J, Kucharic L, Fejdi V, Wójcicki A, Tarkowski R, Uliasz-Misiak B, Sliaupiene R, Brikmane B, Pomeranceva R, Sadrina T, Shogenova A (2007) The EU GeoCapacity project—assessing European capacity for geological storage of carbon dioxide—Czech Republic, Slovakia, Poland, Lithuania, Latvia and Estonia (NE group). CO2NET Annual Seminar, Lisbon, 6–7 Nov

Tarkowski R, Uliasz-Misiak B, Wojcicki A (2008) CO_2 storage capacity of deep aquifers and hydrocarbon fields in Poland—EU GeoCapacity project results. In: Gale J, Herzog H, Braitsch J (eds) Proceedings of the 9th International Conference on Greenhouse Gas Control technologies. Energy Procedia 1:2671–2677

Witherspoon PA (ed) (1991) Geological problems in radioactive waste isolation: a world wide review. Technical Report LBL-29703. Lawrence Berkeley National Laboratory, Berkeley

Witherspoon PA (ed) (1996) Geological problems in radioactive waste isolation: second worldwide review. Technical Report LBNL-38915. Lawrence Berkeley National Laboratory, Berkeley

Witherspoon PA, Bodvarsson GS (eds) (2001) Geological challenges in radioactive waste isolation: third worldwide review. Technical Report LBNL-49767. Lawrence Berkeley National Laboratory, Berkeley

Witherspoon PA, Bodvarsson GS (eds) (2006) Geological challenges in radioactive waste isolation: fourth worldwide review. Technical Report LBNL-59808. Lawrence Berkeley National Laboratory, Berkeley

Wojcicki A (2008) CO_2 storage potential in Poland. 1st EAGE CO_2 Geological Storage Workshop, Budapest, 29–30 Sept

Comparison of the Geological Disposal of Carbon Dioxide and Radioactive Waste in European Russia

Alexey Cherepovitsyn and Alexander Ilinsky

Abstract In this study a review is conducted of natural geological formations in European Russia in terms of their suitability for storage of carbon dioxide (CO_2) and radioactive waste. The geological conditions of European Russia are described, and the regional features and locations suitable for nuclear waste disposal are identified. A scheme is presented of the location of endogenous activity zones (seismic risks, volcanism) and increased radon risk in European Russia. A map showing suitable areas for nuclear waste storage is presented. The clay formations of the St Petersburg region are reviewed as a potential area for radioactive waste disposal. The main characteristics of the geological conditions that have potential as CO_2 storage sites are determined. A conceptual scheme of the CO_2 storage potential in north-west Russia, the most favourable region, is presented. Information about geological structures and depleted oilfields in north-west Russia is provided. The near-term outlook for CO_2 enhanced oil recovery in the oilfields in north-west Russia and the Kaliningrad region is given. A table of comparative assessments of the geological and economic characteristics of radioactive waste and CO_2 storage is also presented in the review.

Keywords Radioactive waste disposal • CO_2 storage • Geological formations • Enhanced oil recovery • Russian Federation

A. Cherepovitsyn (✉)
Department of Organisation and Management, Saint Petersburg State Mining Institute, St Petersburg, Russia
e-mail: alekseicherepov@inbox.ru; aecherepovitsyn@spmi.ru; oandam@spmi.ru

A. Ilinsky
All-Russia Petroleum Research Exploration Institute (VNIGRI), St Petersburg, Russia
e-mail: alex.ilinsky@bk.ru

F.L. Toth (ed.), *Geological Disposal of Carbon Dioxide and Radioactive Waste: A Comparative Assessment*, Advances in Global Change Research 44,
DOI 10.1007/978-90-481-8712-6_16,

1 Introduction

Russia is large enough to have almost every kind of geological structure and geo-dynamic property, every type of geological formation, and rich deposits of gasiform hydrocarbons, liquid hydrocarbons (oil, groundwater) and hard minerals (metallic and non-metallic). Four geographical locations are considered to be possible places of permanent geological disposal of radioactive waste (RW) and/or carbon dioxide (CO_2):

1. The Nizhnekamsk granitoid massif in the Krasnoyarsk region of Siberia;
2. The Murmansk area in north-west Russia;
3. The Kuril Ridge on Simushir Island, Sakhalin, in the Russian Far East;
4. The Novaya Zemlya (New Land) Island Territory between the Kara and Barents Seas, administered by the Arkhangelsk Oblast.

According to official information, there are no radioactive waste repositories in the Russian Federation today, just 20 temporary storehouses.

Officially, there are also no imports of spent nuclear fuel to Russia from abroad; however, according to ecologists, most of the uranium imported for uranium ore-dressing is still in the country. This RW represents an enormous danger to those employed in the atomic energy industry and to local residents exposed to radiation. Exact statistics on the quantity of uranium stored in Russia are not available. The RW that has accumulated in Russia in recent years has now reached enormous quantities, and there is an urgent need for permanent disposal.

Turning to CO_2, the Kyoto Protocol was the first international instrument to use market mechanisms as a basis for addressing global ecological problems related to greenhouse gas (GHG) emissions and climate change. In response to the economic incentives established by the Kyoto Protocol, most developed countries will reduce their levels of GHG emissions, particularly CO_2, by 5% in 2008–2012, compared with the 1990 level. The Ministry of Natural Resources of the Russian Federation has set the goal of achieving an overall decrease in CO_2 emissions, mainly through energy saving and a switch to modern energy efficient technologies. While the Kyoto Protocol requires efforts to reduce emissions and to implement the rational use of energy and heat, this in no way restrains the economic development of Russia. Furthermore, it is very important for Russia to take advantage of the economic incentives of the Protocol.

The Russian Federation ratified the Kyoto Protocol on 22 October 2004. The CO_2 quota for Russia established by the Protocol was 100% that of the 1990 level. The level of CO_2 emissions in 1990 was 2,360 million tonnes (Mt). Currently, total CO_2 emissions in Russia are 1,572 Mt/year, and specific CO_2 emissions in the power sector are 553 g/kWh (Cherepovitsyn and Ilinsky 2006). Actual emissions in the Russian Federation are thus below quota.

Because of the limited potential of other CO_2 emission reduction options, the concept of CO_2 sequestration by capture and storage in underground reservoirs is gaining ground in Western Europe. Increasing attention is also being paid to this option in the Russian Federation.

The Russian oil and gas industry has a great deal of experience regarding the exploration of subsurface reservoirs for use as spare gasholders. The technologies used in exploring and creating these spare gasholders could also be used for CO_2 storage. Furthermore, injection of CO_2 into oil reservoirs could increase the efficiency of oil recovery and enhance gas recovery.

European Russia, particularly the north-west region, has a large number of oil- and gasfields with only a low level of reservoir development. The geological subsurface of this region is indicative of a large number of aquifers, and the region also has an unutilized supply of gasholders that were created for the storage of gas reserves. All these reservoirs could be used for CO_2 storage (Ilinsky 2005, 2006; Cherepovitsyn and Ilinsky 2006). CO_2 capture and storage thus shows great potential in a region that has a concentration of energy-intensive industries. Moreover, north-west Russia is situated relatively close to Finland, Germany, Poland and other European countries that are potential renters of the subsurface reservoirs for CO_2 storage under the Joint Implementation Mechanism of the Kyoto Protocol.

The problems of CO_2 storage and the technology of sequestration development are in the early stages of scientific research in Russia, and only preliminary estimations of storage potential are currently being conducted. Studies to enhance oil recovery by CO_2 injection were carried out as early as 1970 in the Soviet Union, but were not implemented commercially. There are no natural CO_2 deposits in Russia, but because of the economic incentives of the Kyoto Protocol, interest in such projects has now started to grow.

2 Radioactive Waste Disposal

2.1 Sources of Radioactive Waste in Russia

Radioactive pollution in various regions in Russia, and hence the need to develop RW repositories, is due to nuclear technology-based activities. Statistical data on the radioactive materials and waste that have accumulated in Russia as a result of these activities are presented in Table 1.

The European part of Russia has a huge number of industrial, defence and other enterprises that are potential sources of nuclear danger. Their overall number is close to 10,000, with at least one third being connected to a military or industrial undertaking. In the Murmansk and Arkhangelsk areas there are more than 270 nuclear power installation units, representing 18% of all nuclear power installation in operation worldwide. Many of the enterprises using radioactive materials are concentrated in the region. The most important are the nuclear power stations, shipyards, nuclear-powered icebreaker fleet, Northern Navy, and related infrastructure in St Petersburg and in the Kola (Murmansk) area—in total nearly 4,000 enterprises that use radioactive materials and other sources of ionizing radiation.

The main centres of nuclear power use in north-west Russia, along with the regional geological environment, are shown in Fig. 1.

Table 1 Radioactive waste and the materials that have accumulated in Russia as a result of defence and industrial activities (Shishits 1998).

Stage of nuclear cycle, enterprise, type of waste	Type of material, category of radioactive waste	Weight (t) or volume (m^3) of fuel	Total activity (Ci)	Location
Extraction of uranium and thorium ores	Natural radionuclide LAW	5.6×10^7 t	6×10^5	Tailing dump
Production of fuel- and heat-generating products	LAW	1.6×10^6 t	9.3×10^4	Open storehouses
NPP	Liquid LAW	8×10^4 m^3	3.5×10^3	SLAW on NPP territory
	Solid LAW	5×10^4 t	1×10^3	SLAW on NPP territory
	Hardened waste (bituminous compound)	1×10^4 t	2×10^3	SLAW at St Petersburg NPP and Kalinin NPP
	RW of RBMK (MAW and HLW)	5.325×0^3 t	1×10^9	Storehouse for SNF at NPP
	RW of WWER-440 (MAW and HLW)	9.4×10^2 t	–	Storehouse of SNF at NPP
	RW of WWER-1000 (MAW and HLW)	1.1×10^3 t	–	Storehouse of SNF at NPP
Processing of SNF	RW of WWER-400, BN-350, BN-600, transport reactor (MAW and HLW)	3.5×10^3 t	–	Storehouse of SNF at Mayak plant
	Nuclear waste glass liquid HLW from processing fuel of WWER-440	5.5×10^8 m^3	9.5×10^6	Same as above
Waste from defence programmes	Liquid HLW and MAW	n.a.	5.5×10^8	Capacity storehouses at Mayak plant
	Liquid LAW	n.a.	1.25×10^8	Reservoir No. 9 at Mayak plant
	Solid MAW, LAW: equipment, building and other material	n.a.	1.2×10^7	Storehouse of SNF at Mayak plant
	Liquid HLW, MAW, LAW	n.a.	1.26×10^8	Reservoir atSCE (Tomsk)
			4×10^8	Collectors in deep layers at SCE

(continued)

Table 1 (continued)

Stage of nuclear cycle, enterprise, type of waste	Type of material, category of radioactive waste	Weight (t) or volume (m^3) of fuel	Total activity (Ci)	Location
	Liquid HLW, MAW	n.a.	8.4×10^6	Special KMCE storehouse
	Liquid HLW, MAW, LAW	n.a.	5.0×10^8	Collectors in deep layers at KMCE

BN fast neutron reactor, *HLW* high-level waste, *RBMK* high-power channel-type reactor, *KMCE* Krasnoyarsk mining chemical enterprise, *LAW* low-activity waste, *MAW* medium-activity waste, *NPP* nuclear power plant, *RW* radioactive waste, *SCE* Siberian chemical enterprise, *SLAW* storehouse for low-activity waste, *SLNW* storehouse for liquid nuclear waste, *SNF* spent nuclear fuel, *WWER* water-moderated water-cooled power reactor, *n.a.* not available

The sources of radioactive contamination in the area of study are:

1. Nuclear testing in Novaya Zemlya;
2. Underground nuclear explosions for industrial (non-defence-related) purposes;
3. Nuclear waste deposits;
4. Submerged nuclear ships and the nuclear waste on the Kara and Barents Sea beds;
5. Radioactive fallout from the accident at Chernobyl nuclear power station;
6. Transportation of radioactive cargo (Komlev 1998; Tikhonov 2004).

Near the Lovozerskii and Kovdorskii ore mining and processing enterprises on the Kola Peninsula, the ecological situation is complex (and critical) because of the presence of natural radioactive ore, processed raw materials and finished products. Special action thus needs to be taken to prevent serious accidents. The nearby Loviisa nuclear power station in Finland and Ignalina nuclear power station in Lithuania also represent a potential radiation threat to the Karelia and the Pskov areas (Shishits 1998).

2.2 *Geological Disposal Options for Radioactive Waste*

Table 2 shows a classification of the mining characteristics of geological environments on a regional basis for European Russia (see also Fig. 1).

The geomechanical parameters of a rock massif govern the geological surroundings and the underground storage that can be used for RW waste. Other properties related to the different components of the sphere (such as ectoplasms, groundwater, the gas-bearing parts of the massif and its geochemical and physical fields) are not as important in determining the technical and mining characteristics of the massif. Nevertheless, they can significantly affect the exploitation of the underground area. The presence of radon, for example, is a serious adverse factor.

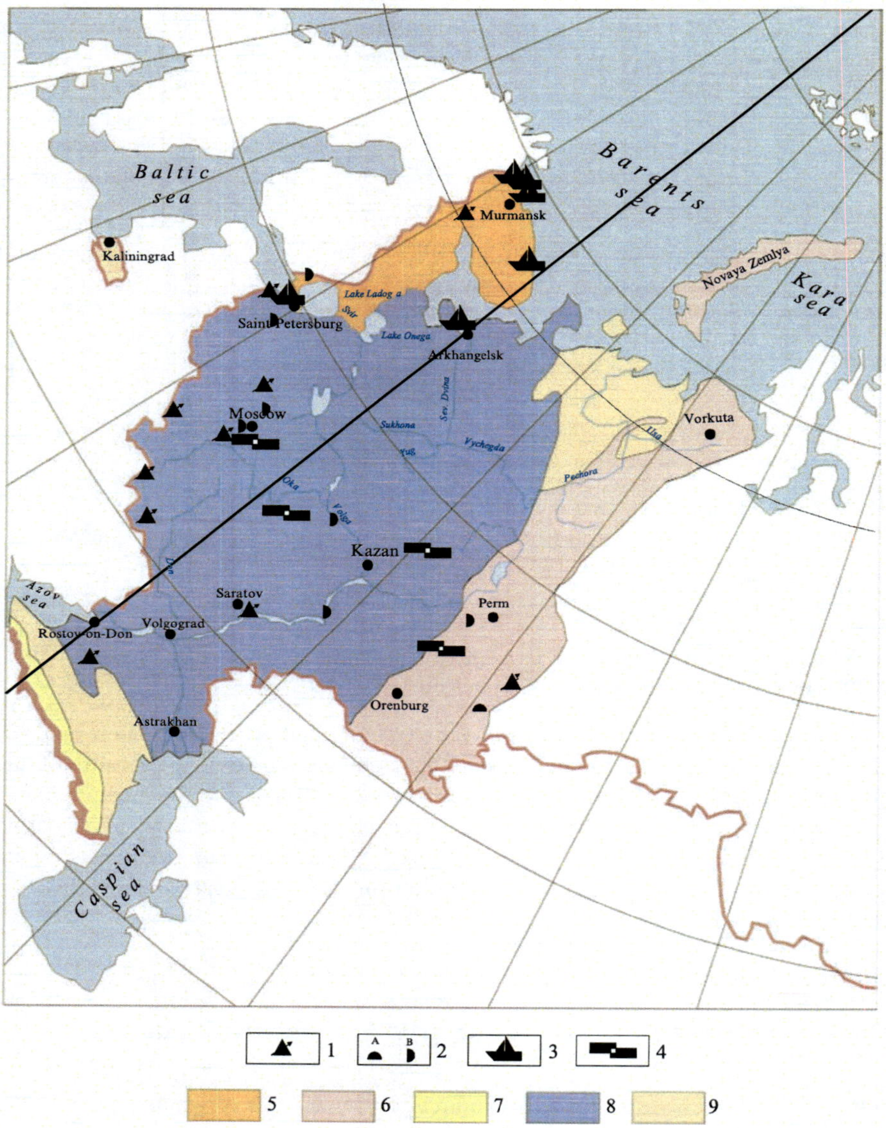

Fig. 1 The main centres of nuclear power use in north-west Russia (1–4) and the regional geological environment (5–9) (*see* Colour Plates). 1 Nuclear power stations. 2 Nuclear reactor: A – technological; B – research. 3 Bases of nuclear fleet. 4 Radiochemical and metallurgical plants. 5 Mountain ranges of Precambrian metamorphic complexes. 6 Folded and magmatic Phanerozoic rocks. 7 Sedimentary and volcanogenic rocks of recent geodynamic active mobile zones. 8 Complexes of lithified sedimentary rocks and vulcanites of ancient platforms. 9 Weakly lithified basic sediments of recent platforms

Table 2 Classification of the mining characteristics of geological environments on a regional basis for European Russia

Engineering geological rock types (pertinent regions in bold)	Mining features geological complex	Evaluation of favourable level of mining-geological complex for RW disposal
Rocky and semi-rocky, in the upper 10–25 m decomposition zones, mainly granitoid **(Republic of Karelia, Murmansk area, part of St Petersburg area, Voronezh area)**	Homogeneous, steady	Suitable
Rocky (in low dislocated zones), semi-rocky, friable non-cohesive (in crush zone), soft cohesive (in Mylonite zones) **(Caucasia, Ciscaucasia)**	Heterogeneous, unsteady	Unsuitable
Rocky (in low dislocated zones), semi-rocky, friable non-cohesive (in crush zone), soft cohesive (in Mylonite zones) **(Ural, Cis-Ural)**	Moderately heterogeneous; moderately steady	Moderately suitable
Friable non-cohesive, soft cohesive, rare semi-rocky **(some central parts of European Russia)**	Heterogeneous, unsteady	Unsuitable

Figure 2 shows the location of endogenous activity zones (seismic/volcanic activity) and increased radon risks. A region's potential for underground storage development is examined in terms of the suitability of conditions, including both internal and external factors. The presence of permafrost in a location is a favourable indicator for nuclear waste storage. Such a location is divided into: (1) the cryolite zone (permafrost); and (2) the area outside the cryolite zone. Within the cryolite zone the impact of negative hydrogeological factors is reduced, but the homogeneity of rocks and their total stability increases. At the same time the danger of radon also decreases. These features need to be taken into account when underground storage development is being considered.

A scheme of the potential for subsurface storage development is presented in Fig. 3.

The most favourable locations for RW disposal are:

1. Areas with cratons of ancient platforms and similar mountain geological complexes at depths of up to 500 m;
2. Outcrops of Precambrian rocks in Phanerozoic folded areas;
3. Areas of platform comprising essentially homogeneous carbonate and clay rocks;
4. Areas of widespread granitoid intrusions with an insignificant display of residual soil (Shishits 1998).

The Baltic Craton is an ideal location for RW disposal. Here, there is a widespread and uniform distribution of granular granite, gneissose granites, migmatite and other formations with a high density and homogeneity, as well as a limited number of recently active breaks. This area's potential is enhanced by its favourable economic-geographical conditions. The slopes of the Baltic Craton's blocked

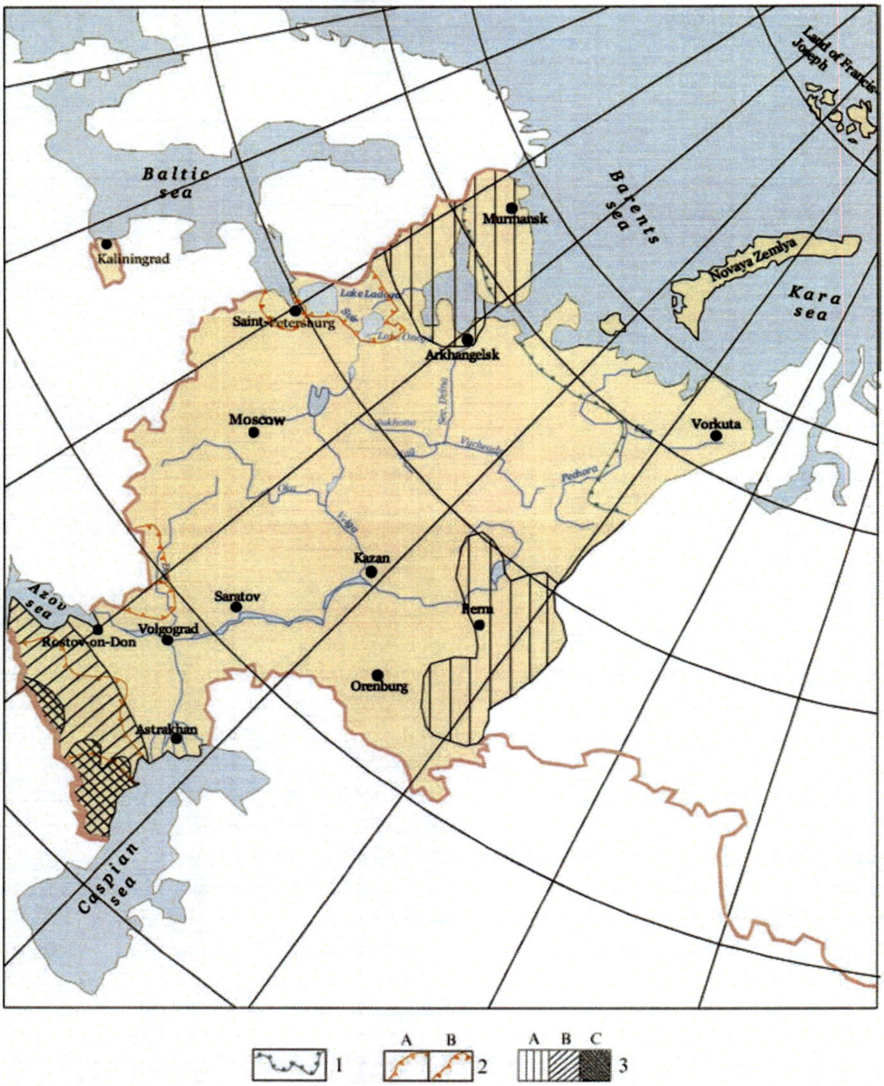

Fig. 2 Locations of endogenous activity zones (seismic risks, volcanism) and increased radon risk. 1 Cryolite zone boundary. 2 Regions of increased (A) and high (B) radon risk. 3 Seismic risk zones: A – with rare random earthquakes with a magnitude up to 4 (according to the Richter Scale); B–C – with constant earthquakes with a magnitude up to 7 (B) and above 7 (C)

homogeneous, terrigenous and clay formations are also quite favourable for RW storage. Complications within the craton and on its slopes are related to areas of incidental earthquakes and the presence of tectonic breaks which interrupt the homogeneity and stability of the rock massif. On the slopes, the hydrogeological features of sandstone are also present (Smyslov 1996).

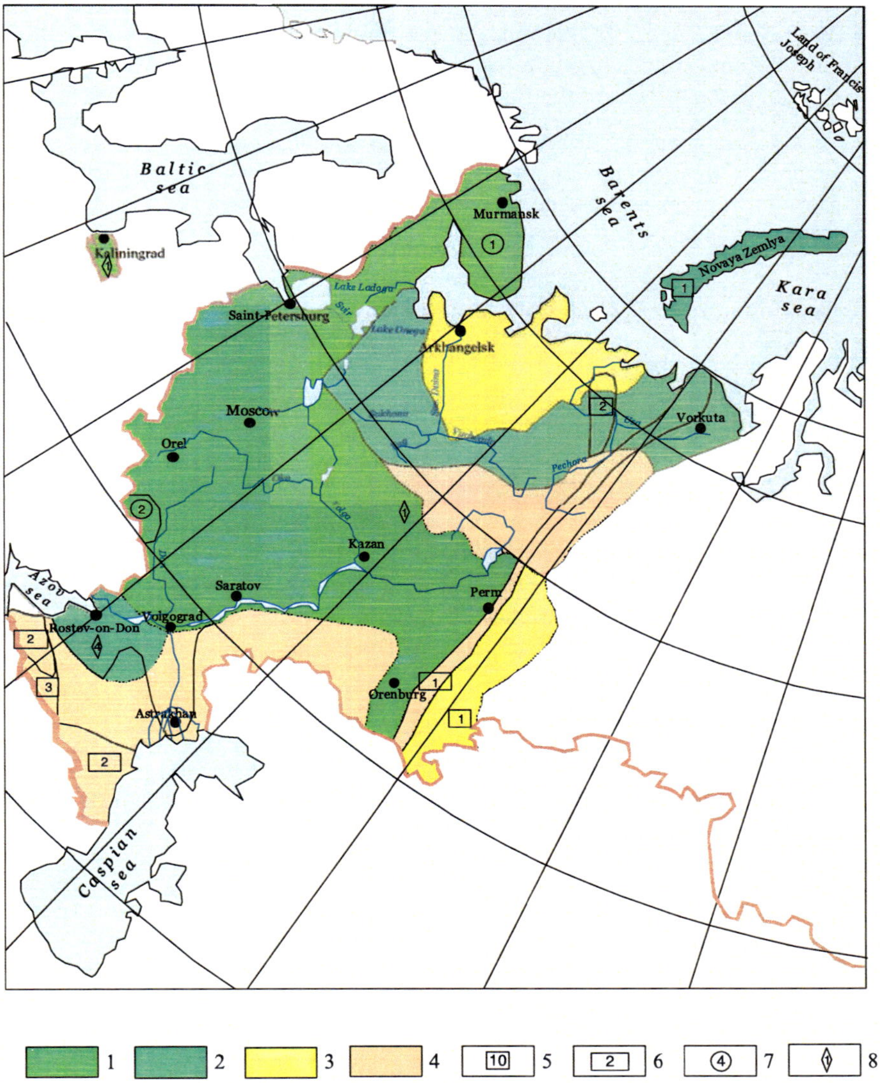

Fig. 3 Map of potential for subsurface disposal site development for radioactive waste (*see* Colour Plates). Coloured areas 1–4, indicating regions: 1 high potential; 2 average potential; 3 suitable areas and regions; 4 low potential. Numbers in geometric shapes: 5 Late Proterozoic Phanerozoic folded areas (*number given in square*) (*1* Urals – Novaya Zemlya; *2* Tieman; *3* Caucasus). 6 Regional deflections (*number given in rectangle*) (*1* attached to Urals; *2* attached to Caucasus). 7 Precambrian folded areas (*number given in circle*) (*1–2* cratons: *1* Baltic; *2* Voronezh Crystal Range). 8 Ancient and recent platforms (*number given in rhombus*) (*4* Skif-Turanic). A dashed line shows the boundary of the respective tectonic structure

The Precambrian crystalline rocks of the Baltic Craton, especially in southern Karelia, are under consideration for underground RW disposal. The following favourable characteristics of the Baltic Craton render it suitable for RW disposal:

- Weak development of surface interstices (pores) of erosion;
- Apparently weak geodynamic activity;
- Low temperatures at the neutral layer level (2–8°C at a depth of 15–30 m), and at greater depths.

In the St Petersburg region, the Lower Cambrian blue clay of the Koporja area can be used for waste disposal. However, this territory is located in one of the country's most active fault zones. Blue clay is an environment with low sorption ability and a high level of vulnerability not only to nuclear irradiation but also to changes in physical, chemical and biochemical conditions. The transformation of these clay sediments under the influence of technogenic factors would adversely affect their isolating potential, allowing the active migration of radioactive nuclides; this, in turn, could cause pollution of the underlying aquiferous stratum used for water supply. As a rule, clay formations and rocks are free from circulating subsurface waters and possess enough plasticity to make them suitable for RW isolation.

Suitable clay formations are abundant in all parts of Russia; their mineralogy, bedding, low permeability and other characteristics make them one of the most promising formations for the construction of RW repositories. The advantages of clay are:

- Insolubility of clay minerals in underground waters;
- Good sorption ability of most clay minerals.

The isolating ability of clay rocks—widely exploited in mining and in the manufacturing of mining equipment—has been widely investigated by the oil and gas industry. However, before storage sites are developed in clay rocks, the following should be considered (Tatarchuk 1997):

- Fluids and hydrated minerals can adversely affect isolation integrity;
- The specific heat conductivity of clay sediments is three to four times lower than that of rock salt. The thermal influence of waste can alter not only the plastic characteristics of clay but also its sorption abilities;
- Clay excavations are difficult to carry out and maintain;
- The volume and circulation rate of fluid passing through the pressure head sites of a clay formation are difficult to determine.

To locate clay formations that can be used as RW repositories, homogeneous clay layers should be sought in favourable mining, geological and tectonic conditions. The most promising formations are deep-water facies pools with homogeneous layers of montmorillonite and montmorillonite-hydromica clay. The main problem is to sustain the capacity of the clay to provide efficient and safe isolation of nuclear waste (Smyslov et al. 2002)

An assessment of the geological criteria fulfilled by the blue clays in the St Petersburg region in terms of suitability for RW disposal are presented in Table 3.

Table 3 Assessment of geological criteria fulfilled by the blue clays in St Petersburg area in terms of suitability for radioactive waste disposal

Level of suitability	Criteria for estimation: Tectonic elements	Conditions of accumulation of deposits	Form of clay rock deposit	Depth of bed roof of clay breeds (m)	Capacity of working thickness (m)	Hydrogeological conditions	Seismicity on the MSK-64 scale
Potentially suitable	Platforms and regional deflections	Deep-water sea and coastal sea pools	Homogeneous regional sustained layers of clay and argillites	from 300 to 700 m	>100	Zones of complicated and extremely slowed-down water exchange	<7
Suitable, with restrictions	Platforms, areas of regional deflections, dividing platform and young orogen	Sea, coastal sea and lagoon phases	Homogeneous layers of clay and argillites, local distribution	>300	>100	Zones of complicated water exchange	<7
			Powerful regional sustained layers of clay and argillites	<300	>200	Zones of complicated water exchange	<7
Practically unsuitable	Platforms and regional deflections, intermountain hollows	Continental adjournment of glaciers, lake marsh phases and barks of aeration	Lenses, moraines, alternation of clay with non-tight differences	<300	<300	Zones of active and complicated water exchange	>7

MSK Medvedev–Sponheuer–Karnik

There are also suitable geological formations, including gneiss and granite dome-shaped reservoirs and massifs of Rapakivi granite, on the northern shore of Lake Ladoga in Karelia.

Large granitoid massifs offer the most stable environment for underground RW disposal. When locations are being sought for underground gas storage, monolithic blocks in geological structures are of particular interest. Investigations in north-west Russia have shown that there are monolithic blocks of this kind in many places (for example, the Kola region) (Smyslov et al. 2002).

The Voronezh crystal massif is also a potentially favourable location for RW disposal. This area is characterized by dense homogeneous metamorphosed formations at technically accessible depths and is capped by carbonate rock massifs. Despite the development of Cretaceous and Jurassic sediments on the boundaries of formations with terrigenous structures, the carbonates are characterized by a high level of homogeneity, which suggests that they are of marine origin.

The geological formations that characterize the East European platform are also favourable for RW disposal. Here, the most important areas are those with predominantly clay, sulphate, halogen, and carbonate formations. Even if their capacity is small, they can be used for underground disposal. In some areas these formations lie at technically and economically accessible depths (Ordovician and Silurian carbonate rock mass in north-west Russia, carboniferous deposits in the central part of the Russian Platform, Permian system in the Cis-Ural region, etc.). Complications can arise if there is karst present, especially if the karst is active, as is the case in the Cis-Ural region, and could become more active if the underground area is developed (Smyslov 1996).

Halogen formations are chemical deposits that have accumulated as a result of the evaporation of large volumes of water containing halogen salts. The deposits are evaporites that have precipitated over time in pools, isolated from oceans. Usually the cycle of evaporation begins with sedimentation of dispersed clay, then of dolomite and anhydrite. Most of the evaporation cycle results in rock salt, which frequently includes layers of potash salts. There are huge saliferous reserves in the Near-Caspian hollow. Hydrochloric formations with a wide seam thickness (more than 2,000 m) are widespread over an extensive area.

The characteristics of the saliferous areas of Russia are presented in Table 4.

In the east and north-east parts of the East European platform there are Permian-Triassic terrigene rock masses, characterized by significant lateral heterogeneity and the presence of sulphides with aggressive formation waters. These are harsh environments for concrete and metal.

When areas in the Russian platform are considered for development of underground disposal, in all but a few specific cases the possible lack of heterogeneity of the geological environment must be taken into account, as must the possible difficulties involved in mapping small amplitude breaks.

The Urals should be considered as a region of low to average utility for geological disposal. The need for RW disposal, especially in the Middle and Southern Urals, is indisputable because of the presence of many nuclear installations with a great deal of radioactively contaminated material. In the Urals, carbonate complexes

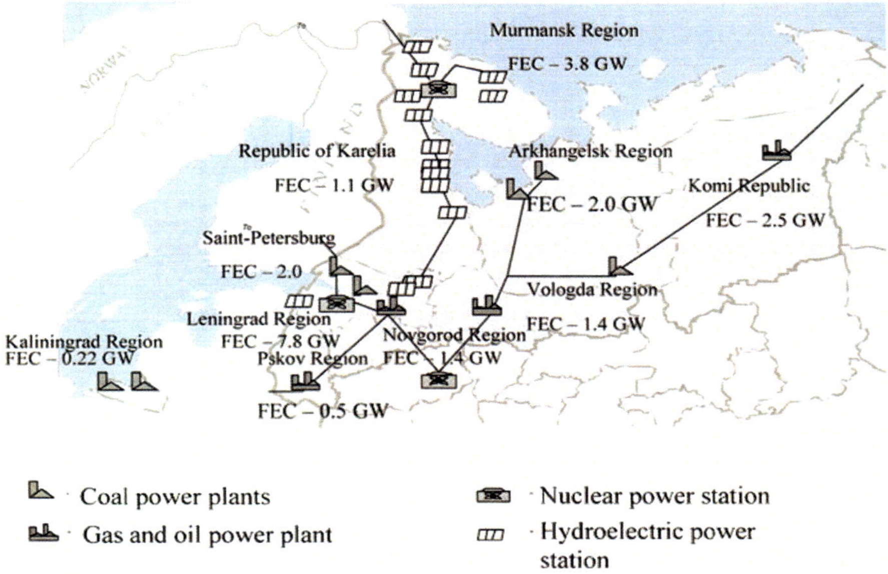

Fig. 6 Energy sector of the north-west region of Russia (Source: Ilinsky 2005), FEC: Full eletric capacity

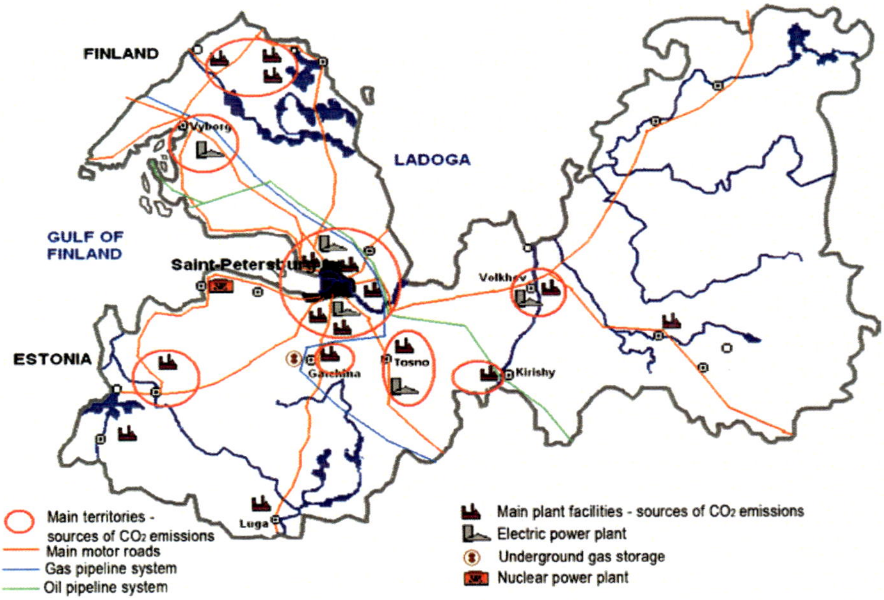

Fig. 7 Main sources of CO_2 emissions in the St Petersburg region

3.2 *Geological Disposal Options for CO_2*

An estimation of the potential capacity of depleted oil and gas reservoirs for storage of CO_2 has been made based on estimates of the cumulative production and proven reserves of oil and natural gas. The overall capacity for the Russian West Siberian Basin (depleted oil- and gasfield capacity combined) is estimated to be around 177 gigatonnes of CO_2 (Gt CO_2) (Zakharova 2004).

The option of sequestration of CO_2 in unmineable coal seams is still at the feasibility study stage worldwide. However, if this option is to be considered for storing CO_2, account must be taken of the fact that about 82% of the country's coal resources are located in the deposits of western and eastern Siberia, which are quite a distance from the main coal consumption areas. This wide geographic distribution of the areas of CO_2 capture and the potential sink areas can substantially increase the cost of using such reservoirs for CO_2 storage (Zakharova 2004).

In Great Britain, Norway and Germany the main sources of GHG emissions are located 200–500 km from the offshore and onshore oil- and gasfields and aquifers (Stevens et al. 2001; Kjärstad and Johnsson 2004). In Russia, however, the main sources of greenhouse emissions are around 2,000–4,000 km from the disposal sites. It is neither economically nor technologically viable to transport CO_2 to western or eastern Siberia from the Central European part of Russia or from Europe, for that matter.

The north-west region (including Kaliningrad) seems to have the most potential in terms of providing suitable underground reservoirs. This territory is not far from countries that may wish to rent underground CO_2 storage capacity, such as Germany and Poland. However, the geological potential of this territory is estimated by many investigators to be only moderate and further detailed geological studies are necessary.

There are potential storage sites in the north-west around St Petersburg (north and south of Ladoga) and in the Murmansk region (the territories near the Shtokmanovskoe offshore gas condensate field).

Unified Energy Systems, a Joint Stock Corporation, plans to start more than 30 projects in response to the Kyoto Protocol economic incentives. These projects aim to reduce GHG emissions (estimated at more than 20 Mt annually) from the company's power plants. A special Energy Carbon Reserve has been set up to implement these projects, some of which are presented in Table 5.

As can be seen from Table 5, there are no CO_2 capture and storage projects currently under consideration. Problems related to CO_2 emission are not on the agenda in Russia, and development of sequestration technologies is not even at the research stage.

A conceptual scheme of storage potential is represented in Fig. 8. There are four different options for CO_2 storage in north-west Russia.

- CO_2 storage in oil and gas reservoirs with a high level of depletion. At present the level of depletion of the oilfields in the region is 27%. From the economic standpoint CO_2 storage could have an additional effect in terms of utilization of CO_2 for enhanced oil recovery (EOR) methods in the oilfields; however, this is not yet used in Russia.

Table 5 Carbon projects chosen for investment by the Energy Carbon Fund of Russia at power stations in the north-west region of Russia

No	Project and initiator	Project title	Project status	Project cost in million €	CO_2 emission reduction in million t/year
1	Kaliningrad CHP-2 JSC UES	Greenfield construction of 900 MW generation capacity	Feasibility study	532	1.0
2	North West CHP JSC UES	Construction of an additional 900 MW generation capacity	Feasibility study	230	n.a.
3	Pskov TPP JSC UES	Construction of an additional 215 MW generation capacity	Feasibility study	32.5	n.a.
4	Kirishi TPP JSC UES	Transition of boilers No. 1 and 2 (co-generation part of Kirishi TPP) from heavy fuel oil to gas	Project proposal	3.76	0.130

For further information, see website of Energy Carbon Fund of Russia (http://www.reeep.ru)
JSC UES Joint Stock Company Unified Energy System of Russia, *CHP* combined heat and power, *TPP* thermal power plant, *n.a.* not available

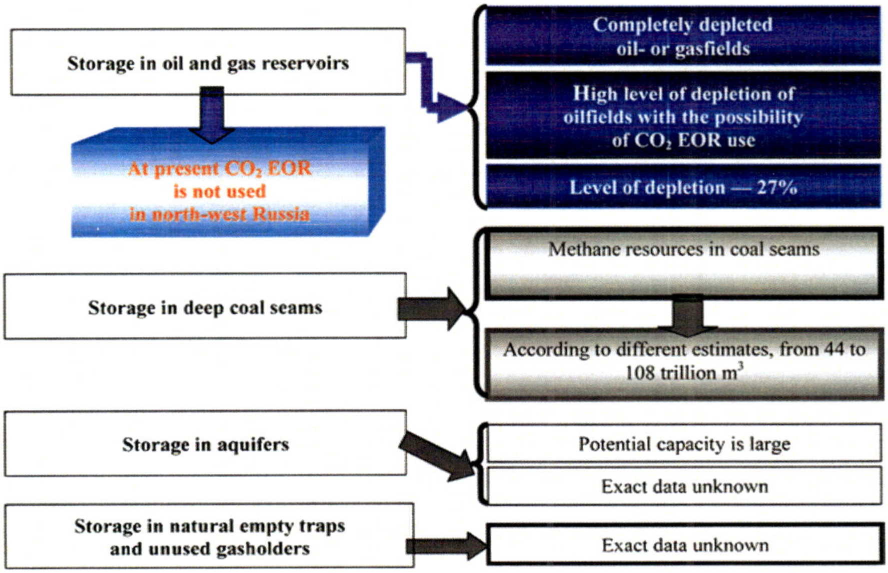

Fig. 8 Conceptual diagram of CO_2 storage potential in north-west Russia

- CO_2 storage in deep coalbeds. The methane resources in coalbeds in the north-west are estimated at 44–108 trillion m^3: this storage method could also be used for enhanced coalbed methane (ECBM) recovery in these coalfields.
- CO_2 storage in aquifers. According to preliminary estimates, this method has a very large potential; however, exact data for Russia are presently unknown.
- CO_2 storage in natural empty traps and unused gasholders. There are a few gasholders in the region, but information regarding their capacity is classified. Thus, currently, the most probable method of CO_2 storage in the north-west region of Russia is storage in oil reservoirs and CO_2 utilization for EOR (Cherepovitsyn 2005, 2006).

According to Cherepovitsyn and Ilinsky (2006), three possible areas for GHG storage in the north-west region of Russia are:

- Timano-Pecherskaya oil and gas province (geological data show initial reserves to be approximately 9.8 Mt oil equivalent (Mtoe));
- Continental shelf area (initial reserves of 3,698 Mtoe);
- Off-shore area (initial reserves are 6,072 Mtoe).

Only sites with an average porosity of not less than 10–15% for normal conditions and not less than 5% for fractured rocks can be used for future underground CO_2 storage. For an aquiferous stratum, the average permeability should be not less than 0.15 μm^2. Permeable beds in quaternary sediments are characterized by considerably better flow capacities.

Aquifers to create underground CO_2 storage repositories require the following qualities:

- Presence and integrity of structural or screened traps;
- Establishment of geological peculiarities of the trap and main characteristics of geological objects, including caprocks, within the exploration area;
- Acquisition of hydrogeological data on all aquifers to determine their sealing properties.

The following geological requirements for reservoirs already exist:

- Collectors should have capping strata of impermeable plastic or hard rocks;
- Caprocks should be homogeneous and their thickness not less than 2–6 m for depths of 600 m and from 4 to 5 m for depths over 600 m;
- To guarantee the long-term operation of CO_2 storage, additional interlayers with sealing properties should be present in the formations;
- Within the calculated contour of the future CO_2 storage there should be no tectonic faults that could lead to a decrease in impermeability of the main and reserve capping rocks;
- Permeability of caprocks should not exceed 10^{-10} mD.

An alternative method of calculating the potential CO_2 storage capacity in the CO_2 fields is the Reidulv Bøe methodology (Bøe et al. 2002).

Table 6 CO_2 storage capacity in oilfields in north-west Russia

Oilfield	Cumulative production (Mt)	Min total capacity of CO_2 storage (kt)	Max total capacity of CO_2 storage (kt)
Komi Republic (onshore)	393,220	271,291	349,176
Nenetsky AA (onshore)	56,850	39,222	50,482
Kaliningrad Region (onshore)	31,030	21,408	27,544
North-west Russia (total)	481,100	331,921	427,202

The primary data required in this case are:

- Initial recoverable resources of oil;
- Initial recoverable resources of natural gas;
- Minimal and maximal values of underground oil density (kg/m^3);
- Minimal and maximal values of underground CO_2 density (kg/m^3).

The CO_2 storage capacity in oilfields in north-west Russia is represented in Table 6. Calculation of reservoir capacity for storage of CO_2 was carried out in this investigation using the Reidulv Bøe methodology.

CO_2 EOR needs to be introduced in Russia for the following reasons (Cherepovitsyn and Ilinsky 2006):

1. To assess the potential for joint implementation in CO_2 EOR processes;
2. To determine the main aspects and size of the CO_2 EOR market, providing details for each oil producing region;
3. To investigate the range of prices for industrially captured CO_2 that can be afforded by oilfield operators, bearing in mind that the price would include not only the volume of CO_2 purchased, but also the distance to the oil basin and the quality of the oilfield;
4. To investigate opportunities for establishing public–private partnerships that would encourage large-scale joint CO_2 EOR and CO_2 storage activities in each of the major oil basins, including policies, incentives, improved CO_2 EOR R&D/in situ demonstration projects and 'zero emission' hydrocarbon processing plants.

Oil and gas reservoirs, aquifers and unmined coalbeds are all geological structures with the potential for CO_2 storage. The geographical distribution of potential storage formations differs from that of RW deposits. However, the priority is given to the north-west region with its rich reserves of hydrocarbons in the Timano-Pechera province and the Kaliningrad area. EOR processes are the only way that there can be large-scale utilization of CO_2 and where CO_2 acquires a positive economic value. The total potential capacity of CO_2 storage sites in north-west Russia (depleted oil- and gasfields) is estimated to range from 331.9 to 427.2 Mt.

4 Comparative Assessment

The main criteria for and a comparative assessment of RW disposal and CO_2 storage are presented in Table 7 in terms of the geological features of European Russia.

4.1 Disposal of Radioactive Waste

The isolation of RW is based on the principle of the creation and use of a natural engineering system that protects against ionizing radiation via a basic barrier that is poorly permeable in a seismically stable geological environment.

Many types of geological formation and rock compositions can be considered as potentially suitable environments for RW isolation. The most important parameters of these environments are: size, homogeneity, thickness, structural and hydraulic characteristics, physical and chemical properties, mineralogical structure and petrological features, physical, thermal and mechanical characteristics, possible geochemical reactions, etc.

The selection of geological formations for RW disposal in European Russia is a complex scientific and practical problem. Many types of waste produce a great deal of heat over long periods of time and possess a large number of aggressive chemical and radiation properties. In an underground disposal site, they can thus fundamentally change over time and, in due course, the original properties of a massif can be transformed or destroyed.

In European Russia there is a large potential for developing underground RW disposal sites.

Assessment on a regional basis of such underground areas is as follows:

- Baltic Craton (Murmansk region, Republic of Karelia, Part of St Petersburg region): favourable;
- Voronezh Crystal Massif: favourable;
- East European (Russian) Platform (west and north-west): favourable;
- Russian Platform (East and North-East): unfavourable;
- Northern Caucasia and Ciscaucasia: unfavourable.

4.2 Disposal of CO_2

The highest potential for geological CO_2 storage can be found in:

- The Timano-Pechera oil and gas province (Komi Republic and Nenets Autonomous Area);
- The Kaliningrad area.

Table 7 Comparative assessment of geological conditions for favourable disposal of CO_2 and radioactive waste

Comparative criteria	CO_2	Radioactive waste
Type of geological formation	Depleted oil- and gasfields, aquifers, deep coalbeds	Many types of rocks: clay formations, halogen formations, sandstones, basalts, tuffs
Disposal capacity	High (in Russia approximately 177 Gt CO_2 in oil and gas reservoir)	High, but Russian nuclear power industry produces only a relatively small amount of radioactive waste
Caprocks	Caprocks should be homogeneous and their thickness no less than 2–6 m at depths of 600 m and 4–5 m at depths over 600 m	Magmatic rocks>500 m Halogen formations 75 m Layers 200 m Domes, clay formations >100 m
Permeability of caprock	→ Min. Permeability of capping rocks should not exceed 10^{-10} mD	→Min.
Hydrogeological condition	Acquisition of hydrogeological data on all aquifers with determination of their sealing properties	Zones with extremely slow water exchange
Depth	In most cases, over 600 m	More than 300 m
Seam pressure	Pressure of ground waters>intrinsic pressure of CO_2	
Tectonic breaks, seismicity	→ Min.	→ Min. seismicity on the MSK-64 scale to be<7 (less than 6.2 on the Richter scale)
Retention time	→ Max., more than 5,000–10,000 years	Max.
Location	Oil and gas province in north-west Russia	Murmansk region, Republic of Karelia, Part of St Petersburg region, Voronezh Crystal Massif
Provision index of transport infrastructure	→ Max. (well-developed network of pipelines)	Depends on development of infrastructure (road, container shipment—development of port area, railway)
Distance from points of emission source to disposal sites	→ Min. (30–1500 km in north-west Russia)	Not so important: depends on safety of transport means
Transport infrastructure	Absent for CO_2 transportation, but good prospects of being developed in the future	Good developed transport infrastructure in European Russia
Storage cost	Worldwide analogue	Exact data are unknown

(continued)

Table 7 (continued)

Comparative criteria	CO_2	Radioactive waste
Economic benefits	Additional economic value with use of CO_2 EOR Project of joint implementation in context of Kyoto economic incentives	Economic incentives of radioactive waste disposal need a legal basis
Public acceptance	Low level of knowledge about CO_2 sequestrations processes	Some civil society organizations occasionally raise issue of nuclear waste disposal and safety

MSK Medvedev–Sponheuer–Karnik

There are many oil- and gasfields situated in these two areas, including ones with a high level of depletion, and possible aquifers. The unmined coal seams of the Pechera coal basin (Komi Republic) have a huge storage potential.

4.3 Viability of Geological Storage in Russia Today

In the north-west and other regions, the participation of enterprises in joint implementation projects, including CO_2 storage, is fairly limited. The main reason is the absence of a Russian state register for GHG emissions. Another reason is the absence of a normative-legal base regulating the economic incentives of the Kyoto Protocol (Ilinsky and Cherepovitsyn 2005).

The economic success of CO_2 storage projects in the north-west region would be dependent on greater use of EOR. Research shows that there are depleted oil- and gasfields available for storage in the Komi Republic and the Kaliningrad area. The distances from sources of industrial emissions range from 30 to 1500 km to the Komi fields and from 20 to 50 km to the Kaliningrad fields.

It must be noted that the new Energy Strategy of Russia to 2030 recommends an accelerated introduction of innovative industrial technologies related to EOR, with gas, water-gas, thermogas and thermal methods being the priorities. Comparative analysis shows that one of the most effective and rational technological processes from the point of view of energy and resource saving, and in terms of increasing the oil recovery ratio, is stimulation of petroleum reservoirs by gas injection, including injecting CO_2.

To transport CO_2 to the injection sites, the opportunity to use existing, dedicated gas pipelines (for example, in north-west Russia) must be considered. If new pipelines have to be built, the cost of a CO_2 sequestration project will multiply.

For economic evaluation of RW disposal, there is a need for institutional and legal regulations to encourage nuclear enterprises to implement RW disposal projects. Private–state partnerships are needed for similar projects. It is envisaged that both economic and ecological risks would be shared among the participants through a risk insurance fund. This will promote RW disposal in geological formations and encourage investment in such projects. However, it is unlikely that a satisfactory method of accounting for risk in nuclear power will be found unless greater social, economic and environmental efficiency is achieved.

Public organizations and the population in general have insufficient information about the problem of CO_2 storage. To date, no CO_2 sequestration projects have been carried out. We would thus anticipate that societal and public acceptance in Russia of CO_2 storage will take a long time to achieve (Ilinsky and Cherepovitsyn 2006).

The problem of RW disposal is discussed on occasion by environmental organizations. It is assumed that Russia imports RW and that RW disposal sites already exist. However, no confirmation of this is forthcoming from state bodies, and there are no discussions held on this issue at a national level.

5 Conclusions

This chapter is one of the first attempts to define the regional geological conditions for RW disposal and CO_2 storage in Russia. Questions regarding RW disposal in geological formations have been studied by Russian scientists for a long time. However, according to official sources there are no permanent RW disposal sites in Russia, just 20 temporary storehouses.

Meanwhile, in European Russia there are many geological formations that could be used for RW disposal. The north-west region is a priority area not only from the geological standpoint but also in view of the region's developed nuclear infrastructure. Possible future research directions would include a study of geological and other conditions in mudstones, permafrost limestones and the typical lithologies of north-west Russia.

The geological structures of the part of the Baltic Craton in the Murmansk area are suitable for RW disposal. These structures, for example, the Pechengskaya area, lie in an area that has a developed transport infrastructure and is close to sources of radioactive pollution. The permafrost rocks of the Novaya Zemlya archipelago are also interesting geographically for RW disposal. Favourable areas for RW disposal are the crystalline rocks of the Voronezh Craton and some of the halogen formations of the East European platform.

No economic information on the problem of RW disposal in geological formations is available. Any financial investment for research work will be initiated by the state. The participation of private investors in private–state partnerships is unlikely. The state will also establish what levels of economic profitability are necessary for RW disposal projects in geological formations. Public acceptance of long-term RW disposal in geological formation is very low. It remains so because information on this issue is inaccessible.

To date, hardly any scientific research on CO_2 storage in geological formations in Russia has been conducted. Oil and gas reservoirs, aquifers and unmined coalbeds are the types of geological structure that can be used for CO_2 storage. The geographical distribution of potential CO_2 storage formations differs from that of RW storage. However, the priority areas for both are the north-west province of the Timano-Pechora region, with its rich reserves of hydrocarbons, and the Kaliningrad area.

At present the actual CO_2 storage capacity is known only for oil- and gasfields. Some regions, like Kaliningrad and the Komi Republic, have a high level of depleted oilfields. CO_2 EOR is not currently used in oilfields of north-west Russia.

The total potential capacity of CO_2 storage sites in north-west Russia (depleted oil- and gasfields) is estimated to range from 331.9 to 427.2 Mt. In view of the current level of CO_2 emissions and existing fuel and transportation costs, CO_2 capture and sequestration technologies are a possible option for reduction of CO_2 emissions in Russia only after 2012–2015.

The development of CO_2 sequestration technologies and innovative CO_2 storage projects will be based on large-scale use of the Joint Implementation Mechanisms of the Kyoto Protocol. If the development of EOR technologies, including the gas methods designated in the Energy Strategy of Russia to 2030, are prioritized, this will stimulate innovative projects of CO_2 storage in north-west Russia.

The amount of information concerning CO_2 sequestration in Russia is very low, with practically no scientific articles or any other literature at all on this issue. There is as yet no indication as to what public opinion might be with regard to CO_2 sequestration. Investment by the government and various scientific funds are needed to foster research into the issue of CO_2 storage.

References

Bøe R, Magnus C, Osmundsen P, Rindstad B (2002) CO_2 point sources and subsurface storage capacities for CO_2 in aquifers in Norway. NGU Report 2002.010. Norsk Geologiske Undersøkelse (Geological Survey of Norway), Trondheim

Cherepovitsyn AE (2005) Economic-technological evaluation of greenhouse mitigation options: Russian federation potential. In Konfliktmanagement bei der Planung von Rohstoffvorhaben: Freiberger Forschungsforum 55. Berg- und Hüttenmännischer Tag. Freib Forsch C 509:228–233

Cherepovitsyn AE (2006) Oil and gas fields as future prospect for CO_2 storage in the north-west of Russian federation. In development, exploration and processing of raw materials: Freiberger Forschungsforum 57. Berg- und Hüttenmännischer Tag. Freib Forsch C 512:71–74

Cherepovitsyn AE, Ilinsky AA (2006) CO_2 storage potential in north-west Russia. In: Røkke N, Bolland O, Gale J (eds) Proceedings of the 8th International Conference on Greenhouse Gas Control Technologies. Elsevier, London, pp 61–62, CD-ROM

Ilinsky AA (2005) Working out innovation strategy of development of energy complex of north-west region and justification directions of environmental technologies. Informational report. Saint Petersburg State Mining Institute (Technical University), Saint Petersburg [in Russian]

Ilinsky AA (2006) The development of public policy foundations in energy complex of north-west region. Informational report. Saint Petersburg State Mining Institute (Technical University), Saint Petersburg [in Russian]

Ilinsky AA, Cherepovitsyn AE (2005) Conceptual approaches to Kyoto Protocol market mechanisms implementation. In: Proceedings of the 4th Nordic Minisymposium on Carbon Dioxide Capture and Storage. Helsinki University of Technology, Helsinki, pp 43–46

Ilinsky AA, Cherepovitsyn AE (2006) Implementation of the Kyoto Protocol market mechanism in the Russian federation. In: Røkke N, Bolland O, Gale J (eds) Proceedings of the 8th International Conference on Greenhouse Gas Control Technologies. Elsevier, London, pp 203–204, CD-ROM

Kjärstad J, Johnsson F (2004) Simulating future paths of European power generation—Applying the Chalmers power plant database to the British and German power generation system. In: Rubin ES, Keith DW, Gilboy CF (eds) Proceedings of the 7th International Conference on Greenhouse Gas Control Technologies, vol 1. Elsevier, London, pp 881–890

Komlev VN (1998) Use of natural materials from northern Russia for the isolation of radioactive wastes and spent nuclear fuel. In: Stenhouse MJ, Kirko VI (eds) Defence Nuclear Waste Disposal in Russia: International Perspective, vol 18, NATO ASI Series: disarmament technologies. Kluwer Academic, Dordrecht, pp 85–98

Shishits IU (1998) The Basis of Geology and Radioecology. Publishing House of Moscow State Mining University, Moscow [in Russian]

Smyslov AA (ed) (1996) Geological Atlas of Russia. Roskomnedra, Moscow [in Russian]

Smyslov AA, Sendek SV, Lir YV, Kozlov AV (2002) The interior of Russia. In: Ecology of Geological Sphere, vol 2. Saint Petersburg State Mining Institute (Technical University), Saint Petersburg [in Russian]

Stevens SH, Kuuskraa VA, Gale J (2001) Sequestration of CO_2 in depleted oil and gas fields: global capacity, costs, and barriers. In: Williams DJ, Durie RA, McMullan P, Paulson CAJ, Smith AY (eds) Greenhouse Gas Control Technologies: Proceedings of the Fifth International Conference on Greenhouse Gas Control Technologies. CSIRO Publishing, Collingwood, pp 278–283

Tatarchuk BJ (1997) Complex researches of regions under geological, natural, economic and social forecasts with the purpose of a choice of sites for detailed investigation of nuclear waste burial grounds and storehouses location. Mining Bull 6:23–27 [in Russian]

Tikhonov MN (2004) Dangerous spots on a map of Russia. Industry information agency PRoAtom [in Russian]. http://www.proatom.ru/modules.php?name=News&file=article&sid=215

Zakharova E (2004) Some aspects of application of CO_2 capture and storage technologies for CO_2 emission reduction in the Russian electric power production sector. In: Rubin ES, Keith DW, Gilboy CF (eds) Proceedings of the 7th International Conference on Greenhouse Gas Control Technologies, vol 1. Elsevier Science, Oxford

Comparison Between Geological Disposal of Carbon Dioxide and Radioactive Waste in China

Ju Wang and Zhonghe Pang

Abstract The large amount of carbon dioxide (CO_2) emissions and the fast development of nuclear power plants in China pose challenges for the safe disposal of CO_2 and high-level waste (HLW). Significant progress has been made in both areas. The disposal of CO_2 has focused on making commercial use of CO_2. Several enhanced oil recovery and enhanced coalbed methane projects have been implemented in China. Seventy disposal sites in 24 major sedimentary basins have been identified for CO_2 disposal. The amount of spent fuel will reach about 82,000 t of heavy metal when all of the planned 58 reactors on the Chinese mainland reach the end of their lifetime. A target to build a national HLW repository in around 2050 has been set. CO_2 disposal and radioactive waste disposal have much in common, but there are also many differences, including disposal principles and capacity, host media, potential sites, site characterization and cost. The site with the most potential for HLW disposal in China is the Beishan granite site in north-western Gansu Province, while most of the potential sites for CO_2 disposal are in the eastern and south-western basins of China. For HLW, only one repository is planned, but for CO_2 disposal, many sites are needed. The disposal of CO_2 and radioactive waste are facing similar scientific and technical challenges, including site selection, monitoring of disposal site, prediction of how disposal systems will work, safety assessments, and social and economic issues. To meet these challenges, the scientists working in these fields need to intensify the exchange of information and increase cooperation.

J. Wang (✉)
Beijing Research Institute of Uranium Geology, China National Nuclear Corporation, 9818, Beijing 100029, China
e-mail: wangju9818@163.com

Z. Pang
Institute of Geology and Geophysics, Chinese Academy of Science, Beijing 100029, China
e-mail: z.pang@mail.iggcas.ac.cn

F.L. Toth (ed.), *Geological Disposal of Carbon Dioxide and Radioactive Waste: A Comparative Assessment*, Advances in Global Change Research 44,
DOI 10.1007/978-90-481-8712-6_17,

Keywords Geological disposal of CO_2 • Geological disposal of radioactive waste • High-level radioactive waste • China

1 Introduction

China has many kinds of energy resources and is especially rich in hydropower and coal. China's reserves of hydropower rank first in the world, and its coal reserves rank third. The total confirmed reserves of coal in China are 184.2 gigatonnes (Gt). The structure of primary energy consumption in China in 2006 was: coal 69.4%, oil 20.4%, natural gas 3%, the rest (hydropower, nuclear power and wind power) accounted for 7.2% (Jiang 2008). The structure of the installed capacity for electric power generation in 2007 was: thermal power (including coal-, oil- and gas-burning power plants) 78%; hydropower 20%; nuclear power 1%; wind and other technologies 1%. These figures show the obvious importance of coal in the energy mix of China. The production of coal in 2007 came to 2.54 Gt. China's carbon dioxide (CO_2) emissions in 2007 amounted to 5.06 Gt (Jiang 2008), making it second to the USA in the list of the world's top CO_2 emitters.

Coal-burning power plants are the main sources of CO_2, sulphur dioxide (SO_2), dust and coal ash, and have significant environmental impacts. However, based on the level of renewable energy development in China, it is foreseen that coal will remain the dominant fuel in China's energy supply for decades to come.

China is also one of the countries adversely affected by global climate change: extreme climate events and disasters have been more frequent in recent years, including heavy flooding in southern China as well as the cold winter storms with icy rain that caused loss of human life and property damage of 150 billion Chinese renminbi yuan (CNY) in early 2008. It has been predicted that climate change will exacerbate the extreme weather conditions in the northern and southern regions of China: north China is expected to experience less rainfall but the south much more. To mitigate global climate change, a significant reduction of CO_2 emissions must be achieved to stabilize atmospheric concentrations of CO_2. Because of the great global concern over climate change, China is facing pressure to reduce its CO_2 emissions.

Although some future energy demands can be reduced by more efficient use of energy and use of renewables, the coal mining capacity will still be increased substantially. Although clean coal technology is being developed, it is not expected to reduce CO_2 emissions by more than 10%. Carbon dioxide capture and disposal (CCD) is part of a portfolio of solutions, with disposal of CO_2 in geological formations being one of the options under consideration.

The Chinese Government has embarked on the development of a higher share of nuclear power in the energy mix as an effective way of reducing CO_2 emissions. In 2007 the State Council approved a medium- and long-term plan for nuclear power development in China (2005–2020), which set the target of 40 GW installed capacity by 2020, with construction work beginning by the same date to produce an additional 18 GW. At the same time, the Government is implementing a new law on renewable energy, passed in early 2006, which encourages the use of renewable and clean energy resources such as wind power, solar power, geothermal and biomass, etc. The Government has also set the target of reducing energy consumption by 20% per unit of GDP by 2012. These measures will certainly contribute to the reduction of CO_2 emissions.

2 Geological Disposal of CO_2 in China

2.1 Principles of Geological Disposal of CO_2

CO_2 produced by the energy sector is the main cause of global warming. Deep reductions in CO_2 emissions are required to meet the goal of the United Nations Framework Convention on Climate Change (UNFCCC) to stabilize anthropogenic greenhouse gas emissions. CCD technology would be used in combination with other mitigation measures (e.g. fuel switching, energy efficiency and use of renewable energy) to achieve these goals.

CO_2 can be captured from a variety of anthropogenic sources such as power plants and large industrial plants and then compressed and transported to a disposal site. In China, for example, the main sources of CO_2 emissions are coal-fired power plants and large steel factories. CO_2 can be captured through: (1) post-combustion; (2) pre-combustion; (3) oxyfuel combustion; (4) other methods (e.g. chemical looping combustion) (IEA GHG 2005).

Geological disposal involves emplacing the captured CO_2 in selected deep geological reservoirs. It is a promising option, capable of achieving deep reductions in emissions in the foreseeable future. A number of geological media are candidates for storage of captured CO_2. These include: (1) depleted and disused oil- and gas-fields; (2) deep saline aquifers; and (3) deep unmineable coal seams. The geological media suitable for CO_2 disposal via various physical and chemical trapping mechanisms must have the necessary disposal capacity and fluid injectivity; they must also have a tough, dense caprock layer to confine the CO_2 and prevent its rapid migration and vertical leakage to more shallow strata, in particular those containing potable groundwater and soils and/or to the atmosphere. As many of these geological traps have already held hydrocarbons or liquids for millions of years, they can be considered as stable and intact under normal conditions. Disposal of gases, including CO_2, in these media has been demonstrated on a field scale by enhanced oil recovery (EOR) and enhanced coalbed methane (ECBM) operations, natural gas disposal and acid gas disposal projects.

After CO_2 is injected into the ground, it can be trapped through: (1) structural trapping; (2) solubility trapping; and (3) mineralization trapping. These processes take time to complete, as models have predicted.

According to the Intergovernmental Panel on Climate Change (IPCC) report (IPCC 2005):

> 'Observations from engineered and natural analogues as well as models suggest that the fraction retained in appropriately selected and managed geological reservoirs is very likely to exceed 99% over 100 years and is likely to exceed 99% over 1,000 years.'

'Very likely' means a 90–99% probability; 'likely' means a 66–90% probability.

The global disposal capacity for the main suitable geological disposal reservoirs has been estimated by the IEA Greenhouse Gas R&D Programme, based on injection costs of up to US $20 per tonne of CO_2 sequestered (excluding cost of capture, conditioning and site closure). The capacities for depleted gasfields, depleted oilfields (using CO_2 EOR), unmineable coal seams (using CO_2 ECBM) and deep saline aquifers are 690, 120, 40 and 400–10,000 Gt CO_2, respectively (Bradshaw et al. 2007). Deep saline aquifers are clearly the main underground space suitable for disposal, and are of no economic significance.

CO_2 capture and geological disposal is an enabling technology that will allow the continued use of fossil fuels, mainly coal, well into this century, for power generation and combustion in industrial processes. With fossil fuels being relatively abundant, cheap, available and widely distributed in China, they will enhance the security and stability of the country's energy supply.

2.2 CO_2 Emissions in China

According to statistics, in 2004 China's total CO_2 emissions amounted to 5.1 Gt, accounting for 13% of total world emissions, with average per capita emissions being 87% of the average world level. Total coal use and coal power generation accounted for 73% and 36% of emissions, respectively. Although the share of coal in primary energy consumption is expected to decline with time, it will still remain at around 45–50% by the year 2050. The share of new and renewable energy sources will increase in the power generation sector; however, coal-burning power plants will still represent around 50% of power generation by 2050. Therefore CCD technology is badly needed in China, and geological disposal of CO_2 (GDC) constitutes the final step in the whole operation.

The main sources of CO_2 emissions in China are coal-fired power plants and large steel factories. These sources are located mainly in the eastern coastal areas of China. Since the sources are relatively concentrated, the emissions can be captured and disposed of in a fairly efficient manner, without transportation over long distances.

2.3 Policies and Plan for the Geological Disposal of CO_2 in China

The Chinese Government has realized the serious consequences of global climate change and is taking strong action to reduce CO_2 emissions in the country. China is one of the founding members of Carbon Sequestration Leadership Forum. China signed the UNFCCC in June 1992 and the Kyoto Protocol in May 1998. The Government has also improved the country's energy consumption structure and taken measures to increase energy efficiency. It has established policies and laws to encourage energy saving and the use of renewable energy resources. China also hosts many projects under the Clean Development Mechanism of the Kyoto Protocol.

The study of CCD as a cutting-edge technology is included in several key national R&D plans. CCD technology has been integrated into the National Medium- and Long-Term Plan for Science and Technology Development towards 2020. In the 11th Five-Year National Plan period (2006–2010), the National High-Tech R&D Program (the 863 Program) supports the development of CCD technology. About CNY 7 billion have been allocated to the Scientific & Technological Actions on Climate Change, in which CCD technology is included.

2.4 R&D Activities Related to Geological Disposal of CO_2 in China

The main research efforts related to GDC in China have been focused on using CO_2 to achieve the highest possible economic returns. Many projects have been implemented and have had fruitful outcomes. These projects have covered a wide range of topics including EOR (Ma et al. 2007) and ECBM (Ye et al. 2007). Fundamental research such as chemical thermodynamic modelling of water-rock-gas systems (e.g. Li 2008) has also shifted towards this field.

Because of the complicated geology of Chinese oilfields and the high value of its oil and natural gas resources, the country's oil recovery technology has progressed very fast and China has become a world leader in this field. The recovery rate is up to 60% in the Daqin oilfield, for example, while the world average is about 35%. Methane recovery through CO_2 injection has resulted in increased methane production: a corresponding field-scale test in Shanxi Province was concluded to be feasible, with the manufacture of a full-scale demonstration unit being recommended.

Use of geothermal energy for heating and cooling has also boomed in China in the last decade as a result of improved living conditions and the need to mitigate climate change. In Tianjin, North China, for example, more than one million people live in houses that are heated by geothermal energy, and another four million people use hot tap water from geothermal wells. The demand for geothermal energy is still increasing, and with the capacity of limestone formations approaching its limit, exploitation is moving increasingly towards sandstone formations. However, the injection rate for sandstone formations is quite low, currently less than 20% of the water produced (from the aquifers exploited) (Pang 2007). To overcome this barrier to geothermal development, a new concept known as enhanced aquifer thermal energy recovery (EATER) using CO_2 as a catalyst has been proposed (Pang et al. 2008; Pang et al. in press). As there is a need to understand the response of these saline aquifers to CO_2 injection, this project has come at a good time, allowing the EATER project and the CO_2 disposal project to be carried out together.

It should be pointed out that various uses of CO_2 in the energy industry have promoted studies of technologies for GDC. However, all three types of project—EOR, ECBM and EATER—are looking to maximize economic returns while minimizing CO_2 loss so that it can be reused. These projects will not achieve the final goal of GDC at a commercial scale. Therefore studies that focus on making a detailed assessment of major target formations, such as deep saline aquifers, are needed since saline aquifers are considered as the main target formations with the largest disposal capacity, as discussed above. A new project is under way to conduct detailed studies on saline aquifers for CO_2 disposal. It is also supported by the Ministry of Science and Technology (MOST) through the 863 Program, to be implemented jointly by institutions of the Chinese Academy of Sciences, universities and the geothermal industry. This project includes a field-scale test with CO_2 injection.

In the meantime, several academic institutions such as Qinghua University will continue their research on carbon capture technologies with the support of the same national 863 high-tech programme. It is hoped that CCD technology will progress as an integrated package of technologies for the mitigation of climate change.

Planning of the GreenGen Program by the China Huaneng Group (CHNG) and eight other big Chinese energy producers is under way. The plan is to construct a demonstration project in the special economic zone in Tianjin to integrate CCD technology for a 250 MW integrated gasification combined cycle (IGCC) power plant. These choices have made Tianjin a focal region for different tests related to GDC and CCD.

2.5 Preliminary Estimates of the Capacity for Geological Disposal of CO_2 in China

Most of the coal-fired power plants are located in the eastern part of China, where most of the population and industry are also located. In eastern China there are many sedimentary basins; their potential for CO_2 disposal has been estimated by Liu et al. (2005a, b) and Li et al. (2006).

There are many large sedimentary basins on the Chinese mainland and its continental shelf, with excellent potential for the disposal of CO_2 (Fig. 1). Li et al. (2006) identified 70 disposal sites in 24 major sedimentary basins in China, with a total disposal capacity of 144 Gt CO_2 in the saline aquifers. The North China Basin, Sichuan Basin, Zhunger Basin and the east coast basins are the main potential sites for GDC.

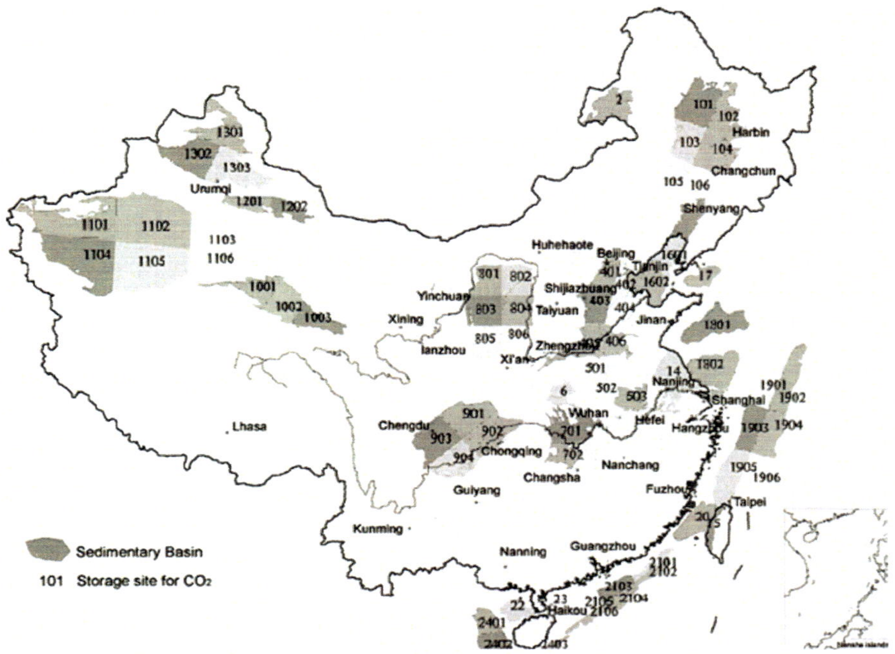

Fig. 1 Distribution map of main sedimentary basins in China and the potential CO_2 disposal sites (Source: Ren 2000)

Most of the large sedimentary basins in China are composed of multi-layered sediment systems; saline aquifers are widely dispersed in these basins and they are a suitable space for CO_2 storage. According to a rough estimate by Li et al. (2006) based on solubility, continental reservoir capacity is 7.738×10^{10} t, accounting for 53.92% of total disposal capacity. The storage capacities for individual basins are: Huabei Basin (including Bohai Sea), about 2.046×10^{10} t; Sichun Basin, about 6.407×10^{8} t; Zhunger Basin, about 4.688×10^{8} t; Tarim Basin, about 2.787×10^{10} t. Continental shelf reservoir capacity is about 6.613×10^{10} t. The total capacity of the major basins listed above is about 1.434×10^{11} t CO_2, which is about 40 times the CO_2 emissions of China in 2003.

Liu et al. (2005a, b) estimate total disposal capacity in unmineable coalbeds and depleted natural gasfields in China. As a result of these and other studies, primary estimates are as follows: (1) 46 oil and gas reservoirs—7.2 Gt CO_2; (2) 68 unmineable coalbeds with methane recovery—12 Gt CO_2; (3) 24 saline aquifers—1,435 Gt CO_2. For stationary CO_2 sources and geological disposal potential the findings are: (1) capacity—1,445.8 Gt, about 400 years (3,300~4,000 million tonnes per year); (2) north-east—one of the most suitable areas, (3) south—further exploration expected; (4) central China and the south-west—mainly in the Sichuan Basin.

Studies in this direction are still continuing, with the aim of achieving more precise estimates with more detailed research.

2.6 Key Challenges and Barriers to the Geological Disposal of CO_2 in China

CO_2 injection into geological media is technologically feasible, as indicated by CO_2 EOR, ECBM and combined CO_2/hydrogen sulphide (H_2S) injection operations, including the more recent CO_2 disposal operations at Sleipner in the North Sea. These operations show that there are no major technological barriers to CO_2 geological disposal at a local scale (Stevens et al. 1999). More tests such as EATER will verify the specific characteristics of local geological variations in the target geological formations.

Most of the risks associated with GDC are common to and comparable with any other industrial activities for which extensive safety and regulatory frameworks are in place. The specific risks of GDC are associated with the injection and the post-injection phases, of which the risks of most concern are those posed by the potential for acute or chronic CO_2 leakage from the disposal formation, the local risks to health and safety, and the environmental impacts on fresh water and air quality, which need to be assessed and managed.

There are concerns that public opinion and public acceptance or rejection of this technology will likely affect the large-scale implementation of GDC. The lack of policy and regulation is presently the most significant barrier. Whether or not policymakers decide to regulate GDC will have an important bearing on the economic viability and the financial risks involved in the adoption of this technology. The timing of such a decision will also affect how quickly (or slowly) the technology is deployed. From the public's perspective there are two basic questions that need to be satisfactorily answered: Will CO_2 leak? And what will happen if it does leak? The public may perceive risk differently from the engineer, who is concerned with likelihood and hazard in a subjective manner that considers more the qualitative nature and characteristics of risks. The decision of the government to implement GDC or not is largely dependent on the cost of the operation and pressure from the international community.

When it comes to large quantities, multiple sites and long-term operation, GDC faces more scientific and technical challenges, such as:

- How to select suitable disposal sites;
- How to evaluate the suitability of disposal sites, and how to estimate the disposal capacity of a site;
- How to monitor the long-term behaviour of the CO_2 disposed of, including the reaction between CO_2 and the components in the host formation, and the migration of CO_2 within host formations;
- How to assess the environmental impacts of a GDC site.

To answer the above-mentioned challenges, several national projects, such as the national 863 high-tech development programme, jointly supported by the Ministry of Science and Technology of China, institutions of the Chinese Academy of Sciences, universities and the geothermal industry, are under way. Studies including laboratory experiments, numerical modelling, as well as the first pilot field test are being implemented.

3 Geological Disposal of Radioactive Waste in China

3.1 Principles of Radioactive Waste Disposal

Radioactive waste (RW) is generated at every phase of the nuclear fuel cycle and as a consequence of the use of radioactive materials in industrial, medical, military and research applications. The main objective of the safe disposal of the waste is the protection of people and the environment in the short and long term. In some countries low-level waste (LLW) and intermediate-level waste (ILW) are disposed of in shallow underground repositories and also in former mines. However, for high-level waste (HLW), spent nuclear fuel (SNF) and also for all long-lived waste categories, deep geological disposal at depths of hundreds of metres is considered worldwide as the safest and most feasible method of protecting humans and the environment for extremely long periods of time (tens of thousands of years) (IAEA 1993). For the last 4 decades, various concepts have been developed in many countries for the disposal of RW in deep geological formations (including rock salt, granite, clay and tuff rocks). To demonstrate the feasibility of the disposal concepts, numerous full-scale experiments have been carried out in underground research laboratories (URLs) under realistic repository conditions (Wang et al. 2006a).

The conceptual design of repositories in different geological formations generally relies on a multi-barrier system, which typically comprises the natural geological barriers provided by the repository host rock and its surroundings and an engineered barrier system (EBS). The ability of natural barriers to isolate radionuclides from the environment on a long-term basis and their long-term stability are the main reasons for the geological disposal of RW. The EBSs are man-made, engineered materials placed within a repository, including the waste form, waste canisters, buffer materials, backfill and seals (NEA 2005). The main functions of EBS components are the following:

1. The waste matrix is designed to provide a stable waste form that is resistant to leaching and gives slow rates of radionuclide release over the long term;
2. The container/overpack is designed to facilitate waste handling, emplacement and retrievability, and to provide containment for up to 1,000 years or longer depending on the waste type;
3. The buffer/backfill is designed to assure the stability of the repository and of the thermo-hydro-mechanical-chemical conditions, and to provide low permeabilities and diffusivities, chemical buffering and long-term retardation;
4. The other EBS components (e.g. seals, plugs) are designed to prevent releases via drifts and shafts and to prevent access to the repository.

3.2 Nuclear Power Plants in China

As a result of the recent rapid economic growth in China, electricity shortages have become a major problem over the past few years in some coastal provinces. This situation has made the Chinese Government change its policy regarding nuclear power plant (NPP) development. In 2003, to meet the strong demand for electricity, the Government decided to pursue the 'active development of nuclear power'. According to the Medium and Long-Term Development Plan for Nuclear Power (2005–2020) that was approved by the State Council in 2007, the installed nuclear power capacity will reach 40 GW by 2020 and the electricity produced by these plants will provide 4% of total electricity production. In the meantime, a further 18 GW nuclear power capacity will be under construction in 2020.

At present, there are 5 NPPs (11 reactors) in operation on the Chinese mainland: Qinshan, Qinshan-II and Qinshan-III in eastern China (Zhejiang Province), Daya Bay and Lin'ao in southern China (Guangdong Province), and Tianwan in eastern China (Jiangsu Province). The total installed nuclear power capacity was 9.10 GW in 2007. This accounted for 1.3% of the total installed electricity capacity in China, while the electricity produced accounted for 1.92%.

3.3 Policies and Plan for High-Level Radioactive Waste Disposal

China has been accumulating HLW from the defence industry since the 1950s, and will generate more from NPPs, which means that there is an urgent need for the safe disposal of HLW.

At present, the 11 reactors on the Chinese mainland generate 370 t of spent fuel each year. Based on the long-term plan for NPPs in China, an estimation of the amount of spent fuel was made, and this revealed that there would be about 10,300 t of heavy metal (tHM) of spent fuel by 2020 and 82,000 tHM when all the reactors reached the end of their lifetime.

In the 1980s the China National Nuclear Corporation (CNNC) proposed a plan for HLW disposal. With the increasing need for and concern about a geological repository programme, the China Atomic Energy Authority (CAEA) instituted a series of procedures and worked out a national strategy and long-term plan for a repository in February 2006 called R&D Guidelines for the Geological Disposal of High Level Radioactive Waste in China (CAEA et al. 2006). It is a three-step strategy (site selection first, then research in an underground laboratory, and finally the building of a repository) and includes a plan consisting of five fields of activities: policies, regulations and management; site selection and characterization; engineering design; safety assessment; and radiochemical studies.

In the R&D Guidelines it is announced that the purpose of the long-term plan is to build a geological repository at a site with stable geological and suitable socio-economic conditions in the mid-twenty-first century; its purpose will be to use engineered and natural retardation barriers to protect public health and the environment against unacceptable harm from HLW by means of the containment. The three-step development strategy described in the Guidelines comprises:

(a) Site selection and comprehensive studies;
(b) In situ tests and demonstration in a URL;
(c) Construction of the HLW repository.

At the same time, comprehensive and supporting studies, including performance assessment and covering such subjects as backfill material, radionuclide migration, natural analogues, and the technology for construction, sealing, and closure of the repository, are to be conducted.

During Step 1, nationwide site screening, regional screening, subregional screening, studies of the deep geological environment, and site characterization will be carried out, with the objective of selecting and confirming the final site. During Step 2, a URL will be built either at the site selected during Step 1 or elsewhere. This URL will serve both as a methodological laboratory and for site evaluation. During Step 3, a final repository will be built at the site. It might also be built based on results from a previous URL. According to the State Council (2007), the construction of a URL for R&D of geological disposal should be complete by 2020, and a national repository will be ready in around 2050.

3.4 Major Progress in Geological Disposal Programme for High-Level Radioactive Waste

3.4.1 Site Selection

Since site selection began in 1985, it has been an important part of China's HLW disposal programme. The whole siting process was divided into four stages: nationwide screening, regional screening, area screening and site confirmation. During the siting process, socio-economic factors and natural factors were considered, including political support, population, economic potential, plant/animal resources, mineral resources, land use, local public acceptance, geological/hydrogeological conditions and engineering geological conditions. Since 1985 the following activities have been conducted in connection with site selection (Wang et al. 2004, 2006b):

1. Nationwide screening (1985–1986): According to the preliminary siting criteria, five regions have been selected as potential regions: south-western China, eastern China, Inner Mongolia, southern China and north-western China.
2. Regional screening (1986–1989): Based on the results from the first stage nationwide screening further investigations were conducted, and 21 candidate areas

were selected. Some clay formations in sedimentary basins were also considered. In north-western China, the Beishan area in Gansu Province is considered as the most important area.

3. Area screening (1990–present): Since 1990 most efforts have been concentrated on the Beishan area in Gansu Province. Studies include the investigation of regional crust stability, study of the tectonic evolution, lithological studies, hydrogeological studies and a preliminary geophysical survey. At the same time, possible host rock types for the repository have also been investigated, with the conclusion being reached that granite is the most suitable for repositories in China.

3.4.2 Site Characterization

Research in the Beishan area since 1990 includes the regional geological setting, crust stability, geological characteristics, hydrogeology and methodological studies for site characterization. The International Atomic Energy Agency (IAEA) has assisted—and continues to assist—China's site characterization programme through its technical cooperation projects (IAEA TC Projects CPR/9/026, CPR/4/024 and CPR/3/008 (ongoing)).

The Beishan area, located in north-west China's Gansu Province, was selected as the area most suitable for an HLW repository. Within the Beishan area, three granitic subareas (Jiujing, Xiangyangshan-Xinchang and Yemaquan) are considered as being the most suitable. During 1999–2008, surface geological, hydrogeological and geophysical surveys and borehole drilling were conducted in the Jiujing, Yemaquan and Xinchang granitic sections. Six deep and eight shallow boreholes were drilled and a series of borehole tests, such as pumping tests, injection tests, borehole televiewer and borehole radar survey, sample taking and geostress measurements were conducted. Favourable findings were obtained, which provided important data that were used to evaluate the suitability of the Jiujing, Xinchang and Yemaquan subareas.

The boreholes in the Beishan area provided deep geological environmental parameters and groundwater samples for geological research, enabled hydrogeological experiments and in situ geostress measurements to be carried out, and allowed rock and groundwater samples to be collected. The suitability of the region was evaluated through a series of site characterization methods, with proven effectiveness. This provided a reference for similar work and the formulation of standards for the future. Thus, valuable experience was gained in terms of evaluating sites in fractured granite media in arid areas.

According to the Chinese programme, once the suitability of the Beishan area has been recognized, a URL will be built there, and more detailed site evaluation, in situ tests and underground experiments conducted. The URL will serve both as a methodological laboratory and as a site confirmation tool, and might also be developed into an actual HLW repository.

3.4.3 Geology of the Beishan Area

The Beishan area in Gansu Province has been selected as the most suitable area to serve as China's HLW repository. The crust in the area has a block structure, with a crust thickness of 47–50 km. The seismic intensity of the area is below grade VI, and, as far as is known, no earthquakes with a magnitude of Ms>4.75 have taken place. The topography of the area is characterized by flatter Gobi Desert terrain and small hills with elevations ranging between 1,000 and 2,000 m above sea level. Since the Tertiary it has been a slowly uplifting area without obvious differential movement; the geological characteristics of the Beishan area show that the crust in the area is stable and that it has a great potential for the construction of an HLW repository.

In the area, eight granite subareas have been selected as potential sites for the future URL and HLW repository.

3.4.4 Hydrogeology in the Beishan Region

The Beishan region is poor in groundwater resources. Pumping tests carried by local teams of geologists in the area in the 1980s showed that, for most of the wells, the outflow rates are less than 50 m^3/day. Beishan groundwater can be divided into three categories: (1) upland rocky fissured units; (2) valley and depression pore-fissure units; and (3) basin pore-fissure units. The upland rocky fissured unit is the most prevalent one in the area.

1. Upland rocky fissured groundwater: This is the most important water type in the Beishan region, occurring in weathered and structural fractures. Groundwater recharge is primarily from precipitation infiltration, with discharge mostly through evaporation and lateral outflows into the water-bearing fracture zones, intermountain areas and valley depressions. The present water table in the potential site area is 28–46 m below the surface.
2. Valley and depression pore-fissured groundwater: In the Beishan region, valley and depression topography generally coincides with the fault zones. This water is commonly more abundant than in other areas. The water table is shallower, with the depth varying between 2 and 8 m below the surface. The water is recharged mainly through infiltration of rainfall and temporary seasonal floods, and the main discharge includes evaporation and run-off towards the basin and the Hexi Corridor.
3. Basin pore-fissured groundwater: This groundwater is mainly distributed among the basins in the north and north-east parts of the area and also among the fault basins of the Hexi Corridor. The basins are mainly composed of Jurassic, Tertiary and Quaternary formations. Groundwater is recharged from the lateral inflows. Well production varies within a wide range (from 10 to 1,000 m^3/day), depending mainly on the conditions of the basin scale, lithology of the aquifer, and structure. In general, the water table is close to the surface. In some areas, the groundwater is artesian.

3.5 Key Challenges

The safe disposal of HLW and SNF is a challenging task in scientific and technological terms because they have to be fully and reliably isolated for hundreds of thousands (even millions) of years. Radionuclides, such as Np, Pu, Am, Tc, etc., are highly radioactive, toxic, and have a long half-life. If any of these radionuclides were to pollute the environment, this would cause tremendous harm to the biosphere. Therefore deep geological repositories are constructed to isolate the HLW and SNF. However, the construction of such repositories involves a number of key issues (Wang et al. 2006a, 2007), including:

- How to select a suitable site, and how to evaluate its suitability;
- How to select engineered barrier materials to effectively isolate HLW/SNF;
- How to design and construct a deep repository;
- How to assess the long-term safety of the disposal system.

To solve the challenging issues mentioned above, many large-scale R&D projects have been carried out around the world (Kickmaier and McKinley 1997; NEA 1999, 2001, 2003; Zhang et al. 2006), including:

(a) Development and testing excavation techniques;
(b) Studies of excavated damage zone;
(c) Site characterization studies;
(d) Hydrogeological tests;
(e) In situ radionuclide migration tests;
(f) Simulation of effects caused by emplacement of RW;
(g) Demonstration of EBSs;
(h) Prototype repository tests;
(i) Natural analogue and anthropogenic analogue studies.

Most of the above R&D projects are carried out to deal with the following key scientific challenges:

1. Precise prediction of the evolution of a repository site: As there are many long half-life radionuclides in HLW/SNF, they need to be isolated for a very long period, as long as $(1 \sim 10) \times 10^5$ years. Therefore the prediction of how a repository site will evolve should be carried out in precise detail, including prediction of the geological stability, regional geological conditions, regional and local groundwater flow, climate changes, landform evolution, geological hazards (volcanism, earthquakes, faulting, etc.).
2. The behaviour of engineering material under coupled conditions: The engineering materials for repositories include waste forms (such as waste glass), canisters (carbon steel, copper, etc.) and buffer and backfill materials. The behaviour of such materials under coupled conditions (intermediate to high temperature, geostress, hydraulic, chemical, biological and radiation processes) is much different from its usual behaviour. The long-term evolution of such materials under coupled conditions has also been a hot topic in recent R&D programmes.

3. The geochemical behaviour of transuranic radionuclides with low concentration and their migration with groundwater: The radionuclides released from the repositories will migrate with groundwater and diffuse into the matrix. The migration behaviour of radionuclides depends on the groundwater flow and on complicated geochemical processes. At present, we have little knowledge regarding the geochemical processes related to radionuclides such as Np, Pu, Am and Tc. The speciation, complexation, colloidal and biological effect of these radionuclides under realistic repository conditions are challenging topics. Some of the radionuclides such as Tc, I and ^3H are difficult to contain; thus selecting suitable materials to retard them is also challenging.
4. Safety assessment of disposal system: The geological disposal system is a complex system, composed of many subsystems (waste forms, canister, buffer material, near field, far field, biosphere, groundwater etc.) which will experience complicated and long-term coupling processes. A detailed safety assessment of the system thus presents a very difficult challenge to current computational and technical abilities.

In China several R&D projects are currently being conducted in response to the above challenges. At the Beishan site in Gansu Province, comprehensive site characterization is under way, one of the purposes being to investigate the geological history of the site and to predict future geological evolution. A mock-up facility is under construction to allow the behaviour of buffer and backfill material (Chinese bentonite, equivalent to GMZ bentonite) under coupled conditions to be better understood. Methodological studies on the geochemical behaviour of transuranic radionuclides at low concentrations and their migration with groundwater, as well as a safety assessment of disposal systems, are also in progress.

4 Comparison of Geological Disposal of CO_2 and Radioactive Waste

The geological disposal of CO_2 (GDC) and RW is one of the largest environmental engineering undertakings in the world. GDC shares many similarities with the geological disposal of RW, but there are also many differences between the two (see Table 1).

The Chinese Government considers disposal of both CO_2 and RW to be important environmental issues that are key to the sustainable development of the country, and has been paying increasing attention to them. Several regulations and standards have been put into force, and development strategies and guidelines have been proposed for both types of disposal. Most importantly, several key R&D projects with concrete financial support have been started and many scientists and engineers are involved in these projects. Thus, significant progress has made both in CO_2 and RW disposal.

The differences between the two are as follows:

Table 1 Comparison between geological disposal of CO_2 and radioactive waste in China

Attribute	CO_2 disposal	Radioactive waste disposal
Philosophy	1. To mitigate global climate change, China has decided to dispose of CO_2 in geological formations 2. In China CO_2 is considered not only as waste but also as a source of profit	1. To ensure the sustainable development of nuclear energy and to protect the environment, China has decided to dispose of HLW in deep geological formations 2. As spent fuel is considered as a resource in China, the spent fuel will be used to recycle plutonium and uranium. The vitrified HLW is considered as final waste
National policies	China has signed the United Nations Framework Convention on Climate Change and the Kyoto Protocol in an effort to mitigate climate change and reduce the emission of CO_2	China has signed the Joint Convention on the Safety of Spent Fuel Management and the Safety of Radioactive Waste Management. For HLW disposal, the policy is that the spent fuel should be reprocessed; a centralized deep geological repository is planned to dispose of HLW in China
National plan	CCD is included in several key national R&D plans, such as the National Medium- and Long-Term Plan for Science and Technology Development towards 2020. In the 11th Five-Year National Plan period (2006–2010), the National High-Tech R&D Program supports the development of CCD technology. In addition, the document on Scientific & Technological Actions on Climate Change has been promulgated by the Ministry of Science and Technology	R&D Guidelines for the Geological Disposal of High Level Radioactive Waste in China were jointly issued by three government ministries. It is planned to build a URL for HLW disposal by 2020, and to construct a geological repository for HLW by 2050
Inventory	In 2007 China's CO_2 emissions amounted to 5.06 Gt; the average emission per person is 87% of the world average. The main sources of CO_2 are coal-fired power plants and steel factories	In 2008 there were 11 reactors in operation, producing 370 t of spent fuel. When the 58 reactors reach the end of their lifetime, total spent fuel will amount to 82,000 tHM

(continued)

Table 1 (continued)

Attribute	CO_2 disposal	Radioactive waste disposal
Disposal concept	CO_2 is captured and injected into selected deep geological porous media (reservoirs) and confined by the host rock through various physical and chemical trapping mechanisms. The host media must have a tough dense caprock layer to confine the CO_2 and to prevent its rapid migration and vertical leakage to more shallow strata	The geological repository is a 500–1,000 m deep underground mine-type facility; the multi-barrier concept is also under consideration, including waste glass, container, buffer and backfill material (bentonite) and host rock. The idea is to isolate the RW permanently
Disposal methods	CO_2 is captured and injected into deep geological porous media (reservoirs). EOR and ECBM operations and injection into saline aquifers are the main methods	Deep geological disposal. HLW is buried in a deep geological repository
Host media	Porous media in sedimentary basins. The potential host media include: (1) depleted and disused oil- and gasfields, (2) deep saline aquifers, and (3) deep unmineable coal seams	Granite mass with good integrity and adequate volume, or clay layers with adequate thickness and stability
Potential sites	There are 70 disposal sites in 24 major sedimentary basins on the Chinese mainland, with a total disposal capacity of 144 Gt CO_2 in saline aquifers. The North China Basin, Sichuan Basin, Zhunger Basin and the east coast basins are the main potential sites for GDC. In addition 46 oil and gas reservoirs, 68 unmineable coalbeds with methane recovery and 24 saline aquifers have been identified for potential storage use	North-west China (granite); southern China (granite), eastern China (tuff and clay), south-west China (granite and clay), Inner Mongolia (granite). The most suitable area is the Beishan site in north-west China
Site characteri-zation methods	Borehole drilling, borehole logging and tests. In situ borehole injection tests	Satellite image and airborne image analysis; surface geological, hydrogeological and geophysical survey; borehole drilling and systematic borehole logging and tests; in situ tests in the URL

(continued)

Table 1 (continued)

Attribute	CO_2 disposal	Radioactive waste disposal
In situ tests	Borehole injection tests and monitoring are the major in situ test methods for evaluating the capacity and injectivity of geological formations. EOR and ECBM tests have been conducted in several oilfields and coal mines	In situ tests are planned in repository sites and the URL. Data from in situ tests will be used for repository design, safety assessment and licensing application
Numerical modelling	Widely used to simulate the mixture process of CO_2 and groundwater and the movement of CO_2. Geochemical models are used to simulate the water–rock reactions	It is a very important tool and often used for predicting the performance of the disposal system and the long-term safety of the repository. Numerous computer codes are currently used in R&D programmes, repository design and safety assessment
Cost	Injection costs up to US$20 per tonne of CO_2 disposed of (excluding the cost for capture, conditioning and site closure). No data on the actual cost of injection or on the economic benefit of EOR are available	The cost of disposing of the 82,000 tHM of spent fuel generated from the 58 reactors is estimated at about 140 billion CNY or 1.2% of the total electricity income
Risk	CO_2 leakage from the disposal formation with local risk to health and safety and impact on fresh water and air quality	RW migrates with groundwater and enters the biosphere
Challenges	Scientific, technological, engineering, social and economic challenges	Scientific, technological, engineering and social challenges

CCD carbon capture and disposal, *CNY* Renminbi yuan, *ECBM* enhanced coalbed methane, *EOR* enhanced oil recovery, *GDC* geological disposal of CO_2, *HLW* high-level waste, *RW* radioactive waste, *tHM* tonnes of heavy metal, *URL* underground research laboratory

1. Philosophy: In China CO_2 is considered not only as waste but also as a source of profit, with CO_2 injection being used to enhance oil and gas recovery and methane production. SNF is also considered as a resource in China, with the SNF destined to be reprocessed for the recycling of plutonium and uranium. The high-level radioactive liquid waste residue remaining after the reprocessing of the SNF will be vitrified into its final form for disposal.
2. National policies: In efforts to mitigate climate change and reduce CO_2 emissions, China has signed the UNFCCC and the Kyoto Protocol. With regard to RW, China signed the Joint Convention on the Safety of Spent Fuel Management and the Safety of Radioactive Waste Management. HLW disposal policy is that

the spent fuel should be reprocessed: a centralized deep geological repository is planned for this purpose in China.

3. National plan: CCD has been included in several key national R&D plans. R&D Guidelines for the Geological Disposal of High Level Radioactive Waste in China have been jointly issued by three government ministries. The building of a URL for HLW disposal is planned by 2020, and the construction of a geological repository for HLW by 2050.
4. Inventory: In 2007 China's CO_2 emissions amounted to 5.06 Gt; the average emissions per person amount to 87% of the average world level. The main sources of CO_2 are coal-fired power plants and steel factories. In 2008, 11 nuclear power reactors were in operation, generating 370 t SNF. When the 58 reactors reach the end of their lifetime, the total SNF will amount to 82,000 tHM.
5. Financial support: To some extent, CCD methods such as EOR, ECBM, EATER will not only be used to dispose of CO_2, but will also produce economic benefits for the companies. The EOR method will enhance oil recovery; ECBM can enhance methane production; EATER can provide heat for houses. Therefore these technologies have gained strong support both from the Government and commercial companies, and are developing very fast. However, as the disposal of RW will not produce profits for nuclear power companies, financial support for it in China at this stage is very limited.
6. Disposal principles and capacity: The principle for CO_2 disposal varies greatly, whereas the principle of RW disposal relies mainly on the multi-barrier concept, i.e. retardation of radionuclides by means of engineered and geological barriers. As, compared with CO_2, the amount of RW is relatively small, one site will probably be enough in most countries to dispose of all of the HLW. For LLW and ILW, several more disposal sites are needed. To dispose of CO_2 in China, many sites are required.
7. Disposal concept: CO_2 is captured and injected into selected deep geological porous media (reservoirs) and confined by the host rock through various physical and chemical trapping mechanisms. The host media must have a tough dense caprock layer to confine the CO_2 and to prevent its rapid migration and vertical leakage to more shallow strata. For HLW, the type of geological repository being considered is a 500–1,000 m deep underground mine-type facility using the multi-barrier concept of waste glass, container, buffer and backfill material (bentonite), and host rock. The idea is to permanently isolate the RW.
8. Host media: For CO_2 disposal, the host media consists of porous media or aquifers in sedimentary basins. The potential host media include: (a) depleted and disused oil- and gasfields, (b) deep saline aquifers, and (c) deep unmineable coal seams. For HLW disposal in China, the major host rocks are granite mass with good integrity and adequate volume, or clay layers with adequate thickness and stability.
9. Potential sites: For CO_2 disposal, there are 70 disposal sites in 24 main sedimentary basins on the Chinese mainland, with a total disposal capacity of 144 Gt CO_2 in saline aquifers. The North China Basin, Sichuan Basin,

Zhunger Basin and the east coast basins are the main potential sites for GDC. In addition, 46 oil and gas reservoirs, 68 unmineable coalbeds with methane recovery and 24 saline aquifers have been identified. For HLW, north-west China (granite), southern China (granite), eastern China (tuff and clay), south-west China (granite and clay) and Inner Mongolia (granite) are the key regions under consideration. However, the most likely one is the Beishan site in north-west China.

10. Site selection: This is an important issue both for CO_2 and RW disposal. However, site selection for RW takes a long time, entails complicated procedures and has to involve many stakeholders. Public acceptance is much more important in terms of RW disposal than of CO_2 disposal. In China the site selection efforts for RW have been concentrated in the north-west area since 1990s, while efforts to select CO_2 disposal sites are mainly concentrated in the east.
11. Site characterization: The sites selected both for CO_2 disposal and RW disposal need to be characterized in a very detailed manner. There are some common characterization methods, such as surface geophysical survey, drilling, borehole logging and tests. However, there are many specific methods used for CO_2 disposal, and also for RW disposal.
12. In situ tests: For CO_2 disposal, small- or large-scale pilot injection plants are needed, and many in situ tests are carried out to verify the disposal site. For RW, especially for HLW, URLs are needed to verify the disposal concept, test the performance of engineered barriers and, most importantly, use the underground facility to win public acceptance. In China there are several test sites for EOR and ECBM, but there is still no URL available for RW disposal.
13. Mathematical modelling: This is very important and is often used both in CO_2 and RW disposal. Some of the mathematical models and codes, for example some hydrogeological models and geochemical models, are used in both fields. However, RW disposal has many specific models, such as a total system performance assessment model, rock mechanical models coupled with thermal and chemical reaction processes, and geochemical models that calculate the decay rate of radionuclides. Modellers in both fields can learn from each other.
14. Cost: Injection costs up to US$20 per tonne of CO_2 disposed of (excluding the cost of capture, conditioning and site closure). However, there are no data on the actual cost of injection or on the economic benefits of EOR. The cost of disposal of the 82,000 tHM of spent fuel generated from the 58 reactors is estimated at about 140 billion CNY, 1.2% of the total electricity income.
15. Risk: CO_2 can leak from the disposal formation and cause risk to local health and safety and impact on fresh water and air quality. In HLW disposal, RW can migrate with groundwater and enter the biosphere.

Table 1 shows a comprehensive comparison between GDC and RW in China.

5 Conclusions

China's total CO_2 emissions in 2007 amounted to 5.06 Gt, making the country the second highest global CO_2 emitter; average emissions per person are 87% of the world average. The main sources of CO_2 are coal-fired power plants and steel factories. This means that China faces an increasing challenge in CCD. The rapid development of NPPs is also challenging in terms of the safe disposal of HLW.

The Chinese Government considers both CO_2 and RW disposal to be important for the sustainable development of the country, and has paid increasing attention to them. Several regulations and standards have come into force, and development strategies and guidelines have been proposed. Several key R&D projects with concrete financial support have been initiated; significant progress has been made both in CO_2 and RW disposal.

On the Chinese mainland 70 storage sites in 24 major sedimentary basins have been identified, with a total storage capacity of 144 Gt CO_2 in the saline aquifers. China has extensive EOR experience, existing ageing oilfields and pure CO_2 sources; CO_2 EOR is the most ready CO_2 disposal option in terms of commercial demonstration in China. ECBM shows a certain potential, but not as good as that of saline aquifers. There are potentially huge disposal capacities in deep saline aquifers in China.

The 11 nuclear power reactors on the Chinese mainland generate 370 t of spent fuel a year, a figure that will reach about 82,000 tHM when all the planned 58 reactors reach the end of their lifetime. For HLW disposal, a long-term target to build a national HLW repository in around 2050 has been set, while site selection has been conducted since 1985.

GDC and the geological disposal of RW have many differences, including disposal principles and capacity, host media, potential sites, site characterization and cost. The site with the most potential for HLW disposal in China is the Beishan granite site in north-western Gansu Province, while most of the potential sites for CO_2 disposal are in basins in eastern and south-western China. For HLW, only one repository is planned, but for CO_2 disposal many sites are needed.

GDC and the geological disposal of RW share many similarities. They also face similar scientific and technical challenges, including site selection and characterization, long-term performance of storage sites, monitoring of storage sites, safety assessment of the disposal system, social and economic issues and public acceptance.

Based on the available information and discussion in this chapter, geological formations and capacities for disposal of CO_2 and RW are available in several regions in China. This implies that China could rely on various combinations of fossil or nuclear sources. This would give the country various choices, depending on the circumstances and driving forces of energy policies.

Acknowledgements The authors would like to express their sincere thanks to Ferenc Toth, the editor of the book, for his encouragement and helpful comments and suggestions on this chapter. Thanks are also due to the three referees of this chapter for their useful reviews.

References

Bradshaw J, Bachu S, Bonijoly D, Burruss R, Holloway S, Christensen NP, Mathiassen OM (2007) CO_2 storage capacity estimation: issues and development of standards. Int J Greenhouse Gas Control 1:62–68

CAEA (China Atomic Energy Authority) (2006) Ministry of Science and Technology of the People's Republic of China and State Environment Protection Administration of China. R&D Guidelines for the Geological Disposal of High Level Radioactive Waste in China. Beijing

IAEA (International Atomic Energy Agency) (1993) Report on Radioactive Waste Disposal. Technical Reports Series No. 349. IAEA, Vienna

IEA GHG (International Energy Agency Greenhouse Gas R&D Programme) (2005) Capture and storage of CO_2. http://www.ieagreen.org.uk/ccs.html

IPCC (Intergovernmental Panel on Climate Change) (2005) IPCC Special Report on Carbon Dioxide Capture and Disposal. Metz B, Davidson OR, de Coninck HC, Loos M, Meyer LA (eds). Cambridge University Press, New York/Cambridge, UK http://www.ipcc.ch/pdf/special-reports/srccs/srccs_wholereport.pdf

Jiang ZM (2008) Reflections on energy issues in China. China Nucl Power 1:98–113

Kickmaier W, McKinley I (1997) A review of research carried out in European rock laboratories. Nucl Eng Des 176:75–81

Li D (2008) Gas-water-salt-rock system phase equilibrium and speciation equilibrium and its application in CO2 geological storage numerical simulation. PhD dissertation, Graduate University of the Chinese Academy of Sciences, Beijing [in Chinese]

Li X, Liu Y, Bai B, Fang Z (2006) Ranking and screening of CO_2 saline aquifer storage zones in China. Chinese J Rock Mech Eng 25(5):963–968 [in Chinese]

Liu Y, Li X, Bai B (2005a) Preliminary estimation of CO_2 storage capacity of coalbeds in China. Chinese J Rock Mech Eng 24(16):2947–2952 [in Chinese]

Liu Y, Li X, Fang Z (2005b) Preliminary estimation of CO_2 storage capacity in gas fields in China. Chinese J Rock Mech Eng 27(12):2277–2281 [in Chinese]

Ma I, Tang D, Jiang P, Ning Q (2007) The status of enhanced oil recovery by CO_2 injection. Oilfield Chem 24(4):379–383 [in Chinese]

NEA (Nuclear Energy Agency) (1999) Geological Disposal of Radioactive Waste: Review of Development in the Last Decade. OECD Nuclear Energy Agency, Paris

NEA (Nuclear Energy Agency) (2001) The Role of Underground Laboratories in Nuclear Waste Disposal Programmes. OECD Publications, Paris

NEA (Nuclear Energy Agency) (2003) Engineered Barrier Systems and the Safety of Deep Geological Repositories, State-of-the-Art Report. OECD Publications, Paris, Report EUR 19964 EN

NEA (Nuclear Energy Agency) (2005) Geological repositories: political and technical progress. Workshop Proceedings Stockholm, Sweden, 7–10 December 2003. OECD Publications, Paris

Pang Z (2007) Geothermal resources development and key technical issues in China. In: Proceedings of the NEET Workshop on Energy Technology Collaboration, Beijing, China: 1–2 November 2007. http://www.iea.org/Textbase/work/2007/neet_Beijing/Pangzhonghe_en.pdf

Pang Z, Duan Z, Yang F, Huang T (2008) Gas–water–rock interactions in sandstone reservoirs: implications for enhanced re-injection into geothermal reservoirs and CO_2 geological sequestration. In: Proceedings of the Workshop for Decision Makers on Direct Heating use of Geothermal Resources in Asia. United Nations University Geothermal Training Programme, Tianjin, pp 214–218

Pang Z, Yang F, Duan Z, Wang J (in press) Geological disposal of CO_2: review and prospect. In: Proceedings of the 2nd National Conference on Underground Disposal of Waste. Chinese Society for Rock Mechanics and Engineering, Beijing [in Chinese]

Ren J (2000) Geotectonics Map of China and its Vicinity (1:5,000,000). Geological Publishing House, Beijing [in Chinese]

State Council of China (2007) Medium and long-term development plan for nuclear power (2005–2020)
Stevens SH, Kuuskraa VA, Taber JJ (1999) Sequestration of CO_2 in depleted oil and gas fields: Barriers to Overcome in implementation of CO_2 capture and storage (disused oil and gas fields). IEA Greenhouse Gas Research and Development Programme Report PH3/22. Advanced resources international, Arlington, VA
Wang J, Fan XH, Xu GQ et al (2004) Geological Disposal of High Level Radioactive Waste in China: Progress in the Last Decade (1991-2000). Atomic Energy Press, Beijing [in Chinese]
Wang J, Chen W, Su R, Guo Y, Jin Y (2006a) Geological disposal of high-level radioactive waste and its key scientific issues. Chinese J Rock Mech Eng 25(4):801–812 [in Chinese]
Wang J, Su R, Chen W, Guo Y, Jin Y, Wen Z, Liu Y (2006b) Deep geological disposal of high-level radioactive waste in China. Chinese J Rock Mech Eng 25(4):649–658
Wang J (2007) Key scientific challenges in geological disposal of high level radioactive waste. Eng Sci 5(4):45–50 [in Chinese]
Ye J, Feng S, Fan Z (2007) Micro-pilot test for enhanced coalbed methane recovery by injecting carbon dioxide in south part of Qinshui Basin. Acta Petrolei Sinica 28(4):77–80 [in Chinese]
Zhang CL, Wang J, Su K (2006) Concepts and tests for disposal of radioactive waste in deep geological formations. Chinese J Rock Mech Eng 25(4):750–767

Geological Disposal of Carbon Dioxide and Radioactive Waste in the Geotectonically Active Country of Japan

Hitoshi Koide and Kinichiro Kusunose

Abstract In Japan, site selection for geological disposal of radioactive waste (RW) and carbon dioxide (CO_2) is very important because of the large regional differences in tectonic activity. Assessment of the long-term stability of geological environments is key to geological RW disposal in Japan. A comprehensive system of long-term prediction of crustal movement and the groundwater regime around the virtual RW disposal sites has been developed in Japan. CO_2 is naturally abundant, but geological disposal of the gigantic volumes of CO_2 may have big impacts on the environment. One of the adverse effects of underground fluid injection is that it may induce earthquakes. Underground carbon microbubble injection accelerates advanced geological disposal mechanisms. The autogenously sealed 'CO_2 capsules' can be formed in large basaltic sheets, ophiolite complex and oceanic crust. Sub-seabed aquifers under the deep sea floor can provide very safe and virtually limitless CO_2 disposal. Different disposal strategies for CO_2 and RW are needed because of the extreme difference in their toxicity and volume. The dispersion and dilution principle is possible for the CO_2 disposal, while RW is strictly contained by the multiple barrier system. The stability of the geological environment is important for both CO_2 and RW disposal.

Keywords Geological prediction • Induced earthquake • Microbubbles • Tectonic stability • Site selection

H. Koide (✉)
CC Geosystem Laboratory, 2-6-1-602 Higashi-Ikebukuro, Toshima, Tokyo 170-0013, Japan
e-mail: MXE02263@nifty.com

K. Kusunose
Geological Survey of Japan, AIST, 1-1-1, Higashi, Tsukuba 305-8567, Japan
e-mail: k.kusunose@aist.go.jp

F.L. Toth (ed.), *Geological Disposal of Carbon Dioxide and Radioactive Waste: A Comparative Assessment*, Advances in Global Change Research 44, DOI 10.1007/978-90-481-8712-6_18,

1 Introduction

Despite a lengthy history of meticulous research into the geological disposal of radioactive waste (RW) and carbon dioxide (CO_2), Japan has yet to successfully implement either geological disposal of high-level radioactive waste or full-scale underground disposal of CO_2. The Japanese, having experienced at first hand natural disasters such as earthquakes and also nuclear disasters, are particularly sensitive to events such as these. Indeed, many Japanese people have major concerns about the safety of radioactive facilities, especially as Japan is an earthquake-prone country with many active faults.

The arcuate islands of Japan were formed along plate boundaries between two major oceanic plates and two continental plates. Many earthquakes, active fault lines and active volcanoes are concentrated in and around the Japanese islands. Subduction of the oceanic plates under the continental plates dominates tectonic deformation on the Japanese islands. Volcanic activity and faults are concentrated along the volcanic fronts, but no active volcanoes and few active fault lines are located between the volcanic front and the subduction zone (Sugimura and Uyeda 1973). Site selection for geological disposal of RW and CO_2 is very important because of the large regional differences in tectonic activity in Japan (see Fig. 1). It is also important to win public confidence in the long-term safety of RW disposal. Accurate assessment of the long-term stability of the geological environment is crucial to the geological disposal of RW and CO_2 in Japan.

2 Current Status and Issues of Geological Disposal of CO_2 in Japan

2.1 Geological Disposal of CO_2 in Saline Aquifers and in Gas- and Oilfields

Japan has a long history of research into carbon capture and storage (CCS) in geological sites. The first paper on CO_2 disposal in saline aquifers appeared in 1990

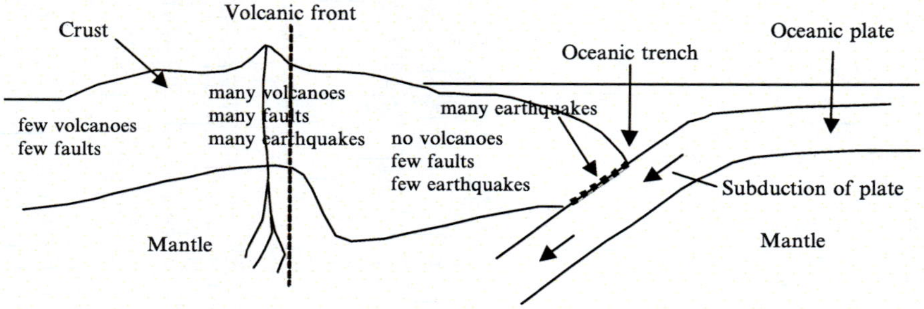

Fig. 1 Schematic distribution of crustal movement features in an island archipelago

(in Japanese). Koide (1990) proposed the pumping up of deep saline groundwater using naturally dissolved methane, separating the methane from the groundwater and then reinjecting the groundwater and anthropogenically produced CO_2 into the aquifers. A voluntary CCS R&D team from national research institutes and some companies led by H. Koide presented the first comprehensive report on the disposal of CO_2 in saline aquifers at the First International Conference on Carbon Dioxide Removal, held in Amsterdam, the Netherlands, in March 1992 (Koide et al. 1992). They suggested that the relevant technologies were generally available, disposal costs were reasonable (about US$22/t CO_2) and disposal capacity was adequate (with at least a 319 gigatonnes of CO_2 (Gt CO_2) disposal capacity in solution in aquifers worldwide) (Koide et al. 1993). Koide (1999) revised his estimate of world subsurface disposal potential in saline aquifers to 3×10^{12} t in CO_2 solution, as he considered his first estimate to be far too conservative.

This voluntary study was expanded to a CCS R&D project within the Engineering Advancement Association of Japan (ENAA). ENAA's CCS team estimated the CO_2 disposal potential of Japan on a systematic basis. It found the disposal capacity of CO_2 supercritical fluid in Japan to be 2 Gt CO_2 in oil/gas deposits (category I) and 1.5 Gt CO_2 in dome structures (category II), and the disposal capacity of CO_2 in solution in Japan to be 16 Gt CO_2 in onshore saline aquifers (category III) and 72 Gt CO_2 in offshore saline aquifers (category IV) (see Fig. 2 and Tanaka et al. 1995). The total aquifer disposal potential in Japan was estimated to be as much as 92 Gt CO_2.

Recently, the Research Institute of Innovative Technology for the Earth (RITE) revised the estimated disposal capacity of CO_2 in Japan (Takahashi et al. 2009). Supercritical CO_2 may be stored as follows: 3.5 Gt CO_2 in oil/gas deposits; 5.2 Gt CO_2 in probable dome structures; and 21.4 Gt CO_2 in possible dome structures, mainly in offshore sedimentary basins around Japan. Additional CO_2 may be stored in solution—as much as 22 Gt CO_2 in onshore saline aquifers and 94 Gt CO_2 in offshore saline aquifers. The total CO_2 aquifer disposal potential was estimated at 146 Gt CO_2; this is enough to accommodate more than 100 years of industrial CO_2 emissions in Japan.

Although the generic potential for CO_2 aquifer disposal is enough to accommodate Japan's anthropogenic CO_2 emissions, Japan needs to refine the concept further to adapt it to its actual and industrial environments. This should be done in the next few years for Japan to fulfil its obligations under the Kyoto Protocol. The refinement of the concept depends on the answers to the following questions: (1) How much CO_2 can be stored in Japan? (2) Where? (3) How? (4) At what cost? and (5) How safely can it be stored? Japan's dense population and urbanization make the transportation of RW and CO_2 on land costly, if not prohibitive. Most of the big CO_2 emitters are near the coast for convenience of transportation. It is expected that the RW and CO_2 will most probably be disposed of geologically underground on the coast or, alternatively, in offshore sub-seabed rocks. Li et al. (2005) rank CO_2 disposal sites in terms of potential capacity and supply of CO_2, both of which significantly affect the economics of disposal (Fig. 2). They preliminarily selected 11 premier-ranking candidate sites recommended for early deployment in five regions: Kanto, Joban, Niigata, Toyama and south-west Hokkaido.

In 2000 RITE began a systematic R&D project on the geological disposal of CO_2 with support from Japan's Ministry of Economy, Trade and Industry (METI)

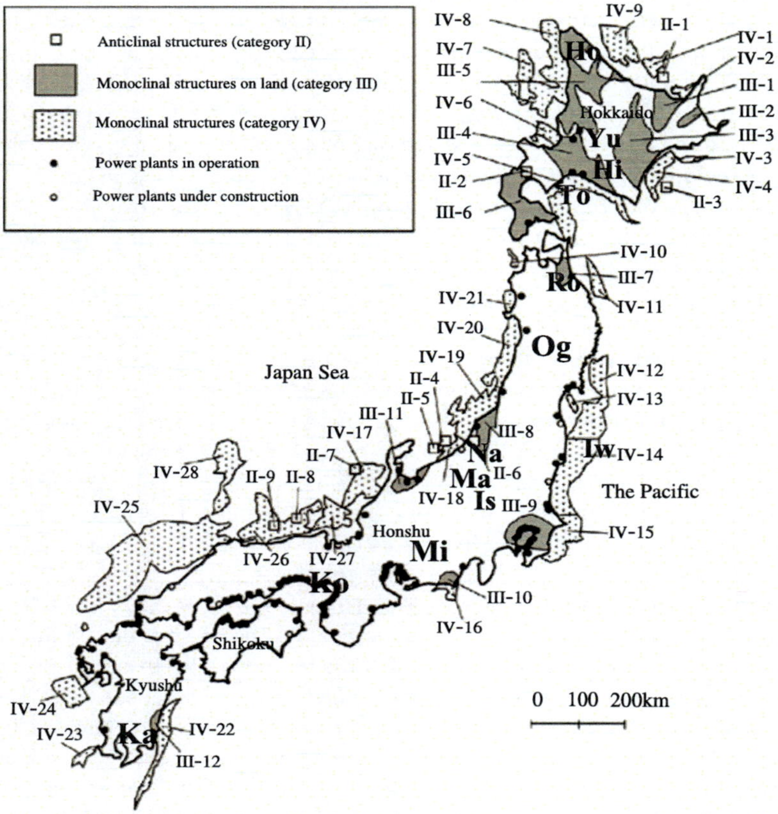

Fig. 2 Simplified location map of fossil fuel-fired power plants and potential storage sites. (Source: Li et al. 2005.) Storage sites of categories II, III and IV are indicated. Test-only sites for radioactive waste disposal in sedimentary rocks (Ho—Horonobe) and in granite (Mi—Mizunami). Ro (Rokkasho): low-level radioactive wastes disposal and vitrified high-level waste storage site. Ko (Kobe): predictions regarding geotectonic stability are carried out. The CO_2 disposal test sites are Na—Nagaoka (saline aquifer), Yu—Yubari (coal), Og—Ogati (granite) and Hi—Hidaka (serpentine). Natural analogue sites are Ma—Matsushiro (earthquake analogue), Is—Isobe (saline aquifer disposal analogue) and Ka—Kagoshima Bay (sub-seabed disposal analogue). The preliminary geological investigation for the second-phase CO_2 disposal test started in 2009 at Iw (Iwaki-oki), To (Tomakomai-oki) and two other sites in preparation

(Kaya et al. 2001). RITE started underground injection of supercritical CO_2 at a depth of 1,100 m into a saline sandy aquifer above the natural gas reservoirs in Nagaoka, Niigata Prefecture, on 7 July 2003. The underground injection ended on 11 January 2005, although monitoring continued up to 2007. During the injection, the Chuetsu earthquake, which registered 6.8 on the Richter scale, occurred 15 km south-east of the test site. The earthquake was caused by the movement of a major active fault system running through the Nagaoka area along a branch fault.

The shock of the earthquake immediately stopped the injection operation, but no other effects were observed in the test wells. The triggering of the Chuetsu earthquake seems to have been unrelated to the underground injection of CO_2, as the active faults near the test site were dormant.

In May 2008, the Japan Petroleum Exploration Co. Ltd. (JAPEX) and other major companies in the country co-founded the Japan Carbon Capture & Sequestration (JCCS) Company Limited. The 32 co-founders of JCCS included 11 major electric power companies, five oil distributors and three oil developers, as of June 2009. In July 2009, JCCS, on behalf of METI, began a field survey for a feasibility project along the planned CO_2 pipeline route connecting the Integrated Gasification Combined Cycle (IGCC) power plant at Iwaki City in the Fukushima Prefecture and the offshore Iwaki-oki gasfield. JCCS also started a preliminary field survey at the offshore Tomakomai-oki site in Hokkaido. METI and JCCS intend to make a preliminary geological investigation at four offshore sites, including Iwaki-oki and Tomakomai-oki, for a larger second-phase underground CO_2 injection test of possibly around 100,000 t/year rate classes.

2.2 *Geological Disposal of CO_2 in Coal Seams*

The General Environmental Technos Co. Ltd., Kansai Electric (KEPCO), Japan Coal Energy Center (JCOAL) and other institutions began the Japan 'CO_2 Geodisposal in Coal Seams' Project in 2002 with support from METI (Koide et al. 2003). From November 2004 to 2007 they injected a total of 884 t CO_2 into a 5.6 m thick coal seam at a depth of 890–895 m and recovered 150,672 normal m^3 of methane from the coal seam. Nitrogen injection effectively enhanced injectivity damaged by coal swelling. The CO_2 disposal capacities of coalfields in Japan were estimated from old geological and experimental data (Koide et al. 2005). The remaining coal seams in old coal mines in Japan are estimated to be able to adsorb about 600 million tonnes of CO_2 (Mt CO_2). Explored but untapped coal seams are suitable for the early application of CO_2 disposal. A preliminary field test on using CO_2 for enhanced coalbed methane (CO_2 ECBM) is being carried out for untapped coal seams in the coalbed methane-rich Ishikari coalfield in Hokkaido. It is possible that the main untapped coalfields in Japan can store 380 Mt CO_2 and produce 84×10^9 m^3 of coalbed methane in total. All the major conventional coalfields in Japan can sequestrate about 1 Gt CO_2 and produce 240×10^9 m^3 of coalbed methane from the coalbeds remaining in old coal mines and in untapped coalfields, combined. Near shore coalfields in the Kyushu district can provide cost-effective disposal for CO_2.

Systematic exploration for oil and natural gas by the Japan National Oil Corporation and other oil companies has revealed that there are huge coal seam volumes in the depths of the Palaeogene and Cretaceous sedimentary basins in central Hokkaido, off the Pacific coast of north-eastern Honshu, and north-west of Kyushu in and around Japan (Koide and Kuniyasu 2006). Deep unmineable coal seams and coaly shale layers provide a possible sink for CO_2 disposal and a source of coalbed methane. In Japan there are unmineable coal seams deeper than 1,200 m and shallower than 3,000 m that exceed 300 Gt capacity and could store 10 Gt CO_2 as well as producing 3×10^{12} m^3 of coalbed methane. Very deep (over 3,000 m) unmineable coal seams in Japan are estimated to exceed 3×10^{12} t capacity and could contain some 24×10^{12} m^3 of coalbed methane. The CO_2 disposal potential in very deep unmineable coal seams (deeper than 3,000 m) is more than 100 Gt in Japan.

The total capacity of deep unmineable coalbeds is between 10 and 100 times greater than the shallow coal reserves that are mineable using conventional drift mining or open-cut mining practices. The huge volumes of deep unmineable coal seams are potential sinks for CO_2 and also enormous untapped energy resources. When pure CO_2 is injected, it at once behaves as a supercritical fluid around injection wells in deep coal seams. Supercritical CO_2-enhanced coal seam gas recovery and in situ fire-free microbial gasification of coal can turn deep unmineable coal seams into CO_2 sinks and methane sources for a CO_2 emission-free closed-circuit power plant. A new type of 'coal mine' using borehole mining is proposed for the development of 'deep unmineable coalbeds' (Koide and Yamazaki 2001).

In June 2009, the Agency for Natural Resources and Energy (ANRE) of METI proposed a new and environmentally friendly coal policy that includes: (1) the 'CoolGen' project for the zero emission coal-fired power plant feasibility study, and (2) the 'Clean Coal for Earth' project for international cooperation by Japan's cutting-edge clean coal technologies.

2.3 *Natural Analogues for Geological Disposal of* CO_2

A natural analogue study for safe disposal of CO_2 is very important in the tectonically active islands of Japan. A series of small earthquakes, the Matsushiro swarm earthquakes, began in the Matsushiro area of Nagano City in central Japan in August 1965, culminating in April 1966. The Matsushiro swarm earthquakes were accompanied by a combined eruption of CO_2 and brine from a double en echelon system of numerous new small cracks. The earthquake activity quickly declined after large volumes of water and CO_2 were released from newly formed surface cracks. The surface cracks increased in number to form a double en echelon system of numerous small cracks filled with CO_2-rich brine. The double en echelon system of cracks indicates an underlying left-lateral strike-slip fault: the Matsushiro earthquake fault zone (Koide 1971). A large amount of CO_2 and brine cascaded from cracks in and around the earthquake fault zone. Mechanical and chemical analyses suggest that CO_2 bubbles caused an increase in fluid pressure in the shallower tips of cracks and that this eventually triggered the swarm earthquakes (see Fig. 3 and Koide et al. 2006). Large underground CO_2 bubbles of supercritical fluid and gas can induce fault instability and small earthquakes under critical tectonic stress states (see Figs. 4 and 5). Although CO_2-driven earthquakes tend to be very small, careful investigation of faults, in situ stresses, crustal deformation and seismicity need to be carried out before a large amount of CO_2 can be stored underground. Mechanical

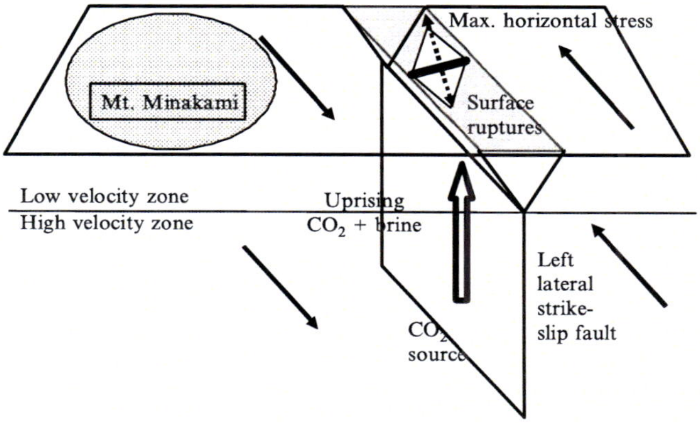

Fig. 3 The Matsushiro earthquake fault zone. The CO_2 bubbles induced the increase in fluid pressure in shallower tips of cracks that eventually triggered the Matsushiro swarm earthquakes

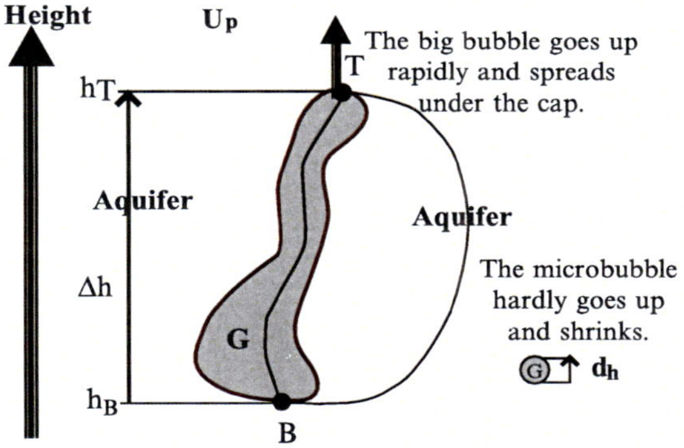

Fig. 4 CO_2 bubble sizes and dynamics in underground rocks.The big bubbles of CO_2 fluid (gas, supercritical fluid, liquid) ascend by buoyant force, while tiny microbubbles have little buoyancy, shrink and do not coalesce into a big bubble. Microbubbles of CO_2 can be emplaced stably into the pores and interstices of underground rocks

analyses indicate that CO_2 solution and CO_2 microbubbles do not induce such a large underground mechanical instability as big bubbles of CO_2 supercritical fluid and CO_2 gas (Koide and Xue 2009).

Although an estimated 100,000 t CO_2 were released with about 10 million tonnes (Mt) of brine from cracks and cascades during the Matsushiro swarm earthquakes, there were no casualties, as the water-rich surface condition buffered the adverse effect of CO_2 release. About 40 years later, shallow fresh groundwater has almost completely covered the deep CO_2-rich brine except for a few spots where a few small bubbles sporadically rise in an area with a thin and shallow groundwater cover. Shallow groundwater covers the deep CO_2-rich brine, almost completely in

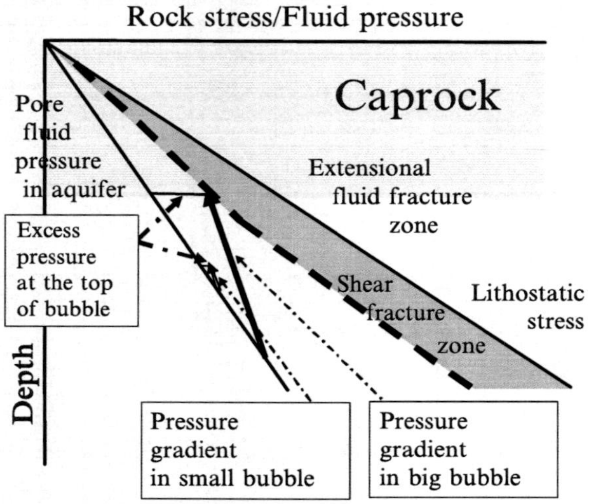

Fig. 5 CO_2 bubble sizes and pressure gradients. Large excess fluid pressure builds up at the top of large bubble of CO_2 fluid because of the difference in density between the inside and outside of bubbles (Fig. 4). The CO_2 fluid (gas or liquid or supercritical fluid) is lighter than the groundwater at normal underground pressure and temperature. Fluid fracturing of caprocks and fluid-induced earthquakes may occur at the top of the large CO_2 bubble. Fluid fracturing and induced earthquakes are unlikely around small bubbles because of the insignificant excess fluid pressure build-up that occurs only at the top of the small bubble

contrast to the monoclinal geological structure of the Isobe natural CO_2 reservoir in Annaka City, Gunma Prefecture, central Japan.

CO_2-rich volcanic gas (72–95% CO_2) gushes from some ten vents on the sea floor at depths of about 78 and 200 m in the northern Kagoshima bay, south-western Japan. The northern part of Kagoshima bay is an almost closed sea basin, divided from the main part of Kagoshima bay by the active Sakura-jima volcano. It was formed mainly by the volcanic Aira caldera. The total amount of CO_2 released is estimated at around 20–100 t CO_2 per day. Most of the CO_2 is in gas bubbles which dissolve into the seawater within several metres above the vents, while CO_2-free gas bubbles appear on the sea surface. The deep seawater is acidified in the summer. However, the deep block of acidic seawater disappears in winter when the convection of seawater becomes active. The shallow water is important for the safety of underground CO_2 disposal, acting as a buffer against the adverse effects of CO_2.

2.4 *Advanced Methods of Geological CO_2 Disposal*

Using advanced methods for swift and secure underground storage of CO_2 is also very important in unstable regions. Small-scale CO_2 injection tests were conducted for the study of underground geochemical CO_2 fixation in a serpentine body in

Hidaka, Hokkaido (Yajima et al. 2005; Okamoto et al. 2006) and in a geothermal granite body in Ogati, in Akita Prefecture (Wakahama et al. 2008). Serpentine (mainly composed of the serpentine $Mg_3Si_2O_5(OH)_4$) is a type of ultramafic rock, which is derived from the Earth's mantle and occurs at collision plate boundaries like the Japanese islands. A preliminary field study showed that the estimated capacity of CO_2 mineralization in serpentine massif all over Japan is about 700 Mt (Yajima et al. 2005). The CO_2 injection makes the groundwater acidic near the injection well, and this dissolves some of the rock-forming minerals in the CO_2-rich acidic groundwater. However, the chemical reaction between the CO_2 and the rocks makes the groundwater alkaline. If CO_2 encounters an alkaline groundwater rich in metallic cations such as calcium ions, magnesium ions, ferrum ions, etc., the carbonate minerals precipitate and clog the pores and fissures in the rocks. Thus, while CO_2 injection induces partial dissolution of rocks, thereby increasing the porosity of rocks near the injection well, it also precipitates as carbonates in pores in alkaline groundwater far from the injection well (see Fig. 6). Autogenous sealing of CO_2 occurs in cation-rich alkaline aquifers, especially around ultramafic and mafic rock bodies such as peridotite, serpentine, basalt, tuff, ophiolite and oceanic crust (Koide 2001).

The in situ geophysical and geochemical observations and related laboratory experiments during the first Japanese project on CO_2 geological disposal at the Nagaoka site revealed some interesting behaviours on the part of deep saline aquifers in Japan (Xue et al. 2009; Mito et al. 2008). The results of time-lapse cross-well seismic tomography indicate an area of P-wave velocity decrease due to CO_2 saturation, while the CO_2-bearing zone near the injection well expands clearly along the formation to the updip direction during CO_2 injection. The presence of

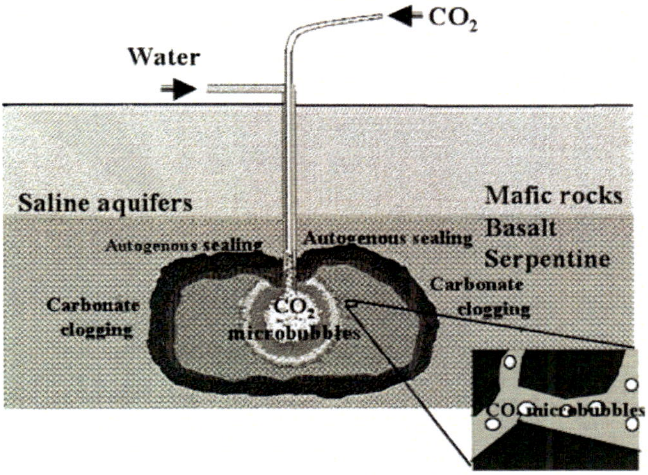

Fig. 6 Carbon microbubble sequestration into incompletely confined aquifers or open rock bodies and the formation of carbonates

CO_2 was also identified by induction, sonic and neutron logging at the observation wells. Changes in the composition of the formation water sampled by the Cased Hole Dynamics Tester (CHDT) tool after the CO_2 injection suggest that solubility trapping is an important mechanism in geological disposal of CO_2. Dissolution of plagioclase and chlorite has great potential for enhancing neutralization of the acidified formation water (Mito et al. 2008).

Numerical simulation using TOUGHREACT software has shown that more than 20 Mt CO_2 can be stored for a period up to 10,000 years in the young sedimentary aquifers of the Tokyo Bay area (Okuyama et al. 2009). Carbonate precipitation occurs extensively on the periphery of expanding areas of carbonated water, forming a shell that encloses the CO_2. The distribution of dawsonite is predicted to be dependent on the dissolution of plagioclase, which is present in abundance in the sandstones of the Tokyo Bay area and in the young sedimentary strata of the Japanese islands, suggesting its potential importance in the mineral trapping of CO_2 (Okuyama et al. 2009).

A novel technology for generating numerous microbubbles of CO_2 and/or other gases in water in underground injection wells (Fig. 6) allows greenhouse gases to be emplaced in the tiny pores of saline aquifers and narrow cracks of fractured rock bodies (Koide and Xue 2009). Carbon microbubbles of less than several 10 μm in diameter tend to shrink and quickly dissolve in water. As these microbubbles have hardly any buoyancy, they do not merge to form large bubbles with a large buoyant force in groundwater. The capillary effect traps many carbon microbubbles as residual gas in rock pores (Figs. 6 and 7). As the CO_2 solution is heavier than the original water, the dispersion of CO_2 into underground rocks as microbubbles prevents the rising upwards of plumes of large CO_2 (gas or supercritical fluid) bubbles in underground aquifers. The injection of underground carbon microbubbles accelerates geological disposal mechanisms such as capillary trapping, solubility trapping, ionization trapping, mineral trapping and microbial trapping by dispersion and quick dissolution (Fig. 7).

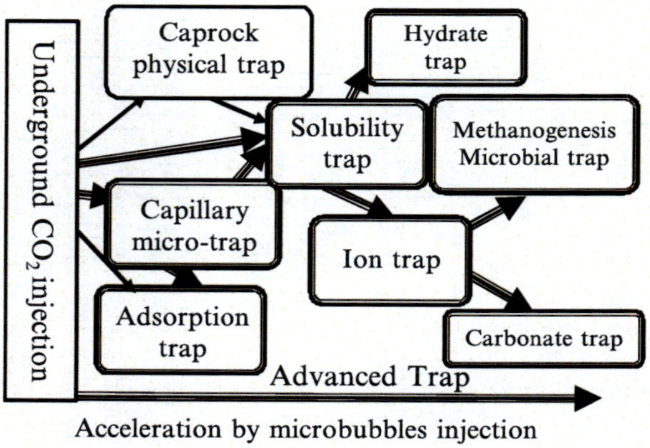

Fig. 7 Acceleration of advanced sequestration of CO_2 through carbon microbubble injection

The combined effect of the very weak buoyant force of carbon microbubbles, heavy CO_2 solution and various trapping mechanisms makes the underground injection of carbon microbubbles stable across a wide geological range, even where there are some structural imperfections such as depleted oil/gas reservoirs with abandoned wells, faulted domes, saline aquifers with incomplete caprocks in young unfolded sedimentary basins, large horizontal aquifers with unproven caprocks, synclinal sedimentary basins, fractured large basalt layers, faulted serpentine bodies, etc. (Koide and Xue 2009). The flexibility of site selection makes source–sink matching much easier for subsurface injection of carbon microbubbles than for conventional direct injection. As suitable sites for sinks can be found near many of the main sources of CO_2, the subsurface injection of carbon microbubbles is a practical energy saving and cost-effective greenhouse gas reduction method in many regions.

CO_2 injection under gas hydrate-filled or permafrost layers may achieve greenhouse gas mitigation and recovery of unused natural gas. Autogenous sealing of CO_2 in deep (>300 m) and cool (<5°C) aquifers assures virtually complete and practically unlimited subsurface containment of CO_2 (Koide et al. 1997). Sediments under the deep sea floor are very cool because the deep oceanic water usually only has a temperature of a few degrees centigrade. CO_2 hydrate is formed in sediments under wide areas of ocean floor deeper than about 300 m. Virtually complete isolation of huge amounts of CO_2 is made possible by deep sub-seabed disposal. Liquid CO_2 with heavy suspension intrudes laterally under light, unconsolidated sediments at sea floor depths of under around 3,700 m. Using the lateral intrusion technique for the super-deep sub-seabed disposal of CO_2 can protect the ecology of the sea floor. Virtually unrestricted volumes of CO_2 can be sequestered in the sediments and oceanic crusts under the deep sea floor around the Japanese islands.

Chemoautotrophic microbes fix CO_2 in deep underground aquifers even in the absence of sunlight. Thermophilic methanogens can convert the CO_2 into methane in anoxic and hot aquifers (Koide 1997). Biogenic restoration of subsurface hydrocarbon deposits may be possible in CO_2-injected aquifers, probably after decades. Microbiological recycling of CO_2 in aquifers is an attractive possibility for energy-poor countries such as Japan (Koide and Yamazaki 2001). The key to underground CO_2 recycling is hydrogen supply, that is, energy sources from microbial decomposition of organic matter, from geochemical water–rock interaction and from deep geothermal activity. In view of the wide diversity of underground microorganisms, an extensive search for favourable species and ecosystems is needed.

The natural gas and/or oil reservoirs in and around Japan are relatively small-scale and distant from major CO_2 emitters. However, the CO_2 disposal potential of saline aquifers has been estimated at as much as 146 Gt CO_2—enough to accommodate more than a century of industrial CO_2 emissions in Japan. The real problem is the geotectonic instability (earthquakes, faults and geothermal activity) of the Japanese islands. Advanced scientific studies are anticipated to develop secure CO_2 disposal technologies and underground microbial carbon recycling.

3 Current Status and Issues of Geological Disposal of Radioactive Waste in Japan

3.1 Geological Disposal of High-Level Radioactive Waste in Japan

Since 1955 Japan has been promoting the use of nuclear energy for strictly peaceful purposes. The establishment of the nuclear fuel cycle is the fundamental policy for nuclear energy development and utilization in Japan, with spent fuel from nuclear power generation being chemically processed at fuel reprocessing plants to recover uranium and plutonium which can be recycled as fuel for electricity generation. After the reprocessing of spent radioactive fuel and the recovery of the uranium and plutonium, the resultant high-level radioactive liquid waste is mixed with raw materials, then vitrified into a solid glass form to ensure physical and chemical stability. The vitrified high-level waste (HLW) is encapsulated in stainless steel containers and placed into temporary disposal for 30–50 years to cool. It is subsequently placed in a deep underground, geological disposal repository, at a depth greater than 300 m.

Japan has a much longer history of research into geological disposal of RW than into CCS. The national R&D project on geological disposal of HLW started along Atomic Energy Commission (AEC) of Japan guidelines published in 1976 and in 1980 (AEC 1976, 1980). The Special Committee of the AEC indicated a stepwise procedure for geological disposal of HLW as outlined below (AEC 1980):

Phase 1 Research on potential geological environments;
Phase 2 Research on suitable geological environments;
Phase 3 In situ experiments with mocked-up vitrified waste forms;
Phase 4 In situ experiments with vitrified waste forms;
Phase 5 Demonstration of disposal.

The Power Reactor and Nuclear Fuel Development Corporation (PNC, later JNC) compiled the R&D achievements of Japan in a first progress report (PNC 1992). In November 1999, the Japan Nuclear Cycle Development Institute (JNC) submitted a second progress report entitled H12: Project to Establish the Scientific and Technical Basis for HLW Disposal in Japan (H12 Report, for short) (JNC 2000a, b, c, d) to the AEC for an official review. The H12 Report consists of the Project Overview Report (JNC 2000a); Supporting Report 1—Geological Environment in Japan (JNC 2000b); Supporting Report 2—Repository Design and Engineering Technology (JNC 2000c); Supporting Report 3—Safety Assessment of the Geological Disposal System (JNC 2000d); and Supplementary Report—Background of Geological Disposal.

The Japanese Government enacted the Specified Radioactive Waste Final Disposal Act (Final Disposal Act) in May 2000 to establish a framework for the final disposal of high-level radioactive waste. The Final Disposal Act established the Nuclear Waste Management Organization of Japan (NUMO) as the responsible organization by private initiative. In the same year the Nuclear Safety Commission (NSC) issued the Basic Policy for Safety Regulations for High-Level Radioactive Waste (First Report). In December 2002, NUMO began an open solicitation process to invite municipalities to volunteer for preliminary investigation sites. Although since December 2002 some ten municipalities have shown interest in being voluntary NUMO 'preliminary investigation sites', none have been able to obtain the consensus of local residents to be able to apply officially. Many Japanese people are deeply concerned about the safety of radioactive facilities from earthquakes and active faults. It is important to gain public confidence in the long-term safety of RW disposal.

Until the H12 Report, research on the long-term stability of the geological environment was conducted mainly to show that there is a wide range of stable geological environments in Japan that are suitable for RW geological disposal. The potential and scale of future events can be discussed based on the trends and frequency of occurrence of natural phenomena in the past. After the H12 Report, in addition to continuing academic research such as nationwide data acquisition and studies on individual events and their mechanisms, the emphasis was placed more on the development of appropriate investigation and assessment technologies. In September 2005, JNC published the report entitled H17: Development and Management of the Technical Knowledge Base for the Geological Disposal of HLW (H17 Report, for short) (JNC 2005a, b, c, d). The H17 Report consists of the Knowledge Management Report (JNC 2005a); Supporting Report 1—Geoscience Study (JNC 2005b); Supporting Report 2—Repository Engineering Technology (JNC 2005c); and Supporting Report 3—Safety Assessment Methods (JNC 2005d).

Disposal technology is a multidisciplinary field about which a wide range of information is needed to develop safety scenarios. Knowledge, both explicit and tacit, obviously increases with time and is essential to all stakeholders, including implementers, regulators, political decision makers and the general public. Although essential for developing a project, a structured approach to assuring that all relevant knowledge is available is particularly critical at times when major project decisions have to be made.

The Japan Atomic Energy Agency (JAEA) was established in October 2005 by merging the Japan Atomic Energy Research Institute (JAERI) and JNC. JAEA carries out geoscientific research into actual deep geological environments at the Mizunami Underground Research Laboratory (granite, hard rock, fresh groundwater) in Gifu Prefecture and at the Horonobe Underground Research Laboratory (sedimentary rock, soft rock, saline groundwater) in Hokkaido.

In 1992, low-level waste (LLW) disposal and, in 1995, vitrified HLW storage began at the Rokkasyo, Aomori Prefecture. The LLW disposal centre has a total capacity of 80,000 m^3 (400,000 drums of 200 l) and an ultimate future capacity of 600,000 m^3. The vitrified waste (HLW) storage centre has a storage capacity of 1,440 canisters and ultimate future capacity of 2,880 canisters.

In Japan, except for eliminating unconsolidated rocks, potential geological formations for HLW disposal have not yet been identified. Both crystalline and sedimentary rocks are being investigated generically. The long-term stability of the geological environment and groundwater regime is being investigated carefully for the long-term isolation of RW (Marui and Kusunose 2008).

3.2 *Prediction of Long-Term Stability of the Geological Environment for the Safe Disposal of Radioactive Waste*

3.2.1 Methods of Geological Prediction

The long-term stability of the geological environment is essential for the safe geological disposal of RW (Koide 1991). Long-term crustal movement must be predicted to evaluate the stability of geological repositories of RW and neighbouring rock mass during an assessment period. As a case study, a numerical analysis method for the prediction of crustal movement in Japan, which was studied using three-dimensional elastic analysis by the finite element method (FEM) for the geological block structure of the Kinki region and the Awaji-Rokko (Kobe) area, is presented (Sasaki et al. 2000). Stability analysis for a disposal cavern was also investigated. Geological modelling, based on geophysical and drilling information, including information on the surface structure, is performed to delineate the subsurface structure (Kouda and Murakami 2002). The Kinki region and the Awaji-Rokko (Kobe) area are not candidate sites for the geological disposal of RW, as they are geotectonically unstable areas with a high earthquake risk. They were studied for convenience of evaluation of the prediction of long-term crustal movement.

There are various methods of predicting crustal movement, such as research and model testing of past crustal movement patterns. In these methods, the establishment of a computer-based model is indispensable for quantitative prediction. The computational techniques involved in simulating the crustal movement for about 10,000 years were proposed and the numerical results compared with the features of surveyed crustal movement.

Geological prediction methods are generally classified as follows (Koide 1991, 1997):

1. Prediction by extrapolation: the fundamental method for geological prediction is especially effective in steady slow phenomena, and is also applicable to periodic and cyclical ones. The prediction should be made based on detailed observations over a sufficiently longer time span than the prediction period, and the possibility of changes in tendency should be checked by other prediction methods.
2. Prediction by analogy: the future prediction is carried out by identifying similar past phenomena. This method, known as the natural analogue method, is an effective approach for crustal movement phenomena, and the only reliable method of validating conceptual and numerical models for long-term prediction for RW disposal.
3. Prediction by experiment: the phenomena are artificially modelled, and the prediction is then carried out using the experimental model. The method is necessary

for understanding the mechanism of geological phenomena, as the conditions can be precisely controlled. Long-term quantitative prediction using this method is difficult, however.

4. Prediction by probability: statistical research is indispensable for assessing the risk of some catastrophic events; however, statistically fine data of geological phenomena are difficult to obtain.
5. Prediction by conceptual model: the plate tectonic model is useful for predicting tectonic movements in a plate boundary region such as the Japanese archipelago.
6. Prediction by numerical simulation model: as past phenomena cannot exactly equate to present ones, a complicated underground environment cannot be simulated, even experimentally. Numerical calculations by computer are necessary to quantitatively predict changes in crustal movement phenomena in which various factors interact. Mechanical stability and seepage analyses are carried out to evaluate the performance of current geological disposal.
7. Prediction by safety assessment model: simplified safety assessment models are useful for analysing the overall performance of a complex disposal system. The safety assessment requires the development of synthetic models, scenario analyses and consequence analyses. Long-term geological prediction should be performed and cross-checked by the different methods.

A systematic approach with several prediction methods, as mentioned above for the prediction of long-term tectonic stability, is more practical.

3.2.2 Outline of Crustal Movement in Japan

The disposal of RW in deep geological formations is a safer, viable technology at present. However, as the Japanese archipelago is close to the plate boundary, crustal movements and associated active faults are more intense than in continental areas (Koide 1992). Thus, volcanoes are not evenly distributed. There is no volcanic activity between the volcanic front and the oceanic trench where the plate sinks, but much volcanic activity concentrated in the continental vicinity of the volcanic front (Sugimura and Uyeda 1973). Uneven distribution by region is also noteworthy in relation to the active faults. There are regions characterized by few active faults between the volcanic front and the oceanic trench (see Fig. 1).

Up to about 10 km underground, earth pressure and temperature, as well as rock strength, all increase with depth. However, deeper than 10 km underground the strength of the rock decreases drastically, as the high geothermal temperature makes the rock ductile. Near the volcanic front, the temperature is high because of magma rising from the mantle. Hence, in the volcanic front region the crust with high rigidity is very thin, while in the region between the oceanic plate and the volcanic front, the crust with high rigidity is thick because of the relatively low temperature caused by plate subduction. As a result, crust deformation is concentrated near the volcanic front region with the thin crust, whereas the crust with high rigidity between the volcanic front and oceanic trench is relatively stable (Fig. 1).

The region where crust upheaval is greatest is concentrated near the plate boundary and volcanic front. In the region of maximum upheaval the rate is about several mm/year on average. In most regions, crustal upheaval or subsidence rates are within 1 mm/year. The average displacement rate is several mm/year, even in an active fault. Such active faults are distributed near the plate boundary, volcanic front and the median tectonic line. The average slip rate of the majority of active faults in other regions is less than 1 mm/year. To evaluate the long-term stability of disposal sites, it is essential to estimate the magnitude of future geological events.

3.2.3 Crustal Movement Prediction Analysis

Japan is located in a particularly complex geological region, being close to four major plates. North-eastern Japan is primarily compressed between two converging plate boundaries. The north-east is characterized by the east–west horizontal compression and the north–south trend of reverse faults. On the other hand, the strike-slip faults represented by the major median tectonic line develop in the part of western Japan that is affected by the oblique subduction of the Philippine Sea Plate at the Tokai-Nankai Trough under the Eurasian Plate. The north–south trend of many short reverse faults in the Kinki district is related to the bends in major strike-slip faults.

In western Japan, the Kinki district has been chosen as a region for large-scale modelling. The Kinki district is divided geologically into several blocks by active faults. A qualitative evaluation by lateral slip, upheaval and subsidence behaviour around the Osaka block is carried out by three-dimensional linear elastic FEM (Sasaki et al. 2000). In the depth direction it is modelled to the lower-limit plane depth (estimated from the hypocentral depth distribution) of the upper crust. The lower crust is treated mechanically as a distribution of springs. The inclination of the block boundary is also considered. The compressive strain in the east–west direction is considered by giving an enforcement displacement (30 m/1,000 years), referring to the horizontal strain of the Kinki district over the past 100 years measured by the Geographical Survey Institute. This result simulates the lateral slip behaviour of the fault, generated over the past 10,000 years.

In addition, a medium-scale model has been developed. The purpose of the medium-scale model is to establish the analytical conditions of the vicinity model around the virtual RW disposal vault. In the medium-scale model, minor active faults that are not considered in the large-scale model have been modelled. The boundary conditions for the enforcement displacement are given, and the value of displacement is obtained from large-scale model analysis. The mechanical properties of the rock mass and faults are assumed to be linear elastic as in the large-scale model. Figure 8 compares the analytical results for the relative slip rate vector with those surveyed. It appears that the analytical results qualitatively match the survey results. This result shows that the strain is concentrated around the Awaji-Kobe-Rokko active faults. The disastrous 1995 Kobe earthquake (which measured 7.2 on the Richter scale) occurred along the active faults with a large slip rate during the

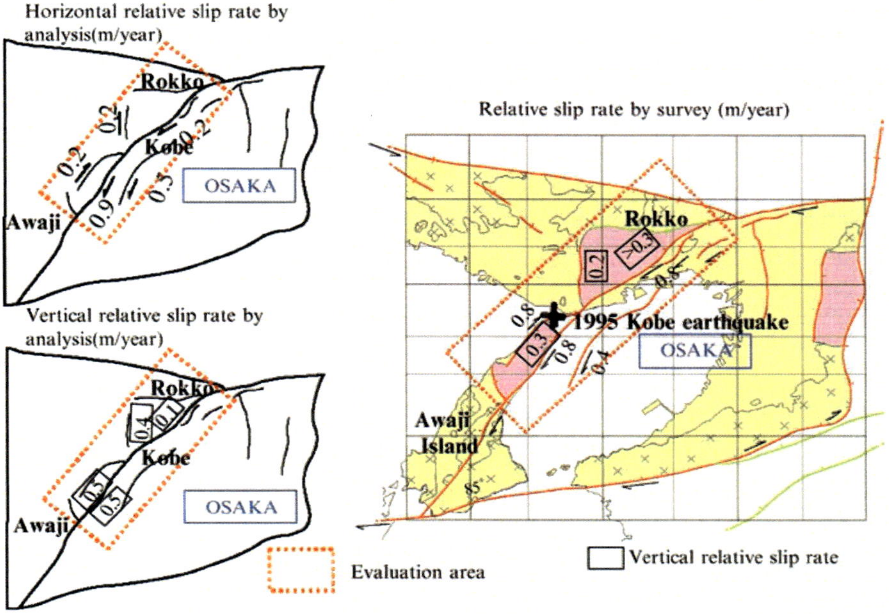

Fig. 8 Comparison of analytical results on the relative slip-rate vectors of the medium-scale model and survey results. (*see* Colour Plates) The 1995 Kobe earthquake (measuring 7.2 on the Richter scale) occurred along active faults with large slip rates (Source: Sasaki et al. 2000)

mechanical analysis (Koide 1997). The analytical results provide useful suggestions on the relation between the fault geometry and earthquake-fault stability.

Research on crustal movement prediction techniques is required for underground RW disposal sites for long periods exceeding 1,000 years. Although there are various prediction methods, computational simulation techniques are necessary to quantitatively predict crustal movement. One of the problems in crustal movement prediction is that there is little information regarding the deep underground. The long-term prediction of crustal movement for geological disposal sites and an example of the procedure and modelling technique using FEM have been proposed. To improve the accuracy of crustal movement prediction, more precise modelling of the fault is required. It is also important to judge synthetically the result of the numerical analysis using geological and rock engineering knowledge, as crustal movement prediction involves many uncertain and complicated factors. The validity of the model must be verified by comparison with crustal movements in the past.

Assessment of the long-term stability of geological environments is the key issue for geological RW disposal in Japan. The comprehensive system of long-term prediction of crustal movement and the groundwater regime around the virtual RW disposal sites has been developed through long and painstaking efforts in Japan.

4 Comparison of CO_2 and Radioactive Waste Geological Disposal

4.1 Extreme Differences in Toxicity and Volume

Although RW is extremely toxic, its volume is much smaller than that of CO_2 (see Table 1). As the radionuclides in RW do not exist in nature, they should not be dispersed into the environment. Hence the disposal strategy for RW is long-term containment. Although the radioactivity of HLW rapidly decreases, it is still highly toxic for up 10,000 years and toxic for up to 100,000 years. LLW has low toxicity after 300 years; however, some LLW may be weakly toxic for up to 10,000 years. That is why RW is stored in carefully mined tunnels and/or vault systems in multiple natural and engineered barrier systems (EBSs).

Almost 800 Gt CO_2 have accumulated in the atmosphere since the industrial revolution—enough to cover the whole earth with an 80 cm thick layer. The annual emission of CO_2 is about 1.3 Gt in Japan and about 27 Gt worldwide. A single major fossil fuel power plant can emit several Mt CO_2 per year. It is virtually impossible to construct a big enough tunnel and/or vault system to accommodate such a gigantic amount of CO_2. Containment by EBSs used in RW disposal is too costly for the disposal of a huge amount of CO_2. Cost-effective CO_2 disposal uses injection of CO_2 through drilled wells into the natural underground rock pore system and natural containment mechanisms. Natural gas and/or oil reservoirs are ideal reservoirs for the CO_2 disposal. However, many natural gas and/or oil reservoirs are still occupied by valuable resources. Some available reservoirs are too far from large CO_2 emission sources to be economically feasible. Disposal in saline aquifers and other advanced options for geological disposal of CO_2 are being investigated to provide a wider range of CO_2 reduction options.

While a CO_2 density in air of under 1% is not toxic to humans, CO_2 that is denser than 0.04% (current level) has too high a greenhouse effect. On the other hand, CO_2 density that is under 0.03% is dangerously low for plants and has too low a greenhouse effect, leading to potential global cooling. The natural density of CO_2 in the air (about 0.03%) is not only harmless but also necessary for human health and for the global climate. Using CO_2 microbubble injection as a strategy to disperse and dilute CO_2 may provide the safest disposal option, in terms of avoiding a large underground accumulation of a high density of CO_2.

4.2 Underground Injection of CO_2 Fluid and Emplacement of Solidified Radioactive Waste

The impact of RW disposal on the environment is limited. While CO_2 is naturally abundant, the geological disposal of CO_2, through the injection of huge volumes of CO_2 directly into the pore system of underground rocks, may have

Table 1 Comparison of CO_2 and radioactive waste disposal[a]

Characteristic or attribute	CO_2 disposal	Radioactive waste disposal
Characteristics of the geological media		
1. Tectonic stability	Tectonic stability preferred: • Avoid active volcanoes • Avoid major active faults • Avoid underlying large unstable faults (so as not to induce large earthquakes) • Avoid rapid uplift zones	Tectonic stability preferable, prediction of long-term stability (>100,000 years) is required: • Avoid volcanic activity • Avoid major active faults • Avoid geothermally active areas • Avoid rapid uplift zones • Avoid plate boundary zones • Avoid continental vicinity of volcanic fronts
2. Past stability	Investigate past seismic activity	Important to understand and demonstrate past physical and chemical stability to increase confidence that such stability will continue into the future
3. Geological environment	• In folded sedimentary basins that: have formations of sufficient porosity (for capacity) and permeability (for injectivity) and domal or other confining structures of impermeable (clay or shale) covers (caprocks) that can trap a buoyant plume of CO_2 gas and/or CO_2 supercritical fluid • In less folded flat or synclinal sedimentary basins that: have formations of sufficient porosity and permeability and less permeable (clay or shale) covers that can prevent the upward flow of saline groundwater, CO_2 solution and residual gas • In sedimentary basins with imperfect caprocks and/or damaged (by drilling and/or extraction subsidence) caprocks that can still prevent the upward flow of saline groundwater, CO_2 solution and residual gas	• Groundwater fluxes at depth are sufficiently low • Sufficient volume of host rock to house repository • Host rock has suitable geotechnical properties for underground construction • Geological complexity is acceptably low • Geotectonically stable to assure no possibility of adverse disturbances of repository over the next hundred thousand years • No underground igneous activity and acceptably low geothermal activity over the next hundred thousand years

(continued)

Table 1 (continued)

Characteristic or attribute	CO_2 disposal	Radioactive waste disposal
	• In coal seams or tuff formations (for CO_2 adsorption) • In cation-rich alkaline brine in deep sedimentary formations and in mafic and ultramafic rocks such as peridotite, serpentine, basalt, etc. (for carbonate precipitation) • In deep (>about 300 m) and cool (<about 5°C) areas under the deep sea floor and in frigid regions (for CO_2 hydrate formation) • In anoxic hydrogen-rich aquifers in deep saline sedimentary formation, in methane hydrate-bearing formation and in continental and oceanic basalt (for CO_2-reduction methanogenesis)	
4. Rock type	• Porous sedimentary rocks (sandstone, carbonate) • Coal, tuff (adsorption) • Oceanic and continental basalt, peridotite, serpentine, ophiolite, etc. (carbonate clogging) • Sediments and rocks under deep sea floor and in frigid regions (hydrate clogging)	Hard (crystalline) rock (granite), sedimentary rocks (mudstones, clays, tuff)
Emplacement characteristics		
5. Toxicity	Concentrated CO_2 is toxic;	The radioactivity of HLW rapidly decreases but still highly toxic > 10,000 years and toxic >100,000 years
	1% > CO_2 >0.04% is not toxic for humans but has too high a greenhouse effect CO_2 <0.03% is dangerously low for plants and has too low a greenhouse effect, leading to potential global cooling	The LLW is hardly toxic after 300 years, but relatively high LLW may be weakly toxic for as long as 10,000 years
6. Mode of disposal	Injection through wells	Emplacement in (and from) tunnel and/or vault systems in multiple natural and EBSs
7. Volume	Very large (large-scale deployment > some Mt/year/project, possibly several Gt/year in Japan, many smaller disposal sites and test projects are also possible)	Comparatively small (high-level vitrified radioactive waste in some hundred thousand canisters in Japan: about 1.3 m high, about 40 cm in diameter, 500 kg total weight)

(continued)

Table 1 (continued)

Characteristic or attribute	CO_2 disposal	Radioactive waste disposal
8. Depth	• CO_2 supercritical fluid: > 800 m • CO_2 solution: >50 m	HLW disposal: >300 m and probably <2,000 m Relatively high LLW: >50 m and <100 m LLW: <50 m
9. Physical state	• Gas at the well top, mostly supercritical fluid at the well bottom, occasionally liquid or gas at the injection points • In some options, CO_2 solution or WAG or mixture of water and small CO_2 bubbles (microbubbles or nanobubbles of CO_2 supercritical fluid or gas or liquid)	Glass (high-level vitrified radioactive waste) in a stainless steel canister
10. Containment mode	• CO_2 supercritical fluid and gas need completely confining caprocks (natural barriers: shale, clay, tuff, coal) • Dissolved CO_2 is likely dispersed and diluted in aquifers under restricting aquitards, groundwater flux and chemistry • Quasi-natural barriers (autogenous sealing: carbonate precipitation clogging, hydrate precipitation clogging)	Multiple natural and engineered barrier system Geological barrier always acting in tandem with an EBS
11. Timescale of interest	Two timescales: • Associated with global warming (greater than centuries) • Associated with local risks posed by injection and possible leakage (decades)	Detailed, quantitative calculations required for 100,000 years, less quantitative for longer, possibly up to 1 million years
12. Containment period	Depends on disposal concept Concentrated CO_2 should be contained for several centuries, up to millennia Small amounts of seepage of diluted CO_2 are harmless	Depends on disposal concept and waste types HLW should be contained for 10^5 years or longer; LLW should be monitored for several hundred years, but some relatively high LLW requires longer evaluation of safety depending on the characteristics of the waste
Effects of emplacement and potential migration from the disposal site		
13. Direct effects of disposal	• Pressure increase (swell, fluid fracturing)	• Thermal effects due to radioactive decay (for heat emitting waste)

(continued)

Table 1 (continued)

Characteristic or attribute	CO_2 disposal	Radioactive waste disposal
	• Thermal effects	• Geochemical reactions and processes in both the near and far fields
	• Well integrity (breakthrough)	• Biochemical processes in both the near and far fields
	• Geochemical reactions with groundwater (pH and Eh change) • Groundwater displacement • Buoyancy effect of CO_2 gas and/or CO_2 supercritical fluids	• Geomechanical effects due to repository construction
14. Effects on the natural barrier	No significant structural modifications to the geological environment caused by the engineered systems (wells), but the CO_2 itself may have adverse effects on barrier integrity • Geochemical reactions in the presence of groundwater and host rocks (dissolution and precipitation of carbonates, weathering effect of silicate minerals in both the near and far fields) • Geomechanical effects as a result of pressure increase and stresses (swell, fault instability, induced earthquakes) • Caprock integrity (breakthrough, fluid fracturing, leaching) • Decomposition and formation of gas hydrates in Arctic regions and under deep sea floors (>300 m)	The construction of the repository and the EBS employed will directly affect the natural barrier (although probably only locally); also, heat emitting waste will directly affect the natural barrier, although any effects will be limited to a few thousand years, at most
15. Transport mechanisms of CO_2 or radionuclides	The CO_2 injection pressure:	Mainly via groundwater (advective and diffusional transport), but to a lesser extent via gas (produced by a variety of geochemical and biochemical processes in the near field and by radioactive decay)
	• Buoyancy effect of bubbles of CO_2 gas and/or CO_2 supercritical fluids (upward rising plume, pressure instability) • Groundwater regime and gravity flow of CO_2-dissolved water	Transport of radionuclides can also take place in colloidal form

(continued)

Table 1 (continued)

Characteristic or attribute	CO_2 disposal	Radioactive waste disposal
16. Return to the biosphere, hydrosphere and atmosphere	There are, in effect, no engineered barriers, and leaky wells, fractures and other local geological features may provide a pathway for the return of CO_2	A considerable amount of the long-lived waste is encapsulated within containers that will remain intact for a considerable time; even in the event of canister failure, the EBS will delay the release of radionuclides for a time
	• Upward rising plume of CO_2 gas and/ or CO_2 supercritical fluids • Breakthrough, fluid-fracturing and/or leaching of caprocks. • 'Bathtub effect' • 'Champagne effect' • Sand wedging and mud volcanoes with earthquakes • Seepage along faults	Geotectonical disturbances (earthquakes, fault activity, igneous or sand dike intrusion, hydrothermal intrusion, etc.) may affect the containment of radionuclides in the long term
Site activities		
17. Site characterization	Currently based only on the existing geological data for oil and gas exploration. Original geological investigations for CO_2 disposal are anticipated.	Very comprehensive and lengthy investigation programme on the test-only sites (Mizunami and Horonobe) and on the generic geological situation of Japan
18. Monitoring	Monitoring (land surface deformation, gas seepage, groundwater regime, groundwater chemistry, pressure, stress, seismic activity, etc.) required for baseline conditions (site selection), during injection, and decreasingly after cessation of injection for site closure and ensuring long-term safety of the system	All disposal concepts have extensive barrier systems; thus, although monitoring will be a requirement, no releases are likely to be detected for a considerable time (i.e. several millennia) after closure
	Monitoring of underground CO_2 behaviour and surface 'watchdog' monitoring	
19. Future access, intrusion or penetration	Penetration by future wells drilled for other purposes (e.g. groundwater and mineral water tapping, geothermal wells, exploration and production of oil, gas and other mineral resources) is quite possible	Sites will be selected only in areas where the intrusion risk is considered low (i.e. no mineral resources)

EBS engineered barrier system
HLW high-level waste
LLW low-level waste
WAG water alternating gas
[a] Only the geologically related issues discussed in this chapter are considered here

huge environmental impacts (Table 1). One of the adverse effects of underground fluid injection is to induce earthquakes (Sminchak et al. 2001).

The rapid underground injection of a large amount of CO_2 may cause the formation of a CO_2 plume, that is, a large underground CO_2 bubble; this penetrates the pore spaces of the reservoir rocks until they are more or less filled by CO_2. The pure CO_2 fluid may be gas, liquid or supercritical fluid, depending on the pressure and temperature conditions in the reservoir. As the CO_2 fluid (gas, liquid or supercritical fluid) is lighter than the groundwater at the normal underground combination of pressure and temperature, a big underground CO_2 bubble acquires a large buoyant force (see Figs. 4 and 5). The plume-like rising upwards of a large CO_2 bubble was observed seismically in the highly permeable sandy aquifer in the first large-scale CO_2 disposal project at the Norwegian Sleipner Field in the North Sea (Arts et al. 2001). For the safe disposal of a large buoyant body of CO_2 fluid, a perfect cover for an underground geological structure such as a caprock dome is necessary (Figs. 4 and 5). Natural gas and oil deposits have long-term geological traps that have contained the buoyant fluids for longer than several million years. It has also been suggested that undetected small gaps of shale layers provided pathways for the CO_2 plume in the Sleipner aquifer (SACS 2003). No effective technologies have yet been developed to check the integrity of caprocks that has not been proven naturally. Moreover, abandoned production wells, exploration boreholes or artificial fractures caused by subsidence resulting from extraction may form pathways for CO_2 leakage, even in naturally proven caprocks.

Because of the difference in density between the inside and outside of the underground bubble, there is a big build-up of excess fluid pressure at the top of it (Fig. 4). The CO_2 fluid (gas, liquid or supercritical fluid) inside the bubble is lighter than the groundwater outside it at normal underground pressure and temperature. Pressure instability is inevitable around the large buoyant bubble. Pressure in the CO_2 reservoir cannot exceed the interfacial threshold pressure for the CO_2 breakthrough of caprocks (Fig. 5). The fluid fracturing of caprocks and fluid-induced earthquakes may occur at the top of a large CO_2 bubble if the excess pressure due to the buoyancy effect reduces the effective stress enough to cause the shear instability of underground rocks (Fig. 5).

Fluid fracturing and induced earthquakes are unlikely around microbubbles, as the very small excess fluid pressure only builds up at the top of the microbubbles. Koide and Xue (2009) proposed a carbon microbubble injection technology that can drastically reduce the size of CO_2 bubbles. The novel carbon microbubble injection technology (Fig. 6) can generate numerous tiny uniform bubbles of CO_2 and/or other greenhouse gases smaller than several 10 μm in diameter in water. The carbon microbubbles of diameters of less than several 10 μm tend to shrink and quickly dissolve in water. The greenhouse gas microbubbles possess hardly any buoyancy and do not tend to join together to form large bubbles that have large buoyant force in groundwater. Tiny CO_2 microbubbles percolate deep into tiny pores and cracks in the aquifer rocks (Fig. 6). Interfacial force and the capillary effect trap many carbon microbubbles in pores of rocks as residual gas (Fig. 7). With the underground high pressure, CO_2 microbubbles rapidly dissolve in the groundwater. As the CO_2

solution is heavier than the primary groundwater, the CO_2 solution tends to flow downwards. The dispersion of CO_2 into underground rocks as microbubbles prevents the plume-like rising upwards of large CO_2 (gas or supercritical fluid) bubbles in partially confined aquifers and even in unconfined aquifers. The carbon microbubble injection accelerates the residual gas (interfacial and capillary effects) trapping, dissolution trapping and ionization trapping of CO_2 for stable saline aquifer disposal of CO_2. The natural methane solution deposits (in Chiba, Niigata and Miyazaki prefectures) and natural CO_2 solution deposits (in the Isobe and Izumi districts) in unconfined saline aquifers indicate that long-term CO_2 disposal in solution is possible in young semi-open unfolded sedimentary basins because of dissolution, ionizing and residual gas trapping (Koide et al. 1992). The disposal of CO_2 in saline aquifers is the most economically favourable option for geological carbon disposal, with a huge disposal potential of three trillion tonnes of CO_2 in solution in sedimentary basins worldwide (Koide 1999). The sequestration of carbon microbubbles in saline aquifers provides an economic option for CCS for many urban and industrial areas.

The extreme differences in toxicity and volume of CO_2 and RW result in different disposal strategies. The dispersion and dilution principle is possible for CO_2 disposal, while RW is strictly contained by multiple barrier systems. The large volume of CO_2 fluid injection may destabilize the underground geological environment, while the effect of RW is restricted to near the repository. The stability of the geological environment is important for both CO_2 and RW disposal. The NIMBY (Not In My Backyard) syndrome, as a manifestation of public wariness with respect to disposal sites for both CO_2 and RW, is also common.

5 Conclusions

The geological disposal of high-level radioactive waste and full-scale underground disposal of CO_2 have not yet been realized in spite of the lengthy and painstaking history of research on geological disposal of RW and CO_2 in Japan. Site selection for geological disposal of RW and CO_2 is very important because of the large regional differences in tectonic activity in Japan.

The natural gas and/or oil reservoirs in and around Japan are relatively small-scale and distant from major CO_2 emitters. However, the CO_2 disposal potential of saline aquifers has been estimated to be as much as 146 Gt CO_2: enough to accommodate more than 100 years of industrial CO_2 emissions in Japan. The real problem is the geotectonic instability (earthquakes, faults and geothermal activities) of the Japanese islands. Assessment of the long-term stability of geological environments is the key issue for geological RW disposal in Japan. The comprehensive system of long-term prediction of crustal movement and the groundwater regime around the virtual RW disposal sites has been developed after long and meticulous efforts in Japan.

Future geological predictions should be cross-checked by several different methods. However, only historical geological evidence can verify long-range predictions for the subterranean containment of waste. Plate interactions form complex geologic structures and cause serious tectonic instability. Frequent earthquakes and numerous active faults clearly indicate severe deformations of the Earth's crust in Japan. Long-term stability of the geological environment is an important factor in the disposal of RW in Japan. The density of active faults and volcanoes shows dramatic regional variations. There are few active faults and no volcanoes in stable regions between the subduction zone and volcanic fronts in the island arc region.

While RW is extremely toxic, it has a much smaller volume than CO_2. RW contains radionuclides not found in nature, but it is stable when placed in solid assemblies and carefully buried in deep repositories. The impact of RW on the environment is restricted. CO_2 is naturally abundant, but injecting huge volumes of CO_2 directly into a pore system of underground rocks may have enormous impacts on the Earth. One of the adverse effects of underground fluid injection is that it may induce earthquakes. The fluid fracturing of caprocks and fluid-induced earthquakes may occur at the top of large CO_2 bubbles if the excess pressure due to the buoyancy effect reduces the effective stress enough to cause shear instability in underground rocks. However, the combined effect of the very weak buoyant force of greenhouse gas microbubbles, the heavy CO_2 solution and various trapping mechanisms makes the subsurface injection of carbon microbubbles stable in a wide variety of geological conditions, even where there are structural imperfections such as depleted oil/gas reservoirs with abandoned wells, barren geological domes with unproven caprocks, faulted anticlines of sedimentary formations, saline aquifers with incomplete caprocks, large horizontal or monoclinal formations, synclinal sedimentary basins, fractured basalt layers, fractured serpentine bodies, fractured igneous bodies, etc. Although a mixed injection of CO_2 microbubbles and water reduces the availability of definite pore spaces, greater varieties of pores become available as a result of the carbon microbubble injection. Carbon microbubble sequestration in saline aquifers provides safe and economic options for CCS for many urban and industrial areas, as closer, shallower and bigger aquifers are available for most CO_2 emission sources.

The underground injection of carbon microbubbles accelerates advanced geological disposal mechanisms such as residual gas trapping, dissolution trapping, ionization trapping, mineral trapping, microbial trapping and methanogenesis by dispersion and quick dissolution into groundwater in deep tiny pores in underground rocks. Autogenously sealed 'CO_2 capsules' can be formed in large basaltic sheets, ophiolite complex and oceanic crust. A cool temperature lower than 5°C and hydrostatic pressure greater than 3 MPa in sediments directly below the seabed at depths over 300 m are sufficient to form CO_2 hydrate in the sub-seabed sediments. Sub-seabed aquifers under the deep sea floor around the Japanese islands can provide very safe and virtually limitless CO_2 disposal.

The selection of geologically stable disposal sites is essential for the extremely toxic but very small volume of RW. On the other hand, the forthcoming geological disposal of huge volumes of CO_2 from widespread numerous sources in the carbon-constrained world may have greater impacts on the environment. The development of advanced environmentally friendly disposal technologies is important for the geological disposal of CO_2.

The extreme difference in toxicity and volume of CO_2 and RW entails different disposal strategies. The dispersion and dilution principle can be used for CO_2 disposal, while RW is strictly contained by multiple barrier systems. The large volume of CO_2 fluid injection may destabilize the underground geological environment, while the possible adverse effect of RW is restricted to the vicinity of the repository. The stability of the geological environment is important for both CO_2 and RW disposal. The NIMBY (Not In My Back Yard) syndrome is common, making public confidence in long-term safety assessments and disposal practices extremely important.

References

AEC (Atomic Energy Commission) (1976) Report on Radioactive Waste Management in Japan. Japan Atomic Energy Commission, Tokyo [in Japanese]

AEC (Atomic Energy Commission) (1980) Promotion of Research and Development on the Disposal of High-Level Radioactive Waste. Special Committee on Radioactive Waste Management. Japan Atomic Energy Commission, Tokyo [in Japanese]

Arts R, Brevik I, Eiken O, Sollie R, Causse E, van der Meer B (2001) Geophysical methods for monitoring marine aquifer CO_2 storage—Sleipner experiences. In: Williams DJ, Durie RA, McMullan P, Paulson CAJ, Smith AY (eds) Greenhouse Gas Control Technologies: Proceedings of the Fifth International Conference on Greenhouse Gas Control Technologies. CSIRO Publishing, Collingwood, pp 366–371

Kaya Y, Koide H, Nakaya S, Ohsumi H, Takita K, Yohimura S (2001) Underground sequestration of carbon dioxide in tectonically active areas. In: Williams DJ, Durie RA, McMullan P, Paulson CAJ, Smith AY (eds) Greenhouse Gas Control Technologies: Proceedings of the Fifth International Conference on Greenhouse Gas Control Technologies. CSIRO Publishing, Collingwood, pp 296–298

Koide H (1971) Fractures aligned en echelon and fracture patterns. In: Takeuchi Y (ed) Proceedings of the IMA-IAGOD Meetings '70. Society of Mining Geologists of Japan, Special Issue No. 3 (IAGOD volume):107–114

Koide H (1990) Underground gas disposal. In: Komiyama H (ed) Global Warming Handbook. Industrial Publishing & Consulting, Tokyo, pp 345–347 [in Japanese]

Koide H (1991) Geologic problems of radioactive waste disposal in Japan. Episodes 14:299–302

Koide H (1992) Prediction of tectonic stability in an island arc region for waste disposal. In: Koide H (ed) Waste Disposal and Geology: Scientific Perspectives (pp. 409–421). Proceedings of the Workshop WC-1 of the 29th International Geological Congress. Committee of the workshop WC-1 of the 29th International Geological Congress, Tokyo

Koide H (1997) Geotechnical challenges to geological hazard and protection of environment, Keynote paper. In: Fujii T (ed) Proceedings of the 8th International Congress on Rock Mechanics. A. A. Balkema, Rotterdam, pp 1005–1012

Koide H (1999) Geological sequestration and microbiological recycling of CO_2 in aquifers. In: Eliasson B, Riemer PWF, Wokaun A (eds) Greenhouse Gas Control Technologies: Proceedings of the 4th International Conference on Greenhouse Gas Control Technologies. Pergamon, Oxford, pp 201–205

Koide H (2001) Subsurface biogeochemical carbon recycling: CO_2-EGR with microbial methanogenesis in geochemically sealed aquifers. In: Williams DJ, Durie RA, McMullan P, Paulson CAJ, Smith AY (eds) Greenhouse Gas Control Technologies: Proceedings of the Fifth International Conference on Greenhouse Gas Control Technologies. CSIRO Publishing, Collingwood, pp 578–582

Koide H, Xue Z (2009) Carbon microbubbles sequestration: a novel technology for stable underground emplacement of greenhouse gases into wide variety of saline aquifers, fractured rocks and tight reservoirs. In: Gale J, Herzog H, Braitsch J (eds) Proceedings of the 9th International Conference on Greenhouse Gas Control Technologies; Energy Procedia 1:3655–3662

Koide H, Kuniyasu M (2006) Deep unmineable coalbeds in Japan: Potential CO_2 sink and untapped energy resource. In: Gale JJ, Røkke N, Zweigel P, Svenson H (eds) Proceedings of the 8th International Conference on Greenhouse Gas Technologies. Elsevier, Oxford, CD-ROM

Koide H, Yamazaki K (2001) Subsurface CO_2 disposal with enhanced gas recovery and biogeochemical carbon recycling. Environ Geosci 8(3):218–224

Koide H, Tazaki Y, Noguchi Y, Nakayama S, Iijima M, Ito K, Shindo Y (1992) Subterranean containment and long-term storage of carbon dioxide in unused aquifers and in depleted natural gas reservoirs. Energy Convers Manage 33(5–8):619–626

Koide H, Tazaki Y, Noguchi Y, Iijima M, Ito K, Shindo Y (1993) Underground storage of carbon dioxide in depleted natural gas reservoirs and unused aquifers. Eng Geol 34:175–179

Koide H, Shindo Y, Tazaki Y, Iijima M, Ito K, Kimura N, Omata K (1997) Deep sub-seabed disposal of CO_2—The most protective storage. Energy Convers Manage 38(Supplement):253–258

Koide H, Nishimura S, Satsumi S, Xue Z, Li X (2003) Carbon sequestration in coal seams in Japan and biogeochemical carbon cycle in tertiary sedimentary basins. In: Gale JJ, Kaya Y (eds) Greenhouse Gas Control Technologies: Proceedings of the 6th International Conference on Greenhouse Gas Control Technologies, vol 1. Elsevier Science, Amsterdam, pp 703–708

Koide H, Yamazaki T, Kuniyasu M, Aso K, Chinju J (2005) Possibility of CO_2 sequestration into coal seams in Japan. In: Wilson M, Morris T, Gale JJ, Thambimuthu K (eds) Greenhouse Gas Control Technologies: Proceedings of the 7th International Conference on Greenhouse Gas Control Technologies, vol 2., pp 2209–2212

Koide H, Yamamoto K, Todaka N, Nakanishi S, Tosha T (2006) Mechanical interaction of CO_2-rich brine and the multiple echelons Matsushiro earthquake fault—Matsushiro swarm earthquakes as natural analogues for geological storage of CO_2. Chikyu Mon 28:801–808

Kouda R, Murakami Y (2002) Subsurface 3D solid modeling for areas of earthquake active faults, active volcanoes, and a hyper-scale subsurface construction of neutrino detectors. Terra Nostra 4:21–23

Li X, Ohsumi T, Koide H, Akimoto K, Kotsubo H (2005) Near-future perspective of CO_2 aquifer storage in Japan: site selection and capacity. Energy 30(11–12):2360–2369

Marui A, Kusunose K (2008) Stability and buffering capacity of the geosphere for long-term isolation of radioactive waste: Application to crystalline rock. In: Marui A, Ahn JS, He Q, Ito N (eds) Proceedings of the 36th Congress of the International Association of Hydrogeologists. International Association of Hydrogeologists, Reading, CD-ROM

Mito S, Xue Z, Ohsumi T (2008) Case study of geochemical reactions at the Nagaoka CO_2 injection site, Japan. In: Busch A, Kühn M, Etheridge D (eds) EGU General Assembly 2007: Advances in CO2 Storage in Geological Systems—EGU (2007). Int J Greenhouse Gas Control 2(3):309–318

JNC (Japan Nuclear Cycle Development Institute) (2000a) H12: Project to establish the scientific and technical basis for HLW disposal in Japan. Project Overview Report. Report JNC TN1410 2000-001. Japan Nuclear Cycle Development Institute, Tokai-mura

JNC (Japan Nuclear Cycle Development Institute) (2000b) H12: Project to establish the scientific and technical basis for HLW disposal in Japan. Supporting Report 1: Geological Environment in Japan. Report JNC TN1410 2000-002. Japan Nuclear Cycle Development Institute, Tokai-mura

JNC (Japan Nuclear Cycle Development Institute) (2000c) H12: Project to establish the scientific and technical basis for HLW disposal in Japan. Supporting Report 2: Repository Design and Engineering Technology. Report JNC TN1410 2000-003. Japan Nuclear Cycle Development Institute, Tokai-mura

JNC (Japan Nuclear Cycle Development Institute) (2000d) H12: Project to establish the scientific and technical basis for HLW disposal in Japan. Supporting Report 3: Safety Assessment of the Geological Disposal System. Report JNC TN1410 2000-004. Japan Nuclear Cycle Development Institute, Tokai-mura

JNC (Japan Nuclear Cycle Development Institute) (2005a) H17: Development and management of the technical knowledge base for the geological disposal of HLW. Knowledge Management Report. Report JNC TN1400 2005-020. Japan Nuclear Cycle Development Institute, Tokai-mura [in Japanese]

JNC (Japan Nuclear Cycle Development Institute) (2005b) H17: Development and management of the technical knowledge base for the geological disposal of HLW. Supporting Report 1: Geoscience Study. Report JNC TN1400 2005-014. Japan Nuclear Cycle Development Institute, Tokai-mura [in Japanese]

JNC (Japan Nuclear Cycle Development Institute) (2005c) H17: Development and management of the technical knowledge base for the geological disposal of HLW. Supporting Report 2: Repository Engineering Technology. Report JNC TN1400 2005-015. Japan Nuclear Cycle Development Institute, Tokai-mura [in Japanese]

JNC (Japan Nuclear Cycle Development Institute) (2005d) H17: Development and management of the technical knowledge base for the geological disposal of HLW. Supporting Report 3: Safety Assessment Method. Report JNC TN1400 2005-016. Japan Nuclear Cycle Development Institute, Tokai-mura [in Japanese]

PNC (Power Reactor and Nuclear Fuel Development Corporation) (1992) Research and development on geological disposal of high-level radioactive waste. First Progress Report (H-3). Report PNC TN 1410 93059. Power Reactor and Nuclear Fuel Development Corporation, Tokyo

Okamoto I, Mieuochi Y, Ninomiya A, Kato T, Yajima T, Ohsumi T (2006) In situ test on CO_2 fixation by serpentinite rock mass in Japan. In: Gale JJ, Røkke N, Zweigel P, Svenson H (eds) Proceedings of the 8th International Conference on Greenhouse Gas Technologies. Elsevier, Oxford, CD-ROM

Okuyama Y, Sasaki M, Nakanishi S, Todaka N, Ajima S (2009) Geochemical CO_2 trapping in open aquifer storage—the Tokyo Bay model. In: Gale J, Herzog H, Braitsch J(eds) Proceedings of the 9th International Conference on Greenhouse Gas Control Technologies; Energy Procedia 1:3253–3258

SACS (Saline Aquifer CO_2 Storage Project) (2003) Saline aquifer CO_2 storage project (SACS) best practise manual. Report Number PH4/21. IEA Greenhouse Gas R&D Programme, Trondheim

Sasaki T, Morikawa S, Tabei K, Koide H, Tashiro T (2000) Prediction of long term crustal movement for geological disposal of radioactive waste. In: Proceedings of the 4th International Conference on Supercomputing in Nuclear Applications (pp 92–101). SNA2000 Office, Japan Atomic Energy Research Institute, Tokyo

Sminchak J, Gupta N, Byrer C, Bergman P (2001) Issues related to seismic activity induced by the injection of CO_2 in deep saline aquifers. In: Proceedings of the First National Conference on Carbon Sequestration. National Energy Technology Laboratory, US Department of Energy. http://www.netl.doe.gov/publications/proceedings/01/carbon_seq/p37.pdf

Sugimura A, Uyeda S (1973) Island Arcs. Elsevier, Amsterdam

Takahashi T, Ohsumi T, Nakayama K, Koide K, Miida H (2009) Estimation of CO_2 aquifer storage potential in Japan. In: Gale J, Herzog H, Braitsch J (eds) Proceedings of the 9th International Conference on Greenhouse Gas Technologies; Energy Procedia 1:2631–2638

Tanaka S, Koide H, Sasagawa A (1995) Possibility of CO_2 underground sequestration in Japan. Energy Convers Manage 36:527–530

Wakahama H, Mito S, Ohsumi T, Ueda A, Yajima T, Satoh H, Sugiyama T, Ozawa A, Ajima S, Todaka N, Sato T, Kato M, Kaji Y, Tokumarui T, Kaieda H, Kubota K (2008) A concept of CO_2 georeactor sequestration at the Ogachi HDR site, NE Japan. In: Gale J, Herzog H, Braitsch J(eds) Proceedings of the 9th International Conference on Greenhouse Gas Control Technologies; Energy Procedia 1:3683–3689

Xue Z, Mito S, Kitamura K, Matsuoka T (2009) Case study: Trapping mechanisms at the pilot-scale CO_2 injection site, Nagaoka, Japan. In: Gale J, Herzog H, Braitsch J (eds) Proceedings of the 9th International Conference on Greenhouse Gas Control Technologies; Energy Procedia 1:2057–2062

Yajima T, Ohsumi T, Mizuochi Y, Ninomiya A, Takayuki K (2005) Field study on CO_2 fixation by serpentinite rock-bed. In: Wilson M, Morris T, Gale JJ, Thambimuthu K (eds) Greenhouse Gas Control Technologies: Proceedings of the 7th International Conference on Greenhouse Gas Control Technologies, vol 2. Elsevier, Oxford, pp 2027–2030

The Geological Storage of Carbon Dioxide and Disposal of Nuclear Waste in South Africa

Anthony D. Surridge, Marthinus Cloete, and Philip J. Lloyd

Abstract South Africa has a coal-based energy economy, and the use of coal is likely to increase as new coal-fired electricity generation stations and coal-to-liquids plants are built. This situation has been exacerbated by the decision of the electricity utility to delay the construction of further nuclear-powered electricity generation stations. Notwithstanding the introduction of renewable energies and energy efficiency measures, the use of fossil fuels is therefore expected to increase. At the South African Climate Change Summit held during March 2009, it was announced that South Africa will increase its carbon dioxide emissions until 2020–2025, plateau for 10 years, and thereafter decrease emissions in real terms. Carbon dioxide capture and storage is being investigated as a measure to mitigate greenhouse gas emissions: a Centre for Carbon Capture and Storage was established on 30 March 2009. Preliminary studies have already indicated some potential, and the completion of a carbon geological storage atlas by mid-2010 is expected to result in more accurate information. Presently South Africa has one nuclear-powered electricity generation station which provides a mere 2.8% of primary energy supply; further stations have been postponed. Currently, high-level radioactive waste is stored on-site at nuclear installations. Low- and intermediate-level waste is buried in a remote, desert-like location. The Government is undertaking an investigation into the handling and final disposal of nuclear waste to cater for current and future nuclear installations. An institute that will manage radioactive waste on a national basis is the subject of the National

A.D. Surridge (✉)
South African National Energy Research Institute, 158 Jan Smuts Ave,
Sandton 011 280 0300, South Africa
e-mail: tonys@saneri.org.za

M. Cloete
Council for Geoscience, 280 Pretoria St, Pretoria 012 841 1911, South Africa
e-mail: mcloete@geoscience.org.za

P.J. Lloyd
Cape Peninsula University of Technology, Symphony Way,
Cape Town 021 959 6911, South Africa
e-mail: lloydp@cput.ac.za

F.L. Toth (ed.), *Geological Disposal of Carbon Dioxide and Radioactive Waste: A Comparative Assessment*, Advances in Global Change Research 44,
DOI 10.1007/978-90-481-8712-6_19, © International Atomic Energy Agency, 2011

Radioactive Waste Disposal Institute Act, 2008 (No. 53 of 2008) that was gazetted on 9 January 2009. Separate institutions address carbon dioxide storage and radioactive waste, and currently there is little interaction between them. This chapter discusses the status of activities regarding carbon capture and storage and radioactive waste disposal in South Africa.

Keywords Carbon capture and storage • Nuclear waste • Carbon storage geological atlas • South Africa

1 Introduction

A coal-based energy economy and an increasing coal-based energy infrastructure bestows on South Africa a high per capita carbon dioxide (CO_2) emission rate. With few other economically exploitable energy resources, and in common with similarly placed countries, such emissions are likely to continue, in spite of renewable energy programmes and energy efficiency measures. Consequently, South Africa is investigating the use of carbon capture and storage (CCS) for mitigation of greenhouse gas emissions—as a transition measure until renewable and nuclear energies can play a greater part in the country's energy economy. This investigation has been reinforced by an announcement during 2008 by Eskom (South Africa's electricity generation utility) that it has postponed plans for its next nuclear electricity generation build. This is likely to lead to the building of more coal-fired power stations in addition to those already planned.

Only 2.8% of South Africa's primary energy supply (~6% of its electricity) is nuclear-sourced, and that percentage is forecast to decrease as more coal-fired electricity generation stations are constructed. As a consequence of the delay in further nuclear build, there is a need to focus on mitigating the emissions from coal-fired stations.

A preliminary study commissioned by the Department of Minerals and Energy of the Council for Scientific and Industrial Research (CSIR) of South Africa indicated that there were not only capturable sources of CO_2 in South Africa but also storage possibilities (Engelbrecht et al. 2004). A workshop of stakeholders during June 2006, commissioned by the Department of Minerals and Energy and organised by the South African Fossil Fuel Foundation, came to two conclusions:

1. CCS should be investigated in South Africa;
2. Storage should be focused in geological sites.

To develop capacity, both human and technical, in this relatively new field, a Centre for Carbon Capture and Storage commenced operations on 30 March 2009 within the South African National Energy Research Institute. The purpose of the work of the Centre is to lead to a state of 'country readiness' for CCS and to the establishment of a demonstration plant by the year 2020.

2 Greenhouse Gas Emissions and Carbon Capture and Storage in South Africa

During the year 2000, total CO_2 emissions were 426 million tonnes (Mt), of which 59% or 249 Mt were deemed to be capturable. Of the latter, 65% was attributed to electricity generation stations (Engelbrecht et al. 2004). However, the most promising source was seen as coming from the synthetic fuel industry, which emits nearly 30 Mt of CO_2 of approximately 95% concentration per year: to a large extent, the capture component has already been done.

2.1 Geological Storage of CO_2 in South Africa

South Africa is considering the following types of geological storage sites for CO_2:

- Depleted oil and gas wells: South Africa has a dearth of oil and gas reserves. The little it has is primarily gas (offshore with some condensate offtake), which is converted into liquid fuels. Some of these wells have come to the end of their lifetime and others are nearing the end of their production.
- Unmineable coal seams: some studies are being undertaken to investigate the storage of CO_2 in unmineable coal seams. However, such an action could sterilise the coal in terms of later technology advances and exploitation.
- Deep saline aquifers: these offer the best option for CO_2 storage in South Africa, and will be addressed in detail in Sect. 2.2 below.

2.2 Preliminary Calculations for Geological Storage of CO_2 in Deep Saline Aquifers

In South Africa very little (if any) exploration has been conducted specifically for deep-seated saline aquifers. Geophysical exploration for deep-seated mineral deposits and the geological structures that host them have been successfully undertaken (e.g. the Witwatersrand gold deposits and the Bushveld Complex); however, this has not been done for either freshwater or saline aquifers. Although, generally speaking, fresh or potable water resources are explored for at depths of less than 100 m, in a few cases (e.g. the Kalahari Basin and the Peninsula Formation in the Western Cape) depths of up to ~500 m have been explored. However, for the safe storage of CO_2, aquifers with the requisite architecture are necessary at depths of greater than 800 m.

South Africa is well endowed with mineral resources, but poorly endowed with oil- and gas- rich reservoirs. Most of these mineral resources, excluding coal, occur in dense crystalline rocks that typically have low porosity and are therefore unlikely

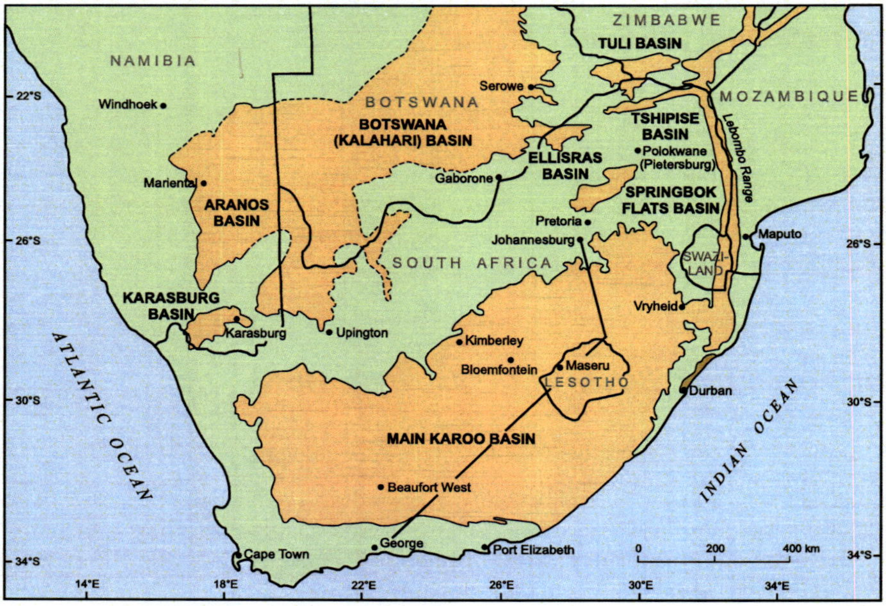

Fig. 1 Site of the Karoo basins in southern Africa (Source: Johnson et al. 2006) (*see* Colour Plates)

candidates for CO_2 storage. Excluding these crystalline (metamorphic and igneous) rock formations, the Karoo Supergroup (Fig. 1) is the only onshore sedimentary basin that is likely to contain viable storage reservoirs and caprock formations.

The majority of the Karoo strata occur in the main basin, which covers an area of approximately 700,000 km^2. Significant deposits of Karoo strata also occur in the smaller Springbok Flats, Ellisras, Tshipese and Tuli satellite basins (Fig. 1). Also noteworthy are the large tracts of Karoo strata that occur in Botswana, much of which is covered by the younger semi-consolidated porous Kalahari Basin sediments. A narrow strip of Karoo rocks also occurs in a linear belt along the eastern margin of South Africa (Johnson et al. 2006).

The Karoo Basin ranges in age from 350 to 180 million years and attains a cumulative thickness of about 2 km in the south-east, whereas to the north it thins out to a thin veneer that eventually disappears only to reveal older crystalline basement rocks. The major stratigraphic units of the Karoo Supergroup crop out concentrically around the main basin. In the southern part of the basin the strata have northerly dips, but elsewhere the rocks are essentially flat-lying with slight centripetal dips.

The pros and cons of three potential CO_2 storage areas are discussed below in the light of information available in the literature. The storage capacities suggested for the respective areas are theoretical estimates (CSLF 2007) which assume that the entire rock volumes mentioned are available to store CO_2 in all their available pore space. These values, which represent maximum upper limits to the capacity estimates, are unrealistic because, in reality, there are always physical, technical, regulatory and economic limitations that prevent full utilisation of such storage capacities. Furthermore, the assessments presented here are based mainly on stratigraphic information available in Johnson et al. (2006). Although the accuracy of the data is good for specific (measured) localities, those of intermediate and subsurface locations are mostly generalised (interpolated from figures and tables in Johnson et al. 2006).

2.2.1 Storage Prospectivity of Area A

Area A (see Fig. 2), previously proposed by Engelbrecht et al. (2004), here referred to as the Vryheid Formation Area, was estimated to have a CO_2 storage capacity of 183 gigatonnes (Gt). This estimate is essentially based on an average sandstone thickness of 350 m, a strike length of 350 km, a down-dip length of 75 km and an effective storage volume of 2%. The burial depth was assumed or calculated to be

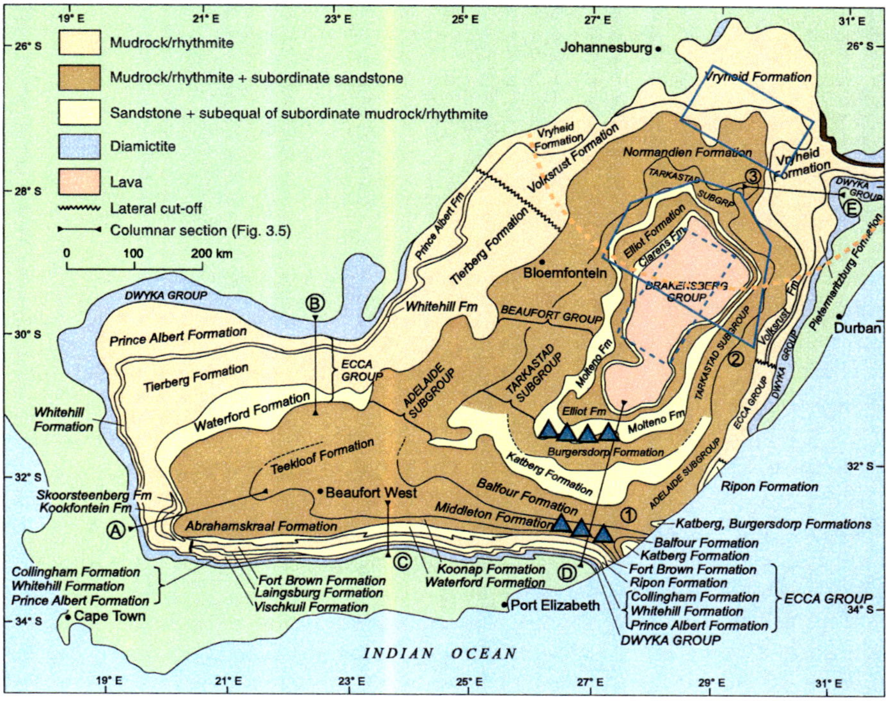

Fig. 2 Schematic map illustrating the distribution of the lithostratigraphic units of the main Karoo Basin (Johnson et al. 2006) and the location of the five (A,B,C,D,E) prospective CO_2 storage areas. The yellow arc depicts an area approximately 300 km from Secunda, in which there are several large CO_2-emitting point sources (*see* Colour Plates)

between 800 and 3,000 m. If this is the case, the estimated storage capacity would probably be a reasonable estimate, taking into account the authors' cautionary remarks.

However, the stratigraphic profile of the Karoo Supergroup in Area A is wedge-shaped, thickening towards the south, and therefore uncertain. The maximum thickness of the Karoo Supergroup in Area A is probably less than 700 m and not the 800–3,000 m suggested above (Engelbrecht et al. 2004). If one were to subtract the suggested average reservoir thickness of 350 m, the depth below surface would be of the order of 300–400 m, which would be insufficient for the safe storage of CO_2.

2.2.2 Storage Prospectivity of Area B

Area B is situated approximately 200 km south-west of Area A, which includes parts of the northern and eastern Free State, the KwaZulu-Natal Highlands and the northern half of Lesotho. The distance of Area B from the major CO_2 point sources would fall roughly around the 300 km mark (see the yellow stippled arc in Fig. 2). Two potential target storage areas exist:

1. The deeper Vryheid Formation sandstones;
2. The shallower Molteno and Clarens Formation sandstones.

Calculations indicate that approximately 80 Gt of storage could be available.

2.2.3 Storage Prospectivity of Area C

If the Vryheid Formation is targeted for storage from areas overlain by at least the Elliot Formation and stratigraphically higher ground, there should be ample caprock to retain injected CO_2. The caprocks would consist of the Volksrust Formation and the entire Beaufort Group, comprising close to 2,000 m of mudrocks and some subordinate sandstones (see Johnson et al. 2006, chapter 22, figure 16).

According to available stratigraphic information, the Vryheid Formation is expected to thin towards the west and south until it finally disappears against a north-west to south-east trending lateral transition zone beneath central Lesotho (Johnson et al. 2006). The lateral transition is a zone in which the Vryheid Formation sandstones cross over into impermeable mudrocks; this is taken to be the south-western storage boundary of Area B.

If requisite CO_2 reservoir storage characteristics exist for the Vryheid Formation beneath the Drakensberg basalts, the target storage zone would be situated at depths of about 2,000–3,000 m. Alternately, if sites inside Lesotho cannot be secured for CO_2 injection, it is estimated that suitable injection could still be achieved from areas within South Africa, but only on land surfaces that are stratigraphically higher than the Elliot Formation. For the latter, storage could still occur nearly 2,000 m below the Elliot Formation.

If we assume an average sandstone thickness of 150 m for the Vryheid Formation within Area B and an aerial extent of 200 km by 140 km (schematically represented by the blue box B in Fig. 2) with an effective storage volume of 2%, the pore space for 1 km^2 would be of the order of 3 million m^3. It follows then that for an area of 28,000 km^2, a total pore space of 84,000 million m^3 could be expected that would accommodate around 80 Gt CO_2.

2.2.4 Storage Prospectivity of Area C

The viability of using the Molteno and Clarens Formations beneath the Lesotho basalts is not certain and will have to be determined by a systematic multidisciplinary evaluation. It is conservatively assumed that the storage prospects for Clarens Formation sandstone storage will be too low in terms of effective porosities, while those of the deeper Molteno Formation with its coarser sandstone beds and of the Elliot Formation caprock will have suitable storage prospects. For the Molteno Formation the presence of a subtle dip and concomitant thickening towards the south, which could be useful for down-dip transfer of the denser, CO_2-laden formational waters, will extend storage capacity.

The total depth below surface of the Molteno Formation may be as much as 2,000 m and over small distances as little as 1,000 m (or less) because of the deeply incised river valleys. The effective pore volumes are estimated to be twice that (4%) of the Vryheid Formation. It is estimated that storage beneath the Elliot Formation caprock may accommodate a plume of 80×80 km^2 and 80×160 km^2 if down-dip flow is modelled. An average thickness of 100 m for the sandstone in the northern part of Area C is estimated, which is expected to increase further south. It is thus deduced that 1 km^2 would have an effective pore volume of 4 million m^3 and that an area of (80×80 km^2) 6,400 km^2 would amount to a storage space of 25,600 million m^3. At the prevailing pressures, such a volume would be able to accommodate storage of ~24 Gt, and if down-dip storage is modelled, ~48 Gt CO_2 could be a possibility. If the down-dip flow of CO_2 is possible, an additional advantage would be that northerly injection sites could be used, limiting the need to extend costly CO_2 pipelines further south.

2.2.5 Storage Prospectivity of Areas D and E

Area D (Fig. 2) is underlain by Molteno Formation sandstones and, in places, sandstones containing (uneconomical) low-grade coal measures following a broad east–west trending belt, also known as the Molteno Coalfield. In a zone that is from ~1,000 to 2,000 m deep, thick sandstone-rich beds of the Katberg Formation occur, and these could have excellent storage prospectivity. Between the Katberg Formation and the outcropping Molteno Formation, ~1,000 m of caprock mudstone of the Burgersdorp Formation are expected. Towards the north, the Katberg Formation sandstone beds are expected to grade into mudrocks of the Tarkastad

Subgroup because of the existence (at depth) of a northerly transition zone (Johnson et al. 2006). The low-grade coalbeds of the Molteno can, nevertheless, present some interesting development opportunities, should CO_2 be available for deep storage.

Area E (Fig. 2), on the other hand, would be much closer to the coastal development nodes and would thus require greatly reduced pipeline transfer costs. The area is also conveniently situated in a region where the Karoo Supergroup attains maximum thicknesses and thus its greatest storage prospectivity. Storage in these parts should target the 600–1,000 m thick Ripon Formation sandstones, while also taking into account that the strata are folded, implying that anticlinal structures would need to be targeted. The stratigraphic columns also indicate the presence of adequate caprock to ensure safe storage. The only problem may be if there is requisite permeability in the sandstones: information that may be found in old oil exploration data or by first-hand drilling.

2.2.6 Storage Prospectivity of the Smaller Karoo Basins

Significant smaller deposits of Karoo strata occur north of the main basin and are known as the Springbok Flats, Ellisras, Tshipise and Tuli basins (Fig. 1). From a CO_2 storage point of view some salient features warrant mentioning. It is evident that none of the smaller basins exceed the 800 m thickness mark and all contain appreciable amounts of coal, especially the Ellisras Basin.

2.3 Summary of CO_2 Storage Capacity

Noting the above discussion and the uncertainties therein, an estimate of the potential for storage of CO_2 in South Africa is summarised in Table 1. Storage capacity estimates for Areas D and E have not been attempted because of geological uncertainties, and are therefore not listed in Table 1.

Omitting the Vryheid Formation, which may be too shallow, the calculated storage capacity for only the deep saline aquifer areas B and C is 104 Gt. If one considers that approximately 10% of emissions (40 Mt/year) could be captured and stored, then a 100-year operation would require a storage capacity of 4 Gt. The capacity

Table 1 Estimated CO_2 storage potential in South Africa for three deep saline aquifer formations

Area	Name	Estimated storage potential Gt CO_2
A	Vryheid Formation	0 (183)
B	Free State, KwaZulu Natal and Lesotho	80
C	Molteno and Clarens Formations	24
Total		~104 (~287) Gt

Note: Numbers in parentheses indicate capacity that may be too shallow for CO_2 storage calculated above exceeds this requirement 25-fold and provides incentive for further investigation.

2.4 CO_2 Geological Storage Atlas

A study is under way to produce a detailed CO_2 geological storage atlas for South Africa. Using currently available data, it will locate and characterise potential storage sites. The study for the atlas will address not only the deep saline aquifer potential discussed above, but also all possible onshore and offshore CO_2 geological storage possibilities.

The atlas is scheduled for completion and publication during mid-2010 and will contain a more accurate and broader calculation of total storage capacity than that addressed above. It is anticipated that, based on the information contained in the atlas, appropriate seismic and drilling operations could be undertaken to identify a specific storage site. Thus, it should be possible to schedule an experimental injection of CO_2 to test the reaction of rock in South Africa to the injected gas.

3 Storage/Disposal of Nuclear Waste in South Africa

3.1 Current Sources of Waste

South Africa's current main sources of nuclear waste are:

1. One nuclear-powered electricity generation station located at Koeberg near Cape Town;
2. One research reactor, Safari, located at Pelindaba near Pretoria;
3. Medical and mining sources;
4. Other small sources such as the iThemba accelerators used for basic and applied research, particle radiotherapy for the treatment of cancer and the supply of isotopes for nuclear medicine and research. The iThemba laboratories are based at two sites: Western Cape and Gauteng.

The nuclear-powered electricity generation station comprises two 900 MWe units supplied by the French company Framatome to a Westinghouse design. It is situated at Koeberg on the Atlantic coast some 27 km almost due north of Cape Town at 33.67°S, 18.43°E. The design is that of the standard pressurised water reactor, and the units came into operation during 1984 and 1985, respectively. The high-level waste comprises spent fuel elements, fully assembled. These are presently stored in 12 m deep pools on-site. There are 157 fuel elements in each reactor, one-third of which are replaced every 17 months on average. The decay heat is such that it will take between 30 and 50 years for the surface of the elements to cool below 100°C. The storage racks were originally designed for a total of 275 elements, but in 2002 a new design gave a revised capacity of 728 elements per pool (Dell 2003). At present (2009) there are about 1,100 tonnes of fuel elements cooling in the ponds at Koeberg.

Low-level waste, mainly contaminated clothing and cleaning materials, is generated during routine operations, and by 2009 amounted to about 1,200 tonnes, including the mass of the steel drums into which it is compacted. These are transported about 600 km to the storage site at Vaalputs (see Sect. 3.4.1). There is a small quantity of intermediate-level waste, mainly failed equipment that has been irradiated. By 2009 this amounted to about 1,400 tonnes including the concrete into which it was cast. It too is transported to Vaalputs.

The Safari-1 research reactor is located at Pelindaba at 25.80°S, 27.92°E. It is a light water cooled tank-type reactor of 20 MWth, constructed to a design by the US firm Allis Chalmers. It began critical operation in 1965. It was originally fuelled with 4.5 kg of 90% highly enriched uranium. Until 1976 the USA supplied the fuel; thereafter the fuel elements were produced in South Africa. During July 2005 it was announced that in future it would be fuelled by low-enriched (19.5% U^{235}) uranium rather than high-enriched uranium. The conversion took place over the period 2006–2008. Safari-1 is key to developing isotopes for nuclear medicines. These isotopes are used in South Africa and exported to more than 50 nations worldwide.

3.2 Nuclear Developments in South Africa

3.2.1 Uranium Mining

Uranium production in South Africa has generally been a by-product of gold or copper mining. During 1951 a company was formed to exploit the uranium-rich slurries from gold mining. In 1967, this function was taken over by Nuclear Fuels Corporation of South Africa (Nufcor), which, in 1998, became a subsidiary of AngloGold Ltd. It produces over 1,000 tonnes of triuranium octoxide (U_3O_8) compound per year from uranium slurries trucked in from various gold mines and from the Palabora copper mine. Most of South Africa's production is sourced from the Moab Khotsong gold mine operated by AngloGold. The Witwatersrand conglomerates and tailings contain up to 80% of South Africa's resources, estimated at 284,000 tonnes of uranium.

The tailings from gold mining operations have long been known to contain uranium, often at economically attractive grades, but the material is not classified as an ore (Lloyd 1980). The tailings contain pyrite and other sulphides, which oxidise on exposure to air to generate sulphuric acid; the sulphuric acid then solubilises the uranium (and other minerals) which are released into the environment. However, air is only able to penetrate a few metres into the surface of the dumps, which are made of very finely ground material. Thus it is only the uranium in those few metres that is released. This finds its way into watercourses and is immobilised during passage through wetlands.

3.2.2 Uranium Refining

The uranium produced in the mines is shipped as 'yellow cake', a mud of ammonium diuranate, or ADU, to the Nufcor plant situated to the south of Randfontein. It is extruded into pellets and calcined to produce a reasonably pure U_3O_8 which is drummed for shipment. While 'reasonably pure', it still contains a number of impurities, primarily strong neutron adsorbers that must be removed before it can be used for fuel preparation.

South Africa has declared its intention to investigate enrichment, and possibly also reprocessing, as part of the total fuel cycle, as discussed in the policy section below. It clearly has the technological capability to address the waste disposal challenges of fuel production; its ability to handle high-level waste arising from reprocessing has, however, not been tested.

3.2.3 Nuclear Reactor Developments

The South African electricity utility Eskom had previously announced its intention of installing 20 GW of nuclear generating capacity by 2025, although the South African government, the only shareholder, has agreed to 6 GW by 2020, and even that is in doubt. During 2008 two proposals to construct 2 GW of new nuclear capacity were considered, and in November of the same year it was announced that, as neither of the bids had proved commercially attractive in the prevailing economic climate, the plans would accordingly be put on hold. Modified plans see 3.2 GW of new nuclear power being operational by 2019.

South Africa is also developing its own high-temperature gas-cooled reactor, the Pebble Bed Modular Reactor (PBMR), a 400 MWth, 165 MWe scale-up of the reactor that was operated for 19 years at Jülich. The reactor is a carbon-moderated, helium-cooled design with the fuel contained in micropellets encased in layers of silicon carbide and graphite, embedded in graphite spheres about 6 cm in diameter. It is intended to build a demonstration plant alongside the Koeberg nuclear power station and a fuel fabrication facility at Pelindaba.

During February 2009 it was announced that the PBMR would be developed to service both the electricity and process heat markets (Parliamentary Portfolio Committee 2009). In the light of these changes, it is not certain when the PBMR will start contributing to the nuclear waste stream in South Africa.

3.2.4 Decommissioning of Facilities

Waste arises in the decommissioning of facilities. The old production and enrichment facility at Pelindaba is being decommissioned and the resultant low- and intermediate-level waste is contributing to the waste stream. Decommissioning work is ongoing and is funded by the Government through annual budget appropriations as well as special allocations from the Department of Minerals and Energy.

An unresolved issue is that of the evaporation pans. These, with a total evaporation area of about 75,000 m^2, were constructed during the late 1970s and early 1980s, for evaporation of effluent generated at Pelindaba. The effluent contained salts in excess of the allowable limit for disposal to the Crocodile River, so it was necessary to store them in the evaporation ponds. The long-term fate of these residues is not yet resolved.

3.3 Nuclear Energy Regulatory Environment in South Africa

There are two main acts of parliament regulating nuclear activities, including nuclear waste disposal in South Africa, namely:

1. The Nuclear Energy Act, 1999 (No. 46 of 1999) provides for the establishment of the South African Nuclear Energy Corporation Limited, a public company wholly owned by the State (the 'Corporation'). It defines the Corporation's functions and powers and its financial and operational accountability; provides for the Corporation's governance and management by a board of directors and a chief executive officer; apportions responsibilities for the implementation and application of the Safeguards Agreement and any additional protocols entered into by the Republic and the International Atomic Energy Agency (IAEA) in support of the Nuclear Non-Proliferation Treaty, which the Republic has ratified; regulates the acquisition and possession of nuclear fuel, certain nuclear-related material and equipment, and their import and export in compliance with the international obligations of the Republic; and prescribes measures regarding the discarding of radioactive waste and the storage of irradiated nuclear fuel;
2. The National Nuclear Regulator Act, 1999 (No. 47 of 1999) provides for the establishment of a National Nuclear Regulator of nuclear activities, its objects and functions, management and staff-related matters; provides for safety standards and regulatory practices for protection of persons, property and the environment against nuclear damage.

In addition, the Hazardous Substances Act, 1973 (No. 15 of 1973) provides for the control of Group IV hazardous substances (radioactive material not at nuclear installations or not part of the nuclear fuel cycle, for example fabricated radioactive sources, medical isotopes) and Group III hazardous substances (involving exposure to ionising radiation emitted from equipment). Radioactive waste arising from activities authorised under this Act falls under the regulation of the Department of Health's Directorate of Radiation Control. In practice, the Department of Health does not regulate naturally occurring radioactive material. The Directorate of Radiation Control also acts as the national competent authority in connection with the IAEA Regulations for the Safe Transport of Radioactive Material.

During 2005 the Department of Minerals and Energy published a policy document (DME 2005) on radioactive waste management, which built on the IAEA principles for safe management of radioactive waste. Specifically, it foresaw the implementation of the following:

1. *Polluter pays principle*: The financial burden for the management of radioactive waste shall be borne by the generator of that waste.
2. *Transparency regarding all aspects of radioactive waste management*: All radioactive waste management activities shall be conducted in an open and transparent manner and the public shall have access to information regarding waste management where this does not infringe on the security of radioactive material.
3. *Sound decision-making based on scientific information, risk analysis and optimisation of resources*: Decision-making shall be based on proven scientific information and the recommendation of competent national and international institutions dealing with radioactive waste management.
4. *Precautionary principle*: Where there are threats of serious irreversible damage, lack of full scientific certainty shall not be used as a reason for postponing cost-effective measures to prevent environmental degradation (Rio Principle 15).
5. *No Import or Export of Radioactive waste:* In principle South Africa will neither import nor export radioactive waste.
6. *Co-operative governance and efficient national co-ordination*: Due to their cross-cutting nature all activities involving radioactive waste management shall be managed in a manner that prevents duplication of effort and maximises coordination.
7. *International cooperation*: The government recognises that it shares a responsibility with other countries for global and regional radioactive waste management issues. Its actions shall follow the principles in this policy and in relevant regional and international agreements.
8. *Public Participation*: Radioactive waste management shall take into account the interests and concerns of all interested and affected, when decisions are being made.
9. *Capacity building and education*: The government shall create opportunities to develop people's understanding, skills and general capacity concerning radioactive waste management' (DME 2005, p. 9).

This has led to the development of legislation to establish an institute that will manage radioactive waste on a national basis as provided for by the National Radioactive Waste Disposal Institute Act, 2008 (No. 53 of 2008).

During June 2008 the South African Department of Minerals and Energy released a Nuclear Energy Policy for the Republic of South Africa (DME 2008). The essence of that policy, which focuses on radioactive waste-related matters, is addressed in this subsection.

The vision for nuclear waste disposal in South Africa is to be guided by a number of principles. It is fully accepted that all activities undertaken will be in a manner that takes into account the environmental impacts and strives to mitigate those impacts as far as possible. Furthermore, DME (2008) states: 'All nuclear energy sector activities shall take place within a *legal regulatory framework consistent with international best practice.*' In that connection, 'Nuclear energy shall be used only for *peaceful purposes* and in conformity with national and international legal obligations and commitments.' The national nuclear energy programme must commit fully to ensuring 'that *nuclear and radiation safety* receives the highest priority to provide for the protection of persons, property and the environment.' DME (2008) continues: 'South Africa shall endeavour to *use* [its] *uranium resources in a sustainable manner* through the recognition of the three interdependent and mutually reinforcing pillars of sustainable development namely economic development, social development and environmental protection.' It goes on to stipulate that technologies chosen for nuclear power plants 'shall… allow for optimum utilisation of uranium resources including the use of recycled uranium.'

The existing nuclear energy governance framework comprises:

(a) White Paper on Energy Policy (1998). The Government undertook that the complete nuclear fuel cycle, in particular the issues of spent nuclear fuel, nuclear fuel procurement and radioactive waste management would be investigated by the Department of Minerals and Energy.
(b) Nuclear Energy Act, 1999 (Act No. 46 of 1999), as detailed in Sect. 3.3, point 1.
(c) National Nuclear Regulator Act, 1999 (Act No. 47 of 1999), as detailed in Sect. 3.3, point 2.
(d) Radioactive Waste Management Policy and Strategy (2005), also described above.

Other legislation having a bearing on nuclear waste disposal includes:

(a) Hazardous Substances Act, 1973 (Act No. 15 of 1973), which specifically excludes nuclear waste.
(b) Non-Proliferation of Weapons of Mass Destruction Act, 1993 (Act No. 87 of 1993), which provides for control of weapons of mass destruction and the establishment of a Council to control and manage matters relating to the proliferation of such weapons in the Republic. Weapons utilising nuclear effects are specifically banned.
(c) National Environmental Management Act (Act No. 107 of 1998), which defines pollution to include radioactive waste, so that its provisions apply to nuclear waste materials.

With regard to spent nuclear fuel and radioactive waste management, the Nuclear Energy Policy states:

(a) 'Radioactive Waste including used nuclear fuel shall be managed in terms of the radioactive waste management policy and strategy', as approved by the Government during 2005, and discussed above.
(b) '[The] Government, through NECSA [the Nuclear Energy Corporation of South Africa] shall investigate the viability of building an indigenous reprocessing

facility' for 'Used (Irradiated) Fuel and Recycling of Fissile Materials'. 'In the short term South Africa shall make use of existing commercial reprocessing facilities in other countries.'

3.4 Current Storage/Disposal Activities

The Radioactive Waste Management Policy and Strategy for the Republic of South Africa (DME 2005) categorises radioactive waste into various classes. These can be broadly divided into:

(a) *Low- and intermediate-level waste*: Two low-level radioactive waste management options are currently used in South Africa:

   i. Above-ground disposal in engineered facilities for the bulk of the mining waste;
   ii. Near-surface disposal for low- and intermediate-level waste at the Vaalputs site in the Northern Cape.

(b) *Spent fuel and high-level waste*: There are two mechanisms (dry and wet storage) for the management of used nuclear fuel and high-level waste:

   i. The used fuel from the Koeberg electricity generation station is stored in authorised used-fuel pools on site as well as in casks designed and constructed for the storage of used fuel;
   ii. The used fuel from the Safari research reactor is stored at an authorised dry storage facility on site as well as in the reactor pool.

Whereas low- and intermediate-level nuclear waste may be disposed of, all high-level waste is merely stored at this stage. The long-term disposal of high-level waste is currently under review. A National Radioactive Waste Management Institute is to be established to address this matter.

Currently, low- and intermediate-level radioactive waste is disposed of in the Vaalputs facility. The site is located approximately 600 km north of Cape Town at 29.29°S, 21.37°E. It covers an area of about 10,000 ha and measures 16.5 km east–west and 6.5 km north–south. The first waste deliveries took place during 1986. All the activities at Vaalputs are conducted in accordance with accepted standards and practices as prescribed by the IAEA and according to the licence conditions imposed by the South African National Nuclear Regulator. Based on the life expectancy of the Koeberg electricity generation station, the operational life of the Vaalputs site is expected to last until 2035.

Spent fuel from Safari-1 and other radioactive waste is stored at Thabana Hill (Radiation Hill) in Pelindaba. The store includes eight trenches, each containing approximately 17 tonnes of uranium, a steel tubular storage facility, a medium-active waste storage chamber for the disposal of activation products with up to 30 years half-life, a fluoride storage facility for the disposal of radioactive sludge, and storage for hazardous chemicals.

3.5 Future Plans

The disposal of low- and intermediate-level waste from current plants is assured at least for the next 2 decades. However, high-level waste disposal still needs to be tackled. A recent document released for public comment by the Department of Minerals and Energy addressed a number of options that are being investigated for long-term disposal of nuclear waste. These options include:

1. *Above-ground storage at a licensed off-site facility*: The benefit of this option is that such waste would be easily accessible should more appropriate waste disposal technologies be developed in the future. The downside is that storing nuclear waste above ground indefinitely may produce an undue burden for future generations.
2. *Reprocessing, conditioning and recycling*: An investigation commissioned by the Department of Minerals and Energy has concluded that this option should remain open. Such processing could take place in South Africa (where there are currently no facilities) or in another country.
3. *Deep geological storage*: This technology is internationally the most pursued option and as such is being carefully considered by South Africa. If it were chosen as a nuclear waste disposal mechanism, then preferably geological disposal should take place in such a manner that the waste could be retrieved, so as not to exclude the possibility of future improved technologies being applied. A particular South African expertise that could facilitate the introduction of this technology is experience in the construction of large-scale chambers at great depths, acquired through gold and uranium mining at depths up to 4,000 m below surface. A further advantage is the tectonically stable nature of the subcontinent. The deep level mines pass through successive facies, many of which are conformable, to reach strata more than 2 billion years old at depth. This means that much of the inherent topology of the subcontinent has been stable for at least that period.
4. *Transmutation*: Although this option would not be researched in South Africa, international developments would be monitored.

3.6 National Radioactive Waste Management Institute

It may appear at first sight that the disposal of high-level radioactive waste is on hold in South Africa. However, as well as ongoing technical investigations, the country's institutional capacity is being strengthened to address the matter.

The National Radioactive Waste Disposal Institute Act, 2008 (No. 53 of 2008) provides for the establishment of a National Radioactive Waste Management Institute. This independent body would be established to manage the disposal of all radioactive waste. The duties of the Institute would be to, inter alia:

(a) Design and implement disposal solutions for all categories of radioactive waste;
(b) Manage, operate and monitor radioactive waste disposal facilities;
(c) Design and construct new facilities, as required.

The implication is that the entire Vaalputs disposal facility, including all staff, would be incorporated into the Institute. Moreover, the establishment of the Institute would provide the impetus for developing a plan for the disposal of high-level radioactive waste.

4 Regional Aspects

South Africa is the 'power house' of Africa – approximately 40% of Africa's current electricity generation capacity is in South Africa. Moreover, South Africa operates the only nuclear electricity generation station in Africa. It is therefore not surprising that CO_2 storage and radioactive waste disposal are not high on the agenda of other southern African countries. However, it would be remiss to state that nothing is being done. For example, a successful conference was held in Botswana on 28 June 2007 that addressed, inter alia, CCS.

A new electricity generation plant in Africa is more likely to be hydro (for example, the Inga options) or fossil fuel-based generation (for example, the proposed Mmamabula project in Botswana and other fossil fuel projects) rather than nuclear. With regard to carbon storage, while there are investigations into coal-fired electricity generation plants in southern Africa and the occasional conference addressing the topic, there is currently no concerted effort to investigate and demonstrate the potential of carbon capture and geological storage.

On the other hand, geological formations in countries neighbouring South Africa could be suitable for CO_2 storage. Although CCS is not high-profile, it is expected that interest on the part of neighbouring countries will be raised when South Africa's CO_2 geological storage atlas is published and with the operationalisation of the South African Centre for Carbon Capture and Storage.

5 Carbon and Nuclear Comparison

As indicated above, there is no significant work being undertaken in the southern African region regarding geological storage of CO_2 and definitely none in nuclear disposal, other than the work in South Africa as outlined in this chapter. Moreover, the teams working in each area tend to do so in isolation from each other. The initial perception is that, apart from the requirement for a stable geology and the long period of times over which the two degrade, there is little similarity between nuclear disposal and carbon storage. CO_2 is a fluid; radioactive waste is generally solid. The latter is preferably buried shallow, the former at depth.

Table 2 Comparison of CO_2 storage and radioactive waste disposal

	Carbon capture and storage	Radioactive waste disposal
Stable geology	Both require geological stability, but CCS is susceptible to the development and propagation of rock cracks during its fluid phase so that it is essential that the CO_2 be retained by an impermeable caprock	Solid radioactive waste is less susceptible to minor cracks
Time frames	Long time frames are required for both	
Regulatory systems	A regulatory system for CCS is not in place, but it is one of the issues for CO_2 storage that could find commonality with radioactive disposal	A recent Act makes provision for the establishment of a National Radioactive Disposal Institute. Regulations (discussions, debate) are advanced.
Transformation	CO_2 may calcify in deep saline aquifers	Radioactive material decays to daughter elements over time
Volume	~40 Mt/year	~55 t/year high-level waste at present (2009)
Depth	>800 m	South Africa's preference is for relatively shallow disposal to facilitate future recycling
Geological characteristics	Porous rock contained by a caprock	Excavated chamber that is usually back-filled
Coal seams	CO_2 adheres to coal, but this sterilises the coal for future use	Coal seams are unsuitable for storage of radioactive waste
Depleted oil and gas mines	Most suitable sites for CO_2 storage	Depleted oil and gas mines are unsuitable for storage of radioactive waste
Exhausted gold mines	Exhausted gold and similar mines are unsuitable for CO_2 storage	Exhausted gold and similar mines are suitable for radioactive waste storage
Pressure	CO_2 is stored under pressure; hence the need for depths >800 m	radioactive waste requires no pressurisation
Injection	CO_2 can be stored at depth via borehole injection	Nuclear material requires handling methods associated with bulk solid materials

Radioactive waste
CCS carbon capture and storage

Table 2 briefly addresses the comparisons and synergies between CO_2 storage and radioactive waste, as viewed in South Africa.

The establishment of institutional capacity is a factor common to both CO_2 storage and radioactive waste disposal in South Africa. The formation of the National Radioactive Waste Disposal Institute is imminent following the gazetting of the National Radioactive Waste Disposal Institute Act, 2008 (No. 53 of 2008). The South

African Centre for Carbon Capture and Storage was established on 30 March 2009. The Centre is essentially a research and development body and has no mandate to issue regulations. However, researching regulatory requirements is seen as one of its functions. Radioactive waste regulations are well advanced, but a regulatory environment for CCS has yet to be initiated. One expects there to be some synergy between the two that may provide for fruitful cooperation between the Institute and Centre.

6 Conclusion

South Africa has an interest in the long-term storage of CO_2 and final disposal of nuclear waste. The former is scheduled to increase faster than the latter as the fossil fuel-driven electricity and synthetic fuel industries expand to cater for increasing demand.

With regard to radioactive waste, high-level waste is currently stored 'on site' awaiting decisions as to its fate. Currently, there is no immediate urgency in this matter as there is sufficient temporary storage at Koeberg to last its planned life. On the other hand, low- and intermediate-level nuclear waste is disposed of in the Vaalputs site, which has a lifetime of at least the life of the Koeberg plant. A National Radioactive Waste Disposal Institute is to be established to address the matter of radioactive waste in South Africa.

CO_2 storage, on the other hand, is less advanced. Currently, there is no geological storage of CO_2 in South Africa. However, the South African Centre for Carbon Capture and Storage was established on 30 March 2009. The vision for the Centre is that a demonstration plant will be operational by 2020. The mission of the Centre is to facilitate a state of 'country readiness' for CCS. Already a preliminary theoretical study has shown that there may be potential for CCS in South Africa. A detailed atlas to identify and characterise potential sites is scheduled to be published in mid-2010.

The disposal of nuclear waste and the storage of CO_2 are currently the subjects of determined investigation, reinforced by preliminary investigations and stable geological structures that are necessary for both challenges. Moreover, there are synergies that may be exploited by the Centre for Carbon Capture and Storage and the National Radioactive Waste Disposal Institute when the latter is operationalised, especially as far as regulatory matters are concerned.

References

Dell T (2003) Project management solutions using HR and QM principles. Management Today, November. Available at http://www.saqi.co.za/Publications_BackIssues_QMS Standards.asp

DME (Department of Minerals and Energy) (2005) Radioactive waste management policy and strategy for the Republic of South Africa 2005. http://www.dme.gov.za/pdfs/energy/nuclear/radwaste_policy_2005.pdf

DME (Department of Minerals and Energy) (2008) Nuclear Energy Policy for the Republic of South Africa. http://www.dme.gov.za/energy/pdf/FINAL%20NUCLEAR%20ENERGY%20POLICY%20JUNE%202008.pdf

Engelbrecht A, Golding A, Hietkamp S, Scholes B (2004) The potential for sequestration of carbon dioxide in South Africa. Contract Report 86DD/HT339. Department of Minerals and Energy. http://www.dme.gov.za/pdfs/energy/coal/carbon_dioxide.pdf

Johnson MR, van Vuuren CJ, Visser JNJ, Cole DI, de Ville Wickens H, Christie ADM, Roberts DL, Brandl G (2006) Sedimentary rocks of the Karoo Supergroup. In: Johnson MR, Anhaeusser C, Thomas RJ (eds) The Geology of South Africa. Geological Society of South Africa/Council for Geoscience, Johannesburg/Pretoria, pp 461–499

Lloyd PJD (1980) Ninety years experience in the preservation of uranium ore dumps. In: Brawner CO (ed) Proceedings of the First International Conference on Uranium Mine Waste Disposal, May 19–21, 1980, Vancouver, British Columbia, Canada, pp 33–42. Society of Mining Engineers of the American Institute of Mining, Metallurgical, and Petroleum Engineers, New York, NY

Parliamentary Portfolio Committee on Public Enterprises (2009) Briefing on the pebble bed modular reactor. Parliament of the Republic of South Africa, Cape Town, 18 February

CSLF (Carbon Sequestration Leadership Forum) (2007) Phase II final report from the task force for review and identification of standards for CO_2 storage capacity estimation: estimation of CO_2 storage capacity in geological media. http://cslforum.org/publications/documents/PhaseIIReportStorageCapacityMeasurementTaskForce.pdf

Assessment of the Geological Disposal of Carbon Dioxide and Radioactive Waste in Brazil, and Some Comparative Aspects of Their Disposal in Argentina

Roberto Heemann, João Marcelo Medina Ketzer, Goro Hiromoto, and Alexandro Scislewski

Abstract Brazil and Argentina have a huge geological capacity for carbon dioxide (CO_2) and radioactive waste (RW) disposal. Projects for carbon capture and disposal in coal seams and depleted oilfields have important economic benefits, significantly enhancing gas and oil productivity through enhanced coalbed methane and enhanced oil recovery, respectively. In Brazil and South America as a whole, saline aquifers have the greatest storage capacity and thus the greatest potential for CO_2 disposal. Despite the costs of CO_2 capture, transport, injection and monitoring in saline aquifers (at present, without direct financial returns), these projects protect the atmosphere by reducing the amount of greenhouse gas emissions. Disposal of RW in deep geological repositories raises the environmental protection issue of preventing RW from nuclear power plants from causing underground (deep aquifers) and surface contamination. In both CO_2 and RW disposal, the long-term (millennial) safety of underground isolation in deep geological repositories must be assured. Thus, in selecting geological sites for permanent CO_2 and RW disposal, the following should be considered: (1) the occurrence of caprocks to prevent leakage; (2) the structural and geological context (stable regions without earthquake hazards); (3) the disposal capacity; and (4) the cost-efficiency of projects. The definition and characterization of disposal sites is a key question for the energy supply and the geopolitical and environmental security of all Latin American developing countries. The need for a clean

R. Heemann (✉) and J.M.M. Ketzer
Center of Excellence in Research and Innovation in Petroleum, Mineral Resources and Carbon Storage (CEPAC), Pontifical Catholic University of Rio Grande do Sul, Porto Alegre, Brazil
e-mail: roberto.heemann@pucrs.br; marcelo.ketzer@pucrs.br

G. Hiromoto
Nuclear Wastes Laboratory, Brazilian Energy and Nuclear Research Institute (IPEN), São Paulo, Brazil
e-mail: hiromoto@ipen.br

A. Scislewski
Caetité District - DICAE, Brazilian National Nuclear Energy Commission (CNEN), Rio de Janeiro, Brazil
e-mail: ascislewski@cnen.gov.br

F.L. Toth (ed.), *Geological Disposal of Carbon Dioxide and Radioactive Waste: A Comparative Assessment*, Advances in Global Change Research 44,
DOI 10.1007/978-90-481-8712-6_20,

diversified energy matrix in a regional context of economic growth, global warming and climate change must be agreed upon by governments and by the public.

Keywords CO_2 geological disposal • Radioactive waste • Deep geological repository • Potential sites • Environmental impacts • Energy supply

1 Introduction

In view of the projected 50% increase in global energy consumption by 2030 (70% in developing countries) and the predicted decline in the supply of oil and natural gas, nuclear power and the application of carbon capture and storage (CCS) to coal-fired power plants in the near future will make an important contribution to the energy matrix in South American developing countries. Against a background of global warming, and with nuclear power plants (NPPs) being among the lowest carbon dioxide (CO_2) emitters of all energy sources, South America not only has enough uranium resources to fuel NPPs, but also a huge potential for CO_2 and radioactive waste (RW) disposal.

The most important South America uranium reserves (around 309,000 t) are located in the Brazilian states of Minas Gerais, Bahia, and Ceará (DNPM 2008). As uranium costs represent only about 5–15% of the total cost of nuclear-generated electricity, global uranium price fluctuations are not of great significance. In Brazil uranium enrichment plants are operated by the state-owned Nuclear Industries of Brazil (INB) located in Resende, state of Rio de Janeiro, where fuel is produced for the two NPPs located in Angra dos Reis. Brazil and Argentina are the only two countries in South America with nuclear reactors. Both countries have plans to build deep geological repositories for the disposal of long-lived RW.

2 CO_2 Emissions and Disposal in Brazil and Argentina

2.1 CO_2 Emissions in Brazil and Argentina

Brazil has a very unusual emissions portfolio, as most of its electricity is produced by hydroelectric power plants. Total CO_2 emissions in Brazil with potential for CCS (i.e. stationary sources) are around 300 million tonnes of CO_2 per year (Mt CO_2/year) (Fig. 1), most of which are produced from biomass combustion (33%), followed by electricity (25%), cement factories (17%), the iron and steel industry (10%) and petroleum refineries (9%). There are minor emissions from plants producing ethylene (4%), ethanol (2%) and ammonia (<1%), according to data from the Ministry of Science and Technology (www.mct.gov.br) and a report published by the OECD International Energy Agency (IEA 2002). Most emissions (around 80%) are concentrated in the south and south-eastern part of Brazil; there are also

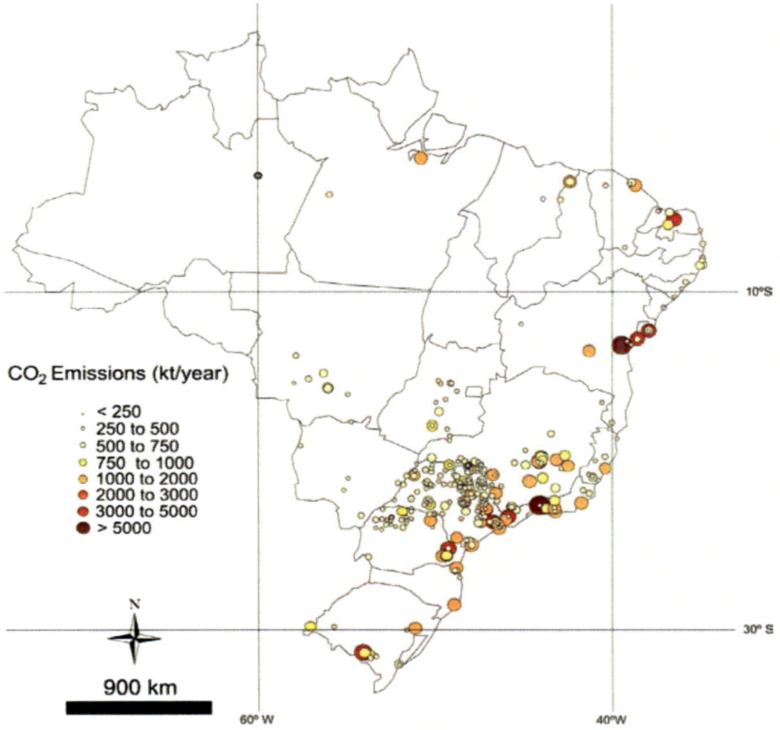

Fig. 1 Large stationary sources of CO_2 emissions in the south-east region of Brazil (kt/year) (Modified from Ketzer et al. 2007) (*see* Colour Plates)

localized emissions from point sources in the north-east of the country (see Fig. 1). There are coal-fired power plants only in the southern part of Brazil, where all the country's domestic coal reserves are located (see Fig. 2). The most important Brazilian coalfields are in the states of Rio Grande do Sul and Santa Catarina (in Brazil's far south). The lower quality coals are high-ash (up to 50%), sub-bituminous and bituminous. The deep coal reserves (>300 m) amount to 11,691 Mt (not including the huge potential offshore resources). New coal gasification projects (syngas generation) are combined with CCS and hydrogen production.

The Brazilian energy matrix is based mainly on hydropower (84% of total electricity consumption and 14% of total energy); nuclear energy provides about 2.3% of the country's electricity (see Fig. 3 for more details). Hydroelectric power systems are directly affected by climatic changes. Brazil and other Latin American countries are improving the proportion of CO_2 free energy sources in the energy matrix. Nuclear power generation and the use of fossil fuels, such as clean coal technologies with CCS, thus appear to be a way of diminishing dependence on hydropower. (The Itaipu Hydro Power Plant alone is responsible for 20% of total power generation.) In this context CO_2 storage and NPPs are of great significance in minimizing global warming and climate change and in assuring energy security. The energy matrix of

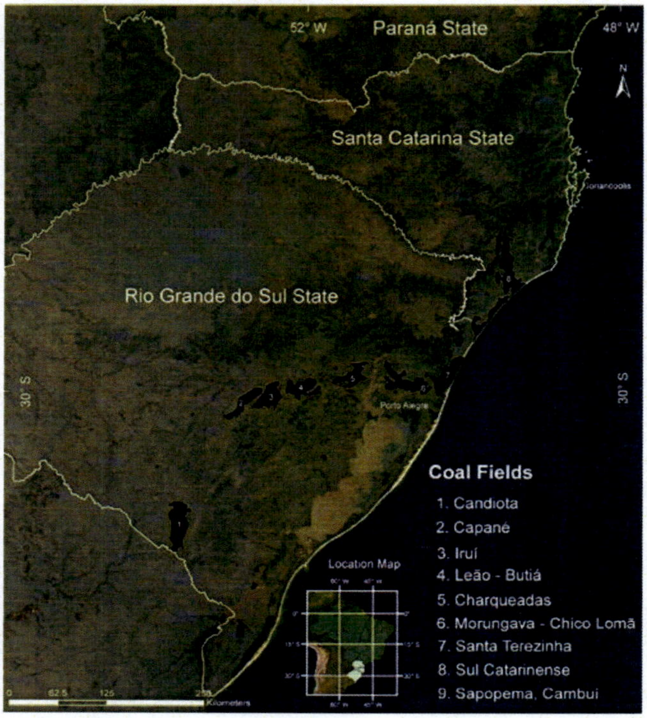

Fig. 2 Location of major coal deposits in Brazilian area of Paraná Basin (RS, SC and PR States). Modified from Süffert (1997) and Aboarrage and Lopes (1986) (*see* Colour Plates)

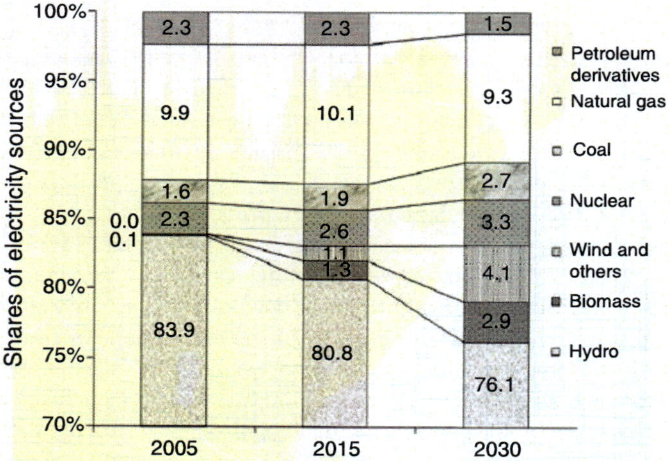

Fig. 3 Electricity matrix of Brazil. (Source: Brazilian Electricity Regulatory Agency (ANEEL 2010) and Brazilian Energy and Mines Ministry (MME 2010) http://www.aneel.gov.br/aplicacoes/capacidadebrasil (accessed in December 2010))

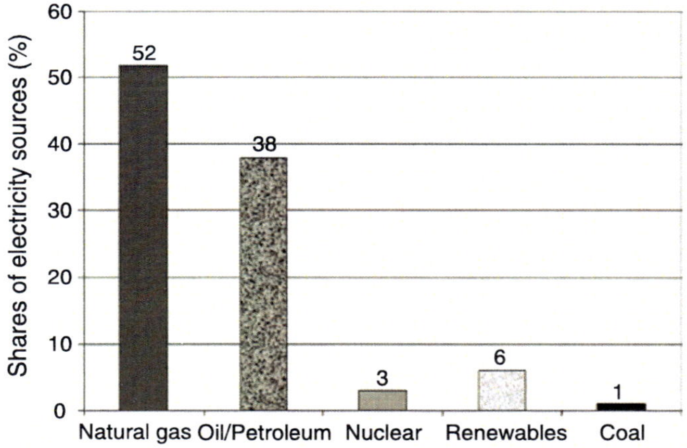

Fig. 4 Electricity matrix of Argentina (Source: World Nuclear Association and Secretaría de Energía-Republíca da Argentina (2010))

Argentina, on the other hand, is highly dependent on natural gas and petroleum (86%), while nuclear represents only 3% of overall power generation (see Fig. 4) in the country.

Primary world energy demand will grow by up to 50% in the next 25 years, and there will be high dependence on fossil fuels like coal, gas and petroleum. The energy policies of Brazil and Argentina thus need to consider CO_2 injection projects to minimize their contribution to global warming during the development process. Energy-related CO_2 emissions in Brazil amount to over 336.7 Mt and in Argentina to about 146 Mt (IEA 2007).

2.2 *CO_2 Disposal Potential in Brazil and Argentina*

Brazil has a very large potential CO_2 storage capacity, as nearly half its territory (around 4.8 million km^2) is occupied by sedimentary basins. The most important potential storage sites can be divided into three categories: saline aquifers, depleted oilfields and coal seams.

Petroleum reservoirs and saline aquifers suitable for CO_2 disposal are located in offshore and onshore basins in Brazil. There are 12 offshore basins located on the northern and eastern margins of Brazil, which came into being as a result of the opening of the Atlantic Ocean. These are the Pelotas, Santos, Campos, Espírito Santo, Bahia Sul, Sergipe-Alagoas, Pernambuco-Paraíba, Potiguar, Ceará, Barreirinhas, Pará-Maranhão and Foz do Amazonas Basins. These offshore basins have a lacustrine rift phase (Barremian) with both clastic and carbonate rocks which are potential reservoirs for CO_2 storage. The most important source rocks for petroleum plays in offshore basins are also related to this rift phase. The rift phase is followed by a transitional restricted marine environment with deposition of a thick (up to 2,000 m) evaporitic sequence (Aptian), which is a regional seal for the rift reservoirs. The open marine phase consists of a transgressive-regressive sequence (Albian to the present) with shelf sandstones and limestones, and deepwater clastic systems, from which around 90% of Brazilian petroleum is produced.

Onshore basins are:

- Intracratonic sags with deposition during the Phanerozoic (predominantly Paleozoic), these being the Paraná, Parnaíba, Amazonas and Solimões Basins;
- Intracratonic rifts of Cretaceous age, these being the Recôncavo, Tucano and Jatobá Basins.

The intracratonic sags in northern Brazil (Amazonas and Solimões) have reservoirs of Early Pennsylvanian age (sandstones), which are regionally sealed by Pennsylvanian Evaporites; these are suitable for CO_2 storage. In the Parnaíba Basin in north-eastern Brazil, reservoirs are sandstones of Givetian and Frasnian age, which are sealed by Famenian mudstones. The onshore Recôncavo and Tucano rift basins contain Late Jurassic fluvial sandstone reservoirs of pre-rift phase, sealed by lacustrine mudstones of pre-rift phase, and Berriasian to Barremian syn-rift sandstone reservoirs, sealed by syn-rift mudstones.

The reservoirs suitable for CO_2 storage in the Paraná Basin (in southern Brazil) are of both Devonian and Early Permian sandstones (Aboarrage and Lopes 1986), which are regionally sealed by Devonian and Kungurian shales, respectively. The Paraná Basin—more specifically in the coal-bearing, Permian Rio Bonito formation—also contains extensive coal deposits (around 37 gigatonnes (Gt)) which can be used for CO_2 storage.

Brazilian coal reserves (Fig. 2) amount to 37.7 Gt, and there are 15 coal production companies located in the southern region of Brazil in the states of Rio Grande do Sul (3), Santa Catarina (11) and Paraná (1), as well as five coal power plants. The installed capacity of the coal power plants is 1,414 MW, which is expected to increase to 4,176 MW in 2012 (see Table 1). Generation capacity of more than 2,764 MW is projected for the new coal-fired power plants (see Table 2).

Brazil's petroleum fields are both onshore and offshore. Most of the oil reserves and production, however, are located in the offshore Campos Basin (>10 billion barrels), followed by the Potiguar, Sergipe-Alagoas, Recôncavo, Espírito Santo and Santos Basins. Gas reserves and production are situated in the Campos (ca 130,000 million m^3), Santos (ca 50,000 million m^3), Solimões (ca 50,000 m^3), and Espírito Santos, Potiguar and Recôncavo Basins (ANP 2005).

Table 1 Coal power plants operating in Brazil

Operating power plants	State	Capacity (MW)
Jorge Lacerda	Santa Catarina – SC	857
Charqueadas	Rio Grande do Sul – RS	72
Presidente Médice	Rio Grande do Sul – RS	446
São Jeronimo	Rio Grande do Sul – RS	20
Figueira	Paraná – PR	20
Total		1,415

Table 2 New coal power plant projects in Brazil

Power plants projects	State	Capacity (MW)
USITESC	Santa Catarina – SC	440
Candiota III	Rio Grande do Sul – RS	350
Jacuí	Rio Grande do Sul – RS	357
Seival	Rio Grande do Sul – RS	500
CTSUL	Rio Grande do Sul – RS	650
Pampa I	Rio Grande do Sul – RS	340
Figueira (upgrade)	Paraná – PR	127
Total		2,764

The total storage capacity of (a) coal deposits and petroleum fields, calculated using the methodology recommended by the Carbon Sequestration Leadership Forum (see Bachu et al. 2007); and (b) saline aquifers, assessed on the basis of porous volumes of sedimentary basins, is estimated to be approximately 2,000 Gt CO_2 (Ketzer et al. 2007). This estimate is of the same order of magnitude as that of Canada and the USA (i.e. ca 3,900 Gt CO_2) (NETL 2007).

A preliminary matching of sources to sinks indicates that most of the stationary CO_2 sources are in the south and south-east and are associated with geological sinks. These potential CO_2 storage sites are: the onshore Paraná Basin (saline aquifers and coal seams); the offshore Campos and Santos Basins (saline aquifers and petroleum fields); and the onshore São Francisco Basin (saline aquifers). The ongoing enhanced oil recovery (EOR) operation also makes the Recôncavo Basin in north-eastern Brazil an important 'sink' candidate (Ketzer et al. 2007). However, there is 'mismatching' between sources and sinks in onshore basins in northern Brazil; for example, in the Solimões and Amazon Basins there is a large storage capacity but few stationary CO_2 sources. Together, the large storage capacity and the projected growth in the production of fossil fuels represent an incentive for the future adoption of CCS in Brazil.

Given the large number of sedimentary basins in Latin America, the CO_2 storage potential on the continent is vast. With the exception of the highly active tectonic areas near the Andean mountains (onshore and Pacific coast), the greatest potential (outside Brazil) is in Venezuela, Bolivia and Argentina, where most of the hydrocarbon production occurs.

The following EOR projects are being conducted in South America:

- Brazil: in the state of Bahia since 1987;
- Argentina: Charco Bayo/Piedras Blancas Field; Entre Lomas Area (Neuquen Basin); and Cuyo Basin (Mendoza);
- Venezuela: Lagomar Field, since 2001;
- Suriname: Tambaredjo, State Oil Company Suriname N.V. (Staatsolie), since 2003.

In Brazil there is only one project under way to inject CO_2 into saline aquifers and one pilot project for disposal of CO_2 in coal seams (Carbon capture and geological storage (CCGS) associated with enhanced coalbed methane production recovery (ECBM)), which is due to start in the state of Rio Grande do Sul in the far south of Brazil in 2011 (Porto Batista CCGS/CBM Pilot Site).

3 Radioactive Waste Disposal Options in Brazil and Argentina

NPPs in Argentina account for 6.7% of all electricity generated. Nuclear energy consumption in Argentina is 5.8 TWh (terawatt hour). Argentina's gross domestic product grew by 7.9% in 2007 (IEA 2007) and Brazil's by 5.9% in 2007.

Brazil is a nuclear weapon-free country by virtue of its 1988 constitution. In 1991 the governments of Brazil and Argentina signed the Agreement on the Exclusively Peaceful Use of Nuclear Energy, in Guadalajara, Mexico. Today, there are only four operational nuclear power reactors in South America, two in Brazil and two in Argentina (for further information, see Tables 3 and 4), which have a net generation capacity of 2,730 MW. The Brazilian nuclear reactors (Angra-1 and Angra-2 power plants) are located in the city of Angra dos Reis (state of Rio de Janeiro) about 130 km south of Rio de Janeiro, 220 km from São Paulo and 350 km from Belo Horizonte. The NPPs began operating in 1982 and 2000, respectively.

New nuclear projects currently being planned in Latin America are:

Brazil

Construction of the 1,224 MW Angra-3 unit at the Angra dos Reis nuclear complex, close to Angra-2, for which the Brazilian Institute for the Environment (IBAMA) issued an environmental licence on 24 July 2008. Angra-3 is planned to be operational in 2014 with a power generation capacity of 10.9 TWh/year. The Brazilian Government plans to construct four further nuclear plants, two of which will be in the

Table 3 Operating nuclear power reactors in Brazil

Operating nuclear power reactors in Brazil

Name	Type	Status	Location	Net	Capacity (MW) Gross	Date Connected
ANGRA-1	PWR	Operational	Rio de Janeiro	520	657	1982/04/01
ANGRA-2	PWR	Operational	Rio de Janeiro	1,275	1,350	2000/07/21

Source: IAEA (2007)
Owned by Eletrobras Termonuclear SA—Eletronuclear
PWR pressurized water reactor

Table 4 Nuclear power reactors operating or under construction in Argentina

Operating nuclear power reactors in Argentina

Name	Type	Status	Location	Net	Capacity (MW) Gross	Date Connected
ATUCHA-1	PHWR	Operational	Buenos Aires	335	357	1974/03/19
ATUCHA-2	PHWR	Under construction	Buenos Aires	692	745	2010/10/01
EMBALSE	PHWR	Operational	Cordoba	600	648	1983/04/25

Source: IAEA (2007)
Owned by Nucleoelectrica Argentina S.A.
PHWR pressurized heavy water reactor

north-east of the country. To date, four Brazilian states (Bahia, Pernambuco, Sergipe and Alagoas) have offered to accommodate these plants.

The Brazilian nuclear power company (Eletronuclear) has developed a solution for storing nuclear waste in steel capsules that guarantees residue safety for over 500 years. There are no deep repositories under construction in Brazil. This is perhaps the key challenge for environmental protection and energy security, not only for the Brazilian nuclear programme but for all countries of South America.

The Brazilian Government has plans to build four more 1,000 MW nuclear plants before 2015 and to add a further 8 GW of nuclear capacity before 2030 (through the construction of small power plants in north-east Brazil). Brazil, with 6% of world total uranium resources (309,000 t of uranium), has big enough uranium reserves to meet its own nuclear fuel needs. There are three main uranium deposits in Brazil: (1) Poços de Caldas (state of Minas Gerais); (2) Lagoa Real or Caetité (state of Bahia); and (3) Itataia (state of Ceará). Uranium conversion, enrichment and fuel fabrication are carried out at a nuclear fuel factory at Resende (65 km north of the Angra facility). The Resende nuclear fuel factory is expected to supply 100% of Brazil's enriched uranium in 2015.

Argentina

The Atucha-2 project with 692 MW at Lima in the province of Buenos Aires. Completion of the country's third reactor is expected by early 2011. CNEA is responsible for radioactive waste management and for Argentine NPPs decommissioning. Low and intermediate-level wastes including used fuel from research reactors are handled at CNEA's Ezeiza facility. The used fuel is stored at each power plant and there is 248 Spent Fuel Dry Storage Silos at Embalse since 1992 (Barrera 2010).

3.1 Radioactive Waste

Brazil's National Nuclear Energy Commission (Comissão Nacional de Energia Nuclear (CNEN)) reports to the Ministry of Science and Technology and, in association with IBAMA, is responsible for licensing and supervising the nuclear facilities. CNEN is also in charge of the management and disposal of low- and intermediate-level RW. Brazilian Law No. 10.308 of 20 November 2001 deals with the final disposal of RW. In Article 37 it states: 'CNEN should initiate studies on the selection, design, construction and licensing of a final repository for radioactive waste on national territory with a view to its entry into operation within the shortest period of time technically feasible' (translation from the Portuguese). Eletronuclear and CNEN are responsible for all fuel cycle and reactor technology, radioisotope production, and related R&D at five nuclear research centres.

The Brazil-Argentina Agency for Accounting and Control of Nuclear Materials (ABACC) and the Brazil-Argentina Agency on Nuclear Energy Applications

(ABAEN) are responsible for the control of nuclear materials and nuclear energy applications (nuclear fuel cycle, nuclear waste and nuclear reactors).

Brazilian nuclear waste is classified into three radioactivity classes, all of which are currently stored in temporary intermediate-level sites:

1. Low-level waste (LLW): NPP waste.
2. Intermediate-level waste (ILW): NPP waste.
3. High-level waste (HLW): by-products formed by the irradiated fuel element of the reactor. The HLW is stored in the spent nuclear fuel (SNF) pool inside the reactor (Angra-2) or outside it (Angra-1). The SNF pool enables the HLW to be submerged up to a depth of 10 m. The Angra-1 SNF pool has a nuclear waste storage capacity of 40 years (equivalent to the lifetime of the NPP). Angra-2 has a storage capacity of 20 years (equivalent to half the lifetime of the NPP).

In Argentina, the nuclear waste generated at country level is classified as shown in Table 5.

Considering the increasing demand for nuclear power and the relatively short lifetime of the SNF pool, Latin American countries need, on a regional basis, to choose sites where HLW can be disposed of in deep geological repositories (Black and Chapman 2001). However, in Brazil there are currently no deep geological repositories in which HLW from SNF can be stored. There is only a nuclear waste disposal project, related to the building of underground concrete structures, that was submitted to the Brazilian Federal Senate and is currently going through the approval process. Any future deep repository project should consider not only the natural isolation of HLW in deep geological repositories, but also how the HLW might be retrieved and reprocessed using new emerging technologies—a proposition which is controversial and which, if realized, will be expensive.

Table 5 Argentinian radioactive waste classification

Class	Type	Characteristics	Technology
B	Low-level Short-lived ($\tau < 30$ years)	Low beta/gamma Radioactivity Insignificant alpha content	Surface engineered System
M	Intermediate-level Short-lived ($\tau < 30$ years)	Intermediate beta/gamma Radioactivity Insignificant alpha content	Monolithic near-surface repository
A	Low-level Long-lived ($\tau > 30$ years)	Low beta/gamma Radioactivity Significant alpha content	
	Intermediate-level Long-lived ($\tau > 30$ years)	Intermediate beta/gamma Radioactivity Significant alpha content	Deep geological Repository
	High-level Long-lived ($\tau > 30$ years)	High beta/gamma Radioactivity Significant alpha content Significant heat emission	

Source: CNEA (2008)

The cities where the disposal sites will be built will receive financial compensation (10% of the disposal costs in Brazil). However, in spite of the financial compensation, some policymakers are against disposal of HLW in deep geological repositories because of the potential for causing public controversy and economic losses (for example, the Angra dos Reis region is a tourist destination).

The basic criteria for and preliminary siting of RW repositories (Beninson et al. 1986) and research into deep geological disposal in Argentina (Ninci 2004) have developed as follows:

- 1980–1990: First study of potential repositories for HLW disposal covering 200 granitic bodies all over the country. Sites chosen for detailed geological, geophysical and hydrogeological studies were Sierra del Medio and Gastre, both of them in the province of Chubut (South Argentina). The first stage report was published in 1990.
- 1992: Official cancellation of the project because of criticism and public concern.

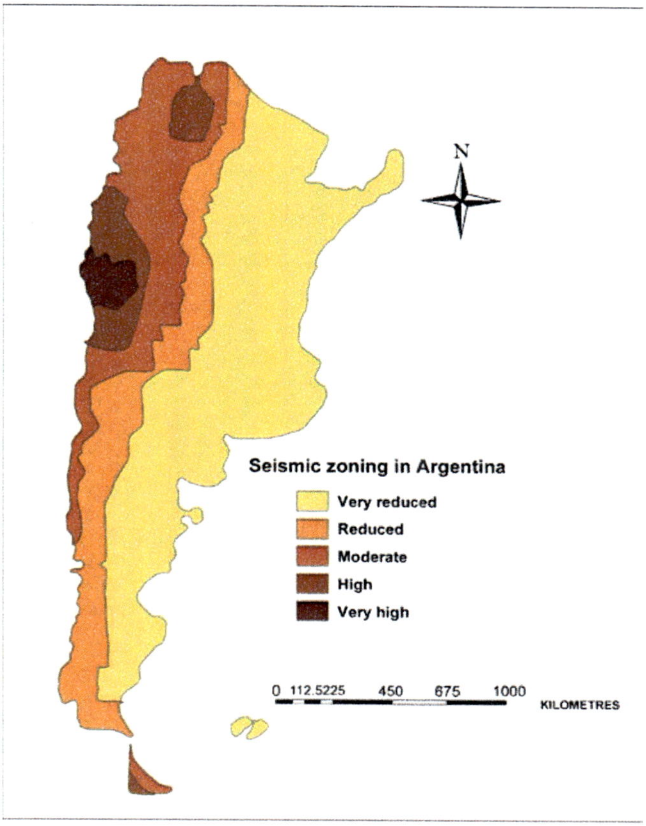

Fig. 5 Seismic zoning[a] in Argentina (Source: Instituto Nacional de Prevención Sísmica (INPRES)) (*see* Colour Plates)

[a] Zones in which the possibility of radioactive waste disposal is excluded are those with high and very high seismicity.

- 1991: The Argentinian National Nuclear Energy Commission (Comisión Nacional de Energía Atómica (CNEA)) starts a new project entitled 'Study of geological formations suitable for the siting of repositories for the disposal of high-, medium- and low-level waste'.
- 1993–1995: Geological investigations regarding the siting of an ILW repository.
- 1996: International Atomic Energy Agency (IAEA) Technical Cooperation project (ARG/4/084) with CNEA on Geology of Repositories for High-Level Waste Disposal (IAEA 1996).
- 1996–2001: Identification of favourable geological formations for final disposal: clay formations, volcaniclastics and evaporates.

The following factors must be taken into account during the selection process for location of deep geological repositories, for example the 'exclusion factors' such as seismic zoning (see Fig. 5) and centres of volcanic activity (details given in Fig. 6). For a classification of regions in Brazil and Argentina with respect to the siting of an HLW repository, see Table 6.

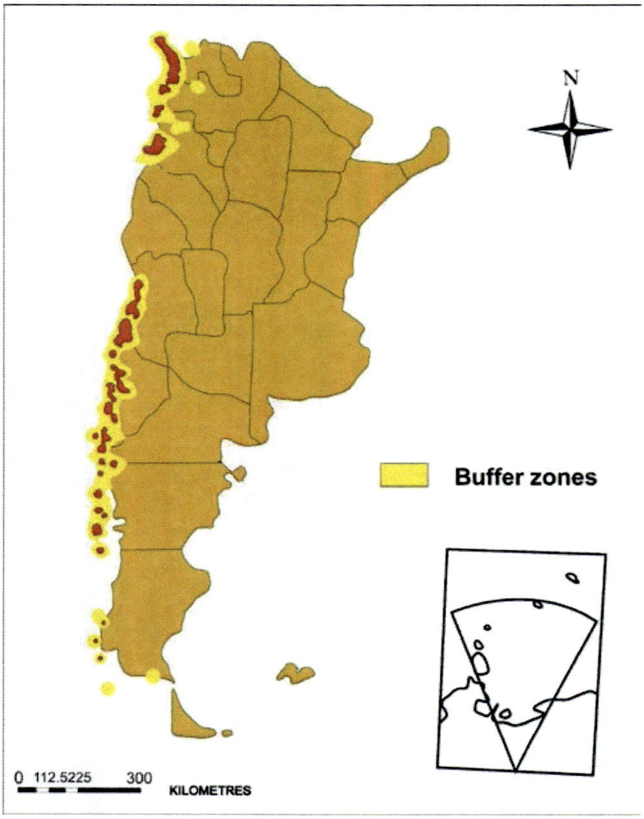

Fig. 6 Volcanic centres in Argentina (Source: Instituto Nacional de Prevención Sísmica (INPRES)) (*see* Colour Plates)

Table 6 Considerations affecting the selection of deep repositories for radioactive waste

Aspects/factors	Brazil	Argentina
Seismicity	Low (Intraplate setting)	High West convergent margins (Andes) (see Fig. 6)
Volcanism	Absent	Andean volcanism (see Fig. 7)
Neotectonism (map of quaternary faults)	Low	Andean margin
Geology	Cratons and onshore sedimentary basins	Cratons and sedimentary basins
Fractures and lineaments (satellite information)	Almost all regions requires detailed studies	Almost all regions requires detailed studies
Geomorphology, slopes	Plateau areas (continental flood basalts), cratons and intracratonic basins	Andean and craton regions, plateau areas and some volcanic regions
Natural resources: petroleum and mining areas	Petroleum, coal and uranium, plus base metals	Petroleum, coal, plus base metals
National and provincial parks	Almost all regions requires detailed studies	Almost all regions requires detailed studies
Cities and population	Southern regions and coastal areas	Central and north regions
Main rivers and hydrology	Many hydrographic basins distributed in south-east regions and central-northern region	Concentrated in north and central-western regions
Hydrogeology	Guarani aquifer (largest fresh water resource) amongst others	Guarani aquifer (largest fresh water resource) amongst others
Geothermal gradients	Low	Low to high (west regions)

No other countries in South America apart from Brazil and Argentina have nuclear power programmes. However, Peru has an RW management programme based at the Peruvian Institute of Nuclear Energy (Instituto Peruano de Energia Nuclear (IPEN)), which operates a research facility (the RACSO Nuclear Research Center) 42 km north-east of Lima. Currently, waste from here is treated and stored pending the licensing of a near-surface repository for final disposal.

4 Comparative Analysis

The geological disposal of CO_2 and that of RW have similarities and differences (see Table 7). The most important environmental issue common to both CO_2 and RW is that both are power plant by-products that can be disposed of underground. Compared with other types of power plants with high CO_2 emission rates, such as modern coal-fired plants, the generation of spent fuel from NPPs represents a

Table 7 Summary of the main differences and similarities between CO_2 and radioactive waste disposal

Type of disposal	CO_2: carbon capture and storage (CCS)	Radioactive waste (RW)
Structural and geological context	Stable regions without earthquake hazards, such as onshore and offshore basins in Brazil and eastern South America (intraplate context; see Fig. 5 for seismic zones). In the western region of South America plate boundary context and volcanic activity must be taken into account Geological trap may be desired	Stable regions without earthquake hazards such as Brazilian shield and Brazilian intracratonic onshore basins (e.g. Paraná and Parnaíba Basins) and eastern South America (intraplate context; see Fig. 5 for seismic zones). In the western region of South America plate boundary context and volcanic activity must be taken into account
Natural analogues to deep geological repositories	Natural analogues such as natural gas reserves, possible natural CO_2 fields in the São Francisco Basin (onshore), high-CO_2 natural gas deposits in deep reservoirs of the Santos Basin (Brazil)	Significantly, deep geological repositories have natural analogues: places on Earth where naturally occurring radioactive materials have been isolated in their geological formation, without human intervention, for millions of years. By adding the benefit of highly engineered containers and barriers, deep geological repositories build on this proven geological phenomenon
Availability and volume of potentially suitable geological formations	Large volume potential (2,000 Gt in Brazil)/many centuries of disposal	
Disposal capacity	High (2,000 Gt CO_2, mainly in saline aquifers in onshore intracratonic basins)	High, but nuclear power is a remarkably clean technology because it produces huge quantities of energy from small quantities of nuclear fuel and correspondingly small amounts of nuclear waste
Deep disposal	Deep repository Safety of long-term (millions of years) underground waste isolation in deep geological formations Saline aquifers and petroleum fields (>800 m CO_2 supercritical) Gondwanic high-ash coal deposits, 300–1,500 m, essentially in the Paraná Basin	Deep repository Safety of long term (millions of years) underground waste isolation in deep geological repository Large shield/basement area formed by granitic massifs (Brazilian shield) Onshore sedimentary basins with thick, regional mudstone seals (e.g. Paraná and Parnaíba Basins)

(continued)

Table 7 (continued)

Type of disposal	CO_2: carbon capture and storage (CCS)	Radioactive waste (RW)
Cavity	CO_2 disposal requires porous geological formation	Nuclear material requires a 'cavern'-like room and pillar system
Pressure	CO_2 is injected into storage site under high pressure (typically >100 bar for aquifer and petroleum and >30 bar for coal)	Nuclear material is usually stored at atmospheric pressure
Location	Onshore/offshore disposal	Offshore disposal is forbidden in Brazil
	Onshore proportion of EOR in Brazil would be very small	Possibility of affecting international waters and other countries
Transformation during period of storage	CO_2 can transform into CO_3^{-2}, HCO_3^-, Ca_2CO_3, Mg_2CO_3, Fe_2CO_3, dawsonite, depending on geochemical conditions, time, etc	Nuclear waste decays into daughter elements and consequently emits radiation
Waste quantities	High quantities (ca 300 Mt/year in Brazil plus ca 500 Mt/year in other South American countries)	Huge quantities of energy from small quantities of nuclear fuel and correspondingly small amounts of nuclear waste 'Clean' technology[a]
Demonstrated technical feasibility of deep geological repositories	Many pilot and demonstration projects to start in the next 5–10 years in saline aquifers, petroleum fields and coal seams. EOR in Brazil since 1987. CCS was to start in 2009	Proceeding in several countries at research laboratories in underground sites that have already been selected As RW is created in many countries, it may prove efficient to seek economies of scale by rationalizing the number of repositories and thus concentrating investment. Such considerations may be especially relevant for countries where establishing the entire stream for waste management (IAEA 2002) and disposal does not represent a sound economic and technological option. The challenge of long-term disposal should not inhibit national decisions to construct new nuclear power reactors

(continued)

Table 7 (continued)

Type of disposal	CO_2: carbon capture and storage (CCS)	Radioactive waste (RW)
Risk assessments	Caprock—injection wells—abandoned wells—geochemistry (CO_2 rock interactions)—aquifer contamination—seismicity—blowout (poor completion)	Percolation—corrosion rates—dissolution—groundwater flux—seismic—igneous event
	Environmental effect on coastlines and marine ecology	Environmental effect on sea coastlines and marine ecology
Distance to main sources of material to be disposed	Source and sink match issue Good matching for south-eastern part where the stationary sources are located	Depends on geological conditions
Competition for space with other uses/resources	Petroleum and coal	–
Potential for retrieval	Possibility for retrievability/production of CO_2 on a commercial basis for industrial uses and EOR/ECBM projects	Possibility for retrieving and reprocessing HLW in the future, using new emerging technologies Repository plans generally allow for retrieval of nuclear waste, at least for an extended period of time. However, safe monitoring of stored waste is conceivable only for a limited period of time, not for an indefinite period
Diversifying energy sources and diminishing dependence on hydro	Yes, certainly from ECBM and EOR, and contributing to sustained fossil fuel consumption	From a national perspective, nuclear power also affords an excellent means of diversifying energy sources, providing baseload electricity cleanly and reliably, while reducing vulnerability to sharp price fluctuations and crippling disruptions in energy supply
Transport risks and costs	Pipeline leakages Cost compatible with petroleum pipelines CO_2 transport by ship between offshore fields in Brazilian basins (environmental effect on marine ecology)	If situated close to provinces directly involved in nuclear fuel cycle, the transport risks and costs are reduced (Tunaboylu et al. 2001)
Social problems and economic losses	Urban regions, natural reserves; most of CO_2 sources near highly populated areas in south-eastern Brazil	Angra dos Reis (Brazil) is a tourism region Better social and public acceptance (NAS 2001)

(continued)

Table 7 (continued)

Type of disposal	CO_2: carbon capture and storage (CCS)	Radioactive waste (RW)
Project compatible costs	High dependence on technology; currently average cost of US $50–100/t CO_2 avoided includes capture, transport and storage monitoring. Possible capture cost reduction through CO_2 separation from natural gas (e.g. new discoveries in the Santos Basin, Brazil)	Financial compensation for disposal of nuclear waste equals 10% of the disposal costs in Brazil In contrast to common perception, the cost of managing and disposing of nuclear waste and SNF represents a very small percentage of the overall cost of producing nuclear energy
Ethical	The deep geological repository is an alternative solution for providing long-term passive protection to future generations against global warming risks	The deep geological repository is the only ethically correct solution, giving passive protection to future generations in the long term

[a]Globally, 2–3 Gt CO_2/year are avoided by nuclear generation according to the Canadian Nuclear Association (CNA 2008)

ECBM enhanced coalbed methane, *EOR* enhanced oil recovery, *HLW* high-level waste, *SNF* spent nuclear fuel (equivalent to *HLW*)

decrease in greenhouse gas emissions. Thus, NPPs can help Latin American countries to reduce their investments in CO_2 sequestration projects related to the use of fossil energy and thus achieve their greenhouse gas emission reduction goals. Considering the high capacity required to replace the non-renewable fossil fuels, nuclear power generation and the location of respective RW disposal and CO_2 injection sites is a key technical and political question for future regional energy supply strategies. Thus, CO_2 and RW disposal are interrelated issues that have a very important role to play in the Latin America environmental and energy matrices, and which will impact the development of the region.

5 Conclusions

In Brazil deep geological sites for the disposal of RW and CO_2 are still at the identification and selection stage. Among sites suitable for CO_2 disposal are mature petroleum fields, coal seams and saline aquifers in onshore intracratonic Paleozoic basins and Cretaceous rift basins and mature petroleum fields and saline aquifers in offshore sedimentary basins (Cretaceous to recent Atlantic Ocean rift and drift phases). The total CO_2 storage capacity of Brazilian reservoirs is large—approximately 2,000 Gt—offering potential storage for domestic emissions for centuries to come. CO_2 storage in Brazilian basins can assist in enhancing the production of oil and natural gas from mature petroleum fields and coal seams, respectively.

RW repository sites are cavities constructed in onshore sedimentary basins such as the Paraná and Parnaíba basins (Brazilian legislation prohibits offshore disposal of RW) and shield/basement areas, such as granitic massifs in the Brazilian Shield. Geologically stable conditions (in the intraplate context) in Brazil and eastern South America are favourable for the disposal of RW and CO_2.

While politicians in Brazil discuss the safety of long-term disposal in 'deep geological repositories', nuclear bureaus and research centres are making efforts to validate deep geological repositories as safe long-term disposal sites for highly radioactive materials (spent nuclear fuel (SNF) and high level waste (HLW)). In Argentina, a site in the Gastre area (Chubut region) was selected, but rejected in 1992.

Before being placed in selected underground deep geological repositories the SNF–HLW and ILW are sealed in a ceramic and glass or concrete matrix, then further sealed in corrosion resistant steel containers to ensure their long-term safe confinement. The nuclear waste (SNF, HLW and ILW) can then be isolated underground for millions of years, with zero impact to the environment (Miller et al. 1999; Hill and Chapman 2001).

Geological formations for use as deep repositories are carefully selected, taking into account all the relevant geological and environmental issues. The social characteristics of target sites are also considered, and there is extensive public consultation and awareness-raising to reduce the possibility of public controversy. Before a decision is made, the nuclear research network, comprising CNEN, the Brazilian Geological Survey, National Environmental Agency, Brazilian Bureau of Mines and the Center of Excellence in Research and Innovation in Petroleum, Mineral Resources and Carbon Storage (CEPAC), conducts detailed viability and safety studies in conformity with IAEA safety procedures. However, in Brazil the Federal Public Ministry must also endorse any decisions made as being in the public and society's interest. These characterization studies are very important in terms of gaining public support and confidence for long-term RW disposal.

RW is isolated in deep geological repositories until its radioactivity decreases to the levels normally found in the Earth's crust. There are naturally occurring analogues of deep geological repositories in some geological formations where radioactive materials have been isolated for millions of years. The most important environmental issue involved in the disposal of SNF–HLW in deep repositories is the need to prevent the RW from contaminating groundwater. A site must therefore be selected for its stable and impermeable geological setting.

New Brazilian investments in NPPs (Angra-3) are dedicated to the development of a pilot project for disposal and monitoring of HLW in geological repositories ('technical demonstration' project). In terms of the overall cost of energy generation from NPPs, the cost of RW disposal in deep geological repositories is proportionately low. Worldwide, the first operation of a Waste Isolation Pilot Plant (WIPP) was in 1999 in a deep stable layer of salt in New Mexico in the USA (Allègre 1999).

Diversification of the Brazilian energy matrix through nuclear power and coal gasification, together with CCS plus hydrogen production, is essential in near-term regional energy planning to reduce dependence on fossil fuels and greenhouse gas emissions.

The need for common repositories in South America should be recognized. Other regions could learn from Brazil and its proposed repositories for CO_2 disposal in the Paraná Basin. The concept of international or multinational repositories for RW could be introduced. This would cover shared (McCombie 2007), well sited and safe facilities, operated for the benefit of a number of users, to make the most effective use of shared resources (McCombie et al. 2001).

Selection criteria for deep repositories for RW and CO_2 in Brazil and Argentina should take into account the following:

- The area should have been stable for millions of years (South America Plate cratons, continental flood basalt areas and intracratonic sedimentary basins) (see Fig. 7) and should lie well away from active fault zones.
- Repository depths should be in the hundreds of metres so that the effects of surface erosion are negligible and any earthquake-related damage is minimized (the effects of earthquakes are less severe at depth).
- Impermeable caprocks are the most appropriate locations for repositories. Sedimentary salt, basalt (e.g. the Serra Geral Formation) or clay/siltstones from

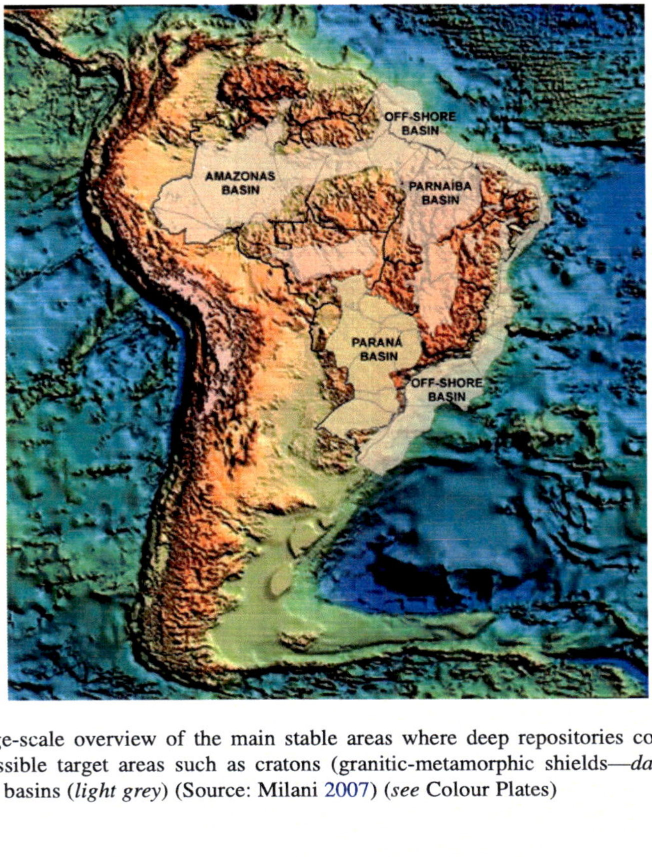

Fig. 7 Large-scale overview of the main stable areas where deep repositories could be sited, showing possible target areas such as cratons (granitic-metamorphic shields—*dark grey*) and sedimentary basins (*light grey*) (Source: Milani 2007) (*see* Colour Plates)

the Paraná Basin are the most suitable formations (see Fig. 8). The Paraná Basin with its large storage capacity (ca 135 Mt CO_2/year) is the best repository for CO_2. The sinks are saline aquifers with a theoretical capacity of ca 462,000 Mt and deep coal seams with a theoretical capacity of ca 200 Mt.

- For RW an engineered barrier can be constructed to complement the geological barrier (see details in Apted et al. 2001a, b).
- A major challenge is the possibility of future retrievability of stored RW for use as NPP fuel or for reprocessing using new technologies to reduce the environmental disposal risk. This, however, is likely to be an expensive operation.
- Regarding HLW disposal in Argentina, the granite site at Gastre in Chubut was selected for HLW disposal, with the earliest start date being 2040. The granite is unfractured and located in a stable region. In Brazil, potential deep repository sites must be better studied and identified; the deposit of thorium (Th) and rare earth elements (REE) (known as the Th-REE ore body) in Morro do Ferro in Minas Gerais is being studied because it is a natural analogue of RW repositories and thus provides an opportunity to model the release and migration of thorium and radium into groundwater (Alexander et al. 2006).
- Ethically speaking, deep repositories are the best solution because: (i) their efficacy can be demonstrated; (ii) site selection is a democratic process (being

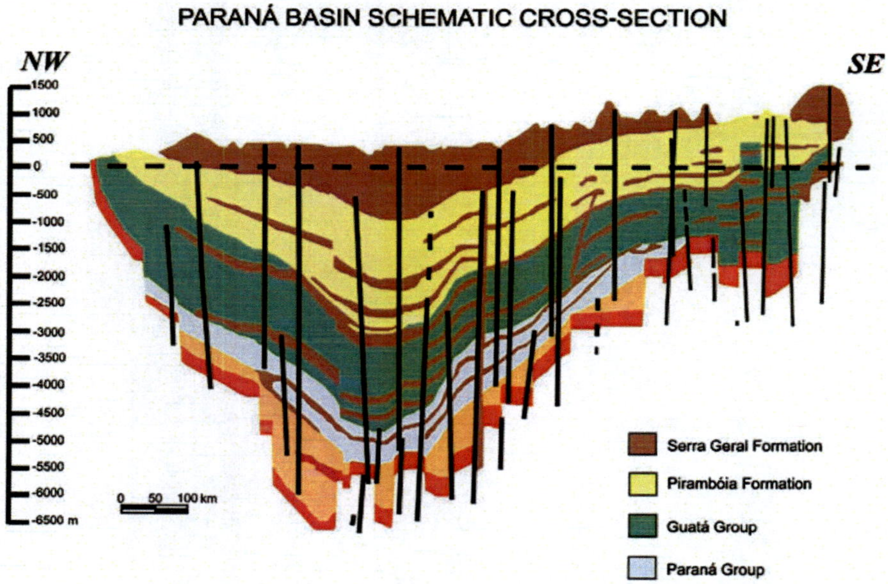

Fig. 8 Target areas for carbon capture and storage and potential sites for deep repositories for nuclear waste disposal in the Paraná Basin, Brazil (*see* Colour Plates)

submitted for public approval in Brazil); and (iii) they do not transfer RW disposal problems, risks and costs to future generations.
- Based on information currently available, geological formations and capacities for safe disposal of CO_2 and RW would appear to be available in several regions in Brazil and Argentina. Thus, these countries could use any combination of fossil and nuclear sources for energy production based on prevailing conditions, particularly the driving forces of energy policies.

Finally, there are important differences between RW and CO_2 disposal: (1) CO_2 can be transported for longer distances than RW with less risk of leakage; (2) deep repositories for multinational use are easier to implement for RW than for CO_2 because of the smaller quantities of RW per unit of energy produced (smaller deep repositories can provide capacity for RW); and (3) high-level RW needs robust, multi-sealed sites plus engineering barriers that take account of the processes of decay and heat transfer of RW.

Among the similarities between RW and CO_2 disposal are the need for detailed caprock characterization and studies on caprock integrity, hydrogeology (e.g. groundwater flow model), seismic survey for geological characterization of the target area (e.g. identification of fault zones) and computer modelling of the geological system (disposal site, caprock efficiency).

Studying the similarities and differences between RW and CO_2 disposal can generate new insights and lessons that are useful in terms of increasing confidence in the selection of sites for disposal activities. An example of this is the recent cooperation carried out by the Brazilian Carbon Storage Research Center (CEPAC) and the French Atomic Energy Commission (CEA) for integrated studies of caprocks integrity. CEPAC expects to work with the Brazilian Energy and Nuclear Research Institute (IPEN) and with the Brazilian National Nuclear Energy Commission (CNEN) in the near future on RW and CO_2 disposal.

Acknowledgements We would like to thank Ferenc L. Toth (IAEA) for the great interest shown in this project and for the constructive comments and criticisms provided in the previous reviews of the draft version of this chapter. Thanks to CNPq (Brazilian Research Council) for research grant and International Atomic Energy Agency (IAEA) for the efforts on this project. We also appreciate the efforts of Mr Giancarlo Caporale in helping us to edit the figures.

References

Aboarrage AM, Lopes R da C (1986) Projeto a borda leste da Bacia do Paraná: Integração geológica e avaliação econômica, vol 18. Departamento Nacional de Produção Mineral (DNPM), Brasilia, and Serviço Geológico do Brasil (CPRM), Porto Alegre. Unpublished [in Portuguese]

Alexander WR, Gieré R, Hidaka H, Yoshida HD (2006) Natural immobilization processes aid the understanding of long-term evolution of deep geological radioactive waste repositories. Geochem Explor Environ Anal 6(1):3–4

Allègre M (1999) The deep geological repository: an unavoidable and ethically correct solution. In: Proceedings of the 24th Annual International Symposium. The Uranium Institute. http://www.world-nuclear.org/sym/1999/allegre.htm

ANEEL (Brazilian Electricity Regulatory Agency), MME (2010) (Brazilian Energy and Mines Ministry). Electricity generation capacities. http://www.aneel.gov.br/aplicacoes/capacidadebrasil (accessed December 2010)

ANP (Agência Nacional do Petróleo, Gás Natural e Biocombustíveis—National Petroleum Agency) (2005) Reservas do Brasil em 31/12/2005 [in Portuguese]. Available at: http://www.anp.gov.br/?pg=17639&m=reservas&t1=&t2=reservas&t3=&t4=&ar=0&ps=1&cachebust=1271251011515

Apted MJ, Smith PA, Chapman NA (2001a) Engineered barrier design strategy for a high isolation repository site. In: Proceedings of the 9th International High-Level Radioactive Waste Management Conference. American Nuclear Society, La Grange Park, CD-ROM

Apted MJ, Smith P, Miller I (2001b) The high-isolation safety concept: a preliminary evaluation of performance, safety and design factors. Pangea Resources International, Baden, Pangea Technical Report PTR-01-04

Bachu S, Bonijoly D, Bradshaw J, Burruss R, Holloway S, Christensen NP, Mathiassen OM (2007) CO_2 storage capacity estimation: methodology and gaps. Int J Greenhouse Gas Control 1(4):430–443

Barrera A (2010) Operating experience and condition assessment of spent fuel dry storage silos and spent fuel pool at embalse NPP. Life extension project – Embalse NPP– Nucleoelectrica Argentina S.A. International Conference on Management of Spent Fuel from Nuclear Power Reactors, IAEA, Vienna

Beninson DJ, González AJ, Palacios E, Ciallella NR (1986) The Argentine radioactive waste repository: basic criteria, preliminary sitting and design conceptual basis. Bull Eng Geol Environ 34(1):5–10

Black JH, Chapman NA (2001) Siting a high-isolation radioactive waste repository: technical approach to identification of potentially suitable regions worldwide. Pangea Resources International, Baden, Pangea Technical Report PTR-0-01

CNA (Canadian Nuclear Association) (2008) World energy: the past and possible futures. CNA, Calgary, http://www.cna.ca/english/pdf/studies/ceri/CNA_CERI07_EN.pdf. Accessed 13 September 2008

CNEA (Comisión Nacional de Energía Atómica—National Atomic Energy Commission) (2008) Joint convention on the safety of spent fuel management and on the safety of radioactive waste management: third national report. Section B4.1: Comisión Nacional de Energía Atómica. http://www.cnea.gov.ar/xxi/residuos/en/THIRD%20NATIONAL%20REPORT%20DV.pdf

DNPM (Departamento Nacional de Produção Mineral—National Department of Mineral Production) (2008) Anuário mineral brasileiro. DNPM, Brasilia [in Portuguese]

Hill MD, Chapman NA (2001) Storage versus disposal of long-lived radioactive wastes. In: Proceedings of the 9th International High-Level Radioactive Waste Management Conference. American Nuclear Society, La Grange Park, CD-ROM

IAEA (International Atomic Energy Agency) (1996) Technical cooperation project ARG/4/084: geology of repositories for high-level waste disposal. IAEA-TC project datasheet available at: http://www-tc.iaea.org/tcweb/default.asp

IAEA (International Atomic Energy Agency) (2002) Joint convention on the safety of spent fuel management and on the safety of radioactive waste management

IAEA (International Atomic Energy Agency) (2007) Power reactor information system—PRIS. http://www.iaea.org/programmes/a2/

IEA (International Energy Agency) (2002) World Energy Outlook 2002. OECD/IEA, Paris

IEA (International Energy Agency) (2007) Tracking Industrial Energy Efficiency and CO_2 Emissions. OECD/IEA, Paris

Instituto Nacional de Prevención Sísmica (INPRES). Publicación Técnica nº 5: Zonificación Sísmica de La República Argentina

Ketzer JM, Villwock JA, Caporale G, Rocha LH, Rockett G, Braum H, Giraffa L (2007) Opportunities for CO_2 capture and geological storage in Brazil: the CARBMAP project.

In: Proceedings of the Sixth Annual Conference on Carbon Capture and Sequestration. National Energy Technology Laboratory, US Department of Energy. http://www.netl.doe.gov/publications/proceedings/07/carbon-seq/data/papers/wed_025.pdf
McCombie C (2007) Multinational cooperation in geological disposal: sharing of concepts, results, research projects and facilities. In: Brewitz W, Kleemann U (eds) Radioactive Waste Disposal in Geological Formations (pp. 106–111). International Conference, Braunschweig ("City of Science 2007"), November 6–9, 2007. RepoSafe 2007—Proceedings. Bundesamt für Strahlenschutz (BfS), Salzgitter, and Gesellschaft für Anlagen- und reaktorsicherheit (GRS) mbH, Braunschweig
McCombie C, Chapman N, Kurzeme M, Stoll R (2001) International repositories: an essential complement to national facilities. In: Proceedings of Geological Problems in Radioactive Waste Management. Lawrence Berkley National Laboratory, Berkeley, CA, Paper available at: http://www.pangea-international.com/mc_berkeley2001.pdf
Milani EJ (2007) Boletim de Geociências da PETROBRAS 15(2). Bacias Sedimentares Brasileiras, Revisão das Cartas Estratigráficas.
Miller I, Black J, McCombie C, Pentz D, Zuidema P (1999) High-isolation sites for radioactive waste disposal: a fresh look at the challenge of locating safe sites for radioactive repositories. In: Proceedings of Waste Management 1999. WM Symposia, Inc, Tucson, CD-ROM
NAS (National Academy of Sciences) (2001) Disposition of High-Level Waste and Spent Nuclear Fuel: The Continuing Societal and Technical Challenges. The National Academies Press, Washington
NETL (National Energy Technology Laboratory) (2007) Carbon sequestration atlas of the United States and Canada. National Energy Technology Laboratory, Office of Fossil Energy, US Department of Energy. http://www.netl.doe.gov/technologies/carbon_seq/refshelf/atlas/index.html
Ninci C (2004) Deep geological disposal research in Argentina. Paper Presented at the Inter-Regional Training Course on Methodologies for Geological Disposal, Berkeley, CA
Süffert T (1997) Carvão nos Estados do Rio Grande do Sul e Santa Catarina. Porto Alegre, CPRM. 39 p. (Série Oportunidades Minerais – Exame Atualizado de Projeto, n°. 21)
Tunaboylu K, Playfair A, Niggli H, Mariapillai N (2001) Waste transport and public safety. Pangea Resources International, Baden, Pangea Technical Report PTR-01-03
World Nuclear Association (WNA) – www.world-nuclear.org. Accessed 08 December 2010

Index

F.L. Toth (ed.), *Geological Disposal of Carbon Dioxide and Radioactive Waste: A Comparative Assessment*, Advances in Global Change Research 44,
DOI 10.1007/978-90-481-8712-6,

F

G

T

U

Colour Plates

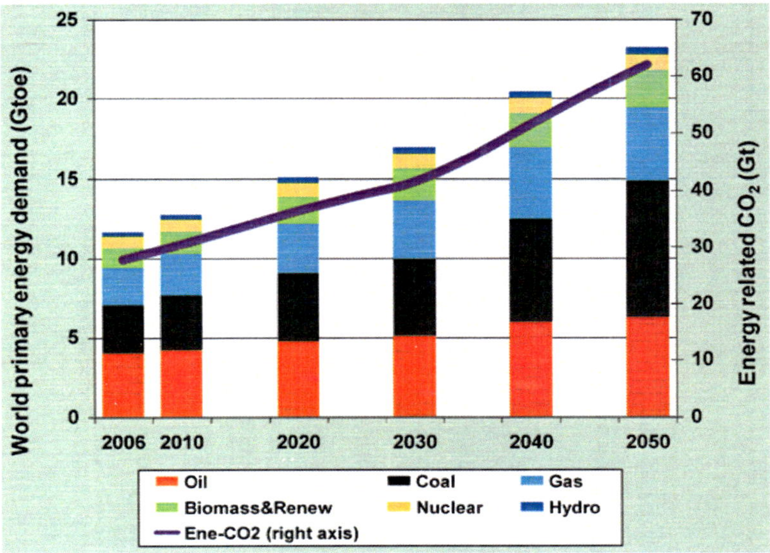

Toth, Fig. 1 Global primary energy sources (*left axis*) and energy-related CO_2 emissions (*right axis*) in the IEA's reference scenarios (Based on IEA 2008a, b)

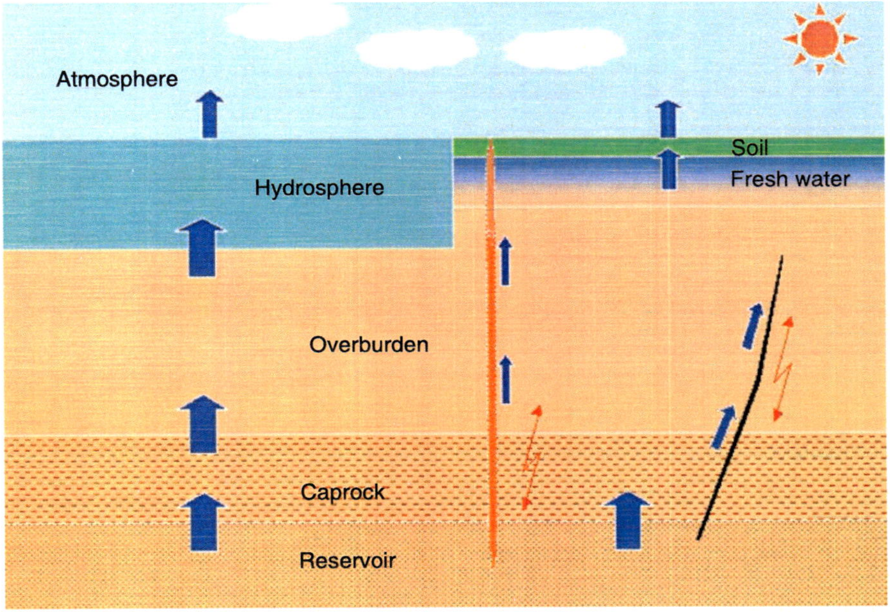

Maul, Fig. 1 Barriers and transport pathways for carbon dioxide

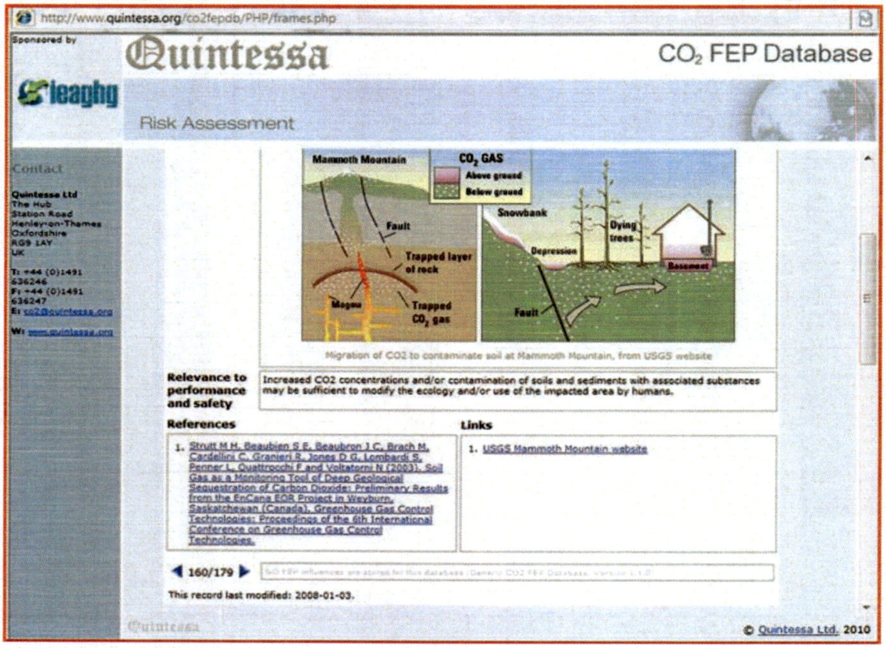

Maul, Fig. 3 An example entry in the generic FEP database

Reiner and Nuttall, Fig. 1 Michael Simonian's Plutonium Memorial concept '24110'. (Images copyright Simonian; see: http://www.designboom.com/eng/cool/simonian.html). The artist imagines a central Washington DC location for a plutonium store just under the Ellipse, a field 1 km in circumference, near the White House, which takes to an extreme the notion that plutonium storage should not be *out of sight and out of mind*

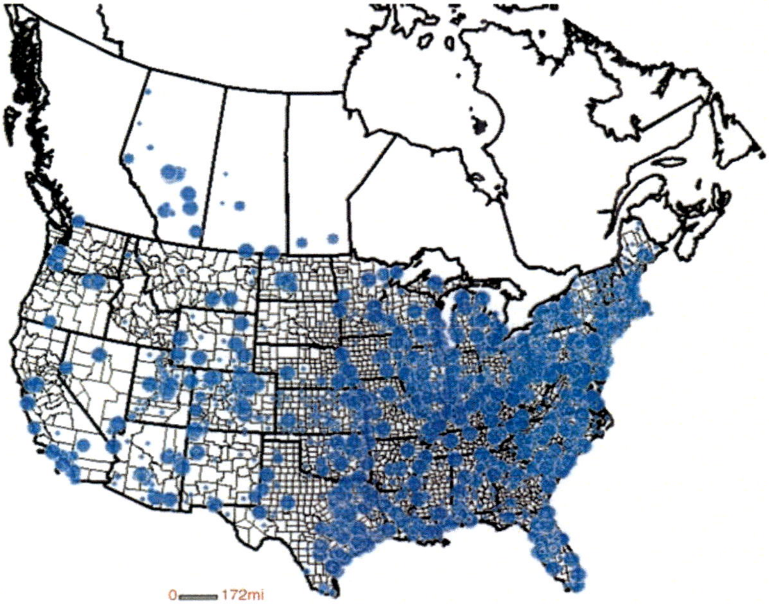

Oldenburg and Birkholzer, Fig. 1 North American CO_2 sources from electricity generation (Source: NATCARB: http://drysdale.kgs.ku.edu/natcarb/eps/natcarb_alpha_content.cfm)

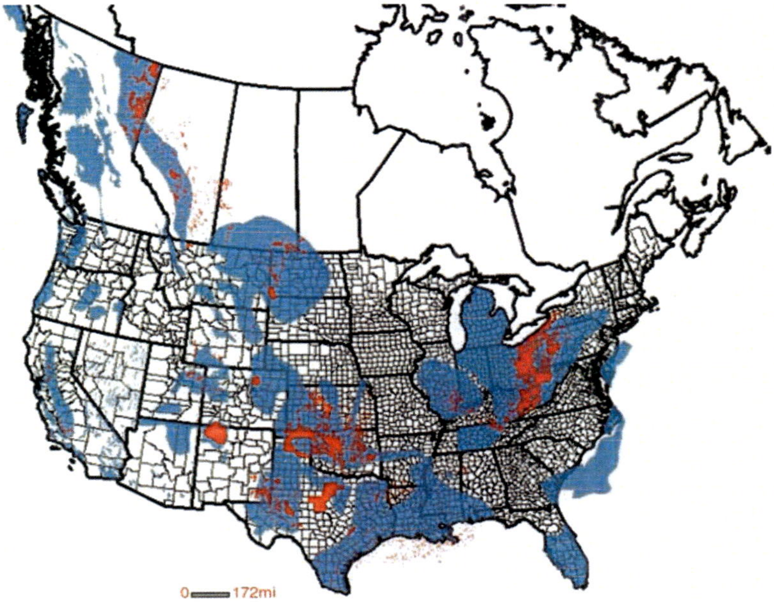

Oldenburg and Birkholzer, Fig. 2 North American deep saline reservoirs (*blue*) and oil and gas reservoirs (*red*) potentially available for GCS (Source: NATCARB: http://drysdale.kgs.ku.edu/natcarb/eps/natcarb_alpha_content.cfm)

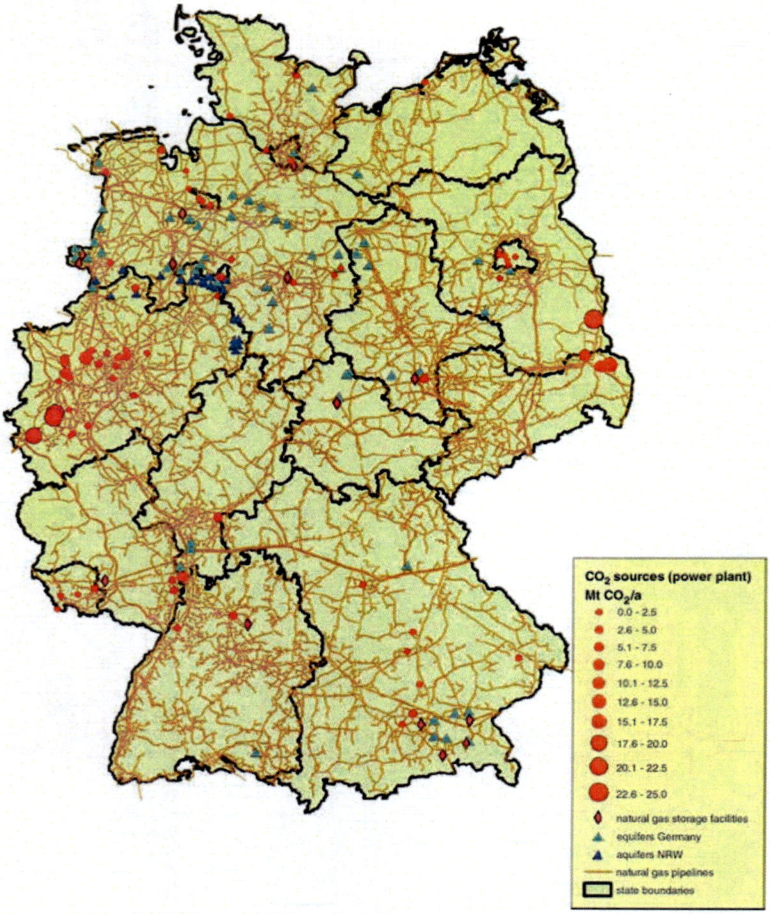

Toth et al., Fig. 2 Major stationary sources of CO_2 (*power plants*), potential disposal in saline aquifers and natural gas storage facilities, and the existing gas pipeline network in Germany (Source: Fischedick et al. 2007)

Toth et al., Fig. 3 Nuclear power plants and storage facilities in Germany (Source: Sailer 2008)

Toth et al., Fig. 4 Major sources of CO_2 emissions (Data taken from the Registre français des emissions polluantes 2009)

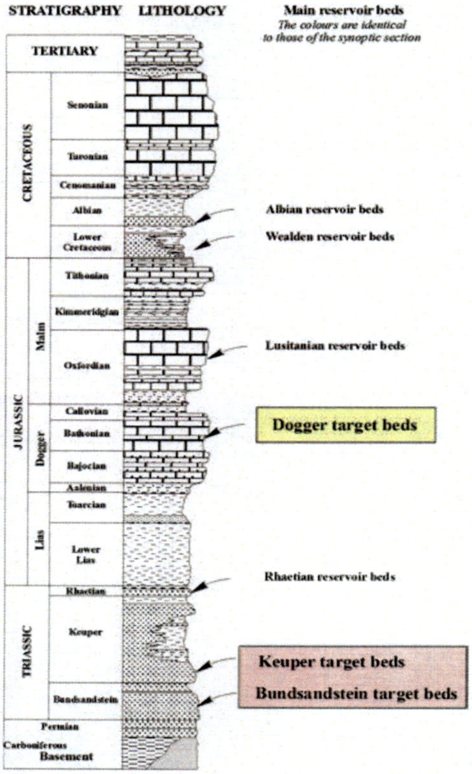

Toth et al., Fig. 6a Geological formations and main CO_2 reservoirs in the Paris Basin (Source: Bonijoly et al. 2003) Panel **a**. Synoptic log of sedimentary formations in the Paris Basin

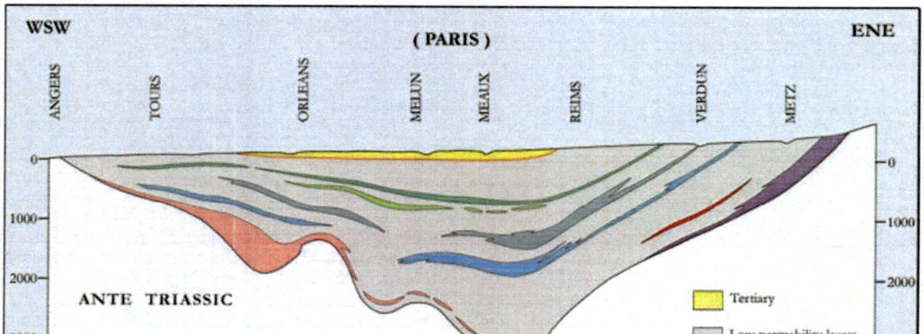

Toth et al., Fig. 6b (continued) Panel **b**. Main reservoirs identified in the Paris Basin

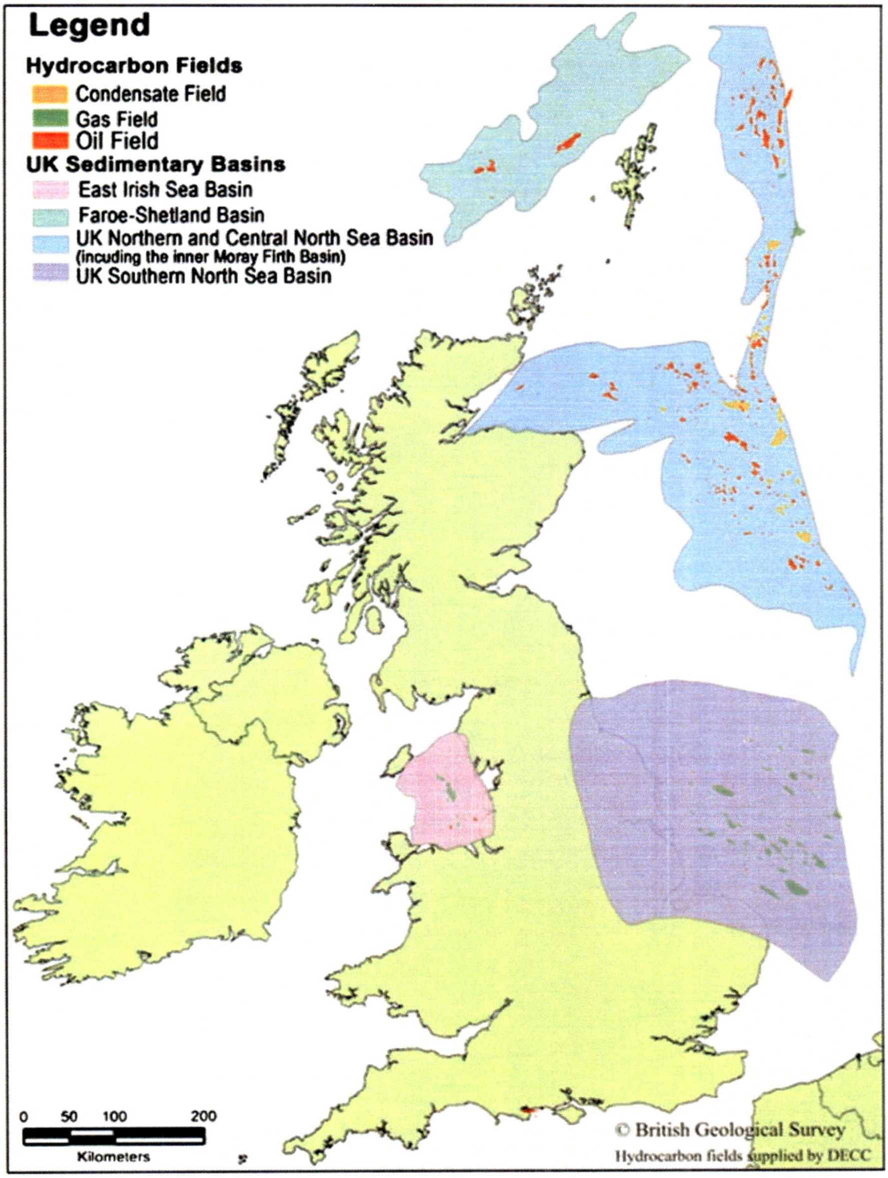

Toth et al., Fig. 9 Offshore hydrocarbon fields and the major oil- and gas-bearing sedimentary basins (Source: Holloway et al. 2006)

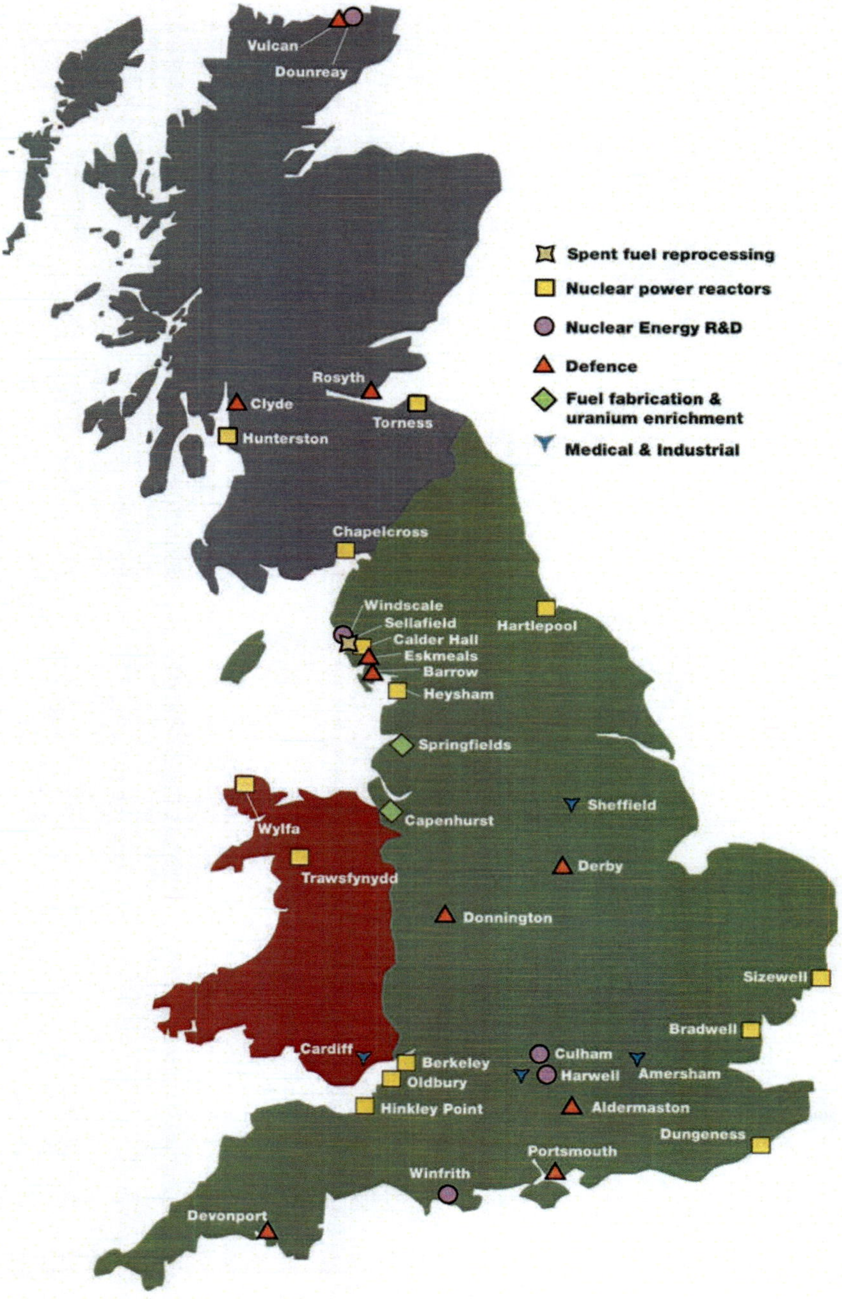

Toth et al., Fig. 10 Locations of major UK radioactive waste producers (Source: Defra 2008)

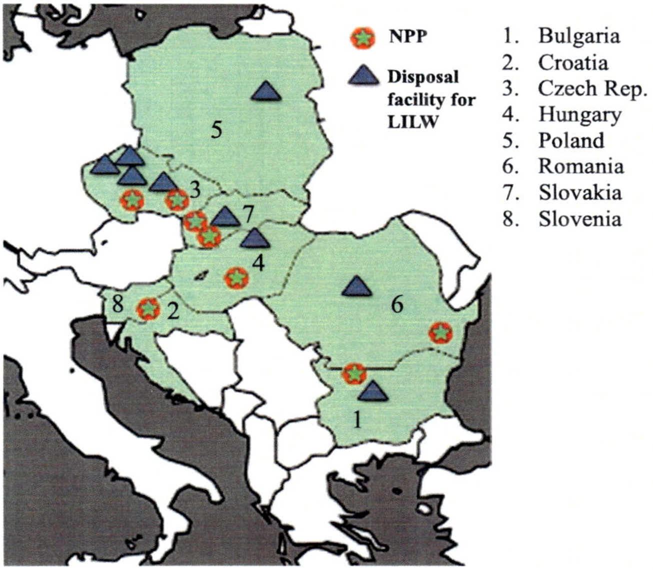

Hódossyné Hauszmann et al., Fig. 3 Location of existing disposal facilities and nuclear power plants in the Central and Eastern European region

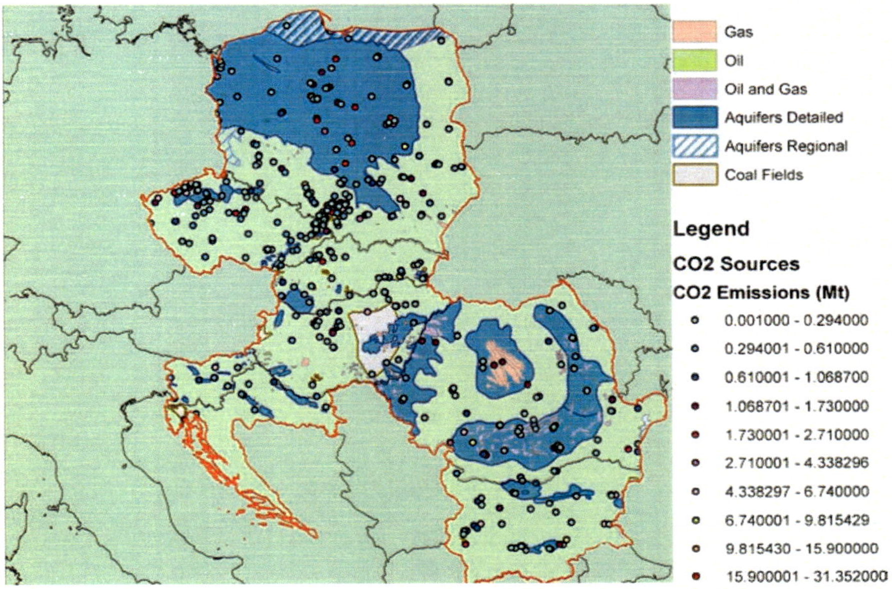

Hódossyné Hauszmann et al., Fig. 4 The study area with larger emission point sources (*circles*), depleted hydrocarbon fields, saline aquifer and unmineable coal reserve sites

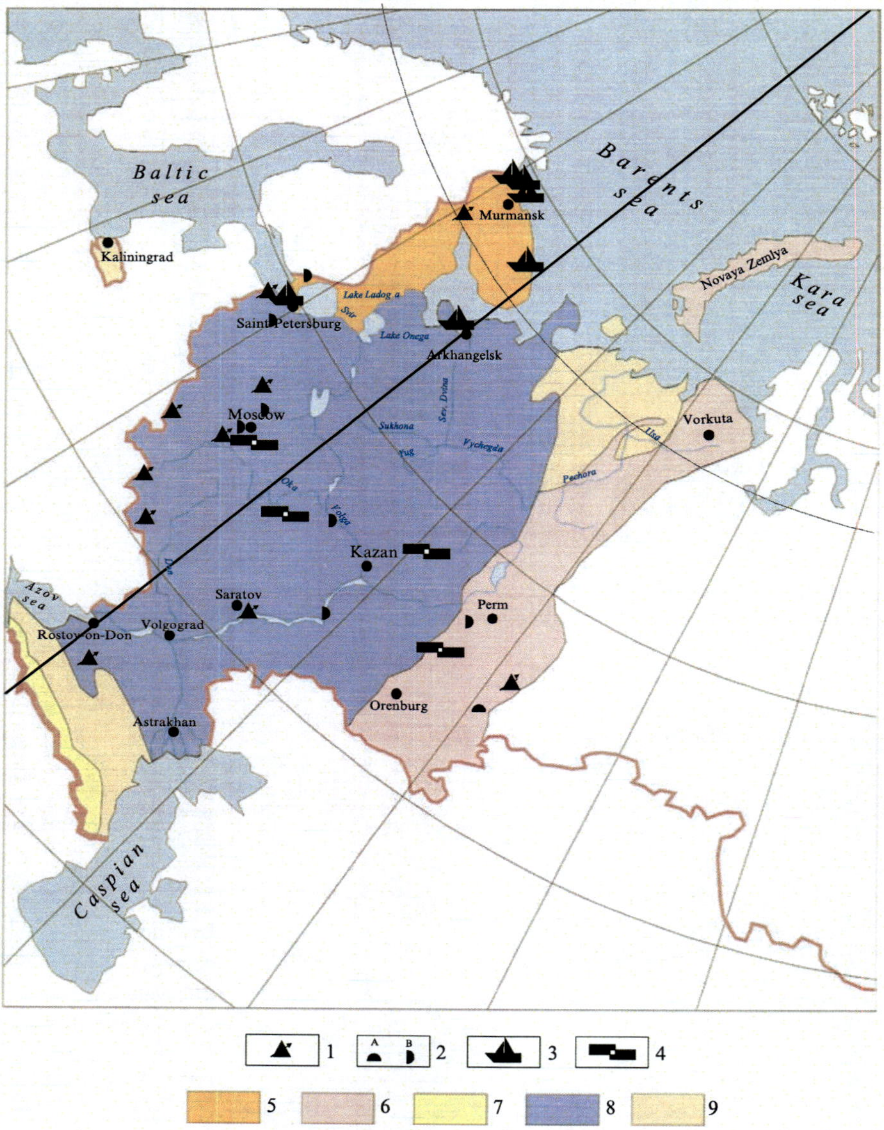

Cherepovitsyn and Ilinsky, Fig. 1 The main centres of nuclear power use in north-west Russia (1–4) and the regional geological environment (5–9). 1 Nuclear power stations. 2 Nuclear reactor: A – technological; B – research. 3 Bases of nuclear fleet. 4 Radiochemical and metallurgical plants. 5 Mountain ranges of Precambrian metamorphic complexes. 6 Folded and magmatic Phanerozoic rocks. 7 Sedimentary and volcanogenic rocks of recent geodynamic active mobile zones. 8 Complexes of lithified sedimentary rocks and vulcanites of ancient platforms. 9 Weakly lithified basic sediments of recent platforms

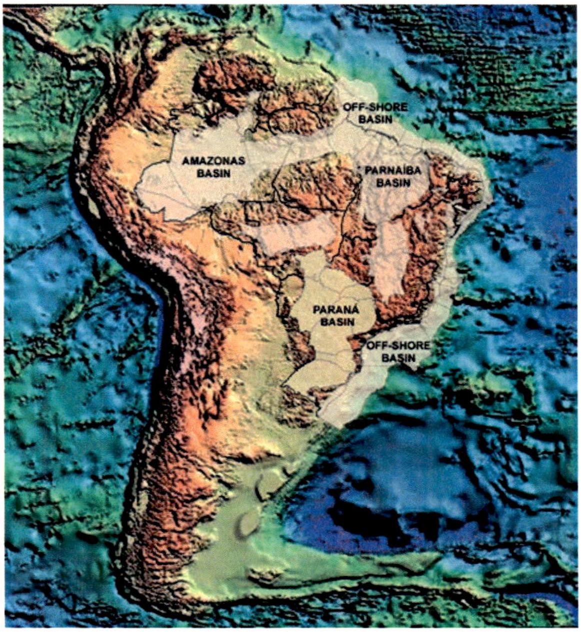

Heemann et al., Fig. 7 Large-scale overview of the main stable areas where deep repositories could be sited, showing possible target areas such as cratons (granitic-metamorphic shields—*dark grey*) and sedimentary basins (*light grey*) (Source: Milani 2007)

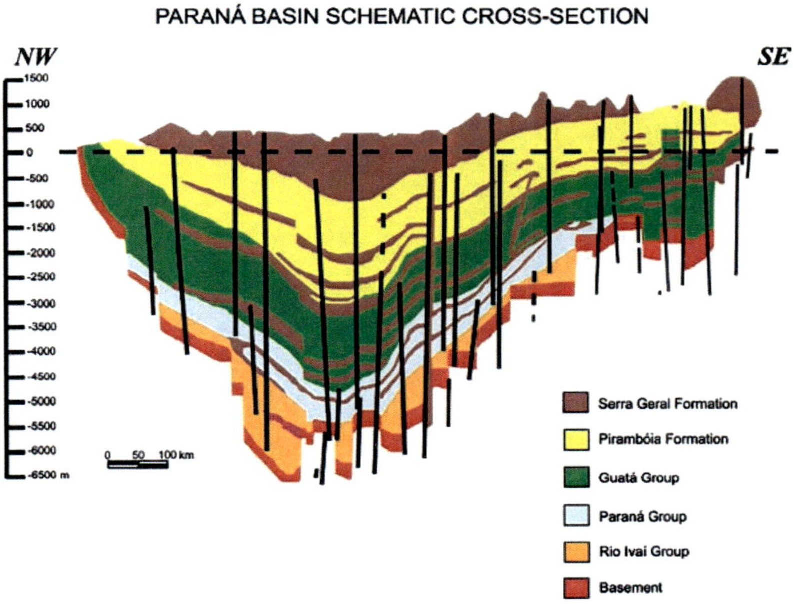

Heemann et al., Fig. 8 Target areas for carbon capture and storage and potential sites for deep repositories for nuclear waste disposal in the Paraná Basin, Brazil

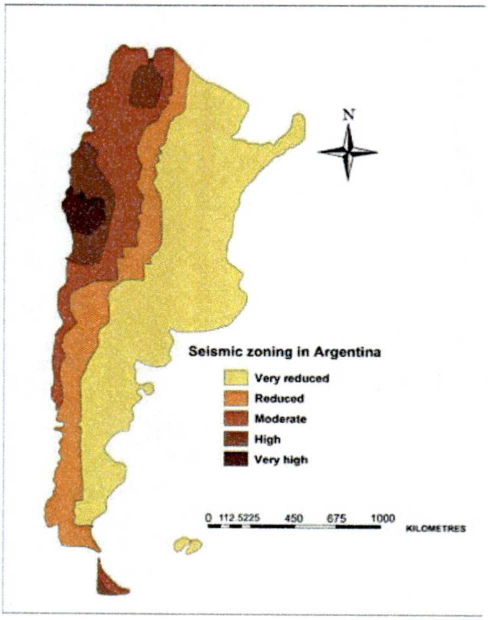

Heemann et al., Fig. 5 Seismic zoning[a] in Argentina (Source: Instituto Nacional de Prevención Sísmica (INPRES))

[a]Zones in which the possibility of radioactive waste disposal is excluded are those with high and very high seismicity.

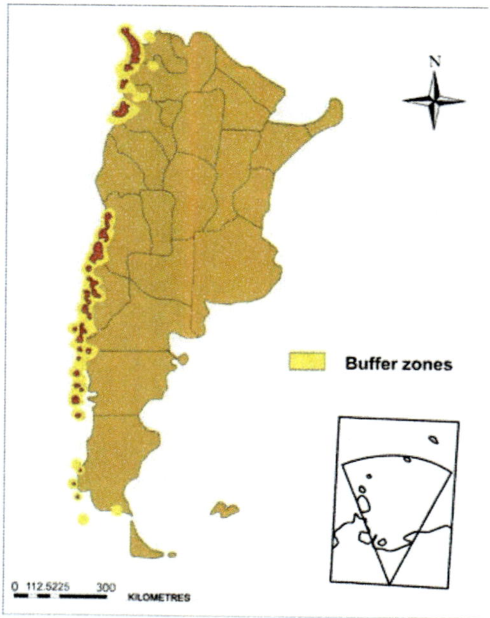

Heemann et al., Fig. 6 Volcanic centres in Argentina (Source: Instituto Nacional de Prevención Sísmica (INPRES))

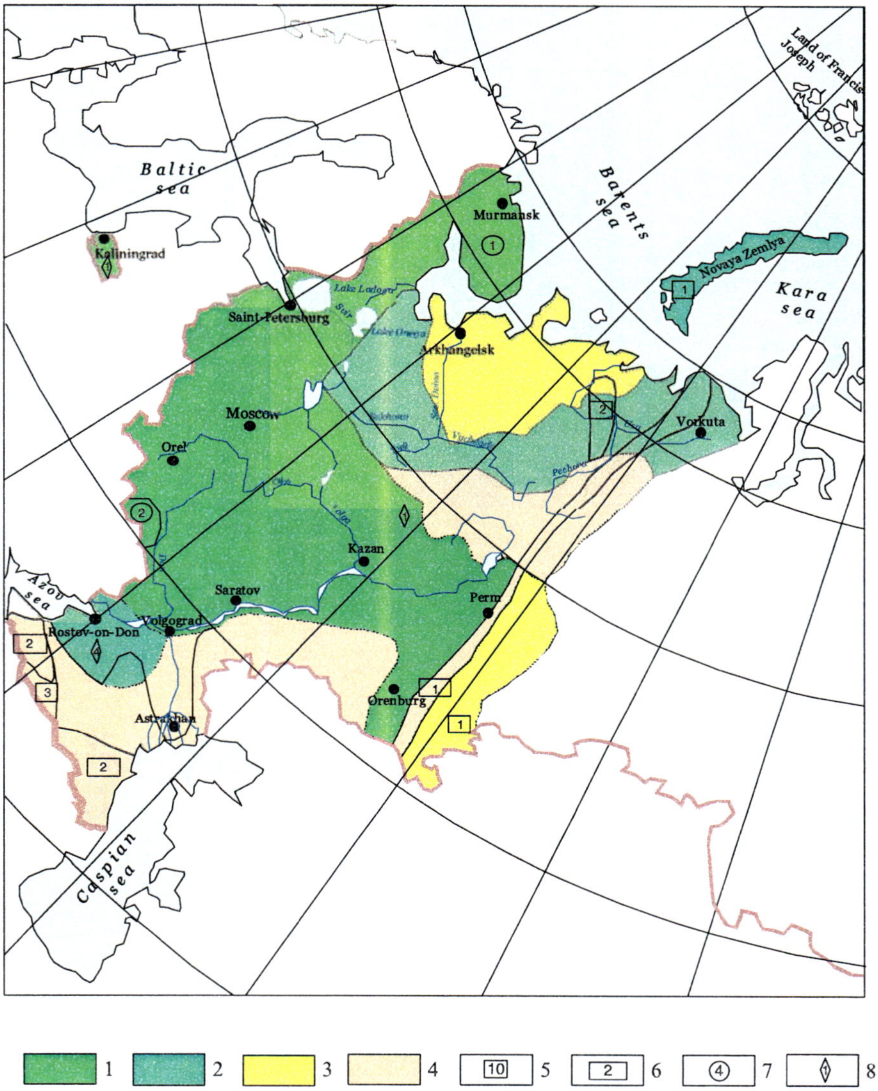

Cherepovitsyn and Ilinsky Fig. 3 Map of potential for subsurface disposal site development for radioactive waste. Coloured areas 1–4, indicating regions: 1 high potential; 2 average potential; 3 suitable areas and regions; 4 low potential. Numbers in geometric shapes: 5 Late Proterozoic Phanerozoic folded areas (*number given in square*) (*1* Urals – Novaya Zemlya; *2* Tieman; *3* Caucasus). 6 Regional deflections (*number given in rectangle*) (*1* attached to Urals; *2* attached to Caucasus). 7 Precambrian folded areas (*number given in circle*) (*1–2* cratons: *1* Baltic; *2* Voronezh Crystal Range). 8 Ancient and recent platforms (*number given in rhombus*) (*4* Skif-Turanic). A dashed line shows the boundary of the respective tectonic structure

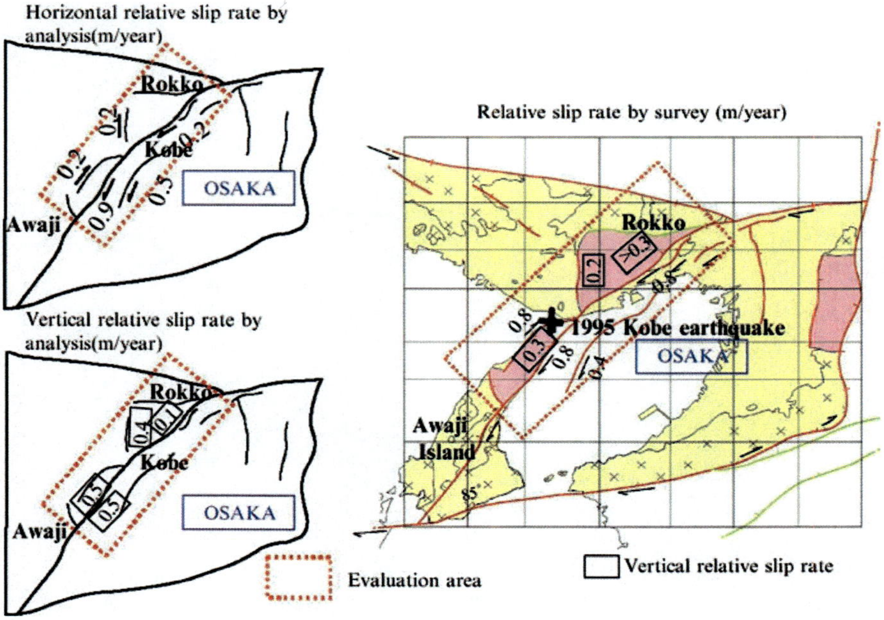

Koide and Kusunose, Fig. 8 Comparison of analytical results on the relative slip-rate vectors of the medium-scale model and survey results. The 1995 Kobe earthquake (measuring 7.2 on the Richter scale) occurred along active faults with large slip rates (Source: Sasaki et al. 2000)

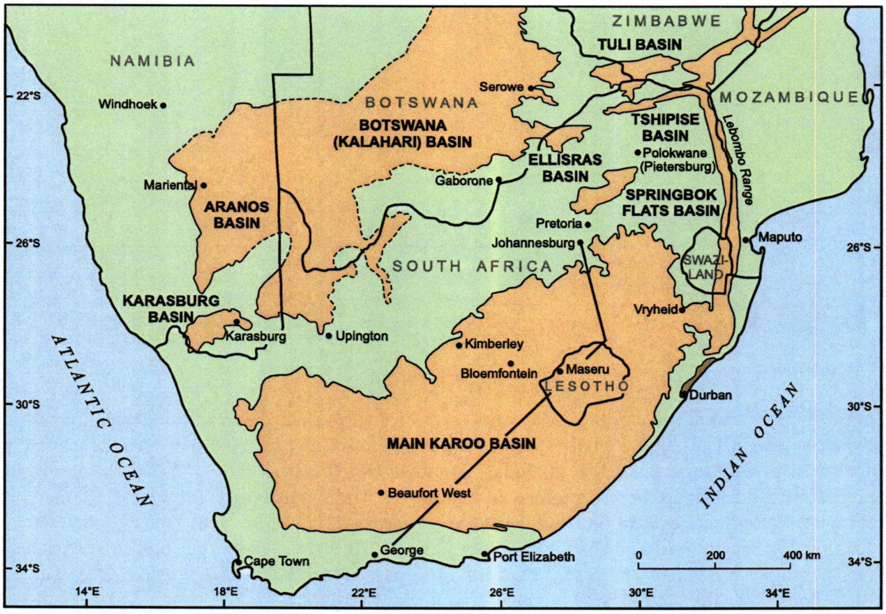

Surridge et al., Fig. 1 Site of the Karoo basins in southern Africa (Source: Johnson et al. 2006)

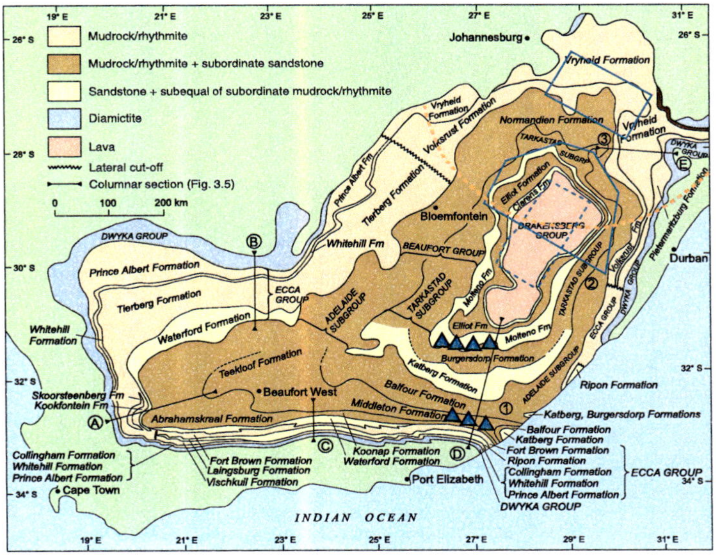

Surridge et al., Fig. 2 Schematic map illustrating the distribution of the lithostratigraphic units of the main Karoo Basin (Johnson et al. 2006) and the location of the five (A,B,C,D,E) prospective CO_2 storage areas. The yellow arc depicts an area approximately 300 km from Secunda, in which there are several large CO_2-emitting point sources

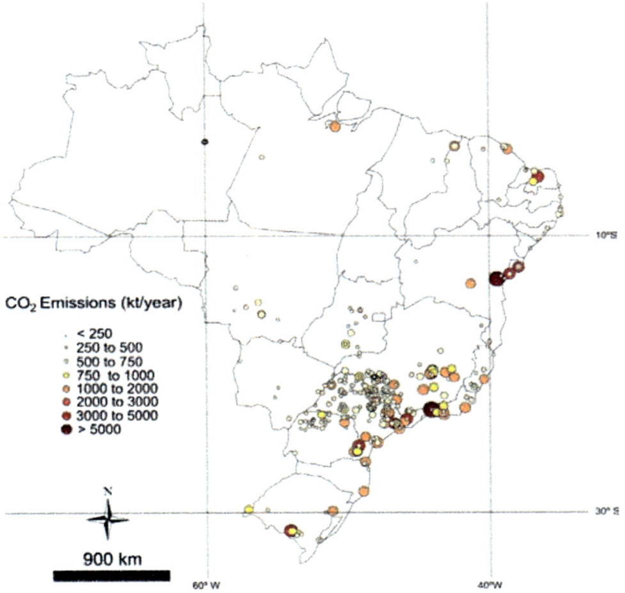

Heemann et al., Fig. 6 Large stationary sources of CO_2 emissions in the south-east region of Brazil (kt/year) (Modified from Ketzer et al. 2007)

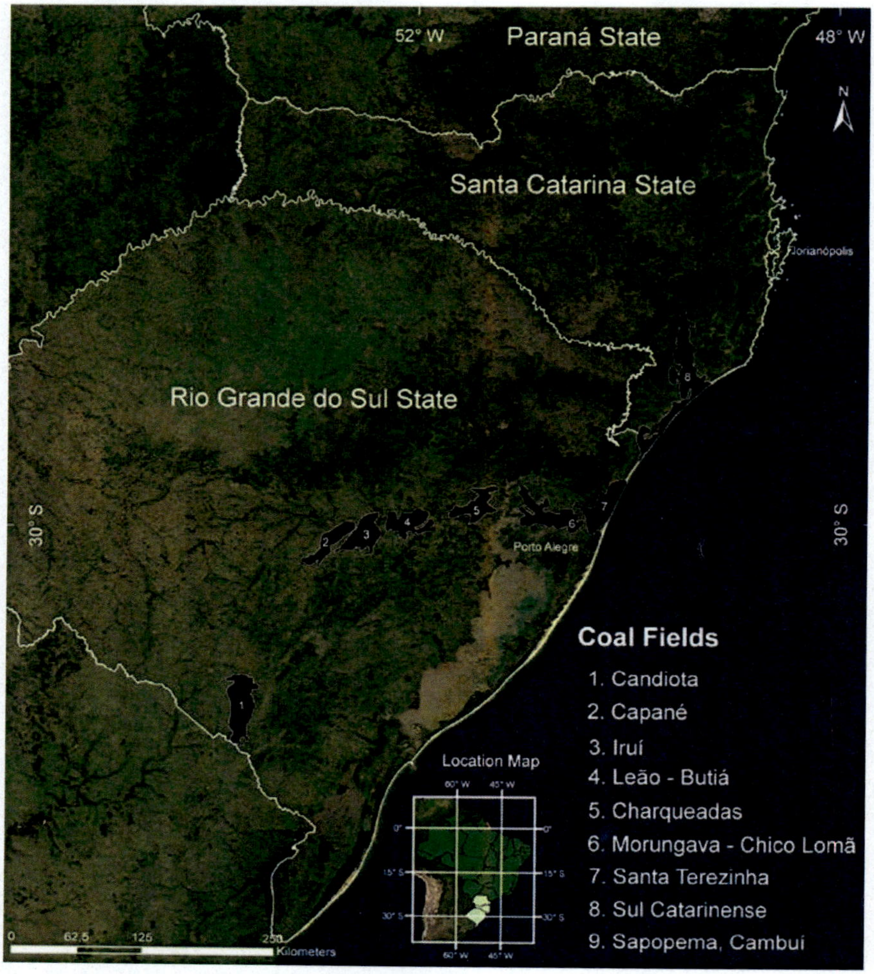

Heemann et al., Fig. 2 Location of major coal deposits in Brazilian area of Paraná Basin (RS, SC and PR States). Modified from Süffert (1997) and Aboarrage and Lopes (1986)